STUDENT'S SOLUTIONS MANUAL

John R. Martin
Tarrant County College

Michelle Boué

BASIC TECHNICAL MATHEMATICS

WITH CALCULUS

SI Version

Tenth Edition

Allyn J. Washington
Dutchess Community College

Michelle Boué

Toronto

Editor-in-Chief: Michelle Sartor
Acquisitions Editor: Cathleen Sullivan
Marketing Manager: Michelle Bish
Developmental Editor: Mary Wat
Project Manager: Kim Blakey
Production Editor: Lila Campbell
Compositor: Cenveo
Cover Designer: Alex Li
Interior Designer: Cenveo
Cover Image: Gencho Petkov/Shutterstock

Credits and acknowledgments for material borrowed from other sources and reproduced, with permission, in this textbook appear on the appropriate page within the text.

Original edition published by Pearson Education, Inc., Upper Saddle River, New Jersey, USA.

10 9 8 7 6 5 4 3 2 1 [WC]

Library and Archives Canada Cataloguing in Publication

Boué, Michelle, author

Basic technical mathematics with calculus : SI version, tenth edition. Student's solutions manual, Allyn J. Washington, Dutchess Community College / Michelle Boué.

Supplement to: Basic technical mathematics with calculus.
ISBN 978-0-13-398276-3 (pbk.)

1. Mathematics--Handbooks, manuals, etc. 2. Calculus--Handbooks, manuals, etc. I. Title.

QA37.2.W372 2014 Suppl. 510.76 C2014-905639-7

ISBN 978-0-13-398276-3

CONTENTS

CHAPTER 5 SYSTEMS OF LINEAR EQUATIONS; DETERMINANTS

CHAPTER 6 FACTORING AND FRACTIONS

CHAPTER 7 QUADRATIC EQUATIONS

CHAPTER 8 TRIGONOMETRIC FUNCTIONS OF ANY ANGLE

CHAPTER 9 VECTORS AND OBLIQUE TRIANGLES

CHAPTER 10 GRAPHS OF THE TRIGONOMETRIC FUNCTIONS

CHAPTER 11 EXPONENTS AND RADICALS

CHAPTER 12 COMPLEX NUMBERS

CHAPTER 13 EXPONENTIAL AND LOGARITHMIC FUNCTIONS

CHAPTER 23 THE DERIVATIVE

CHAPTER 24 APPLICATIONS OF THE DERIVATIVE

CHAPTER 25 INTEGRATION

CHAPTER 26 APPLICATIONS OF INTEGRATION

CHAPTER 31 DIFFERENTIAL EQUATIONS

CHAPTER 1

BASIC ALGEBRAIC OPERATIONS

1.1 Numbers

1. The numbers –3 and 14 are integers. They are also rational numbers since they can be written as $\frac{-3}{1}$ and $\frac{14}{1}$.

5. 3 is an integer, rational $\left(\frac{3}{1}\right)$, and real.

$\sqrt{-4}$ is imaginary.

$-\frac{\pi}{6}$ is irrational (because π is an irrational number) and real.

9. $6 < 8$; 6 is to the left of 8.

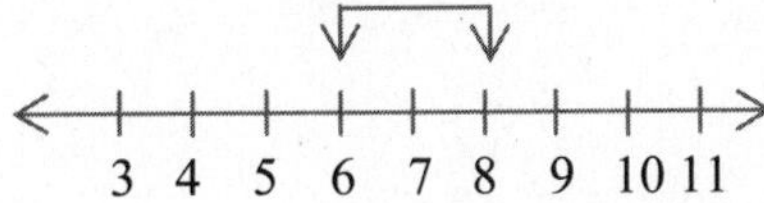

13. $-4 < -|-3|$; –4 is to the left of $-|-3|$, $(-|-3| = -(3) = -3)$.

17. The reciprocal of 3 is $\frac{1}{3}$.

The reciprocal of $-\frac{4}{\sqrt{3}}$ is $-\frac{1}{\frac{4}{\sqrt{3}}} = -\frac{\sqrt{3}}{4}$.

The reciprocal of $\frac{y}{b}$ is $\frac{1}{\frac{y}{b}} = \frac{b}{y}$.

21. An absolute value is not always positive, $|0| = 0$ which is not positive.

25. A rational number can be expressed as a fraction of integers. So, if the numerator in the fraction must be 3, then find the integer x, so that $0.13 < \frac{3}{x} < 0.14$.

So $0.13 = \frac{13}{100}$, and we can find the equivalent fraction of $\frac{13}{100}$ that has 3 as a numerator by rearranging and solving the following equation:

$$\frac{13}{100} = \frac{3}{x}$$
$$13(x) = 100 \times 3$$
$$x = \frac{300}{13}$$
$$x = 23.077$$
$$0.13 = \frac{3}{23.077}$$

However, since x must be an integer and less than the answer above to make our fraction with the numerator of 3 greater than 0.13, we assign 23 to x, making $\frac{3}{23} = 0.1304$, which is rational.

$0.13 < 0.1304 < 0.14$.

29. If a and b are positive integers and $b > a$, then

(a) $b - a$ is a positive integer.

(b) $a - b$ is a negative integer.

(c) $\frac{b-a}{b+a}$, the numerator and denominator are both positive, but the numerator is less than the denominator, so the answer is a positive rational number than is less than 1.

33. **(a)** If $x > 0$, then x is a positive number located to the right of zero on the number line.

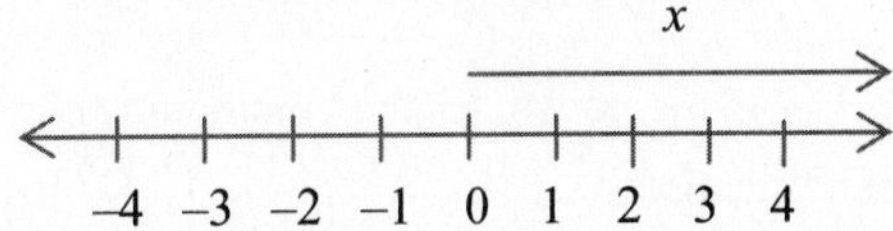

(b) If $x < -4$, then x is a negative number located to the left of –4 on the number line.

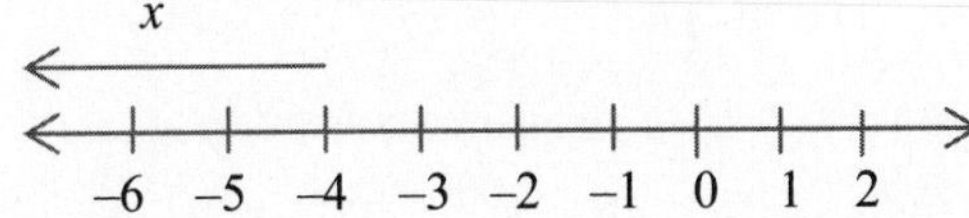

37. $a+bj=a+b\sqrt{-1}$ is a real number when $\sqrt{-1}$ is eliminated, which is when $b=0$. So $a+bj$ is a real number for all real values of a and $b=0$.

41. $N=\frac{a\text{ bits}}{\text{bytes}}\times\frac{1000\text{ bytes}}{1\text{ kilobyte}}\times n\,\text{kilobytes}$

$N=1000\,an$ bits

1.2 Fundamental Operations of Algebra

1. $16-2\times(-3)=16-(-6)=16+6=22$

5. $8+(-4)=8-4=4$

9. $-19-(-16)=-19+16=-3$

13. $-7(-5)=+(7\times5)=35$

17. $-2(4)(-5)=-8(-5)=40$

21. $-9-|2-10|=-9-|-8|=-9-8=-17$

25. $8-3(-4)=8+12=20$

29. $30(-6)(-2)\div(0-40)=30(12)\div(-40)$

$=360\div(-40)=-9$

33. $-7-\frac{|-14|}{2(2-3)}-3|6-8|=-7-\frac{14}{2(-1)}-3|-2|$

$=-7-\frac{14}{-2}-3(2)$

$=-7-(-7)-6$

$=-7+7-6$

$=-6$

37. $6(7)=(7)6$ demonstrates the commutative law of multiplication.

41. $3+(5+9)=(3+5)+9$ demonstrates the associative law of addition.

45. $-a+(-b)=-a-b$, which is expression (d).

49. **(a)** The sign of a product of an even number of negative numbers is positive.
Example: $-3(-6)=18$

(b) The sign of a product of an odd number of negative numbers is negative.
Example: $-5(-4)(-2)=-40$

53. **(a)** $-xy=1$ is true for values of x and y that are negative reciprocals of each other or $y=-\frac{1}{x}$, providing that the number x in the denominator is not zero. So if $x=12$, then $y=-\frac{1}{12}$ and $-xy=-(12)\left(-\frac{1}{12}\right)=1$.

(b) $\frac{x-y}{x-y}=1$ is true for all values of x and y, providing that $x\neq y$ to prevent division by zero.

57. The average temperature for the week is:

$$T_{avg}=\frac{7+3+(-2)+(-3)+(-1)+4+6}{7}\ ^\circ\text{C}$$

$$T_{avg}=\frac{7+3-2-3-1+4+6}{7}\ ^\circ\text{C}$$

$$T_{avg}=\frac{14}{7}\ ^\circ\text{C}=2.0\ ^\circ\text{C}$$

61. The oil drilled by the first well is $100\text{ m}+200\text{ m}=300\text{ m}$ which equals the depth drilled by the second well $200\text{ m}+100\text{ m}=300\text{ m}$. $100\text{ m}+200\text{ m}=200\text{ m}+100\text{ m}$ demonstrates the commutative law of addition.

1.3 Measurement, Calculation, and Approximate Numbers

1. 0.390 has three significant digits since the zero is after the decimal. The zero is not necessary as a placeholder and should not be written unless it is significant.

5. 1 megahertz = 1 MHz = 1 000 000 Hz

9. 1 kV = 1 kilovolt = 1000 volts

13. $1\text{ km}\left(\frac{1000\text{ m}}{1\text{ km}}\right)\cdot\left(\frac{100\text{ cm}}{1\text{ m}}\right)=100\ 000\text{ cm}$

17. $250\text{ mm}^2\left(\dfrac{1\text{ m}}{1\ 000\text{ mm}}\right)^2 = 0.000\ 25\text{ m}^2$

21. $45.0\text{ m/s}\left(\dfrac{100\text{ cm}}{1\text{ m}}\right) = 4500\text{ cm/s}$
(with 3 significant digits)

25. $25\text{ h}\left(\dfrac{3600\text{ s}}{1\text{ h}}\right)\cdot\left(\dfrac{1000\text{ ms}}{1\text{ s}}\right) = 90\ 000\ 000\text{ ms}$
(with 2 significant digits)

29. The distance around the orbit should be
$d = 2\pi r = 2\pi(150\ 000\ 000\text{ km}) = 942\ 477\ 796.1\text{ km}.$

$1\text{ y}\left(\dfrac{365.25\text{ d}}{1\text{ y}}\right)\cdot\left(\dfrac{24\text{ h}}{1\text{ d}}\right) = 8766\text{ h}$

Orbital speed is the ratio of distance travelled to time elapsed:

$$v = \frac{d}{t} = \frac{942\ 477\ 796.1\text{ km}}{8766\text{ h}} = 107\ 515\text{ km/h}$$
$$= 110\ 000\text{ km/h}$$

33. $6800\ \dfrac{\text{m}}{\text{s}}\left(\dfrac{1\text{ km}}{1000\text{ m}}\right)\cdot\left(\dfrac{3600\text{ s}}{1\text{ h}}\right) = 24\ 480\text{ km/h}$
$= 24\ 000\text{ km/h}$

37. $1000\ \dfrac{\text{kg}}{\text{m}^3}\left(\dfrac{1000\text{ g}}{1\text{ kg}}\right)\cdot\left(\dfrac{1\text{ m}^3}{1000\text{ L}}\right) = 1000\text{ g/L}$

41. $1.35\ \dfrac{\text{kW}}{\text{m}^2}\left(\dfrac{1000\text{ W}}{1\text{ kW}}\right)\cdot\left(\dfrac{1\text{ J/s}}{1\text{ W}}\right)\cdot\left(\dfrac{1\text{ m}}{100\text{ cm}}\right)^2$
$= 0.135\text{ J/(s}\cdot\text{cm}^2)$

45. 8 cylinders is exact because they can be counted. 55 km/h is approximate since it is measured.

49. 107 has 3 significant digits; 3004 has 4 significant digits.

53. 3000 has 1 significant digit; 3000.1 has 5 significant digits.

57. **(a)** Both 0.1 and 78.0 have the same precision as they have the same number of decimal places.
(b) 78.0 is more accurate because it has more significant digits (3) than 0.1, which has 1 significant digit.

61. **(a)** 4.936 rounded to 3 significant digits is 4.94.
(b) 4.936 rounded to 2 significant digits is 4.9.

65. **(a)** 9545 rounded to 3 significant digits is 9540 since 4 is the nearest even to 4.5
(b) 9549 rounded to 2 significant digits is 9500.

69. **(a)** Estimate: $13 + 1 - 2 = 12$
(b) Calculator: $12.78 + 1.0495 - 1.633 = 12.1965$, which is 12.20 to 0.01 precision

73. **(a)** Estimate: $\dfrac{20\times 0.02}{10-8} = 0.2$
(b) Calculator: $\dfrac{23.962\times 0.01537}{10.965 - 8.249} = 0.135\ 602\ 334$, which is 0.1356 to 4 significant digits

77. $0.9788 + 14.9 = 15.8788$ since the least precise number in the question has 4 decimal places.

81. With a frequency listed as 2.75 MHz, the least possible frequency is 2.745 MHz, and the greatest possible frequency is 2.755 MHz. Any measurements between those limits would round to 2.75 MHz.

85. **(a)** $2.2 + 3.8\times 4.5 = 2.2 + (3.8\times 4.5) = 19.3$
(b) $(2.2 + 3.8)\times 4.5 = 6.0\times 4.5 = 27$

89. Pick any six digit integer for $x = 231\ 465$ and rearrange those digits for $y = 164\ 352$.
$(x - y)\div 9 = (231\ 465 - 164\ 352)\div 9 = 7457$.
A smaller *integer* number results.

93. **(a)** $1\div 3 = 0.333...$ It is a rational number since it is a repeating decimal.
(b) $5\div 11 = 0.454545...$ It is a rational number since it is a repeating decimal.
(c) $2\div 5 = 0.400...$ It is a rational number since it is a repeating decimal (0 is the repeating part).

97. $1\text{ K} = 1024$ bytes

$256\text{ K}\cdot\left(\dfrac{1024\text{ bytes}}{1\text{ K}}\right) = 262\ 144\text{ bytes}$

1.4 Exponents

1. $(-x^3)^2 = \left[(-1)x^3\right]^2 = (-1)^2(x^3)^2 = (1)x^6 = x^6$

5. $x^3x^4 = x^{3+4} = x^7$

9. $\dfrac{m^5}{m^3} = m^{5-3} = m^2$

13. $\left(P^2\right)^4 = P^{2(4)} = P^8$

17. $\left(aT^2\right)^{30} = a^{30}T^{2(30)} = a^{30}T^{60}$

21. $\left(\frac{x^2}{2}\right)^4 = \frac{x^{2(4)}}{(2)^4} = \frac{x^8}{16}$

25. $-3x^0 = -3(1) = -3$

29. $\frac{1}{R^{-2}} = R^2$

33. $(2v^2)^{-6} = (2)^{-6}v^{2(-6)} = \frac{v^{-12}}{(2)^6} = \frac{1}{64v^{12}}$

37. $\frac{2v^4}{(2v)^4} = \frac{2v^4}{(2)^4(v^4)} = \frac{2v^4}{16v^4} = \frac{1}{8}$

41. $(\pi^0x^2a^{-1})^{-1} = \pi^{0(-1)}x^{2(-1)}a^{-1(-1)} = \pi^0x^{-2}a^1 = \frac{a}{x^2}$

45. $\left(\frac{4x^{-1}}{a^{-1}}\right)^{-3} = \frac{(4)^{-3}x^{-1(-3)}}{a^{-1(-3)}} = \frac{x^3}{64a^3}$

49. $7(-4)-(-5)^2 = -28-25 = -53$

53. $\frac{3.07(-1.86)}{(-1.86)^4+1.596} = \frac{-5.7102}{11.96883216+1.596}$

$= \frac{-5.7102}{13.56483216} = -0.420956185$

which gets rounded to 3 significant digits: –0.421.

57. $\left(\frac{1}{x^{-1}}\right)^{-1} = \frac{1^{-1}}{x^{-1(-1)}} = \frac{1}{x}$, which is the reciprocal of x.

61. $(x^a \cdot x^{-a})^5 = (x^{a-a})^5 = (x^o)^5 = x^{0(5)} = x^0 = 1$, provided that $x \neq 0$.

65. $\pi\left(\frac{r}{2}\right)^3\left(\frac{4}{3\pi r^2}\right) = \pi\left(\frac{r^3}{8}\right)\left(\frac{4}{3\pi r^2}\right) = \frac{4r}{24} = \frac{r}{6}$

1.5 Scientific Notation

1. $8.06\times10^3 = 8060$

5. $2.01\times10^{-3} = 0.002\ 01$

9. $1.86\times10 = 18.6$

13. $0.0087 = 8.7\times10^{-3}$

17. $0.063 = 6.3\times10^{-2}$

21. $28\ 000(2\ 000\ 000\ 000) = 2.8\times10^4(2\times10^9) = 5.6\times10^{13}$

25. $2\times10^{-35}+3\times10^{-34} = 0.2\times10^{-34}+3\times10^{-34} = 3.2\times10^{-34}$

29. $1280(865\ 000)(43.8) = 4.849\ 536\times10^{10}$

which gets rounded to 4.85×10^{10}.

33. $(3.642\times10^{-8})(2.736\times10^5) = 9.964\ 512\times10^{-3}$

which gets rounded to 9.965×10^{-3}.

37. $2\ 000\ 000\ \text{kW} = 2\times10^6\ \text{kW}$

41. $2\ 000\ 000\ 000\ \text{Hz} = 2\times10^9\ \text{Hz}$

45. $2\ 000\ 000\ \text{kW} = 2\times10^6\ \text{kW}$

$= 2\times10^6\times10^3\ \text{W}$

$= 2\ \times10^9\ \text{W}$

$= 2\ \text{GW}$

49. $2\ 000\ 000\ 000\ \text{Hz} = 2\times10^9\ \text{Hz}$

$= 2\ \text{GHz}$

53. googol $= 10^{100}$, so to find the ratio $\frac{10^{100}}{10^{79}}$

$= 10^{100-79} = 10^{21}$

A googol is 10^{21} times larger than the number of electrons in the universe.

57. **(a)** $1\ \text{day}\times\frac{24\ \text{h}}{\text{day}}\times\frac{60\ \text{min}}{\text{h}}\times\frac{60\ \text{s}}{\text{min}} = 86400\text{s} = 8.64\times10^4\ \text{s}$

(b) $100\ \text{year}\times\frac{365.25\ \text{day}}{\text{year}}\times\frac{24\ \text{h}}{\text{day}}\times\frac{60\ \text{min}}{\text{h}}\times\frac{60\ \text{s}}{\text{min}}$

$= 3\ 155\ 760\ 000\ \text{s} = 3.155\ 760\ 0\times10^9\ \text{s}$

61. $\frac{1.496\times10^8\text{km}}{\text{AU}}\times\frac{\text{AU}}{4.99\times10^2\text{s}}$
$= 2.997\ 995\ 991\ 98\times10^5\text{ km/s} = 2.998\times10^5\text{ km/s}$
This is the same speed mentioned in Question 56 as the speed of radio waves.

1.6 Roots and Radicals

1. $-\sqrt[3]{64} = -\sqrt[3]{(4)^3} = -4$

5. $\sqrt{81} = \sqrt{9^2} = 9$

9. $-\sqrt{49} = -\sqrt{7^2} = -7$

13. $\sqrt[3]{125} = \sqrt[3]{5^3} = 5$

17. $\left(\sqrt{5}\right)^2 = \sqrt{5}\times\sqrt{5} = 5$

21. $\left(-\sqrt[4]{53}\right)^4 = (-1)^4\left(\sqrt[4]{53}\right)^4 = (1)(53) = 53$

25. $2\sqrt{84} = 2\sqrt{(4)(21)} = 2\times\sqrt{4}\times\sqrt{21} = 2\times2\times\sqrt{21} = 4\sqrt{21}$

29. $\sqrt[3]{8^2} = \sqrt[3]{64} = \sqrt[3]{4^3} = 4$

33. $\sqrt{36+64} = \sqrt{100} = \sqrt{10^2} = 10$

37. $\sqrt{85.4} = 9.24121204171$, which is rounded to 9.24

41. **(a)** $\sqrt{1296+2304} = \sqrt{3600} = 60$, which is expressed as 60.00
(b) $\sqrt{1296}+\sqrt{2304} = 36+48 = 84$, which is expressed as 84.00

45. $\sqrt{207s} = \sqrt{(207)(46)} = \sqrt{9522}$
$= 97.5807358037 = 98\text{ km/h}$

49. $\sqrt{w^2+h^2} = \sqrt{(93.0\text{ cm})^2+(52.1\text{ cm})^2}$
$= \sqrt{8649\text{ cm}^2+2714.41\text{ cm}^2}$
$= \sqrt{11363.41\text{ cm}^2}$
$= 106.599296432\text{ cm}$
$= 107\text{ cm}$

53. **(a)** $\sqrt[3]{2140} = 12.8865874254$, which is rounded to 12.9
(b) $\sqrt[3]{-0.214} = -0.59814240297$, which is rounded to -0.598

```
³√(2140)
        12.88658743
³√(-0.214
        -.598142403
```

1.7 Addition and Subtraction of Algebraic Expressions

1. $3x+2y-5y = 3x-3y$

5. $5x+7x-4x = 8x$

9. $2F-2T-2+3F-T = 5F-3T-2$

13. $s+(3s-4-s) = s+(2s-4) = s+2s-4 = 3s-4$

17. $2-3-(4-5a) = -1-4+5a = 5a-5$

21. $-(t-2u)+(3u-t) = -t+2u+3u-t = -2t+5u$

25. $-7(6-3j)-2(j+4) = -42+21j-2j-8 = 19j-50$

29. $2[4-(t^2-5)] = 2[4-t^2+5]$
$= 2[-t^2+9]$
$= -2t^2+18$

33. $aZ-[3-(aZ+4)] = aZ-[3-aZ-4]$
$= aZ-[-aZ-1]$
$= aZ+aZ+1$
$= 2aZ+1$

37. $5p-(q-2p)-[3q-(p-q)] = 5p-q+2p-[3q-p+q]$
$= 5p-q+2p-[4q-p]$
$= 7p-q-4q+p$
$= 8p-5q$

41. $5V^2-(6-(2V^2+3)) = 5V^2-(6-2V^2-3)$
$= 5V^2-(-2V^2+3)$
$= 5V^2+2V^2-3$
$= 7V^2-3$

45. $-4[4R-2.5(Z-2R)-1.5(2R-Z)]$
$=-4[4R-2.5Z+5R-3R+1.5Z]$
$=-4[6R-Z]$
$=-24R+4Z$

49. $\left[\left(B+\frac{4}{3}\alpha\right)+2\left(B-\frac{2}{3}\alpha\right)\right]-\left[\left(B+\frac{4}{3}\alpha\right)-\left(B-\frac{2}{3}\alpha\right)\right]$
$=\left[B+\frac{4}{3}\alpha+2B-\frac{4}{3}\alpha\right]-\left[B+\frac{4}{3}\alpha-B+\frac{2}{3}\alpha\right]$
$=[3B]-\left[\frac{6}{3}\alpha\right]$
$=3B-2\alpha$

53. (a) $(2x^2-y+2a)+(3y-x^2-b)$
$=2x^2-y+2a+3y-x^2-b$
$=x^2+2y+2a-b$

(b) $(2x^2-y+2a)-(3y-x^2-b)$
$=2x^2-y+2a-3y+x^2+b$
$=3x^2-4y+2a+b$

1.8 Multiplication of Algebraic Expressions

1. $2s^3(-st^4)^3(4s^2t)=2s^3(-1)^3s^3t^{12}(4s^2t)$
$=-2s^6t^{12}(4s^2t)$
$=-8s^8t^{13}$

5. $(a^2)(ax)=a^3x$

9. $(2ax^2)^2(-2ax)=(2ax^2)(2ax^2)(-2ax)$
$=(4a^2x^4)(-2ax)$
$=-8a^3x^5$

13. $-3s(s^2-5t)=(-3s)(s^2)+(-3s)(-5t)$
$=-3s^3+15st$

17. $3M(-M-N+2)=(3M)(-M)+(3M)(-N)+(3M)(2)$
$=-3M^2-3MN+6M$

21. $(x-3)(x+5)=(x)(x)+(x)(5)+(-3)(x)+(-3)(5)$
$=x^2+5x-3x-15$
$=x^2+2x-15$

25. $(2a-b)(3a-2b)=(2a)(3a)+(2a)(-2b)$
$+(-b)(3a)+(-b)(-2b)$
$=6a^2-4ab-3ab+2b^2$
$=6a^2-7ab+2b^2$

29. $(x^2-1)(2x+5)=(x^2)(2x)+(x^2)(5)+(-1)(2x)+(-1)(5)$
$=2x^3+5x^2-2x-5$

33. $2(a+1)(a-9)=2[(a)(a)+(a)(-9)+(1)(a)+(-9)(1)]$
$=2[a^2-9a+a-9]$
$=2[a^2-8a-9]$
$=2a^2-16a-18$

37. $2L(L+1)(4-L)=2L[(L)(4)+(L)(-L)+(1)(4)+(1)(-L)]$
$=2L[-L^2+4L-L+4]$
$=2L[-L^2+3L+4]$
$=-2L^3+6L^2+8L$

41. $(x_1+3x_2)^2=(x_1+3x_2)(x_1+3x_2)$
$=(x_1)(x_1)+(x_1)(3x_2)+(3x_2)(x_1)+(3x_2)(3x_2)$
$=x_1^2+3x_1x_2+3x_1x_2+9x_2^2$
$=x_1^2+6x_1x_2+9x_2^2$

45. $2(x+8)^2=2[(x+8)(x+8)]$
$=2[(x)(x)+(x)(8)+(8)(x)+(8)(8)]$
$=2[x^2+8x+8x+64]$
$=2[x^2+16x+64]$
$=2x^2+32x+128$

49. $3T(T+2)(2T-1)$
$=3T[(T)(2T)+(T)(-1)+(2)(2T)+(2)(-1)]$
$=3T[2T^2-T+4T-2]$
$=3T[2T^2-T+4T-2]$
$=3T[2T^2+3T-2]$
$=6T^3+9T^2-6T$

53. $(4)^2-1=16-1=15=(3)(5)$
If we let x equal an integer between 1 and 9, $1<x<9$, then x^2-1 can be factored to $(x-1)(x+1)$:
$(x-1)(x+1)=(x)(x)+(1)(x)+(-1)(x)+(1)(-1)$
$=x^2-x+x-1$
$=x^2-1$
(x–1) is the number before x, and (x+1) is the number after x.

57. $P(1+0.01r)^2 = P(1+0.01r)(1+0.01r)$

$= P[(1)(1)+(1)(0.01r)+(0.01r)(1)$
$\qquad +(0.01r)(0.01r)]$

$= P[1+0.01r+0.01r+0.0001r^2]$

$= 0.0001r^2P+0.02rP+P$

61. Number of switches $= n^2$

$= (n+100)^2$

$= (n+100)(n+100)$

$= (n)(n)+(n)(100)+(100)(n)$
$\qquad +(100)(100)$

$= n^2+100n+100n+10\ 000$

$= n^2+200n+10\ 000$

1.9 Division of Algebraic Expressions

1. $\dfrac{-6a^2xy^2}{-2a^2xy^5} = \left(\dfrac{-6}{-2}\right)\dfrac{a^{2-2}x^{1-1}}{y^{5-2}} = \dfrac{3}{y^3}$

5. $\dfrac{8x^3y^2}{-2xy} = -4x^{3-1}y^{2-1} = -4x^2y$

9. $\dfrac{(15x^2)(4bx)(2y)}{30bxy} = \dfrac{120x^3by}{30bxy} = 4x^{3-1}b^{1-1}y^{1-1} = 4x^2$

13. $\dfrac{3a^2x+6xy}{3x} = \dfrac{3a^2x}{3x}+\dfrac{6xy}{3x} = \dfrac{3a^2x^{1-1}}{3}+\dfrac{6x^{1-1}y}{3} = a^2+2y$

17. $\dfrac{4pq^3+8p^2q^2-16pq^5}{4pq^2} = \dfrac{4pq^3}{4pq^2}+\dfrac{8p^2q^2}{4pq^2}-\dfrac{16pq^5}{4pq^2}$

$= p^{1-1}q^{3-2}+2p^{2-1}q^{2-2}-4p^{1-1}q^{5-2}$

$= -4q^3+2p+q$

21. $\dfrac{3ab^2-6ab^3+9a^2b^2}{9a^2b^2} = \dfrac{3ab^2}{9a^2b^2}-\dfrac{6ab^3}{9a^2b^2}+\dfrac{9a^2b^2}{9a^2b^2}$

$= \dfrac{b^{2-2}}{3a^{2-1}}-\dfrac{2b^{3-2}}{3a^{2-1}}+a^{2-2}b^{2-2}$

$= \dfrac{1}{3a}-\dfrac{2b}{3a}+1$

25.
$$\begin{array}{r} 2x+1 \\ x+3\overline{)2x^2+7x+3} \\ \underline{2x^2+6x\quad} \\ x+3 \\ \underline{x+3} \\ 0 \end{array}$$

$\dfrac{2x^2+7x+3}{x+3} = 2x+1$

29.
$$\begin{array}{r} 4x^2-x-1 \\ 2x-3\overline{)8x^3-14x^2+x+0} \\ \underline{8x^3-12x^2\qquad\quad} \\ -2x^2+x\quad \\ \underline{-2x^2+3x}\quad \\ -2x+0 \\ \underline{-2x+3} \\ -3 \end{array}$$

$\dfrac{8x^3-14x^2+x}{2x-3} = 4x^2-x-1-\dfrac{3}{2x-3}$

33.
$$\begin{array}{r} x^2+x-6 \\ x+2\overline{)x^3+3x^2-4x-12} \\ \underline{x^3+2x^2\qquad\qquad} \\ x^2-4x\qquad \\ \underline{x^2+2x}\qquad \\ -6x-12 \\ \underline{-6x-12} \\ 0 \end{array}$$

$\dfrac{x^3+3x^2-4x-12}{x+2} = x^2+x-6$

37.
$$\begin{array}{r} x^2-2x+4 \\ x+2\overline{)x^3+0x^2+0x+8} \\ \underline{x^3+2x^2\qquad\qquad} \\ -2x^2+0x\qquad \\ \underline{-2x^2-4x}\qquad \\ 4x+8 \\ \underline{4x+8} \\ 0 \end{array}$$

$\dfrac{x^3+8}{x+2} = x^2-2x+4$

41. $x+y-z\,\overline{)\,x^2-y^2+0xy+0xz+2yz-z^2}$ with quotient $x-y+z$

$$\begin{array}{l} x^2 \quad +xy-xz \\ \hline -y^2-xy+xz+2yz-z^2 \\ -y^2-xy \quad +yz \\ \hline xz+yz-z^2 \\ xz+yz-z^2 \\ \hline 0 \end{array}$$

$$\frac{x^2-y^2+2yz-z^2}{x+y-z}=x-y+z$$

45. $x+1\,\overline{)\,x^4+0x^3+0x^2+0x+1}$ with quotient x^3-x^2+x-1

$$\begin{array}{l} x^4+x^3 \\ \hline -x^3+0x^2 \\ -x^3-x^2 \\ \hline x^2+0x \\ x^2+x \\ \hline -x+1 \\ -x-1 \\ \hline 2 \end{array}$$

$$\frac{x^4+1}{x+1}=x^3-x^2+x-1+\frac{2}{x+1}\neq x^3$$

49.
$$\frac{GMm[(R+r)-(R-r)]}{2rR}=\frac{GMm[R+r-R+r]}{2rR}=\frac{GMm[2r]}{2rR}=\frac{GMm\,[\cancel{2r}]}{\cancel{2r}R}=\frac{GMm}{R}$$

1.10 Solving Equations

1. (a)
$$x-3=-12$$
$$x-3+3=-12+3$$
$$x=-9$$

(b)
$$x+3=-12$$
$$x+3-3=-12-3$$
$$x=-15$$

(c)
$$\frac{x}{3}=-12$$
$$3\left(\frac{x}{3}\right)=3(-12)$$
$$x=-36$$

(d)
$$3x=-12$$
$$\frac{3x}{3}=\frac{-12}{3}$$
$$x=-4$$

5.
$$x-2=7$$
$$x=7+2$$
$$x=9$$

9.
$$\frac{t}{2}=-5$$
$$t=2(-5)$$
$$t=-10$$

13.
$$3t+5=-4$$
$$3t=-4-5$$
$$t=\frac{-9}{3}$$
$$t=-3$$

17.
$$3x+7=x$$
$$x-3x=7$$
$$-2x=7$$
$$x=-\frac{7}{2}$$

21.
$$6-(r-4)=2r$$
$$2r=6-r+4$$
$$2r+r=10$$
$$3r=10$$
$$r=\frac{10}{3}$$

25.
$$0.1x-0.5(x-2)=2$$
$$x-5(x-2)=2(10)$$
$$x-5x+10=20$$
$$-4x=20-10$$
$$x=\frac{10}{-4}=-\frac{5}{2}$$

29. $\dfrac{4x-2(x-4)}{3}=8$

$$4x-2x+8=3(8)$$
$$2x=24-8$$
$$x=\frac{16}{2}$$
$$x=8$$

33. $5.8-0.3(x-6.0)=0.5x$

$$0.5x=5.8-0.3x+1.8$$
$$0.5x+0.3x=7.6$$
$$0.8x=7.6$$
$$x=\frac{7.6}{0.8}$$
$$x=9.5$$

37. $\dfrac{x}{2.0}=\dfrac{17}{6.0}$

$$x=2.0\left(\frac{17}{6.0}\right)$$
$$x=5.6666666...$$
$$x=5.7$$

41. (a) $2x+3=3+2x$

$$2x+3=2x+3$$

Is an identity, since it is true for all values of x.

(b) $2x-3=3-2x$

$$4x=6$$
$$x=\frac{6}{4}=\frac{3}{2}$$

Is conditional as x has one answer only.

45. $2.0v+40=2.5(v+5.0)$

$$2.0v+40=2.5v+12.5$$
$$40-12.5=2.5v-2.0v$$
$$27.5=0.5v$$
$$v=\frac{27.5}{0.5}$$
$$v=55\text{ km/h}$$

49. $0.14n+0.06(2000-n)=0.09(2000)$

$$0.14n+120-0.06n=180$$
$$0.14n-0.06n=180-120$$
$$0.08n=60$$
$$n=\frac{60}{0.08}$$
$$n=750\text{ L}$$

1.11 Formulas and Literal Equations

1. $v=v_0+at$

$$v-v_0=at$$
$$a=\frac{v-v_0}{t}$$

5. $E=IR$

$$R=\frac{E}{I}$$

9. $Q=SLd^2$

$$L=\frac{Q}{Sd^2}$$

13. $A=\dfrac{Rt}{PV}$

$$Rt=APV$$
$$t=\frac{APV}{R}$$

17. $T=\dfrac{c+d}{v}$

$$c+d=Tv$$
$$d=Tv-c$$

21. $a=\dfrac{2mg}{M+2m}$

$$a(M+2m)=2gm$$
$$aM+2am=2gm$$
$$aM=2gm-2am$$
$$M=\frac{2gm-2am}{a}$$

25. $N=r(A-s)$

$$N=Ar-rs$$
$$rs+N=Ar$$
$$rs=Ar-N$$
$$s=\frac{Ar-N}{r}$$

29. $Q_1=P(Q_2-Q_1)$

$$Q_1=PQ_2-PQ_1$$
$$PQ_2=Q_1+PQ_1$$
$$Q_2=\frac{Q_1+PQ_1}{P}$$

33.
$$L=\pi(r_1+r_2)+2x_1+2x_2$$
$$L=\pi r_1+\pi r_2+2x_1+2x_2$$
$$\pi r_1=L-\pi r_2-2x_1-2x_2$$
$$r_1=\frac{L-\pi r_2-2x_1-2x_2}{\pi}$$

37.
$$C=\frac{2eAk_1k_2}{d(k_1+k_2)}$$
$$Cd(k_1+k_2)=2eAk_1k_2$$
$$e=\frac{Cd(k_1+k_2)}{2Ak_1k_2}$$

41.
$$\eta=\frac{T_2}{T_1+T_2}$$
$$\eta(T_1+T_2)=T_2$$
$$\eta T_1+\eta T_2=T_2$$
$$\eta T_1=T_2-\eta T_2$$
$$T_1=\frac{T_2-\eta T_2}{\eta}$$
$$T_1=\frac{875\text{ K}-0.450(875\text{ K})}{0.450}$$
$$T_1=\frac{875\text{ K}-393.75\text{ K}}{0.450}$$
$$T_1=\frac{481.25\text{ K}}{0.450}$$
$$T_1=1069.444444\text{ K}$$
$$T_1=1070\text{ K}$$

45.
$$V_1=\frac{VR_1}{R_1+R_2}$$
$$V_1(R_1+R_2)=VR_1$$
$$R_1+R_2=\frac{VR_1}{V_1}$$
$$R_2=\frac{VR_1}{V_1}-R_1$$
$$R_2=\frac{(12.0\text{ V})(3.56\ \Omega)}{6.30\text{ V}}-(3.56\ \Omega)$$
$$R_2=6.780952381\ \Omega-3.56\ \Omega$$
$$R_2=3.220952381\ \Omega$$
$$R_2=3.22\ \Omega$$

1.12 Applied Word Problems

1. Let x = the number of 1.5 Ω resistors.
Let $34-x$ = the number of 2.5 Ω resistors.

$$1.5x+2.5(34-x)=56$$
$$1.5x+85-2.5x=56$$
$$-x=56-85$$
$$-x=-29$$
$$x=29$$

There are 29 of the 1.5 Ω resistors and (34 – 29) = 5 of the 2.5 Ω resistors.

Check:
$$29(1.5\ \Omega)+5(2.5\ \Omega)=56\ \Omega$$
$$43.5\ \Omega+12.5\ \Omega=56\ \Omega$$
$$56\ \Omega=56\ \Omega$$

5. Let x = the cost of the car 6 years ago.
Let x + \$5000 = the cost of the car model today.

$$x+(x+\$5000)=\$49\ 000$$
$$2x=\$44\ 000$$
$$x=\frac{\$44\ 000}{2}$$
$$x=\$22\ 000$$

The cost of the car 6 years ago was \$22 000, and the cost of the today's model is (\$22 000 + 5000) = \$27 000.

Check:
$$\$22\ 000+(\$22\ 000+\$5000)=\$49\ 000$$
$$\$22\ 000+\$27\ 000=\$49\ 000$$
$$\$49\ 000=\$49\ 000$$

9. Let x = the number hectares of land leased for \$200 per hectare.
Let $140-x$ = the number of hectares of land leased for \$300 per hectare.

$$\$200/\text{hectare}\ x+\$300/\text{hectare}(140\text{ hectares}-x)$$
$$=\$37\ 000$$
$$-\$100/\text{hectare}\ x=-\$5\ 000$$
$$x=\frac{-\$5000}{-\$100/\text{hectare}}$$
$$x=50\text{ hectares}$$

There are 50 hectares leased at \$200 per hectare and (140 hectares – 50 hectares) = 90 hectares leased for \$300 per hectare.

Check:
$$\$200/\text{hectare}\ (50\text{ hectares})+\$300/\text{hectare}$$
$$(140\text{ hectares}-(50\text{ hectares}))$$
$$=\$37\ 000$$
$$\$10\ 000+\$27\ 000=\$37\ 000$$
$$\$37\ 000=\$37\ 000$$

13. Let x = the number of 18-m girders needed.
Let $x + 4$ = the number of 15-m girders needed.

$$(18\text{ m})x = (15\text{ m})(x+4)$$
$$(18\text{ m})x = (15\text{ m})x + 60\text{ m}$$
$$(3\text{ m})x = 60\text{ m}$$
$$x = \frac{60\text{ m}}{3\text{ m}}$$
$$x = 20\text{ girders}$$

There would be 20 18-m girders needed or (20 girders + 4 girders) = 24 15-m girders needed.

Check:
$$(18\text{ m})20 = (15\text{ m})(20+4)$$
$$360\text{ m} = 360\text{ m}$$

17. Let x = the length of the first pipeline in km.
Let $x + 2.6$ km = the length of the 3 other pipelines.

$$x + 3(x + 2.6\text{ km}) = 35.4\text{ km}$$
$$x + 3x + 7.8\text{ km} = 35.4\text{ km}$$
$$4x = 27.6\text{ km}$$
$$x = \frac{27.6\text{ km}}{4}$$
$$x = 6.9\text{ km}$$

The first pipeline is 6.9 km long, and the other three pipelines are each (6.9 km + 2.6 km) = 9.5 km long.

Check:
$$6.9\text{ km} + 3(6.9\text{ km} + 2.6\text{ km}) = 35.4\text{ km}$$
$$6.9\text{ km} + 3(9.5\text{ km}) = 35.4\text{ km}$$
$$6.9\text{ km} + 28.5\text{ km} = 35.4\text{ km}$$
$$35.4\text{ km} = 35.4\text{ km}$$

21. Let x = the amount of time the skier spends on the ski lift in minutes.
Let 24 minutes $-x$ = the amount of time the skier spends skiing down the hill in minutes.

$$(50\text{ m/min})x = (150\text{ m/min})(24\text{ min} - x)$$
$$(50\text{ m/min})x = 3600\text{ m} - (150\text{ m/min})x$$
$$(200\text{ m/min})x = 3600\text{ m}$$
$$x = \frac{3600\text{ m}}{200\text{ m/min}}$$
$$x = 18\text{ min}$$

The length of the slope is 18 minutes × 50 m/minute = 900 m.

Check:
$$(50\text{ m/min})18\text{ min} = (150\text{ m/min})(24\text{ min} - 18\text{ min})$$
$$900\text{ m} = 3600\text{ m} - (150\text{ m/min})(18\text{ min})$$
$$900\text{ m} = 3600\text{ m} - 2700\text{ m}$$
$$900\text{ m} = 900\text{ m}$$

25. Let $x - 30.0$ s = time since the first car started moving in the race in seconds.
Let x = time since the second car started the race in seconds.
The distance travelled by each car will be the same at the point where the first car overtakes the second car.
Distance = speed × time.

$$79.0\text{ m/s}(x - 30.0\text{ s}) = 73.0\text{ m/s}(x)$$
$$(79.0\text{ m/s})x - (79.0\text{ m/s})(30.0\text{ s}) = (73.0\text{ m/s})$$
$$(79.0\text{ m/s})x - 2370\text{ m} = (73.0\text{ m/s})x$$
$$(6.0\text{ m/s})x = 2370\text{ m}$$
$$x = \frac{2370\text{ m}}{6.0\text{ m/s}}$$
$$x = 395\text{ s}$$

The first car will overtake the second car after 395 s.
The first car travels 79 m/s × (395 s – 30 s) = 28 835 m by this point. 8 laps around the track is 4.36 km/lap.
8 laps × 1000 m/km = 34 880 m, so the first car will already be in the lead at the end of the 8th lap.

Check:
$$79.0\text{ m/s}(395\text{ s} - 30.0\text{ s}) = 73.0\text{ m/s}(395\text{ s})$$
$$79.0\text{ m/s}(365\text{ s}) = 73.0\text{ m/s}(395\text{ s})$$
$$28\,835\text{ m} = 28\,835\text{ m}$$

29.

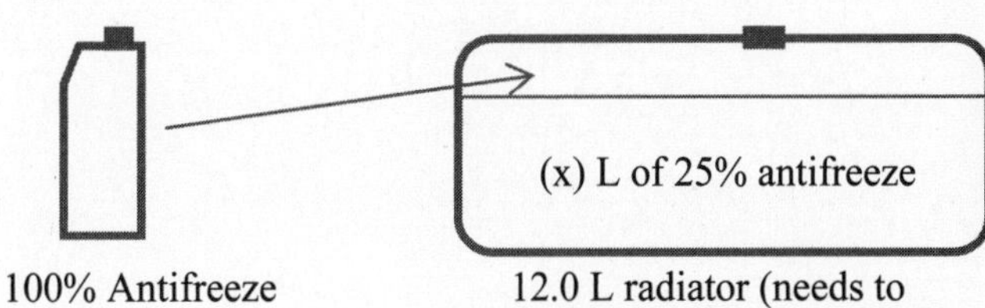

Let x = the amount in L of 25% antifreeze left in radiator
Let 12.0 L $- x$ = the amount of 100% antifreeze added in L.

$$0.25(x) + 1.00(12.0\text{ L} - x) = 0.5(12.0\text{ L})$$
$$0.25(x) + 12.0\text{ L} - 1.00(x) = 6.0\text{ L}$$
$$-0.75(x) = -6.0\text{ L}$$
$$x = \frac{-6.0\text{ L}}{-0.75}$$
$$x = 8.0\text{ L}$$

There needs to be 8L of 25% antifreeze left in radiator, so (12.0 L – 8.0 L) = 4.0 L must be drained.

Check:
$$0.25(8.0\text{ L}) + 1.00(12.0\text{ L} - 8.0\text{ L}) = 0.5(12.0\text{ L})$$
$$2\ .0\text{L} + 1.00(4.0\text{ L}) = 6.0\text{ L}$$
$$2.0\text{ L} + 4.0\text{ L} = 6.0\text{ L}$$
$$6.0\text{ L} = 6.0\text{ L}$$

Review Exercises

1. $(-2)+(-5)-3=-2-5-3=-10$

5. $-5-|2(-6)|+\frac{-15}{3}=-5-|-12|+(-5)=-5-12-5=-22$

9. $\sqrt{16}-\sqrt{64}=\sqrt{(4)(4)}-\sqrt{(8)(8)}=4-8=-4$

13. $(-2rt^2)^2=(-2)^2r^2t^{2\times2}=4r^2t^4$

17. $\frac{-16N^{-2}(NT^2)}{-2N^0T^{-1}}=\frac{8N^{-2+1}T^{2+1}}{(1)}=\frac{8N^{-1}T^3}{(1)}=\frac{8T^3}{N}$

21. 8840 has 3 significant digits. Rounded to 2 significant digits, it is 8800.

25.
$$\begin{aligned}37.3-16.92(1.067)^2&=37.3-16.92(1.138489)\\&=37.3-19.26323388\\&=18.03676612\end{aligned}$$
which rounds to 18.0.

29. $a-3ab-2a+ab=-2ab-a$

33.
$$\begin{aligned}(2x-1)(x+5)&=(2x)(x)+(2x)(5)+(-1)(x)+(-1)(5)\\&=2x^2+10x-x-5\\&=2x^2+9x-5\end{aligned}$$

37.
$$\begin{aligned}\frac{2h^3k^2-6h^4k^5}{2h^2k}&=\frac{2h^3k^2}{2h^2k}-\frac{6h^4k^5}{2h^2k}\\&=h^{3-2}k^{2-1}-3h^{4-2}k^{5-1}\\&=-3h^2k^4+hk\end{aligned}$$

41.
$$\begin{aligned}2xy-\{3z-[5xy-(7z-6xy)]\}&=2xy-\{3z-[5xy-7z+6xy)]\}\\&=2xy-\{3z-[11xy-7z]\}\\&=2xy-\{3z-11xy+7z\}\\&=2xy-\{10z-11xy\}\\&=2xy-10z+11xy\\&=13xy-10z\end{aligned}$$

45.
$$\begin{aligned}-3y(x-4y)^2&=-3y(x-4y)(x-4y)\\&=-3y[(x)(x)+(x)(-4y)\\&\quad+(-4y)(x)+(-4y)(-4y)]\\&=-3y[x^2-4xy-4xy+16y^2]\\&=-3y[x^2-8xy+16y^2]\\&=-3x^2y+24xy^2-48y^3\end{aligned}$$

49.
$$\begin{aligned}\frac{12p^3q^2-4p^4q+6pq^5}{2p^4q}&=\frac{12p^3q^2}{2p^4q}-\frac{4p^4q}{2p^4q}+\frac{6pq^5}{2p^4q}\\&=\frac{6q^{2-1}}{p^{4-3}}-\frac{2\cancel{p^4q}}{\cancel{p^4q}}+\frac{3q^{5-1}}{p^{4-1}}\\&=\frac{3q^4}{p^3}+\frac{6q}{p}-2\end{aligned}$$

53.
$$\begin{array}{r}x^2-2x+3\\3x-1\overline{)3x^3-7x^2+11x-3}\\\underline{3x^3-x^2\quad\quad\quad}\\-6x^2+11x\quad\\\underline{-6x^2+2x\quad}\\9x-3\\\underline{9x-3}\\0\end{array}$$

57.
$$\begin{aligned}-3\{(r+s-t)-2[(3r-2s)-(t-2s)]\}&=-3\{r+s-t-2[3r-2s-t+2s)]\}\\&=-3\{r+s-t-2[3r-t]\}\\&=-3\{r+s-t-6r+2t]\}\\&=-3\{-5r+s+t\}\\&=15r-3s-3t\end{aligned}$$

61.
$$\begin{aligned}3x+1&=x-8\\2x&=-9\\x&=-\frac{9}{2}\end{aligned}$$

65.
$$\begin{aligned}6x-5&=3(x-4)\\6x-5&=3x-12\\3x&=-7\\x&=-\frac{7}{3}\end{aligned}$$

69.
$$\begin{aligned}3t-2(7-t)&=5(2t+1).\\3t-14+2t&=10t+5\\5t-14&=10t+5\\-5t&=19\\t&=-\frac{19}{5}\end{aligned}$$

73. **(a)** 60 000 000 000 bytes = 6×10^{10} bytes

(b) 60 000 000 000 bytes = 60×10^9 bytes = 60 gigabytes

77. **(a)** 4.05×10^{13} km = 40 500 000 000 000 km

(b) 4.05×10^{13} km $= 40.5\times10^{12}$ km

$= 40.5\times10^{15}$ m

= 40.5 Pm

(Note that the symbol P stands for peta, which is the SI prefix associated with the multiple 10^{15}.)

81. **(a)** 1.5×10^{-1} Bq/L = 0.15 Bq/L

(b) 1.5×10^{-1} Bq/L $= 150\times10^{-3}$ mBq/L

85.
$$P = \frac{\pi^2 EI}{L^2}$$
$$L^2 P = \pi^2 EI$$
$$E = \frac{L^2 P}{\pi^2 I}$$

89.
$$d = (n-1)A$$
$$d = An - A$$
$$d + A = An$$
$$n = \frac{d+A}{A}$$

93.
$$R = \frac{A(T_2 - T_1)}{H}$$
$$HR = AT_2 - AT_1$$
$$AT_2 = HR + AT_1$$
$$T_2 = \frac{HR + AT_1}{A}$$

97. $\dfrac{5.25\times10^{10}\text{ bytes}}{6.4\times10^{4}\text{ bytes}} = 82\ 0312.5$

which rounds to 8.2×10^5. The newer computer's memory is 8.2×10^5 larger.

101.
$$\frac{R_1R_2}{R_1+R_2} = \frac{(0.0275\ \Omega)(0.0590\ \Omega)}{0.0275\ \Omega + 0.0590\ \Omega}$$
$$= \frac{0.0016225\ \Omega^2}{0.0865\ \Omega}$$
$$= 0.018757225\ \Omega$$

which rounds to 0.0188 Ω. The combined electric resistance is 0.0188 Ω.

105.
$$4(t+h) - 2(t+h)^2$$
$$= 4t + 4h - 2(t+h)(t+h)$$
$$= 4t + 4h - 2[(t)(t) + (t)(h) + (h)(t) + (h)(h)]$$
$$= 4t + 4h - 2[t^2 + 2ht + h^2]$$
$$= 4t + 4h - 2t^2 - 4ht - 2h^2$$
$$= -2t^2 - 2h^2 - 4ht + 4t + 4h$$

109.
$$x - (3 - x) = 2x - 3$$
$$x - 3 + x = 2x - 3$$
$$2x - 3 = 2x - 3$$

The equation is valid for all values of the unknown, so the equation is an identity.

113. $\dfrac{8\times10^{-3}}{2\times10^{4}} = 4\times10^{-7}$

117. Let $2x$ = the amount of oxygen produced in cm^3 by the first reaction.

Let x = the amount of oxygen produced in cm^3 by the second reaction.

Let $4x$ = the amount of oxygen produced in cm^3 by the third reaction.

$$2x + x + 4x = 560\text{ cm}^3$$
$$7x = 560\text{ cm}^3$$
$$x = \frac{560\text{ cm}^3}{7}$$
$$x = 80\text{ cm}^3$$

The first reaction produces (2 × 80 cm^3) = 160 cm^3 of oxygen, the second reaction produces 80 cm^3 of oxygen, and the third reaction produces (4 × 80 cm^3) = 320 cm^3 of oxygen.

Check: 160 cm^3 + 80 cm^3 + 320 cm^3 = 560 cm^{3*}

121. Let x = the time taken in hours for the crew to build 250 m of road.

The crew works at a rate of 450 m/12 h, which is 37.5 m/h. Time = distance/speed.

$$x = \frac{250\text{ m}}{37.5\text{ m/h}}$$
$$x = 6.666666667\text{ h}$$

which rounds to 6.7 h.

125. Let x = the number of litres of 0.50% grade oil used.

Let 1000L − x the number of litres of 0.75% grade oil used.

$$0.005(x)+0.0075(1000\text{ L}-x)=0.0065(1000\text{ L})$$
$$0.005(x)+7.5\text{ L}-0.0075(x)=6.5\text{ L}$$
$$-0.0025(x)=-1.0\text{ L}$$
$$x=\frac{-1.0\text{ L}}{-0.0025}$$
$$x=400\text{ L}$$

It will take 400 L of the 0.50% grade oil and (1000 L – 400 L) = 600 L of the 0.75% grade oil to make 1000 L of 0.65% grade oil.

Check:

$$0.005(400\text{ L})+0.0075(1000\text{ L}-400\text{ L})=0.0065(1000\text{ L})$$
$$2\text{ L}+4.5\text{ L}=6.5\text{ L}$$
$$6.5\text{ L}=6.5\text{ L}$$

129.

$$P=P_0+P_0rt$$
$$P-P_0=P_0rt$$
$$r=\frac{P-P_o}{P_0t}$$
$$r=\frac{\$7625-\$6250}{\$6250(4.000\text{ years})}$$
$$r=\frac{\$1375}{25\ 000}$$
$$r=0.055$$

The rate is equal to 5.500%.

On the calculator type:

$(7625-6250)/(6250\times4.000)$

CHAPTER 2

GEOMETRY

2.1 Lines and Angles

1. $\angle ABE = 90^\circ$ because it is a vertically opposite angle to $\angle CBD$ which is also a right angle.

5. $\angle EBD$ and $\angle DBC$ are acute angles (i.e., $< 90^\circ$).

9. The complement of $\angle CBD = 65^\circ$ is $\angle DBE$

$$\angle CBD + \angle DBE = 90^\circ$$
$$65^\circ + \angle DBE = 90^\circ$$
$$\angle DBE = 90^\circ - 65^\circ$$
$$\angle DBE = 25^\circ$$

13. $\angle AOB = \angle AOE + \angle EOB$

but $\angle AOE = 90^\circ$ because it is vertically opposite to $\angle DOF$ a given right angle,

and $\angle EOB = 50^\circ$ because it is vertically opposite to $\angle COF$ a given angle of 50°,

so $\angle AOB = 90^\circ + 50^\circ = 140^\circ$

17. $\angle 1$ is supplementary to 145°, so

$$\angle 1 = 180^\circ - 145^\circ = 35^\circ$$
$$\angle 2 = \angle 1 = 35^\circ$$

$\angle 4$ is vertically opposite to $\angle 2$, so

$$\angle 4 = \angle 2$$
$$\angle 4 = 35^\circ$$

21. $\angle 6 = 90 - 62^\circ$ since they are complementary angles

$$\angle 6 = 28^\circ$$

$\angle 3$ is an alternate-interior angle to $\angle 6$, so

$$\angle 3 = \angle 6$$
$$\angle 3 = 28^\circ$$

25. $\angle CBE = \angle BAD = 44^\circ$ because they are corresponding angles $\angle DEB$ and $\angle CBE$ are alternate interior angles, so

$$\angle DEB = \angle CBE$$
$$\angle DEB = 44^\circ$$

29. Using Eq. (2.1),

$$\frac{a}{4.75} = \frac{3.05}{3.20}$$
$$a = 4.75 \cdot \frac{3.05}{3.20}$$
$$a = 4.53 \text{ m}$$

33. $\angle BCE = 47^\circ$ since those angles are alternate interior angles.

$\angle BCD$ and $\angle BCE$ are supplementary angles

$$\angle BCD + \angle BCE = 180^\circ$$
$$\angle BCD = 180^\circ - 47^\circ$$
$$\angle BCD = 133^\circ$$

37. $\angle 1 + \angle 2 + \angle 3 = 180^\circ$, because $\angle 1$, $\angle 2$, and $\angle 3$ form a straight angle.

2.2 Triangles

1. $\angle 5 = 45^\circ$

$\angle 3 = 45^\circ$ since $\angle 3$ and $\angle 5$ are alternate interior angles.

$\angle 1$, $\angle 2$, and $\angle 3$ make a stright angle, so

$$\angle 1 + \angle 2 + \angle 3 = 180^\circ$$
$$70^\circ + \angle 2 + 45^\circ = 180^\circ$$
$$\angle 2 = 65^\circ$$

5.
$$\angle A + \angle B + \angle C = 180^\circ$$
$$\angle A + 40^\circ + 84^\circ = 180^\circ$$
$$\angle A = 56^\circ$$

9.
$$A = \frac{1}{2}bh$$
$$A = \frac{1}{2}(7.6)(2.2)$$
$$A = 8.4 \text{ m}^2$$

13. One leg can represent the base, the other leg the height.

$$A=\frac{1}{2}bh$$

$$A=\frac{1}{2}(3.46)(2.55)$$

$$A=4.41\text{ cm}^2$$

17. We add the lengths of the sides to get

$$p=205+322+415$$

$$p=942\text{ cm}$$

21. $c^2=a^2+b^2$

$$c=\sqrt{a^2+b^2}$$

$$c=\sqrt{13.8^2+22.7^2}$$

$$c=26.6\text{ mm}$$

25. All interior angles in a triangle add to 180°

$$23^\circ+\angle B+90^\circ=180^\circ$$

$$\angle B=180^\circ-90^\circ-23^\circ$$

$$\angle B=67^\circ$$

29.

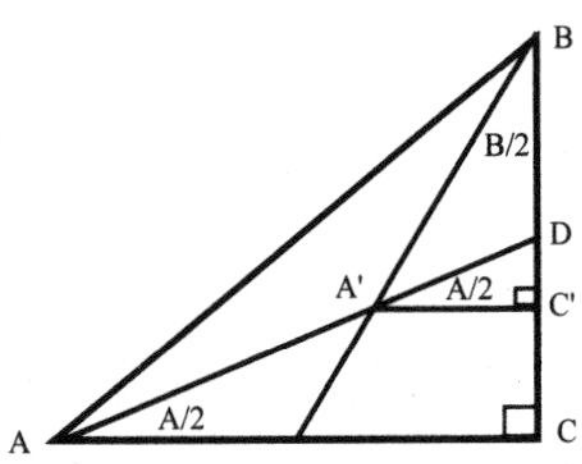

$$\Delta ADC\sim\Delta A'DC'$$

$$\angle DA'C'=A/2$$

$\angle BA'D=\angle$ between bisectors

From $\Delta BA'C'$, and all angles in a triangle must sum to 180°

$$\frac{B}{2}+(\angle BA'D+A/2)+90^\circ=180^\circ$$

$$\angle BA'D=90^\circ-\left(\frac{A}{2}+\frac{B}{2}\right)$$

$$\angle BA'D=90^\circ-\left(\frac{A+B}{2}\right)$$

But ΔABC is a right triangle, and all angles in a triangle must sum to 180°, so $A+B=90^\circ$

$$\angle BA'D=90^\circ-\frac{90^\circ}{2}$$

$$\angle BA'D=45^\circ$$

33.

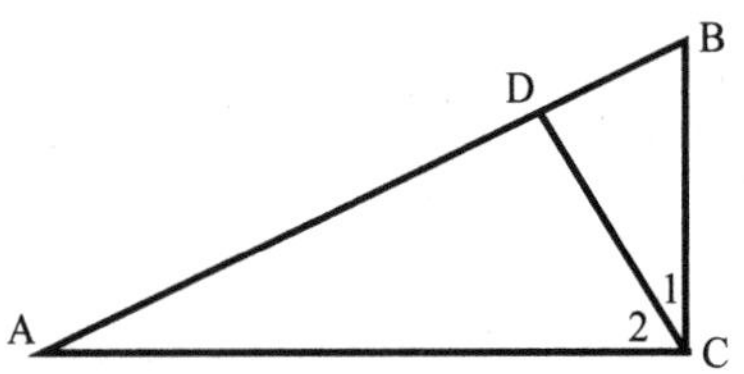

$$\angle A+\angle B=90^\circ$$

$$\angle 1+\angle B=90^\circ$$

$$\angle A=\angle 1$$

redraw ΔBDC as

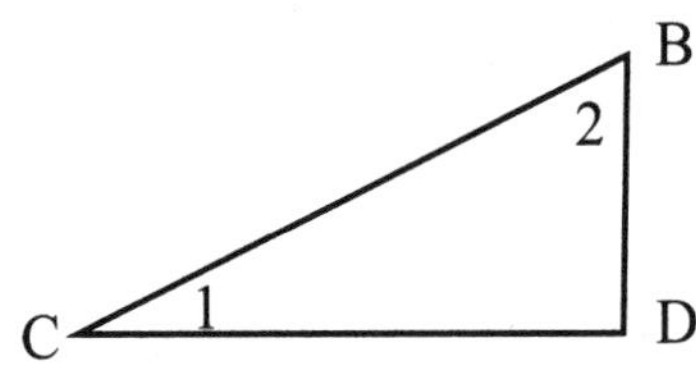

$$\angle 1+\angle 2=90^\circ$$

$$\angle 1+\angle B=90^\circ$$

$$\angle 2=\angle B$$

and ΔADC as

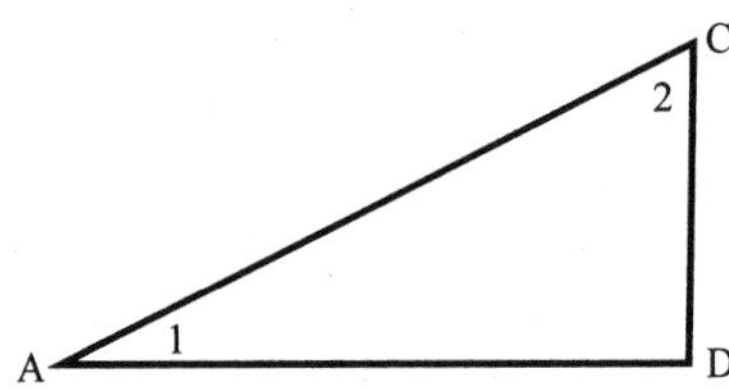

ΔBDC and ΔADC are similar.

37.

$$KM=KN-MN$$

$$KM=15-9$$

$$KM=6$$

Since $\Delta MKL\sim\Delta MNO$

$$\frac{LM}{KM}=\frac{OM}{MN}$$

$$\frac{LM}{6}=\frac{12}{9}$$

$$LM=\frac{(6)(12)}{9}$$

$$LM=8$$

41. $s = \frac{p}{2} = \frac{2(76.6)+30.6}{2} = 91.9 \text{ cm}$

By Hero's formula,

$$A = \sqrt{s(s-a)(s-b)(s-c)}$$

$$A = \sqrt{91.9(91.9-76.6)(91.9-76.6)(91.6-30.6)}$$

$$A = 1150 \text{ cm}^2$$

45.

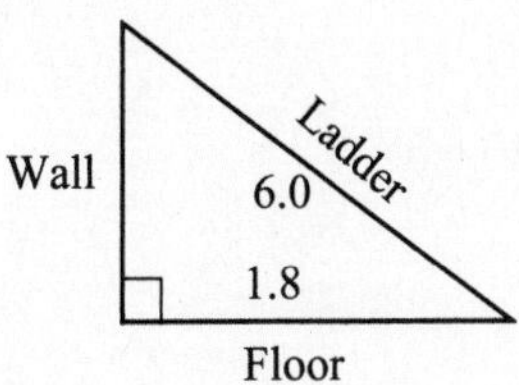

$$c^2 = a^2 + b^2$$

$$b = \sqrt{c^2 - a^2}$$

$$b = \sqrt{6.0^2 - 1.8^2}$$

$$b = 5.7 \text{ m}$$

49.

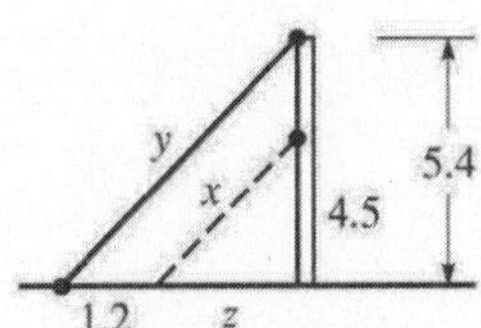

By Eq. (2.1),

$$\frac{z}{4.5} = \frac{1.2}{0.9}$$

$$z = \frac{(4.5)(1.2)}{0.9}$$

$$z = 6.0 \text{ m}$$

$$x^2 = z^2 + 4.5^2$$

$$x = \sqrt{56.25} \text{ m}$$

$$x = 7.5 \text{ m}$$

$$y^2 = (1.2+6)^2 + 5.4^2$$

$$x = \sqrt{81.0} \text{ m}$$

$$y = 9.0 \text{ m}$$

53. Redraw ΔBCP as

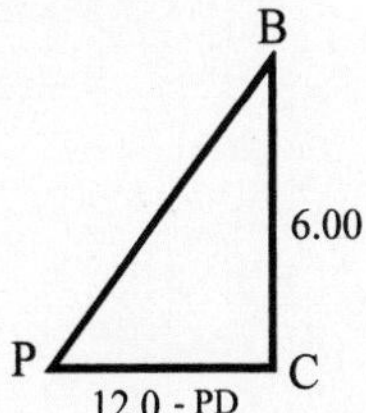

ΔAPD is

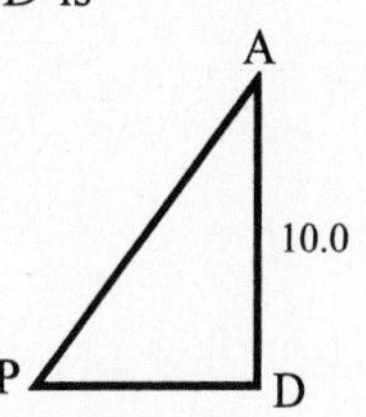

since $\Delta BCP \sim \Delta ADP$

$$\frac{6.00}{12.0-PD} = \frac{10.0}{PD}$$

$$6PD = 120 - 10PD$$

$$16PD = 120$$

$$PD = 7.50 \text{ km}$$

$$PC = 12.0 - PD$$

$$PC = 4.50 \text{ km}$$

$$l = PB + PA$$

$$l = \sqrt{4.50^2 + 6.00^2} + \sqrt{7.50^2 + 10.0^2}$$

$$l = 7.50 + 12.5$$

$$l = 20.0 \text{ km}$$

2.3 Quadrilaterals

1.

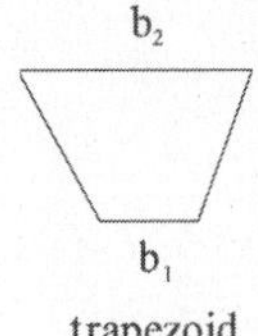

trapezoid

5. $p = 4s = 4(65) = 260 \text{ m}$

9. $p = 2l + 2w = 2(3.7) + 2(2.7) = 12.8 \text{ m}$

13. $A = s^2 = 2.7^2 = 7.3 \text{ mm}^2$

17. $A = bh = 3.7(2.5) = 9.2 \text{ m}^2$

21. $p = 2b + 4a$

25. The parallelogram is a rectangle.

29. The diagonal always divides the rhombus into two congruent triangles. All outer sides are always equal.

33.

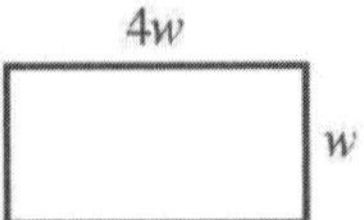

If width increases by 1500 mm and length decreases by 4500 mm the dimensions will be equal (a square).

$w + 1500 = 4w - 4500$

$6000 = 3w$

$w = 2000$ mm

$4w = 8000$ mm

37.

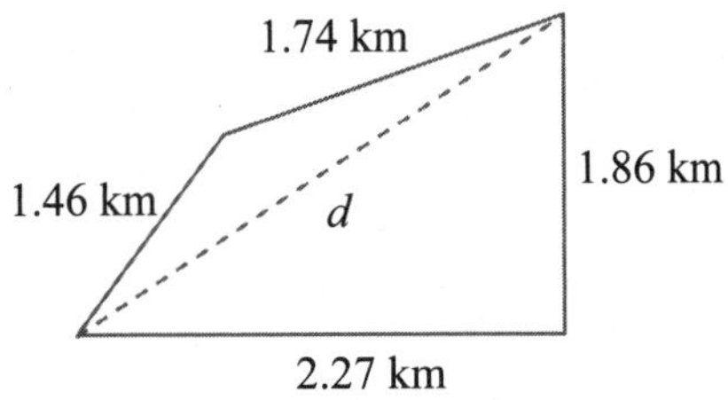

$d = \sqrt{2.27^2 + 1.86^2}$

$d = 2.934706$ km

For the right triangle,

$A = \frac{1}{2}bh$

$A = \frac{1}{2}(2.27)(1.86)$

$A = 2.1111 \text{ km}^2$

For obtuse triangle,

$s = \dfrac{1.46 + 1.74 + d}{2}$

$s = \dfrac{1.46 + 1.74 + 2.934706}{2}$

$s = 3.06735$ km

$A = \sqrt{s(s-1.46)(s-d)(s-1.74)}$

$A = \sqrt{3.06735(3.06735-1.46)(3.06735-2.934706)(3.06735-1.74)}$

$A = 0.931707 \text{ km}^2$

$A_{\text{quadrilateral}}$ = Sum of areas of two triangles

$A = 2.1111 \text{ km}^2 + 0.931707 \text{ km}^2$

$A = 3.04 \text{ km}^2$

2.4 Circles

1. $\angle OAB + OBA + \angle AOB = 180^\circ$

$\angle OAB + 90^\circ + 72^\circ = 180^\circ$

$\angle OAB = 18^\circ$

5. (a) AD is a secant line.

(b) AF is a tangent line.

9. $c = 2\pi r = 2\pi(275) = 1730$ cm

13. $A = \pi r^2 = \pi(0.0952)^2 = 0.0285 \text{ km}^2$

17. $\angle CBT = 90^\circ - \angle ABC = 90^\circ - 65^\circ = 25^\circ$

21. $\overset{\frown}{BC} = 2(60^\circ) = 120^\circ$

25. $022.5^\circ = 022.5^\circ\left(\dfrac{\pi \text{ rad}}{180^\circ}\right) = 0.393$ rad

29. Perimeter $= \dfrac{1}{4}(2\pi r) + 2r = \dfrac{\pi r}{2} + 2r$

33. All are on the same diameter.

37. $C = 2\pi r = 2\pi(6375) = 40\ 060$ km

41. $c = 112$

$c = \pi d$

$d = c / \pi$

$= 112 / \pi$

$= 35.7$ cm

45. A of room = A of rectangle $+ \dfrac{3}{4} A$ of circle

$A = 8100(12\ 000) + \dfrac{3}{4}\pi(320)^2$

$A = 9.7 \times 10^7 \text{ mm}^2$

49. $s = \theta r$

$s = (2.8)\left(\dfrac{450}{2} \text{ km}\right)$

$s = 630$ km

2.5 Measurement of Irregular Areas

1. The use of smaller intervals improves the approximation since the total omitted area or the total extra area is smaller. Also, since the number of intervals would be 10 (an even number) Simpson's Rule could be employed to achieve a more accurate estimate.

5. $A_{\text{trap}} = \frac{h}{2}[y_0 + 2y_1 + 2y_2 + ... + 2y_{n-1} + y_n]$

$$A_{\text{trap}} = \frac{2.0}{2}[0.0 + 2(6.4) + 2(7.4) + 2(7.0) + 2(6.1) + 2(5.2) + 2(5.0) + 2(5.1) + 0.0]$$

$A_{\text{trap}} = 84.4 = 84 \text{ m}^2$ to two significant digits

9. $A_{\text{trap}} = \frac{h}{2}[y_0 + 2y_1 + 2y_2 + ... + 2y_{n-1} + y_n]$

$$A_{\text{trap}} = \frac{0.5}{2}[0.6 + 2(2.2) + 2(4.7) + 2(3.1) + 2(3.6) + 2(1.6) + 2(2.2) + 2(1.5) + 0.8]$$

$$A_{\text{trap}} = 9.8 \text{ km}^2$$

13. $A_{\text{trap}} = \frac{h}{2}[y_0 + 2y_1 + 2y_2 + ... + 2y_{n-1} + y_n]$

$$A_{\text{trap}} = \frac{2.0}{2}[0 + 2(5.2) + 2(14.1) + 2(19.9) + 2(22.0) + 2(23.4) + 2(23.6) + 2(22.5) + 2(17.9) + 2(16.5) + 2(13.5) + 2(9.1) + 0]$$

$$A_{\text{trap}} = 375.4 \text{ km}^2 = 380 \text{ km}^2$$

17. $A_{\text{trap}} = \frac{h}{2}[y_0 + 2y_1 + 2y_2 + ... + 2y_{n-1} + y_n]$

$$A_{\text{trap}} = \frac{0.500}{2}[0.0 + 2(1.732) + 2(2.000) + 2(1.732) + 0.0]$$

$$A_{\text{trap}} = 2.73 \text{ cm}^2$$

This value is less than 3.14 cm^2 because all of the trapezoids are inscribed.

2.6 Solid Geometric Figures

1. $V_1 = lwh$

$$V_2 = (2l)(w)(2h)$$
$$V_2 = 4lwh$$
$$V_2 = 4V_1$$

The volume increases by a factor of 4.

5. $V = s^3$

$$V = (7.15 \text{ cm})^3$$
$$V = 366 \text{ cm}^3$$

9. $V = \frac{4}{3}\pi r^3$

$$V = \frac{4}{3}\pi(0.877 \text{ m})^3$$
$$V = 2.83 \text{ m}^3$$

13. $V = \frac{1}{3}Bh$

$$V = \frac{1}{3}(76 \text{ cm})^2(130 \text{ cm})$$
$$V = 250\ 293 \text{ cm}^3$$
$$V = 2.5\times10^5 \text{ cm}^3$$

17. $V = \frac{1}{2}\left(\frac{4}{3}\pi r^3\right)$

$$V = \frac{2\pi}{3}\left(\frac{d}{2}\right)^3$$
$$V = \frac{2\pi}{3}\left(\frac{0.83 \text{ cm}}{2}\right)^3$$
$$V = 0.14969 \text{ cm}^3$$
$$V = 0.15 \text{ cm}^3$$

21. $V = \frac{4}{3}\pi r^3$

$$V = \frac{4}{3}\pi\left(\frac{d}{2}\right)^3$$
$$V = \frac{4}{3}\pi\frac{d^3}{8}$$
$$V = \frac{1}{6}\pi d^3$$

25. $\frac{A_{final}}{A_{original}} = \frac{4\pi(2r)^2}{4\pi r^2}$

$$\frac{A_{final}}{A_{original}} = \frac{16\pi r^2}{4\pi r^2}$$
$$\frac{A_{final}}{A_{original}} = 4$$

29. $V = \pi r^2 h$

$$V = \pi\left(\tfrac{d}{2}\right)^2 h$$
$$V = \frac{\pi}{4}(0.76 \text{ m})^2(540\ 000 \text{ m})$$

$V = 244\ 969\ \text{m}^3$

$V = 2.4 \times 10^5\ \text{m}^3$

33. $V = \frac{4}{3}\pi r^3$

$V = \frac{4}{3}\pi\left(\frac{d}{2}\right)^3$

$V = \frac{4}{3}\pi\left(\frac{50.3}{2}\right)^3$

$V = 66\ 635\ \text{m}^3$

$V = 66\ 600\ \text{m}^3$

37. $c = 2\pi r$

$75.7 = 2\pi r$

$r = \frac{75.7}{2\pi}$

$V = \frac{4}{3}\pi r^3$

$V = \frac{4}{3}\pi\left(\frac{75.7}{2\pi}\right)^3$

$V = 7330\ \text{cm}^3$

Review Exercises

1. $\angle CGH$ and given angle 148° are corresponding angles, so

$\angle CGH = 148^\circ$

$\angle CGE$ and $\angle CGH$ are supplementary angles so

$\angle CGE + \angle CGH = 180^\circ$

$\angle CGE = 180^\circ - 148^\circ$

$\angle CGE = 32^\circ$

5. $c^2 = a^2 + b^2$

$c = \sqrt{9^2 + 40^2}$

$c = \sqrt{1681}$

$c = 41$

9. $c^2 = a^2 + b^2$

$c = \sqrt{6.30^2 + 3.80^2}$

$c = \sqrt{54.13}$

$c = 7.357309291$

$c = 7.36$

13. $p = 3s$

$p = 3(8.5\ \text{mm})$

$p = 25.5\ \text{mm}$

17. $c = 2\pi r$

$c = \pi d$

$c = \pi(98.4\ \text{mm})$

$c = 309.1327171\ \text{mm}$

$c = 309\ \text{mm}$

21. $V = Bh$

$V = \frac{1}{2}bl \cdot h$

$V = \frac{1}{2}(26.0\ \text{cm} \times 34.0\ \text{cm})(14.0\ \text{cm})$

$V = 6188\ \text{cm}^3$

$V = 6190\ \text{cm}^3$

25. $A = 6s^2$

$A = 6(0.520\ \text{m})^2$

$A = 1.6224\ \text{m}^2$

$A = 1.62\ \text{m}^2$

29. $\angle BTA = \frac{50^\circ}{2} = 25^\circ$

33. $\angle ABE$ and $\angle ADC$ are corresponding angles since $\Delta ABE \sim \Delta ADC$

$\angle ABE = \angle ADC$

$\angle ABE = 53^\circ$

37. p = base of triangle + hypotenuse of triangle + semicircle perimeter

$p = b + \sqrt{b^2 + (2a)^2} + \frac{1}{2}\pi(2a)$

$p = b + \sqrt{b^2 + 4a^2} + \pi a$

41. A square is a rectangle with four equal sides.
A rectangle is a parallelogram with perpendicular intersecting sides so a square is a parallelogram.
A rhombus is a parallelogram with four equal sides and since a square is a parallelogram,
a square is a rhombus.

45.

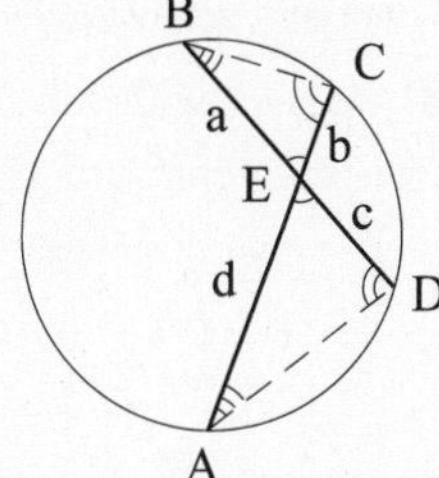

$\angle BEC = \angle AED$, since they are vertically opposite angles

$\angle BCA = \angle ADB$, both are inscribed in $\overset{\frown}{AB}$

$\angle CBD = \angle CAD$, both are inscribed in $\overset{\frown}{CD}$

which shows $\Delta AED \sim \Delta BEC$

$$\frac{a}{d} = \frac{b}{c}$$

49.

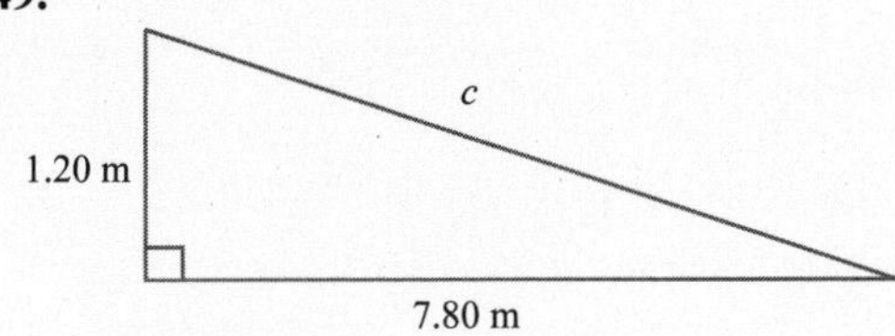

$$c^2 = a^2 + b^2$$

$$c = \sqrt{(1.20 \text{ m})^2 + (7.80 \text{ m})^2}$$

$$c = \sqrt{62.28 \text{ m}^2}$$

$$c = 7.891767863 \text{ m}$$

$$c = 7.89 \text{ m}$$

53.

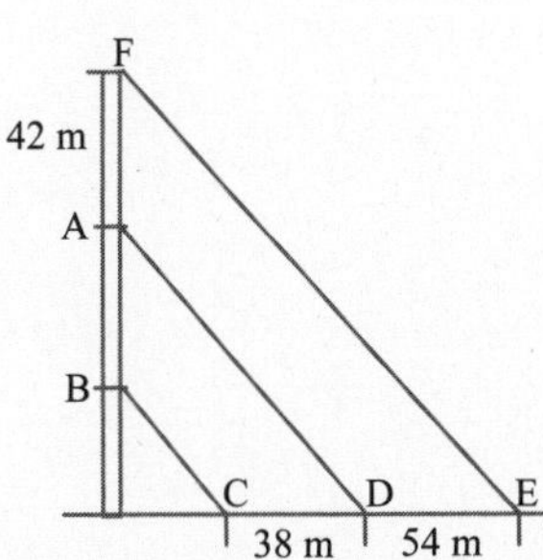

Since line segments BC, AD, and EF are parallel, the segments AB and CD are proportional to AF and DE

$$\frac{AB}{CD} = \frac{AF}{DE}$$

$$\frac{AB}{38 \text{ m}} = \frac{42 \text{ m}}{54 \text{ m}}$$

$$AB = \frac{38 \text{ m}(42 \text{ m})}{54 \text{ m}}$$

$$AB = 29.55555556 \text{ m}$$

$$AB = 30 \text{ m}$$

57. The longest distance between points on the photograph is

$$c^2 = a^2 + b^2$$

$$c = \sqrt{(20.0 \text{ cm})^2 + (25.0 \text{ cm})^2}$$

$$c = \sqrt{1025 \text{ cm}^2}$$

$$c = 32.01562119 \text{ cm}$$

Find the distance in km represented by the longest measure on the map

$$x = (32.01562119 \text{ cm})\left(\frac{18\,450}{1}\right)\left(\frac{1 \text{ m}}{100 \text{ cm}}\right)\left(\frac{1 \text{ km}}{1000 \text{ m}}\right)$$

$$x = 5.906882 \text{ km}$$

$$x = 5.91 \text{ km}$$

61. Area of the drywall is the area of the rectangle subtract the two circular cutouts.

$$A = lw - 2\left(\pi r^2\right)$$

$$A = lw - 2\left(\frac{\pi d^2}{4}\right)$$

$$A = lw - \frac{\pi d^2}{2}$$

$$A = (1200 \text{ mm})(2400 \text{ mm}) - \frac{\pi(350 \text{ mm})^2}{2}$$

$$A = 2\,687\,577.45 \text{ mm}^2$$

$$A = 2.7 \times 10^6 \text{ mm}^2$$

65. $V = \pi r^2 h$

$$V = \frac{\pi d^2}{4} h$$

$$V = \frac{\pi(4.32 \text{ m})^2}{4}(13.2 \text{ m})$$

$$V = 193.47787 \text{ m}^3$$

$$V = 193 \text{ m}^3$$

69. $V = V_{cylinder} + V_{dome}$

$$V = \pi r^2 h + \frac{1}{2}\left(\frac{4}{3}\pi r^3\right)$$

Note the height if the cylinder is the total height less the radius of the hemisphere.

$$V = \pi(0.380 \text{ m})^2 (2.05 \text{ m} - 0.380 \text{ m}) + \frac{2}{3}\pi(0.380 \text{ m})^3$$

$$V = 0.872512433 \text{ m}^3$$

Convert m^3 to L,

$$V = 0.872512433 \text{ m}^3 \left(\frac{1000 \text{ L}}{\text{m}^3}\right)$$

$$V = 872.512433 \text{ L}$$

$$V = 873 \text{ L}$$

73.

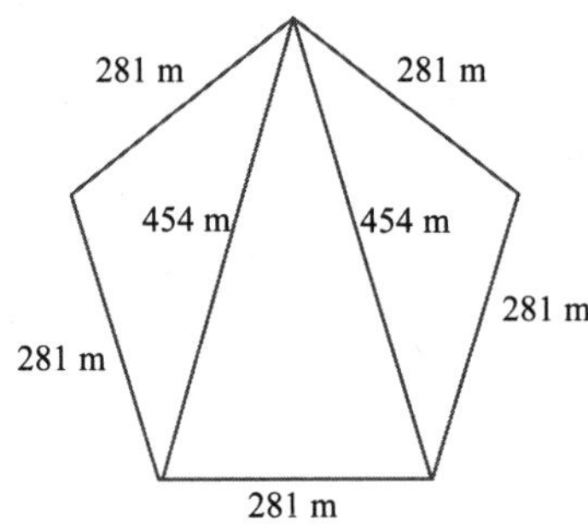

The area is the sum of the areas of three triangles, one with sides 454, 454, and 281 and two with sides 281, 281, and 454. The semi-perimeters are given by

$$s_1 = \frac{281 + 281 + 454}{2} = 508$$

$$s_2 = \frac{454 + 454 + 281}{2} = 594.5$$

$$A = 2\sqrt{508(508 - 281)(508 - 281)(508 - 454)}$$

$$+ \sqrt{594.5(594.5 - 454)(594.5 - 454)(594.5 - 281)}$$

$$A = 136\ 000 \text{ m}^2$$

CHAPTER 3

FUNCTIONS AND GRAPHS

3.1 Introduction to Functions

1. $f(x) = 3x - 7$

$f(-2) = 3(-2) - 7 = -13$

5. **(a)** $A(r) = \pi r^2$

(b) $A(d) = \pi\left(\dfrac{d}{2}\right)^2 = \dfrac{\pi d^2}{4}$

9. $A(s) = s^2$

$\sqrt{A} = \sqrt{s^2}$

$s(A) = \sqrt{A}$

13. $f(x) = 2x + 1$

$f(1) = 2(1) + 1 = 3$

$f(-1) = 2(-1) + 1 = -1$

17. $\phi(x) = \dfrac{6 - x^2}{2x}$

$\phi(2\pi) = \dfrac{6 - (2\pi)^2}{2(2\pi)} = \dfrac{6 - 4\pi^2}{4\pi} = \dfrac{3 - 2\pi^2}{2\pi} = -2.66$

$\phi(-2) = \dfrac{6 - (-2)^2}{2(-2)} = \dfrac{2}{-4} = -\dfrac{1}{2}$

21. $K(s) = 3s^2 - s + 6$

$K(-s) = 3(-s)^2 - (-s) + 6 = 3s^2 + s + 6$

$K(2s) = 3(2s)^2 - (2s) + 6 = 12s^2 - 2s + 6$

25. $f(x) = 5x^2 - 3x$

$f(3.86) = 5(3.86)^2 - 3(3.86) = 62.918 = 62.9$

$f(-6.92) = 5(-6.92)^2 - 3(-6.92) = 260.192$

$= 2.60 \times 10^2$

29. $f(x) = x^2 + 2$

Square the value of the independent variable and add 2 to the result.

33. $R(r) = 3(2r + 5) - 1$

Mulitply the value of the independent variable and then add 5.

Multiply this result by 3, then subtract 1.

37. $A = 5s^2$

$f(s) = 5s^2$

41. $s = f(t) = 17.5 - 4.9t^2$

$f(12) = 17.5 - 4.9(1.2)^2 = 10.4$ m

45. $d = rate \times t = 55t$

$d(t) = 55t$

3.2 More about Functions

1. $f(x) = -x^2 + 2$ is defined for all real values of x.

Domain: all real numbers $\mathbb{R}$ or $(-\infty, \infty)$

Since x^2 cannot be negative, the maximum value of $f(x)$ is 2.

Range: all real numbers $f(x) \le 2$, or $(-\infty, 2]$

5. $f(x) = x + 5$

Domain: all real numbers $\mathbb{R}$ or $(-\infty, \infty)$

Range: all real numbers $\mathbb{R}$ or $(-\infty, \infty)$

9. $f(s) = \dfrac{2}{s^2}$

$f(s)$ is not defined at $s = 0$ since it gives a division by zero error.

Domain: all real values $s \ne 0$, or $(-\infty, 0)$ and $(0, \infty)$

Since $\dfrac{2}{s^2}$ is always positive in this restricted domain

Range: all real numbers $f(s) > 0$, or $(0, \infty)$

13. $y = |x - 3|$

Domain: all real numbers $\mathbb{R}$ or $(-\infty, \infty)$

Absolute value is never negative, so

Range: all real numbers $y \ge 0$, or $[0, \infty)$

17. $f(D)=\dfrac{D}{D-2}+\dfrac{4}{D+4}$

Division by zero is undefined, so the domain must be restricted to exclude any value for which $D-2,\ D+4$ are equal to zero.

In this case, $D\neq 2,\ -4$.

Domain: all real numbers $D\neq 2,\ -4$, or $(-\infty,-4),\ (-4,\ 2)$, and $(2,\ \infty)$

21. $f(x)=\begin{cases} x+1 & \text{for } x<1 \\ \sqrt{x+3} & \text{for } x\geq 1 \end{cases}$

$f(1)=\sqrt{1+3}=\sqrt{4}=2$ (since $1\geq 1$)

$f(-0.25)=-0.25+1=0.75$ (since $-0.25<1$)

25. Weight w is expressed in Mg,

$w(t)=5500-2t$

29. For any length up to 50 m in length, it costs \$500. For every additional metre, \$5 is charged. But with the domain restriction that $l>50$ m,

If cost C is in dollars,

$C(l)=500+5(l-50)$

$C(l)=5l+250$

33. Profit is expressed in dollars.

$p=$ Profit from cell phones + Profit from DVD players

$2750=15x+25y$

$25y=2750-15x$

$y(x)=\dfrac{2750-15x}{25}$

37. The distance d from the helicopter is a function of the height h of the helicopter, related by the Pythagorean theorem.

$h^2+120^2=d^2$

$d(h)=\sqrt{h^2+14\ 400}$

since distance above ground is nonnegative,

Domain: all real values $h\geq 0$, or $(0,\ \infty)$

Distance d is a minium of 120 m since d is 120 m when h is 0, and when h increases, d increases.

Range: all real values $d\geq 120$ m, or $(120,\ \infty)$

41. $f=\dfrac{1}{2\pi\sqrt{C}}$

To avoid a division by zero error, $C\neq 0$.

And to avoid a negative in a square root error, $C>0$.

Domain: all real values $C>0$, or $(0,\ \infty)$

45.

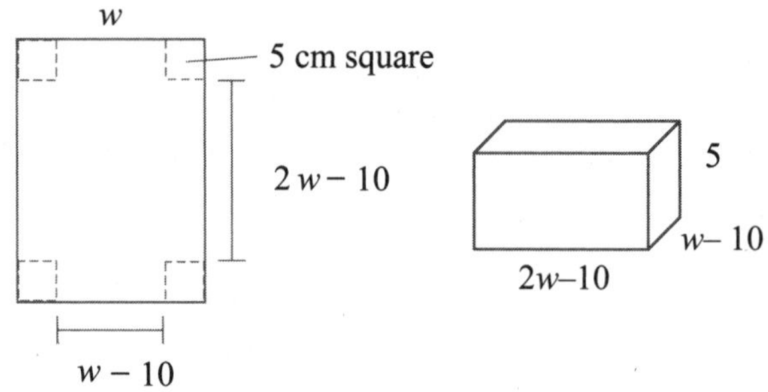

(a) $V=lwh$

$V(w)=(2w-10)(w-10)(5)$

$V(w)=\left(2w^2-20w-10w+100\right)(5)$

$V(w)=\left(2w^2-30w+100\right)(5)$

$V(w)=10w^2-150w+500$

(b) The width has to be larger than 10 cm to allow two 5-cm cutout squares to be removed from its length and still produce a box.

Domain: $w>10$ cm, or (10 cm, ∞)

49. $f(x)=|x|+|x-2|$,

Since $|x|\geq 0$ for all real x, and $|x-2|\geq 2$ for all real x,

$f(x)\geq 0+2=2$

Range: all real values $f(x)\geq 2$, or $[2,\ \infty)$

3.3 Rectangular Coordinates

1. $A(-1,-2),\ B(4,-2),\ C(4,1)$

The vertices of base AB both have y-coordinates of -2, which means the base CD which must be parallel, has the same y-coordinates for its vertices. Since at point C the y-coordinate is 1, then at D, y must also be 1.

In the same way, the x-coordinates of the left side must both be -1.

Therefore the fourth vertex is $D(-1, 1)$.

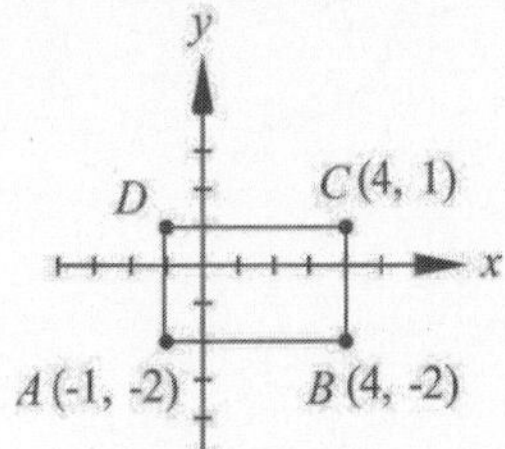

5.

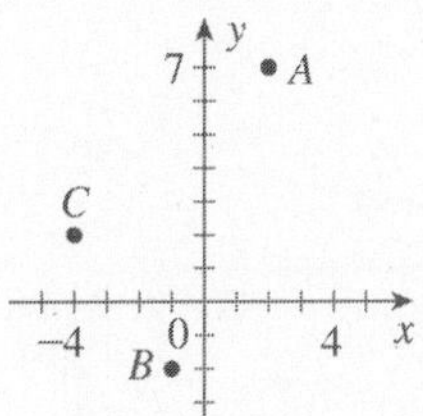

9. Figure $ABCD$ forms a rectangle.

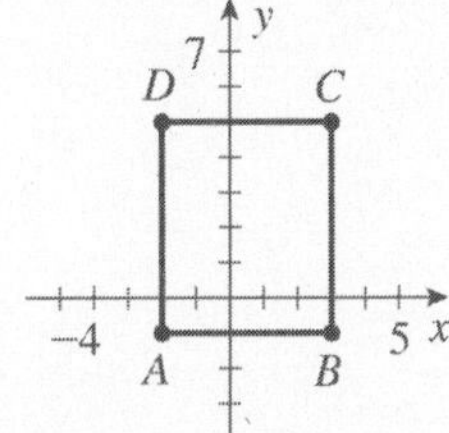

13. In order for the x-axis to be the perpendicular bisector of the line segment joining $P(3, 2)$ and $Q(x, y)$, Q must be located the same distance on the opposite side of the x-axis from P, and at the same x-coordinate so that the segment PQ is vertical (perpendicular to the x-axis). Therefore the point is $Q(3, -2)$.

17. All points with y-coordinate of 3 are on a horizontal line 3 units above the x-axis, passing through $(0, 3)$. The equation of this horizontal line is $y = 3$.

21. The x-coordinate of all points on the y-axis is zero, since all points on that line have form $(0, y)$, where y can be any real number.

25. All points for which $x < -1$ are in Quadrant II and Quadrant III to the left of the line $x = -1$, which is parallel to the y-axis, one unit to the left of the y-axis.

29. If $xy = 0$ then the product of the coordinates x and y must be zero.
Therefore, either coordinate may be zero.
For $x = 0$ points lie on the y-axis, and for $y = 0$ points lie on the x-axis.
So, all points where $xy = 0$ lie on the x- or y-axis.

33. (a) Distance between $(3, -2)$ and $(-5, -2)$ can be found because these two points have the same y-coordinate, they are at the same vertical position. The distance between the points will then just be the difference in the x-coordinates of the points.

$$d = 3-(-5) = 8$$

(b) Distance between $(3, -2)$ and $(3, 4)$ can be found because these two points have the same x-coordinate, they are at the same horizontal position. The distance between the points will then just be the difference in the y-coordinates of the points.

$$d = 4-(-2) = 6$$

3.4 The Graph of a Function

1. $f(x) = 3x + 5$

x	y
−3	−4
−2	−1
−1	2
0	5
1	8

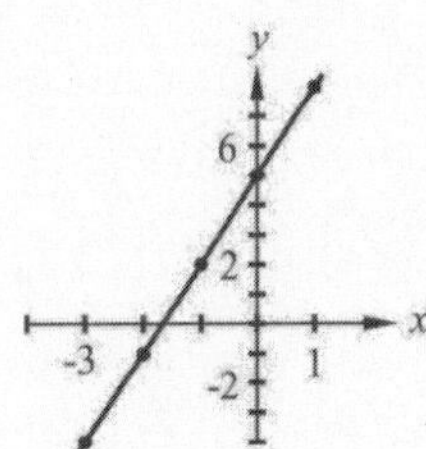

5. $y = 3x$

x	y
−1	−3
0	0
1	3

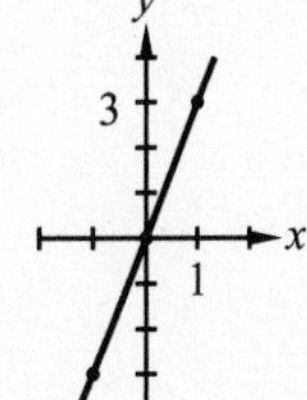

9. $s = 7 - 2t$

s	t
–1	9
0	7
1	5

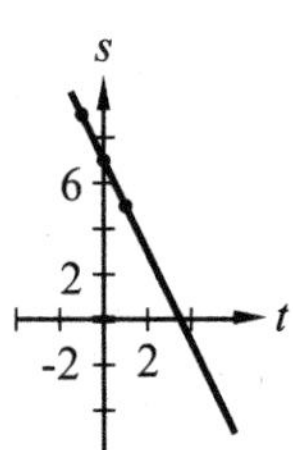

13. $y = x^2$

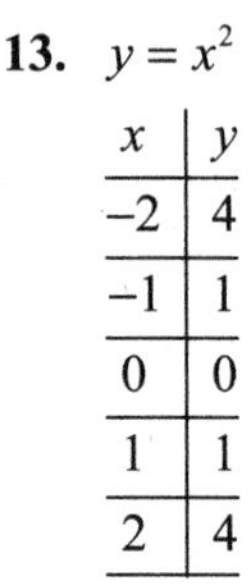

x	y
–2	4
–1	1
0	0
1	1
2	4

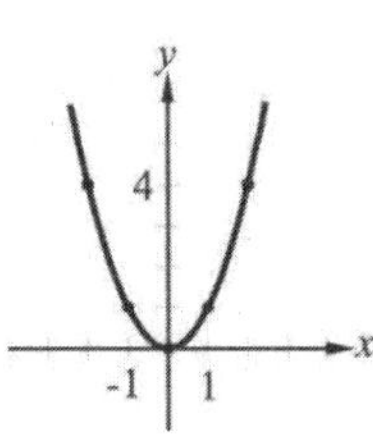

17. $y = \frac{1}{2}x^2 + 2$

x	y
–4	10
–2	4
0	2
2	4
4	10

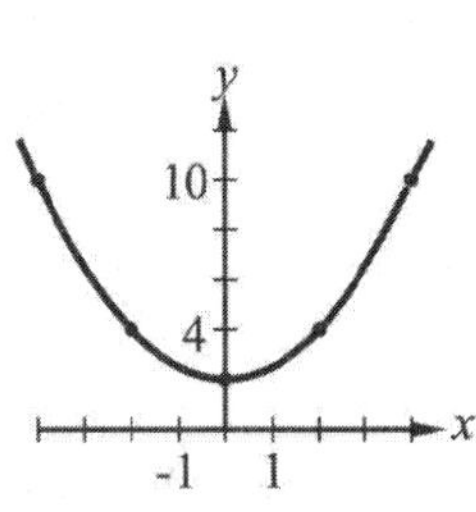

21. $y = x^2 - 3x + 1$

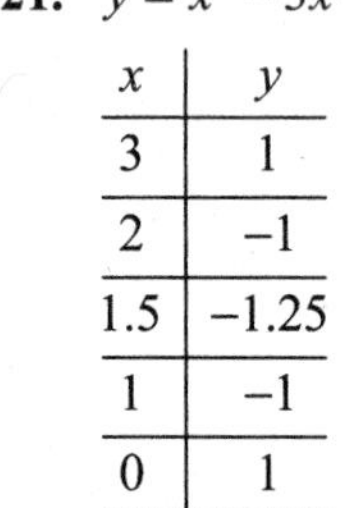

x	y
3	1
2	–1
1.5	–1.25
1	–1
0	1

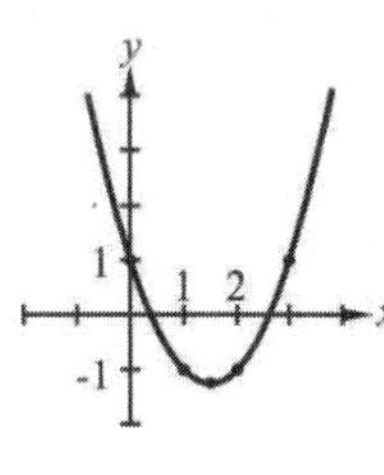

25. $y = x^3 - x^2$

x	y
–2	–12
–1	–2
0	0
2/3	–4/27
1	0
2	4

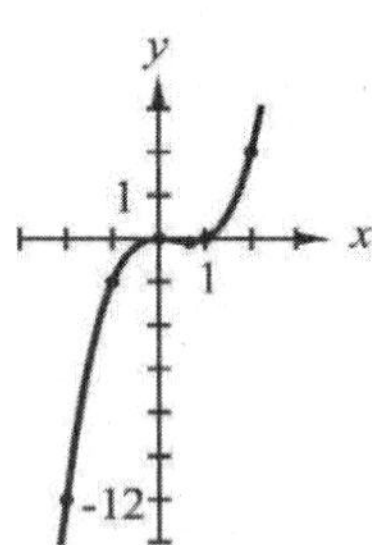

29. $P = \frac{1}{V} + 1$

V	P
–3	2/3
–2	1/2
–1	0
1	2
2	3/2
3	4/3

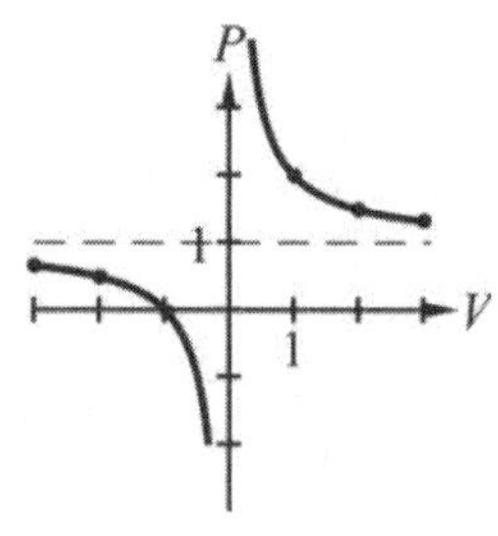

33. $y = \sqrt{x}$

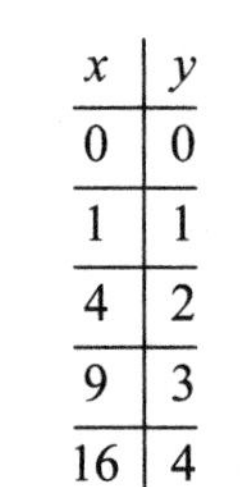

x	y
0	0
1	1
4	2
9	3
16	4

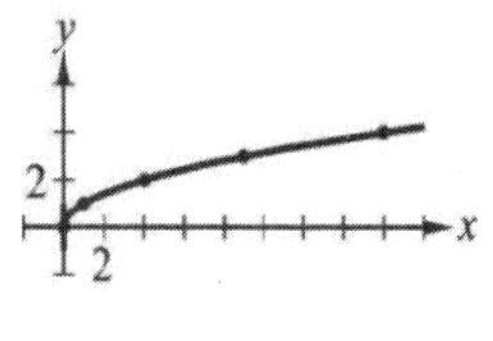

37. $n = 0.40m$

m (L)	n (L)
10	4
50	20
80	32

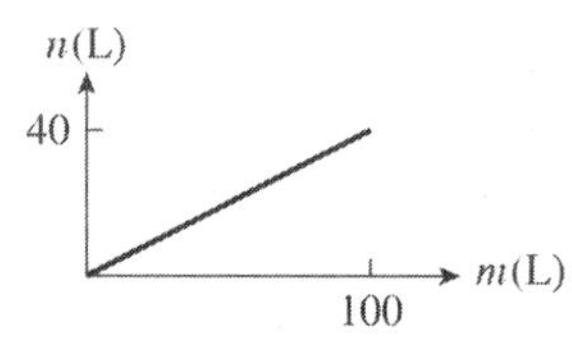

41. $H = 240I^2$

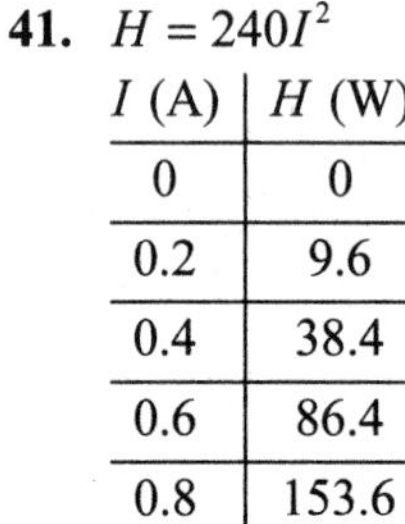

I (A)	H (W)
0	0
0.2	9.6
0.4	38.4
0.6	86.4
0.8	153.6

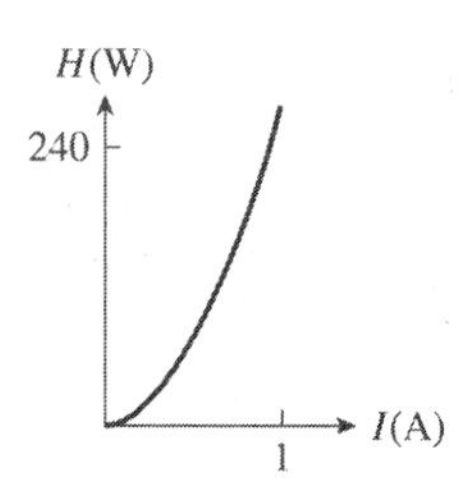

45. $P = 0.004v^3$

v (m/s)	P (W)
0	0
5	0.5
10	4.0
15	13.5
20	32.0

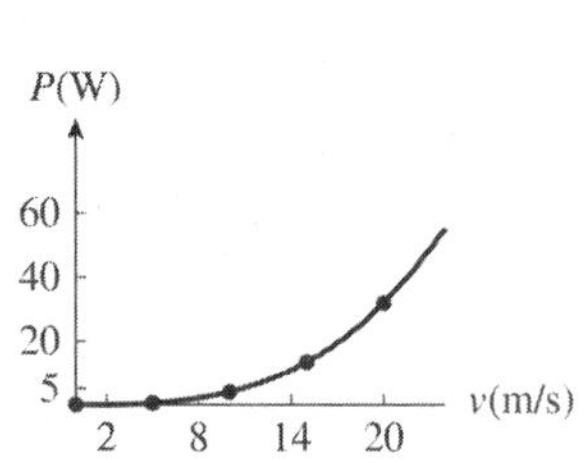

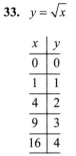

49. $P = 2l + 2w$

$200 = 2l + 2w$

$l = \dfrac{200 - 2w}{2}$

$l = 100 - w$

$A = lw$

$A = (100 - w)w$

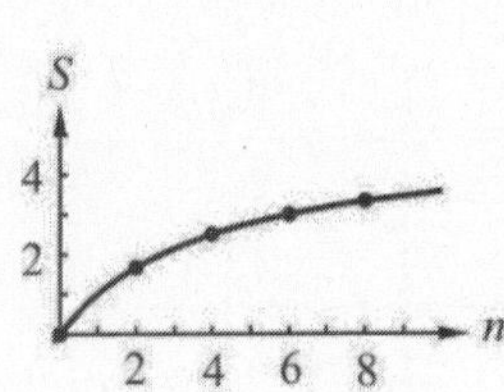

$A(w) = 100w - w^2$ for $30 \le w \le 70$

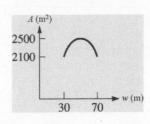

w (m)	30	40	50	60	70
A (m^2)	2100	2400	2500	2400	2100

53. $S = \dfrac{5n}{4+n}$

n	S
0	0
2	5/3
4	5/2
6	3
8	10/3

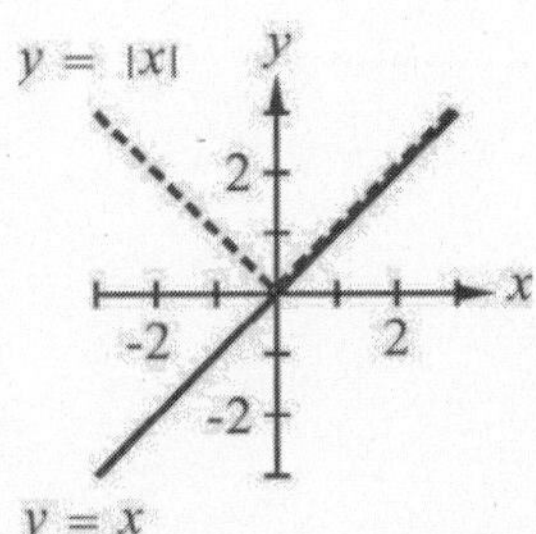

57.

x	y
−2	2
−1	1
0	0
1	1
2	2

$y = x$ is the same as $y = |x|$ for $x \ge 0$.

$y = |x|$ is the same as $y = -x$ for $x < 0$.

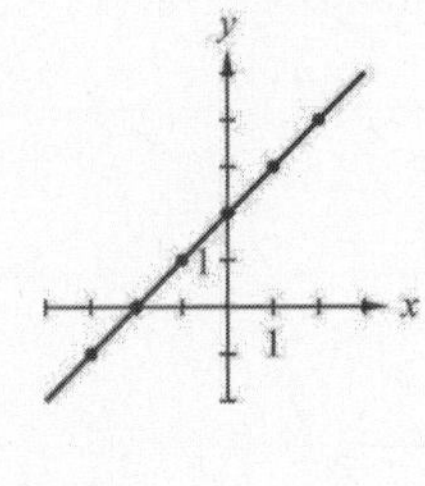

For negative values of x, $y = |x|$ becomes $y = -x$.

61. **(a)** $y = x + 2$

x	y
−3	−1
−2	0
−1	1
0	2
1	3
2	4

(b) $y = \dfrac{x^2 - 4}{x - 2}$

x	y
−3	−1
−2	0
−1	1
0	2
1	3
2	undefined

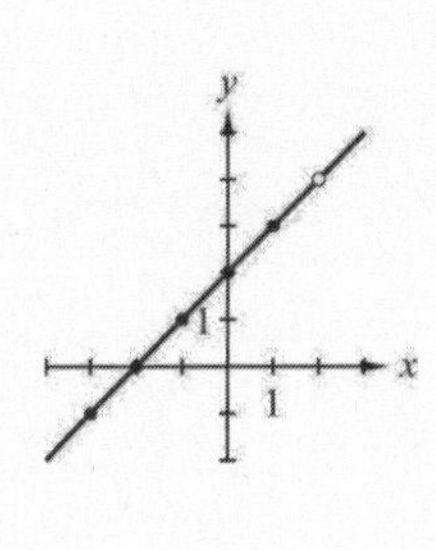

The graphs of $y = x + 2$ and $y = \dfrac{x^2 - 4}{x - 2}$ are identical except the second curve is undefined at a single point $x = 2$. It turns out that if you factor the second function (see Chapter 6), it will reduce to the first function everywhere except at $x = 2$ where it is undefined.

65. No. Some vertical lines will intercept the graph at multiple points when $x > 0$. The graph is that of a relation.

3.5 More about Graphs

1. $x^2 + 2x = 1$

$x^2 + 2x - 1 = 0$

Graph the following function, and estimate solutions.

$y = x^2 + 2x - 1$

x	y
−4	7
−3	2
−2	−1
−1	−2
0	−1
1	2
2	7

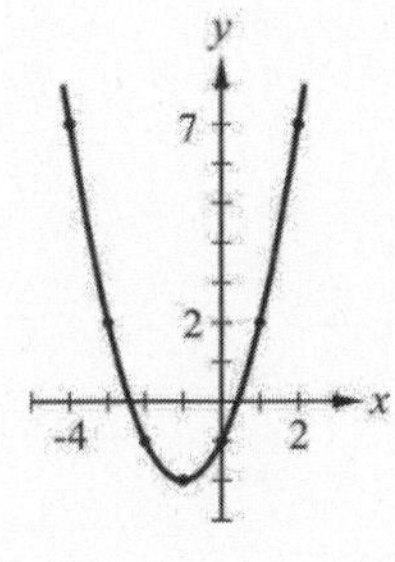

$x = -2.4,\ x = 0.4$

5. $y = 3x - 1$

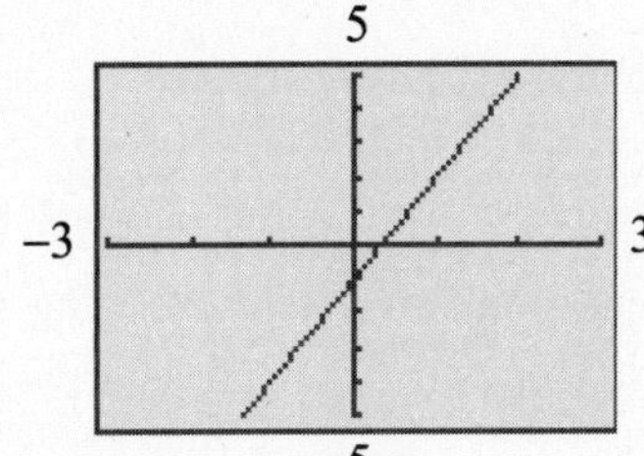

9. $y = 6 - x^3$

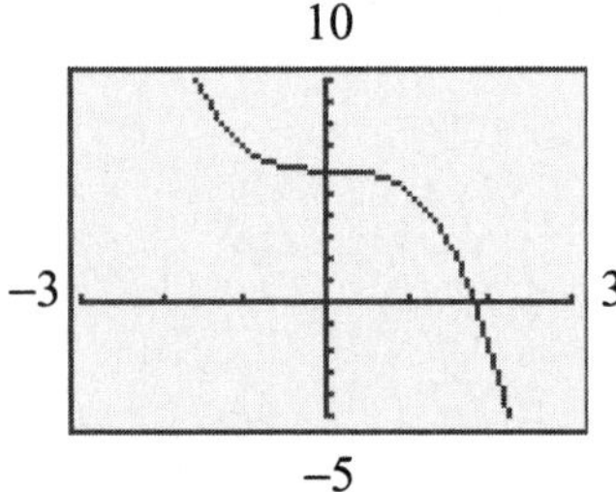

13. $y = \dfrac{2x}{x-2}$

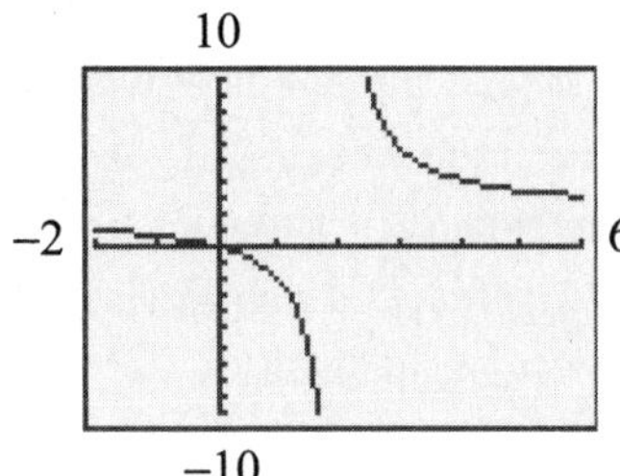

17. $y = 3 + \dfrac{2}{x}$

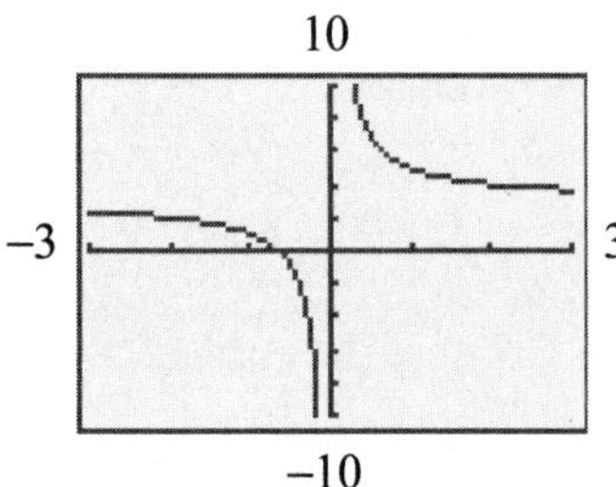

21. To solve $x^3 - 3 = 3x$ or $x^3 - 3x - 3 = 0$, graph $y = x^3 - 3x - 3$ and use the zero feature to solve.

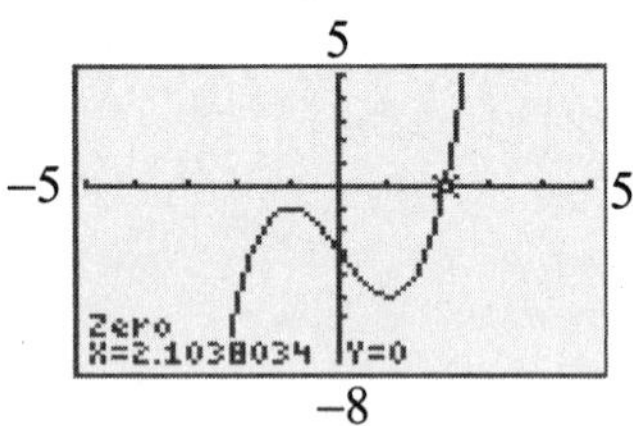

Solution $x = 2.104$

25. To solve $\sqrt{5R+2} = 3$ or $\sqrt{5R+2} - 3 = 0$ graph $y = \sqrt{5x+2} - 3$ and use the zero feature to solve.

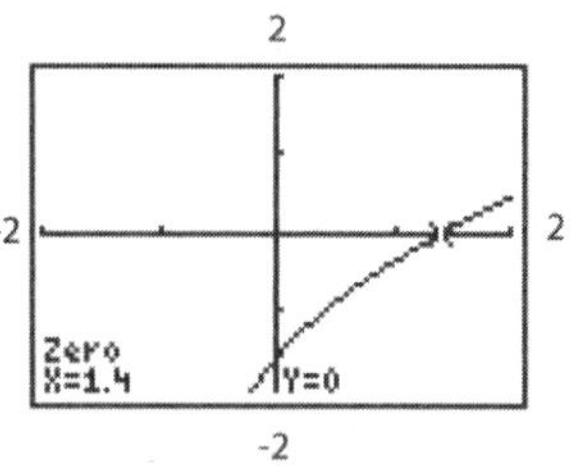

Solution $R = 1.400$

29. From the graph, $y = \dfrac{4}{x^2 - 4}$ has

Range: all real values $y > 0$ when $x < -2$ or $x > 2$

Range: all real values $y \le -1$ when $-2 < x < 2$

$x = \pm 2$ are vertical asymptotes for the function.

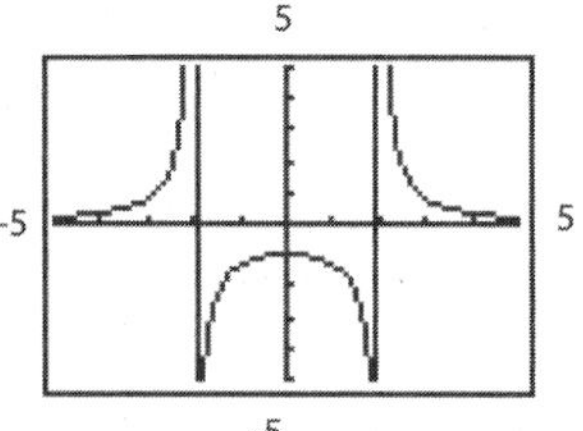

33. To find range of $Y(y) = \dfrac{y+1}{\sqrt{y-2}}$

graph $y = \dfrac{x+1}{\sqrt{x-2}}$ on the graphing calculator and use the minimum feature.

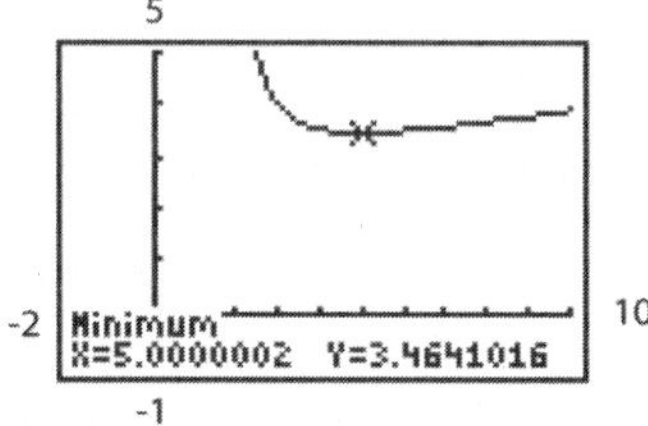

Range: all real values $Y(y) \ge 3.464$

$x = 2$ is a vertical asymptote for the function.

37. function: $y = 3x$

function shifted up 1: $y = 3x + 1$

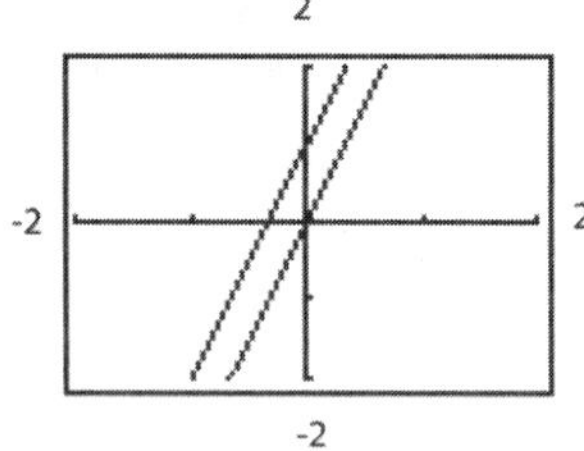

41. function: $y = -2x^2$

function shifted down 3, left 2: $y = -2(x+2)^2 - 3$

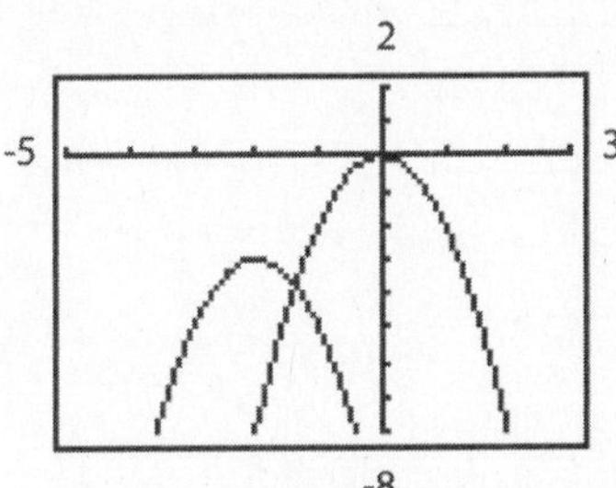

45. $i = 0.01v - 0.06$, graph $y = 0.01x - 0.06$

We need to solve for $i = 0$, so use the zero feature to solve.

From the graph, $v = 6.00$ V for $i = 0$.

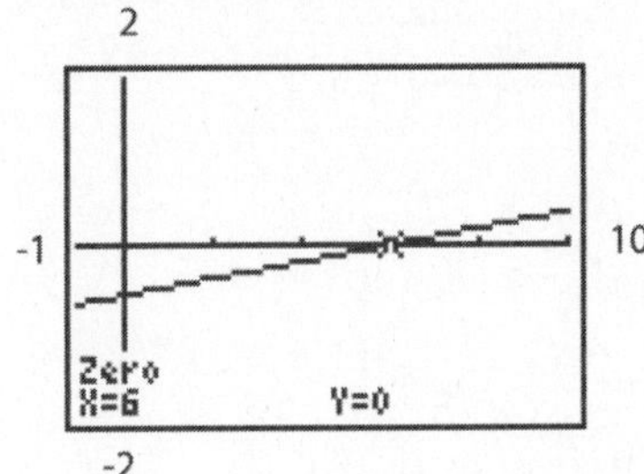

49. Let w = width of the panel

Let $l = w + 12$

$$A = lw$$

$$520 = (w+12)w$$

$$520 = w^2 + 12w$$

$$w^2 + 12w - 520 = 0$$

So, graph $y = x^2 + 12x - 520$ and use the zero feature to solve.

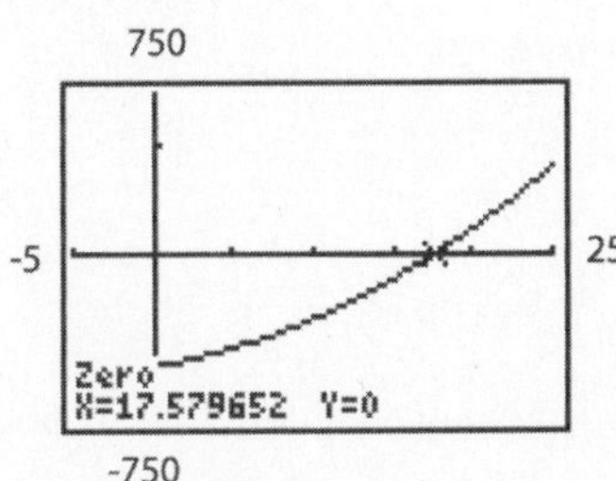

The approximate dimensions are $w = 17.6$ cm, $l = 29.6$ cm.

53. Graph $y = x^3 - 2.0000x^2 - 30.000$ and use the zero feature to solve.

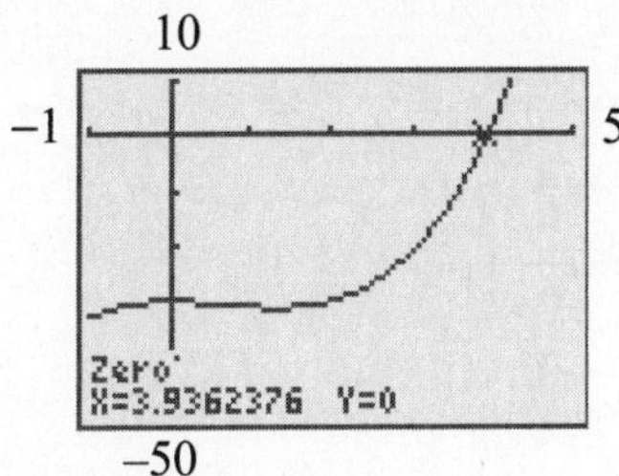

From the graph $x = 3.9362$.

The dimensions are

3.9362 cm, 3.9362 cm, and 1.9362 cm.

$V = (3.9362 \text{ cm})(3.9362 \text{ cm})(1.9362 \text{ cm})$

$V = 29.999 \text{ cm}^3$

57. (a) Since V is increasing at a constant rate and $V = \frac{4}{3}\pi r^3$ it seems reasonable that r^3 should be increasing at a constant rate.

Thus, r should be proportional to the cube root of time, $r = k\sqrt[3]{t}$. The graph will look similar to the graph shown below.

(b) $r = \sqrt[3]{3t}$ would be a typical situation.

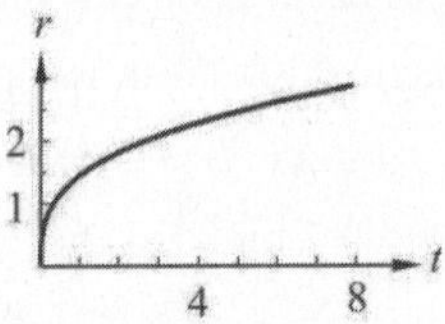

3.6 Graphs of Functions Defined by Tables of Data

1.

Week	Production
1	765
2	780
3	840
4	850
5	880
6	840
7	760
8	820

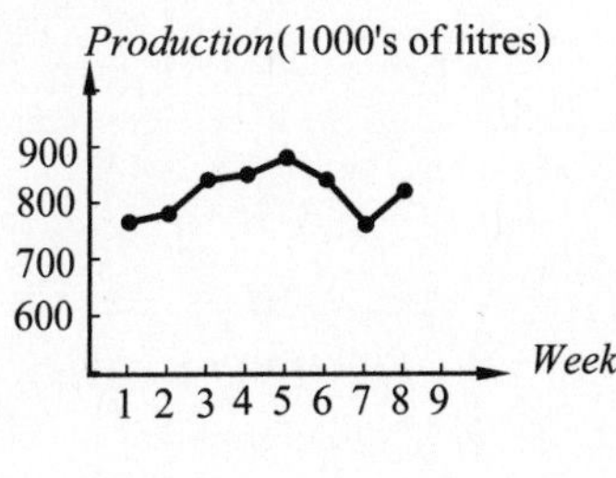

5.

Dist. (cm)	M. ind. (H)
0.0	0.77
2.0	0.75
4.0	0.61
6.0	0.49
8.0	0.38
10.0	0.25
12.0	0.17

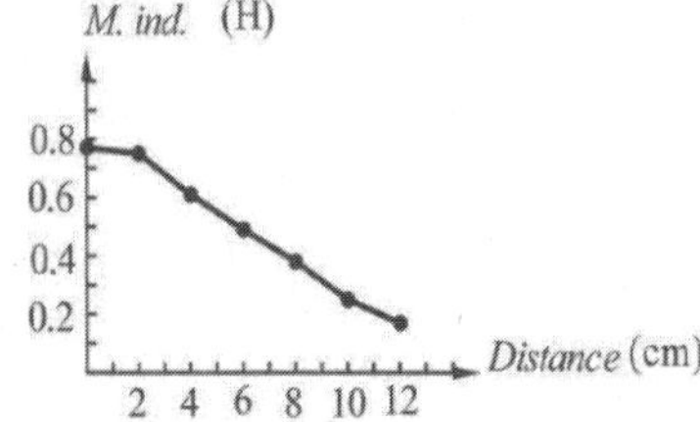

9. **(a)** Estimate 0.3 of the interval between 4 and 5 on the t-axis and mark it.

Draw a line vertically from this point to the graph. From the point where it intersects the graph, draw a horizontal line to the T-axis which it crosses at 132°C.

Therefore for $t = 4.3$ min, $T = 132^\circ\text{C}$

(b) Draw a line horizontally from $T = 145.0^\circ\text{C}$ to the point where it intersects the graph.

Draw a vertical line from this point to the t-axis, which it crosses between 0 and 1.

Estimate $t = 0.7$ min. So, for $T = 145.0^\circ\text{C}$, $t = 0.7$ min

13.

$$2\left[\begin{array}{ll} 1.2\left[\begin{array}{ll} 8.0 & 0.38 \\ 9.2 & ? \end{array}\right] x \\ \quad 10.0 \quad 0.25 \end{array}\right] -0.13$$

$$\frac{1.2}{2} = \frac{x}{-0.13}$$

$$x = \frac{(1.2)(-0.13)}{2}$$

$$x = -0.078$$

$$M.ind. = 0.38 + (-0.078)$$

$$M.ind. = 0.30\ \text{H}$$

17.

Height (cm)	Rate (m^3/s)
0	0
50	1.0
100	1.5
200	2.2
300	2.7
400	3.1
600	3.5

(a) For $R = 2.0\ \text{m}^3/\text{s}$, $H = 170$ cm

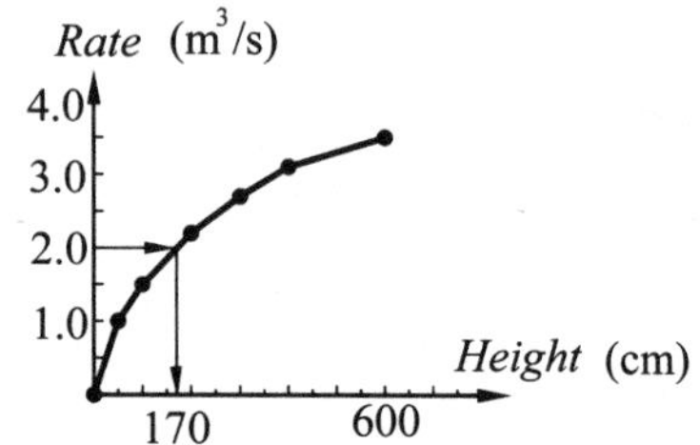

(b) For $H = 240$ cm, $R = 2.4\ \text{m}^3/\text{s}$

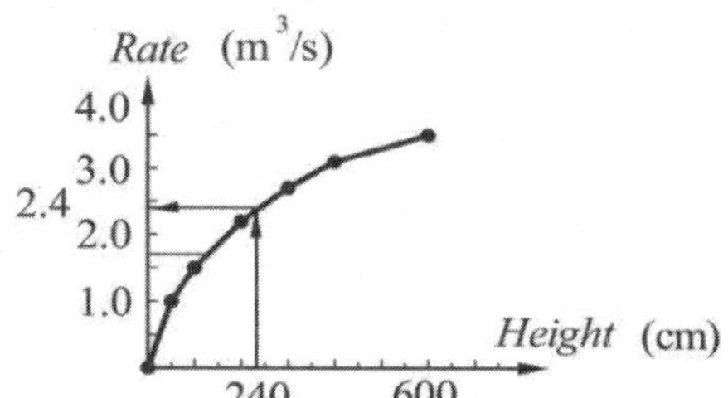

21.

$$10\left[\begin{array}{ll} 6\left[\begin{array}{ll} 30 & 0.30 \\ 36 & ? \end{array}\right] x \\ \quad 40 \quad 0.37 \end{array}\right] 0.07$$

$$\frac{6}{10} = \frac{x}{0.07}$$

$$x = \frac{6(0.07)}{10}$$

$$x = 0.042$$

$$f = 0.30 + 0.042$$

$$f = 0.34$$

25. The graph is extended using a straight line segment.

$T \approx 130.3^\circ\text{C}$ for $t = 5.3$ min

Review Exercises

1. $A = \pi r^2$

If $r = 2 \text{ m/s} \times t$

$A(t) = \pi(2t)^2$

$A(t) = 4\pi t^2$

5. $f(x) = 7x - 5$

$f(3) = 7(3) - 5 = 21 - 5 = 16$

$f(-6) = 7(-6) - 5 = -42 - 5 = -47$

9. $F(x) = x^3 + 2x^2 - 3x$

$F(3+h) - F(3) = (3+h)^3 + 2(3+h)^2 - 3(3+h) - ((3)^3 + 2(3)^2 - 3(3))$

$F(3+h) - F(3) = (3+h)(9+6h+h^2) + 2(9+6h+h^2) - 9 - 3h - (27 + 2(9) - 9)$

$F(3+h) - F(3) = (27+18h+3h^2+9h+6h^2+h^3) + 18+12h+2h^2-9-3h-27-18+9$

$F(3+h) - F(3) = h^3 + 11h^2 + 36h$

13. $f(x) = 8.07 - 2x$

$f(5.87) = 8.07 - 2(5.87)$

$= 8.07 - 11.74 = -3.67$

$f(-4.29) = 8.07 - 2(-4.29)$

$= 8.07 + 8.58 = 16.65 = 16.6$

17. $f(x) = x^4 + 1$

There are no restrictions to the value of x, so

Domain: all real numbers, or $(-\infty, \infty)$

Because x^4 is always positive, the minimum value of $f(x)$ is 1.

Range: all real numbers $f(x) \ge 1$, or $[1, \infty)$

21. $f(n) = 1 + \dfrac{2}{(n-5)^2}$

To avoid a division by zero error, $n \ne 5$, so

Domain: all real numbers $n \ne 5$, or $(-\infty, 5)$ and $(5, \infty)$

Since $(n-5)^2 > 0$, for all $n \ne 5$, no value of n will produce $f(n) \le 1$

Range: all real numbers $f(n) > 1$, or $(1, \infty)$

25. The graph of $s = 4t - t^2$ is

t	s
−1	−5
0	0
1	3
2	4
3	3
4	0
5	−5

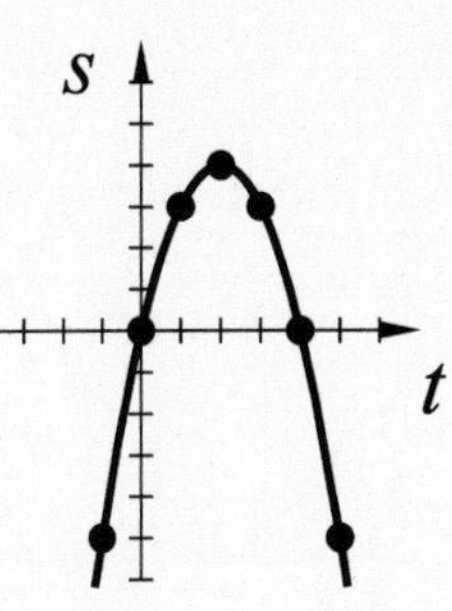

29. The graph of $A = 2 - s^4$ is

A	s
−2	−14
−1	1
0	2
1	1
0	−14

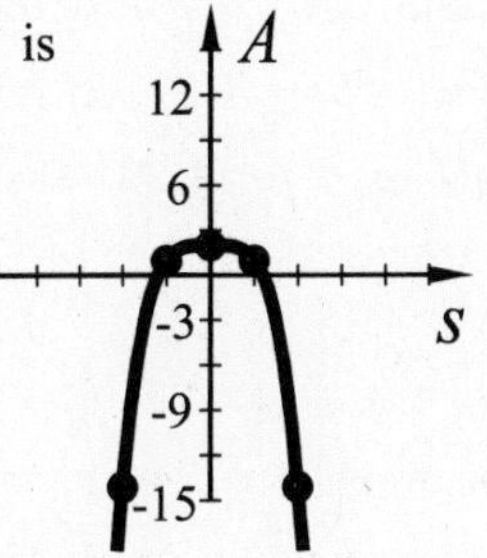

33. To solve $7x - 3 = 0$, graph $y = 7x - 3$ and use the zero feature to solve.

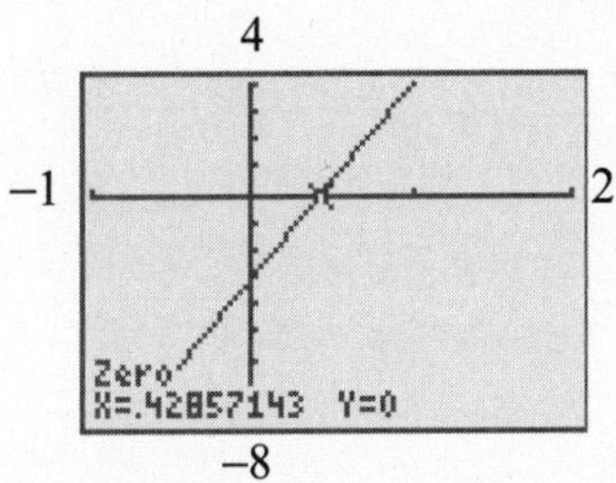

Solution $x = 0.43$

37. $x^3 - x^2 = 2 - x$

$x^3 - x^2 + x - 2 = 0$

Graph $y = x^3 - x^2 + x - 2$ and use the zero feature to solve.

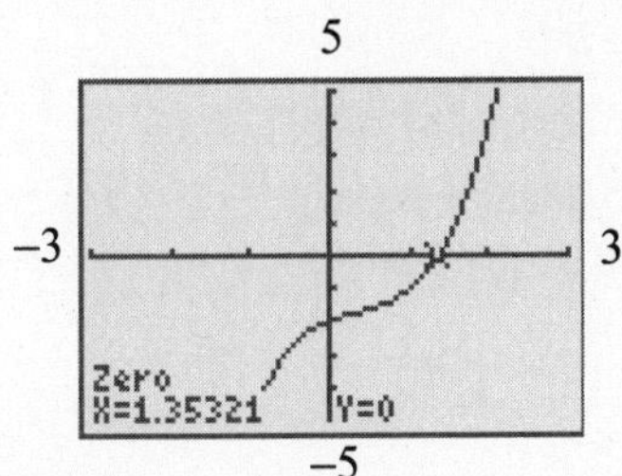

Solution $x = 1.35$

41. Graph $y = x^4 - 5x^2$ and use the minimum feature,
Range: all real values $y \geq -6.25$ or $[-6.25, \infty)$

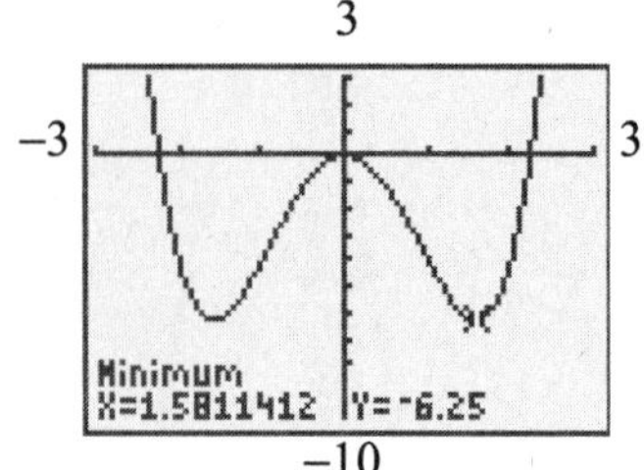

45. When a and b have opposite signs, then $A(a, b)$ and $B(b, a)$ will be in different quadrants.
If $A(a, b) = A(2, -3)$ it is in Quadrant IV but $B(b, a) = B(-3, 2)$ is in Quadrant II.

49. $\left|\dfrac{y}{x}\right| > 0$ for $\dfrac{y}{x} \neq 0$, so $y \neq 0$ to make the fraction nonzero, and $x \neq 0$ to avoid a division by zero error.
$\left|\dfrac{y}{x}\right| > 0$ for all values of (x, y) that are not on x-axis or y-axis.

53. $y = \sqrt{x-1}$ shifted left 2 and up 1 is
$y = \sqrt{(x+2)-1} + 1$
$y = \sqrt{x+1} + 1$

57. $(2, 3), (5, -1), (-1, 3)$
None of these points are vertically above one another (i.e. none have the same x-coordinate with different y-coordinates. Therefore this passes the vertical line test, and the three points could all lie on the same function.

61. $A = 8.0 + 12t^2 - 2t^3$
Graph $y = 8.0 + 12x^2 - 2x^3$ for $0 \leq x \leq 6$ and use the maximum feature.
The maximum angle is 72°.

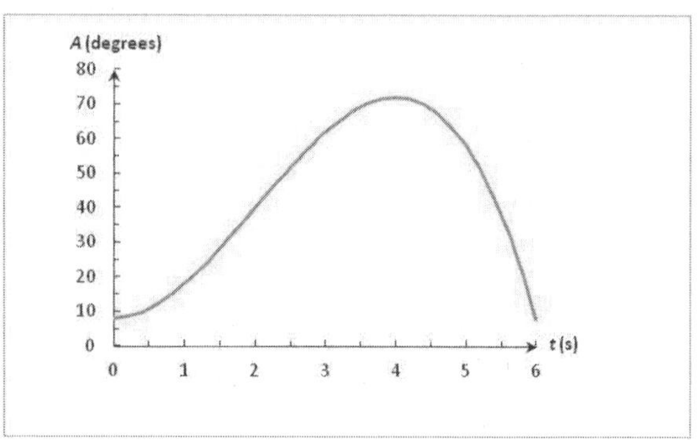

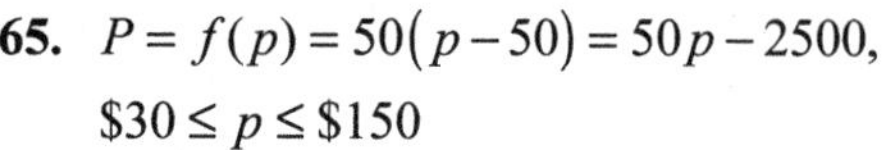

65. $P = f(p) = 50(p-50) = 50p - 2500$,
$\$30 \leq p \leq \$150$

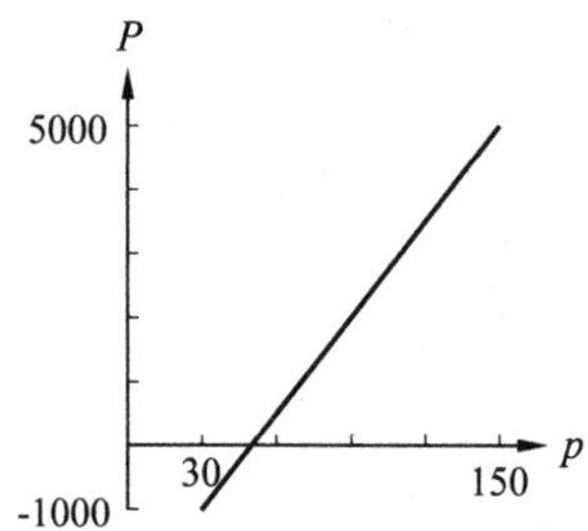

69. $\varepsilon = f(T) = \dfrac{100(T^4 - 307^4)}{307^4}$

$f(309) = \dfrac{100(309^4 - 307^4)}{307^4}$

$f(309) = \dfrac{100(233\ 747\ 360)}{8\ 882\ 874\ 001}$

$f(309) = 2.631438428$

$f(309) = 2.63\%$

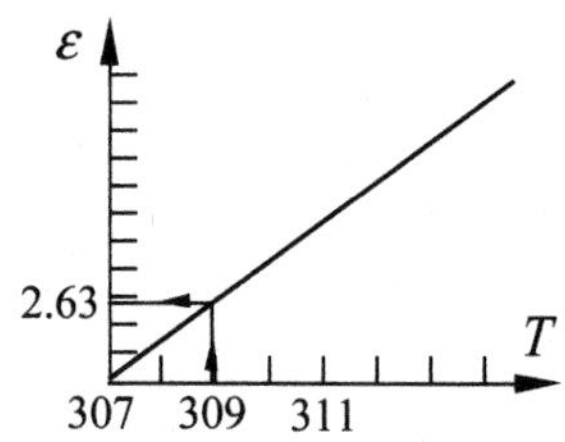

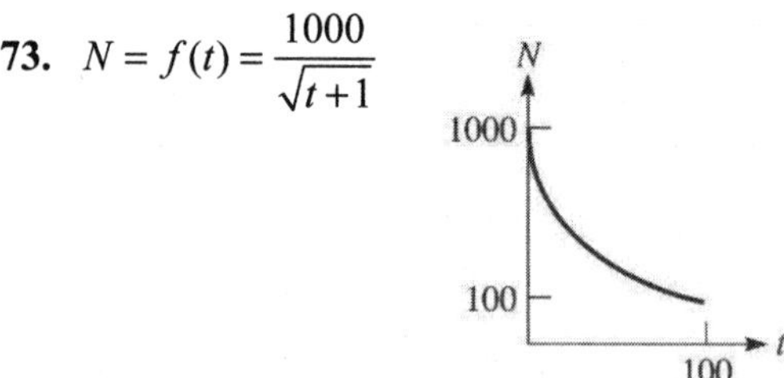

73. $N = f(t) = \dfrac{1000}{\sqrt{t+1}}$

77. Since $f(14)$ lies between the 10 °C and 20 °C

The distance between 10 and 20 is 10.

The distance between 10 and 14 is 4.

The distance between the corresponding distances is $3.57 - 3.38 = 0.19$

$$10\begin{bmatrix} 4\begin{bmatrix} 10 & 3.38 \\ 14 & ? \end{bmatrix} x \\ 20 \quad 3.57 \end{bmatrix} 0.19$$

$$\text{so, } \frac{x}{4} = \frac{0.19}{10}$$

$$x = \frac{(0.19)(4)}{10}$$

$$x = 0.076$$

$$f(14) = 3.38 + 0.076 = 3.46 \text{ m}$$

81.

$$s = 135 + 4.9T + 0.19T^2$$

$$500 = 135 + 4.9T + 0.19T^2$$

$$0.19T^2 + 4.9T - 365 = 0$$

Graph $y = 0.19x^2 + 4.9x - 365$ and use the zero feature to solve.

Then $T = 32.8$ °C, the negative temperature is likely too low to be physical.

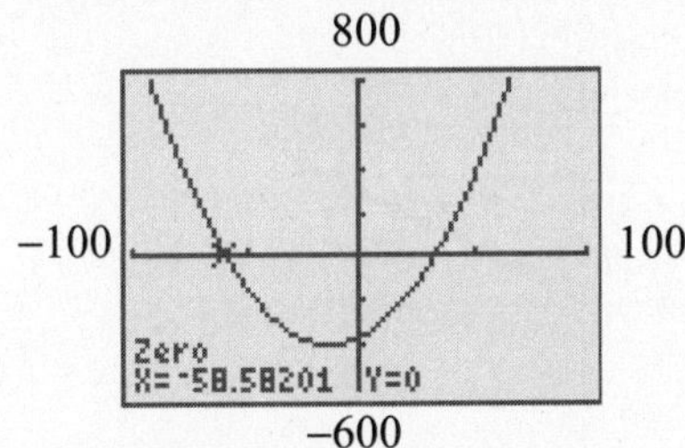

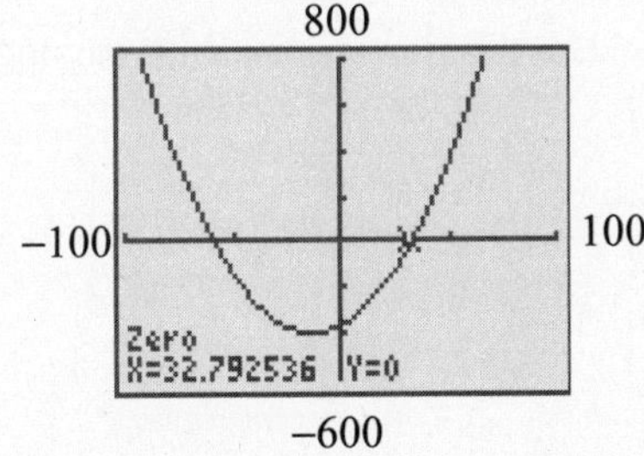

85.

$$T = \frac{4t^2}{t+2} - 20,\ t \geq 0$$

$$0 = \frac{4t^2}{t+2} - 20$$

$$4t^2 - 20(t+2) = 0$$

$$4t^2 - 20t - 40 = 0$$

$$t^2 - 5t - 10 = 0$$

Graph $y = x^2 - 5x - 10$ for $x \geq 0$ and use the zero feature to solve.

Solution $t = 6.53$ hours

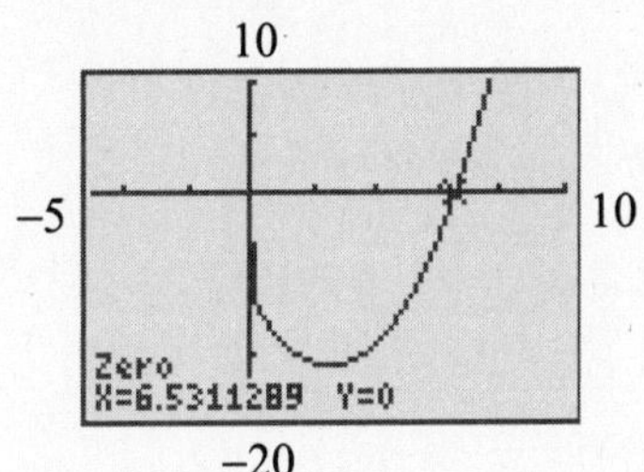

CHAPTER 4

THE TRIGONOMETRIC FUNCTIONS

4.1 Angles

1. $145.6^\circ + 2(360^\circ) = 865.6^\circ$, or

$145.6^\circ - 2(360^\circ) = -574.4^\circ$

5. $60^\circ, 120^\circ, -90^\circ$

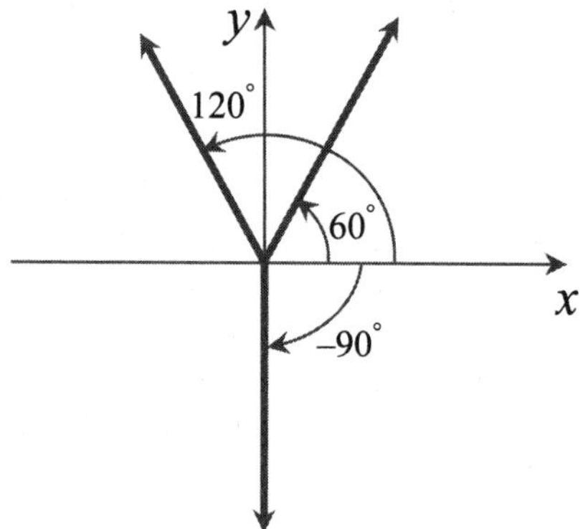

9. positive: $45^\circ + 360^\circ = 405^\circ$

negative: $45^\circ - 360^\circ = -315^\circ$

13. positive: $70^\circ 30' + 360^\circ = 430^\circ 30'$

negative: $70^\circ 30' - 360^\circ = -289^\circ 30'$

17. $0.265 \text{ rad } = 0.265 \text{ rad}\left(\dfrac{180^\circ}{\pi \text{ rad}}\right) = 15.18^\circ$

21. With calculator in Degree mode, enter 0.329^r (ANGLE menu #3 on TI-83+) $0.329 \text{ rad} = 18.85^\circ$

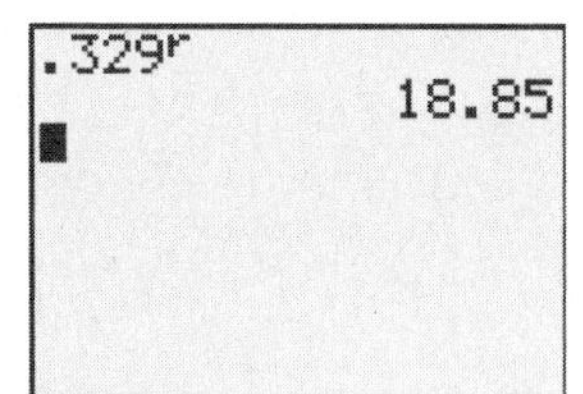

25. With calculator in Radian mode, enter 56.0° (ANGLE menu #1 on TI-83+) $56.0^\circ = 0.977$ rad

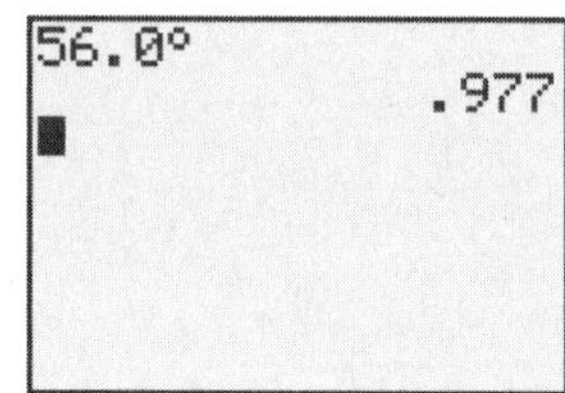

29. $47.50^\circ = 47^\circ + 0.50^\circ\left(\dfrac{60'}{1^\circ}\right) = 47^\circ + 30' = 47^\circ 30'$

33. $15^\circ 12' = 15^\circ + 12'\left(\dfrac{1^\circ}{60'}\right) = 15^\circ + 0.20^\circ = 15.20^\circ$

37. Angle in standard position terminal side passing through $(4, 2)$.

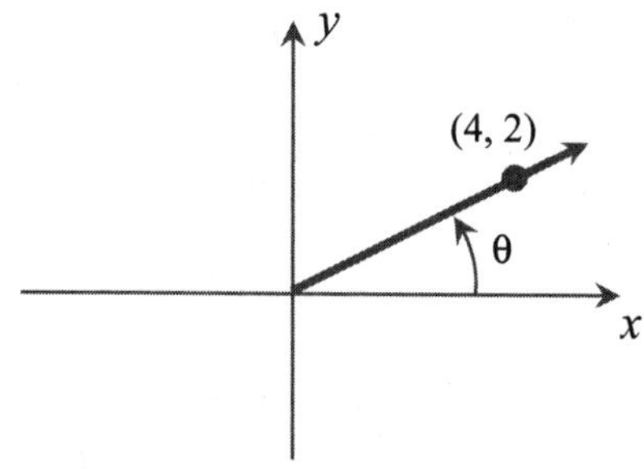

41. Angle in standard position terminal side passing through $(-7, 5)$.

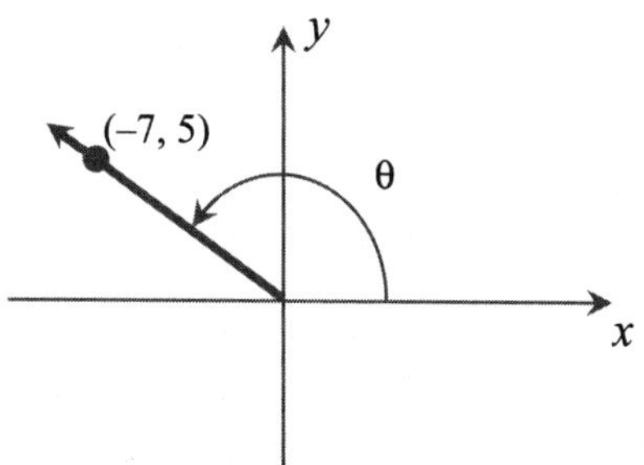

45. 31° lies in Quadrant I, so it is a first-quadrant angle. 310° lies in Quadrant IV, so it is a fourth-quadrant angle.

49. $1 \text{ rad} = 57.3^\circ$ lies in Quadrant I, so it is a first-quadrant angle.
$2 \text{ rad} = 114.6^\circ$ lies in Quadrant II, so it is a second-quadrant angle.

53. $21^\circ 42' 36'' = 21^\circ + 42'\left(\frac{1^\circ}{60'}\right) + 36''\left(\frac{1'}{60''}\right)\left(\frac{1^\circ}{60'}\right)$

$21^\circ 42' 36'' = 21^\circ + 0.7^\circ + 0.01^\circ$

$21^\circ 42' 36'' = 21.710^\circ$

4.2 Defining the Trigonometric Functions

1. For a point (4, 3) on the terminal side,

$x = 4,\ y = 3$

$r = \sqrt{4^2 + 3^2} = \sqrt{25} = 5$

$\sin\theta = \frac{y}{r} = \frac{3}{5}$ $\quad \csc\theta = \frac{r}{y} = \frac{5}{3}$

$\cos\theta = \frac{x}{r} = \frac{4}{5}$ $\quad \sec\theta = \frac{r}{x} = \frac{5}{4}$

$\tan\theta = \frac{y}{x} = \frac{3}{4}$ $\quad \cot\theta = \frac{x}{y} = \frac{4}{3}$

5. For a point (15, 8) on the terminal side,

$x = 15,\ y = 8$

$r = \sqrt{15^2 + 8^2} = \sqrt{289} = 17$

$\sin\theta = \frac{y}{r} = \frac{8}{17}$ $\quad \csc\theta = \frac{r}{y} = \frac{17}{8}$

$\cos\theta = \frac{x}{r} = \frac{15}{17}$ $\quad \sec\theta = \frac{r}{x} = \frac{17}{15}$

$\tan\theta = \frac{y}{x} = \frac{8}{15}$ $\quad \cot\theta = \frac{x}{y} = \frac{15}{8}$

9. For a point $(1, \sqrt{15})$ on the terminal side,

$x = 1,\ y = \sqrt{15}$

$r = \sqrt{1^2 + \left(\sqrt{15}\right)^2} = \sqrt{16} = 4$

$\sin\theta = \frac{y}{r} = \frac{\sqrt{15}}{4}$ $\quad \csc\theta = \frac{r}{y} = \frac{4}{\sqrt{15}}$

$\cos\theta = \frac{x}{r} = \frac{1}{4}$ $\quad \sec\theta = \frac{r}{x} = 4$

$\tan\theta = \frac{y}{x} = \sqrt{15}$ $\quad \cot\theta = \frac{x}{y} = \frac{1}{\sqrt{15}}$

13. For a point (50, 20) on the terminal side,

$x = 50,\ y = 20$

$r = \sqrt{50^2 + 20^2} = \sqrt{2900} = 10\sqrt{29}$

$\sin\theta = \frac{y}{r} = \frac{20}{10\sqrt{29}} = \frac{2}{\sqrt{29}}$ $\quad \csc\theta = \frac{r}{y} = \frac{\sqrt{29}}{2}$

$\cos\theta = \frac{x}{r} = \frac{50}{10\sqrt{29}} = \frac{5}{\sqrt{29}}$ $\quad \sec\theta = \frac{r}{x} = \frac{\sqrt{29}}{5}$

$\tan\theta = \frac{y}{x} = \frac{20}{50} = \frac{2}{5}$ $\quad \cot\theta = \frac{x}{y} = \frac{5}{2}$

17. $\cos\theta = \frac{x}{r} = \frac{12}{13}$

Use $x = 12,\ r = 13$, so

$y = \sqrt{r^2 - x^2} = \sqrt{13^2 - 12^2} = \sqrt{169 - 144}$

$= \sqrt{25} = 5$

$\sin\theta = \frac{y}{r} = \frac{5}{13}$

$\cot\theta = \frac{x}{y} = \frac{12}{5}$

21. $\sin\theta = \frac{y}{r} = \frac{0.750}{1}$

Use $y = 0.750$ and $r = 1$

$x = \sqrt{r^2 - y^2} = \sqrt{1^2 - 0.750^2} = \sqrt{0.4375}$

$\cot\theta = \frac{x}{y} = \frac{\sqrt{0.4375}}{0.750} = 0.882$

$\csc\theta = \frac{r}{y} = \frac{1}{0.750} = 1.33$

25. For $(3, 4)$, $x = 3,\ y = 4,\ r = \sqrt{3^2 + 4^2} = \sqrt{25} = 5,$

$\sin\theta = \frac{y}{r} = \frac{4}{5}$ and $\tan\theta = \frac{y}{x} = \frac{4}{3}$

For $(6, 8)$, $x = 6,\ y = 8,\ r = \sqrt{6^2 + 8^2} = \sqrt{100} = 10,$

$\sin\theta = \frac{y}{r} = \frac{8}{10} = \frac{4}{5}$ and $\tan\theta = \frac{y}{x} = \frac{8}{6} = \frac{4}{3}$

For $(4.5, 6)$, $x = 4.5,\ y = 6,\ r = \sqrt{4.5^2 + 6^2}$

$= \sqrt{56.25} = 7.5,$

$\sin\theta = \frac{y}{r} = \frac{6}{7.5} = \frac{4}{5}$ and $\tan\theta = \frac{y}{x} = \frac{6}{4.5} = \frac{4}{3}$

29. If $\tan\theta = \frac{y}{x} = \frac{3}{4}$

Use $y = 3$, $x = 4$, $r = \sqrt{x^2 + y^2} = \sqrt{4^2 + 3^2}$
$= \sqrt{25} = 5$

$$\sin\theta = \frac{y}{r} = \frac{3}{5} \quad \text{and} \quad \cos\theta = \frac{x}{r} = \frac{4}{5}$$

$$\sin^2\theta + \cos^2\theta = \left(\frac{3}{5}\right)^2 + \left(\frac{4}{5}\right)^2$$

$$\sin^2\theta + \cos^2\theta = \frac{9}{25} + \frac{16}{25}$$

$$\sin^2\theta + \cos^2\theta = \frac{25}{25}$$

$$\sin^2\theta + \cos^2\theta = 1$$

33. If the points are on the same terminal side, then all trigonometric function ratios are the same, since the angle is the same. Choosing $\tan\theta$

$$\tan\theta = \frac{y}{x} = \frac{4}{x+1} = \frac{6}{-2}$$

$$4(-2) = 6(x+1)$$

$$6x + 6 = -8$$

$$6x = -14$$

$$x = -\frac{7}{3}$$

4.3 Values of the Trigonometric Functions

1. $\sin\theta = 0.3527$

$\theta = \sin^{-1}(0.3527)$

$\theta = 20.65^\circ$

5.

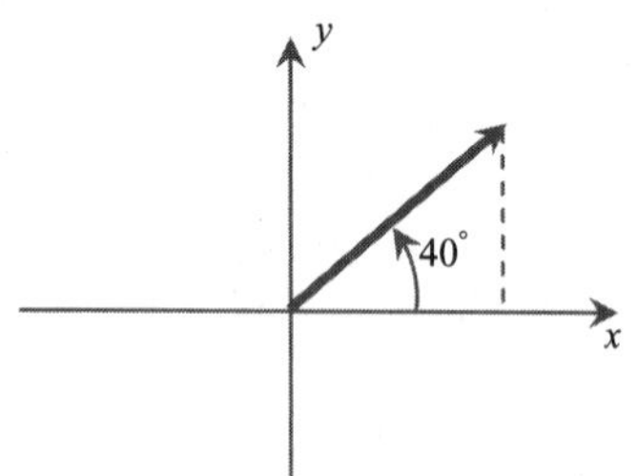

Answers may vary. One set of measurements gives $x = 7.6$ and $y = 6.5$.

$\sin 40^\circ = \frac{6.5}{10} = 0.65$ $\quad \csc 40^\circ = \frac{10}{6.5} = 1.5$

$\cos 40^\circ = \frac{7.6}{10} = 0.76$ $\quad \sec 40^\circ = \frac{10}{7.6} = 1.3$

$\tan 40^\circ = \frac{6.5}{7.6} = 0.86$ $\quad \cot 40^\circ = \frac{7.6}{6.5} = 1.2$

9. $\sin 22.4^\circ = 0.381$

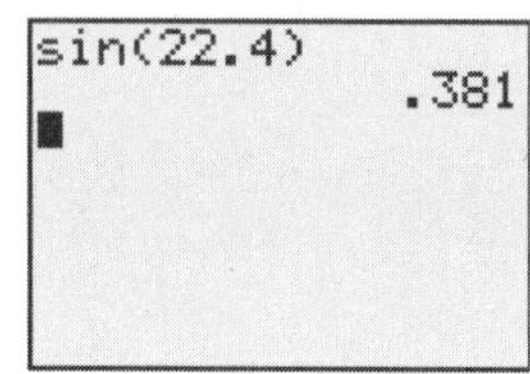

13. $\cos 15.71^\circ = 0.9626$

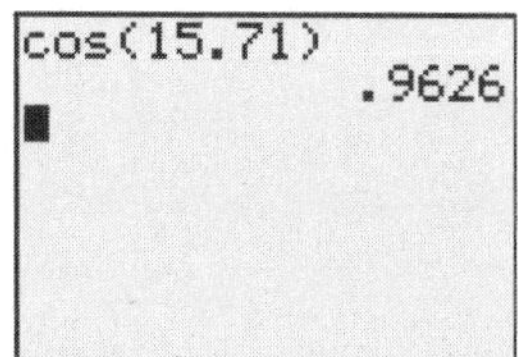

17. $\cot 67.78^\circ = 0.4085$

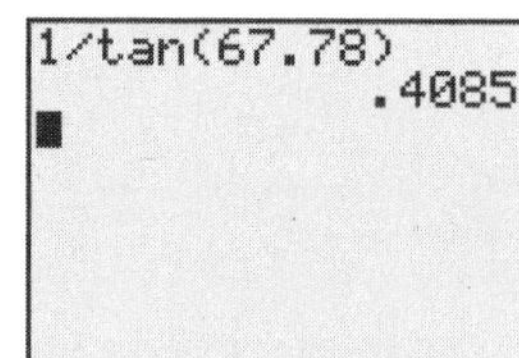

21. $\csc 0.4900^\circ = 116.9$

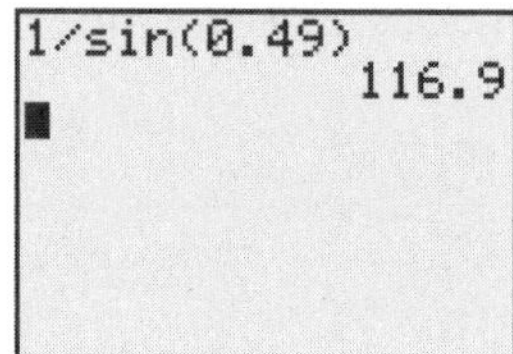

25. $\cos\theta = 0.3261$

$\theta = \cos^{-1}(0.3261)$

$\theta = 70.97^\circ$

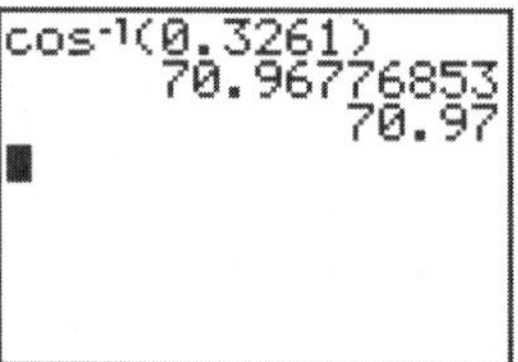

29. $\tan\theta = 0.207$

$\theta = \tan^{-1}(0.207)$

$\theta = 11.7^\circ$

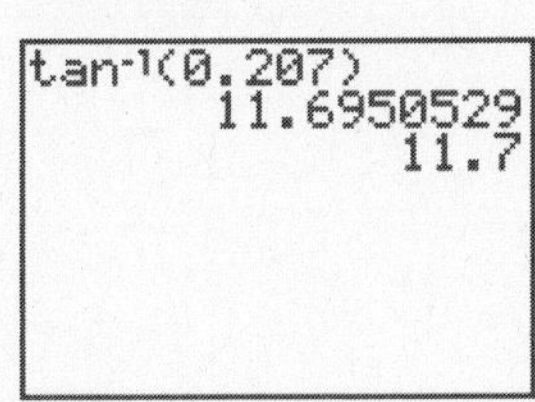

33. $\csc\theta = 1.245$

$\sin\theta = 1/1.245$

$\theta = \sin^{-1}(1/1.245)$

$\theta = 53.44^\circ$

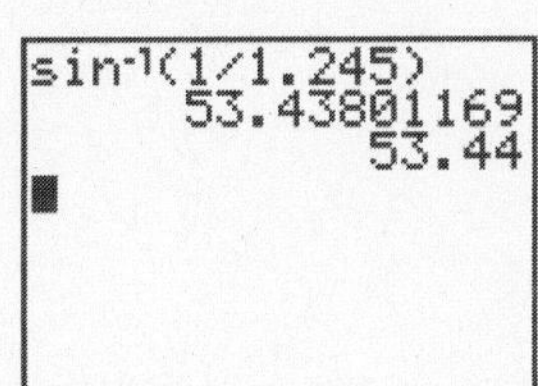

37. $\sec\theta = 3.65$

$\cos\theta = 1/3.65$

$\theta = \cos^{-1}(1/3.65)$

$\theta = 74.1^\circ$

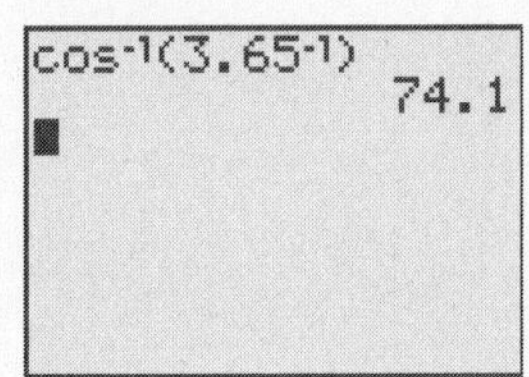

41. $\dfrac{\sin 43.7^\circ}{\cos 43.7^\circ} = \tan 43.7^\circ$

$0.956 = 0.956$

```
sin(43.7)/cos(43
.7)
      .9556207567
tan(43.7)
      .9556207567
```

45. Since $\sin\theta = \dfrac{y}{r}$, and since y is always less than or equal to r,

$y \le r$

$\dfrac{y}{r} \le 1$

$\sin\theta \le 1$

The minimum value of y is 0, so

$\sin\theta \ge 0$

Together,

$0 \le \sin\theta \le 1$

49. Given $\tan\theta = 1.936$

$\theta = \tan^{-1} 1.936$

$\sin\theta = \sin\left(\tan^{-1} 1.936\right)$

$\sin\theta = 0.8885$

53.

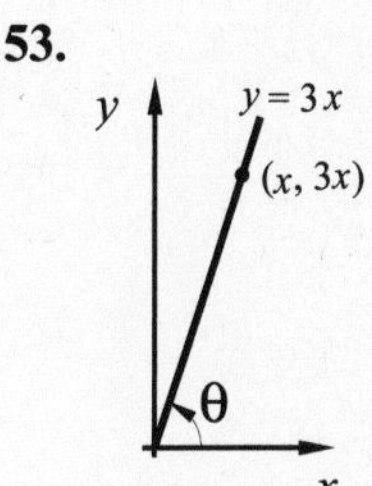

For every variable x, the y-coordinate will be $3x$

$\tan\theta = \dfrac{3x}{x}$

$\theta = \tan^{-1}\dfrac{3x}{x}$

$\theta = \tan^{-1} 3$

$\theta = 71.6^\circ$

57. $\sin\theta = \dfrac{1.50\lambda}{d}$

$\sin\theta = \dfrac{1.5(200\text{ m})}{400\text{ m}}$

$\theta = \sin^{-1}\dfrac{1.5(200)}{400}$

$\theta = 48.6^\circ$

4.4 The Right Triangle

1. $\sin A = \dfrac{7}{\sqrt{65}} = 0.868$

$\cos A = \dfrac{4}{\sqrt{65}} = 0.496$

$\tan A = \dfrac{7}{4} = 1.75$

$\sin B = \dfrac{4}{\sqrt{65}} = 0.496$

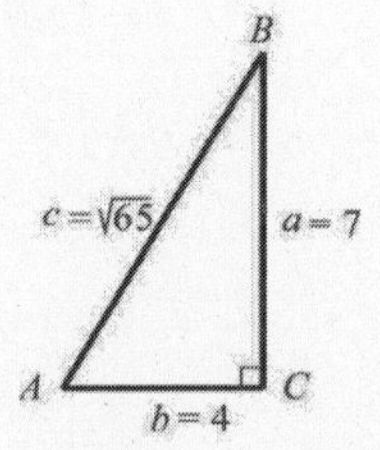

$\cos B = \dfrac{4}{\sqrt{65}} = 0.868$

$\tan B = \dfrac{4}{7} = 0.571$

5. A 60° angle between sides of 3 cm and 6 cm determines the unique triangle shown in the figure below.

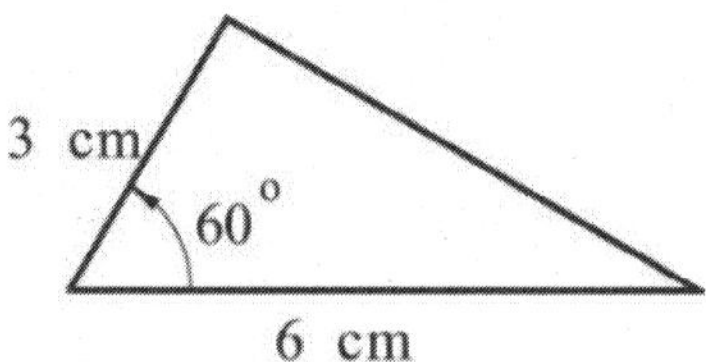

9. Given $A = 77.8^\circ$, $a = 6700$

$$\sin A = \frac{a}{c}$$
$$c = \frac{a}{\sin A}$$
$$c = \frac{6700}{\sin 77.8^\circ}$$
$$c = 6850$$

$$\tan A = \frac{a}{b}$$
$$b = \frac{a}{\tan A}$$
$$b = \frac{6700}{\tan 77.8^\circ}$$
$$b = 1450$$
$$B = 90^\circ - 77.8^\circ$$
$$B = 12.2^\circ$$

13. Given $B = 32.1^\circ$, $c = 23.8$

$$\sin B = \frac{b}{c}$$
$$b = c\sin B$$
$$b = 23.8\sin 32.1^\circ$$
$$b = 12.6$$

$$\cos B = \frac{a}{c}$$
$$a = c\cos B$$
$$a = 23.8\cos 32.1^\circ$$
$$a = 20.2$$
$$A = 90^\circ - 32.1^\circ$$
$$A = 57.9^\circ$$

17. Given $A = 32.10^\circ$, $c = 56.85$

$$\sin A = \frac{a}{c}$$
$$a = c\sin A$$
$$a = 56.85\sin 32.10^\circ$$
$$a = 30.21$$

$$\cos A = \frac{b}{c}$$
$$b = c\cos A$$
$$b = 56.85\cos 32.10^\circ$$
$$b = 48.16$$
$$B = 90^\circ - 32.10^\circ$$
$$B = 57.90^\circ$$

21. Given $B = 37.5^\circ$, $a = 0.862$

$$\cos B = \frac{a}{c}$$
$$c = \frac{a}{\cos B}$$
$$c = \frac{0.862}{\cos 37.5^\circ}$$
$$c = 1.09$$

$$\tan B = \frac{b}{a}$$
$$b = a\tan B$$
$$b = 0.862\tan 37.5^\circ$$
$$b = 0.661$$
$$A = 90^\circ - 37.5^\circ$$
$$A = 52.5^\circ$$

25. Given $a = 591.87$, $b = 264.93$

$$c = \sqrt{a^2 + b^2}$$
$$c = \sqrt{591.87^2 + 264.93^2}$$
$$c = 648.46$$

$$\tan A = \frac{a}{b}$$
$$A = \tan^{-1}\left(\frac{591.87}{264.93}\right)$$
$$A = 65.883^\circ$$

$$\tan B = \frac{b}{a}$$
$$B = \tan^{-1}\left(\frac{264.93}{591.87}\right)$$
$$B = 24.117^\circ$$

29. Given $B = 9.56^\circ$, $c = 0.0973$

$$\sin B = \frac{b}{c}$$
$$b = c\sin B$$
$$b = 0.0973\sin 9.56^\circ$$
$$b = 0.0162$$
$$\cos B = \frac{a}{c}$$
$$a = c\cos B$$
$$a = 0.0973\cos 9.56^\circ$$
$$a = 0.0959$$
$$A = 90^\circ - 9.56^\circ$$
$$A = 80.44^\circ$$

33. $\sin 61.7^\circ = \frac{3.92}{x}$

$$x = \frac{3.92}{\sin 61.7^\circ}$$
$$x = 4.45$$

37.

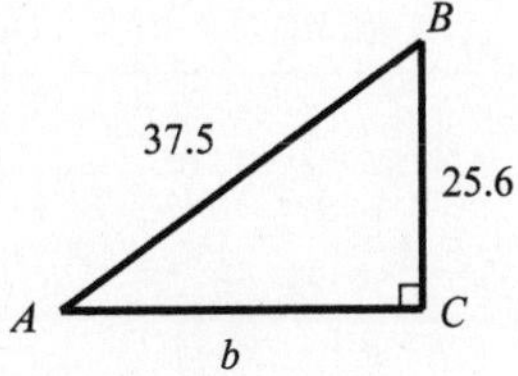

$$\sin A = \frac{25.6}{37.5}$$
$$A = \sin^{-1}\left(\frac{25.6}{37.5}\right)$$
$$A = 43.1^\circ$$
$$B = 90^\circ - 43.1^\circ$$
$$B = 47.9^\circ$$

A is the smaller acute angle.

41. $\sin A = \frac{a}{c}$

$$a = c\sin A$$
$$\cos A = \frac{b}{c}$$
$$b = c\cos A$$
$$B = 90^\circ - A$$

4.5 Applications of Right Triangles

1. $\frac{565 \text{ m}}{d} = \sin 62.1^\circ$

$$d = \frac{565 \text{ m}}{\sin 62.1^\circ}$$
$$d = 639 \text{ m}$$

5.

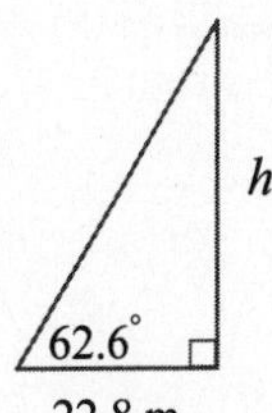

$$\tan 62.6^\circ = \frac{h}{22.8}$$
$$h = 22.8 \tan 62.6^\circ$$
$$h = 44.0 \text{ m}$$

9.

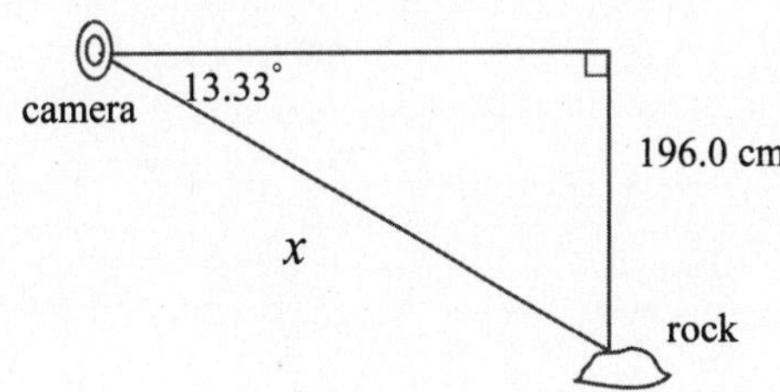

$$\sin 13.33^\circ = \frac{196.0 \text{ cm}}{x}$$
$$x = \frac{196.0 \text{ cm}}{\sin 13.33^\circ}$$
$$x = 850.1 \text{ cm}$$

13.

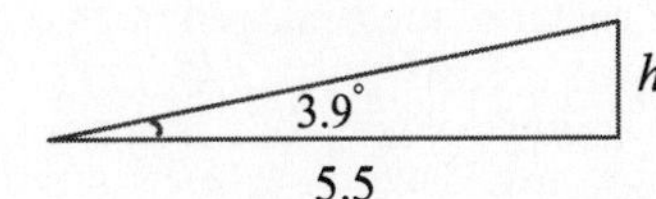

$$\tan 3.9^\circ = \frac{h}{5.5 \text{ km}}$$
$$h = 5.5 \text{ km}\left(\tan 3.9^\circ\right)$$
$$h = 0.37 \text{ km} = 370 \text{ m}$$

17. $\tan\theta = \dfrac{6.0\text{ m}}{100\text{ m}}$

$$\theta = \tan^{-1} 0.060$$

$$\theta = 3.4^{\circ}$$

21.

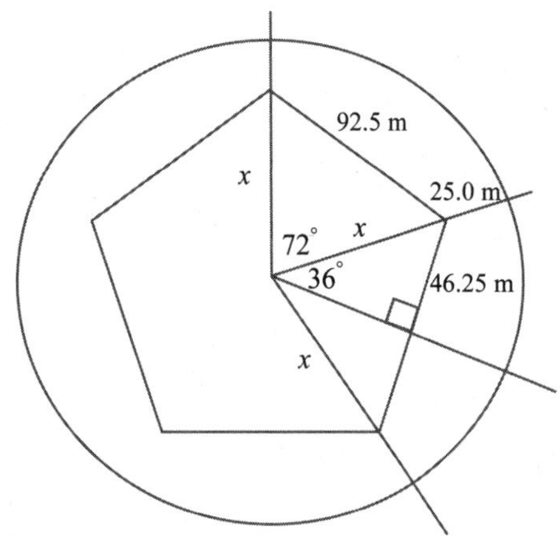

$$\sin\theta = \frac{12.0\text{ m}}{85.0\text{ m}}$$

$$\theta = \sin^{-1}\frac{12.0}{85.0}$$

$$\theta = 8.12^{\circ}$$

25.

Each of the five triangles in the pentagon has a central angle $\frac{360^{\circ}}{5}$ or 72°. Radii drawn from the center of the pentagon (which is also the center of the circle) through adjacent vertices of the pentagon form an isosceles triangle with base 92.5 m and equal sides x. A perpendicular bisector from the center of the pentagon to the base of this isosceles triangle forms a right triangle with hypotenuse x and base 46.25 m. The central angle of this right triangle is $\frac{72^{\circ}}{2}$ or 36°. Thus,

$$\sin 36^{\circ} = \frac{46.25}{x}$$

$$x = \frac{46.25}{\sin 36^{\circ}}$$

$$x = 78.685\text{ m}$$

and $C = 2\pi(x+25)$

$$C = 2\pi(78.685+25)$$

$$C = 651\text{ m}$$

29.

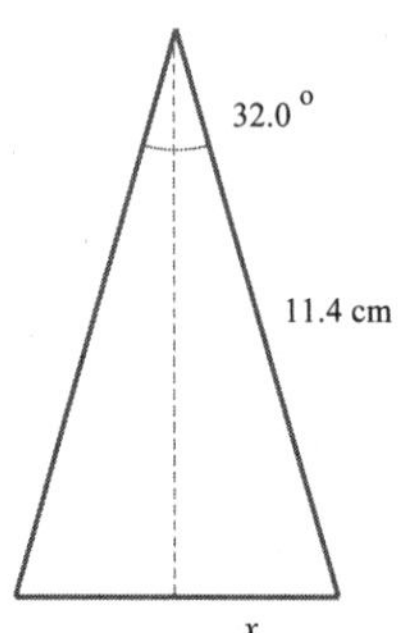

$$\sin 16.0^{\circ} = \frac{x}{11.4\text{ cm}}$$

$$x = 3.14\text{ cm}$$

If d is the distance between the ends

$$d = 2x$$

$$d = 6.28\text{ cm}$$

33.

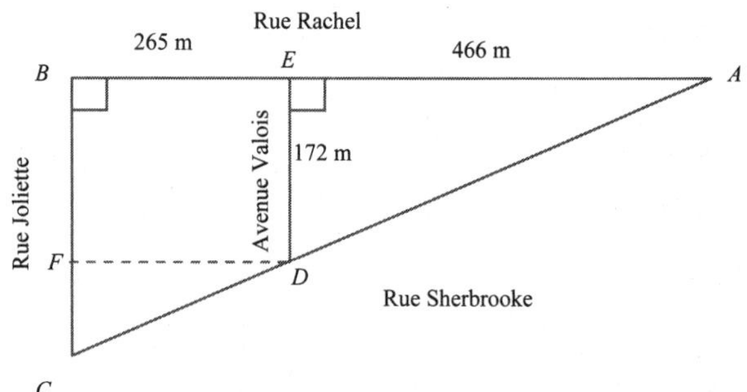

Part (a)

Triangles ΔABC and ΔAED are similar.

$$\frac{172}{466} = \frac{BC}{466+265}$$

$$BC = \frac{172(466+265)}{466}$$

$$BC = 270\text{ m}$$

Part (b)

$$\tan\angle EAD = \frac{172}{466}$$

$$\angle EAD = \tan^{-1}\left(\frac{172}{466}\right)$$

$$\angle EAD = 20.259^{\circ}$$

$\angle EAD$ and $\angle FDC$ are corresponding angles

$$\angle FDC = 20.259^\circ$$
$$\cos \angle FDC = \frac{265}{CD}$$
$$CD = \frac{265}{\cos 20.259^\circ}$$
$$CD = 282 \text{ m}$$

37.

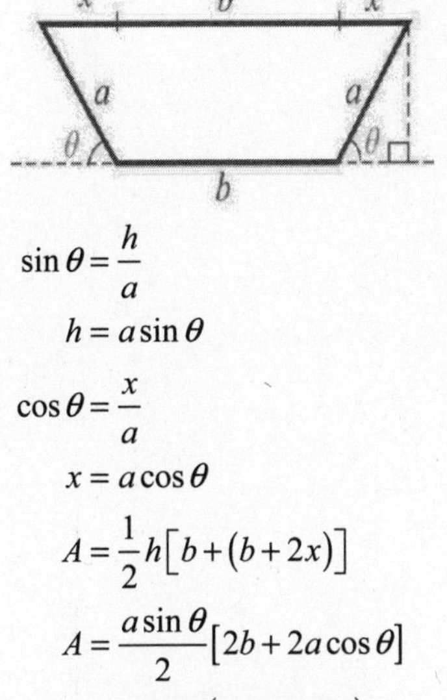

$$\sin\theta = \frac{h}{a}$$
$$h = a\sin\theta$$
$$\cos\theta = \frac{x}{a}$$
$$x = a\cos\theta$$
$$A = \frac{1}{2}h[b + (b + 2x)]$$
$$A = \frac{a\sin\theta}{2}[2b + 2a\cos\theta]$$
$$A = a\sin\theta(b + a\cos\theta)$$

Review Exercises

1. positive: $17.0^\circ + 360^\circ = 377.0^\circ$
negative: $17.0^\circ - 360^\circ = -343.0^\circ$

5. $31^\circ 54' = 31^\circ + 54'\left(\frac{1^\circ}{60'}\right) = 31^\circ + 0.90^\circ = 31.90^\circ$

9. $17.5^\circ = 17^\circ + 0.5^\circ\left(\frac{60'}{1^\circ}\right) = 17^\circ 30'$

13. $x = 24, \ y = 7$

$$r = \sqrt{x^2 + y^2} = \sqrt{24^2 + 7^2} = \sqrt{576 + 49} = \sqrt{625} = 25$$

$\sin\theta = \frac{y}{r} = \frac{7}{25}$ $\quad$ $\csc\theta = \frac{r}{y} = \frac{25}{7}$

$\cos\theta = \frac{x}{r} = \frac{24}{25}$ $\quad$ $\sec\theta = \frac{r}{x} = \frac{25}{24}$

$\tan\theta = \frac{y}{x} = \frac{7}{24}$ $\quad$ $\cot\theta = \frac{x}{y} = \frac{24}{7}$

17. $\sin\theta = \frac{y}{r} = \frac{5}{13}$, so $y = 5, \ r = 13$

$$x = \sqrt{r^2 - y^2} = \sqrt{13^2 - 5^2} = \sqrt{169 - 25}$$
$$= \sqrt{144} = 12$$
$$\cos\theta = \frac{x}{r} = \frac{12}{13} = 0.923$$
$$\cot\theta = \frac{x}{y} = \frac{12}{5} = 2.40$$

21. $\sin 72.1^\circ = 0.952594403 = 0.952$

25. $\sec 18.4^\circ = \frac{1}{\cos 18.4^\circ} = 1.053878471 = 1.05$

29. $\cos\theta = 0.950$

$$\theta = \cos^{-1}(0.950)$$
$$\theta = 18.2^\circ$$

33. $\csc\theta = \frac{1}{\sin\theta} = 4.713$

$$\sin\theta = \frac{1}{4.713}$$
$$\theta = \sin^{-1}\left(\frac{1}{4.713}\right)$$
$$\theta = 12.25^\circ$$

37. $\cot\theta = \frac{1}{\tan\theta} = 7.117$

$$\tan\theta = \frac{1}{7.117}$$
$$\theta = \tan^{-1}\left(\frac{1}{7.117}\right)$$
$$\theta = 7.998^\circ$$

41. Given $A = 17.0^\circ$, $b = 6.00$

$$B = 90.0^\circ - 17.0^\circ$$
$$B = 73.0^\circ$$
$$\tan A = \frac{a}{b}$$
$$a = b\tan A$$
$$a = (6.00)\tan 17.0^\circ$$
$$a = 1.83$$

$$\cos A = \frac{b}{c}$$
$$c = \frac{b}{\cos A}$$
$$c = \frac{6.00}{\cos 17.0^\circ}$$
$$c = 6.27$$

45. Given $A = 37.5^\circ$, $a = 12.0$

$$B = 90^\circ - 37.5^\circ$$
$$B = 52.5^\circ$$

$$\tan A = \frac{a}{b}$$
$$b = \frac{a}{\tan A}$$
$$b = \frac{12.0}{\tan 37.5^\circ}$$
$$b = 15.6$$

$$\sin A = \frac{a}{c}$$
$$c = \frac{a}{\sin A}$$
$$c = \frac{12.0}{\sin 37.5^\circ}$$
$$c = 19.7$$

49. Given $A = 49.67^\circ$, $c = 0.8253$

$$B = 90^\circ - 49.67^\circ$$
$$B = 40.33^\circ$$

$$\sin A = \frac{a}{c}$$
$$a = c \sin A$$
$$a = 0.8253 \sin 49.67^\circ$$
$$a = 0.6292$$

$$\cos A = \frac{b}{c}$$
$$b = c \cos A$$
$$b = 0.8253 \cos 49.67^\circ$$
$$b = 0.5341$$

53.

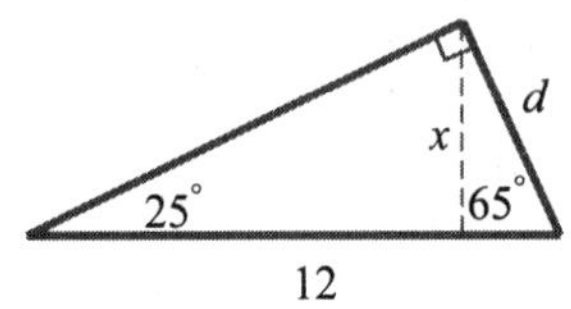

$$\sin 25^\circ = \frac{d}{12}$$
$$d = 12 \sin 25^\circ$$
$$d = 5.0714$$
$$90^\circ - 25^\circ = 65^\circ$$
$$\sin 65^\circ = \frac{x}{d}$$
$$x = d \sin 65^\circ$$
$$x = 5.0714 \sin 65^\circ$$
$$x = 4.6$$

57.

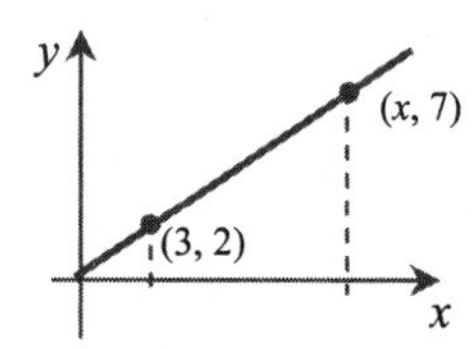

$$\frac{x}{3} = \frac{7}{2}$$
$$x = \frac{3(7)}{2}$$
$$x = 10.5$$

61.

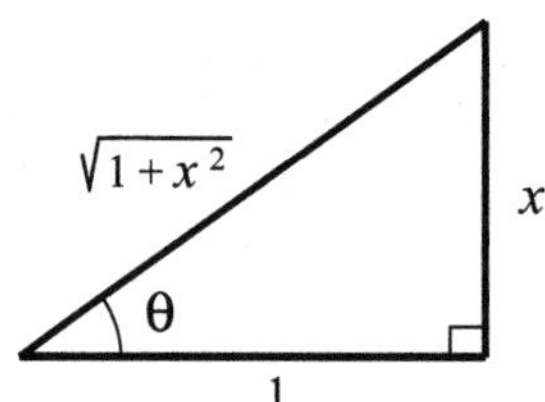

$$\tan \theta = x = \frac{x}{1}$$

Side opposite $o = x$

Side adjacent $a = 1$

Hypotenuse $h = \sqrt{x^2 + 1}$

$$\csc \theta = \frac{h}{o} = \frac{\sqrt{x^2+1}}{x}$$

65. $e = E \cos \alpha$

$$\cos \alpha = \frac{e}{E}$$
$$\alpha = \cos^{-1}\left(\frac{56.9}{339}\right)$$
$$\alpha = 80.3^\circ$$

69.

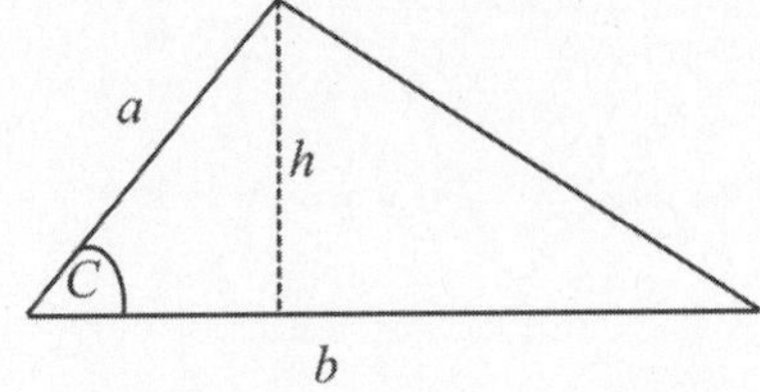

(a) A triangle with angle C included between sides a and b, has an altitude h

$$\sin C = \frac{h}{a}$$

$$h = a\sin C$$

The area is $A = \frac{1}{2}bh$

$$A = \tfrac{1}{2}b(a\sin C)$$

$$A = \tfrac{1}{2}ab\sin C$$

(b) The area of the tract is

$$A = \tfrac{1}{2}(31.96\text{ m})(47.25\text{ m})\sin 64.09^\circ$$

$$A = 679.2\text{ m}^2$$

73.

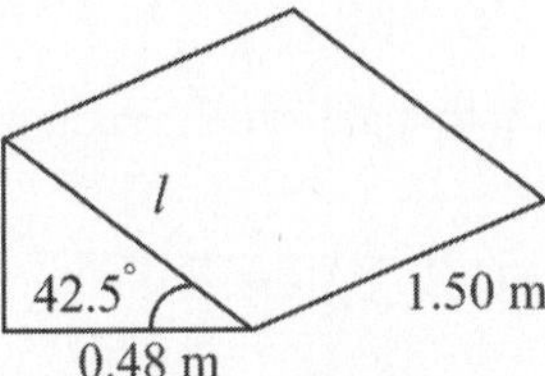

$$\cos 42.5^\circ = \frac{0.480\text{ m}}{l}$$

$$l = \frac{0.480\text{ m}}{\cos 42.5^\circ}$$

$$A = lw$$

$$A = \frac{0.480\text{ m}}{\cos 42.5^\circ}(1.5\text{ m})$$

$$A = 0.977\text{ m}^2$$

77.

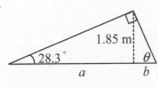

$$\tan 28.3^\circ = \frac{1.85}{a}$$

$$a = \frac{1.85}{\tan 28.3^\circ}$$

$$\theta = 90.0^\circ - 28.3^\circ$$

$$\theta = 61.7^\circ$$

$$\tan 61.7^\circ = \frac{1.85}{b}$$

$$b = \frac{1.85}{\tan 61.7^\circ}$$

$$d = a + b$$

$$d = \frac{1.85}{\tan 28.3^\circ} + \frac{1.85}{\tan 61.7^\circ}$$

$$d = 4.43\text{ m}$$

81.

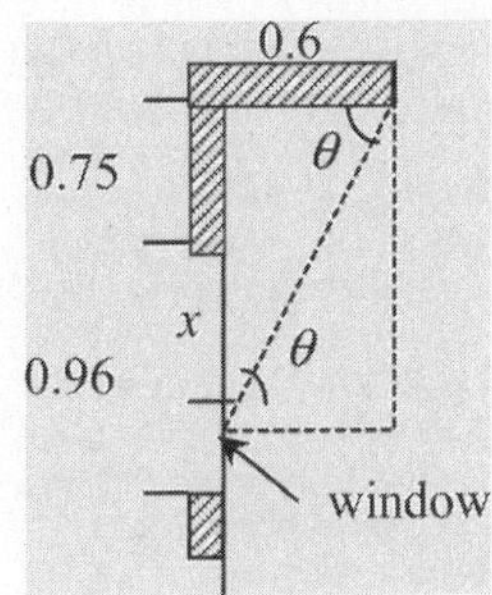

Let x = length of window through which sun does not shine.

$$\tan 65^\circ = \frac{x + 0.75}{0.6}$$

$$x = 0.6\tan 65^\circ - 0.75$$

$$x = 0.53670\text{ m}$$

To find the fraction f of the window shaded,

$$f = \frac{0.53670\text{ m}}{0.96\text{ m}}$$

$$f = 0.55907$$

$$f = 56\%$$

85. I. Line of sight perpendicular to end of span

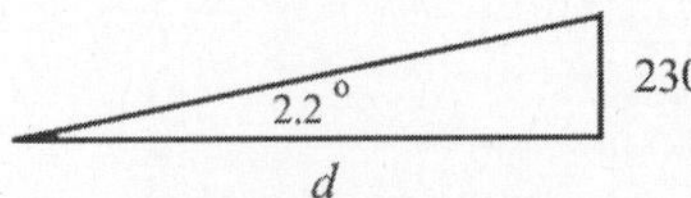

d = distance from helicopter to end of span

$$\tan 2.2^\circ = \frac{230\text{ m}}{d}$$

$$d = \frac{230\text{ m}}{\tan 2.2^\circ}$$

$$d = 5987.1\text{ m}$$

$$d = 6.0\text{ km}$$

II. Line of sight perpendicular to middle of span

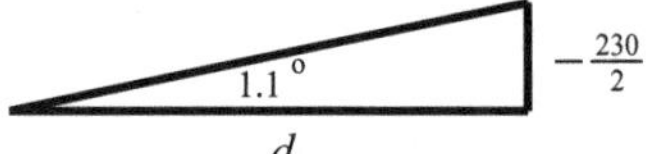

d = distance from helicopter to middle of span

$$\tan 1.1^\circ = \frac{\frac{230\text{ m}}{2}}{d}$$

$$d = \frac{115\text{ m}}{\tan 1.1^\circ}$$

$$d = 5989.3\text{ m}$$

$$d = 6.0\text{ km}$$

89.

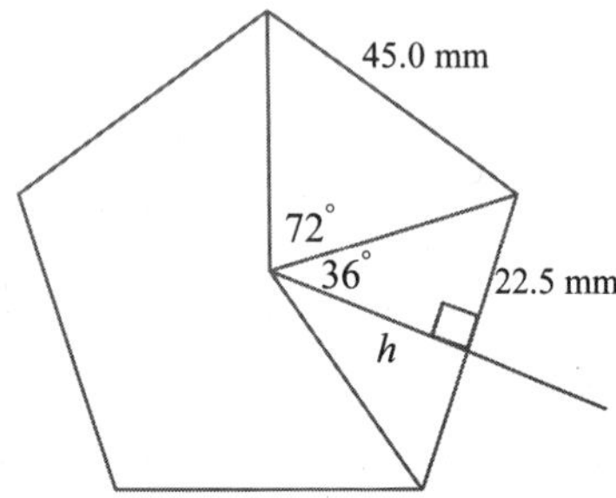

Each of the five triangles in the pentagon has a central angle $\frac{360^\circ}{5}$ or 72°. Radii drawn from the center of the pentagon through adjacent vertices of the pentagon form an isosceles triangle with base 45.0 mm. A perpendicular bisector from the center of the pentagon to the base of this isosceles triangle forms a right triangle with height h and base 22.5 mm. The central angle of this right triangle is $\frac{72^\circ}{2}$ or 36°. Thus,

$$\tan 36^\circ = \frac{22.5}{h}$$

$$h = \frac{22.5}{\tan 36^\circ}$$

$$h = 30.969\text{ mm}$$

The area of each triangle in the pentagon is

$$A = \tfrac{1}{2}bh$$

$$A = \tfrac{1}{2}(45.0\text{ mm})(30.969\text{ mm})$$

$$A = 696.79\text{ mm}^2$$

Each pentagon has five triangles, and there are 12 pentagons on the ball, so the surface area of all the pentagons is

$$A_{\text{pentagons}} = (12)(5)(696.79\text{ mm}^2)$$

$$A_{\text{pentagons}} = 41808\text{ mm}^2$$

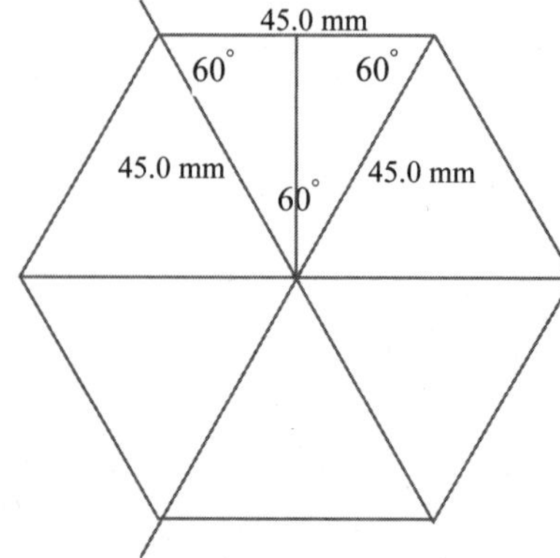

Each of the six triangles in the hexagon has a central angle $\frac{360^\circ}{6}$ or 60°. Radii drawn from the center of the pentagon through adjacent vertices of the pentagon form an equilateral triangle with sides 45.0 mm.

$$\tan 30^\circ = \frac{22.5}{h}$$

$$h = \frac{22.5}{\tan 30^\circ}$$

$$h = 38.971\text{ mm}$$

The area of each triangle in the pentagon is

$$A = \tfrac{1}{2}bh$$

$$A = \tfrac{1}{2}(45.0\text{ mm})(38.971\text{ mm})$$

$$A = 876.85\text{ mm}^2$$

Each hexagon has six triangles, and there are 20 hexagons on the ball, so the surface area of all the hexagons is

$$A_{\text{hexagons}} = (20)(6)(876.85\text{ mm}^2)$$

$$A_{\text{hexagons}} = 105222\text{ mm}^2$$

Total surface area of ball

$$A = 41\ 808\text{ mm}^2 + 105\ 222\text{ mm}^2$$

$$A = 147\ 000\text{ mm}^2$$

Since this is the area of a flat surface it approximates the area of the spherical soccer ball which is given by

$$A = 4\pi r^2$$

$$A = 4\pi\cdot\left(\frac{222\text{ mm}}{2}\right)^2$$

$$A = 155\ 000\text{ mm}^2$$

The flat surface approximation does not account for the curved surface of the soccer ball.

93.

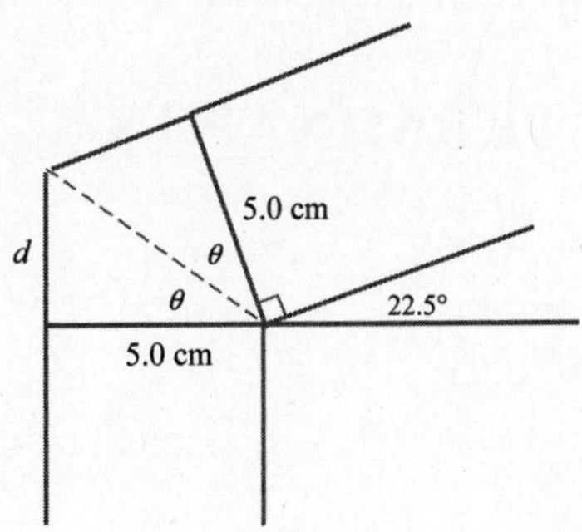

$$2\theta + 90^\circ + 22.5^\circ = 180^\circ$$
$$2\theta = 67.5^\circ$$
$$\theta = 33.75^\circ$$
$$\tan 33.75^\circ = \frac{d}{5.0\text{ cm}}$$
$$d = 5.0\text{ cm} \cdot \tan 33.75^\circ$$
$$d = 3.3409\text{ cm}$$
$$l = 5.0\text{ cm} + 65.0\text{ cm} + 3.3409\text{ cm}$$
$$l = 73.3\text{ cm}$$

CHAPTER 5

SYSTEMS OF LINEAR EQUATIONS; DETERMINANTS

5.1 Linear Equations

1. $x-\frac{y}{6}+z-4w=7$

$x-\frac{1}{6}y+z-4w=7$ is linear, since all terms contain only one variable to the first power (or are constant).

5. $5x+2y=1$

The coordinates of the point $(0.2, -1)$ do not satisfy the equation since

$5(0.2)+2(-1)=1-2=-1\neq 1$

The coordinates of the point $(1, -2)$ do satisfy the equation since

$5(1)+2(-2)=5-4=1$

9. $-5x+6y=60$

If $x=-10$

$$\begin{aligned}-5(-10)+6y&=60\\6y&=60-50\\6y&=10\\y&=\frac{5}{3}\end{aligned}$$

If $x=8$

$$\begin{aligned}-5(8)+6y&=60\\6y&=60+40\\6y&=100\\y&=\frac{50}{3}\end{aligned}$$

13. $2.4y-4.5x=-3.0$

If $x=-0.4$

$$\begin{aligned}2.4y-4.5(-0.4)&=-3.0\\2.4y+1.8&=-3.0\\2.4y&=-4.8\\y&=-2.0\end{aligned}$$

If $x=2.0$

$$\begin{aligned}2.4y-4.5(2.0)&=-3.0\\2.4y-9.0&=-3.0\\2.4y&=6.0\\y&=2.5\end{aligned}$$

17. $-30x+5y=1$

$6x-3y=-4$

If the values $x=\frac{1}{3}$ and $y=2$ satisfy both equations, they are a solution.

$-30\left(\frac{1}{3}\right)+5(2)=-10+10=0\neq 1$

$6\left(\frac{1}{3}\right)-3(2)=2-6=-4$

The first equation is not satisfied, therefore the given values are not a solution.

21. $s-7t=-3.2$

$2s+t=2.5$

If the values $s=-1.1$ and $t=0.3$ satisfy both equations, they are a solution.

$-1.1-7(0.3)=-1.1-2.1=-3.2$

$2(-1.1)+0.3=-2.2+0.3=-1.9\neq 2.5$

The second equation is not satisfied, so they are not a solution.

25. $3x+b=0$

If $x=-2$ is a root,

$$\begin{aligned}3(-2)+b&=0\\b&=6\end{aligned}$$

29. If $F_1=45$ N and $F_2=28$ N are solutions, then both equations should be satisfied.

$$0.80F_1+0.50F_2=0.80(45)+0.50(28)=36+14=50$$

$$0.60F_1-0.87F_2=0.60(45)-0.87(28)=27-24.36=2.64\neq 12$$

The second equation is not satisfied, so the given forces are not a solution.

5.2 Graphs of Linear Functions

1. By taking $(x_2, y_2) = (3, -1)$ and $(x_1, y_1) = (-1, -2)$

$$m = \frac{y_2 - y_1}{x_2 - x_1}$$

$$m = \frac{-1-(-2)}{3-(-1)}$$

$$m = \frac{-1+2}{3+1} = \frac{1}{4}$$

The line rises 1 unit for each 4 units in going from left to right.

5. By taking $(x_2, y_2) = (3, 8)$ and $(x_1, y_1) = (1, 0)$

$$m = \frac{y_2 - y_1}{x_2 - x_1}$$

$$m = \frac{8-0}{3-1} = \frac{8}{2} = 4$$

9. By taking $(x_2, y_2) = (-2, -5)$ and $(x_1, y_1) = (5, -3)$

$$m = \frac{y_2 - y_1}{x_2 - x_1}$$

$$m = \frac{-5-(-3)}{-2-5} = \frac{-5+3}{-7} = \frac{2}{7}$$

13. $m = 2, b = -1$

Plot the y-intercept point $(0, -1)$.

Since the slope is 2/1, from this point go 1 unit to the right and up 2 units, and plot a second point.

Sketch a line passing through these two points.

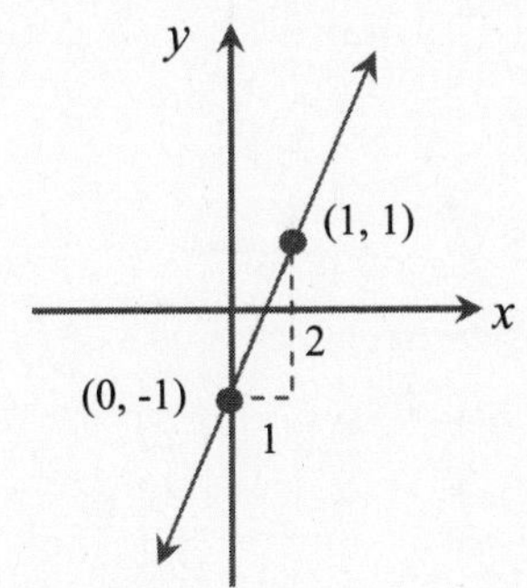

17. $m = \frac{1}{2}, (0, 0)$

Plot the y-intercept point $(0, 0)$.

Since the slope is 1/2, from this point go right 2 units and up 1 unit, and plot a second point.

Sketch a line passing through these two points.

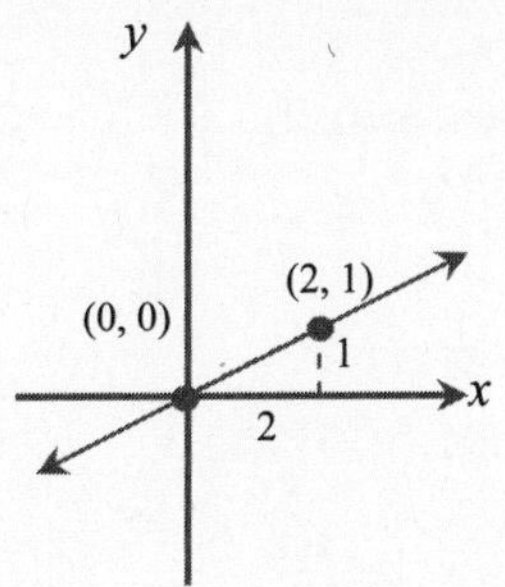

21. $y = -2x + 1$, compare to $y = mx + b$

$m = -2, b = 1$

Plot the y-intercept point $(0, 1)$.

Since the slope is $-2/1$, from this point go right 1 unit and down 2 units, and plot a second point.

Sketch a line passing through these two points.

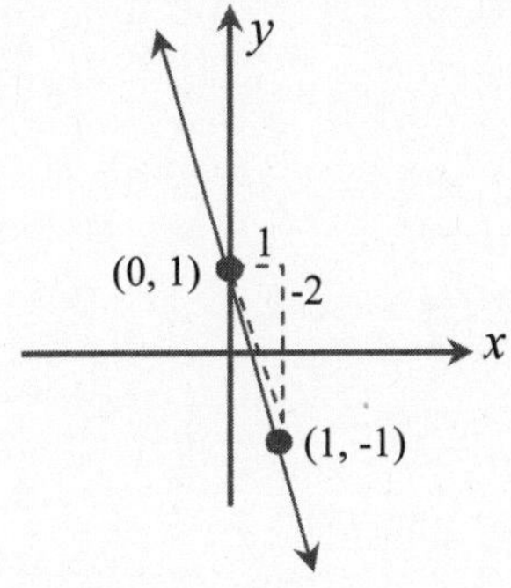

25. $5x - 2y = 40$

$2y = 5x - 40$

$y = \frac{5}{2}x - 20$, compare to $y = mx + b$

$m = \frac{5}{2}, b = -20$

Plot the y-intercept point $(0, -20)$.

Since the slope is 5/2, from this point go right 2 units and up 5 units, and plot a second point.

Sketch a line passing through these two points.

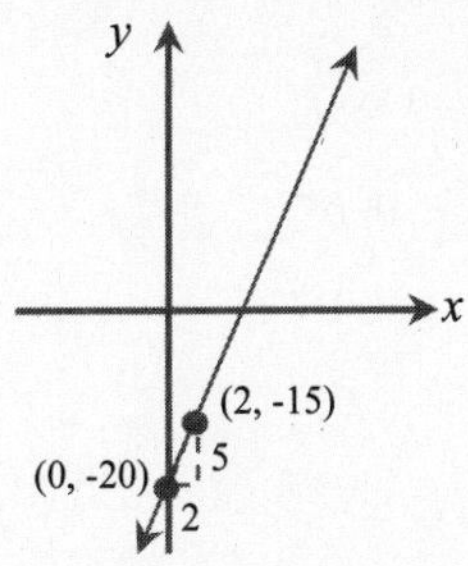

29. $x+2y=4$

For y-int, set $x=0$

$$0+2y=4$$
$$2y=4$$
$$y=2 \quad y\text{-int is } (0, 2)$$

For x-int, set $y=0$

$$x+0=4$$
$$x=4 \quad x\text{-int is } (4, 0)$$

Plot the x-intercept point $(4, 0)$ and the y-intercept point $(0, 2)$.

Sketch a line passing through these two points.

A third point is found as a check.

Let $x=2$

$$2+2y=4$$
$$2y=2$$
$$y=1.$$

Therefore the point $(2, 1)$ should lie on the line.

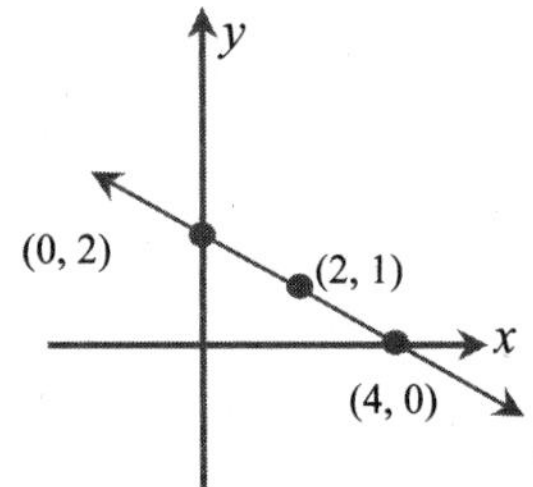

33. $y=3x+6$

For y-int, set $x=0$

$$y=0+6$$
$$y=6 \quad y\text{-int is } (0, 6)$$

For x-int, set $y=0$

$$0=3x+6$$
$$x=-2 \quad x\text{-int is } (-2, 0)$$

Plot the x-intercept point $(-2,0)$ and the y-intercept point $(0,6)$.

Sketch a line passing through these two points.

A third point is found as a check.

Let $x=-1$

$$y=3(-1)+6$$
$$y=3$$

Therefore the point $(-1, 3)$ should lie on the line.

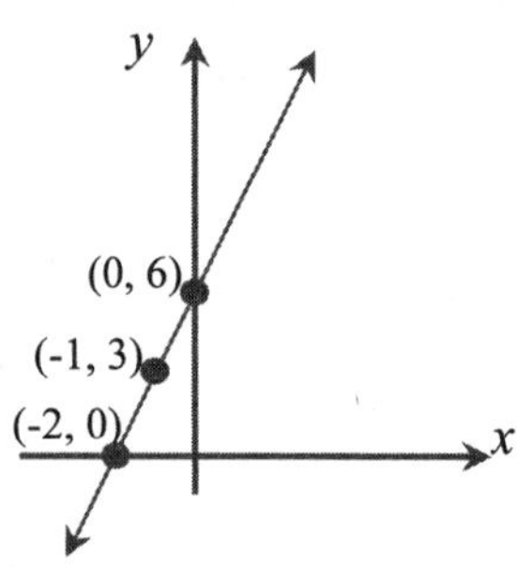

37. $kx-2y=9$

$$2y=kx-9$$
$$y=\frac{k}{2}x-\frac{9}{2}, \text{ compare to } y=mx+b$$
$$m=\frac{k}{2}, \text{ but slope} = 3, \text{ so}$$
$$3=\frac{k}{2}$$
$$k=6$$

41. $4I_1-5I_2=2$

$$5I_2=4I_1-2$$
$$I_2=\tfrac{4}{5}I_1-\tfrac{2}{5}, \text{ compare to } y=mx+b$$
$$m=\tfrac{4}{5},\ b=-\tfrac{2}{5}$$

Plot the I_2-intercept point $\left(0, -\tfrac{2}{5}\right)$.

Since the slope is $\tfrac{4}{5}$, from this point go right 1 units and up $\tfrac{4}{5}$ units, and plot a second point.

Sketch a linepassing through these two points.

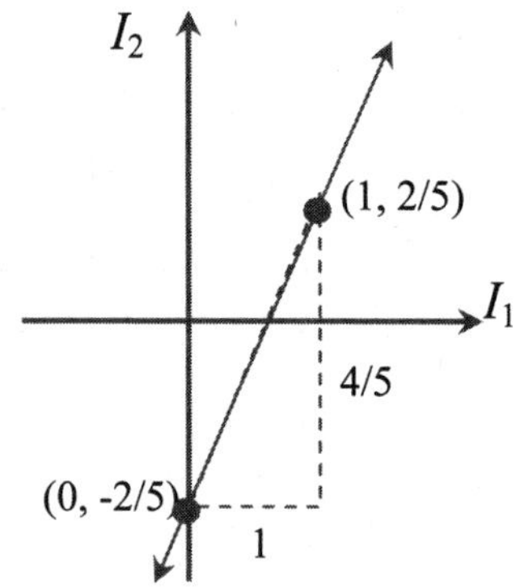

5.3 Solving Systems of Two Linear Equations in Two Unknowns Graphically

1. For the line $2x+5y=10$

Let $x=0$ to find the y-int

$5y=10$

$y=2$ y-int is $(0, 2)$

Let $y=0$ to find the x-int

$2x=10$

$x=5$ x-int is $(5, 0)$

Let $x=1$ to find a third point

$2(1)+5y=10$

$5y=8$

$y=\frac{8}{5}$ A third point is $\left(1, \frac{8}{5}\right)$

For the line $3x+y=6$

Let $x=0$ to find the y-int

$y=6$ y-int is $(0, 6)$

Let $y=0$ to find the x-int

$3x=6$

$x=2$ x-int is $(2, 0)$

Let $x=1$ to find a third point

$3(1)+y=6$

$y=3$ A third point is $(1, 3)$

From the graph the solution is approximately

$x=1.5,\ y=1.4$

Checking both equations,

$2x+5y=10$	$3x+y=6$
$2(1.5)+5(1.4)=10$	$3(1.5)+1.4=6$
$10=10$	$5.9\approx 6$

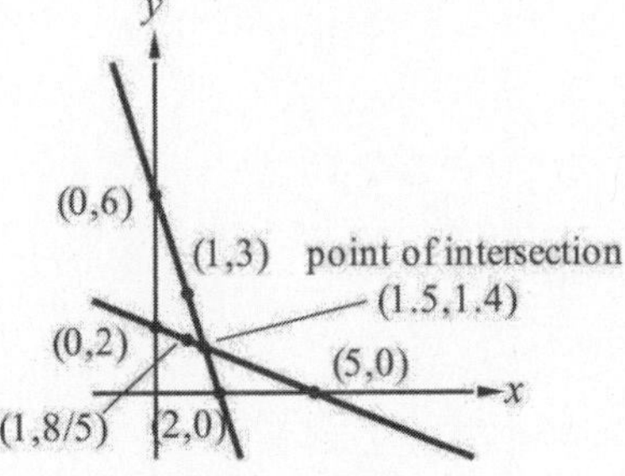

5. $y=2x-6$ and $y=-\frac{1}{3}x+1$

The slope of the first line is 2, and the y-intercept is -6.

The slope of the second line is $-1/3$ and the y-intercept is 1.

From the graph, the point of intersection is $(3, 0)$.

Therefore, the solution of the system of equations is $x=3.0,\ y=0.0$.

Check:

$y=2x-6$	$y=-\frac{1}{3}x+1$
$0.0=2(3.0)-6$	$0.0=-\frac{1}{3}(3.0)+1$
$0.0=0.0$	$0.0=0.0$

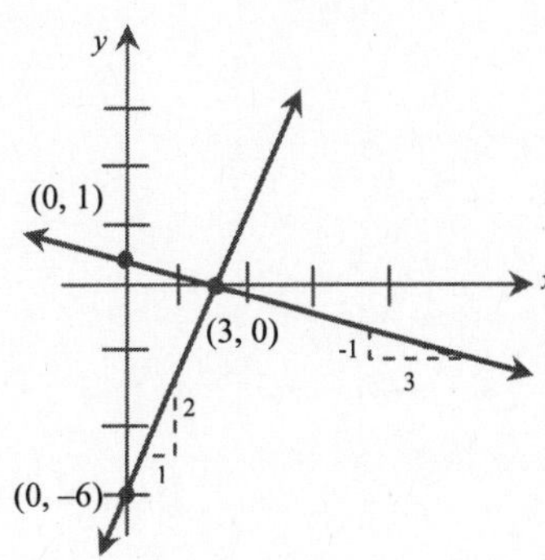

9. For line $2x-5y=10$

Let $x=0$ to find the y-int

$-5y=10$

$y=-2$ y-int is $(0, -2)$

Let $y=0$ to find the x-int

$2x=10$

$x=5$ x-int is $(5, 0)$

Let $x=2.5$ to find a third point

$2(2.5)-5y=10$

$-5y=5$

$y=-1$ A third point is $(2.5, -1)$

For line $3x+4y=-12$

Let $x=0$ to find the y-int

$4y=-12$

$y=-3$ y-int is $(0, -3)$

Let $y=0$ to find the x-int

$3x=-12$

$x=-4$ x-int is $(-4, 0)$

Let $x=-2$ to find a third point

$3(-2)+4y=-12$

$4y=-6$

$y=-\frac{3}{2}$ A third point is $\left(-2, -\frac{3}{2}\right)$

From the graph the solution is approximately $(-0.9, -2.3)$

$$x = -0.9,\ y = -2.3$$

Checking both equations,

$$2x - 5y = 10 \qquad 3x + 4y = -12$$
$$2(-0.9) - 5(-2.3) = 10 \qquad 3(-0.9) + 4(-2.3) = -12$$
$$9.7 \approx 10 \qquad -11.9 \approx -12$$

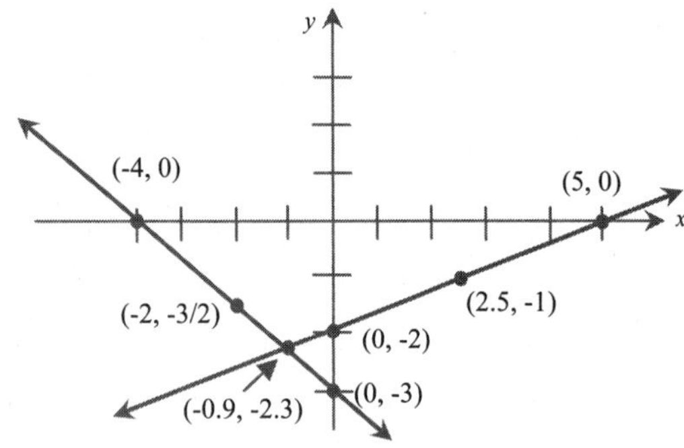

13. $y = -x + 3$ and $y = -2x + 3$

The slope of the first line is -1, and the y-intercept is 3.

The slope of the second line is -2, and the y-intercept is 3.

From the graph, the point of intersection is $(0, 3)$.

Therefore, the solution of the system of equations is

$x = 0,\ y = 3$

Check:

$$y = -x + 3 \qquad y = -2x + 3$$
$$3 = 0 + 3 \qquad 3 = 0 + 3$$
$$3 = 3 \qquad 3 = 3$$

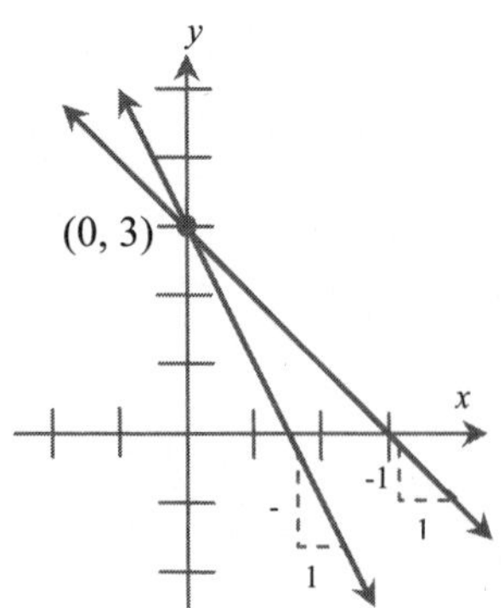

17. For the line $-2r_1 + 2r_2 = 7$

Let $r_1 = 0$ to find the r_2-int

$2r_2 = 7$

$r_2 = \frac{7}{2}$ $\quad r_2$-int is $\left(0, \frac{7}{2}\right)$

Let $r_2 = 0$ to find the r_1-int

$-2r_1 = 7$

$r_1 = -\frac{7}{2}$ $\quad r_1$-int is $\left(-\frac{7}{2}, 0\right)$

Let $r_1 = 1$ to find a third point

$-2(1) + 2r_2 = 7$

$2r_2 = 9$

$r_2 = \frac{9}{2}$ $\quad$ A third point is $\left(1, \frac{9}{2}\right)$

For the line $4r_1 - 2r_2 = 1$

Let $r_1 = 0$ to find the r_2-int

$-2r_2 = 1$

$r_2 = -\frac{1}{2}$ $\quad r_2$-int is $\left(0, -\frac{1}{2}\right)$

Let $r_2 = 0$ to find the r_1-int

$4r_1 = 1$

$r_1 = \frac{1}{4}$ $\quad r_1$-int is $\left(\frac{1}{4}, 0\right)$

Let $r_1 = 1$ to find a third point

$4(1) - 2r_2 = 1$

$-2r_2 = -3$

$r_2 = \frac{3}{2}$ $\quad$ A third point is $\left(1, \frac{3}{2}\right)$

From the graph the solution is approximately $(4, 7.5)$

$r_1 = 4.0,\ r_2 = 7.5$

Checking both equations,

$$-2r_1 + 2r_2 = 7 \qquad 4r_1 - 2r_2 = 1$$
$$-2(4.0) + 2(7.5) = 7 \qquad 4(4.0) - 2(7.5) = 1$$
$$7.0 = 7 \qquad 1.0 = 1$$

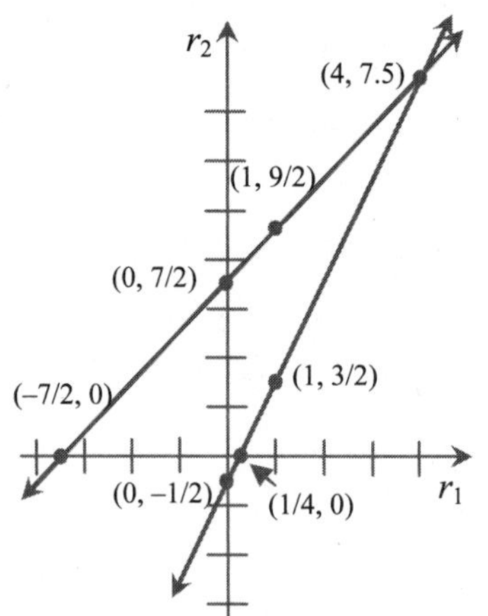

21. $x=4y+2$ and $3y=2x+3$

$y=\frac{x-2}{4}$ $\quad y=\frac{2x+3}{3}$

On a graphing calculator let

$y_1=\frac{x-2}{4}$ and $y_2=\frac{2x+3}{3}$

Using the intersect feature, the point of intersection is $(-3.6,-1.4)$, and the solution of the system of equations is

$x=-3.600$

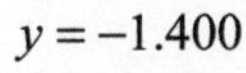
$y=-1.400$

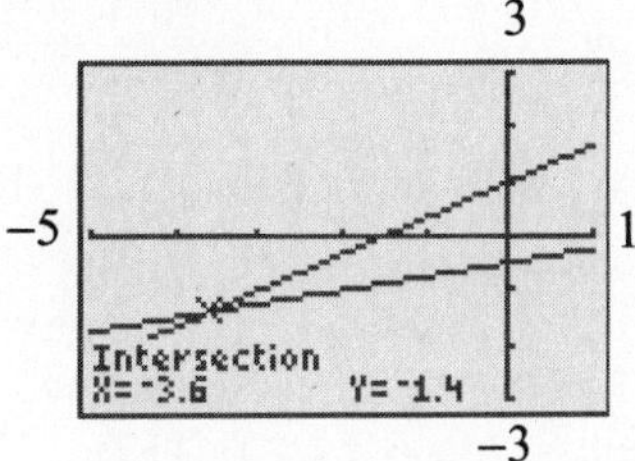

25. $x-5y=10$ and $2x-10y=20$

$y=\frac{x-10}{5}$ $\quad y=\frac{x-10}{5}$

On a graphing calculator let

$y_1=\frac{x-10}{5}$ and $y_2=\frac{x-10}{5}$

From the graph the lines are the same.

The system is dependent.

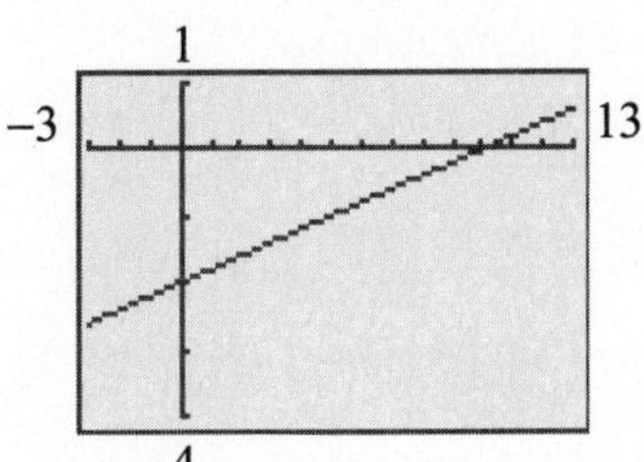

29. $5x-y=3$ and $4x=2y-3$

$y=5x-3$ $\quad y=\frac{4x+3}{2}$

On a graphing calculator let

$y_1=5x-3$ and $y_2=2x+1.5$

Using the intersect feature, the point of intersection is $(1.5, 4.5)$, and the solution to the system of equations is

$x=1.500$

$y=4.500$

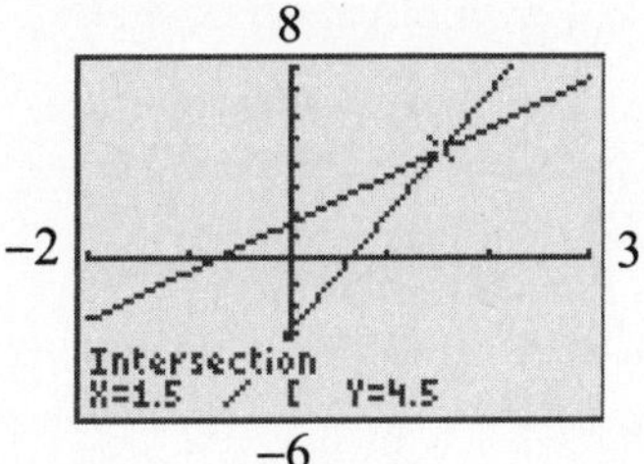

33. $0.8T_1-0.6T_2=12$ and $0.6T_1+0.8T_2=68$

$T_2=\frac{0.8T_1-12}{0.6}$ $\quad T_2=\frac{68-0.6T_1}{0.8}$

On a graphing calculator use

$x=T_1$ and $y=T_2$ and

let $y_1=\frac{0.8x-12}{6}$ and $y_2=\frac{68-0.6x}{0.8}$.

Using the intersect feature, the point of intersection is $(50, 47)$.

The tensions (to the nearest 1 N) are

$T_1=50$ N

$T_2=47$ N

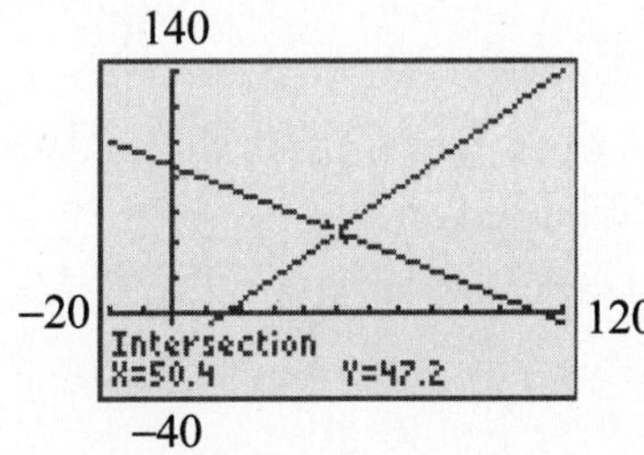

37. Let x = mass of Alloy 1 (70% lead, 30% zinc)

Let y = mass of Alloy 2 (40% lead, 60% zinc)

$x+y=120$ and $0.7x+0.4y=0.5(120)$

$y=-x+120$ $\quad y=-\frac{7}{4}x+150$

The slope of the first line is -1, and the y-intercept is 120.

The slope of the second line is $-7/4$, and the y-intercept is 150.

From the graph, the point of intersection is approximately$(40,80)$.

Therefore, the solution of the system of equations is

$x=40$ kg, mass of Alloy 1

$y=80$ kg, mass of Alloy 2

Check:

$x+y=120$ $\quad 0.7x+0.4y=60$

$80+40=120$ $0.7(40)+0.4(80)=60$

$120=120$ $60=60$

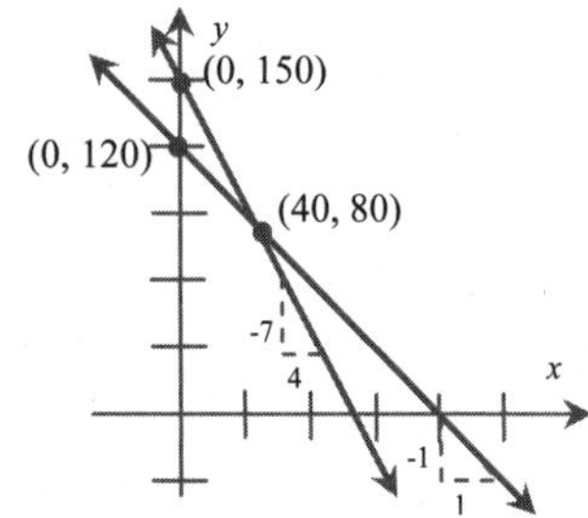

5.4 Solving Systems of Two Linear Equations in Two Unknowns Algebraically

Note to students: For all questions, substitution of the solutions into ***both*** equations serves as a check.

1. $x-3y=6$

$x=3y+6$ (A)

$2x-3y=3$ (B)

$2(3y+6)-3y=3$ substitute x from (A) into (B)

$6y+12-3y=3$

$3y=-9$

$y=-3$

$x=3(-3)+6$ substitute -3 for y into (A)

$x=-3$

The solution to the system is

$x=-3,\ y=-3$

5. $x=y+3$ Equation (A)

$x-2y=5$ Equation (B)

$(y+3)-2y=5$ substitute x from (A) into (B)

$-y=2$

$y=-2$

$x=-2+3$ substitute -2 for y into (A)

$x=1$

The solution to the system is

$x=1,\ y=-2$

9. $x+y=-5$

$y=-x-5$ Equation (A)

$2x-y=2$ Equation (B)

$2x-(-x-5)=2$ substitute y from (A) into (B)

$3x=-3$

$x=-1$

$y=-(-1)-5$ substitute -1 for x into (A)

$y=-4$

The solution to the system is

$x=-1,\ y=-4$

13. $33x+2y=34$

$y=\dfrac{34-33x}{2}$ Equation (A)

$40y=9x+11$

$40y-9x=11$ Equation (B)

$40\left(\dfrac{34-33x}{2}\right)-9x=11$ substitute y from (A) into (B)

$680-660x-9x=11$

$-669x=-669$

$x=1$

$y=\dfrac{34-33(1)}{2}$ substitute 1 for x into (A)

$y=\frac{1}{2}$

The solution to the system is

$x=1,\ y=\frac{1}{2}$

17. $2x-3y=4$ Equation (A)

$2x+y=-4$ Equation (B)

If we subtract Eq. (A) − Eq. (B)

$$\begin{array}{r} 2x-3y=\ \ 4 \\ -\quad 2x+\ \ y=-4 \\ \hline -4y=8 \end{array}$$

$y=-2$

$2x+(-2)=-4$ substitute -2 for y into (B)

$2x=-2$

$x=-1$

The solution to the system is

$x=-1,\ y=-2$

21. $v+2t=7$ Equation (A)

$2v+4t=9$ Equation (B)

If we subtract 2× Eq. (A) − Eq. (B)

$$\begin{array}{r} 2v+4t=14 \\ -\quad 2v+4t=\ \ 9 \\ \hline 0=5 \end{array}$$

The system of equations is inconsistent.

25. $2x - y = 5$ Equation (A)

$6x + 2y = -5$ Equation (B)

If we add $2\times$Eq. (A) + Eq. (B)

$$\begin{array}{r} 4x - 2y = 10 \\ +\ \ 6x + 2y = -5 \\ \hline 10x \quad = 5 \end{array}$$

$x = \frac{1}{2}$

$2\left(\frac{1}{2}\right) - y = 5$ substitute $\frac{1}{2}$ for x into (A)

$y = 1 - 5$

$y = -4$

The solution to the system is

$x = \frac{1}{2},\ y = -4$

29. $15x + 10y = 11$ Equation (A)

$20x - 25y = 7$ Equation (B)

If we add $5\times$Eq. (A) + $2\times$Eq. (B)

$$\begin{array}{r} 75x + 50y = 55 \\ +\ \ 40x - 50y = 14 \\ \hline 115x \quad = 69 \end{array}$$

$x = \frac{69}{115} = \frac{3}{5}$

$15\left(\frac{3}{5}\right) + 10y = 11$ substitute $\frac{3}{5}$ for x into (A)

$10y = 2$

$y = \frac{1}{5}$

The solution to the system is

$x = \frac{3}{5},\ y = \frac{1}{5}$

33. $44A = 1 - 15B$

$A = \dfrac{1-15B}{44}$ Equation (A)

$5B = 22 + 7A$ Equation (B)

$5B = 22 + 7\left(\dfrac{1-15B}{44}\right)$ substitute A from (A) into (B)

$5B = 22 + \dfrac{7}{44} - \dfrac{105}{44}B$

$\dfrac{220+105}{44}B = \dfrac{968+7}{44}$

$325B = 975$

$B = 3$

$A = \dfrac{1-15(3)}{44}$ substitute 3 for B into (A)

$A = -1$

The solution to the system is

$A = -1,\ B = 3$

37. $0.3x - 0.7y = 0.4$ multiply every term by 10

$3x - 7y = 4$ Equation (A)

$0.2x + 0.5y = 0.7$ multiply every term by 10

$2x + 5y = 7$ Equation (B)

If we subtract $2\times$Eq. (A) $- 3\times$Eq. (B)

$$\begin{array}{r} 6x - 14y = 8 \\ -\ \ 6x + 15y = 21 \\ \hline -29y = -13 \end{array}$$

$y = \frac{13}{29}$

$3x - 7\left(\frac{13}{29}\right) = 4$ substitute $\frac{13}{29}$ for y into (A)

$3x = \frac{116+91}{29} = \frac{207}{29}$

$x = \frac{69}{29}$

The solution to the system is

$x = \frac{69}{29},\ y = \frac{13}{29}$

41. $V_1 + V_2 = 15$

$V_2 = 15 - V_1$ Equation (A)

$V_1 - V_2 = 3$ Equation (B)

$V_1 - (15 - V_1) = 3$ substitute V_2 from (A) into (B)

$2V_1 = 18$

$V_1 = 9$

$V_2 = 15 - 9$ substitute 9 for V_1 into (A)

$V_2 = 6$

The solution to the system is

$V_1 = 9\text{ V},\ V_2 = 6\text{ V}$

45. Let x = number of regular email messages

Let y = number of spam messages

$x + y = 79$ Equation (A)

$y = 2x + 4$ Equation (B)

$x + (2x + 4) = 79$ substitute y from (B) into (A)

$3x = 75$

$x = 25$ regular messages

$y = 2(25) + 4$ substitute 25 for x into (B)

$y = 54$ spam messages

49. Let t_1 = time of flight for rocket 1

Let t_2 = time of light for heat-seeking rocket 2

The rockets will meet at the same distance travelled by both, and distance is velocity$\times$time,

$(d = vt)$

$600t_1 = 960t_2$ Equation (A)

The time elapsed by the first rocket will be larger

$t_1 = t_2 + 12$ Equation (B)

$600(t_2 + 12) = 960t_2$ substitute t_1 from (B) into (A)

$7200 = 360t_2$

$t_2 = 20$ s time elapsed for heat-seeking rocket 2

$t_1 = 20 + 12$ substitute 20 for t_2 into (B)

$t_1 = 32$ s time elapsed for rocket 1

53. Let x = windmill power capacity (in kW)

Let y = gas generator power capacity (in kW)

Energy produced = power × time

In a 10-day period, there are 240 h.

For the first 10-day period:

$$(0.450x)240 + y(240) = 3010$$

$$108x + 240y = 3010$$

$$y = \frac{3010 - 108x}{240} \quad \text{Equation (A)}$$

For the second 10-day period:

$$(0.720x)240 + y(240 - 60) = 2900$$

$$172.8x + 180y = 2900 \quad \text{Equation (B)}$$

$$172.8x + 180\left(\frac{3010 - 108x}{240}\right) = 2900 \quad \text{substitute } y \text{ from (A) into (B)}$$

$$172.8x + 2257.5 - 81x = 2900$$

$$91.8x = 642.5$$

$$x = 7.00 \text{ kW} \quad \text{substitute 7.00 for } x \text{ into (A)}$$

$$y = \frac{3010 - 108(7.00)}{240}$$

$$y = 9.39 \text{ kW}$$

The windmill power capacity is 7.00 kW, and the gas generator capacity is 9.39 kW.

57. $ax + y = c$

$bx + y = d$

In order to create a unique solution, the lines must have different slopes. Putting both equations into slope-intercept form,

$y = -ax + c$ has slope $-a$

$y = -bx + d$ has slope $-b$

To make the slopes different, $a \neq b$.

5.5 Solving Systems of Two Linear Equations in Two Unknowns by Determinants

Note to students: In all questions where solving a system of equations is required, substitution of the solutions into *both* equations serves as a check.

1. $\begin{vmatrix} 4 & -6 \\ 3 & 17 \end{vmatrix} = 4(17) - 3(-6) = 68 + 18 = 86$

5. $\begin{vmatrix} 2 & 4 \\ 3 & 1 \end{vmatrix} = (2)(1) - (3)(4) = 2 - 12 = -10$

9. $\begin{vmatrix} 8 & -10 \\ 0 & 4 \end{vmatrix} = (8)(4) - (0)(-10) = 32 - 0 = 32$

13. $\begin{vmatrix} 0.75 & -1.32 \\ 0.15 & 1.18 \end{vmatrix} = 0.75(1.18) - (0.15)$

$(-1.32) = 0.885 + 0.198 = 1.083$

17. $x + 2y = 5$

$x - 2y = 1$

$$x = \frac{\begin{vmatrix} 5 & 2 \\ 1 & -2 \end{vmatrix}}{\begin{vmatrix} 1 & 2 \\ 1 & -2 \end{vmatrix}} = \frac{5(-2) - 1(2)}{1(-2) - 1(2)} = \frac{-12}{-4} = 3$$

$$y = \frac{\begin{vmatrix} 1 & 5 \\ 1 & 1 \end{vmatrix}}{\begin{vmatrix} 1 & 2 \\ 1 & -2 \end{vmatrix}} = \frac{1(1) - 1(5)}{-4} = \frac{-4}{-4} = 1$$

21. Write both equations in standard form.

$12t + 9y = 14$

$6t - 7y = -16$

$$t = \frac{\begin{vmatrix} 14 & 9 \\ -16 & -7 \end{vmatrix}}{\begin{vmatrix} 12 & 9 \\ 6 & -7 \end{vmatrix}} = \frac{14(-7) - (-16)(9)}{12(-7) - 6(9)} = \frac{46}{-138} = -\frac{1}{3}$$

$$y = \frac{\begin{vmatrix} 12 & 14 \\ 6 & -16 \end{vmatrix}}{\begin{vmatrix} 12 & 9 \\ 6 & -7 \end{vmatrix}} = \frac{12(-16) - 6(14)}{-138} = \frac{-276}{-138} = 2$$

25. Rewrite both equations in standard form.

$2x-3y=4$

$3x-2y=-2$

$$x=\frac{\begin{vmatrix}4 & -3\\ -2 & -2\end{vmatrix}}{\begin{vmatrix}2 & -3\\ 3 & -2\end{vmatrix}}=\frac{4(-2)-(-2)(-3)}{2(-2)-3(-3)}=-\frac{14}{5}$$

$$y=\frac{\begin{vmatrix}2 & 4\\ 3 & -2\end{vmatrix}}{\begin{vmatrix}2 & -3\\ 3 & -2\end{vmatrix}}=\frac{2(-2)-3(4)}{5}=-\frac{16}{5}$$

29. $40s-30t=60$

$20s-40t=-50$

$$s=\frac{\begin{vmatrix}60 & -30\\ -50 & -40\end{vmatrix}}{\begin{vmatrix}40 & -30\\ 20 & -40\end{vmatrix}}=\frac{60(-40)-(-50)(-30)}{40(-40)-20(-30)}$$

$$=\frac{-3900}{-1000}=\frac{39}{10}$$

$$t=\frac{\begin{vmatrix}40 & 60\\ 20 & -50\end{vmatrix}}{\begin{vmatrix}40 & -30\\ 20 & -40\end{vmatrix}}=\frac{40(-50)-20(60)}{-1000}$$

$$=\frac{-3200}{-1000}=\frac{16}{5}$$

33. Write both equations in standard form.

$8.4x+1.2y=-10.8$

$3.5x+4.8y=-12.9$

$$x=\frac{\begin{vmatrix}-10.8 & 1.2\\ -12.9 & 4.8\end{vmatrix}}{\begin{vmatrix}8.4 & 1.2\\ 3.5 & 4.8\end{vmatrix}}=\frac{-10.8(4.8)-(-12.9)(1.2)}{8.4(4.8)-3.5(1.2)}$$

$$=\frac{-36.36}{36.12}=-1.0$$

$$y=\frac{\begin{vmatrix}8.4 & -10.8\\ 3.5 & -12.9\end{vmatrix}}{\begin{vmatrix}8.4 & 1.2\\ 3.5 & 4.8\end{vmatrix}}=\frac{8.4(-12.9)-3.5(-10.8)}{36.12}$$

$$=\frac{-70.56}{36.12}=-2.0$$

37. If $a=kb$, $c=kd$

$$\begin{vmatrix}a & b\\ c & d\end{vmatrix}=\begin{vmatrix}kb & b\\ kd & d\end{vmatrix}=kb(d)-kd(b)=kbd-kbd=0$$

41. $x+y=144$

$0.250x+0.375y=44.8$

$$x=\frac{\begin{vmatrix}144 & 1\\ 44.8 & 0.375\end{vmatrix}}{\begin{vmatrix}1 & 1\\ 0.250 & 0.375\end{vmatrix}}=\frac{144(0.375)-44.8(1)}{0.375-0.250}$$

$$=\frac{9.20}{0.125}=73.6\text{ L}$$

$$y=\frac{\begin{vmatrix}1 & 144\\ 0.250 & 44.8\end{vmatrix}}{\begin{vmatrix}1 & 1\\ 0.250 & 0.375\end{vmatrix}}=\frac{1(44.8)-0.250(144)}{0.125}$$

$$=\frac{8.80}{0.125}=70.4\text{ L}$$

45. x = number of phones

y = number of detectors

$x+y=320$

$110x+160y=40\ 700$

$$x=\frac{\begin{vmatrix}320 & 1\\ 40700 & 160\end{vmatrix}}{\begin{vmatrix}1 & 1\\ 110 & 160\end{vmatrix}}=\frac{320(160)-40700}{160-110}$$

$$=\frac{10500}{50}=210$$

$$y=\frac{\begin{vmatrix}1 & 320\\ 110 & 40700\end{vmatrix}}{\begin{vmatrix}1 & 1\\ 110 & 160\end{vmatrix}}=\frac{40700-110(320)}{50}$$

$$=\frac{5500}{50}=110$$

49. Let t_1 = time taken by drug boat

Let t_2 = time taken by Coast Guard

$$24\text{ min}=24\text{ min}\left(\frac{1\text{ h}}{60\text{ min}}\right)=0.40\text{ h}$$

We know the drug boat had a 0.40 h head start

$t_1=t_2+0.40$

$t_1-t_2=0.40$

Remember $d = vt$, and the total distance travelled by each boat is the same

$$63t_1 = 75t_2$$
$$63t_1 - 75t_2 = 0$$

$$t_1 = \frac{\begin{vmatrix} 0.4 & -1 \\ 0 & -75 \end{vmatrix}}{\begin{vmatrix} 1 & -1 \\ 63 & -75 \end{vmatrix}} = \frac{0.4(-75)-0}{-75-(-63)} = \frac{-30}{-12} = 2.5 \text{ h}$$

$$t_2 = \frac{\begin{vmatrix} 1 & 0.4 \\ 63 & 0 \end{vmatrix}}{\begin{vmatrix} 1 & -1 \\ 63 & -75 \end{vmatrix}} = \frac{0-63(0.4)}{-12} = \frac{-25.2}{-12} = 2.1 \text{ h}$$

5.6 Solving Systems of Three Linear Equations in Three Unknowns Algebraically

Note to students: In all questions where solving a system of equations is required, substitution of the solutions into *all three* equations serves as a check.

1.

(1)	$4x + y + 3z = 1$	
(2)	$2x - 2y + 6z = 12$	
(3)	$-6x + 3y + 12z = -14$	
(4)	$8x + 2y + 6z = 2$	(1) multiplied by 2
(2)	$2x - 2y + 6z = 12$	add
(5)	$10x + 12z = 14$	
(6)	$12x + 3y + 9z = 3$	(1) multiplied by 3
(3)	$-6x + 3y + 12z = -14$	subtract
(7)	$18x - 3z = 17$	
(8)	$72x - 12z = 68$	(7) multiplied by 4
(5)	$10x + 12z = 14$	add
(9)	$82x = 82$	
	$x = 1$	
(11)	$18(1) - 3z = 17$	substituting $x = 1$ into (7)
	$-3z = -1$	
	$z = \frac{1}{3}$	
(12)	$4(1) + y + 3\left(\frac{1}{3}\right) = 1$	substitute $x = 1$ and $z = \frac{1}{3}$ into (1)
	$4 + y + 1 = 1$	
	$y = -4$	

The solution is $x = 1$, $y = -4$, $z = \frac{1}{3}$

5.

(1)	$2x + 3y + z = 2$	
(2)	$-x + 2y + 3z = -1$	
(3)	$-3x - 3y + z = 0$	
(4)	$5x + 6y = 2$	Subtract (1) − (3)
(5)	$6x + 9y + 3z = 6$	multiply (1) by 3
(2)	$-x + 2y + 3z = -1$	subtract
(6)	$7x + 7y = 7$	
(7)	$5x + 5y = 5$	multiply (6) by 5/7
(4)	$5x + 6y = 2$	subtract
	$-y = 3$	
	$y = -3$	
(8)	$7x + 7(-3) = 7$	substitute −3 for y into (6)
	$7x = 28$	
	$x = 4$	
(9)	$-3(4) - 3(-3) + z = 0$	substitute 4 for x, and −3 for y into (3)
	$z = 3$	

The solution is $x = 4$, $y = -3$, $z = 3$.

9.

(1)	$2x - 2y + 3z = 5$	
(2)	$2x + y - 2z = -1$	
(3)	$4x - y - 3z = 0$	
(4)	$6x - 5z = -1$	add (2) and (3)
(5)	$4x + 2y - 4z = -2$	multiply (2) by 2
(1)	$2x - 2y + 3z = 5$	add
(6)	$6x - z = 3$	
(4)	$6x - 5z = -1$	subtract
	$4z = 4$	
	$z = 1$	
(7)	$6x - 1 = 3$	substitute 1 for z into (6)
	$x = \frac{4}{6} = \frac{2}{3}$	
(8)	$2\left(\frac{2}{3}\right) + y - 2(1) = -1$	substitute $\frac{2}{3}$ for x, and 1 for z into (2)
	$\frac{4}{3} + y - \frac{6}{3} = -\frac{3}{3}$	
	$y = -\frac{1}{3}$	

The solution is $x = \frac{2}{3}$, $y = -\frac{1}{3}$, $z = 1$.

13.

(1)	$10x+15y-25z=35$	
(2)	$40x-30y-20z=10$	
(3)	$16x-2y+8z=6$	
(4)	$20x+30y-50z=70$	multiply (1) by 2
(2)	$40x-30y-20z=10$	add
(5)	$60x-70z=80$	
(6)	$240x-30y+120z=90$	multiply (3) by 15
(2)	$40x-30y-20z=10$	subtract
(7)	$200x+140z=80$	
(8)	$120x-140z=160$	multiply (5) by 2, then add to (7)
(9)	$320x=240$	
	$x=\frac{3}{4}$	
(10)	$60\left(\frac{3}{4}\right)-70z=80$	substitute $\frac{3}{4}$ for x into (5)
	$-70z=35$	
	$z=-\frac{1}{2}$	
(11)	$16\left(\frac{3}{4}\right)-2y+8\left(-\frac{1}{2}\right)=6$	substitute $\frac{3}{4}$ for x, and $-\frac{1}{2}$ for z into (3)
	$12-2y-4=6$	
	$-2y=-2$	
	$y=1$	

The solution is $x=\frac{3}{4}$, $y=1$, $z=-\frac{1}{2}$.

17. $Ax+By+Cz=D$

(1)	$2A+4B+4C=12$	
(2)	$3A-2B+8C=12$	
(3)	$-A+8B+6C=12$	
(4)	$-2A+16B+12C=24$	multiply (3) by 2
(1)	$2A+4B+4C=12$	add
(5)	$20B+16C=36$	
(6)	$5B+4C=9$	divide (5) by 4
(7)	$-3A+24B+18C=36$	multiply (3) by 3
(2)	$3A-2B+8C=12$	add
(8)	$22B+26C=48$	
(9)	$11B+13C=24$	divide (8) by 2
(10)	$55B+44C=99$	multiply (6) by 11
(11)	$55B+65C=120$	multiply (9) by 5, subtract from (10)
	$-21C=-21$	
	$C=1$	
(12)	$5B+4(1)=9$	substitute 1 for C into (6)
	$5B=5$	
	$B=1$	
(13)	$-A+8(1)+6(1)=12$	substitute 1 for C and B into (3)
	$A=2$	

The solution is $A=2$, $B=1$, $C=1$.

$2x+y+z=12$

21.

(1)	$0.707F_1-0.800F_2=0$	
(2)	$0.707F_1+0.600F_2-F_3=10.0$	
(3)	$3.00F_2-3.00F_3=20.0$	
(4)	$-1.400F_2+F_3=-10.0$	subtract (1) − (2)
(5)	$-4.200F_2+3F_3=-30.0$	multiply (4) by 3
(3)	$3.00F_2-3.00F_3=20.0$	add
(6)	$-1.200F_2=-10.0$	
	$F_2=8.33$	
(7)	$3.00(8.33)-3.00F_3=20.0$	substitute 8.33 for F_2 into (3)
	$-3.00F_3=-5.00$	
	$F_3=1.67$	
(9)	$0.707F_1-0.800(8.33)=0$	substitute 8.33 for F_2 into (1)
	$0.707F_1=6.67$	
	$F_1=9.43$	

The solution is $F_1=9.43$ N, $F_2=8.33$ N, $F_3=1.67$ N.

25. $\theta = at^3 + bt^2 + ct$

$$
\begin{array}{lll}
(1) & 1.00a+1.00b+1.00c=19.0 & \\
(2) & 27.0a+9.00b+3.00c=30.9 & \\
(3) & 125a+25.0b+5.00c=19.8 & \\
\hline
(4) & 3.00a+3.00b+3.00c=57.0 & \text{multiply (1) by 3} \\
(2) & 27.0a+9.00b+3.00c=30.9 & \text{subtract} \\
\hline
(5) & -24.0a-6.00b=26.1 & \\
\\
(6) & 5.00a+5.00b+5.00c=95.0 & \text{multiply (1) by 5} \\
(3) & 125a+25.0b+5.00c=19.8 & \text{subtract} \\
\hline
(7) & -120a-20.0b=75.2 & \\
(8) & 120a+30.0b=-130.5 & \text{multiply (5) by } -5\text{, add to (7)} \\
\hline
(9) & 10.0b=-55.3 & \\
 & b=-5.53 & \\
\\
 & -24.0a-6.00(-5.53)=26.1 & \text{substitute } -5.53 \text{ for } b \text{ into (5)} \\
 & -24.0a=-7.08 & \\
 & a=0.295 & \\
\\
(10) & 0.295-5.53+c=19.0 & \text{substitute } a \text{ and } b \text{ into (1)} \\
 & c=24.235 &
\end{array}
$$

The solution is $a = 0.295,\ b = -5.53,\ c = 24.2.$

$$\theta = 0.295t^3 - 5.53t^2 + 24.2t$$

29.

$$
\begin{array}{lll}
(1) & x-2y-3z=2 & \\
(2) & x-4y-13z=14 & \\
(3) & -3x+5y+4z=0 & \\
\hline
(4) & 2y+10z=-12 & \text{subtract (1)} - \text{(2)} \\
(5) & y+5z=-6 & \text{divide (4) by 2} \\
\\
(6) & 3x-6y-9z=6 & \text{multiply (1) by 3} \\
(3) & -3x+5y+4z=0 & \text{add} \\
\hline
(7) & -y-5z=6 & \\
(8) & y+5z=-6 & \text{divide (7) by } -1 \\
(5) & y+5z=-6 & \text{subtract} \\
\hline
 & 0=0 &
\end{array}
$$

The system is dependent, there are an infinite number of solutions.
One possible solution, if we let $z = 0$,
$y = -6$ from Eq. (8) or Eq. (5).
Substituting into Eq. (1),

$x - 2(-6) - 0 = 2$

$x = -10$

A possible solution is $x = -10,\ y = -6,\ z = 0$

5.7 Solving Systems of Three Linear Equations in Three Unknowns by Determinants

Note to students: In all questions where solving a system of equations is required, substitution of the solutions into *all* equations serves as a check.

1. $\begin{vmatrix} -2 & 3 & -1 \\ 1 & 5 & 4 \\ 2 & -1 & 5 \end{vmatrix} \begin{matrix} -2 & 3 \\ 1 & 5 \\ 2 & -1 \end{matrix}$

$= -2(5)(5) + 3(4)(2) + (-1)(1)(-1) - 2(5)(-1) - (-1)(4)(-2) - 5(1)(3)$

$= -50 + 24 + 1 + 10 - 8 - 15$

$= -38$

This is the same determinant as that of Example 1. Interchanging a single pair of rows alters the determinant in sign only.

5. $\begin{vmatrix} 8 & 9 & -6 \\ -3 & 7 & 2 \\ 4 & -2 & 5 \end{vmatrix} \begin{matrix} 8 & 9 \\ -3 & 7 \\ 4 & -2 \end{matrix}$

$= 280 + 72 + (-36) - (-168) - (-32) - (-135) = 651$

9. $\begin{vmatrix} 4 & -3 & -11 \\ -9 & 2 & -2 \\ 0 & 1 & -5 \end{vmatrix} \begin{matrix} 4 & -3 \\ -9 & 2 \\ 0 & 1 \end{matrix}$

$= -40 + 0 + 99 - 0 - (-8) - (-135) = 202$

13. $\begin{vmatrix} 0.1 & -0.2 & 0 \\ -0.5 & 1 & 0.4 \\ -2 & 0.8 & 2 \end{vmatrix} \begin{matrix} 0.1 & -0.2 \\ -0.5 & 1 \\ -2 & 0.8 \end{matrix}$

$= 0.2 + 0.16 + 0 - 0 - 0.032 - 0.2 = 0.128$

17.

$$
\begin{aligned}
x + y + z &= 2 \\
x \quad - z &= 1 \\
x + y \quad &= 1
\end{aligned}
$$

$$x = \frac{\begin{vmatrix} 2 & 1 & 1 \\ 1 & 0 & -1 \\ 1 & 1 & 0 \end{vmatrix} \begin{matrix} 2 & 1 \\ 1 & 0 \\ 1 & 1 \end{matrix}}{\begin{vmatrix} 1 & 1 & 1 \\ 1 & 0 & -1 \\ 1 & 1 & 0 \end{vmatrix} \begin{matrix} 1 & 1 \\ 1 & 0 \\ 1 & 1 \end{matrix}}$$

$$= \frac{0 + (-1) + 1 - 0 - (-2) - 0}{0 + (-1) + 1 - 0 - (-1) - 0} = \frac{2}{1} = 2$$

$$y=\frac{\left|\begin{array}{ccc|cc}1 & 2 & 1 & 1 & 2\\ 1 & 1 & -1 & 1 & 1\\ 1 & 1 & 0 & 1 & 1\end{array}\right|}{1}$$

$$=\frac{0+(-2)+1-1-(-1)-0}{1}=\frac{-1}{1}=-1$$

$$z=\frac{\left|\begin{array}{ccc|cc}1 & 1 & 2 & 1 & 1\\ 1 & 0 & 1 & 1 & 0\\ 1 & 1 & 1 & 1 & 1\end{array}\right|}{1}$$

$$=\frac{0+1+2-0-1-1}{1}=\frac{1}{1}=1$$

Solution: $x=2,\ y=-1,\ z=1.$

21. $5l+6w-3h=6$

$4l-7w-2h=-3$

$3l+\ w-7h=-1$

$$l=\frac{\left|\begin{array}{ccc|cc}6 & 6 & -3 & 6 & 6\\ -3 & -7 & -2 & -3 & -7\\ 1 & 1 & -7 & 1 & 1\end{array}\right|}{\left|\begin{array}{ccc|cc}5 & 6 & -3 & 5 & 6\\ 4 & -7 & -2 & 4 & -7\\ 3 & 1 & -7 & 3 & 1\end{array}\right|}$$

$$=\frac{294+(-12)+9-21-(-12)-126}{245+(-36)+(-12)-63-(-10)-(-168)}$$

$$=\frac{156}{312}=\frac{1}{2}$$

$$w=\frac{\left|\begin{array}{ccc|cc}5 & 6 & -3 & 5 & 6\\ 4 & -3 & -2 & 4 & -3\\ 3 & 1 & -7 & 3 & 1\end{array}\right|}{312}$$

$$=\frac{105+(-36)+(-12)-27-(-10)-(-168)}{312}$$

$$=\frac{208}{312}=\frac{2}{3}$$

$$h=\frac{\left|\begin{array}{ccc|cc}5 & 6 & 6 & 5 & 6\\ 4 & -7 & -3 & 4 & -7\\ 3 & 1 & 1 & 3 & 1\end{array}\right|}{312}$$

$$=\frac{-35+(-54)+24-(-126)-(-15)-24}{312}$$

$$=\frac{52}{312}=\frac{1}{6}$$

Solution: $l=\frac{1}{2},\ w=\frac{2}{3},\ h=\frac{1}{6}$

25. $3x-7y+3z=6$

$3x+3y+6z=1$

$5x-5y+2z=5$

$$x=\frac{\left|\begin{array}{ccc|cc}6 & -7 & 3 & 6 & -7\\ 1 & 3 & 6 & 1 & 3\\ 5 & -5 & 2 & 5 & -5\end{array}\right|}{\left|\begin{array}{ccc|cc}3 & -7 & 3 & 3 & -7\\ 3 & 3 & 6 & 3 & 3\\ 5 & -5 & 2 & 5 & -5\end{array}\right|}$$

$$=\frac{36+(-210)+(-15)-45-(-180)-(-14)}{18+(-210)+(-45)-45-(-90)-(-42)}$$

$$=\frac{-40}{-150}=\frac{4}{15}$$

$$y=\frac{\left|\begin{array}{ccc|cc}3 & 6 & 3 & 3 & 6\\ 3 & 1 & 6 & 3 & 1\\ 5 & 5 & 2 & 5 & 5\end{array}\right|}{-150}$$

$$=\frac{6+180+45-15-90-36}{-150}$$

$$=\frac{90}{-150}=-\frac{3}{5}$$

$$z=\frac{\left|\begin{array}{ccc|cc}3 & -7 & 6 & 3 & -7\\ 3 & 3 & 1 & 3 & 3\\ 5 & -5 & 5 & 5 & -5\end{array}\right|}{-150}$$

$$=\frac{45+(-35)+(-90)-90-(-15)-(-105)}{-150}$$

$$=\frac{-50}{-150}=\frac{1}{3}$$

Solution: $x=\frac{4}{15},\ y=-\frac{3}{5},\ z=\frac{1}{3}$

29. $3.0x+4.5y-7.5z=10.5$

$4.8x-3.6y-2.4z=\ 1.2$

$4.0x-0.5y+2.0z=\ 1.5$

$$x=\frac{\left|\begin{array}{ccc|cc}10.5 & 4.5 & -7.5 & 10.5 & 4.5\\ 1.2 & -3.6 & -2.4 & 1.2 & -3.6\\ 1.5 & -0.5 & 2.0 & 1.5 & -0.5\end{array}\right|}{\left|\begin{array}{ccc|cc}3.0 & 4.5 & -7.5 & 3.0 & 4.5\\ 4.8 & -3.6 & -2.4 & 4.8 & -3.6\\ 4.0 & -0.5 & 2.0 & 4.0 & -0.5\end{array}\right|}$$

$$x=\frac{-75.6+(-16.2)+4.5-40.5-12.6-10.8}{-21.6+(-43.2)+18.0-108-3.6-43.2}$$

$$=\frac{-151.2}{-201.6}=\frac{3}{4}$$

$$y=\frac{\begin{vmatrix}3.0 & 10.5 & -7.5\\ 4.8 & 1.2 & -2.4\\ 4.0 & 1.5 & 2.0\end{vmatrix}\begin{matrix}3.0 & 10.5\\ 4.8 & 1.2\\ 4.0 & 1.5\end{matrix}}{-201.6}$$

$$y=\frac{7.2+(-100.8)+(-54.0)-(-36.0)-(-10.8)-100.8}{-201.6}$$

$$=\frac{-201.6}{-201.6}=1$$

$$z=\frac{\begin{vmatrix}3.0 & 4.5 & 10.5\\ 4.8 & -3.6 & 1.2\\ 4.0 & -0.5 & 1.5\end{vmatrix}\begin{matrix}3.0 & 4.5\\ 4.8 & -3.6\\ 4.0 & -0.5\end{matrix}}{-201.6}$$

$$z=\frac{-16.2+21.6+(-25.2)-(-151.2)-(-1.8)-32.4}{-201.6}$$

$$=\frac{100.8}{-201.6}=-\frac{1}{2}$$

Solution: $x=\frac{3}{4}$, $y=1$, $z=-\frac{1}{2}$.

33. $\begin{vmatrix}4 & 2 & 1\\ 7 & 8 & 6\\ 7 & 9 & 8\end{vmatrix}=256+84+63-56-216-112=19$

Adding a multiple of one row to another does not change the value of the determinant.

37.
$$s_0+2v_0+2a=20$$
$$s_0+4v_0+8a=54$$
$$s_0+6v_0+18a=104$$

$$s_0=\frac{\begin{vmatrix}20 & 2 & 2\\ 54 & 4 & 8\\ 104 & 6 & 18\end{vmatrix}}{\begin{vmatrix}1 & 2 & 2\\ 1 & 4 & 8\\ 1 & 6 & 18\end{vmatrix}}$$

$$=\frac{1440+1664+648-832-960-1944}{72+16+12-8-48-36}$$

$$=\frac{16}{8}=2$$

$$v_0=\frac{\begin{vmatrix}1 & 20 & 2\\ 1 & 54 & 8\\ 1 & 104 & 18\end{vmatrix}}{8}$$

$$=\frac{972+160+208-108-832-360}{8}$$

$$=\frac{40}{8}=5$$

$$a=\frac{\begin{vmatrix}1 & 2 & 20\\ 1 & 4 & 54\\ 1 & 6 & 104\end{vmatrix}}{8}$$

$$=\frac{416+108+120-80-324-208}{8}$$

$$=\frac{32}{8}=4$$

Solution: $s_0=2.00$ m, $v_0=5.00$ m/s, $a=4.00$ m/s^2

41. Let x = percent of nickel
Let y = percent of iron
Let z = percent of molybdenum

$$x+y+z=100$$
$$x-5y=-1$$
$$y-3z=1$$

$$x=\frac{\begin{vmatrix}100 & 1 & 1\\ -1 & -5 & 0\\ 1 & 1 & -3\end{vmatrix}}{\begin{vmatrix}1 & 1 & 1\\ 1 & -5 & 0\\ 0 & 1 & -3\end{vmatrix}}$$

$$=\frac{1500+0+(-1)-(-5)-0-3}{15+0+1-0-0-(-3)}$$

$$=\frac{1501}{19}=79\%\text{ nickel}$$

$$y=\frac{\begin{vmatrix}1 & 100 & 1\\ 1 & -1 & 0\\ 0 & 1 & -3\end{vmatrix}}{19}$$

$$=\frac{3+0+1-0-0-(-300)}{19}$$

$$=\frac{304}{19}=16\%\text{ iron}$$

$$z=\frac{\begin{vmatrix}1 & 1 & 100\\ 1 & -5 & -1\\ 0 & 1 & 1\end{vmatrix}}{19}$$

$$=\frac{-5+0+100-0-(-1)-1}{19}$$

$$=\frac{95}{19}=5\%\text{ molybdenum}$$

45.
$$F_1 + F_2 + F_3 = 125$$
$$-0.174F_1 - 0.985F_2 + 0.927F_3 = 0$$
$$0.985F_1 - 0.174F_2 - 0.375F_3 = 0$$

$$F_1 = \frac{\begin{vmatrix} 125 & 1 & 1 \\ 0 & -0.985 & 0.927 \\ 0 & -0.174 & -0.375 \end{vmatrix}}{\begin{vmatrix} 1 & 1 & 1 \\ -0.174 & -0.985 & 0.927 \\ 0.985 & -0.174 & -0.375 \end{vmatrix}}$$

$$F_1 = \frac{46.171875 + 0 + 0 - 0 - (-20.16225)}{0.369375 + 0.913095 + 0.030276 - (-0.970225) - (-0.161298) - 0.06525}$$

$$F_1 = \frac{66.334125}{2.379019} = 27.9 \text{ N}$$

$$F_2 = \frac{\begin{vmatrix} 1 & 125 & 1 \\ -0.174 & 0 & 0.927 \\ 0.985 & 0 & -0.375 \end{vmatrix}}{2.379019}$$

$$= \frac{0 + 114.136875 + 0 - 0 - 0 - 8.15625}{2.379019}$$

$$F_2 = \frac{105.980625}{2.379019} = 44.5 \text{ N}$$

$$F_3 = \frac{\begin{vmatrix} 1 & 1 & 125 \\ -0.174 & -0.985 & 0 \\ 0.985 & -0.174 & 0 \end{vmatrix}}{2.379019}$$

$$= \frac{0 + 0 + 3.7845 - (-121.278125) - 0 - 0}{2.379019}$$

$$F_3 = \frac{125.062625}{2.379019} = 52.6 \text{ N}$$

Review Exercises

1. $\begin{vmatrix} -2 & 5 \\ 3 & 1 \end{vmatrix} = (-2)(1) - (3)(5) = -2 - 15 = -17$

5. Find the slope that passes through points $(2,0)$, $(4,-8)$

$$m = \frac{y_2 - y_1}{x_2 - x_1} = \frac{-8 - 0}{4 - 2} = \frac{-8}{2} = -4$$

9. Comparing $y = -2x + 4$ to $y = mx + b$ gives a slope of -2 and y-intercept of 4.

Plot (0,4), then move right one unit and down 2 units and plot a second point.

Draw a line passing through both points.

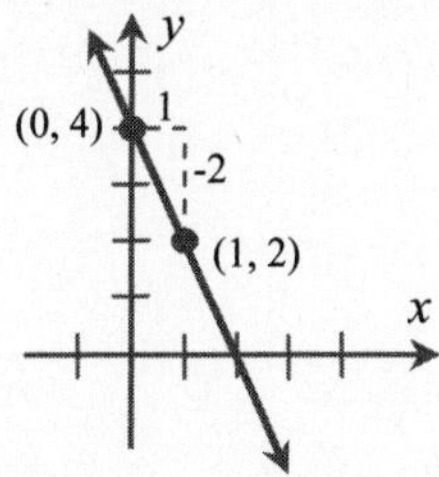

13. $y = 2x - 4$ and $y = -\frac{3}{2}x + 3$

The slope of the first line is 2, and the y-intercept is -4.

The slope of the second line is $-\frac{3}{2}$ and the y-intercept is 3.

From the graph, the point of intersection is approximately at $(2, 0)$.

Therefore, the solution of the system of equations is $x = 2$, $y = 0$.

Check:

$y = 2x - 4$	$y = -\frac{3}{2}x + 3$
$0 = 2(2) - 4$	$0 = -\frac{3}{2}(2) + 3$
$0 = 0$	$0 = 0$

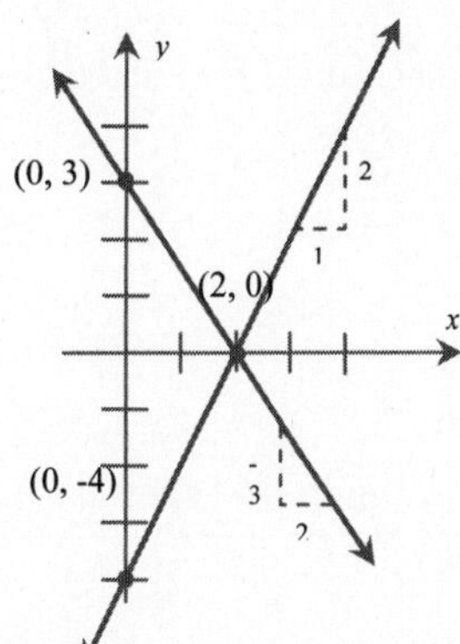

17. $7x = 2y + 14$ $\quad y = -4x + 4$

$2y = 7x - 14$

$y = \frac{7}{2}x - 7$

The slope of the first line is $\frac{7}{2}$, and the y-intercept is -7.

The slope of the second line is -4 and the y-intercept is 4.

From the graph, the point of intersection is approximately at $(1.5, -1.9)$.

Therefore, the solution of the system of equations is

$x = 1.5,\ y = -1.9$

Check:

$$7x = 2y + 14 \qquad y = -4x + 4$$
$$7(1.5) = 2(-1.9) + 14 \qquad -1.9 = -4(1.5) + 4$$
$$10.5 \approx 10.2 \qquad -1.9 \approx 2.0$$

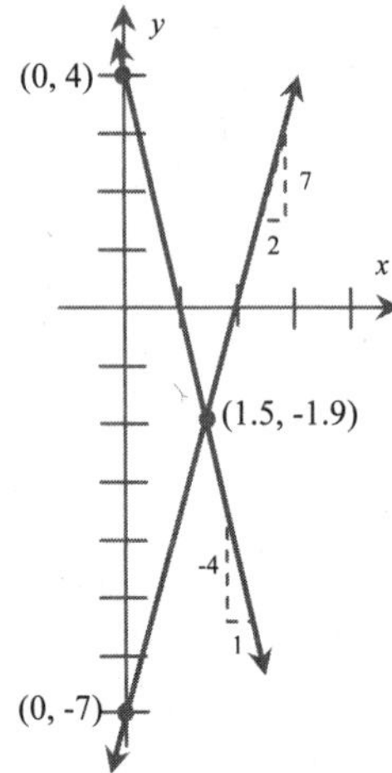

21. (1) $x + 2y = 5$

(2) $x + 3y = 7$ subtract

$$-y = -2$$
$$y = 2$$
$$x + 2(2) = 5 \quad \text{substitute 2 for } y \text{ into (1)}$$
$$x = 1$$

The solution is $x = 1,\ y = 2$.

25. (1) $10i - 27v = 29$

(2) $40i + 33v = 69$

(3) $40i - 108v = 116$ multiply (1) by 4, subtract

(4) $141v = -47$

$$v = -\tfrac{47}{141}$$
$$v = -\tfrac{1}{3}$$

(5) $10i - 27\left(-\tfrac{1}{3}\right) = 29$ substitute $-\tfrac{1}{3}$ for v into (1)

$$10i + 9 = 29$$
$$10i = 20$$
$$i = 2$$

The solution is $i = 2,\ v = -\tfrac{1}{3}$.

29. (1) $90x - 110y = 40$

(2) $30x - 15y = 25$

(3) $90x - 45y = 75$ multiply (2) by 3

(1) $90x - 110y = 40$ subtract

$$65y = 35$$
$$y = \tfrac{35}{65} = \tfrac{7}{13}$$

(4) $30x - 15\left(\tfrac{7}{13}\right) = 25$ substitute $\tfrac{7}{13}$ for y into (2)

$$30x = \tfrac{325+105}{13}$$
$$x = \tfrac{430}{390}$$
$$x = \tfrac{43}{39}$$

The solution is $x = \tfrac{43}{39},\ y = \tfrac{7}{13}$.

33. Solve for $4x + 3y = -4$ and $y = 2x - 3$

Expressing both in standard form:

$4x + 3y = -4$

$2x - y = 3$

$$x = \frac{\begin{vmatrix} -4 & 3 \\ 3 & -1 \end{vmatrix}}{\begin{vmatrix} 4 & 3 \\ 2 & -1 \end{vmatrix}} = \frac{-4(-1) - 3(3)}{4(-1) - 2(3)} = \frac{4 - 9}{-4 - 6} = \frac{-5}{-10} = \frac{1}{2}$$

$$y = \frac{\begin{vmatrix} 4 & -4 \\ 2 & 3 \end{vmatrix}}{-10} = \frac{4(3) - 2(-4)}{-10} = \frac{12 + 8}{-10} = \frac{20}{-10} = -2$$

The solution is $x = \tfrac{1}{2},\ y = -2$.

37. Solve for $7x = 2y - 6$ and $7y = 12 - 4x$

Expressing both in standard form:

$7x - 2y = -6$

$4x + 7y = 12$

$$x = \frac{\begin{vmatrix} -6 & -2 \\ 12 & 7 \end{vmatrix}}{\begin{vmatrix} 7 & -2 \\ 4 & 7 \end{vmatrix}} = \frac{-6(7) - 12(-2)}{7(7) - 4(-2)}$$

$$= \frac{-42 + 24}{49 + 8} = \frac{-18}{57} = -\frac{6}{19}$$

$$y = \frac{\begin{vmatrix} 7 & -6 \\ 4 & 12 \end{vmatrix}}{57} = \frac{7(12) - 4(-6)}{57}$$

$$= \frac{84 + 24}{57} = \frac{108}{57} = \frac{36}{19}$$

The solutions are $x = -\tfrac{6}{19},\ y = \tfrac{36}{19}$.

41. Exercise 33 is most easily solved by substitution because the second equation is already solved for y.

45. $\begin{vmatrix} 4 & -1 & 8 \\ -1 & 6 & -2 \\ 2 & 1 & -1 \end{vmatrix} \begin{matrix} 4 & -1 \\ -1 & 6 \\ 2 & 1 \end{matrix}$

$= -24 + 4 + (-8) - 96 - (-8) - (-1) = -115$

49.

(1)	$2x + y + z = 4$	
(2)	$x - 2y - z = 3$	
(3)	$3x + 3y - 2z = 1$	
(4)	$3x - y = 7$	Add (1) + (2)
(5)	$4x + 2y + 2z = 8$	multiply (1) by 2
(3)	$3x + 3y - 2z = 1$	add
(6)	$7x + 5y = 9$	
(7)	$15x - 5y = 35$	multiply (4) by 5
(6)	$7x + 5y = 9$	add
	$22x = 44$	
	$x = 2$	
(8)	$3(2) - y = 7$	substitute 2 for x into (4)
	$y = -1$	
(9)	$2(2) + (-1) + z = 4$	substitute for x and y into (1)
	$z = 1$	

The solution is $x = 2$, $y = -1$, $z = 1$.

53.

(1)	$3.6x + 5.2y - z = -2.2$	
(2)	$3.2x - 4.8y + 3.9z = 8.1$	
(3)	$6.4x + 4.1y + 2.3z = 5.1$	
(4)	$14.04x + 20.28y - 3.9z = -8.58$	multiply (1) by 3.9
(2)	$3.2x - 4.8y + 3.9z = 8.1$	add
(5)	$17.24x + 15.48y = -0.48$	
(6)	$8.28x + 11.96y - 2.3z = -5.06$	multiply (1) by 2.3
(3)	$6.4x + 4.1y + 2.3z = 5.1$	add
(7)	$14.68x + 16.06y = 0.04$	
(8)	$253.0832x + 227.2464y = -7.0464$	multiply (5) by 14.68
(9)	$253.0832x + 276.8744y = 0.6896$	multiply (7) by 17.24, subtract
	$-49.628y = -7.736$	
	$y = 0.15588$	
(10)	$14.68x + 16.06(0.15588) = 0.04$	substitute for y into (7)
	$14.68x = -2.463429$	
	$x = -0.16781$	
(11)	$3.6(-0.16781) + 5.2(0.15588) - z = -2.2$	substitute for x and y into (1)
	$-z = -2.40646$	
	$z = 2.40646$	

The solution is $x = -0.168$, $y = 0.156$, $z = 2.41$.

57. $2r+s+2t=8$
$3r-2s-4t=5$
$-2r+3s+4t=-3$

$$r=\frac{\begin{vmatrix}8&1&2\\5&-2&-4\\-3&3&4\end{vmatrix}}{\begin{vmatrix}2&1&2\\3&-2&-4\\-2&3&4\end{vmatrix}}$$
$$=\frac{-64+12+30-12-(-96)-20}{-16+8+18-8-(-24)-12}$$
$$=\frac{42}{14}=3$$

$$s=\frac{\begin{vmatrix}2&8&2\\3&5&-4\\-2&-3&4\end{vmatrix}}{14}$$
$$=\frac{40+64+(-18)-(-20)-24-96}{14}$$
$$=\frac{-14}{14}=-1$$

$$t=\frac{\begin{vmatrix}2&1&8\\3&-2&5\\-2&3&-3\end{vmatrix}}{14}$$
$$=\frac{12+(-10)+72-32-30-(-9)}{14}$$
$$=\frac{21}{14}=\frac{3}{2}$$

Solution is $r=3,\ s=-1,\ t=\frac{3}{2}$.

61. $\begin{vmatrix}2&5\\1&x\end{vmatrix}=3$
$2x-5=3$
$2x=8$
$x=4$

65. Let $u=\frac{1}{x}$ and $v=\frac{1}{y}$

(1) $\frac{1}{x}-\frac{1}{y}=\frac{1}{2}$ first original equation

(2) $u-v=\frac{1}{2}$

(3) $\frac{1}{x}+\frac{1}{y}=\frac{1}{4}$ second original equation

(4) $u+v=\frac{1}{4}$

---

(5) $2u=\frac{3}{4}$ adding (2)+(4)

(6) $u=\frac{3}{8}$

(7) $\frac{3}{8}+v=\frac{1}{4}$ substitute u into (4)

(8) $v=-\frac{1}{8}$

$x=\frac{8}{3}$ reciprocal of (6)

$y=-8$ reciprocal of (8)

The solution is $\left(\frac{8}{3},-8\right)$.

69. To be dependent, the equations must be the same.

(1) $3x-ky=6$

(2) $x+2y=2$

(3) $3x+6y=6$ Multiplying (2) by 3

Comparing Eq. (1) with Eq. (3), we have $-k=6$ or $k=-6$ to make the system dependent.

73. $F_1+2.0F_2=26000$
$0.87F_1-F_3=0$
$3.0F_1-4.0F_2=54000$

$$F_1=\frac{\begin{vmatrix}26\,000&2.0&0\\0&0&-1\\54\,000&-4.0&0\end{vmatrix}}{\begin{vmatrix}1&2.0&0\\0.87&0&-1\\3.0&-4.0&0\end{vmatrix}}$$
$$=\frac{0+(-108000)+0-0-104000-0}{0+(-6)+0-0-4-0}$$
$$=\frac{-212\,000}{-10}=21\,000\text{ N}$$

$$F_2=\frac{\begin{vmatrix}1&26\,000&0\\0.87&0&-1\\3.0&54\,000&0\end{vmatrix}}{-10}$$
$$=\frac{0+(-78000)+0-0-(-54000)-0}{-10}$$
$$=\frac{-24\,000}{-10}=2400\text{ N}$$

$$F_3 = \frac{\begin{vmatrix} 1 & 2.0 & 26\,000 \\ 0.87 & 0 & 0 \\ 3.0 & -4 & 54\,000 \end{vmatrix}}{-10}$$

$$= \frac{0+0+(-90480)-0-0-93960}{-10}$$

$$= \frac{-184\,440}{-10} = 18\,000 \text{ N}$$

77. Let x = monthly sales (in \$)

Let I = income derived from payment scheme

(1)	$I = 0.10x$	payment scheme 1
(2)	$I = 2400 + 0.04x$	payment scheme 2
(3)	$0.10x = 2400 + 0.04x$	substitute (1) into (2)
	$0.06x = 2400$	
	$x = 40000$	
	$I = 0.10(40000)$	substitute x into (1)
	$I = 4000$	

\$40 000 is the monthly sales that produces

\$4 000 in income with either payment scheme.

81. $T = \dfrac{a}{x+100} + b$

(1)	$\dfrac{a}{0+100} + b = 14$	for measurement 1
(2)	$0.01a + b = 14$	rewriting (1)
(3)	$\dfrac{a}{900+100} + b = 10$	for measurement 2
(4)	$0.001a + b = 10$	rewriting (3)
(5)	$0.009a = 4$	subtract (2) − (4)
	$a = 444 \text{ m}\cdot{}^\circ\text{C}$	
(6)	$0.01(444) + b = 14$	substitute a into (2)
	$b = 9.56\ {}^\circ\text{C}$	

$$T(x) = \frac{444}{x+100} + 9.56$$

85. $P = I^2R$ for each resistor.

Let R_1 = magnitude of resistance 1

Let R_2 = magnitude of resistance 2

(1)	$(1.0)^2 R_1 + (3.0)^2 R_2 = 14.0$	for measurement 1
(2)	$1.0R_1 + 9.0R_2 = 14.0$	rewriting (1)
(3)	$(3.0)^2 R_1 + (1.0)^2 R_2 = 6.0$	for measurement 2
(4)	$9.0R_1 + 1.0R_2 = 6.0$	rewriting (3)
(5)	$9.0R_1 + 81.0R_2 = 126$	multiply (2) by 9, subtract
	$-80.0R_2 = -120$	
	$R_2 = 1.50\ \Omega$	
(6)	$1.0R_1 + 9.0(1.50) = 14.0$	substitute R_2 into (2)
	$R_1 = 0.500\ \Omega$	

89.

(1)	$A + B + C = 180$	true for all triangles
(2)	$A = 2B - 55$	
(3)	$C = B - 25$	
(4)	$(2B-55) + B + (B-25) = 180$	substituting (2) and (3) into (1)
	$4B = 260$	
	$B = 65$	
	$C = 65 - 25$	substitute B into (3)
	$C = 40$	
	$A = 2(65) - 55$	substitute B into (2)
	$A = 75$	

The angles are $A = 75.0^\circ$, $B = 65.0^\circ$, $C = 40.0^\circ$

93. x = weight of gold in air

y = weight of silver in air

$$x + y = 6.0$$

$$0.947x + 0.9y = 5.6$$

$$x = \frac{\begin{vmatrix} 6.0 & 1 \\ 5.6 & 0.9 \end{vmatrix}}{\begin{vmatrix} 1 & 1 \\ 0.947 & 0.9 \end{vmatrix}} = \frac{5.4 - 5.6}{0.9 - 0.947} = \frac{-0.2}{-0.047} = 4.26 \text{ N}$$

$$y = \frac{\begin{vmatrix} 1 & 6.0 \\ 0.947 & 5.6 \end{vmatrix}}{\begin{vmatrix} 1 & 1 \\ 0.947 & 0.9 \end{vmatrix}} = \frac{5.6 - 5.682}{-0.047} = \frac{-0.082}{-0.047} = 1.74 \text{ N}$$

CHAPTER 6

FACTORING AND FRACTIONS

6.1 Special Products

1. $(3r-2s)(3r+2s)=(3r)^2-(2s)^2$
$=9r^2-4s^2$

5. $40(x-y)=40x-40y$

9. $(T+6)(T-6)=T^2-6^2$
$=T^2-36$

13. $(4x-5y)(4x+5y)=(4x)^2-(5y)^2$
$=16x^2-25y^2$

17. $(5f+4)^2=(5f)^2+2(5f)(4)+4^2$
$=25f^2+40f+16$

21. $(L^2-1)^2=(L^2)^2-2\cdot L^2\cdot 1+1^2$
$=L^4-2L^2+1$

25. $(0.6s-t)^2=(0.6s)^2-2(0.6s)(t)+t^2$
$=0.36s^2-1.2st+t^2$

29. $(3+C^2)(6+C^2)=18+(3C^2+6C^2)+(C^2)^2$
$=18+9C^2+C^4$

33. $(10v-3)(4v+15)=40v^2+(150-12)v-45$
$=40v^2+138v-45$

37. $2(x-2)(x+2)=2(x^2-4)$
$=2x^2-8$

41. $6a(x+2b)^2=6a(x^2+4bx+4b^2)$
$=6ax^2+24abx+24ab^2$

45. $[(2R+3r)(2R-3r)]^2=[4R^2-9r^2]^2$
$=16R^4-72R^2r^2+81r^4$

49. $(3-x-y)^2=[3-(x+y)]^2$
$=9-6(x+y)+(x+y)^2$
$=9-6x-6y+x^2+2xy+y^2$

53. $(3L+7R)^3=(3L)^3+3(3L)^2(7R)+3(3L)(7R)^2+(7R)^3$
$=27L^3+189L^2R+441LR^2+343R^3$

57. $(x+2)(x^2-2x+4)=x^3-2x^2+4x+2x^2-4x+8$
$=x^3+8$

61. $(x+y)^2(x-y)^2=(x^2+2xy+y^2)(x^2-2xy+y^2)$
$=x^4-2x^3y+x^2y^2+2x^3y-4x^2y^2$
$+2xy^3+x^2y^2-2xy^3+y^4$
$=x^4-2x^2y^2+y^4$

65. $4(p+DA)^2=4(p^2+2pDA+D^2A^2)$
$=4p^2+8pDA+4D^2A^2$

69. $\frac{L}{6}(x-a)^3=\frac{L}{6}[x^3-3x^2a+3xa^2-a^3]$
$=\frac{L}{6}x^3-\frac{L}{2}ax^2+\frac{L}{2}a^2x-\frac{L}{6}a^3$

73. $(49)(51)=(50-1)(50+1)$
$=50^2-1^2$
$=2500-1$
$=2499$

77. **(a)** $A=(x+y)^2$
$A=x^2+2xy+y^2$
(b) $A=x^2+xy+xy+y^2$
$A=x^2+2xy+y^2$
This is special product 6.3 shown geometrically.

6.2 Factoring: Common Factor and Difference of Squares

1. $4ax^2 - 2ax = 2ax\left(\frac{4ax^2}{2ax} - \frac{2ax}{2ax}\right)$
$= 2ax(2x-1)$

5. $6x + 6y = 6(x+y)$

9. $3x^2 - 9x = 3x(x-3)$

13. $288n^2 + 24n = 24n(12n+1)$

17. $3ab^2 - 6ab + 12ab^3 = 3ab(b - 2 + 4b^2)$

21. $2a^2 - 2b^2 + 4c^2 - 6d^2 = 2(a^2 - b^2 + 2c^2 - 3d^2)$

25. $100 - 9A^2 = (10)^2 - (3A)^2$
$100 - 9A^2 = (10+3A)(10-3A)$

29. $162s^2 - 50t^2 = 2(81s^2 - 25t^2)$
$= 2(9s+5t)(9s-5t)$

33. $(x+y)^2 - 9 = ((x+y)+3)((x+y)-3)$
$= (x+y+3)(x+y-3)$

37. $300x^2 - 2700z^2 = 300(x^2 - 9z^2)$
$= 300(x+3z)(x-3z)$

41. $x^4 - 16 = (x^2+4)(x^2-4)$
$= (x^2+4)(x+2)(x-2)$

45. Solve $2a - b = ab + 3$ for a.
$2a - ab = b + 3$
$a(2-b) = b + 3$
$a = \frac{b+3}{2-b}$

49. $(x+2k)(x-2) = x^2 + 3x - 4k$
$x^2 - 2x + 2kx - 4k = x^2 + 3x - 4k$
$2kx = 5x$
$k = \frac{5}{2}$

53. $a^2 + ax - ab - bx = (a^2 + ax) - (ab + bx)$
$= a(a+x) - b(a+x)$
$= (a+x)(a-b)$

57. $x^2 - y^2 + x - y = (x^2 - y^2) + (x - y)$
$= (x+y)(x-y) + (x-y)$
$= (x-y)(x+y+1)$

61. $n^2 + n = n(n+1)$, this the product of two consecutive integers of which one must be even. Therefore, the product is even.

65. $Rv + Rv^2 + Rv^3 = Rv(1 + v + v^2)$

69.

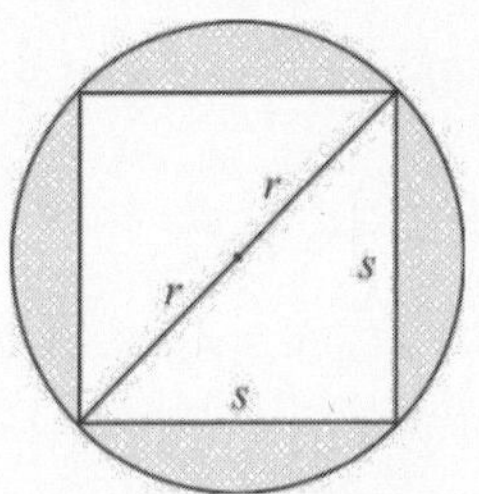

Using Pythagoras' Theorem
$s^2 + s^2 = (2r)^2$
$2s^2 = 4r^2$
$s^2 = 2r^2$
Area of square $= s^2 = 2r^2$
Area of circle $= \pi r^2$
Area left = Area of circle − Area of square
Area left $= \pi r^2 - 2r^2$
Area left $= r^2(\pi - 2)$

73. $3BY + 5Y = 9BS$
$3BY - 9BS = -5Y$
$B(3Y - 9S) = -5Y$
$B = \frac{-5Y}{3Y - 9S}$
$B = \frac{-5Y}{-3(-Y + 3S)}$
$B = \frac{5Y}{3(3S - Y)}$

6.3 Factoring Trinomials

1. $x^2+4x+3=(x+3)(x+1)$

5. $2x^2+6x-36=2(x^2+3x-18)$
$=2(x+6)(x-3)$

9. $s^2-s-42=(s-7)(s+6)$

13. $x^2+2x+1=(x+1)(x+1)$
$=(x+1)^2$

17. $3x^2-5x-2=(3x+1)(x-2)$
(because $-5x=-6x+x$)

21. $2s^2+13s+11=(2s+11)(s+1)$
(because $13s=2s+11s$)

25. $2t^2+7t-15=(2t-3)(t+5)$
(because $7t=10t-3t$)

29. $4x^2-3x-7=(4x-7)(x+1)$
(because $-3x=4x-7x$)

33. $4m^2+20m+25=(2m+5)(2m+5)$
$=(2m+5)^2$
(because $20m=10m+10m$)

37. $9t^2-15t+4=(3t-4)(3t-1)$
(because $-15t=-3t-12t$)

41. $4p^2-25pq+6q^2=(4p-q)(p-6q)$
(because $-25pq=-24pq-pq$)

45. $2x^2-14x+12=2(x^2-7x+6)$
$=2(x-1)(x-6)$

49. $ax^3+4a^2x^2-12a^3x=ax(x^2+4ax-12a^2)$
$=ax(x+6a)(x-2a)$

53. $25a^2-25x^2-10xy-y^2=25a^2-(25x^2+10xy+y^2)$
$=25a^2-(5x+y)^2$
$=(5a+5x+y)(5a-5x-y)$

57. $16t^2-80t+64=16(t^2-5t+4)$
$=16(t-4)(t-1)$

61. $200n^2-2100n-3600=100(2n^2-21n-36)$
$=100(2n+3)(n-12)$

65. $wx^4-5wLx^3+6wL^2x^2=wx^2(x^2-5Lx+6L^2)$
$=wx^2(x-3L)(x-2L)$

69. Write $4x^2+4x-k$ as a perfect square trinomial

$4x^2+4x-k=(2x)^2+2(2x)(1)+(-k)$
and comparing to
$=(2x)^2+2(2x)(1)+1^2$,
which is a perfect square,
gives $-k=1^2$
$k=-1$ from which
$4x^2+4x-k=4x^2+4x+1$
$=(2x+1)^2$

73. $36x^2+9=9(4x^2+1)$

6.4 The Sum and Difference of Cubes

1. $x^3-8=x^3-2^3$
$=(x-2)(x^2+2x+2^2)$
$=(x-2)(x^2+2x+4)$

5. $8-t^3=2^3-t^3$
$=(2-t)(4+2t+t^2)$

9. $4x^3+32=4(x^3+8)$
$=4(x+2)(x^2-2x+4)$

13. $162x^3y-6x^3y^4=6x^3y(27-y^3)$
$=6x^3y(3-y)(9+3y+y^2)$

17. $3a^6-3a^3=3a^3(a^3-1)$
$=3a^3(a-1)(a^2+2a+1)$

21. $27L^6+216L^3=27L^3\left(L^3+8\right)$
$=27L^3(L+2)\left(L^2-2L+4\right)$

25. $64-x^6=4^3-\left(x^2\right)^3$
$=\left(4-x^2\right)\left(16+4x^2+x^4\right)$
$=(2+x)(2-x)\left(16+4x^2+x^4\right)$

29. $D^4-d^3D=D\left(D^3-d^3\right)$
$=D(D-d)\left(D^2+Dd+d^2\right)$

33.

$$\begin{array}{r}
x^4+x^3y+x^2y^2+xy^3+y^4 \\
x-y\overline{)x^5 \qquad\qquad\qquad\qquad -y^5} \\
\underline{x^5-x^4y} \qquad\qquad\qquad\quad \\
x^4y \qquad\qquad\qquad\quad \\
\underline{x^4y-x^3y^2} \qquad\qquad\quad \\
x^3y^2 \qquad\qquad \\
\underline{x^3y^2-x^2y^3} \qquad\quad \\
x^2y^3 \qquad \\
\underline{x^2y^3-xy^4} \quad \\
xy^4-y^5 \\
\underline{xy^4-y^5}
\end{array}$$

$\left(x^5-y^5\right)\div(x-y)=x^4+x^3y+x^2y^2+xy^3+y^4$
$x^5-y^5=(x-y)\left(x^4+x^3y+x^2y^2+xy^3+y^4\right)$
$\left(x^7-y^7\right)\div(x-y)=x^6+x^5y+x^4y^2+x^3y^3+x^2y^4+xy^5+y^6$
$x^7-y^7=(x-y)\left(x^6+x^5y+x^4y^2+x^3y^3+x^2y^4+xy^5+y^6\right)$

37. $n^3+1=(n+1)\left(n^2-n+1\right)$
For $n=1$, this becomes
$1^3+1=(1+1)\left(1^2-1+1\right)$
$2=2\cdot 1$ which is prime
For $n=2,3,4,\ldots,$
$n^3+1=(n+1)\left(n^2-n+1\right)$
means that $(n+1)$ is a factor of
n^3+1 so n^3+1 is not prime.
For example
if $n=4$
$4^3+1=(4+1)\left(4^2-4+1\right)$
$65=(5)(13)$ which is not a prime number.

6.5 Equivalent Fractions

1. $\dfrac{18abc^6}{24ab^2c^5}=\dfrac{6abc^5(3c)}{6abc^5(4b)}$
$=\dfrac{3c}{4b}$ where $a,b,c\neq 0$

5. $\dfrac{2}{3}=\dfrac{2}{3}\cdot\dfrac{7}{7}=\dfrac{14}{21}$

9. $\dfrac{2}{(x+3)}=\dfrac{2(x-2)}{(x+3)(x-2)}=\dfrac{2x-4}{x^2+x-6}$

13. $\dfrac{28}{44}=\dfrac{\frac{28}{4}}{\frac{44}{4}}=\dfrac{7}{11}$

17. $\dfrac{2(R-1)}{(R-1)(R+1)}=\dfrac{\frac{2(R-1)}{(R-1)}}{\frac{(R-1)(R+1)}{(R-1)}}=\dfrac{2}{R+1}$ where $R\neq 1$

21. $\dfrac{A}{6y^2}=\dfrac{3x}{2y}\cdot\dfrac{3y}{3y}$
$\dfrac{A}{6y^2}=\dfrac{9xy}{6y^2}$
$A=9xy$

25. $\dfrac{A}{x^2-1}=\dfrac{2x^3+2x}{x^4-1}$
$\dfrac{A}{x^2-1}=\dfrac{2x\left(x^2+1\right)}{\left(x^2+1\right)\left(x^2-1\right)}$
$A=\dfrac{2x\left(x^2+1\right)}{\left(x^2+1\right)}$
$A=2x$

29. $\dfrac{2a}{8a}=\dfrac{2a}{2a\cdot 4}=\dfrac{1}{4}$ where $a\neq 0$

33. $\dfrac{a+b}{5a^2+5ab}=\dfrac{(a+b)}{5a(a+b)}=\dfrac{1}{5a}$ where $a+b\neq 0$

37. $\dfrac{4x^2+1}{4x^2-1}=\dfrac{4x^2+1}{(2x+1)(2x-1)}$
Since no cancellations can be made the fraction cannot be reduced further.

41. $\frac{3+2y}{4y^3+6y^2}=\frac{(2y+3)}{2y^2(2y+3)}=\frac{1}{2y^2}$ where $y\neq -\frac{3}{2}$

45. $\frac{2w^4+5w^2-3}{w^4+11w^2+24}=\frac{(2w^2-1)(w^2+3)}{(w^2+8)(w^2+3)}=\frac{2w^2-1}{w^2+8}$

49.
$$\frac{N^4-16}{8N-16}=\frac{(N^2+4)(N^2-4)}{8(N-2)}$$
$$=\frac{(N^2+4)(N+2)(N-2)}{8(N-2)}$$
$$=\frac{(N^2+4)(N+2)}{8} \text{ where } N\neq 2$$

53.
$$\frac{(x-1)(3+x)}{(3-x)(1-x)}=\frac{(x-1)(3+x)}{-(3-x)(x-1)}$$
$$=\frac{3+x}{-(3-x)} \text{ where } x\neq 1$$
$$=\frac{x+3}{x-3}$$

57.
$$\frac{x^3+x^2-x-1}{x^3-x^2-x+1}=\frac{(x^3+x^2)-(x+1)}{(x^3-x^2)-(x-1)}$$
$$=\frac{x^2(x+1)-(x+1)}{x^2(x-1)-(x-1)}$$
$$=\frac{(x+1)(x^2-1)}{(x-1)(x^2-1)}$$
$$=\frac{x+1}{x-1} \text{ where } x\neq \pm 1$$

61.
$$\frac{x^3+y^3}{2x+2y}=\frac{(x+y)(x^2-xy+y^2)}{2(x+y)}$$
$$=\frac{x^2-xy+y^2}{2} \text{ where } x\neq -y$$

65. (a) $\frac{x^2(x+2)}{x^2+4}$ will not reduce further since x^2+4 does not factor.

(b)
$$\frac{x^4+4x^2}{x^4-16}=\frac{x^2(x^2+4)}{(x^2+4)(x^2-4)}=\frac{x^2}{x^2-4}$$
$$=\frac{x^2}{(x+2)(x-2)}$$

will not reduce further since there are no more common factors.

69.
$$\frac{mu^2-mv^2}{mu-mv}=\frac{m(u^2-v^2)}{m(u-v)}$$
$$=\frac{(u+v)(u-v)}{(u-v)}$$
$$=u+v \text{ where } u\neq v$$

6.6 Multiplication and Division of Fractions

1.
$$\frac{4x+6y}{(x-y)^2}\times\frac{(x^2-y^2)}{6x+9y}=\frac{2(2x+3y)(x+y)(x-y)}{(x-y)(x-y)\cdot 3(2x+3y)}$$
$$=\frac{2(x+y)}{3(x-y)} \text{ where } x\neq y, -\tfrac{3y}{2}$$

5. $\frac{3}{8}\times\frac{2}{7}=\frac{3}{4(2)}\times\frac{2}{7}=\frac{3}{28}$

(divide out a common factor of 2)

9. $\frac{2}{9}\div\frac{4}{7}=\frac{2}{9}\times\frac{7}{4}=\frac{2(7)}{9(2)(2)}=\frac{7}{9(2)}=\frac{7}{18}$

(divide out a common factor of 2)

13.
$$\frac{4x+12}{5}\times\frac{15t}{3x+9}=\frac{4(x+3)(5)(3)t}{5(3)(x+3)}$$
$$=\frac{4t}{1}=4t \text{ where } x\neq -3$$

(divide out common factors of $5(3)(x+3)$)

17.
$$\frac{2a+8}{15}\div\frac{a^2+8a+16}{125}=\frac{2(a+4)}{3\times 5}\times\frac{5\times 5\times 5}{(a+4)(a+4)}$$
$$=\frac{50}{3(a+4)} \text{ where } a\neq -4$$

(divide out a common factor of $5(a+4)$)

21.
$$\frac{3ax^2-9ax}{10x^2+5x}\times\frac{2x^2+x}{a^2x-3a^2}$$
$$=\frac{3ax(x-3)}{5x(2x+1)}\times\frac{x(2x+1)}{a^2(x-3)}$$
$$=\frac{3x}{5a} \text{ where } x\neq 0, 3, -\tfrac{1}{2} \text{ and } a\neq 0$$

(divide out a common factor of $ax(2x+1)(x-3)$)

25. $\dfrac{\frac{x^2+ax}{2b-cx}}{\frac{a^2+2ax+x^2}{2bx-cx^2}} = \dfrac{x(x+a)}{(2b-cx)} \times \dfrac{x(2b-cx)}{a^2+2ax+x^2}$

$= \dfrac{x^2(a+x)(2b-cx)}{(2b-cx)(a+x)(a+x)}$

$= \dfrac{x^2}{a+x}$ where $x \neq -a,\ b \neq \frac{cx}{2}$

(divide out a common factor of $(a+x)(2b-cx)$)

29. $\dfrac{x^2-6x+5}{4x^2-17x-15} \times \dfrac{6x+21}{2x^2+5x-7}$

$= \dfrac{(x-5)(x-1)(3)(2x+7)}{(4x+3)(x-5)(2x+7)(x-1)}$

$= \dfrac{3}{4x+3}$ where $x \neq 5, 1, -\frac{7}{2}$

(divide out common factor $(x-5)(x-1)(2x+7)$)

33. $\dfrac{7x^2}{3a} \div \left(\dfrac{a}{x} \times \dfrac{a^2x}{x^2}\right) = \dfrac{7x^2}{3a} \times \left(\dfrac{x^3}{a^3x}\right)$

$= \dfrac{7x^5}{3a^4x}$

$= \dfrac{7x^4}{3a^4}$ where $x \neq 0$

(divide out a common factor of x)

37. $\dfrac{x^3-y^3}{2x^2-2y^2} \times \dfrac{y^2+2xy+x^2}{x^2+xy+y^2}$

$= \dfrac{(x-y)(x^2+xy+y^2)(y+x)(y+x)}{2(x^2-y^2)(x^2+xy+y^2)}$

$= \dfrac{(x-y)(x^2+xy+y^2)(x+y)(x+y)}{2(x+y)(x-y)(x^2+xy+y^2)}$

$= \dfrac{x+y}{2}$

(divide out common factors of $(x-y), (x+y)$, (x^2+xy+y^2))

so $x \neq \pm y$ and $x^2+xy+y^2 \neq 0$

41. $\dfrac{x}{2x+4} \times \dfrac{x^2-4}{3x^2} = \dfrac{x(x+2)(x-2)}{2(x+2)(3x^2)}$

$= \dfrac{x-2}{6x}$ where $x \neq -2, 0$

45. $\dfrac{d}{2} \div \dfrac{v_1d+v_2d}{4v_1v_2} = \dfrac{d}{2} \times \dfrac{4v_1v_2}{d(v_1+v_2)}$

$= \dfrac{2v_1v_2}{v_1+v_2}$ where $d \neq 0$

(divide out a common factor $2d$)

6.7 Addition and Subtraction of Fractions

1. $4a^2b = 2 \cdot 2 \cdot a \cdot a \cdot b$

$6ab^3 = 2 \cdot 3 \cdot a \cdot b \cdot b \cdot b$

$4a^2b^2 = 2 \cdot 2 \cdot a \cdot a \cdot b \cdot b$

$L.C.D. = 2^2 \cdot 3 \cdot a^2 \cdot b^3 = 12a^2b^3$

5. $\dfrac{3}{5} + \dfrac{6}{5} = \dfrac{3+6}{5} = \dfrac{9}{5}$

9. $\dfrac{1}{2} + \dfrac{3}{4} = \dfrac{1(2)+3}{4} = \dfrac{2+3}{4} = \dfrac{5}{4}$

13. $\dfrac{a}{x} - \dfrac{b}{x^2} = \dfrac{a(x)-b}{x^2} = \dfrac{ax-b}{x^2}$

17. $\dfrac{2}{5a} + \dfrac{1}{a} - \dfrac{a}{10} = \dfrac{2(2)+1(10)-a(a)}{10a} = \dfrac{14-a^2}{10a}$

21. $\dfrac{3}{2x-1} + \dfrac{1}{4x-2} = \dfrac{3}{(2x-1)} + \dfrac{1}{2(2x-1)}$

$= \dfrac{3(2)+1}{2(2x-1)}$

$= \dfrac{7}{2(2x-1)}$

25. $\dfrac{s}{2s-6} + \dfrac{1}{4} - \dfrac{3s}{4s-12} = \dfrac{s}{2(s-3)} + \dfrac{1}{4} - \dfrac{3s}{4(s-3)}$

$= \dfrac{s(2)+1(s-3)-3s}{4(s-3)}$

$= \dfrac{2s+s-3-3s}{4(s-3)}$

$= \dfrac{-3}{4(s-3)}$

29. $$\frac{3}{x^2-8x+16}-\frac{2}{4-x}=\frac{3}{(x-4)^2}+\frac{2}{x-4}$$
$$=\frac{3+2(x-4)}{(x-4)^2}$$
$$=\frac{3+2x-8}{(x-4)^2}$$
$$=\frac{2x-5}{(x-4)^2}$$

33. $$\frac{x-1}{3x^2-13x+4}-\frac{3x+1}{4-x}=\frac{(x-1)}{(3x-1)(x-4)}+\frac{(3x+1)}{(x-4)}$$
$$=\frac{x-1+(3x+1)(3x-1)}{(3x-1)(x-4)}$$
$$=\frac{x-1+9x^2-1}{(3x-1)(x-4)}$$
$$=\frac{9x^2+x-2}{(3x-1)(x-4)}$$

37. $$\frac{1}{w^3+1}+\frac{1}{w+1}-2=\frac{1}{(w+1)(w^2-w+1)}+\frac{1}{w+1}-2$$
$$=\frac{1+1(w^2-w+1)-2(w+1)(w^2-w+1)}{(w+1)(w^2-w+1)}$$
$$=\frac{1+w^2-w+1-2(w^3+1)}{(w+1)(w^2-w+1)}$$
$$=\frac{w^2-w+2-2w^3-2}{(w+1)(w^2-w+1)}$$
$$=\frac{-2w^3+w^2-w}{(w+1)(w^2-w+1)}$$
$$=\frac{-w(2w^2-w+1)}{(w+1)(w^2-w+1)}$$

41. $$\frac{\frac{x}{y}-\frac{y}{x}}{1+\frac{y}{x}}=\frac{\frac{x(x)-y(y)}{xy}}{\frac{x+y}{x}}$$
$$=\frac{x^2-y^2}{xy}\times\frac{x}{x+y}$$
$$=\frac{x(x+y)(x-y)}{xy(x+y)}$$
$$=\frac{x-y}{y}\quad\text{where } x\neq 0,-y$$

45. $$f(x)=\frac{x}{x+1}$$
$$f(x+h)-f(x)=\frac{x+h}{x+h+1}-\frac{x}{x+1}$$
$$f(x+h)-f(x)=\frac{(x+h)(x+1)-x(x+h+1)}{(x+h+1)(x+1)}$$
$$f(x+h)-f(x)=\frac{x^2+x+hx+h-x^2-xh-x}{(x+1)(x+h+1)}$$
$$f(x+h)-f(x)=\frac{h}{(x+1)(x+h+1)}$$

49. $$(\tan\theta)(\cot\theta)+(\sin\theta)^2-\cos\theta=\frac{y}{x}\cdot\frac{x}{y}+\left(\frac{y}{r}\right)^2-\frac{x}{r}$$
$$=1+\frac{y^2}{r^2}-\frac{x}{r}$$
$$=\frac{1(r^2)+y^2-x(r)}{r^2}$$
$$=\frac{r^2+y^2-rx}{r^2}$$

53. $$f(x)=x-\frac{2}{x}$$
$$f(a+1)=a+1-\frac{2}{a+1}$$
$$=\frac{(a+1)^2-2}{a+1}$$
$$=\frac{a^2+2a+1-2}{a+1}$$
$$=\frac{a^2+2a-1}{a+1}$$

57. $$\frac{y^2-x^2}{y^2+x^2}=\frac{\left(\frac{mn}{m-n}\right)^2-\left(\frac{mn}{m+n}\right)^2}{\left(\frac{mn}{m-n}\right)^2+\left(\frac{mn}{m+n}\right)^2}$$
$$=\frac{\frac{m^2n^2(m+n)^2-m^2n^2(m-n)^2}{(m-n)^2(m+n)^2}}{\frac{m^2n^2(m+n)^2+m^2n^2(m-n)^2}{(m-n)^2(m+n)^2}}$$
$$=\frac{m^2n^2(m+n)^2-m^2n^2(m-n)^2}{m^2n^2(m+n)^2+m^2n^2(m-n)^2}$$
$$=\frac{m^2n^2\left((m+n)^2-(m-n)^2\right)}{m^2n^2\left((m+n)^2+(m-n)^2\right)}$$
$$=\frac{m^2+2mn+n^2-(m^2-2mn+n^2)}{m^2+2mn+n^2+(m^2-2mn+n^2)}$$
$$=\frac{4mn}{2m^2+2n^2}$$
$$=\frac{2mn}{m^2+n^2}$$

61. $$\frac{2n^2-n-4}{2n^2+2n-4}+\frac{1}{n-1}=\frac{2n^2-n-4}{2(n^2+n-2)}+\frac{1}{n-1}$$
$$=\frac{2n^2-n-4}{2(n+2)(n-1)}+\frac{1}{n-1}$$
$$=\frac{2n^2-n-4+1(2)(n+2)}{2(n-1)(n+2)}$$
$$=\frac{2n^2-n-4+2n+4}{2(n-1)(n+2)}$$
$$=\frac{2n^2+n}{2(n-1)(n+2)}$$
$$=\frac{n(2n+1)}{2(n-1)(n+2)}$$

65. $$\frac{\frac{L}{C}+\frac{R}{sC}}{sL+R+\frac{1}{sC}}=\frac{\frac{Ls+R}{sC}}{\frac{(sL+R)sC+1}{sC}}$$
$$=\frac{\frac{Ls+R}{sC}}{\frac{CLs^2+CRs+1}{sC}}$$
$$=\frac{Ls+R}{sC}\times\frac{sC}{CLs^2+CRs+1}$$
$$=\frac{Ls+R}{CLs^2+CRs+1}$$

6.8 Equations Involving Fractions

1. $$\frac{x}{2}-\frac{1}{b}=\frac{x}{2b}$$
$$\frac{x(2b)}{2}-\frac{1(2b)}{b}=\frac{x(2b)}{2b}$$
$$xb-2=x$$
$$x(b-1)=2$$
$$x=\frac{2}{b-1}$$

5. $$\frac{x}{2}+6=2x$$
$$\frac{x(2)}{2}+6(2)=2x(2)$$
$$x+12=4x$$
$$-3x=-12$$
$$x=4$$

9. $$1-\frac{t-5}{6}=\frac{3}{4}$$
$$1(12)-\frac{(t-5)(12)}{6}=\frac{3(12)}{4}$$
$$12-2(t-5)=9$$
$$12-2t+10=9$$
$$-2t=-13$$
$$t=\frac{13}{2}$$

13. $$\frac{3}{T}+2=\frac{5}{3}$$
$$\frac{3(3T)}{T}+2(3T)=\frac{5(3T)}{3}$$
$$9+6T=5T$$
$$T=-9$$

17. $$\frac{2y}{y-1}=5$$
$$\frac{2y(y-1)}{y-1}=5(y-1)$$
$$2y=5y-5$$
$$-3y=-5$$
$$y=\frac{5}{3}$$

21. $$\frac{5}{2x+4}+\frac{3}{6x+12}=2$$
$$\frac{5}{2(x+2)}+\frac{3}{6(x+2)}=2$$
$$\frac{5(6)(x+2)}{2(x+2)}+\frac{3(6)(x+2)}{6(x+2)}=2(6)(x+2)$$
$$15+3=12(x+2)$$
$$18=12x+24$$
$$-12x=6$$
$$x=-\frac{1}{2}$$

25. $$\frac{1}{4x}+\frac{3}{2x}=\frac{2}{x+1}$$
$$\frac{1(4x)(x+1)}{4x}+\frac{3(4x)(x+1)}{2x}=\frac{2(4x)(x+1)}{x+1}$$
$$(x+1)+6(x+1)=8x$$
$$x+1+6x+6=8x$$
$$-x=-7$$
$$x=7$$

29.
$$\frac{1}{x^2-x}-\frac{1}{x}=\frac{1}{x-1}$$
$$\frac{1}{x(x-1)}-\frac{1}{x}=\frac{1}{x-1}$$
$$\frac{1(x)(x-1)}{x(x-1)}-\frac{1(x)(x-1)}{x}=\frac{1(x)(x-1)}{x-1}$$
$$1-(x-1)=x$$
$$1-x+1=x$$
$$-2x=-2$$
$$x=1$$
Substitution reveals a zero in the denominator for the first and third fractions in the original equation. Therefore no solution exists.

33.
$$2-\frac{1}{b}+\frac{3}{c}=0,\ \text{for } c$$
$$2bc-\frac{1bc}{b}+\frac{3bc}{c}=0bc$$
$$2bc-c+3b=0$$
$$c(2b-1)=-3b$$
$$c=\frac{3b}{1-2b}$$

37.
$$\frac{s-s_0}{t}=\frac{v+v_0}{2},\ \text{for } v$$
$$\frac{(s-s_0)(2t)}{t}=\frac{(v+v_0)(2t)}{2}$$
$$2(s-s_0)=t(v+v_0)$$
$$2(s-s_0)=tv+tv_0$$
$$2(s-s_0)-tv_0=tv$$
$$v=\frac{2(s-s_0)-tv_0}{t}$$

41.
$$z=\frac{1}{g_m}-\frac{jX}{g_mR},\ \text{for } R$$
$$zg_mR=\frac{1g_mR}{g_m}-\frac{jXg_mR}{g_mR}$$
$$g_mRz=R-jX$$
$$g_mRz-R=-jX$$
$$R(g_mz-1)=-jX$$
$$R=\frac{jX}{1-g_mz}$$

45.
$$\frac{1}{C}=\frac{1}{C_2}+\frac{1}{C_1+C_3}$$
$$\frac{1CC_2(C_1+C_3)}{C}=\frac{1CC_2(C_1+C_3)}{C_2}+\frac{1CC_2(C_1+C_3)}{C_1+C_3}$$
$$C_2(C_1+C_3)=C(C_1+C_3)+CC_2$$
$$C_1C_2+C_2C_3=CC_1+CC_3+CC_2$$
$$C_1(C_2-C)=CC_3+CC_2-C_2C_3$$
$$C_1=\frac{CC_3+CC_2-C_2C_3}{C_2-C}$$

49. If Pump 1 empties tank of volume V in 5.00 h, its rate of emptying is $\frac{V}{5.00}$.
If Pump 2 empties tank of volume V in 8.00 h, its rate of emptying is $\frac{V}{8.00}$.
If both are operating, then in time elapsed t, to empty the volume
$$\frac{V}{5.0}\times t+\frac{V}{8.0}\times t=V$$
Note how rate (in cubic metres per hour) multiplied by time (in hours) gives volume (in cubic metres). Now solving for t:
$$\frac{Vt(40)}{5.0}+\frac{Vt(40)}{8.0}=V(40)$$
$$8.0Vt+5.0Vt=40V$$
$$13Vt=40V$$
$$t=\frac{40}{13}$$
$$t=3.08\text{ h}$$

53.
$$d=2.00t_1\ \text{for trip up}$$
$$d=2.20t_2\ \text{for trip down}$$
$$t_1+t_2+90.0=5.00(60)$$
$$\frac{d}{2.00}+\frac{d}{2.20}+90.0=300$$
$$\frac{d(4.4)}{2.0}+\frac{d(4.4)}{2.2}+90(4.4)=300(4.4)$$
$$2.2d+2d+396=1320$$
$$4.2d=924$$
$$d=\frac{924}{4.2}=220\text{ m}$$

57.
$$\frac{V}{R_1}+\frac{V}{R_2}=i$$
$$\frac{V}{2.7}+\frac{V}{6.0}=1.2$$
$$\frac{V}{2.7}(16.2)+\frac{V}{6.0}(16.2)=1.2(16.2)$$
$$6V+2.7V=19.44$$
$$8.7V=19.44$$
$$V=2.23\text{ V}$$

Review Exercises

1. $3a(4x+5a)=12ax+15a^2$

5. $(2a+1)^2=4a^2+4a+1$

9. $(2x+5)(x-9)=2x^2-13x-45$

13. $3s+9t=3(s+3t)$

17. $W^2b^{x+2}-144b^x=b^x(W^2b^2-144)$
$$=b^x(Wb+12)(Wb-12)$$

21. $36t^2-24t+4=4(9t^2-6t+1)$
$$=4(3t-1)(3t-1)$$
$$=4(3t-1)^2$$

25. $x^2+x-56=(x+8)(x-7)$

29. $2k^2-k-36=(2k-9)(k+4)$

33. $10b^2+23b-5=(5b-1)(2b+5)$

37. $250-16y^6=2(125-8y^6)$
$$=2\left(5^3-(2y^2)^3\right)$$
$$=2(5-2y^2)(25+10y^2+4y^4)$$

41. $ab^2-3b^2+a-3=b^2(a-3)+(a-3)$
$$=(a-3)(b^2+1)$$

45. $\frac{48ax^3y^6}{9a^3xy^6}=\frac{3axy^6(16x^2)}{3axy^6(3a^2)}$
$$=\frac{16x^2}{3a^2}\quad\text{where } a,x,y\neq 0$$

49. $\frac{4x+4y}{35x^2}\times\frac{28x}{x^2-y^2}=\frac{4(x+y)}{7x(5x)}\times\frac{7x(4)}{(x+y)(x-y)}$
$$=\frac{16}{5x(x-y)}\quad\text{where } x\neq 0,-y$$

53. $\frac{\frac{3x}{7x^2+13x-2}}{\frac{6x^2}{x^2+4x+4}}=\frac{3x}{7x^2+13x-2}\times\frac{x^2+4x+4}{6x^2}$
$$=\frac{3x}{(7x-1)(x+2)}\times\frac{(x+2)(x+2)}{3x(2x)}$$
$$=\frac{x+2}{2x(7x-1)}\quad\text{where } x\neq 0,-2$$

57. $\frac{4}{9x}-\frac{5}{12x^2}=\frac{4(4x)-5(3)}{36x^2}$
$$=\frac{16x-15}{36x^2}$$

61. $\frac{a+1}{a+2}-\frac{a+3}{a}=\frac{(a+1)(a)-(a+2)(a+3)}{a(a+2)}$
$$=\frac{a^2+a-a^2-5a-6}{a(a+2)}$$
$$=\frac{-4a-6}{a(a+2)}$$
$$=\frac{-2(2a+3)}{a(a+2)}$$

65. $\frac{3x}{2x^2-2}-\frac{2}{4x^2-5x+1}=\frac{3x}{2(x+1)(x-1)}$
$$-\frac{2}{(4x-1)(x-1)}$$
$$=\frac{3x(4x-1)-2(2)(x+1)}{2(4x-1)(x+1)(x-1)}$$
$$=\frac{12x^2-3x-4x-4}{2(4x-1)(x+1)(x-1)}$$
$$=\frac{12x^2-7x-4}{2(4x-1)(x+1)(x-1)}$$

69. $x^2-5=\left(x+\sqrt{5}\right)\left(x-\sqrt{5}\right)$

73. $x+y=x\left(1+\frac{y}{x}\right)$

77.
$$\frac{x}{2}-3=\frac{x-10}{4}$$
$$\left(\frac{x}{2}-3\right)(4)=\left(\frac{x-10}{4}\right)(4)$$
$$2x-12=x-10$$
$$x=2$$

81.
$$\frac{2x}{2x^2-5x}-\frac{3}{x}=\frac{1}{4x-10}$$
$$\frac{2x}{x(2x-5)}-\frac{3}{x}=\frac{1}{2(2x-5)}$$
$$\frac{2x(2x)(2x-5)}{x(2x-5)}-\frac{3(2x)(2x-5)}{x}=\frac{1(2x)(2x-5)}{2(2x-5)}$$
$$4x-6(2x-5)=x$$
$$4x-12x+30=x$$
$$9x=30$$
$$x=\frac{10}{3}$$

85. **(a)** Changing an odd number of signs changes the sign of the fraction.

(b) Changing an even number of signs leaves the sign of the fraction unchanged.

89. $2zS(S+1)=2zS^2+2zS$

93.
$$cT_2-cT_1+RT_2-RT_1=c(T_2-T_1)+R(T_2-T_1)$$
$$=(T_2-T_1)(c+R)$$

97.
$$(n+1)^3(2n+1)^3=\left(n^3+3n^2+3n+1\right)\left(8n^3+12n^2+6n+1\right)$$
$$=8n^6+12n^5+6n^4+n^3+24n^5+36n^4$$
$$+18n^3+3n^2+24n^4+36n^3+18n^2$$
$$+3n+8n^3+12n^2+6n+1$$
$$=8n^6+36n^5+66n^4+63n^3+33n^2+9n+1$$

101.
$$V_{change}=V_2-V_1$$
$$=(x+4)^3-x^3$$
$$=x^3+12x^2+48x+64-x^3$$
$$=12x^2+48x+64$$
$$=4\left(3x^2+12x+16\right)$$

105.
$$\frac{\frac{\pi ka}{2}\left(R^4-r^4\right)}{\pi ka\left(R^2-r^2\right)}=\frac{\frac{\pi ka}{2}\left(R^2+r^2\right)\left(R^2-r^2\right)}{\pi ka\left(R^2-r^2\right)}$$
$$=\frac{\frac{1}{2}(R^2+r^2)}{1}$$
$$=\frac{(R^2+r^2)}{2}\quad \text{where } R\neq r,-r$$

109.
$$\frac{4k-1}{4k-4}+\frac{1}{2k}=\frac{4k-1}{4(k-1)}+\frac{1}{2k}$$
$$=\frac{(4k-1)k+1(2)(k-1)}{4k(k-1)}$$
$$=\frac{4k^2-k+2k-2}{4k(k-1)}$$
$$=\frac{4k^2+k-2}{4k(k-1)}$$

113.
$$\frac{\frac{u^2}{2g}-x}{\frac{1}{2gc^2}-\frac{u^2}{2g}+x}=\left(\frac{\frac{u^2}{2g}-x}{\frac{1}{2gc^2}-\frac{u^2}{2g}+x}\right)\cdot\frac{2gc^2}{2gc^2}$$
$$=\frac{u^2c^2-2gc^2x}{1-u^2c^2+2gc^2x}$$
$$=\frac{c^2(u^2-2gx)}{1-u^2c^2+2gc^2x}$$

117.
$$R=\frac{wL}{H(w+L)}$$
$$RH(w+L)=wL$$
$$RHw+RHL=wL$$
$$wL-RHL=RHw$$
$$L(w-RH)=RHw$$
$$L=\frac{RHw}{w-RH}$$

121.
$$s^2+\frac{cs}{m}+\frac{kL^2}{mb^2}=0$$
$$mb^2\left(s^2+\frac{cs}{m}+\frac{kL^2}{mb^2}\right)=0\cdot mb^2$$
$$s^2b^2m+csb^2+kL^2=0$$
$$csb^2=-s^2b^2m-kL^2$$
$$c=-\frac{(s^2b^2m+kL^2)}{sb^2}$$

125. Let $\frac{1}{4}t$ = the amount of the car's battery depleted by the lights in 1 hour

Let $\frac{1}{24}t$ = the amount of the car's battery depleted by the radio in 1 hour

$$\frac{1}{4}t+\frac{1}{24}t=1$$
$$\frac{6}{24}t+\frac{1}{24}t=1$$
$$\frac{7}{24}t=1$$
$$t=\frac{24}{7}$$
$$t=3.43 \text{ h}$$

It will take 3.43 hours for the battery to go dead with both the lights and the radio on.

129. $sg=\frac{w_a}{w_a-w_w}$

$$sg=\frac{1.097w_w}{1.097w_w-w_w}$$
$$sg=\frac{w_w(1.097)}{w_w(1.097-1)}$$
$$sg=\frac{1.097}{(1.097-1)}$$
$$sg=\frac{1.097}{0.097}$$
$$sg=11.3$$

The relative density of lead is 11.3.

133. $\frac{\left(1+\frac{1}{s}\right)\left(1+\frac{1}{s/2}\right)}{3+\frac{1}{s}+\frac{1}{s/2}}=\frac{\left(\frac{s+1}{s}\right)\left(1+\frac{2}{s}\right)}{\left(3+\frac{1}{s}+\frac{2}{s}\right)}$

$$=\frac{\left(\frac{s+1}{s}\right)\left(\frac{s+2}{s}\right)}{\left(\frac{3s+1+2}{s}\right)}$$
$$=\left(\frac{s+1}{s}\right)\left(\frac{s+2}{s}\right)\times\left(\frac{s}{3s+3}\right)$$
$$=\frac{s(s+1)(s+2)}{3s^2(s+1)}$$
$$=\frac{s+2}{3s}$$

When you cancel, the basic operation being performed is division.

CHAPTER 7

QUADRATIC EQUATIONS

7.1 Quadratic Equations; Solution by Factoring

1. $2N^2 - 7N - 4 = 0$

$(2N+1)(N-4) = 0$ factor

$2N+1=0$ or $N-4=0$

$2N=-1$ $\quad N=4$

$N=-\frac{1}{2}$ $\quad N=4$

The roots are $N=-\frac{1}{2}$ and $N=4$.

Checking in the original equation:

$2\left(-\frac{1}{2}\right)^2 - 7\left(-\frac{1}{2}\right) - 4 = 0 \qquad 2(4)^2 - 7(4) - 4 = 0$

$\frac{1}{2} + \frac{7}{2} - 4 = 0 \qquad 32 - 28 - 4 = 0$

$0 = 0 \qquad 0 = 0$

The roots are $-\frac{1}{2}$, 4.

5. $x^2 = (x+2)^2$

$x^2 = x^2 + 4x + 4$

$4x+4=0$, no x^2 term so it is not quadratic

9. $x^2 - 4 = 0$

$(x+2)(x-2) = 0$

$x+2=0$ or $x-2=0$

$x=-2$ $\quad x=2$

13. $x^2 - 8x - 9 = 0$

$(x-9)(x+1) = 0$

$x-9=0$ or $x+1=0$

$x=9$ $\quad x=-1$

17. $40x - 16x^2 = 0$

$2x^2 - 5x = 0$

$x(2x-5) = 0$

$2x-5=0$ or $x=0$

$2x=5$

$x=\frac{5}{2}$

21. $3x^2 - 13x + 4 = 0$

$(3x-1)(x-4) = 0$

$3x-1=0$ or $x-4=0$

$3x=1$ $\quad x=4$

$x=\frac{1}{3}$

25. $6x^2 = 13x - 6$

$6x^2 - 13x + 6 = 0$

$(3x-2)(2x-3) = 0$

$3x-2=0$ or $2x-3=0$

$3x=2$ $\quad 2x=3$

$x=\frac{2}{3}$ $\quad x=\frac{3}{2}$

29. $6y^2 + by = 2b^2$

$6y^2 + by - 2b^2 = 0$

$(2y-b)(3y+2b) = 0$

$2y-b=0$ or $3y+2b=0$

$y=\frac{b}{2}$ $\quad y=\frac{-2b}{3}$

33. $(x+2)^3 = x^3 + 8$

$x^3 + 6x^2 + 12x + 8 = x^3 + 8$

$6x^2 + 12x = 0$

$6x(x+2) = 0$

$6x=0$ or $x+2=0$

$x=0$ $\quad x=-2$

37.
$$x^2 + 2ax = b^2 - a^2$$
$$x^2 + 2ax + a^2 - b^2 = 0$$
$$(x+a)^2 - b^2 = 0$$
$$((x+a)+b)((x+a)-b) = 0$$
$$x+a+b = 0$$
$$\text{or} \quad x+a-b = 0$$
$$x = -a-b$$
$$x = b-a$$

41.
$$V = \alpha I + \beta I^2$$
$$2I + 0.5I^2 = 6$$
$$I^2 + 4I - 12 = 0$$
$$(I+6)(I-2) = 0$$
$$I+6 = 0 \quad \text{or} \quad I-2 = 0$$
$$I = -6 \qquad I = 2$$
The current is -6.00 A or 2.00 A.

45.
$$x^3 - x = 0$$
$$x(x^2 - 1) = 0$$
$$x(x+1)(x-1) = 0$$
$$x+1 = 0 \quad \text{or} \quad x-1 = 0 \quad \text{or} \quad x = 0$$
$$x = -1 \qquad x = 1$$
The three roots are -1, 0, 1.

49.
$$\frac{1}{2x} - \frac{3}{4} = \frac{1}{2x+3}$$
$$\frac{1(4)(2x)(2x+3)}{2x} - \frac{3(4)(2x)(2x+3)}{4} = \frac{1(4)(2x)(2x+3)}{2x+3} \quad \text{multiply by LCD}$$
$$8x + 12 - 12x^2 - 18x = 8x$$
$$-12x^2 - 18x + 12 = 0$$
$$-6(2x-1)(x+2) = 0$$
$$2x - 1 = 0$$
$$\text{or} \quad x+2 = 0$$
$$2x = 1$$
$$x = -2$$
$$x = \frac{1}{2}$$

53. For 120 km round trip, each leg consists of 60 km

$v_1 = \dfrac{60}{t_1}$ going

$v_2 = \dfrac{60}{t_2}$ returning

The total time taken was
$$t_1 + t_2 = 3.5$$
$$t_2 = 3.5 - t_1$$
And we know that the first leg was 10 km/h slower than the return trip
$$\frac{60}{t_1} + 10 = \frac{60}{t_2}$$
$$\frac{60}{t_1} + 10 = \frac{60}{3.5 - t_1} \quad \text{multiply by LCD}$$
$$\frac{60t_1(3.5 - t_1)}{t_1} + 10t_1(3.5 - t_1) = \frac{60t_1(3.5 - t_1)}{3.5 - t_1}$$
$$210 - 60t_1 + 35t_1 - 10t_1^2 = 60t_1$$
$$-10t_1^2 - 85t_1 + 210 = 0$$
$$-5(2t_1^2 + 17t_1 - 42) = 0$$
$$(2t_1 + 21)(t_1 - 2) = 0$$
$$t_1 - 2 = 0$$
$$\text{or} \quad 2t_1 + 21 = 0$$
$$t_1 = 2$$
$$t_1 = \frac{-21}{2} \quad \text{(ignore since } t > 0\text{)}$$
$$v_1 = \frac{60}{2} = 30 \text{ km/h going}$$
$$v_2 = \frac{60}{3.5 - 2} = 40 \text{ km/h returning}$$

7.2 Completing the Square

1.
$$x^2 + 6x - 8 = 0$$
$$x^2 + 6x = 8$$
$$x^2 + 6x + 9 = 8 + 9$$
$$(x+3)^2 = 17$$
$$x + 3 = \pm\sqrt{17}$$
$$x = -3 \pm \sqrt{17}$$

5. $x^2 = 7$

$x = \pm\sqrt{7}$

9. $(x+3)^2 = 7$

$x+3 = \pm\sqrt{7}$

$x = -3 \pm \sqrt{7}$

13. $D^2 + 3D + 2 = 0$

$D^2 + 3D = -2$

$D^2 + 3D + \frac{9}{4} = -2 + \frac{9}{4}$

$\left(D + \frac{3}{2}\right)^2 = \frac{1}{4}$

$D + \frac{3}{2} = \pm\sqrt{\frac{1}{4}}$

$D = -\frac{3}{2} \pm \frac{1}{2}$

$D = -2$ or $D = -1$

17. $v(v+2) = 15$

$v^2 + 2v = 15$

$v^2 + 2v + 1 = 15 + 1$

$(v+1)^2 = 16$

$v + 1 = \pm\sqrt{16}$

$v = -1 \pm 4$

$v = -5$ or $v = 3$

21. $3y^2 = 3y + 2$

$3y^2 - 3y = 2$

$y^2 - y = \frac{2}{3}$

$y^2 - y + \frac{1}{4} = \frac{2}{3} + \frac{1}{4}$

$\left(y - \frac{1}{2}\right)^2 = \frac{11}{12}$

$y - \frac{1}{2} = \pm\sqrt{\frac{11}{12}}$

$y = \frac{1}{2} \pm \sqrt{\frac{11}{4(3)}} \cdot \frac{\sqrt{3}}{\sqrt{3}}$

$y = \frac{1}{2} \pm \frac{\sqrt{33}}{6}$

25. $5T^2 - 10T + 4 = 0$

$T^2 - 2T + \frac{4}{5} = 0$

$T^2 - 2T = -\frac{4}{5}$

$T^2 - 2T + 1 = -\frac{4}{5} + 1$

$(T-1)^2 = \frac{1}{5}$

$T - 1 = \pm\sqrt{\frac{1}{5}}$

$T = 1 \pm \sqrt{\frac{1}{5}} \cdot \frac{\sqrt{5}}{\sqrt{5}}$

$T = 1 \pm \frac{\sqrt{5}}{5}$

29. $x^2 + 2bx + c = 0$

$x^2 + 2bx = -c$

$x^2 + 2bx + b^2 = b^2 - c$

$(x+b)^2 = b^2 - c$

$x + b = \pm\sqrt{b^2 - c}$

$x = -b \pm \sqrt{b^2 - c}$

33. $V = 4.0T - 0.2T^2 = 15$

$0.2T^2 - 4.0T + 15 = 0$

$0.2(T^2 - 20T + 75) = 0$

$T^2 - 20T = -75$

$T^2 - 20T + 100 = -75 + 100$

$(T-10)^2 = 25$

$T - 10 = \pm 5$

$T = 10 \pm 5$

$T = 5$ or $T = 15$

The voltage is 15.0 V when the temperature is 5.0°C or 15°C.

7.3 The Quadratic Formula

1. $x^2+5x+6=0;\ a=1,\ b=5,\ c=6$

$$x=\frac{-5\pm\sqrt{5^2-4(1)(6)}}{2(1)}$$

$$x=\frac{-5\pm\sqrt{1}}{2}$$

$$x=\frac{-5\pm1}{2}$$

$$x=\frac{-5+1}{2} \text{ or } x=\frac{-5-1}{2}$$

$$x=-2 \qquad x=-3$$

5. $x^2+2x-8=0;\ a=1,\ b=2,\ c=-8$

$$x=\frac{-2\pm\sqrt{(2)^2-4(1)(-8)}}{2(1)}$$

$$x=\frac{-2\pm\sqrt{4-(-32)}}{2(1)}$$

$$x=\frac{-2\pm\sqrt{36}}{2}$$

$$x=\frac{-2\pm6}{2}$$

$$x=2 \text{ or } x=-4$$

9. $x^2-4x+2=0;\ a=1,\ b=-4,\ c=2$

$$x=\frac{-(-4)\pm\sqrt{(-4)^2-4(1)(2)}}{(2)}$$

$$x=\frac{4\pm\sqrt{16-8}}{2}$$

$$x=\frac{4\pm\sqrt{8}}{2}$$

$$x=\frac{4\pm2\sqrt{2}}{2}$$

$$x=2\pm\sqrt{2}$$

13. $2s^2+5s=3$

$2s^2+5s-3=0;\ a=2,\ b=5,\ c=-3$

$$s=\frac{-5\pm\sqrt{(5)^2-4(2)(-3)}}{2(2)}$$

$$s=\frac{-5\pm\sqrt{25-(-24)}}{4}$$

$$s=\frac{-5\pm\sqrt{49}}{4}$$

$$s=\frac{-5\pm7}{4}$$

$$s=-3 \text{ or } s=\frac{1}{2}$$

17. $y+2=2y^2$

$2y^2-y-2=0;\ a=2,\ b=-1,\ c=-2$

$$y=\frac{-(-1)\pm\sqrt{(-1)^2-4(2)(-2)}}{2(2)}$$

$$y=\frac{1\pm\sqrt{1-(-16)}}{4}$$

$$y=\frac{1\pm\sqrt{17}}{4}$$

21. $8t^2+61t=-120$

$8t^2+61t+120=0;\ a=8,\ b=61,\ c=120$

$$t=\frac{-61\pm\sqrt{(61)^2-4(8)(120)}}{2(8)}$$

$$t=\frac{-61\pm\sqrt{3721-3840}}{16}$$

$$t=\frac{-61\pm\sqrt{-119}}{16} \text{ (imaginary roots)}$$

25. $25y^2=121$

$25y^2-121=0;\ a=25,\ b=0,\ c=-121$

$$y=\frac{-0\pm\sqrt{(0)^2-4(25)(-121)}}{2(25)}$$

$$y=\frac{-0\pm\sqrt{0-(-12\,100)}}{50}$$

$$y=\frac{\pm\sqrt{12\,100}}{50}$$

$$y=\frac{\pm110}{50}$$

$$y=\frac{11}{5} \text{ or } y=-\frac{11}{5}$$

29. $x^2-0.200x-0.400=0;$

$a=1,\ b=-0.200,\ c=-0.400$

$$x=\frac{-(-0.20)0\pm\sqrt{(-0.200)^2-4(1)(-0.400)}}{2(1)}$$

$$x=\frac{0.200\pm\sqrt{0.0400-(-1.6)}}{2}$$

$$x=\frac{0.200\pm\sqrt{1.64}}{2}$$

$x=-0.540$ or $x=0.740$

33. $x^2+2cx-1=0;\ a=1,\ b=2c,\ c=-1$

$$x=\frac{-(2c)\pm\sqrt{(2c)^2-4(1)(-1)}}{2(1)}$$

$$x=\frac{-2c\pm\sqrt{4c^2-(-4)}}{2}$$

$$x=\frac{-2c\pm2\sqrt{c^2+1}}{2}$$

$$x=-c\pm\sqrt{c^2+1}$$

37. $2x^2-7x=-8$

$2x^2-7x+8=0;\ a=2;\ b=-7;\ c=8$

$$D=\sqrt{(-7)^2-4(2)(8)}=\sqrt{-15},$$

unequal imaginary roots

41. $x^2+4x+k=0$ will have a double root if $b^2-4ac=0$

$4^2-4(1)(k)=0$

$k=4$

45. For $D=3.625$

$$D_0^2-DD_0-0.250D^2=0$$

$$D_0^2-3.625D_0-0.25(3.625)^2=0$$

$$D_0^2-3.625D_0-3.28515625=0$$

$a=1,\ b=-3.625,\ c=-3.28515625$

$$D_0=\frac{-(-3.625)\pm\sqrt{(-3.625)^2-4(1)(-3.28515625)}}{2}$$

$D_0=4.38$ cm or $D_0=-0.751$ cm, reject since $D_0>0$.

49. $Lm^2+Rm+\frac{1}{C}=0;\ a=L,\ b=R,\ c=\frac{1}{C}$

$$m=\frac{-R\pm\sqrt{R^2-4(L)\left(\frac{1}{C}\right)}}{2(L)}$$

$$m=\frac{-R\pm\sqrt{R^2-\frac{4L}{C}}}{2L}$$

53. $A=l\times w=262$

$(w+12.8)\times w=262$

$w^2+12.8w-262=0$

$a=1,\ b=12.8,\ c=-262$

$$r=\frac{-12.8\pm\sqrt{12.8^2-4(1)(-262)}}{2(1)}$$

$w=-3.545$ or $w=0.045$

$w=11.0$ m or $w=-24$, reject since $w>0$.

$l=w+12.8=23.8$ m.

The dimensions of the rectangle are $l=23.8$ m and $w=11.0$ m

57. v = truck speed

$v+20$ = car speed

From $d=vt$

$120=(v+20)t$, for the car, or

$$t=\frac{120}{v+20}$$

$$120=v\left(t+\frac{18}{60}\right),\ \text{truck}$$

$$120=v\left(\frac{120}{v+20}+\frac{18}{60}\right)$$

now multiply by LCD

$$(60)(v+20)120=v(60)(120)+18v(v+20)$$

$$7200v+144000=7200v+18v^2+360v$$

$$18v^2+360v-144000=0$$

$$18(v^2+20v-8000)=0$$

$a=1,\ b=20,\ c=-8000$

$$v=\frac{-20\pm\sqrt{400-4(1)(-8000)}}{2(1)}$$

$$t=\frac{-20\pm180}{2}$$

$v=-100,\ 80$

(use positive solution)

The truck speed is 80.0 km/h and the car speed is 100.0 km/h.

7.4 The Graph of the Quadratic Function

1. $y = 2x^2 + 8x + 6;\ a = 2,\ b = 8,\ c = 6$

$x\text{-coordinate of vertex} = \dfrac{-b}{2a}$

$= \dfrac{-8}{2(2)} = -2$

$y\text{-coordinate of vertex} = 2(-2)^2 + 8(-2) + 6$

$= -2$

The vertex is $(-2, -2)$ and since $a > 0$, it is a minimum. Since $c = 6$, the y-intercept is (0, 6) and the check is:

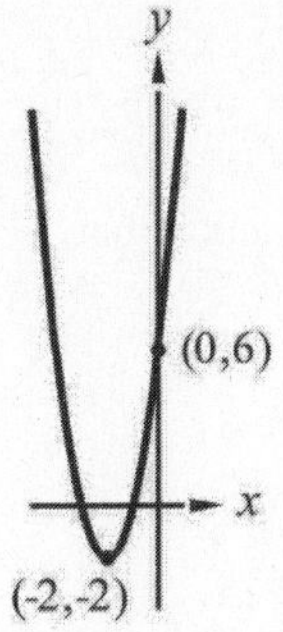

5. $y = -3x^2 + 10x - 4$, with $a = -3$, $b = 10$, $c = -4$. This means that the x-coordinate of the extreme is

$$\frac{-b}{2a} = \frac{-10}{2(-3)} = \frac{10}{6} = \frac{5}{3}$$

and the y-coordinate is

$$y = -3\left(\frac{5}{3}\right)^2 + 10\left(\frac{5}{3}\right) - 4 = \frac{13}{3}.$$

Thus the extreme point is $\left(\frac{5}{3}, \frac{13}{3}\right)$.

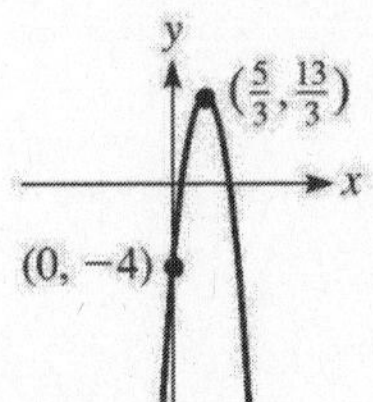

Since $a < 0$, it is a maximum point.
Since $c = -4$, the y-intercept is $(0, -4)$. Use the maximum point $\left(\frac{5}{3}, \frac{13}{3}\right)$, and the y-intercept $(0, -4)$, and the fact that the graph is a parabola, to sketch the graph.

9. $y = x^2 - 4 = x^2 + 0x - 4;\ a = 1,\ b = 0,\ c = -4$
The x-coordinate of the extreme point is $\dfrac{-b}{2a} = \dfrac{-0}{2(1)} = 0$, and the y-coordinate is $y = 0^2 - 4 = -4$.
The extreme point is $(0, -4)$.

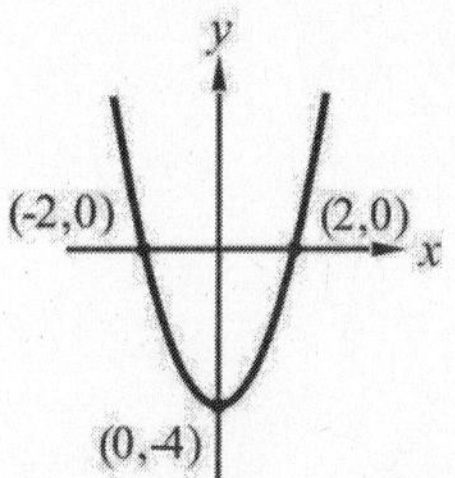

Since $a > 0$, it is a minimum point.
Since $c = -4$, the y-intercept is $(0, -4)$.
$x^2 - 4 = 0,\ x^2 = 4,\ x = \pm 2$ are the x-intercepts.
Use the minimum points and intercepts to sketch the graph.

13. $y = 2x^2 + 3 = 2x^2 + 0x + 3;\ a = 2,\ b = 0,\ c = 3$
The x-coordinate of the extreme point is $\dfrac{-b}{2a} = \dfrac{-0}{2(2)} = 0$, and the y-coordinate is $y = 2(0)^2 + 3 = 3$.
The extreme point is $(0, 3)$. Since $a > 0$ it is a minimum point.

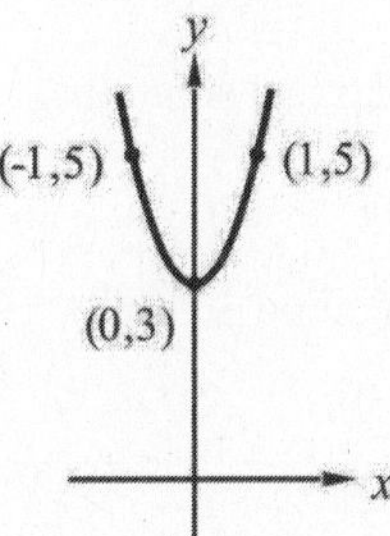

Since $c = 3$, the y-intercept is (0, 3) there are no x-intercepts, $b^2 - 4ac = -24$. $(-1, 5)$ and $(1, 5)$ are on the graph. Use the three points to sketch the graph.

17. $2x^2 - 3 = 0.$

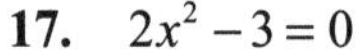

Graph $y = 2x^2 - 3$ and use the zero feature to find the roots.

$x = -1.22$ and $x = 1.22.$

10

−3 3

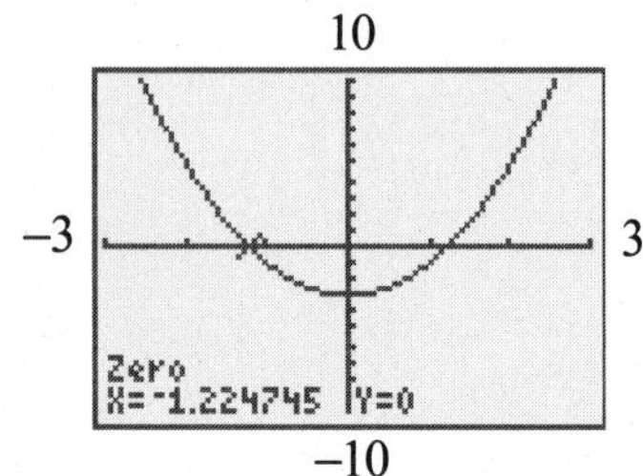

−10

10

−3 3

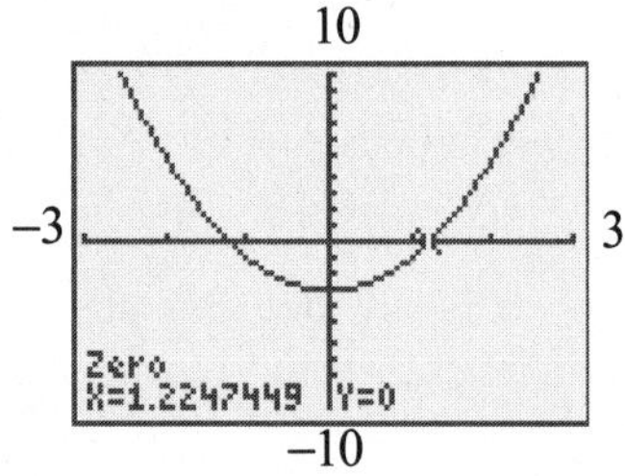

−10

21. $x(2x-1) = -3$

Graph $y_1 = x(2x-1)+3$ and use the zero feature.

10

−3 3

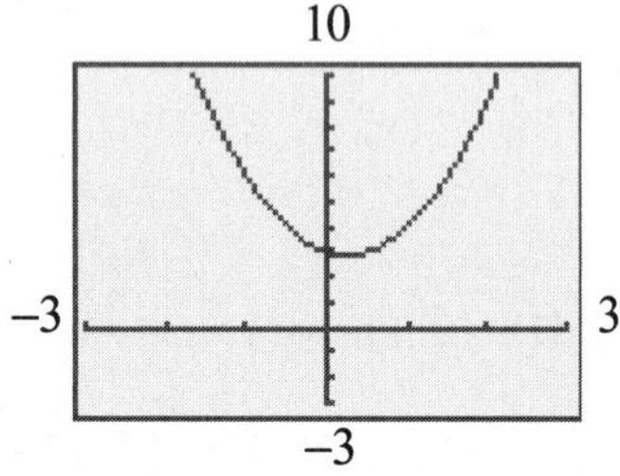

−3

As the graph shows, there are no real solutions.

25. **(a)** $y = x^2$ **(b)** $y = x^2 + 3$ **(c)** $y = x^2 - 3$

The parabola $y = x^2 + 3$ is shifted up $+3$ units (minimum point $(0,\ 3)$).

The parabola $y = x^2 - 3$ is shifted down -3 units (minimum point $(0,\ -3)$).

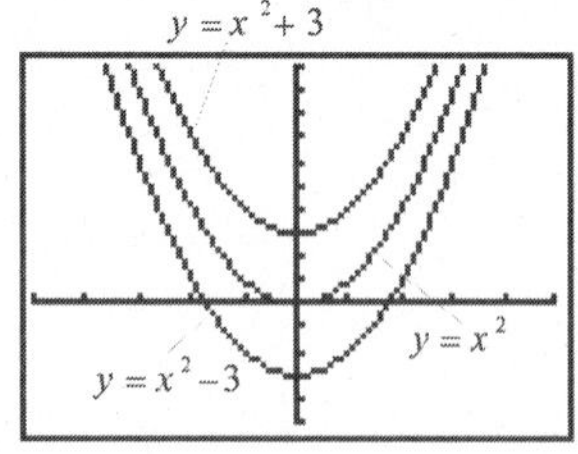

29. **(a)** $y = x^2$ **(b)** $y = 3x^2$ **(c)** $y = \frac{1}{3}x^2$

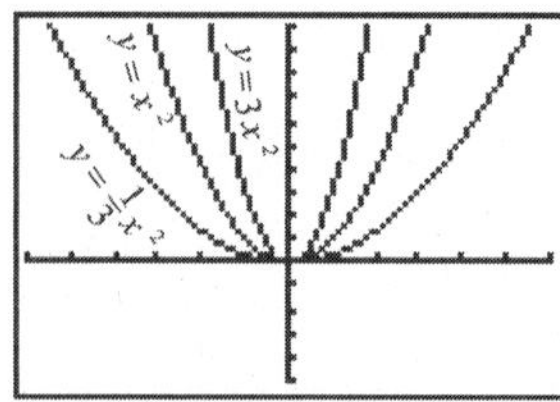

The graph of $y = 3x^2$ is the graph of $y = x^2$ narrowed. The graph of $y = \frac{1}{3}x^2$ is the graph of $y = x^2$ broadened.

33. $y = 2x^2 - 4x - c$ will have two real roots if

$$b^2 - 4ac > 0$$
$$(-4)^2 - 4(2)(-c) \geq 0$$
$$16 + 8c \geq 0$$
$$c \geq -2$$

-2 is the smallest integral value of c such that $y = 2x^2 - 4x - c$ has two real roots.

37. $A = w(8-w)$ for $0 > w > 8.$

w-intercepts occur at

$w = 0$ or $w = 8$

since

$A = 8w - w^2$, $a < 0$ so the curve opens downward.

20

−3 10

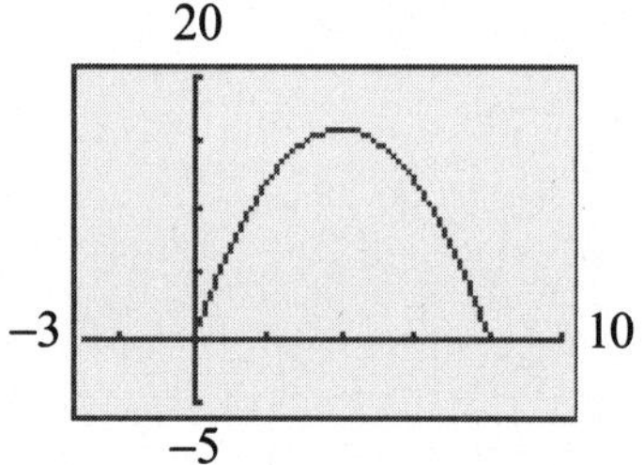

−5

41. **(a)** Graph $s = 50 + 90t - 4.9t^2$

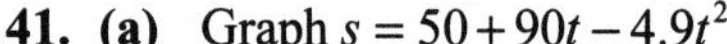

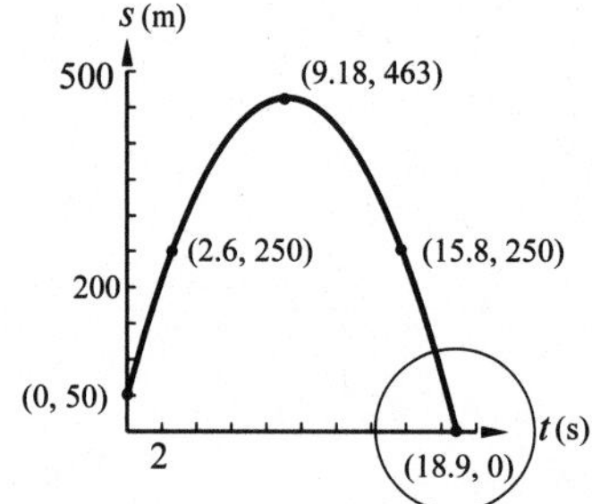

It hits the ground after 18.9s.

(b)

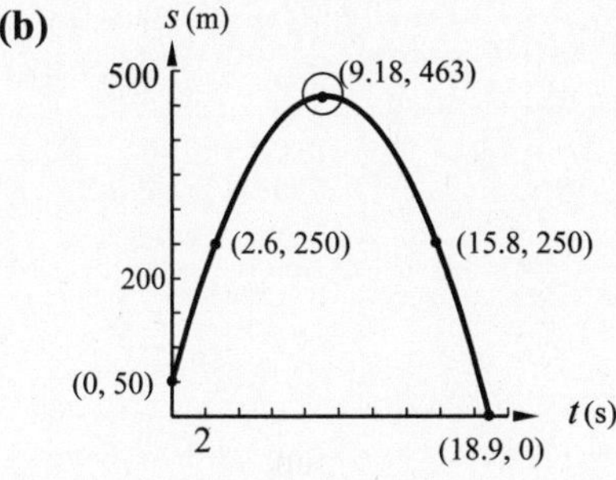

It reaches a height of 463m.

(c)

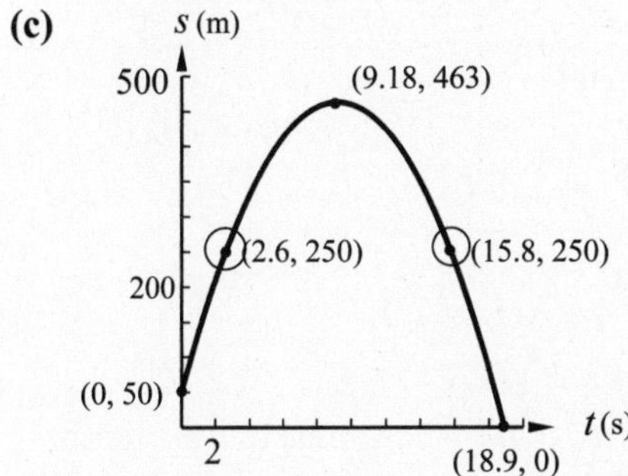

It reaches a height of 250m after 2.6s and again after 15.8s.

45. $A = xy = 2000,$

$$y = \frac{2000}{x}$$

$$\text{cost} = 90x + 60y = 7500$$

$$90x + 60\left(\frac{2000}{x}\right) = 7500$$

$$90x^2 - 7500x + 120\,000 = 0$$

The quadratic formula gives,

$$x = \frac{-(-7500) \pm \sqrt{(-7500)^2 - 4(90)(120\ 000)}}{2(90)}$$

$x = 21.6,\ 61.7$

$$y = \frac{2000}{x},$$

$y = 92.6,\ 32.4$

The dimensions are 21.6 m by 92.6 m, or 32.4 m by 61.7 m.

Review Exercises

1. $x^2 + 3x - 4 = 0$

$$(x+4)(x-1) = 0$$

$$x+4 = 0 \quad \text{or} \quad x-1 = 0$$

$$x = -4 \qquad\qquad x = 1$$

5. $3x^2 + 11x = 4$

$$3x^2 + 11x - 4 = 0$$

$$(3x-1)(x+4) = 0$$

$$3x - 1 = 0 \quad \text{or} \quad x + 4 = 0$$

$$3x = 1 \qquad\qquad x = -4$$

$$x = \frac{1}{3}$$

9. $6s^2 = 25s$

$$6s^2 - 25s = 0$$

$$s(6s - 25) = 0$$

$$6s - 25 = 0 \quad \text{or} \quad s = 0$$

$$6s = 25$$

$$s = \frac{25}{6}$$

13. $x^2 - x - 110 = 0;\ a = 1;\ b = -1;\ c = -110$

$$x = \frac{-(-1) \pm \sqrt{(-1)^2 - 4(1)(-110)}}{2(1)}$$

$$x = \frac{1 \pm \sqrt{1 - (-440)}}{2}$$

$$x = \frac{1 \pm \sqrt{441}}{2}$$

$$x = \frac{1 \pm 21}{2}$$

$$x = -10.0 \quad \text{or} \quad x = 11.0$$

17. $2x^2 - x = 36$

$$2x^2 - x - 36 = 0;\ a = 2;\ b = -1;\ c = -36$$

$$x = \frac{-(-1) \pm \sqrt{(-1)^2 - 4(2)(-36)}}{2(2)}$$

$$x = \frac{1 \pm \sqrt{1 - (-288)}}{4}$$

$$x = \frac{1 \pm \sqrt{289}}{4}$$

$$x = \frac{1 \pm 17}{4}$$

$$x = \frac{9}{2} \quad \text{or} \quad x = -4$$

21. $2.1x^2+2.3x+5.5=0;\ a=2.1;\ b=2.3;\ c=5.5$

$$x=\frac{-2.3\pm\sqrt{(2.3)^2-4(2.1)(5.5)}}{2(2.1)}$$
$$x=\frac{-2.3\pm\sqrt{5.29-46.2}}{4.2}$$
$$x=\frac{-2.3\pm\sqrt{-40.91}}{4.2}$$
(imaginary roots)

25. $x^2+4x-4=0;\ a=1;\ b=4;\ c=-4$

$$x=\frac{-4\pm\sqrt{(4)^2-4(1)(-4)}}{2(1)}$$
$$x=\frac{-4\pm\sqrt{16-(-16)}}{2}$$
$$x=\frac{-4\pm\sqrt{32}}{2}$$
$$x=\frac{-4\pm\sqrt{16\times2}}{2}$$
$$x=\frac{-4\pm4\sqrt{2}}{2}$$
$$x=-2\pm2\sqrt{2}$$

29. $4v^2=v+5$

$$4v^2-v-5=0$$
$$(v+1)(4v-5)=0$$
$$4v-5=0 \quad\text{or}\quad v+1=0$$
$$4v=5 \qquad v=-1$$
$$v=\frac{5}{4}$$

33. $a^2x^2+2ax+2=0;\ a=a^2;\ b=2a;\ c=2$

$$x=\frac{-(2a)\pm\sqrt{(2a)^2-4(a^2)(2)}}{2(a^2)}$$
$$x=\frac{-2a\pm\sqrt{4a^2-8a^2}}{2a^2}$$
$$x=\frac{-2a\pm\sqrt{-4a^2}}{2a^2}$$
$$x=\frac{-2a\pm2a\sqrt{-1}}{2a^2}$$
$$x=\frac{-1\pm\sqrt{-1}}{a}\ (a\neq0,\ \text{imaginary roots})$$

37. $x^2-x-30=0$

$$x^2-x=30$$
$$x^2-x+\frac{1}{4}=30+\frac{1}{4}$$
$$\left(x-\frac{1}{2}\right)^2=\frac{121}{4}$$
$$x-\frac{1}{2}=\pm\sqrt{\frac{121}{4}}$$
$$x-\frac{1}{2}=\pm\frac{11}{2}$$
$$x=\frac{1}{2}\pm\frac{11}{2}$$
$$x=-5 \text{ or } x=6$$

41. $\frac{x-4}{x-1}=\frac{2}{x},\ (x\neq1,0)$

$$x(x-4)=2(x-1)$$
$$x^2-4x=2x-2$$
$$x^2-6x+2=0;\ a=1;\ b=-6;\ c=2$$
$$x=\frac{-(-6)\pm\sqrt{(-6)^2-4(1)(2)}}{2(1)}$$
$$x=\frac{6\pm\sqrt{36-8}}{2}$$
$$x=\frac{6\pm\sqrt{28}}{2}$$
$$x=\frac{6\pm\sqrt{4\times7}}{2}$$
$$x=\frac{6\pm2\sqrt{7}}{2}$$
$$x=3\pm\sqrt{7}$$

45. $y=2x^2-x-1;\ a=2,\ b=-1,\ c=-1$

$c=-1$

y-intercept $=-1$

$$2x^2-x-1=0$$
$$(x-1)(2x+1)=0$$
$x=1$ and $x=-\frac{1}{2}$

are the x-intercepts

$$x\text{ vertex}=\frac{-b}{2a}=\frac{-(-1)}{2(2)}=\frac{1}{4}$$
$$y\text{ vertex}=2\left(\frac{1}{4}\right)^2-\left(\frac{1}{4}\right)-1=-\frac{9}{8}$$

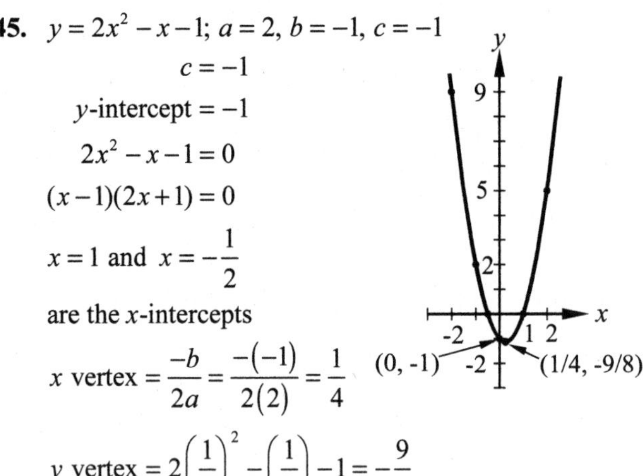

49. Graph $y_1 = 2x^2 + x - 4$ and use the zero feature.

$x = -1.69$ and

$x = 1.19$

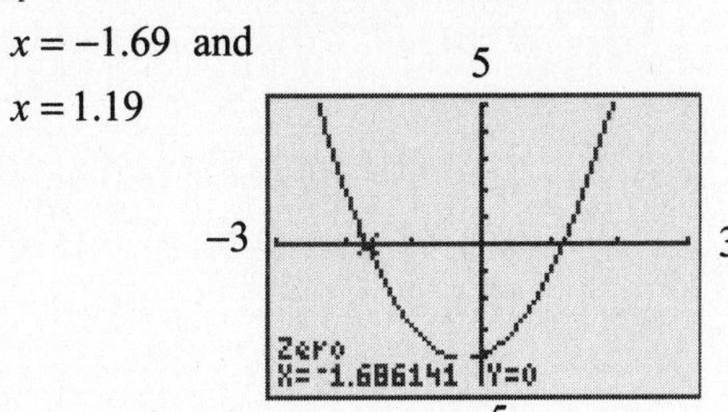

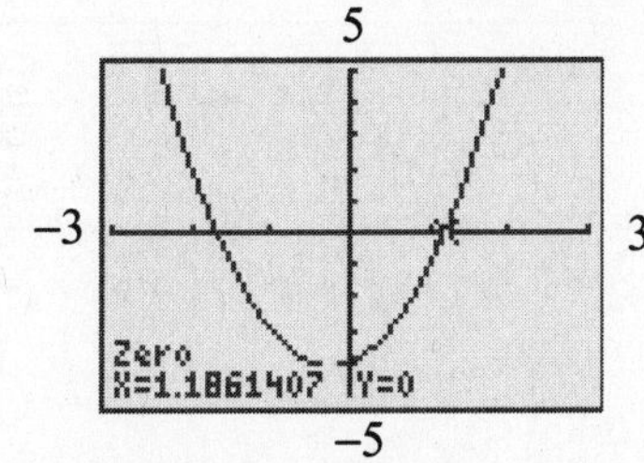

53. The roots are equally spaced on either side of

$x = -1$

$$\frac{x_1 + x_2}{2} = -1$$

$x_1 + x_2 = -2$

$2 + x_2 = -2$

$x_2 = -4$ is the other solution.

57. $0.1x^2 + 0.8x + 7 = 50$

$0.1x^2 + 0.8x - 43 = 0$; $a = 0.1$, $b = 0.8$, $c = -43$

$$x = \frac{-b \pm \sqrt{b^2 - 4ac}}{2a}$$

$$x = \frac{-(0.8) \pm \sqrt{(0.8)^2 - 4(0.1)(-43)}}{2(0.1)}$$

$$x = \frac{-0.8 \pm \sqrt{17.84}}{0.2}$$

$x = 17.1$ units or -25.1 units

17 units can be made for \$50

61. $h = vt \sin\theta - 4.9t^2$

$6.00 = 15.0t \sin 65.0° - 4.9t^2$

$4.9t^2 - 13.59462t + 6.00 = 0$ from which

$a = 4.9, b = -13.59462, c = 6.00$

$$t = \frac{-b \pm \sqrt{b^2 - 4ac}}{2a}$$

$$t = \frac{-(-13.59462) \pm \sqrt{(-13.59462)^2 - 4(4.9)(6.00)}}{2(4.9)}$$

$$t = \frac{13.59462 \pm \sqrt{67.2136}}{9.8}$$

$t = 2.22$ or $t = 0.551$

The height $h = 6.0$ m is reached when $t = 0.55$ s and 2.22 s.

It reaches that height twice (once on the way up, and once on the way down).

65. $A = 2\pi r^2 + 2\pi rh$

$2\pi r^2 + 2\pi hr - A = 0$

$a = 2\pi, b = 2\pi h, c = -A$

$$r = \frac{-b \pm \sqrt{b^2 - 4ac}}{2a}$$

$$r = \frac{-2\pi h \pm \sqrt{(2\pi h)^2 - 4(2\pi)(-A)}}{2(2\pi)}$$

$$r = \frac{-2\pi h + \sqrt{4\pi^2 h^2 + 8\pi A}}{4\pi}$$

$$r = \frac{-2\pi h + \sqrt{4(\pi^2 h^2 + 2\pi A)}}{4\pi}$$

$$r = \frac{-2\pi h + 2\sqrt{\pi^2 h^2 + 2\pi A}}{4\pi}$$

$$r = \frac{-\pi h + \sqrt{\pi^2 h^2 + 2\pi A}}{2\pi}$$

69. $p = 0.090t - 0.015t^2$

$p = -0.015t^2 + 0.090t$

$a = -0.015$, $b = 0.090$, $c = 0$

$y\text{-int} = (0, 0)$

$-0.015t^2 + 0.090t = 0$

$0.015t(-t + 6) = 0$

$t = 0$ or $t = 6$ are the x-intercepts

$$t \text{ vertex} = \frac{-b}{2a} = \frac{-0.090}{2(-0.015)} = 3$$

p vertex $= 0.090(3) - 0.015(3)^2 = 0.135$

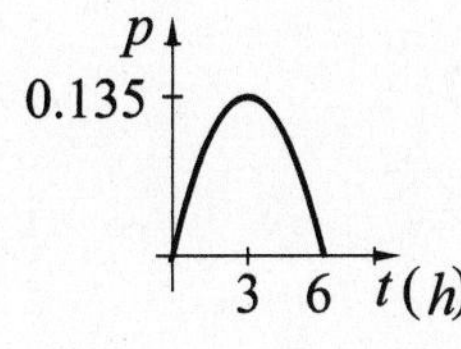

73. Original volume

$V = x^3$

New volume

$$V - 29 = (x - 0.1)^3$$

$$V - 29 = x^3 + 3x^2(-0.1) + 3x(-0.1)^2 + (-0.1)^3$$

$$x^3 - 29 = x^3 - 0.3x^2 + 0.03x - 0.001$$

$$0.3x^2 - 0.03x - 28.999 = 0$$

$a = 0.3,\ b = -0.03,\ c = -28.999$

$$x = \frac{-b \pm \sqrt{b^2 - 4ac}}{2a}$$

$$x = \frac{-(-0.03) \pm \sqrt{(-0.03)^2 - 4(0.3)(-28.999)}}{2(0.3)}$$

$$x = \frac{0.03 \pm \sqrt{34.7997}}{0.6}$$

$x = 9.88$ or $x = -9.78$ (reject since $x > 0$)

$x = 9.88$ cm

77.

$$(h + 14.5)^2 + h^2 = 68.6^2$$

$$h^2 + 2(14.5)h + 14.5^2 + h^2 = 68.6^2$$

$$2h^2 + 29h - 4495.71 = 0$$

$a = 2,\ b = 29,\ c = -4495.71$

$$h = \frac{-b \pm \sqrt{b^2 - 4ac}}{2a}$$

$$h = \frac{-(29) \pm \sqrt{(29)^2 - 4(2)(-4495.71)}}{2(2)}$$

$$h = \frac{-29 \pm \sqrt{36806.68}}{4}$$

$h = 40.7$ or $h = -55.2$ (reject since $h > 0$)

$h = 40.7$ cm

$h + 14.5 = 55.2$ cm

The dimensions of the screen are $40.7 \text{ cm} \times 55.2$ cm.

81. $p = 0.001\,74(10 + 24h - h^2)$

$$p = -0.00174h^2 + 0.04176h + 0.0174$$

If $p = 0.205$ ppm

$$0.205 = -0.00174h^2 + 0.04176h + 0.0174$$

$$0 = -0.00174h^2 + 0.04176h - 0.1876$$

$a = -0.00174,\ b = 0.04176,\ c = -0.1876$

$$h = \frac{-b \pm \sqrt{b^2 - 4ac}}{2a}$$

$$h = \frac{-0.04176 \pm \sqrt{0.04176^2 - 4(-0.00174)(-0.1876)}}{2(-0.00174)}$$

$h = 5.98$ and $h = 18.0$

From the graph $p = 0.205$ at 6 h and 18 h.

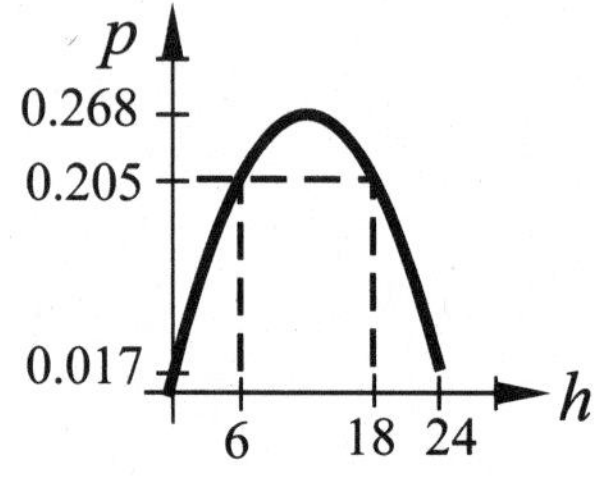

CHAPTER 8

TRIGONOMETRIC FUNCTIONS OF ANY ANGLE

8.1 Signs of the Trigonometric Functions

1. **(a)** $\sin(150^\circ + 90^\circ) = \sin 240^\circ$
which is in Quandrant III is $-$
$\cos(290^\circ + 90^\circ) = \cos 380^\circ$
which is in Quandrant I is $+$
$\tan(190^\circ + 90^\circ) = \tan 280^\circ$
which is in Quandrant IV is $-$
$\cot(260^\circ + 90^\circ) = \cot 350^\circ$
which is in Quandrant IV is $-$
$\sec(350^\circ + 90^\circ) = \sec 440^\circ$
which is in Quandrant I is $+$
$\csc(100^\circ + 90^\circ) = \csc 190^\circ$
which is in Quandrant III is $-$

(b) $\sin(300^\circ + 90^\circ) = \sin 390^\circ$
which is in Quandrant I is $+$
$\cos(150^\circ + 90^\circ) = \cos 240^\circ$
which is in Quandrant III is $-$
$\tan(100^\circ + 90^\circ) = \tan 190^\circ$
which is in Quandrant III is $+$
$\cot(300^\circ + 90^\circ) = \cot 390^\circ$
which is in Quandrant I is $+$
$\sec(200^\circ + 90^\circ) = \sec 290^\circ$
which is in Quandrant IV is $+$
$\csc(250^\circ + 90^\circ) = \csc 340^\circ$
which is in Quandrant IV is $-$

5. $\csc 98^\circ$ is positive since 98° is in Quadrant II, where $\csc\theta$ is positive.
$\cot 82^\circ$ is positive since 82° is in Quadrant I, where $\cot\theta$ is positive.

9. $\cos 348^\circ$ is positive since 348° is in Quadrant IV, where $\cos\theta$ is positive.
$\csc 238^\circ$ is negative since 238° is in Quadrant III, where $\csc\theta$ is negative.

13. $\cot(-2^\circ)$ is negative since -2° is in Quadrant IV, where $\cot\theta$ is negative.
$\cos 710^\circ$ is positive since 710° is coterminal with $710^\circ - 360^\circ = 350^\circ$ which is in Quadrant IV, where $\cos\theta$ is positive.

17. Point $(-2, -3)$, $x = -2$, $y = -3$

$$r = \sqrt{x^2 + y^2}$$
$$r = \sqrt{(-2)^2 + (-3)^2}$$
$$r = \sqrt{13}$$
$$\sin\theta = \frac{y}{r} = \frac{-3}{\sqrt{13}}$$
$$\cos\theta = \frac{x}{r} = \frac{-2}{\sqrt{13}}$$
$$\tan\theta = \frac{y}{x} = \frac{-3}{-2} = \frac{3}{2}$$
$$\csc\theta = \frac{r}{y} = \frac{\sqrt{13}}{-3} = -\frac{\sqrt{13}}{3}$$
$$\sec\theta = \frac{r}{x} = \frac{\sqrt{13}}{-2} = -\frac{\sqrt{13}}{2}$$
$$\cot\theta = \frac{x}{y} = \frac{-2}{-3} = \frac{2}{3}$$

21. Point $(20, -8)$, $x = 20$, $y = -8$

$$r = \sqrt{x^2 + y^2}$$
$$r = \sqrt{20^2 + (-8)^2}$$
$$r = \sqrt{464} = \sqrt{16(29)}$$
$$r = 4\sqrt{29}$$
$$\sin\theta = \frac{y}{r} = \frac{-8}{4\sqrt{29}} = \frac{-2}{\sqrt{29}}$$
$$\cos\theta = \frac{x}{r} = \frac{20}{4\sqrt{29}} = \frac{5}{\sqrt{29}}$$
$$\tan\theta = \frac{y}{x} = \frac{-8}{20} = -\frac{2}{5}$$
$$\csc\theta = \frac{r}{y} = \frac{4\sqrt{29}}{-8} = -\frac{\sqrt{29}}{2}$$
$$\sec\theta = \frac{r}{x} = \frac{4\sqrt{29}}{20} = \frac{\sqrt{29}}{5}$$
$$\cot\theta = \frac{x}{y} = \frac{20}{-8} = -\frac{5}{2}$$

25. $\tan\theta = 1.50$

$\tan\theta = \frac{y}{x}$. Since $\tan\theta$ is positive, x and y must have the same sign, and the terminal side of the angle must lie in either Quadrant I or Quadrant III.

29. $\sin\theta$ is positive and $\cos\theta$ is negative
$\sin\theta$ is positive in Quadrant I and Quadrant II.
$\cos\theta$ is negative in Quadrant II and Quadrant III.
The terminal side of θ must lie in Quadrant II to meet both conditions.

33. $\csc\theta$ is negative and $\tan\theta$ is negative
$\csc\theta$ is negative in Quadrant III and Quadrant IV.
$\tan\theta$ is negative in Quadrant II and Quadrant IV.
The terminal side of θ must lie in Quadrant IV to meet both conditions.

37. $\sin\theta$ is positive and $\cot\theta$ is negative
$\sin\theta$ is positive in Quadrant I and Quadrant II.
$\cot\theta$ is negative in Quadrant II and Quadrant IV.
The terminal side of θ must lie in Quadrant II to meet both conditions.

41. For (x, y) in Quadrant IV, x is $(+)$ and y is $(-)$

$$\frac{y}{x} = \frac{(-)}{(+)} = (-)$$

8.2 Trigonometric Functions of Any Angle

1. $\sin 200^\circ = -\sin(200^\circ - 180^\circ) = -\sin 20^\circ = -0.342$

$\tan 150^\circ = -\tan(180^\circ - 150^\circ) = -\tan 30^\circ = -0.577$

$\cos 265^\circ = -\cos(265^\circ - 180^\circ) = -\cos 85^\circ = -0.0872$

$\cot 300^\circ = -\cot(360^\circ - 300^\circ) = -\cot 60^\circ$

$= -\frac{1}{\tan 60^\circ} = -0.577$

$\sec 344^\circ = \sec(360^\circ - 344^\circ) = \sec 16^\circ = \frac{1}{\cos 16^\circ} = 1.04$

$\sin 397^\circ = \sin(397^\circ - 360^\circ) = \sin 37^\circ = 0.602$

5. $\sin 160^\circ = \sin(180^\circ - 160^\circ) = \sin 20^\circ$

$\cos 220^\circ = -\cos(220^\circ - 180^\circ) = -\cos 40^\circ$

9.

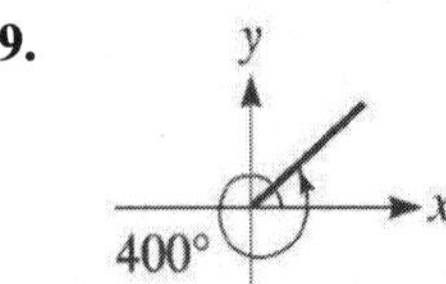

$\cos 400^\circ = \cos(400^\circ - 360^\circ) = \cos 40^\circ$

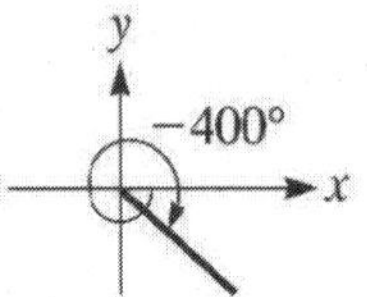

$\tan(-400^\circ) = \tan(-400^\circ + 360^\circ) = \tan(-40^\circ) = -\tan 40^\circ$

13. $\cos 106.3^\circ = -\cos(180^\circ - 106.3^\circ) = -\cos(73.7^\circ)$

$= -0.2807$

17. $\tan(-31.5^\circ) = -\tan 31.5^\circ = -0.613$

21. $\sin 310.36^\circ = -0.76199$

25. $\cos(-72.61^\circ) = 0.2989$

29. $\cos\theta = 0.4003$

$\theta_{ref} = \cos^{-1} 0.4003$

$\theta_{ref} = 66.40^\circ$

Since $\cos\theta$ is positive, θ must lie in Quadrant I or Quadrant IV.

Therefore, $\theta_1 = 66.40^\circ$

or $\theta_4 = 360^\circ - 66.40^\circ$

$\theta_4 = 293.60^\circ$

33. $\sin\theta = 0.870$

$\theta_{ref} = \sin^{-1} 0.870$

$\theta_{ref} = 60.5^\circ$

Since $\sin\theta$ is positive, θ must lie in Quadrant I or Quadrant II.

If $\cos\theta$ is negative, θ must lie in Quadrant II or Quadrant III.

To satisfy both conditions, θ must lie in Quadrant II.

$\theta_2 = 180^\circ - 60.5^\circ$

$\theta_2 = 119.5^\circ$

37. $\tan\theta = -1.366$

$\theta_{\text{ref}} = \tan^{-1}1.366$

$\theta_{\text{ref}} = 53.79^\circ$

Since $\tan\theta$ is negative, θ must lie in Quadrant II or Quadrant IV.

If $\cos\theta$ is positive, θ must lie in Quadrant I or Quadrant IV.

To satisfy both conditions, θ must lie in Quadrant IV.

$\theta_4 = 360^\circ - 53.79^\circ$

$\theta_4 = 306.21^\circ$

41. $\sin\theta = -0.5736$

$\theta_{\text{ref}} = \sin^{-1}0.5736$

$\theta_{\text{ref}} = 35.00^\circ$

Since $\sin\theta$ is negative, θ must lie in Quadrant III or Quadrant IV.

If $\cos\theta$ is positive, θ must lie in Quadrant I or Quadrant IV.

To satisfy both conditions, θ must lie in Quadrant IV.

$\theta_4 = 360^\circ - 35.00^\circ$

$\theta_4 = 325.00^\circ$

In general, the angle could be any solution

$\theta = 325.00^\circ + k\times 360^\circ$ where $k = 0, \pm 1, \pm 2, \cdots$

but all evaluations of the trigonometric functions will be identical for any integer number of rotations from the Quadrant IV solution.

$\tan\theta = \tan 325.00^\circ$

$\tan\theta = -0.7002$

45. $\sin 90^\circ = 1$, and

$2\sin 45^\circ = 2\times\frac{\sqrt{2}}{2} = \sqrt{2}$

$1 < \sqrt{2}$

$\sin 90^\circ < 2\sin 45^\circ$

49. $\theta = 195^\circ$ has $\theta_{\text{ref}} = 15^\circ$

$\cos 195^\circ = -\cos(195^\circ - 180^\circ)$

$\cos 195^\circ = -\cos 15^\circ$

15° and 75° are complementary, so their cofunctions are equivalent

$\cos 195^\circ = -\sin 75^\circ$

$\cos 195^\circ = -0.9659$

53. $i = i_m \sin\theta$

$i = (0.0259\text{ A})\sin 495.2^\circ$

$i = 0.0183\text{ A}$

8.3 Radians

1. $2.80 = (2.80)\left(\frac{180^\circ}{\pi}\right) = 160^\circ$

5. $15^\circ = 15^\circ\left(\frac{\pi}{180^\circ}\right) = \frac{\pi}{12}$

$150^\circ = 150^\circ\left(\frac{\pi}{180^\circ}\right) = \frac{5\pi}{6}$

9. $210^\circ = 210^\circ\left(\frac{\pi}{180^\circ}\right) = \frac{7\pi}{6}$

$27^\circ = 27^\circ\left(\frac{\pi}{180^\circ}\right) = \frac{3\pi}{20}$

13. $\frac{2\pi}{5} = \frac{2\pi}{5}\left(\frac{180^\circ}{\pi}\right) = 72^\circ$

$\frac{3\pi}{2} = \frac{3\pi}{2}\left(\frac{180^\circ}{\pi}\right) = 270^\circ$

17. $\frac{7\pi}{18} = \frac{7\pi}{18}\left(\frac{180^\circ}{\pi}\right) = 70^\circ$

$\frac{5\pi}{3} = \frac{5\pi}{3}\left(\frac{180^\circ}{\pi}\right) = 300^\circ$

21. $23.0^\circ = 23.0^\circ\left(\frac{\pi\text{ rad}}{180^\circ}\right) = 0.401\text{ rad}$

25. $333.5^\circ = 333.5^\circ\left(\frac{\pi\text{ rad}}{180^\circ}\right) = 5.821\text{ rad}$

29. $0.750 = 0.750\left(\frac{180^\circ}{\pi}\right) = 43.0^\circ$

33. $12.6 = 12.6\left(\frac{180^\circ}{\pi}\right) = 722^\circ$

37. $\sin\frac{\pi}{4} = \sin\left[\left(\frac{\pi}{4}\right)\left(\frac{180^\circ}{\pi}\right)\right] = \sin 45^\circ = 0.7071$

41. $\cos\frac{5\pi}{6} = \cos\left[\left(\frac{5\pi}{6}\right)\left(\frac{180^\circ}{\pi}\right)\right] = \cos 150^\circ = -0.8660$

45. $\tan 0.7359 = 0.9056$

49. $\sec 2.07 = \frac{1}{\cos 2.07} = -2.09$

53. $\sin\theta = 0.3090$

$\theta_{ref} = \sin^{-1} 0.3090$

$\theta_{ref} = 0.3141$

Since $\sin\theta$ is positive, θ must lie in Quadrant I or Quadrant II.

$\theta_1 = 0.3141$

$\theta_2 = \pi - 0.3141$

$\theta_2 = 2.827$

57. $\cos\theta = 0.6742$

$\theta_{ref} = \cos^{-1} 0.6742$

$\theta_{ref} = 0.8309$

Since $\cos\theta$ is positive, θ must lie in Quadrant I or Quadrant IV.

$\theta_1 = 0.8309$

$\theta_4 = 2\pi - 0.8309$

$\theta_4 = 5.452$

61. $\frac{5\pi}{8}$ has $\theta_{\text{ref}} = \pi - \frac{5\pi}{8} = \frac{3\pi}{8}$

$\frac{5\pi}{8}$ is in Quadrant II where $\cos\theta$ is negative.

$\cos\frac{5\pi}{8} = -\cos\frac{3\pi}{8}$

$\frac{3\pi}{8}$ and $\frac{\pi}{2} - \frac{3\pi}{8}$ are complementary, so their cofunctions are equivalent.

$\cos\frac{5\pi}{8} = -\sin\left(\frac{\pi}{2} - \frac{3\pi}{8}\right)$

$\cos\frac{5\pi}{8} = -\sin\frac{\pi}{8}$

$\cos\frac{5\pi}{8} = -0.3827$

65. $34.4^\circ = 34.4^\circ\left(\frac{1 \text{ circumference}}{360^\circ}\right)\left(\frac{1 \text{ mil}}{\frac{1}{6400} \text{ circumference}}\right)$

$= 612 \text{ mil}$

69. $V = \frac{1}{2}Wb\theta^2$

$V = \frac{1}{2}(8.75 \text{ N})(0.75 \text{ m})\left(5.5^\circ \frac{\pi}{180^\circ}\right)^2$

$V = 0.030 \text{ N}\cdot\text{m}$

$V = 0.030 \text{ J}$

8.4 Applications of Radian Measure

1. $s = \theta r$

$s = \left(\frac{\pi}{4}\right)(3.00 \text{ cm})$

$s = 2.36 \text{ cm}$

5. $s = \theta r = \left(\frac{\pi}{3}\right)(3.30 \text{ cm}) = 3.46 \text{ cm}$

9. $\theta = \frac{s}{r} = \frac{0.3913 \text{ km}}{0.9449 \text{ km}} = 0.4141 \text{ rad}$

$A = \frac{1}{2}\theta r^2 = \frac{1}{2}(0.4141)(0.9449 \text{ km})^2 = 0.1849 \text{ km}^2$

13. $\theta = 326.0^\circ\left(\frac{\pi}{180^\circ}\right) = 5.690 \text{ rad}$

$A = \frac{1}{2}\theta r^2$

$r = \sqrt{\frac{2A}{\theta}}$

$r = \sqrt{\frac{2(0.0119 \text{ m}^2)}{5.690}}$

$r = 0.0647 \text{ m}$

17. $r = \frac{s}{\theta} = \frac{0.203 \text{ km}}{\frac{3}{4}(2\pi)}$

$r = 0.0431 \text{ km}\left(\frac{1000 \text{ m}}{1 \text{ km}}\right)$

$r = 43.1 \text{ m}$

21. If time t is measured in hours from noon, The hour hand rotates at the rate of a full revolution (2π rad) every 12 hours,

Hour hand $\omega = \frac{2\pi \text{ rad}}{12 \text{ h}}$

$\theta_H = \omega t$

$\theta_H = \frac{2\pi}{12}t$

The minute hand rotates at the rate of a full revolution (2π rad) every 1 hour,

Minute hand $\omega = \dfrac{2\pi \text{ rad}}{1 \text{ h}}$

$$\theta_M = 2\pi t$$

If the angle between the minute hand and hour hand is π rad,

$$\theta_H + \pi = \theta_M$$

$$\frac{2\pi}{12}t + \pi = 2\pi t$$

$$\pi = \left(\frac{12\pi}{6} - \frac{\pi}{6}\right)t$$

$$t = \frac{\pi}{11\pi/6}$$

$$t = \frac{6}{11}\text{ h}\left(\frac{60 \text{ min}}{1 \text{ h}}\right) = 32.727 \text{ min}$$

$$t = 32 \text{ min} + 0.727 \text{ min}\left(\frac{60 \text{ s}}{1 \text{ min}}\right)$$

$$t = 32 \text{ min} + 44 \text{ s}$$

The clock will read 12:32:44 when the hour and minute hands will be at 180° apart.

25. $\omega = \dfrac{\theta}{t} = \dfrac{\pi \text{ rad}}{6.0 \text{ s}} = 0.52 \text{ rad/s}$

29.

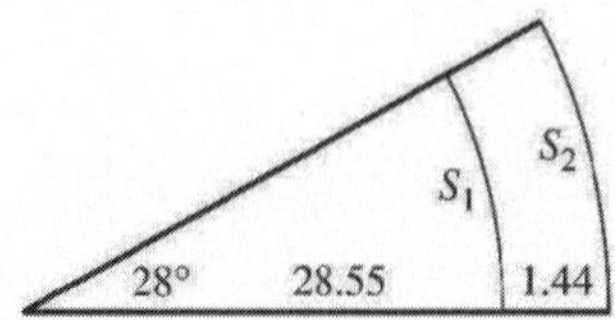

$$\theta = 28.0^\circ\left(\frac{\pi \text{ rad}}{180^\circ}\right) = 0.489 \text{ rad}$$

From $s = \theta r$

$$s_1 = 0.489(28.55 \text{ m}) = 13.952 \text{ m}$$

$$s_2 = 0.489(28.55 \text{ m} + 1.44 \text{ m}) = 14.656 \text{ m}$$

$$s_2 - s_1 = 0.704 \text{ m}$$

Outer rail is 0.704 m longer.

33. $\theta = 15.6^\circ\left(\dfrac{\pi}{180^\circ}\right) = 0.272 \text{ rad}$

$$A = \frac{1}{2}\theta r^2$$

$$r_1 = 285.0 \text{ m}$$

$$r_2 = 285.0 \text{ m} + 15.2 \text{ m} = 300.2 \text{ m}$$

$$A_{road} = A_2 - A_1$$

$$A_{road} = \frac{1}{2}(0.272)\left(300.2^2 - 285.0^2\right)\text{m}^2$$

$$A_{road} = 1210 \text{ m}^2$$

Volume = Area × thickness

$$V = At$$

$$V = 1210 \text{ m}^2(0.305 \text{ m})$$

$$V = 369 \text{ m}^3$$

37. $\omega = 20.0 \text{ r/min}\left(\dfrac{2\pi \text{ rad}}{1 \text{ r}}\right)\left(\dfrac{1 \text{ min}}{60 \text{ s}}\right) = 2.09 \text{ rad/s}$

$$v = \omega r = (2.09 \text{ rad/s})(2.59 \text{ m}) = 5.42 \text{ m/s}$$

41. $\omega = \dfrac{1 \text{ r}}{2.88 \text{ d}}\left(\dfrac{2\pi \text{ rad}}{1 \text{ r}}\right)\left(\dfrac{1 \text{ d}}{24 \text{ h}}\right) = 0.0909 \text{ rad/h}$

$$v = \omega r = (0.0909 \text{ rad/h})(5\,600\,000 \text{ km}) = 509\,000 \text{ km/h}$$

45.

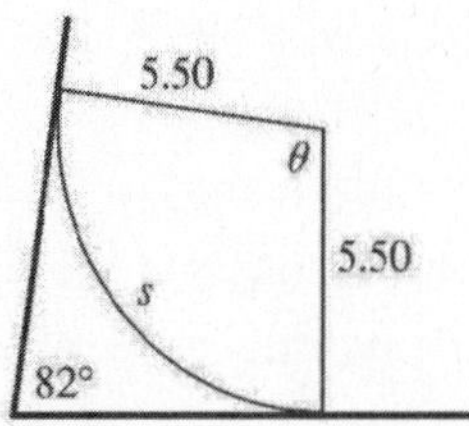

The intersection and the centre of the circle enclose a quadrilateral, all the interior angles should sum to 360°.

$$82.0^\circ + 2(90^\circ) + \theta = 360^\circ$$

$$\theta = 98.0^\circ\left(\frac{\pi \text{ rad}}{180^\circ}\right) = 1.71 \text{ rad}$$

$$s = \theta r$$

$$s = 1.71(5.50 \text{ m})$$

$$s = 9.41 \text{ m}$$

49. $\omega = \dfrac{v}{r} = \dfrac{\frac{1}{4}(6.50 \text{ m/s})}{3.75 \text{ m}} = 0.433 \text{ rad/s}$

53. $y = \sqrt{85.0^2 - 27.4^2} = \sqrt{6474.24} = 80.5 \text{ m}$

$$\cos\theta = \frac{27.4}{85.0}$$

$$\theta = \cos^{-1}\left(\frac{27.4}{85.0}\right) = 71.2^\circ$$

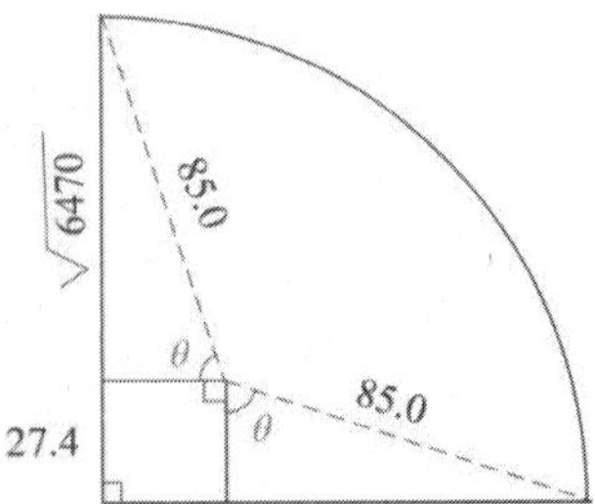

Interior angle in the circular sector is ϕ

$360^\circ = 90^\circ + 2\theta + \phi$

$\phi = 360^\circ - 90^\circ - 2(71.2^\circ) = 127.6^\circ$

$\phi = 127.6^\circ \left(\frac{\pi \text{ rad}}{180^\circ}\right) = 2.23 \text{ rad}$

$A = A_{\text{infield}} + 2A_{\text{triangle}} + A_{\text{sector}}$

$A = x^2 + 2\left(\frac{1}{2}bh\right) + \frac{1}{2}\theta r^2$

$A = 27.4^2 + 2\left(\frac{1}{2}\right)(27.4)(80.5) + \frac{1}{2}(2.23)(85.0)^2$

$A = 11\,001 \text{ m}^2 = 11.0 \times 10^3 \text{ m}^2$

57.

θ	$\sin\theta/\theta$	$\tan\theta/\theta$
0.0001	0.999 999 998 3	1.000 000 003
0.001	0.999 999 833 3	1.000 000 333
0.01	0.999 983 333 4	1.000 033 335
0.01	0.998 334 166 5	1.003 346 721

The sequences both converge on 1.

For small θ in rad, $\theta \approx \sin\theta \approx \tan\theta$.

Review Exercises

1. Point (6, 8), $x = 6$, $y = 8$

$r = \sqrt{6^2 + 8^2} = 10$

$\sin\theta = \frac{y}{r} = \frac{8}{10} = \frac{4}{5}$

$\cos\theta = \frac{x}{r} = \frac{6}{10} = \frac{3}{5}$

$\tan\theta = \frac{y}{x} = \frac{8}{6} = \frac{4}{3}$

$\csc\theta = \frac{r}{y} = \frac{5}{4}$

$\sec\theta = \frac{r}{x} = \frac{5}{3}$

$\cot\theta = \frac{x}{y} = \frac{3}{4}$

5. 132° is in Quadrant II, where $\cos\theta$ is negative.

$\cos 132^\circ = -\cos(180^\circ - 132^\circ) = -\cos 48^\circ$

194° is in Quadrant III, where $\tan\theta$ is positive.

$\tan 194^\circ = +\tan(194^\circ - 180^\circ) = \tan 14^\circ$

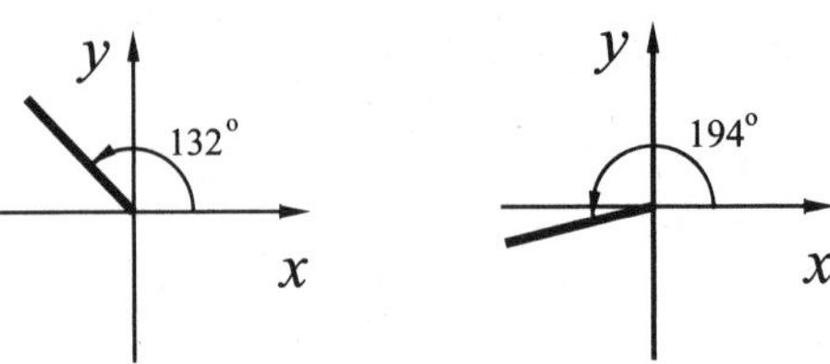

9. $40^\circ = 40^\circ\left(\frac{\pi}{180^\circ}\right) = \frac{2\pi}{9}$

$153^\circ = 153^\circ\left(\frac{\pi}{180^\circ}\right) = \frac{17\pi}{20}$

13. $\frac{7\pi}{5} = \frac{7\pi}{5}\left(\frac{180^\circ}{\pi}\right) = 252^\circ$

$\frac{13\pi}{18} = \frac{13\pi}{18}\left(\frac{180^\circ}{\pi}\right) = 130^\circ$

17. $0.560 = 0.560\left(\frac{180^\circ}{\pi}\right) = 32.1^\circ$

21. $102^\circ = 102^\circ\left(\frac{\pi \text{ rad}}{180^\circ}\right) = 1.78 \text{ rad}$

25. $262.05^\circ = 262.05^\circ\left(\frac{\pi \text{ rad}}{180^\circ}\right) = 4.5736 \text{ rad}$

29. $\cos 245.5^\circ = -0.4147$

33. $\csc 247.82^\circ = -1.0799$

37. $\tan 301.4^\circ = -1.638$

41. $\sin\frac{9\pi}{5} = -0.5878$

45. $\sin 0.5906 = 0.5569$

49. $\tan\theta = 0.1817,\ 0 \le \theta < 360^\circ$

$\theta_{ref} = \tan^{-1}(0.1817) = 10.30^\circ$

Since $\tan\theta$ is positive, θ must lie in Quadrant I or Quadrant III.

$\theta_1 = 10.30^\circ$
$\theta_3 = 180^\circ + 10.30^\circ = 190.30^\circ$

53. $\cos\theta = 0.8387,\ 0 \le \theta < 2\pi$
$\theta_{ref} = \cos^{-1}(0.8387) = 0.5759$
Since $\cos\theta$ is positive, θ must lie in Quadrant I or Quadrant IV.
$\theta_1 = 0.5759$
$\theta_4 = 2\pi - 0.5759 = 5.707$

57. $\cos\theta = -0.7222,\ 0^\circ \le \theta < 360^\circ$
$\theta_{ref} = \cos^{-1}(0.7222) = 43.76^\circ$
Since $\cos\theta$ is negative, θ must lie in Quadrant II or Quadrant III.
Since $\sin\theta$ is negative, θ must lie in Quadrant III or Quadrant IV.
To satisfy both conditions, θ must lie in Quadrant III.
$\theta_3 = 180^\circ + 43.76^\circ = 223.76^\circ$

61. $\theta = 107.5^\circ\left(\dfrac{\pi \text{ rad}}{180^\circ}\right) = 1.876 \text{ rad}$
$r = \dfrac{s}{\theta} = \dfrac{20.3 \text{ cm}}{1.876} = 10.8 \text{ cm}$

65. $A = \dfrac{1}{2}\theta r^2$
$\theta = \dfrac{2A}{r^2} = \dfrac{2(32.8 \text{ m}^2)}{(4.62 \text{ m})^2} = 3.07 \text{ rad}$
$s = \theta r = 3.07(4.62 \text{ m}) = 14.2 \text{ m}$

69.

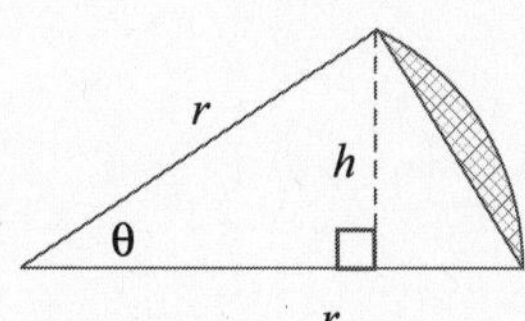

$\sin\theta = \dfrac{h}{r}$
$h = r\sin\theta$
$A_{\text{segment}} = A_{\text{sector}} - A_{\text{triangle}}$
$A_{\text{segment}} = \dfrac{1}{2}\theta r^2 - \dfrac{1}{2}bh$
$A_{\text{segment}} = \dfrac{1}{2}\theta r^2 - \dfrac{1}{2}r(r\sin\theta)$
$A_{\text{segment}} = \dfrac{1}{2}r^2\theta - \dfrac{1}{2}r^2\sin\theta$
$A_{\text{segment}} = \dfrac{1}{2}r^2(\theta - \sin\theta)$

73. $P = P_m \sin^2 377t$
$P = (0.120 \text{ W}) \sin^2(377 \cdot 2\times10^{-3})$
$P = 0.0562 \text{ W}$

77. $v = 5.60 \text{ km/h}\left(\dfrac{1000 \text{ m}}{\text{km}}\right)\left(\dfrac{1 \text{ h}}{60 \text{ min}}\right) = 93.3 \text{ m/min}$
$\omega = \dfrac{v}{r} = \dfrac{93.3 \text{ m/min}}{\frac{1}{2}(1.36 \text{ m})} = 137 \text{ r/min}$

81.

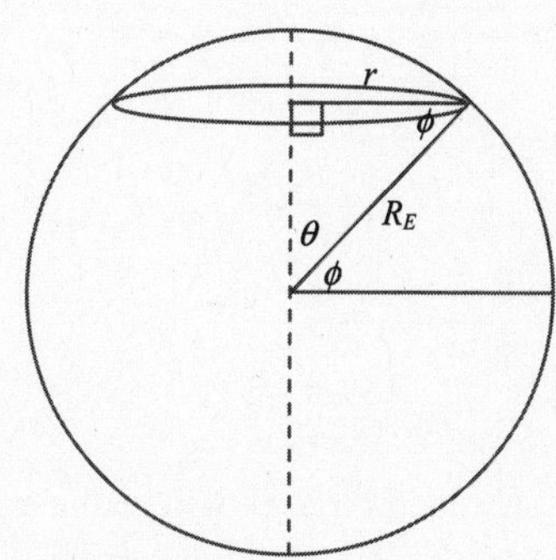

Latitude $\phi = 60^\circ$

(a) The angular distance from each city to the pole is $\theta = 90^\circ - \phi = 30^\circ$, making the over-pole angular distance $2\theta = 60^\circ$ or $\pi/3$ radians.

$s = (2\theta)R_E = \dfrac{\pi}{3}(6370 \text{ km}) = 6670 \text{ km}$

(b) The radius of the circle of latitude is found through

$\cos\phi = \dfrac{r}{R_E}$

At a latitude of 60°, this gives
$r = 6370 \text{ km} \cos 60^\circ$
$r = 3185 \text{ km}$
The net interior angle along the circle of latitude between 135°W and 30°E is 165° or 2.88 radians
$s = 2.88(3185 \text{ km}) = 9170 \text{ km}$
The results show that the distance over the north pole is shorter.

85. $\theta = 80^\circ\left(\dfrac{\pi \text{ rad}}{180^\circ}\right) = \dfrac{4\pi}{9} \text{ rad}$

$A_{\text{circle}} = \pi \cdot 1.08^2$
$A_{\text{hole}} = \pi \cdot 0.25^2$

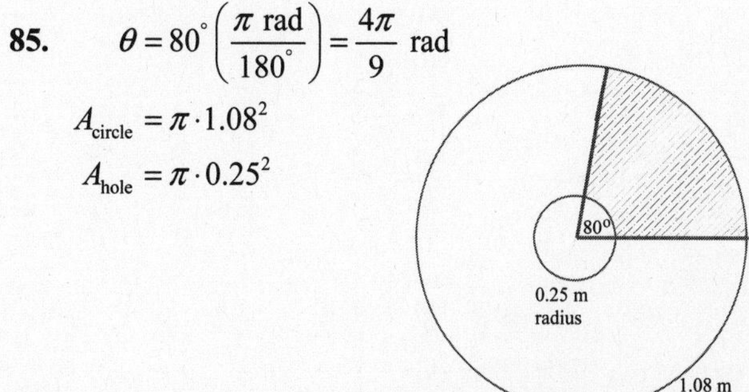

$$A_{\text{sector(without hole)}} = \frac{1}{2}\left(\frac{4\pi}{9}\right)\left(1.08^2 - 0.25^2\right)$$

$$A_{\text{hood}} = A_{\text{circle}} - A_{\text{hole}} - A_{\text{sector(without hole)}}$$

$$A_{\text{hood}} = \pi \cdot 1.08^2 - \pi \cdot 0.25^2 - \frac{1}{2}\left(\frac{4\pi}{9}\right)\left(1.08^2 - 0.25^2\right)$$

$$A_{\text{hood}} = 2.70\text{ m}^2$$

89.

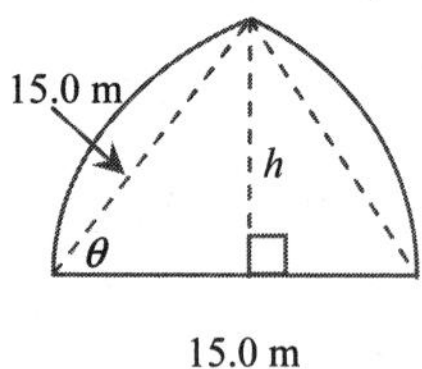

Area in the arch will be the area of one sector, plus the segment area of one side. This segment area is the difference between the sector area and the equilateral triangle in the sector. Since the interior triangle has all sides 15.0 m long, it is equilateral, all internal angles are 60°.

$$\theta = 60^\circ = \frac{\pi}{3}\text{ rad}$$

$$h = (15.0\text{ m})\sin 60^\circ = 15.0\text{ m}\frac{\sqrt{3}}{2}$$

$$A = A_{\text{sector}} + A_{\text{segment}}$$

$$A = \frac{1}{2}\theta r^2 + \left(\frac{1}{2}\theta r^2 - \frac{1}{2}bh\right)$$

$$A = \theta r^2 - \frac{1}{2}bh$$

$$A = \frac{\pi}{3}(15.0\text{ m})^2 - \frac{1}{2}(15.0\text{ m})\left(15.0\text{ m}\frac{\sqrt{3}}{2}\right)$$

$$A = 138\text{ m}^2$$

93. Convert the angular velocity of 1 revolution every 1.95 h into rad/h. Then find the total radius of the orbit by adding the lunar radius to the altitude. Use the equation $v = \omega r$ to solve for the velocity.

$$\omega = \frac{1\text{ r}}{1.95\text{ h}}\left(\frac{2\pi\text{ rad}}{1\text{ r}}\right) = 3.222\text{ rad/h}$$

$$r = 1740\text{ km} + 113\text{ km} = 1853\text{ km}$$

$$v = \omega r = (3.222\text{ rad/h})(1853\text{ km}) = 5970\text{ km/h}$$

CHAPTER 9

VECTORS AND OBLIQUE TRIANGLES

9.1 Introduction to Vectors

1.

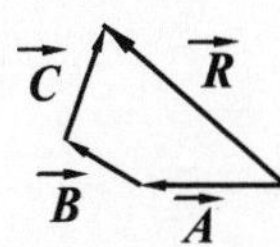

5. **(a)** scalar, no direction given

   (b) vector, magnitude and direction both given

9.

13.

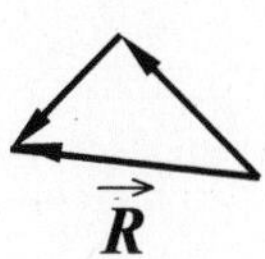

17. 4.3 cm, 156°

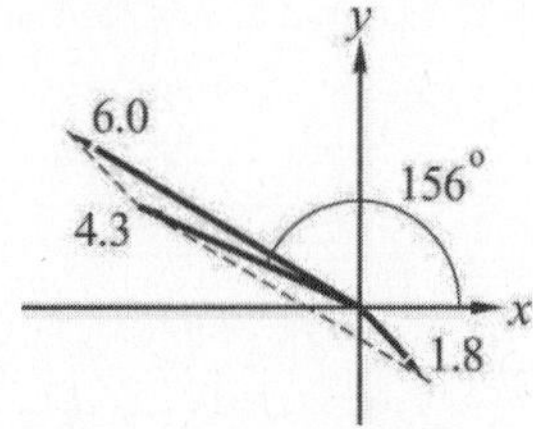

21.

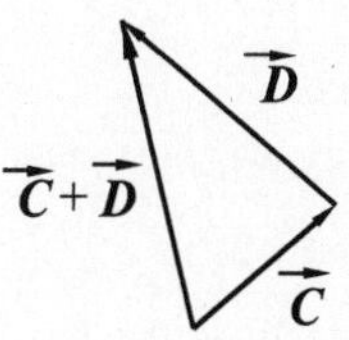

25.

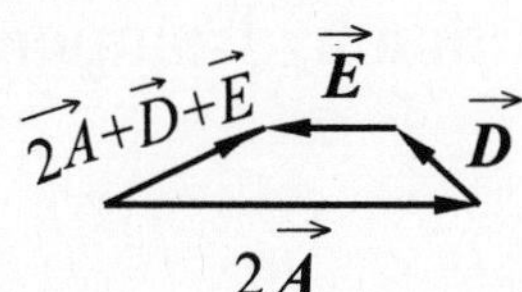

29.

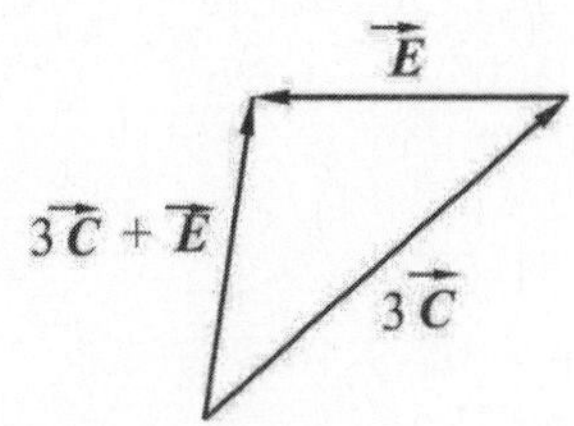

33.

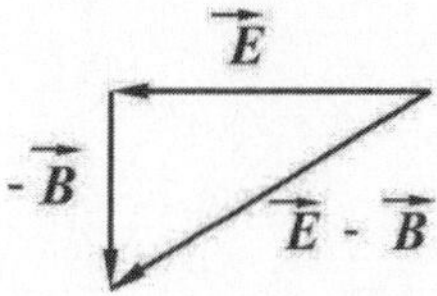

37.

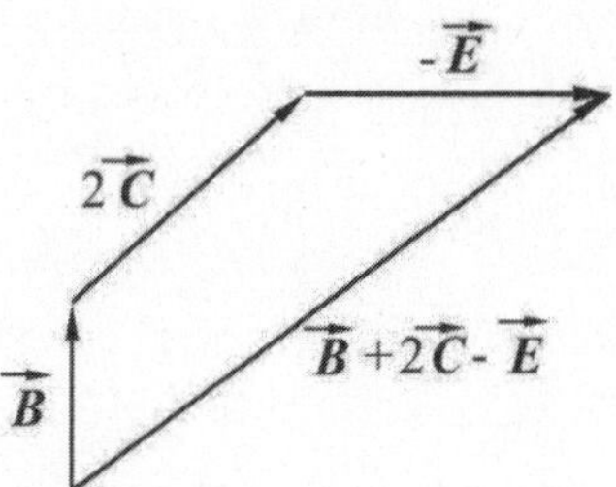

41.

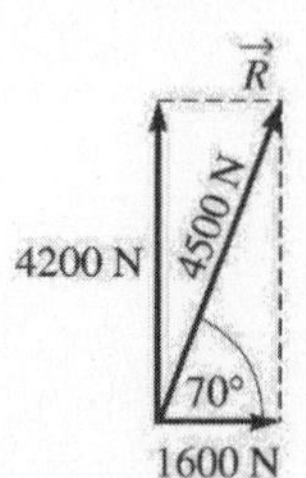

From the drawing, $\vec{R}$ is approximately 4500 N at 70°.

45.

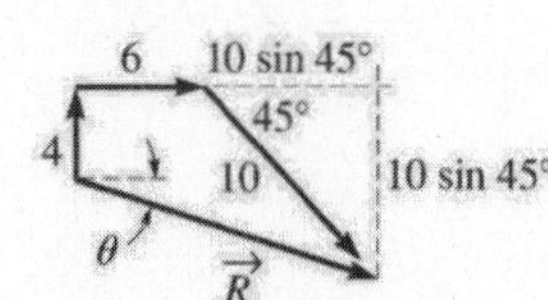

From the drawing,

$R = 13$ km

$\theta = 13^\circ$

9.2 Components of Vectors

1.

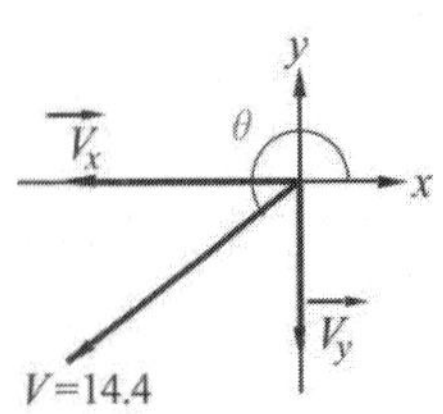

$V_x = V\cos\theta$

$V_x = 14.4\cos 216^\circ = -11.6$

$V_y = V\sin\theta$

$V_y = 14.4\sin 216^\circ = -8.46$

5. $V_x = 750\cos 28^\circ = 662$

$V_y = 750\sin 28^\circ = 352$

9. $V_x = -750$

$V_y = 0$

13. Let $V = 76.8$

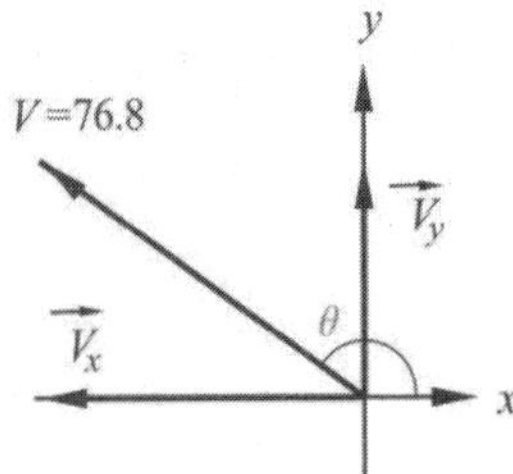

$V_x = V\cos 145.0^\circ = 76.8(-0.819) = -62.9$ m/s

$V_y = V\sin 145.0^\circ = 76.8(0.574) = 44.1$ m/s

17. Let $V = 2.65$

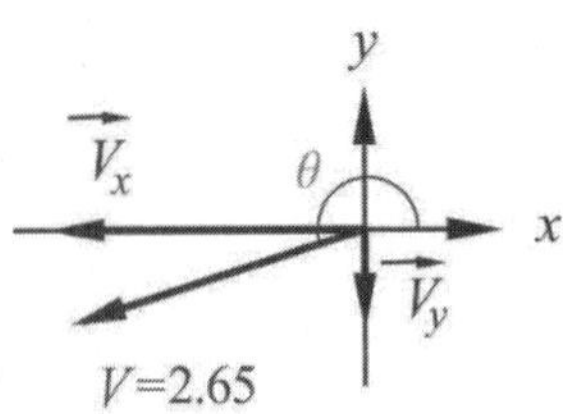

$V_x = V\cos 197.3^\circ = 2.65(-0.955) = -2.53$ mN

$V_y = V\sin 197.3^\circ = 2.65(-0.297) = -0.788$ m/N

21.

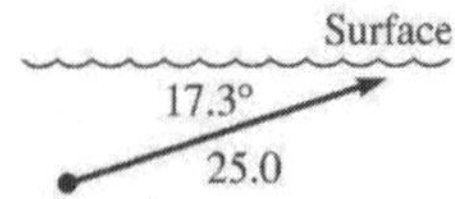

$V_x = 25.0\cos 17.3^\circ = 23.9$ km/h

$V_y = 25.0\sin 17.3^\circ = 7.43$ km/h

25.

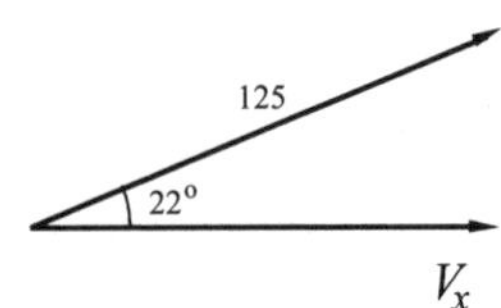

$V_x = 125\cos 22.0^\circ = 116$ km/h

29.

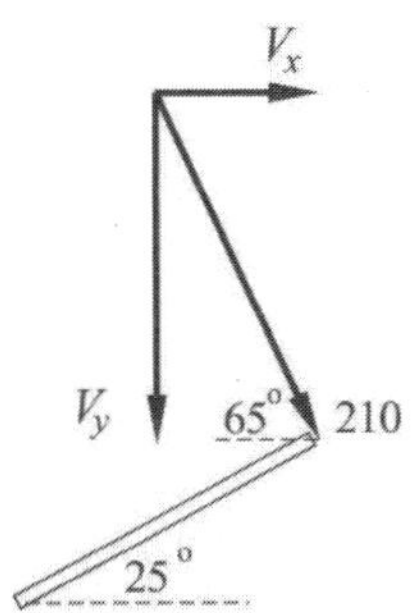

$V_x = 210\cos 65^\circ = 89$ N

$V_y = -210\sin 65^\circ = -190$ N

33.

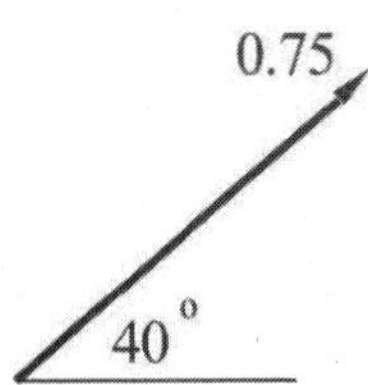

horizontal component $= 0.75\cos 40^\circ$

$= 0.57$ (km/h)/m

vertical component $= 0.75\sin 40^\circ$

$= 0.48$ (km/h)/m

9.3 Vector Addition by Components

1. $A = 1200 = A_x$, $B = 1750$

$Ay = 0$

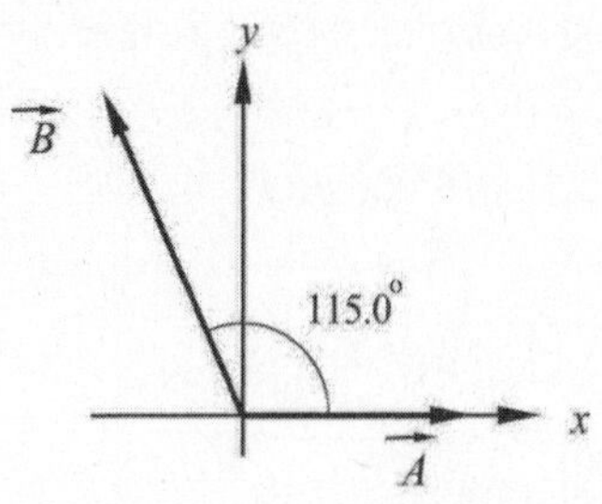

$R_x = A_x + B_x = 1200 + 1750\cos 115^\circ = 460.4$

$R_y = A_y + B_y = 0 + 1750\sin 115^\circ = 1586$

$R = \sqrt{R_x^2 + R_y^2} = \sqrt{460.4^2 + 1586^2} = 1650$

$\theta = \tan^{-1}\dfrac{R_y}{R_x} = \tan^{-1}\dfrac{1586}{460.4} = 73.8^\circ$

5.

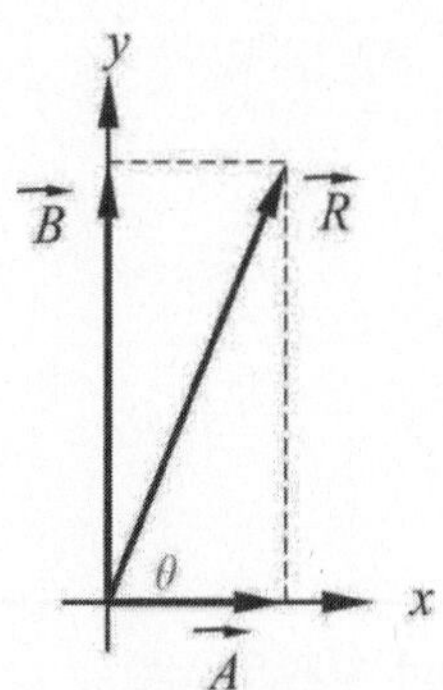

$R = \sqrt{3.086^2 + 7.143^2} = \sqrt{60.54} = 7.781$

$\tan\theta = \dfrac{7.143}{3.086} = 2.315$

$\theta = 66.63^\circ$ (with $\vec{A}$)

9.

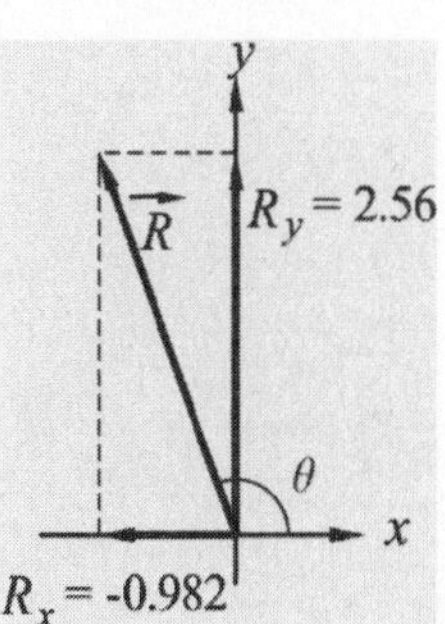

$R_x = -0.982, R_y = 2.56$

$R = \sqrt{R_x^2 + R_y^2} = \sqrt{(-0.982)^2 + 2.56^2} = 2.74$

$\tan\theta_{\text{ref}} = \left|\dfrac{2.56}{-0.982}\right| = 2.61$

$\theta_{\text{ref}} = 69.0^\circ$

$\theta = 180^\circ - 69.0^\circ = 111.0^\circ$

(θ is in Quad II since R_x is negative and R_y is positive)

13.

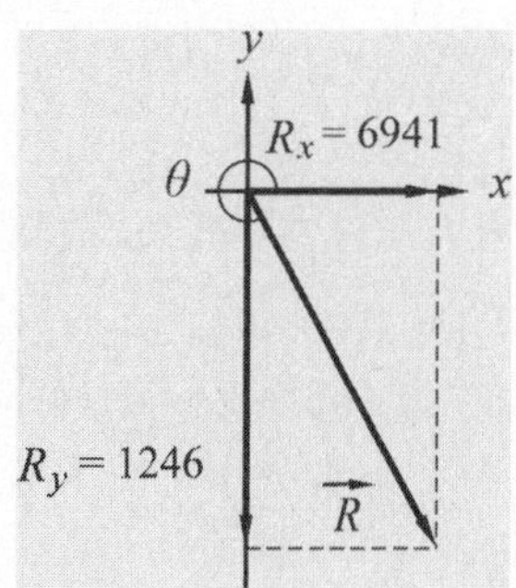

$R_x = 6941,\ R_y = -1246$

$R = \sqrt{6941^2 + (-1246)^2} = 7052$

$\tan\theta_{\text{ref}} = \left|\dfrac{-1246}{6941}\right| = 0.1795$

$\theta_{\text{ref}} = 10.18^\circ$

$\theta = 360^\circ - 10.18^\circ = 349.82^\circ$

(θ is in QIV since R_x is positive and R_y is negative)

17.

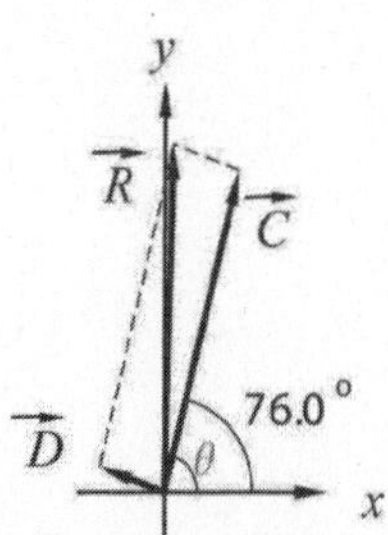

$C = 5650, \theta_C = 76.0^\circ$

$C_x = 5650\cos 76.0^\circ = 1370$

$C_y = 5650\sin 76.0^\circ = 5480$

$D = 1280, \theta_D = 160.0^\circ$

$D_x = 1280\cos 160.0^\circ = -1200$

$D_y = 1280\sin 160.0^\circ = 438$

$R_x = 1370 - 1200 = 170$

$R_y = 5480 + 438 = 5920$

$R = \sqrt{170^2 + 5920^2} = 5920$

$\tan\theta = \dfrac{R_y}{R_x} = \dfrac{5920}{170}, \theta = 88.4^\circ$

21.

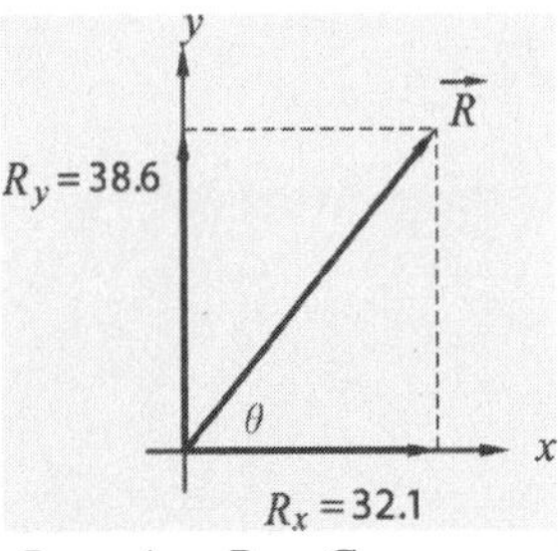

$R_x = A_x + B_x + C_x$

$R_x = 21.9\cos 236.2^\circ + 96.7\cos 11.5^\circ + 62.9\cos 143.4^\circ$

$R_y = A_y + B_y + C_y$

$R_y = 21.9\sin 236.2^\circ + 96.7\sin 11.5^\circ + 62.9\sin 143.4^\circ$

$R = \sqrt{R_x^2 + R_y^2} = 50.2$

$\theta = \tan^{-1}\dfrac{R_y}{R_x} = 50.3^\circ$

25.

Vector	Magnitude	Ref. Angle
A	318	67.5°
B	245	73.7°

	x-component	y-component
	$-318\cos 67.5^\circ = -121.7$	$318\sin 67.5^\circ = 293.8$
	$245\cos 73.7^\circ = 68.76$	$245\sin 73.7^\circ = 235.2$
R	-52.9	529

$\theta_{\text{ref}} = \tan^{-1}\left|\dfrac{529}{-52.9}\right| = 84.3^\circ$

$\theta = 180^\circ - 84.3^\circ = 95.7^\circ$

$R = \sqrt{(-52.9)^2 + 529^2} = 532$

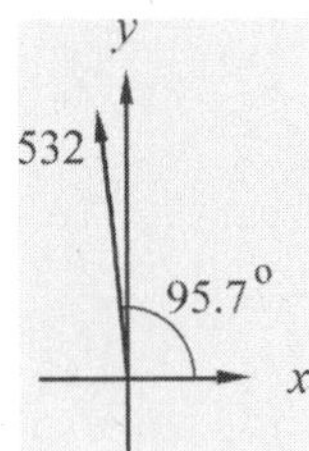

29.

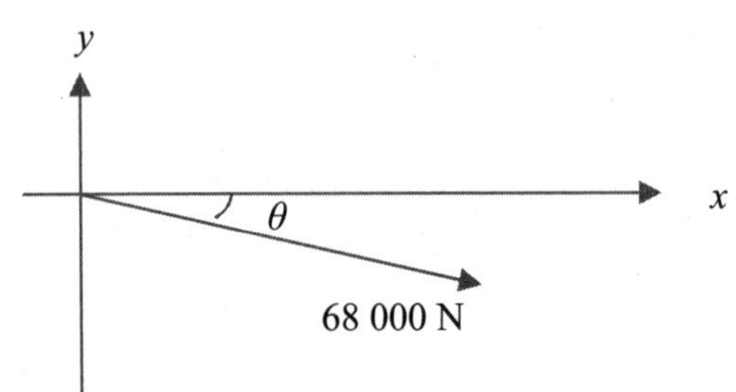

$R_x = 25\ 000 + 29\ 000\cos 15.5^\circ + 16\ 000\cos 37.7^\circ$

$= 65\ 600$

$R_y = -29\ 000\sin 15.5^\circ - 16\ 000\sin 37.7^\circ$

$= -17\ 500$

$R = \sqrt{65\ 600^2 + (-17\ 500)^2} = 68\ 000\ \text{N}$

$\theta_{\text{ref}} = \tan^{-1}\left|\dfrac{-17\ 500}{65\ 600}\right| = 14.9^\circ$

$\theta = 345.1^\circ$

33.

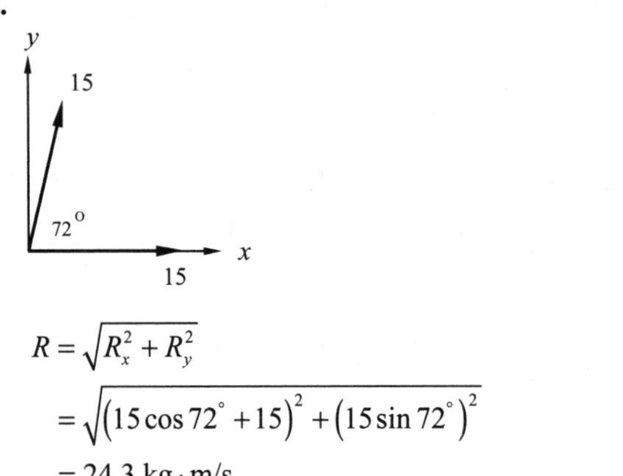

$R = \sqrt{R_x^2 + R_y^2}$

$= \sqrt{(15\cos 72^\circ + 15)^2 + (15\sin 72^\circ)^2}$

$= 24.3\ \text{kg} \cdot \text{m/s}$

9.4 Applications of Vectors

1.

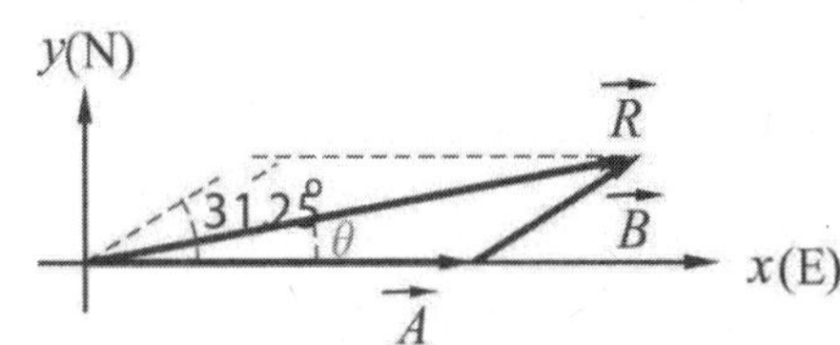

$R_x = A + B_x$

$= 32.50 + 16.18\cos 31.25^\circ$

$= 46.33$

$R_y = 16.18\sin 31.25^\circ$

$= 8.394$

$R = \sqrt{46.33^2 + 8.394^2}$

$= 47.08\ \text{km}$

$\theta = \tan^{-1}\dfrac{8.394}{46.33} = 10.27^\circ$

The ship is 47.08 km from start in direction 10.27° N of E.

5.

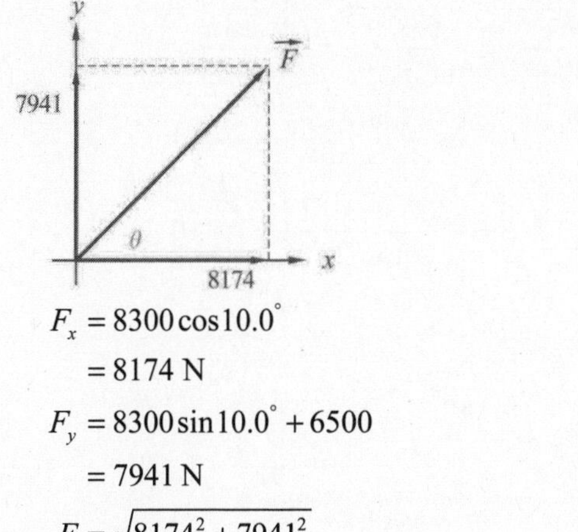

$F_x = 8300\cos 10.0^\circ$

$= 8174 \text{ N}$

$F_y = 8300\sin 10.0^\circ + 6500$

$= 7941 \text{ N}$

$F = \sqrt{8174^2 + 7941^2}$

$= 11000 \text{ N}$

$\tan\theta = \dfrac{7941}{8174} = 0.97, \theta = 44^\circ$ above horizontal

9.

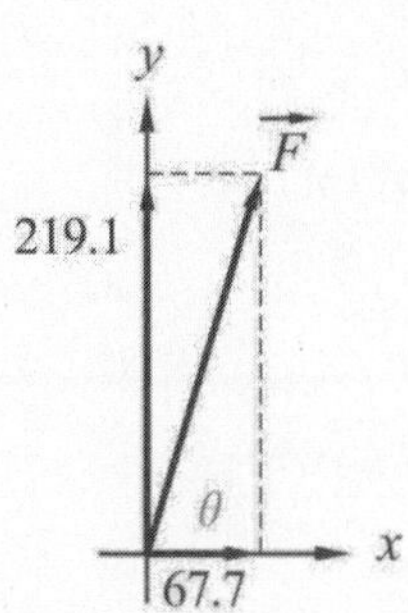

$F_x = 358.2\cos 37.72^\circ - 215.6 = 67.74$

$F_y = 358.2\sin 37.72^\circ = 219.1$

$F = \sqrt{67.7^2 + 219.1^2} = 229.4 \text{ m}$

$\theta = \tan^{-1}\dfrac{219.1}{67.74} = 72.82^\circ$ N of E

13.

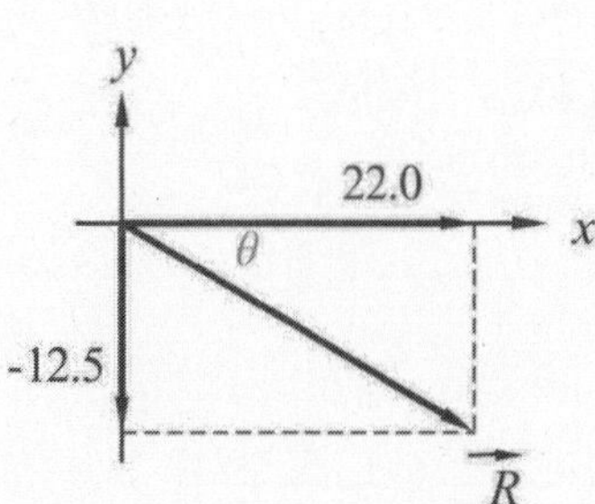

$R = \sqrt{22.0^2 + 12.5^2} = 25.3 \text{ km/h}$

$\theta = \tan^{-1}\dfrac{12.5}{22.0} = 29.6^\circ$ south of east

17.

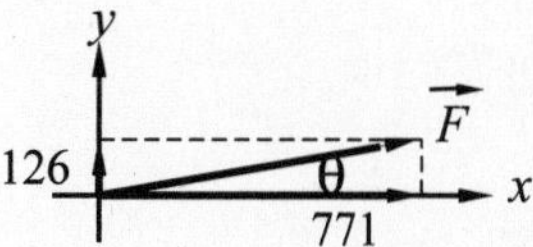

$F_x = 425 + 368\cos 20.0^\circ = 771$

$F_y = 368\sin 20.0^\circ = 126$

$F = \sqrt{771^2 + 126^2} = 781 \text{ N}$

$\tan\theta = \dfrac{126}{771}$

$\theta = 9.3^\circ$, above horizontal and to the right.

21. $v_{sx} = 29\ 370 - 190\cos 5.20^\circ = 29\ 180$

$v_{sy} = 190\sin 5.20^\circ = 17.2$

$v = \sqrt{29\ 180^2 + 17.2^2}$

$= 29\ 180 \text{ km/h}$

$\tan\theta = \dfrac{17.2}{29\ 180}$

$\theta = 0.03^\circ$ from direction of shuttle

25. Assume that as the smoke pours out of the funnel it immediately takes up the velocity of the wind. Let $\vec{w}$ = velocity of wind, $\vec{u}$ = velocity of boat, $\vec{v}$ = velocity of smoke as seen by passenger.

$$w\cos 45^\circ + v\cos 15^\circ = 32$$

$$w\sin 45^\circ = v\sin 15^\circ$$

$$v = \frac{w\sin 45^\circ}{\sin 15^\circ}$$

$$w\cos 45^\circ + \frac{w\sin 45^\circ}{\sin 15^\circ}\cos 15^\circ = 32$$

$$w = 9.3 \text{ km/h}$$

29. $r = \dfrac{d}{2} = \dfrac{8.20}{2} = 4.10 \text{ cm}$

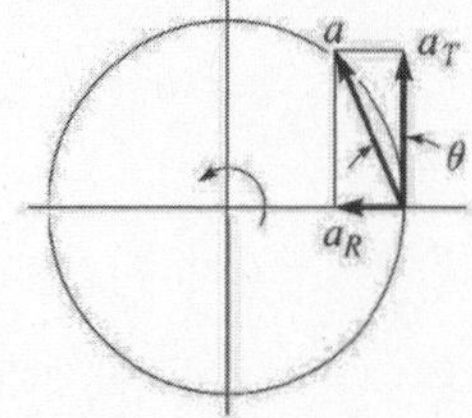

$$a = \sqrt{(a_T)^2 + (a_R)^2}$$
$$= \sqrt{(\alpha r)^2 + (\omega^2 r)^2}$$
$$= \sqrt{(318(4.10))^2 + (212^2 \cdot 4.10)^2}$$
$$= 184\ 000 \text{ cm/min}^2$$
$$\theta = \tan^{-1}\frac{a_R}{a_T}$$
$$= \tan^{-1}\frac{\omega^2 r}{\alpha r}$$
$$= \tan^{-1}\frac{212^2}{318}$$
$$= 89.6^\circ$$

33. top view of the plane:

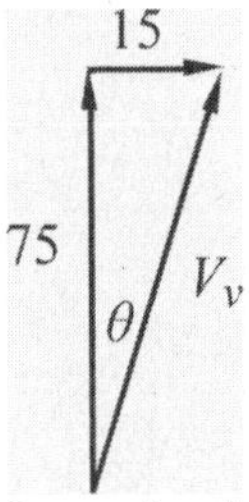

Let v_H be the horizontal component of the package's velocity. It is given by the sum of the velocity of the plane and the ejection velocity.

$$v_H = \sqrt{75.0^2 + 15.0^2} = 76.5$$
$$\theta = \tan^{-1}\frac{15.0}{75.0} = 11.3^\circ$$
$$v_v = 9.80(2.00) = 19.6$$

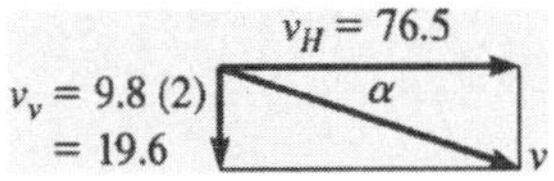

$$v = \sqrt{76.5^2 + 19.6^2} = 79.0 \text{ m/s}$$
$$\alpha = \tan^{-1}\frac{19.6}{76.5} = 14.4^\circ, 75.6^\circ \text{ from vertical}$$

9.5 Oblique Triangles, the Law of Sines

1.

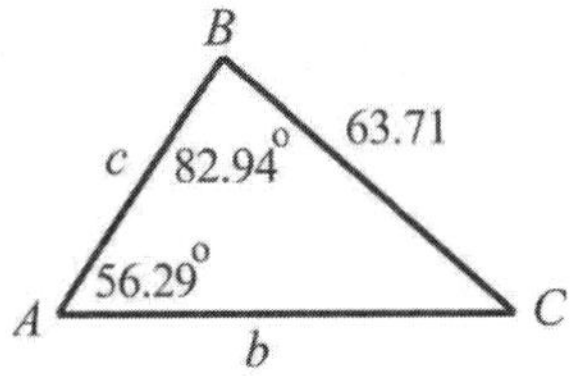

$$C = 180^\circ - (56.29^\circ - 82.94^\circ) = 40.77^\circ$$
$$\frac{b}{\sin 82.94^\circ} = \frac{63.71}{\sin 56.29^\circ} = \frac{c}{\sin 40.77^\circ}$$
$$b = \frac{63.71 \sin 82.94^\circ}{\sin 56.29^\circ} = 76.01$$
$$c = \frac{63.71 \sin 40.77^\circ}{\sin 56.29^\circ} = 50.01$$

5.

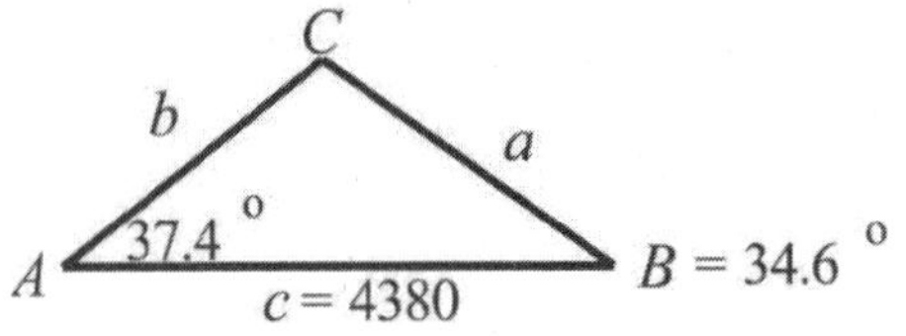

$$c = 4380, A = 37.4^\circ, B = 34.6^\circ$$
$$C = 180.0^\circ - (37.4^\circ + 34.6^\circ) = 108.0^\circ$$
$$\frac{b}{\sin B} = \frac{c}{\sin C}; \frac{b}{\sin 34.6^\circ} = \frac{4380}{\sin 108.0^\circ}$$
$$b = \frac{4380 \sin 34.6^\circ}{\sin 108.0^\circ} = 2620$$
$$\frac{a}{\sin A} = \frac{c}{\sin C}; \frac{a}{\sin 37.4^\circ} = \frac{4380}{\sin 108.0^\circ}$$
$$a = \frac{4380 \sin 37.4^\circ}{\sin 108.0^\circ} = 2800$$

9.

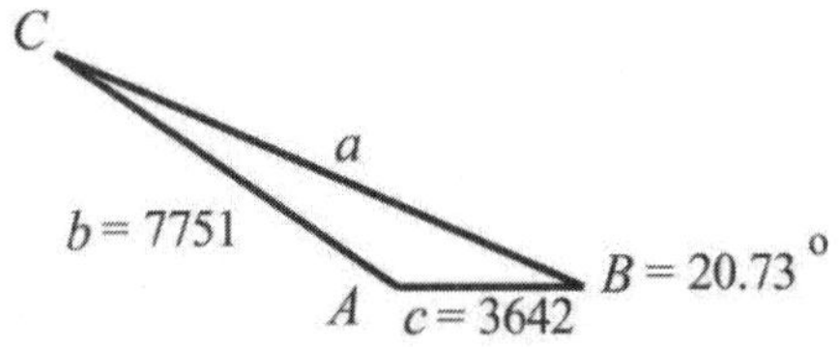

Since 7751 > 3642, the longer side is opposite the known angle, and we have one solution.

$$b = 7751, c = 3642, B = 20.73^\circ$$
$$\frac{b}{\sin B} = \frac{c}{\sin C}; \frac{7751}{\sin 20.73^\circ} = \frac{3642}{\sin C}$$
$$\sin C = \frac{3642 \sin 20.73^\circ}{7751} = 0.1663$$
$$C = 9.574^\circ$$
$$A = 180.0^\circ - (20.73^\circ + 9.574^\circ) = 149.7^\circ$$
$$\frac{a}{\sin A} = \frac{b}{\sin B}; \frac{a}{\sin 149.7^\circ} = \frac{7751}{\sin 20.73^\circ}$$
$$a = \frac{7751 \sin 149.7^\circ}{\sin 20.73^\circ} = 11\,050$$

13.

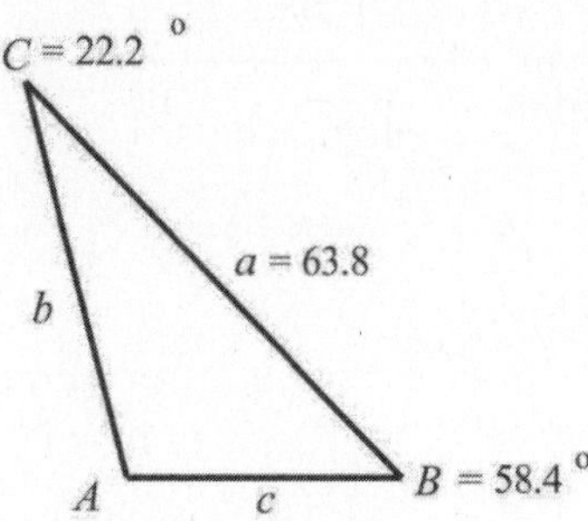

$a = 63.8, B = 58.4°, C = 22.2°$

$A = 180.0° - 58.4° - 22.2° = 99.4°$

$$\frac{a}{\sin A} = \frac{b}{\sin B}; \frac{63.8}{\sin 99.4°} = \frac{b}{\sin 58.4°}$$

$$b = \frac{63.8 \sin 58.4°}{\sin 99.4°} = 55.1$$

$$\frac{a}{\sin A} = \frac{c}{\sin C}; \frac{63.8}{\sin 99.4°} = \frac{c}{\sin 22.2°}$$

$$c = \frac{63.8 \sin 22.2°}{\sin 99.4°} = 24.4$$

17.

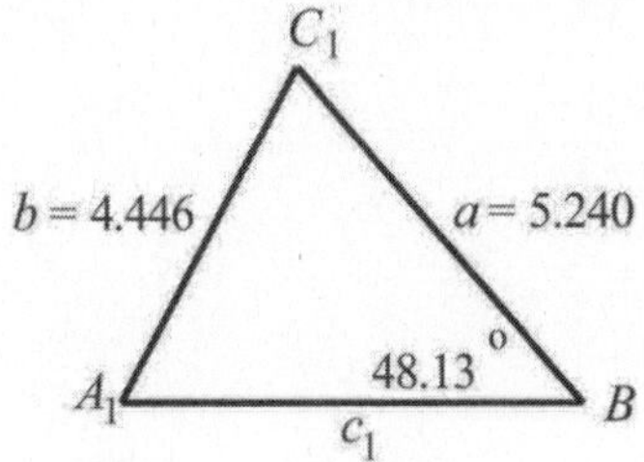

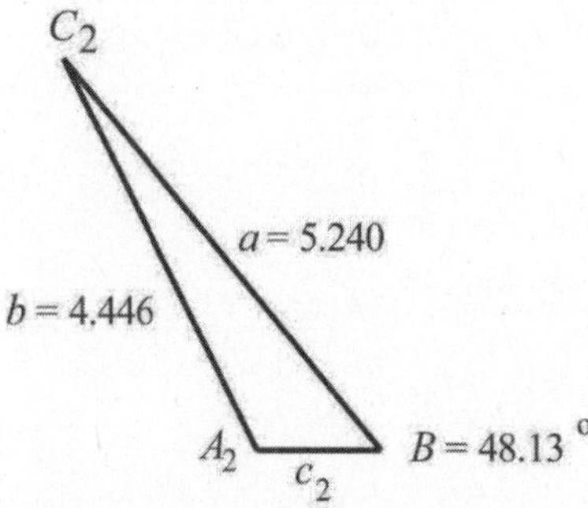

Since 4.446 < 5.240, the shorter side is opposite the known angle, and we have two solutions.

$a = 5.240,\ b = 4.446,\ B = 48.13°$

$$\frac{a}{\sin A} = \frac{b}{\sin B}; \frac{5.240}{\sin A} = \frac{4.446}{\sin 48.13°}$$

$$\sin A = \frac{5.240 \sin 48.13°}{4.446} = 0.8776$$

$A_1 = 61.36°$

$C_1 = 180.0° - 48.13° - 61.36° = 70.51°$

or

$A_2 = 180.0° - 61.36° = 118.64°$

and

$C_2 = 180.0° - 48.13° - 118.64° = 13.23°$

$$\frac{b}{\sin B} = \frac{c_1}{\sin C_1}; \frac{4.446}{\sin 48.13°} = \frac{c}{\sin 70.51°}$$

$$c_1 = \frac{4.446 \sin 70.51°}{\sin 48.13°} = 5.628$$

or

$$\frac{b}{\sin B} = \frac{c_2}{\sin C_2}; \frac{4.446}{\sin 48.13°} = \frac{c_2}{\sin 13.23°}$$

$$c_2 = \frac{4.446 \sin 13.23°}{\sin 48.13°} = 1.366$$

21.

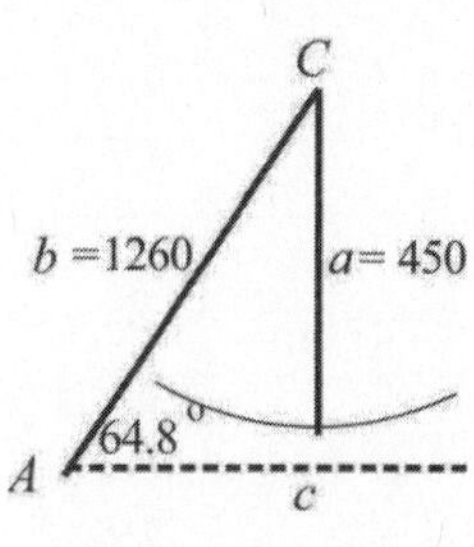

$a = 450,\ b = 1260,\ A = 64.8°$

$$\frac{a}{\sin A} = \frac{b}{\sin B}; \frac{450}{\sin 64.8°} = \frac{1260}{\sin B}$$

$$\sin B = \frac{1260 \sin 64.8°}{450} = 2.53 \text{ (not } \leq 1)$$

Therefore, no solution.

Indeed, $450 < 1260 \sin 64.8° = 1140$, so there is no solution.

25.

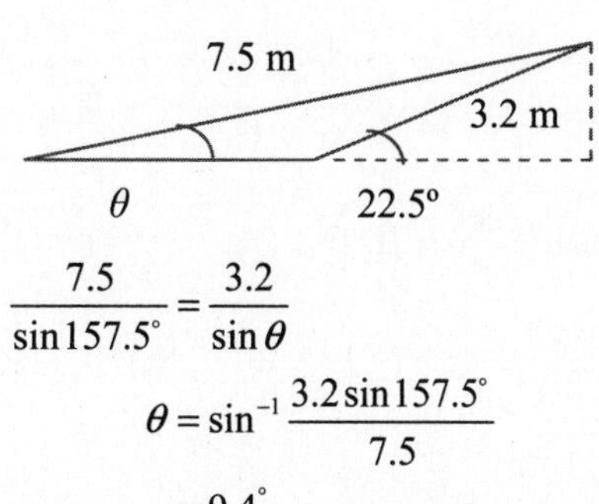

$$\frac{7.5}{\sin 157.5°} = \frac{3.2}{\sin \theta}$$

$$\theta = \sin^{-1} \frac{3.2 \sin 157.5°}{7.5}$$

$$= 9.4°$$

29.

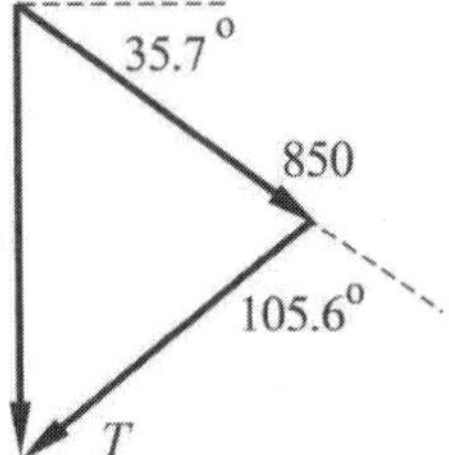

$$\frac{T}{\sin 54.3^\circ} = \frac{850}{\sin 51.3^\circ}; T = \frac{850\sin 54.3^\circ}{\sin 51.3^\circ} = 880 \text{ N}$$

33.

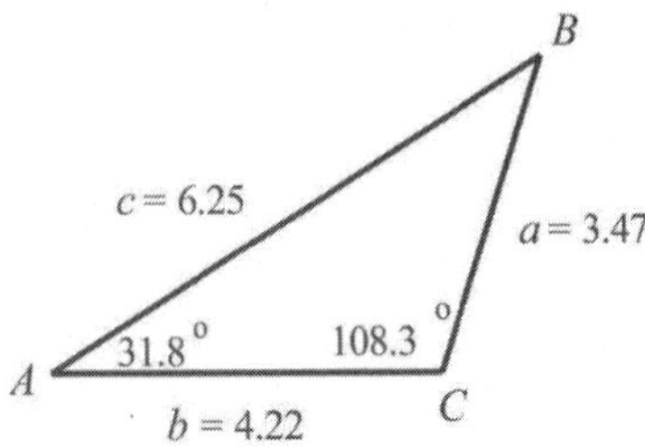

$$\frac{6.25}{\sin 108.3^\circ} = \frac{a}{\sin 31.8^\circ}$$

$$a = \frac{6.25\sin 31.8^\circ}{\sin 108.3^\circ} = 3.47 \text{ cm}$$

$$B = 180^\circ - 108.3^\circ - 31.8^\circ = 39.9^\circ$$

$$\frac{6.25}{\sin 108.3^\circ} = \frac{b}{\sin 39.9^\circ}$$

$$b = \frac{6.25\sin 39.9^\circ}{\sin 108.3^\circ} = 4.22 \text{ cm}$$

Perimeter $= 6.25 + 3.47 + 4.22 = 13.94$ cm

37.

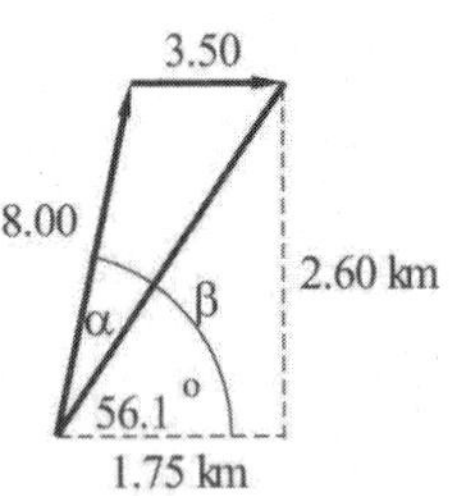

$$\theta = \tan^{-1}\frac{2.60}{1.75} = 56.1^\circ$$

$$\frac{8.00}{\sin 56.1^\circ} = \frac{3.50}{\sin\alpha}$$

$$\alpha = 21.3^\circ,$$

$$\beta = 56.1^\circ + 21.3^\circ = 77.4^\circ$$

with bank downstream

41.

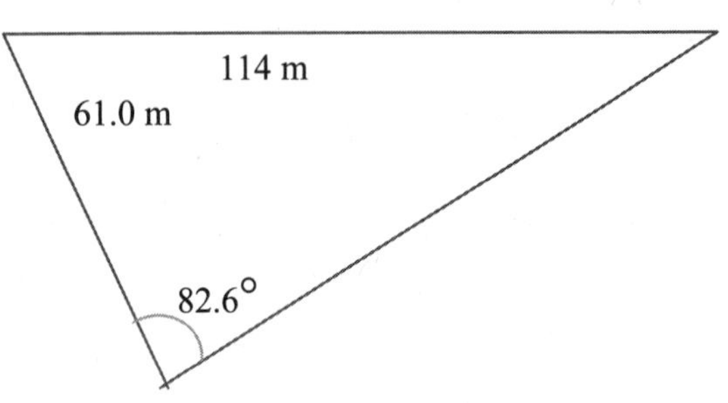

$$\frac{114}{\sin 82.6^\circ} = \frac{61.0}{\sin B} = \frac{c}{\sin C}$$

$$\sin B = \frac{61.0\sin 82.6^\circ}{114}$$

$$B = 32.0^\circ$$

$$C = 180^\circ - (82.6^\circ + 32.0^\circ) = 65.4^\circ$$

$$\frac{114}{\sin 82.6^\circ} = \frac{c}{\sin 65.4^\circ}$$

$$c = 105 \text{ m}$$

9.6 The Law of Cosines

1.

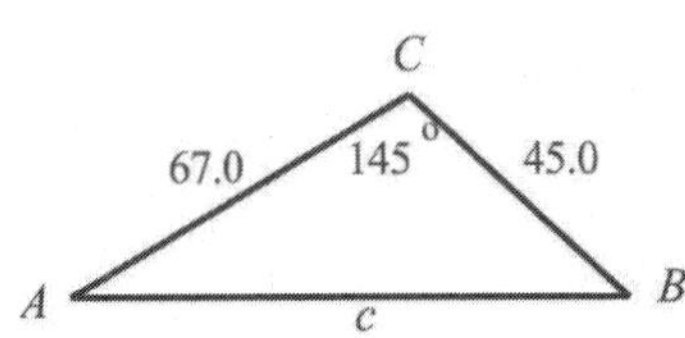

$$c = \sqrt{45.0^2 + 67.0^2 - 2(45.0)(67.0)\cos 145^\circ}$$

$$= 107$$

$$\frac{45.0}{\sin A} = \frac{c}{\sin 145^\circ} = \frac{67.0}{\sin B}$$

$$A = \sin^{-1}\frac{45.0\sin 145^\circ}{c} = 14.0^\circ$$

$$B = \sin^{-1}\frac{67.0\sin 145^\circ}{c} = 21.0^\circ$$

5.

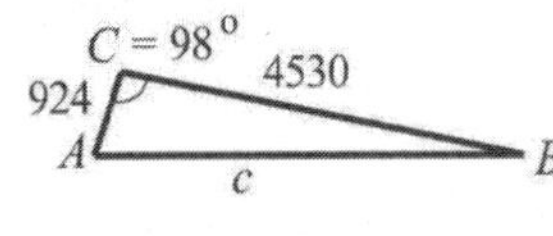

$$a = 4530, b = 924, C = 98.0^\circ$$

$$c = \sqrt{4530^2 + 924^2 - 2(4530)(924)(\cos 98.0^\circ)}$$

$$= 4750$$

$$\frac{c}{\sin C}=\frac{b}{\sin B};\frac{4750}{\sin 98.0^\circ}=\frac{924}{\sin B}$$

$$\sin B=\frac{924\sin 98.0^\circ}{4750}=-0.193$$

$$B=11.1^\circ$$

$$A=180^\circ-98.0^\circ-11.1^\circ=70.9^\circ$$

9.

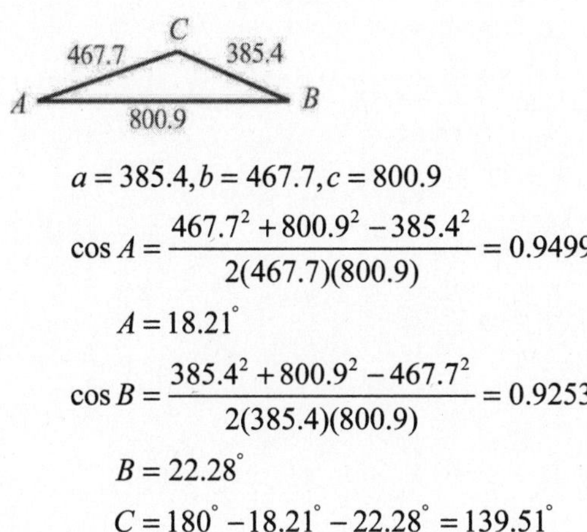

$$a=385.4, b=467.7, c=800.9$$

$$\cos A=\frac{467.7^2+800.9^2-385.4^2}{2(467.7)(800.9)}=0.9499$$

$$A=18.21^\circ$$

$$\cos B=\frac{385.4^2+800.9^2-467.7^2}{2(385.4)(800.9)}=0.9253$$

$$B=22.28^\circ$$

$$C=180^\circ-18.21^\circ-22.28^\circ=139.51^\circ$$

13.

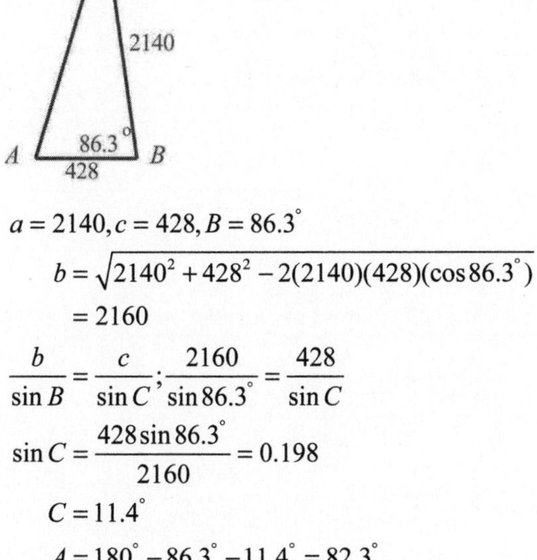

$$a=2140, c=428, B=86.3^\circ$$

$$b=\sqrt{2140^2+428^2-2(2140)(428)(\cos 86.3^\circ)}$$
$$=2160$$

$$\frac{b}{\sin B}=\frac{c}{\sin C};\frac{2160}{\sin 86.3^\circ}=\frac{428}{\sin C}$$

$$\sin C=\frac{428\sin 86.3^\circ}{2160}=0.198$$

$$C=11.4^\circ$$

$$A=180^\circ-86.3^\circ-11.4^\circ=82.3^\circ$$

17.

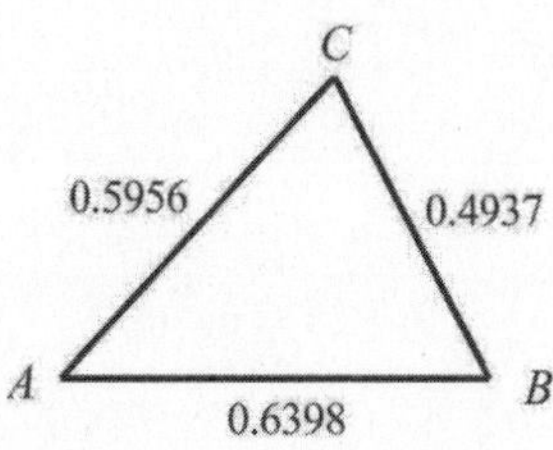

$$a=0.4937, b=0.5956, c=0.6398$$

$$\cos A=\frac{0.5956^2+0.6398^2-0.4937^2}{2(0.5956)(0.6398)}=0.6827$$

$$A=46.94^\circ$$

$$\cos B=\frac{0.4937^2+0.6398^2-0.5956^2}{2(0.4937)(0.6398)}=0.4723$$

$$B=61.82^\circ$$

$$C=180^\circ-46.94^\circ-61.82^\circ=71.24^\circ$$

21.

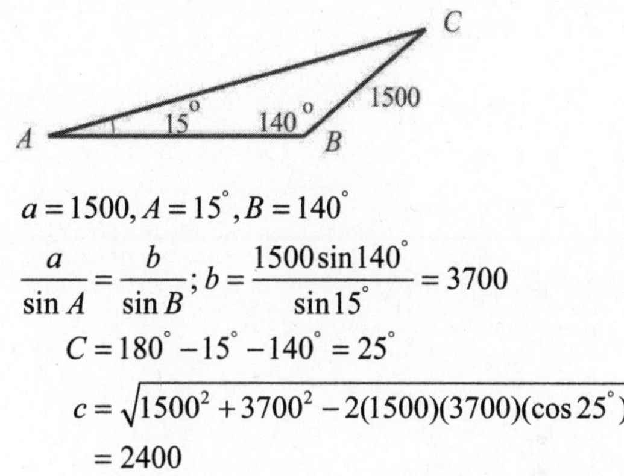

$$a=1500, A=15^\circ, B=140^\circ$$

$$\frac{a}{\sin A}=\frac{b}{\sin B}; b=\frac{1500\sin 140^\circ}{\sin 15^\circ}=3700$$

$$C=180^\circ-15^\circ-140^\circ=25^\circ$$

$$c=\sqrt{1500^2+3700^2-2(1500)(3700)(\cos 25^\circ)}$$
$$=2400$$

25. Case 3: Two sides and the included angle always determine a unique triangle which will have a solution. The triangle may be constructed by drawing the angle and then measuring the two sides along the sides of the angle. The third side is then uniquely determined and the triangle may be solved. Case 4: Three sides determines a unique triangle provided the sum of the lengths of any two sides is greater than the length of the third side. The triangle may be constructed by drawing one side as the horizontal base and the swinging circular arcs from each end. The intersection of these arcs determines the other two sides of the triangle.

29. $48^2=53^2+64^2-2(53)(64)\cos\theta$

$$\theta=47^0$$

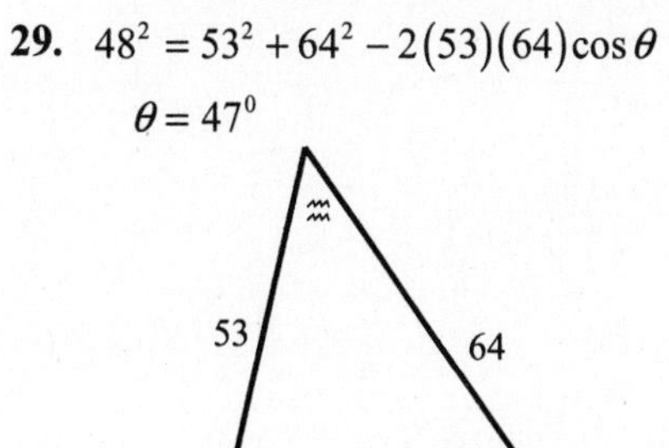

33. $x=\sqrt{9.53^2+9.53^2-2(9.53)(9.53)(\cos 120^\circ)}$

$=16.5$ mm

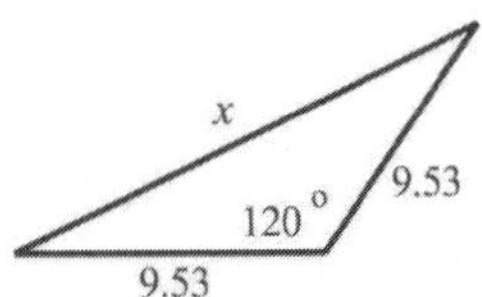

37. $A=26.4^\circ-12.4^\circ=14.0^\circ$

$a=\sqrt{15.8^2+32.7^2-2(15.8)(32.7)\cos 14.0^\circ}$

$=17.8$ km

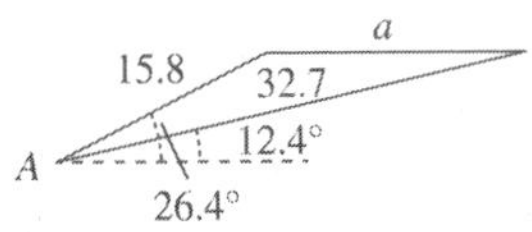

Review Exercises

1.

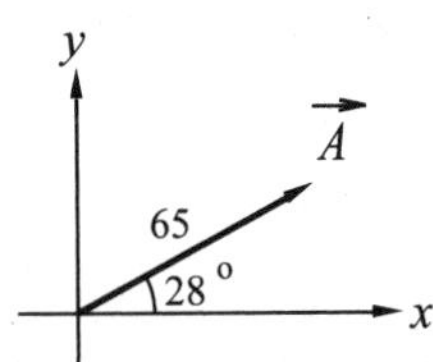

$A_x=65.0\cos 28.0^\circ=57.4$

$A_y=65.0\sin 28.0^\circ=30.5$

5.

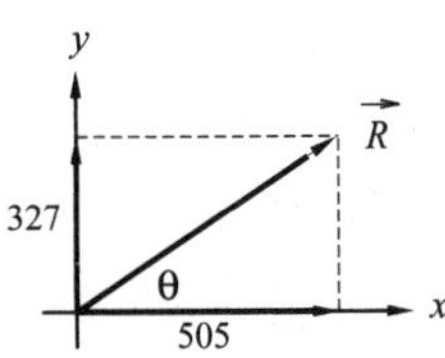

$R=\sqrt{327^2+505^2}=602$

$\theta=\tan^{-1}\dfrac{327}{505}=32.9^\circ$

9.

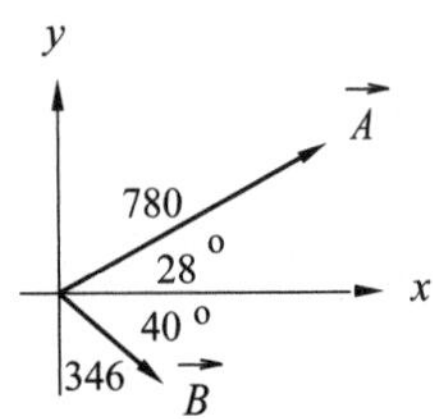

$\vec{R}_y$ $\vec{R}=965$ $R=8.58^\circ$ $\vec{R}_x$

$R_x=780\cos 28.0^\circ+346\cos 40.0^\circ$

$R_y=780\sin 28.0^\circ-346\sin 40.0^\circ$

$R=\sqrt{R_x^2+R_y^2}=\sqrt{954^2+144^2}=965$

$\theta_R=\tan^{-1}\dfrac{144}{954}=8.58^\circ$

13.

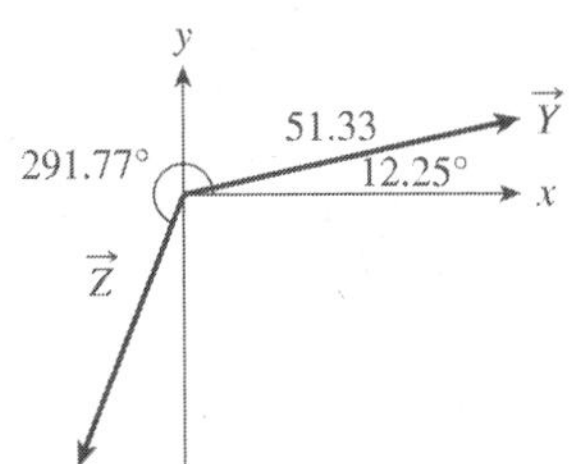

$Y_x=51.33\cos 12.25^\circ=50.16$

$Y_y=51.33\sin 12.25^\circ=10.89$

$Z_x=42.61\cos 291.77^\circ=15.80$

$Z_y=42.61\sin 291.77^\circ=-39.57$

$R_x=50.16+15.80=65.96$

$R_y=10.89-39.57=-28.68$

$R=\sqrt{R_x^2+R_y^2}=\sqrt{65.96^2+(-28.68)^2}=71.94$

$\theta_{ref}=\tan^{-1}\left|\dfrac{-28.68}{65.96}\right|=23.50^\circ$

$\theta_R=360^\circ-23.50^\circ=336.5^\circ$

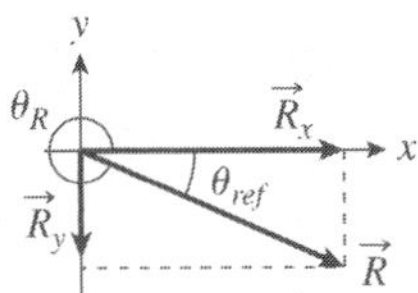

17.

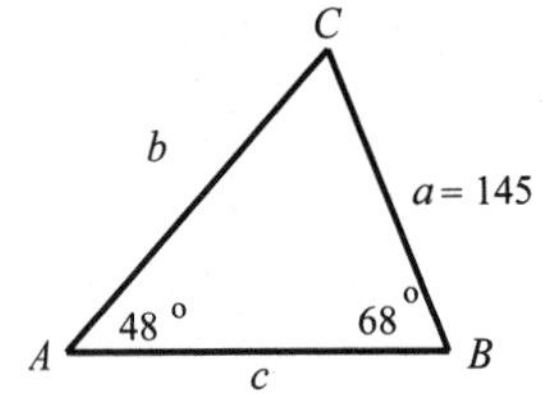

Given: two angles and one side (Case 1).

$C = 180^\circ - 48.0^\circ - 68.0^\circ = 64.0^\circ$

$$\frac{145}{\sin 48.0^\circ} = \frac{b}{\sin 68.0^\circ} = \frac{c}{\sin 64.0^\circ}$$

$$b = \frac{145 \sin 68.0^\circ}{\sin 48.0^\circ} = 181$$

$$c = \frac{145 \sin 64.0^\circ}{\sin 48.0^\circ} = 175$$

21.

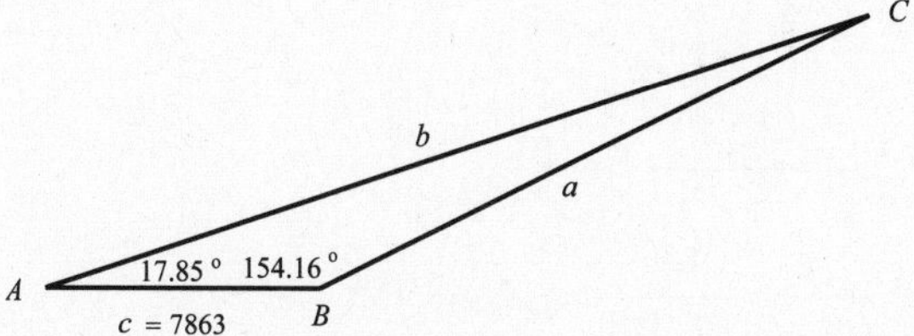

Given: two angles and one side (Case 1).

$C = 180.0^\circ - 17.85^\circ - 154.16^\circ = 7.99^\circ$

$$\frac{a}{\sin 17.85^\circ} = \frac{b}{\sin 154.16^\circ} = \frac{7863}{\sin 7.99^\circ}$$

$$b = \frac{7863 \sin 154.16^\circ}{\sin 7.99^\circ} = 24660$$

$$a = \frac{7863 \sin 17.85^\circ}{\sin 7.99^\circ} = 17340$$

25.

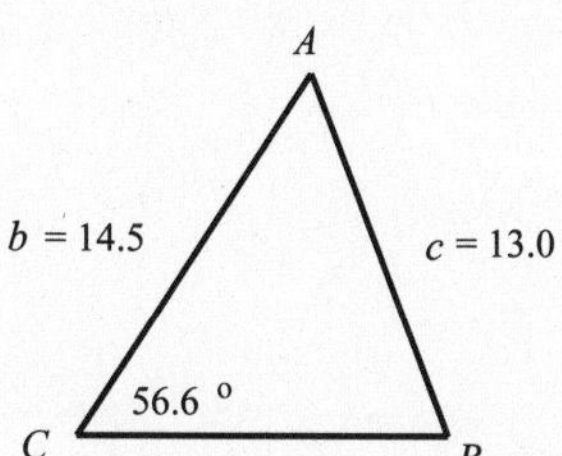

Given: two sides and the angle opposite the shorter side (Case 2, two solutions).

$$\frac{a}{\sin A} = \frac{14.5}{\sin B} = \frac{13.0}{\sin 56.6^\circ}$$

$$\sin B = \frac{14.5 \sin 56.6^\circ}{13.0}$$

$B = 68.6^\circ$ or 111.4°

Solution 1:

$B = 68.6^\circ$

$A = 180^\circ - 68.6^\circ - 56.6^\circ = 54.8^\circ$

$$\frac{a}{\sin 54.8^\circ} = \frac{13.0}{\sin 56.6^\circ}$$

$$a = \frac{13 \sin 54.8^\circ}{\sin 56.6^\circ} = 12.7$$

Solution 2:

$B = 111.4^\circ$

$A = 180^\circ - 111.4^\circ - 56.6^\circ = 12.0^\circ$

$$\frac{a}{\sin 12.0^\circ} = \frac{13.0}{\sin 56.6^\circ}$$

$$a = \frac{13 \sin 12.0^\circ}{\sin 56.6^\circ} = 3.24$$

29.

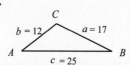

$a = 7.86$ B
C
(2.5°) $b = 2.45$ A

Given: two sides and the included angle (Case 3).

$c^2 = 7.86^2 + 2.45^2 - 2(7.86)(2.45)\cos 2.5^\circ$

$c = 5.413386814$

$c = 5.41$

$7.86^2 = 2.45^2 + c^2 - 2(2.45)(c)\cos A$

(We use c without rounding for more precision since the angle is so close to 180°).

$A = 176.4^\circ$

$B = 180^\circ - 2.5^\circ - 176.4^\circ$

$= 1.1^\circ$

33.

C
$b = 12$ $a = 17$
A B
$c = 25$

Given: three sides (Case 4).

$17^2 = 12^2 + 25^2 - 2(12)(25)\cos A$

$A = 37^\circ$

$12^2 = 17^2 + 25^2 - 2(17)(25)\cos B$

$B = 25^\circ$

$C = 180^\circ - 37^\circ - 25^\circ = 118^\circ$

37.

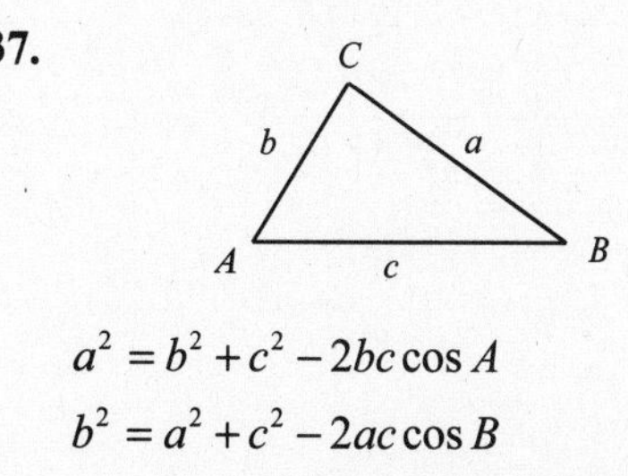

$a^2 = b^2 + c^2 - 2bc\cos A$

$b^2 = a^2 + c^2 - 2ac\cos B$

$\underline{c^2 = a^2 + b^2 - 2ab\cos C}$ add

$$a^2+b^2+c^2=2a^2+2b^2+2c^2-2bc\cos A$$
$$-2ac\cos B-2ab\cos C$$
$$a^2+b^2+c^2=2bc\cos A+2ac\cos B+2ab\cos C$$
$$\frac{a^2+b^2+c^2}{2abc}=\frac{\cos A}{a}+\frac{\cos B}{b}+\frac{\cos C}{c}$$

41.

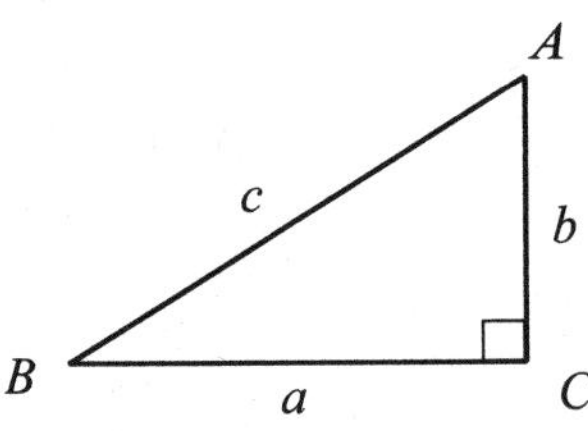

$$\frac{a}{\sin A}=\frac{b}{\sin B}\Rightarrow b=\frac{a\sin B}{\sin A}$$
$$A_t=\frac{1}{2}ab=\frac{1}{2}\cdot a\cdot\frac{a\sin B}{\sin A}$$
$$A_t=\frac{a^2\sin B}{2\sin A}$$

45.

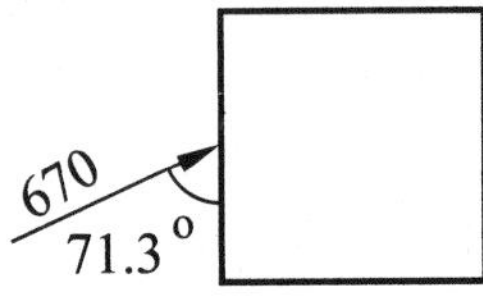

$$v_\perp=670\sin 71.3^\circ$$
$$=630\text{ m/s}$$

We can also obtain $v_\perp$ if we find its direction in standard position, so that $\theta=90^\circ-71.3^\circ=18.7^\circ$. Then

$$v_\perp=670\cos 18.7^\circ$$
$$=630\text{ m/s}$$

49.

vector	x-component	y-component
1300	$-1300\cos 54^\circ$	$1300\sin 54^\circ$
3200	$3200\sin 32^\circ$	$3200\cos 32^\circ$
2100	$-2100\cos 35^\circ$	$-2100\sin 35^\circ$
	-788.6	2561

$$R=\sqrt{(-788.6)^2+(2561)^2}=2700\text{ N}$$
$$\theta_{\text{ref}}=\tan^{-1}\left|\frac{2561}{-788.6}\right|=73^\circ$$
$$\theta=180^\circ-\theta_{\text{ref}}=107^\circ$$

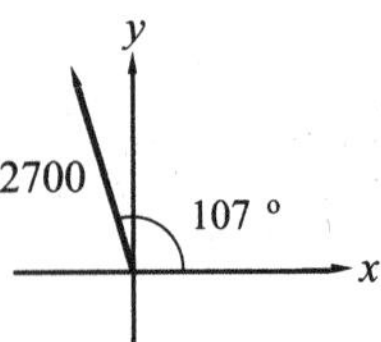

53. upward force $=0.15\sin 22.5^\circ+0.20\sin 15.0^\circ$
$=0.11\text{ N}$

57.

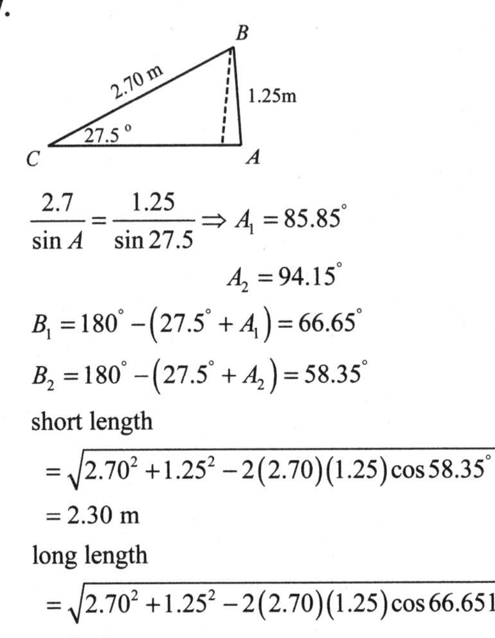

$$\frac{2.7}{\sin A}=\frac{1.25}{\sin 27.5}\Rightarrow A_1=85.85^\circ$$
$$A_2=94.15^\circ$$
$$B_1=180^\circ-\left(27.5^\circ+A_1\right)=66.65^\circ$$
$$B_2=180^\circ-\left(27.5^\circ+A_2\right)=58.35^\circ$$

short length

$$=\sqrt{2.70^2+1.25^2-2(2.70)(1.25)\cos 58.35^\circ}$$
$$=2.30\text{ m}$$

long length

$$=\sqrt{2.70^2+1.25^2-2(2.70)(1.25)\cos 66.651^\circ}$$
$$=2.49\text{ m}$$

61.

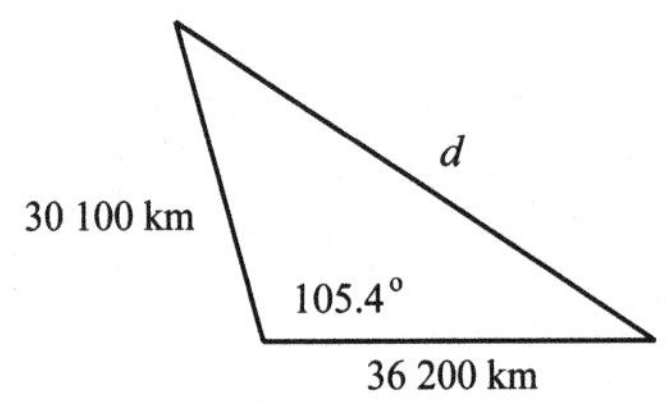

$$d=\sqrt{30\ 100^2+36\ 200^2-2(30\ 100)(36\ 200)\cos 105.4^\circ}$$
$$=52\ 900\text{ km}$$

65.

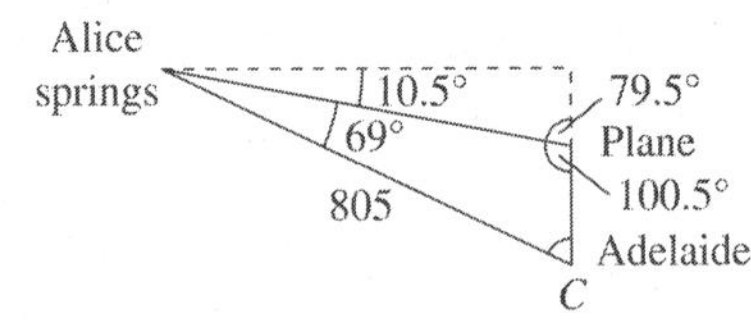

$$C = 180^\circ - \left(\left(69^\circ - 10.5^\circ\right) - 100.5\right)$$
$$= 21^\circ$$
$$\frac{805}{\sin 100.5^\circ} = \frac{c}{\sin 21^\circ}$$
$$c = \frac{805 \sin 21^\circ}{\sin 100.5^\circ}$$
$$= 293 \text{ km}$$

69.

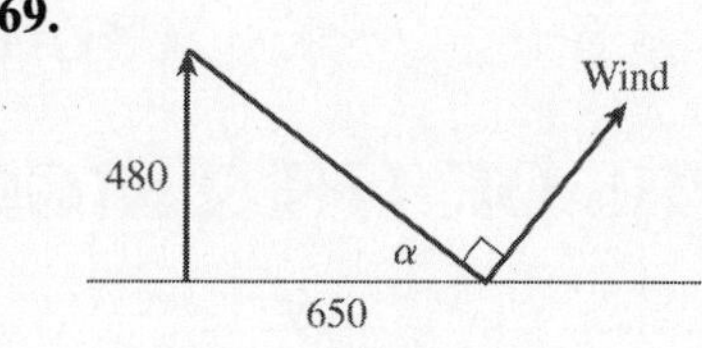

$$\tan \alpha = \frac{480}{650}$$
$$\alpha = 36^\circ \text{ N of E}$$
$$F = \sqrt{F_x^2 + F_y^2} = \sqrt{650^2 + 480^2}$$
$$= 810 \text{ N}$$

CHAPTER 10

GRAPHS OF THE TRIGONOMETRIC FUNCTIONS

10.1 Graphs of $y = a \sin x$ and $y = a \cos x$

1. $y = 3\cos x$

x	0	$\frac{\pi}{6}$	$\frac{\pi}{3}$	$\frac{\pi}{2}$	$\frac{2\pi}{3}$	$\frac{5\pi}{6}$
y	3	2.6	1.5	0	−1.5	−2.6

x	π	$\frac{7\pi}{6}$	$\frac{4\pi}{3}$	$\frac{3\pi}{2}$	$\frac{5\pi}{3}$	$\frac{11\pi}{6}$	2π
y	−3	−2.6	−1.5	0	1.5	2.6	3

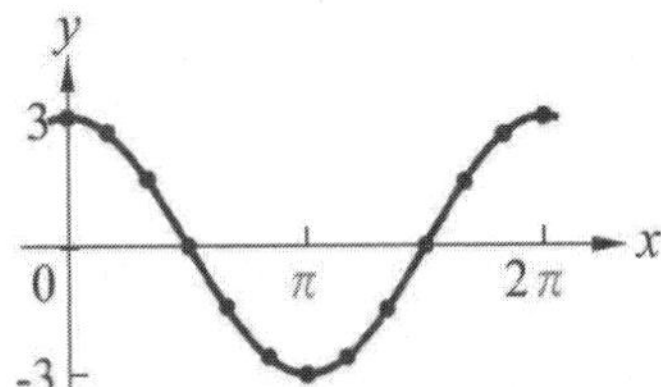

5. $y = 3\cos x$

x	$-\pi$	$-\frac{3\pi}{4}$	$-\frac{\pi}{2}$	$-\frac{\pi}{4}$	0	$\frac{\pi}{4}$	$\frac{\pi}{2}$	$\frac{3\pi}{4}$	π
y	−3	−2.1	0	2.1	3	2.1	0	−2.1	−3

x	$\frac{5\pi}{4}$	$\frac{3\pi}{4}$	$\frac{7\pi}{4}$	2π	$\frac{9\pi}{4}$	$\frac{5\pi}{2}$	$\frac{11\pi}{4}$	3π
y	−2.1	0	2.1	3	2.1	0	−2.1	−3

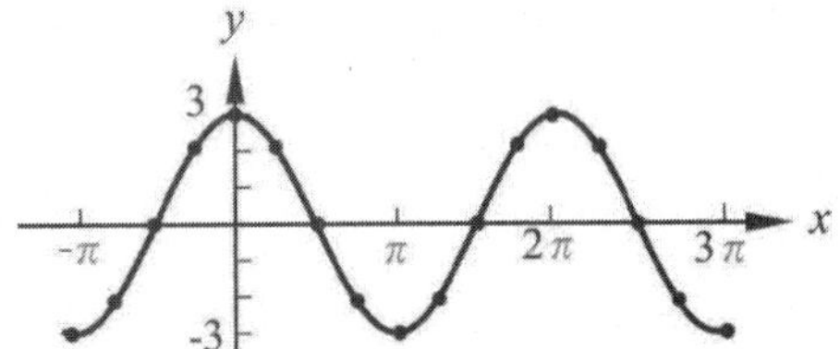

9. $y = \frac{5}{2}\sin x$ has amplitude $\frac{5}{2}$.

The table for key values between 0 and 2π is

x	0	$\frac{\pi}{2}$	π	$\frac{3\pi}{2}$	2π
y	0	$\frac{5}{2}$	0	$-\frac{5}{2}$	0
		max		min	

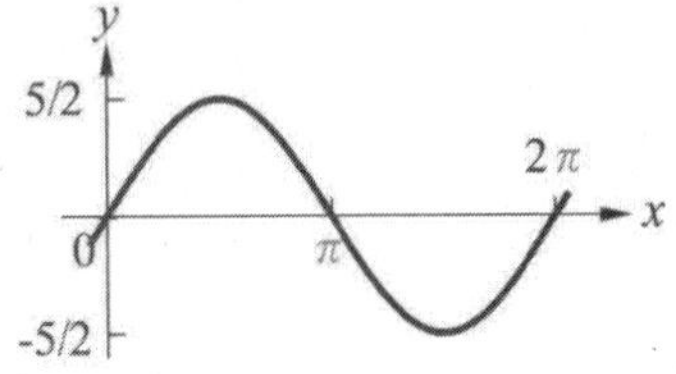

13. $y = 0.8\cos x$ has amplitude 0.8.

The table for key values between 0 and 2π is

x	0	$\frac{\pi}{2}$	π	$\frac{3\pi}{2}$	2π
y	0.8	0	−0.8	0	0.8
	max		min		max

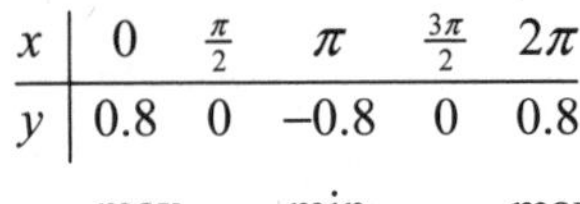

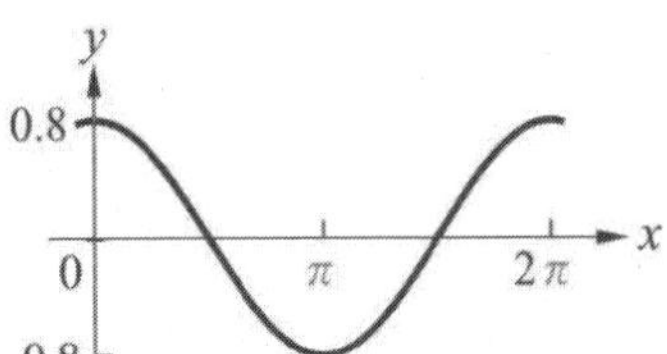

17. $y = -1500\sin x$ has amplitude 1500.

The negative sign will invert the graph, so the table for key values between 0 and 2π is

x	0	$\frac{\pi}{2}$	π	$\frac{3\pi}{2}$	2π
y	0	−1500	0	1500	0
		min		max	

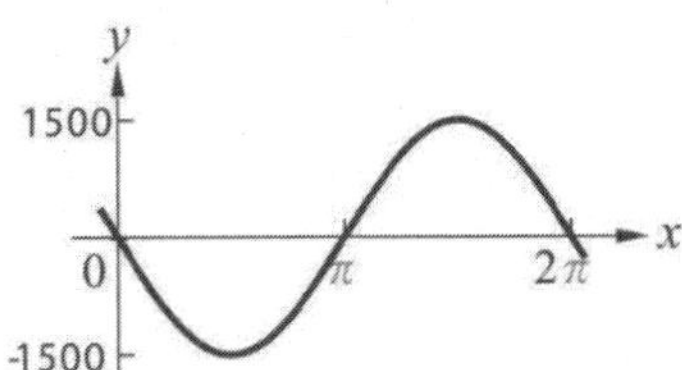

21. $y = -50\cos x$ has amplitude 50.

The negative sign inverts the graph, so the table for key values between 0 and 2π is

x	0	$\frac{\pi}{2}$	π	$\frac{3\pi}{2}$	2π
y	−50	0	50	0	−50
	min		max		min

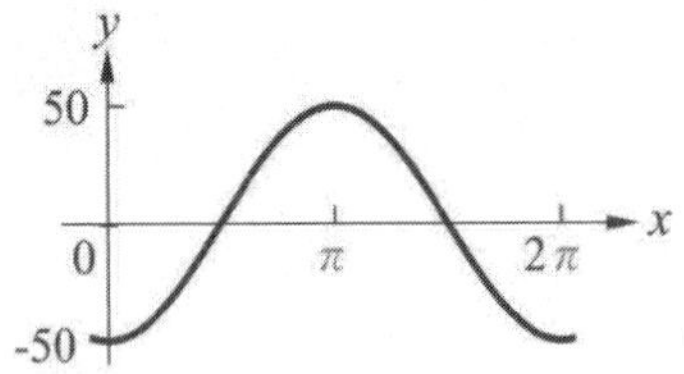

25. Sketch $y = 12\cos x$ for $x = 0, 1, 2, 3, 4, 5, 6, 7$

x	0	1	2	3	4
$12\cos x$	12	6.48	−4.99	−11.9	−7.84

x	5	6	7
$12\cos x$	3.40	11.5	9.05

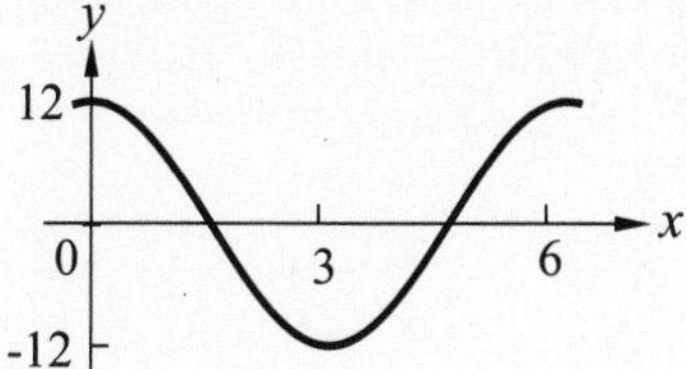

29. $y = a\cos x,\ (\pi, 2)$

$2 = a\cos\pi$

$2 = a(-1)$

$a = -2$

$y = -2\cos x$

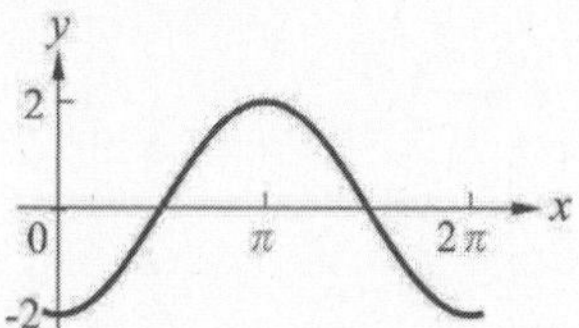

33. The graph has zeros at $x = 0,\ \pi,$ and, 2π, so it is a sine function. Its amplitude is 4, and it has not been inverted. Hence the function is $y = 4\sin x$.

37. If amplitude is 2.50, the function has to be of the form $y = \pm 2.50\sin x$ or $y = \pm 2.50\cos x$. We evaluate each one at $x = 0.67$:

x	$\pm 2.50\sin x$	$\pm 2.50\cos x$
0.67	±1.55	±1.96

The function is $y = -2.50\sin x$.

10.2 Graphs of $y = a\sin bx$ and $y = a\cos bx$

1. $y = 3\sin 6x$, amplitude = 3, period = $\frac{2\pi}{6} = \frac{\pi}{3}$,

key values at multiples of $\frac{1}{4}\left(\frac{\pi}{3}\right) = \frac{\pi}{12}$.

x	0	$\frac{\pi}{12}$	$\frac{\pi}{6}$	$\frac{\pi}{4}$	$\frac{\pi}{3}$	$\frac{5\pi}{12}$	$\frac{\pi}{2}$	$\frac{7\pi}{12}$	$\frac{2\pi}{3}$	$\frac{3\pi}{4}$	$\frac{5\pi}{6}$	$\frac{11\pi}{12}$	π
y	0	3	0	−3	0	3	0	−3	0	3	0	−3	0

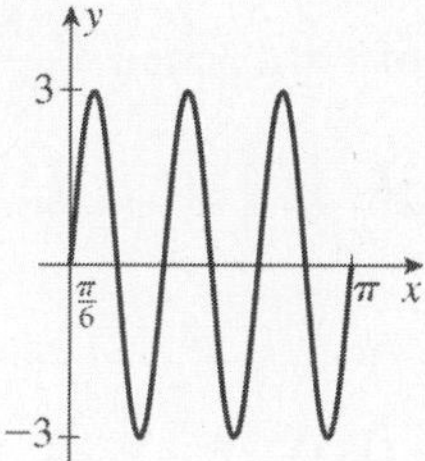

5. Since $\cos bx$ has period $\frac{2\pi}{b}$, $y = 3\cos 8x$ has a period of $\frac{2\pi}{8}$, or $\frac{\pi}{4}$.

9. $y = -\cos 16x$ has period of $\frac{2\pi}{16}$, or $\frac{\pi}{8}$.

13. $y = 3\cos 4\pi x$ has period of $\frac{2\pi}{4\pi}$, or $\frac{2}{4} = \frac{1}{2}$.

17. $y = -\frac{1}{2}\cos\frac{2}{3}x$ has period of $\frac{2\pi}{\frac{2}{3}} = \frac{2\pi}{1}\times\frac{3}{2} = 3\pi$.

21. $y = 3.3\cos\pi^2 x$ has period of $\frac{2\pi}{\pi^2} = \frac{2}{\pi}$.

25. $y = 3\cos 8x$ has amplitude of 3 and period $\frac{\pi}{4}$, with key values at multiples of $\frac{1}{4}\left(\frac{\pi}{4}\right) = \frac{\pi}{16}$.

x	0	$\frac{\pi}{16}$	$\frac{\pi}{8}$	$\frac{3\pi}{16}$	$\frac{\pi}{4}$
y	3	0	−3	0	3

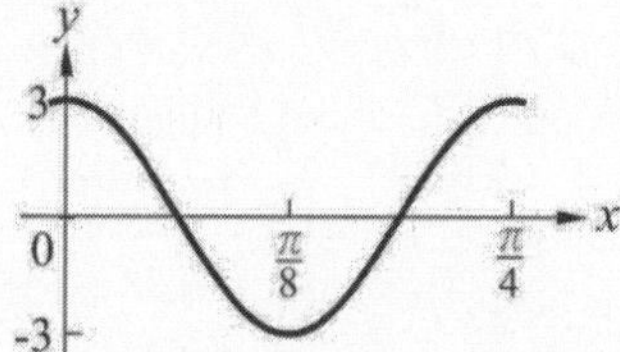

29. $y = -\cos 16x$ has amplitude of $|-1| = 1$, and period of $\frac{\pi}{8}$, with key values at multiples of $\frac{1}{4}\left(\frac{\pi}{8}\right) = \frac{\pi}{32}$.

x	0	$\frac{\pi}{32}$	$\frac{\pi}{16}$	$\frac{3\pi}{32}$	$\frac{\pi}{8}$
y	−1	0	1	0	−1

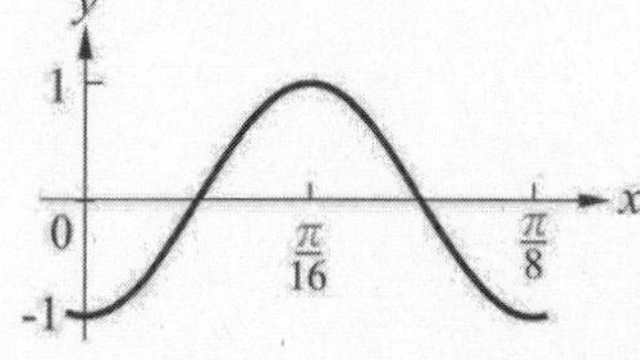

33. $y = 3\cos 4\pi x$ has amplitude of 3 and period of $\frac{1}{2}$, with key values at multiples of $\frac{1}{4}\left(\frac{1}{2}\right) = \frac{1}{8}$.

x	0	$\frac{1}{8}$	$\frac{1}{4}$	$\frac{3}{8}$	$\frac{1}{2}$
y	3	0	−3	0	3

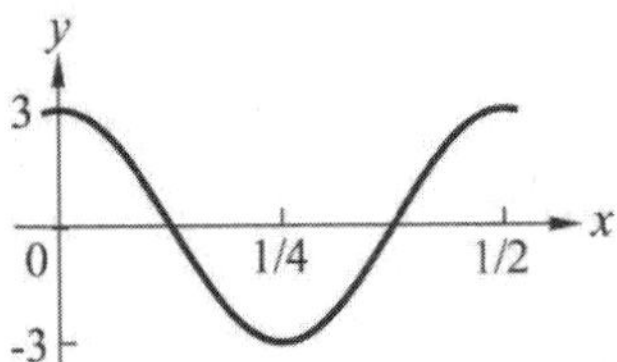

37. $y = -\frac{1}{2}\cos\frac{2}{3}x$ has amplitude of $\left|-\frac{1}{2}\right| = \frac{1}{2}$, and period of 3π, with key values at multiples of $\frac{1}{4}\cdot 3\pi = \frac{3\pi}{4}$.

x	0	$\frac{3\pi}{4}$	$\frac{3\pi}{2}$	$\frac{9\pi}{4}$	3π
y	$-\frac{1}{2}$	0	$\frac{1}{2}$	0	$-\frac{1}{2}$

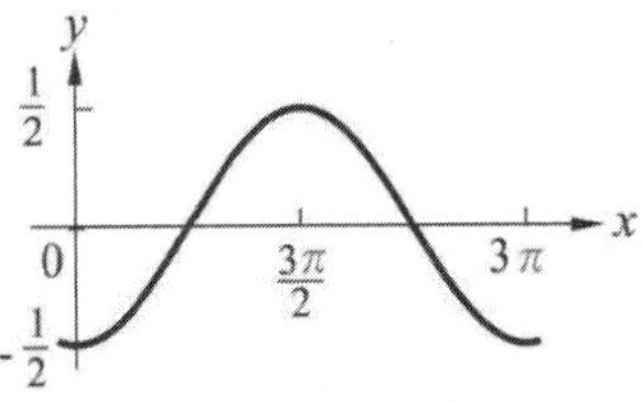

41. $y = 3.3\cos\pi^2 x$ has amplitude of 3.3 and period of $\frac{2}{\pi}$, with key values at multiples of $\frac{1}{4}\left(\frac{2}{\pi}\right) = \frac{1}{2\pi}$.

x	0	$\frac{1}{2\pi}$	$\frac{1}{\pi}$	$\frac{3}{2\pi}$	$\frac{2}{\pi}$
$\pi^2 x$	0	$\frac{\pi}{2}$	π	$\frac{3\pi}{2}$	2π
$\cos\pi^2 x$	1	0	−1	0	1
$3.3\cos\pi^2 x$	3.3	0	−3.3	0	3.3

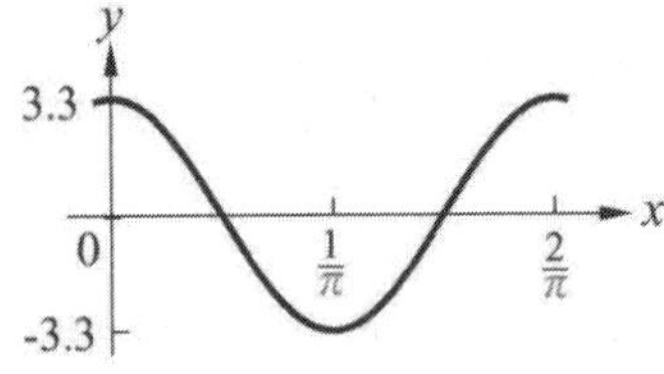

45. $b = \frac{2\pi}{1/3} = 6\pi;\ y = \sin 6\pi x$

49. $y = 8\left|\cos\frac{\pi}{2}x\right|$

This function has amplitude 8. Moreover, $\cos(\pi x/2)$ has period $\frac{2\pi}{\frac{\pi}{2}} = 4$ so that key values are at multiples of $4/4 = 1$.
Also, the absolute value makes the y values positive whenever they are negative and leaves positive values intact, so that the negative parts of the curve are reflected with respect to the x axis. Note that the function repeats itself every 2 units, so that the absolute value changes the period from 4 to 2.

x	0	1	2	3	4
y	8	0	8	0	8

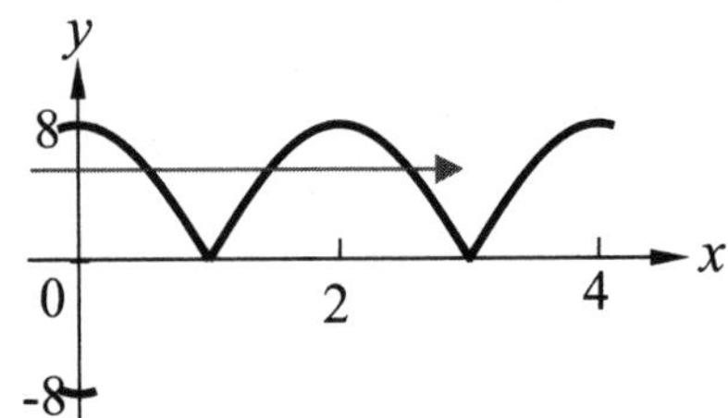

53. $y = -2\sin bx, \left(\frac{\pi}{4}, -2\right), b > 0$

$-2 = -2\sin b\cdot\frac{\pi}{4}$

$\sin\frac{b\pi}{4} = 1$

$\frac{b\pi}{4} = \frac{\pi}{2} + 2\pi n$

$b = 2 + 8n,$ of which the smallest is $b = 2$

$y = -2\sin 2x$ is the function.
It has amplitude 2, period $2\pi/2 = \pi$, with key values at multiples of $\pi/4$

x	0	$\pi/4$	$\pi/2$	$3\pi/4$	π
y	0	−2	0	2	0

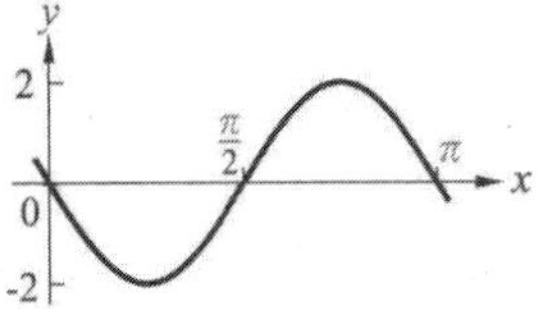

57. $V = 170\sin 120\pi t$

V has amplitude 170 and period $\frac{2\pi}{120\pi} = \frac{1}{60}$ with key values at multiples of 1/240. Between 0 and 0.05, the function completes three cycles $\left(\frac{0.05}{\frac{1}{60}} = 3\right)$.

t	0	1/240	1/120	1/80	1/60	1/48	1/40
V	0	170	0	−170	0	170	0

t	7/240	1/30	3/80	11/24	11/240	1/20 = 0.05
V	−170	0	170	0	−170	0

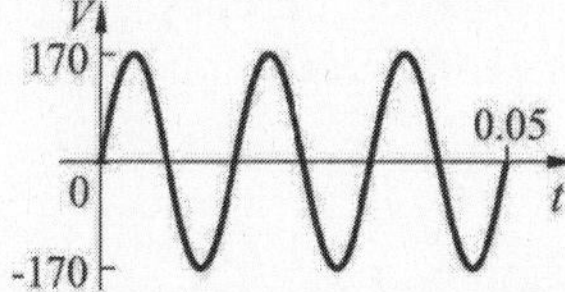

61. The function has amplitude 0.5, period π, with a maximum at $x = 0$ (so it is a cosine function).
$b = 2\pi/\pi$
$b = 2$
Hence the function is $y = \frac{1}{2}\cos 2x$

10.3 Graphs of $y = a\sin(bx + c)$ and $y = a\cos(bx + c)$

1. $y = -\cos\left(2x - \frac{\pi}{6}\right)$

(1) the amplitude is 1

(2) the period is $\frac{2\pi}{2} = \pi$

(3) the displacement is $-\frac{\frac{-\pi}{6}}{2} = \frac{\pi}{12}$

One-fourth period is $\pi/4$, so key values for one full cycle start at $\pi/12$, end at $13\pi/12$, and are found $\pi/4$ units apart. The table of key values is

x	$\frac{\pi}{12}$	$\frac{\pi}{3}$	$\frac{7\pi}{12}$	$\frac{5\pi}{6}$	$\frac{13\pi}{12}$
y	−1	0	1	0	−1

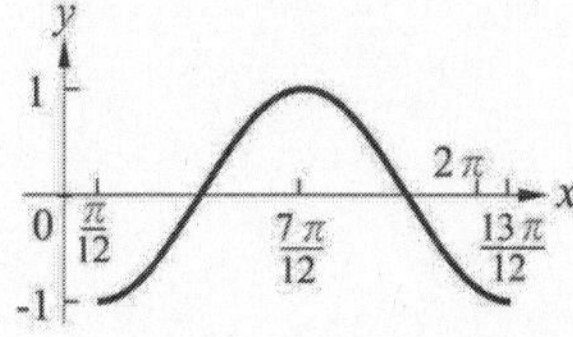

5. $y = \cos\left(x + \frac{\pi}{6}\right)$; $a = 1$, $b = 1$, $c = \frac{\pi}{6}$

Amplitude is $|a| = 1$; period is $\frac{2\pi}{b} = 2\pi$;

displacement is $-\frac{c}{b} = -\frac{\pi}{6}$

One-fourth period is $\pi/2$, so key values for one full cycle start at $-\pi/6$, end at $11\pi/6$, and are found $\pi/2$ units apart. The table of key values is

x	$-\frac{\pi}{6}$	$\frac{\pi}{3}$	$\frac{5\pi}{6}$	$\frac{4\pi}{3}$	$\frac{11\pi}{6}$
y	0	1	0	−1	0

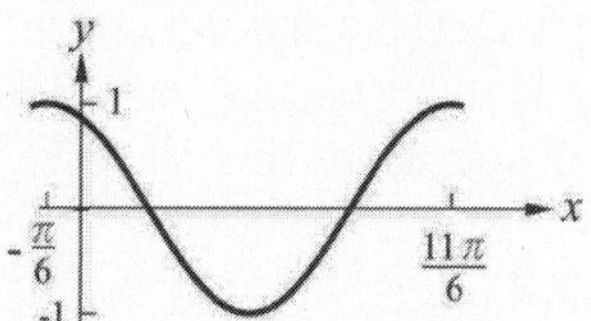

9. $y = -\cos(2x - \pi)$; $a = -1$, $b = 2$, $c = -\pi$

Amplitude is $|a| = 1$; period is $\frac{2\pi}{b} = \frac{2\pi}{2} = \pi$;

displacement is $-\frac{c}{b} = -\left(\frac{-\pi}{2}\right) = \frac{\pi}{2}$

One-fourth period is $\pi/4$, so key values for one full cycle start at $\pi/2$, end at $3\pi/2$, and are found $\pi/4$ units apart. The table of key values is

x	$\frac{\pi}{2}$	$\frac{3\pi}{4}$	π	$\frac{5\pi}{4}$	$\frac{3\pi}{2}$
y	−1	0	1	0	−1

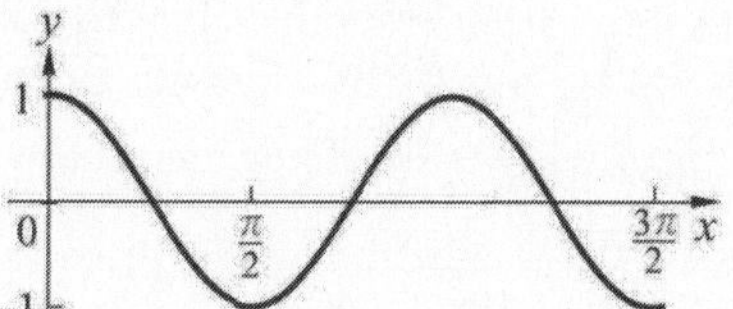

13. $y = 30\cos\left(\frac{1}{3}x + \frac{\pi}{3}\right)$; $a = 30$, $b = \frac{1}{3}$, $c = \frac{\pi}{3}$

Amplitude is $|a| = 30$; period is $\frac{2\pi}{b} = \frac{2\pi}{1/3} = 6\pi$;

displacement is $-\frac{c}{b} = \frac{-\pi/3}{1/3} = -\pi$

One-fourth period is $3\pi/2$, so key values for one full cycle start at $-\pi$, end at 5π, and are found $3\pi/2$ units apart. The table of key values is

x	$-\pi$	$\frac{\pi}{2}$	2π	$\frac{7\pi}{2}$	5π
y	30	0	−30	0	30

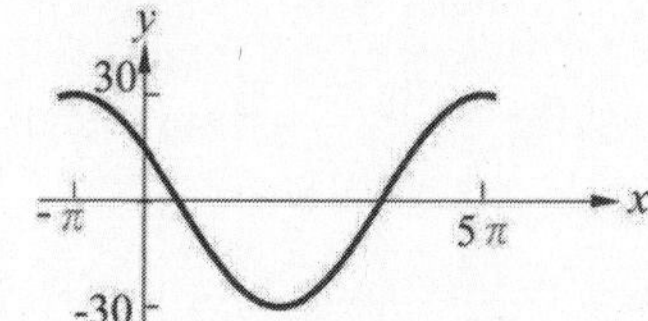

17. $y = 0.08\cos\left(4\pi x - \frac{\pi}{5}\right)$; $a = 0.08$, $b = 4\pi$, $c = -\frac{\pi}{5}$

Amplitude is $|a| = 0.08$; period is $\frac{2\pi}{b} = \frac{2\pi}{4\pi} = \frac{1}{2}$;

displacement is $-\frac{c}{b} = -\left(-\frac{\pi/5}{4\pi}\right) = \frac{1}{20}$

One-fourth period is 1/8, so key values for one full cycle start at 1/20, end at 11/20, and are found 1/8 units apart. The table of key values is

x	$\frac{1}{20}$	$\frac{7}{40}$	$\frac{3}{10}$	$\frac{17}{40}$	$\frac{11}{20}$
y	0.08	0	−0.08	0	0.08

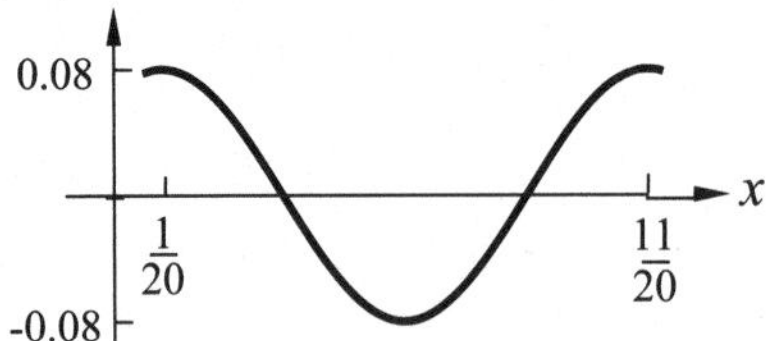

21. $y = 40\cos(3\pi x + 2)$; $a = 40$, $b = 3\pi$, $c = 2$

Amplitude is $|a| = 40$; period is $\frac{2\pi}{b} = \frac{2\pi}{3\pi} = \frac{2}{3}$;

displacement is $-\frac{c}{b} = -\frac{2}{3\pi}$

One-fourth period is 1/6, so key values for one full cycle start at $-2/(3\pi)$, end at $2/3 - 2/(3\pi)$, and are found 1/6 units apart. The table of key values is

x	$-\frac{2}{3\pi}$	$\frac{1}{6}-\frac{2}{3\pi}$	$\frac{1}{3}-\frac{2}{3\pi}$	$\frac{1}{2}-\frac{2}{3\pi}$	$\frac{2}{3}-\frac{2}{3\pi}$
y	40	0	−40	0	40

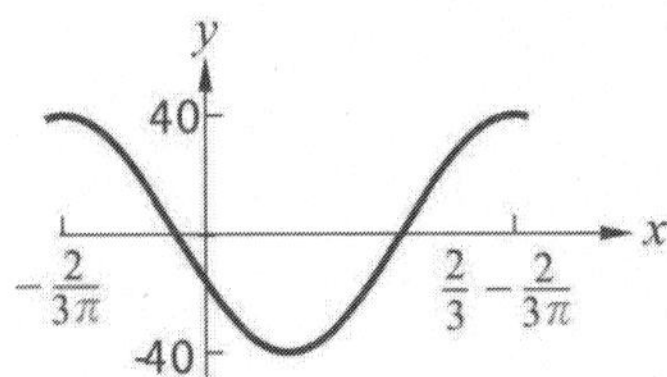

25. $y = -\frac{3}{2}\cos\left(\pi x - \frac{\pi^2}{6}\right)$; $a = -\frac{3}{2}$, $b = \pi$, $c = -\frac{\pi^2}{6}$

Amplitude is $|a| = \frac{3}{2}$; period is $\frac{2\pi}{b} = \frac{2\pi}{\pi} = 2$;

displacement is $-\frac{c}{b} = -\frac{\pi^2/6}{\pi} = -\frac{\pi}{6}$

One-fourth period is 1/2, so key values for one full cycle start at $-\pi/6$, end at $2 - \pi/6$, and are found 1/2 units apart. The table of key values is

x	$-\frac{\pi}{6}$	$-\frac{\pi}{6}+\frac{1}{2}$	$-\frac{\pi}{6}+1$	$-\frac{\pi}{6}+\frac{3}{2}$	$-\frac{\pi}{6}+2$
y	$-\frac{3}{2}$	0	$\frac{3}{2}$	0	$-\frac{3}{2}$

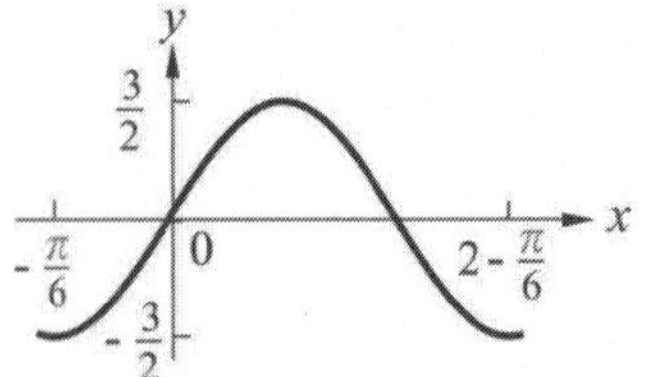

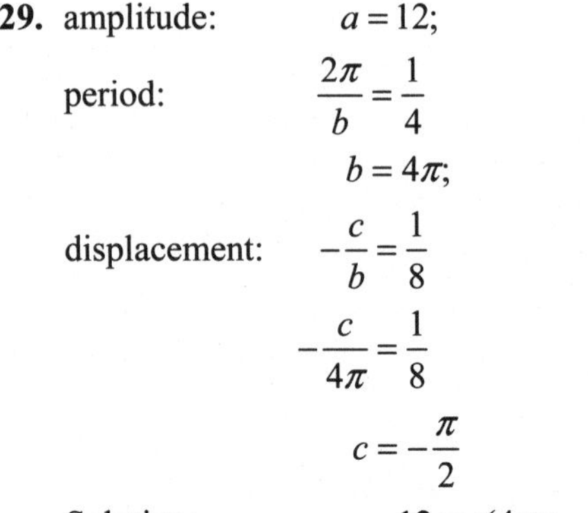

29. amplitude: $a = 12$;

period: $\frac{2\pi}{b} = \frac{1}{4}$

$b = 4\pi$;

displacement: $-\frac{c}{b} = \frac{1}{8}$

$-\frac{c}{4\pi} = \frac{1}{8}$

$c = -\frac{\pi}{2}$

Solution: $y = 12\cos(4\pi x - \frac{\pi}{2})$

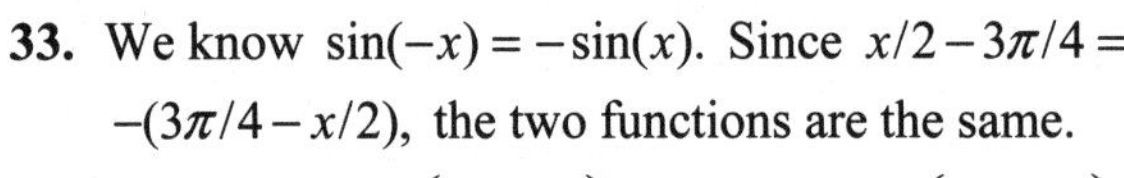

33. We know $\sin(-x) = -\sin(x)$. Since $x/2 - 3\pi/4 = -(3\pi/4 - x/2)$, the two functions are the same.

Graph $y_1 = \sin\left(\frac{x}{2} - \frac{3\pi}{4}\right)$ and $y_2 = -\sin\left(\frac{3\pi}{4} - \frac{x}{2}\right)$.

Graphs are the same.

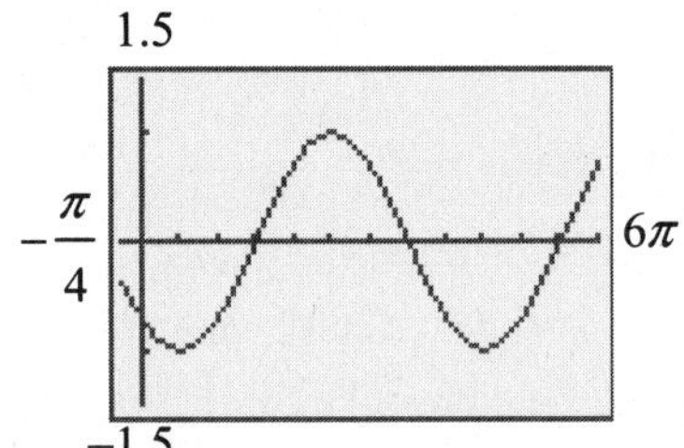

37. $y = 2.00\sin 2\pi\left(\frac{t}{0.100} - \frac{5.00}{20.0}\right)$; $a = 2.00$,

$b = \frac{2\pi}{0.100}$, $c = \frac{-5.00(2\pi)}{20.0}$

Amplitude $= |a| = 2.00$,

period $= \frac{2\pi}{b} = 0.100$,

displacement $= -\frac{c}{b} = 0.025$

One-fourth period is 0.025, so key values for one full cycle start at 0.025, end at 0.125, and are found 0.025 units apart. Three cycles end at 0.325. The table of key values in the first cycle is

t	0.025	0.05	0.075	0.1	0.125
y	0	2.00	0	−2.00	0

The curve repeats after that.

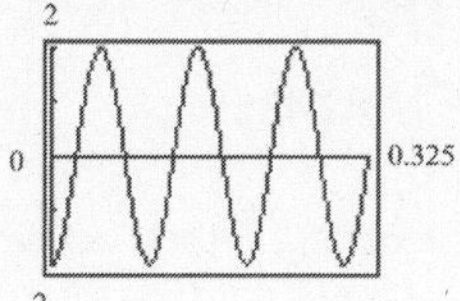

41. The maximum is at $y = 5$, so amplitude is 5. A full cycle takes place between -1 and 15, so the period is 16. The graph has a zero at $x = -1$, so displacement of the sine function is -1. We thus have

amplitude: $a = 5$;

period: $\frac{2\pi}{b} = 16$

$b = \frac{\pi}{8}$;

displacement: $-\frac{c}{b} = -1$

$-\frac{c}{\frac{\pi}{8}} = -1$

$c = \frac{\pi}{8}$

Solution: $y = 5\sin\left(\frac{\pi}{8}x + \frac{\pi}{8}\right)$

45. We determine the values of *a*, *b*, *c*, and *d* as follows:

1. The amplitude is half the distance between the maximum and the minimum daylight hours, so $a = \frac{14.4 - 9.90}{2} = 2.25$.
2. The period is 365 days. Therefore, $\frac{2\pi}{b} = 365$ and $b = \frac{2\pi}{365}$.
3. The maximum of the sine function occurs at one-fourth of the period, which is $\frac{365}{4} = 91.25$. Since the maximum for these data occurs at $t = 355$ (Dec. 21), we must shift the sine function to the right $355 - 91.25 = 263.75$ days. This gives $-\frac{c}{b} = 263.75$, so that $c = -263.75 \times \frac{2\pi}{365} = -4.54$.
4. Finally, the range is [9.90, 14.4] instead of, $[-2.25, 2.25]$ so we must shift the sine function $14.4 - 2.25 = 12.2$ hours vertically by setting $d = 12.2$.

   The sine model is

   $h = 2.25\sin\left(\frac{2\pi}{365}t - 4.54\right) + 12.2$

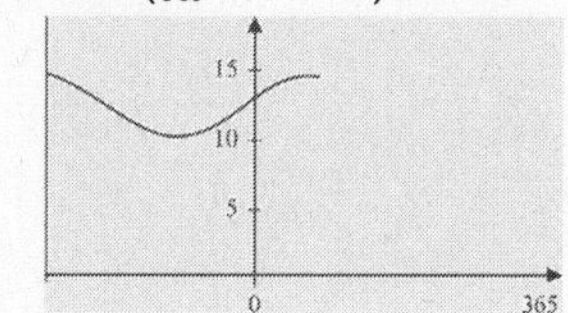

10.4 Graphs of $y = \tan x$, $y = \cot x$, $y = \sec x$, $y = \csc x$

1. $y = 5 \cot 2x$.

Since the period of $y = \cot x$ is π, the period of this function is $\pi/2$. We have the following table of key values:

x	0	$\frac{\pi}{4}$	$\frac{\pi}{2}$	$\frac{3\pi}{4}$	π
y	*	0	*	0	*

(* = asymptote)

Since $a = 5$, the function increases much faster than $y = \cot x$.

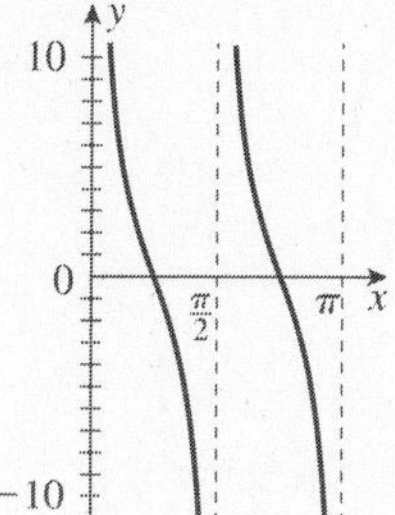

5.

x	$-\frac{\pi}{2}$	$-\frac{\pi}{3}$	$-\frac{\pi}{4}$	$-\frac{\pi}{6}$	0	$\frac{\pi}{6}$	$\frac{\pi}{4}$
sec x	*	2	1.4	1.2	1	1.2	1.4

x	$\frac{\pi}{3}$	$\frac{\pi}{2}$	$\frac{2\pi}{3}$	$\frac{3\pi}{4}$	$\frac{5\pi}{6}$	π
sec x	2	*	−2	−1.4	−1.2	−1

(* = asymptote)

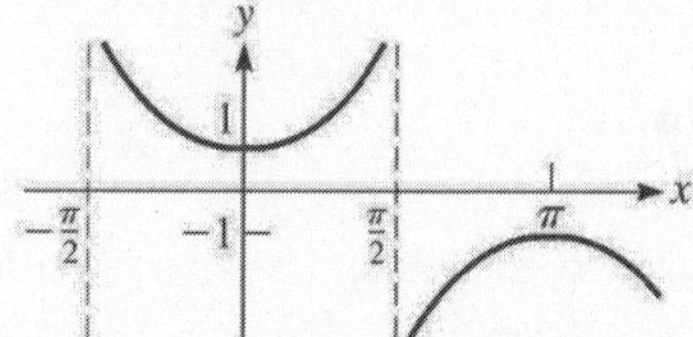

9. For $y = \frac{1}{2}\sec x$, first sketch the graph of $y = \sec x$, then multiply the y − values of the secant function by $\frac{1}{2}$ and graph.

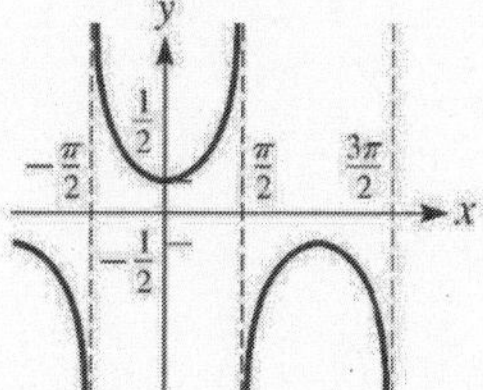

13. For $y = -3 \csc x$, sketch the graph of $y = \csc x$, then multiply the y – values by -3, and resketch the graph. It will be inverted.

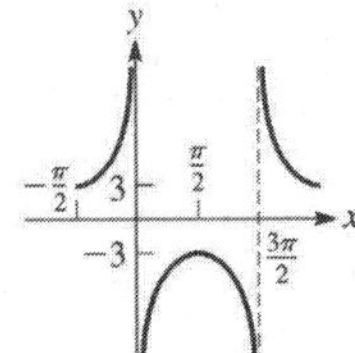

17. Since the period of $\sec x$ is 2π, the period of $y = \frac{1}{2} \sec 3x$ is $\frac{2\pi}{3}$. Graph $y_1 = 0.5(\cos 3x)^{-1}$.

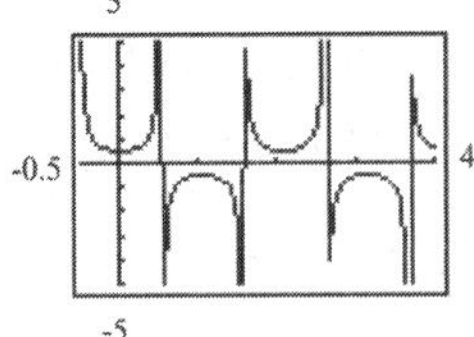

21. Since the period of $\csc x$ is 2π, the period of $y = 18 \csc\left(3x - \frac{\pi}{3}\right)$ is $\frac{2\pi}{3}$. The displacement is $-\left(-\frac{\pi/3}{3}\right) = \frac{\pi}{9}$. Graph $y_1 = 18\left(\sin\left(3x - \frac{\pi}{3}\right)\right)^{-1}$.

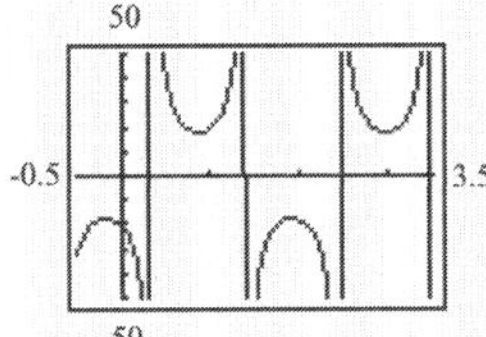

25. If displacement is zero, the secant function is of the form $y = a\sec(bx)$. Since the period is 4π, $2\pi/b = 4\pi$. Therefore, $b = 1/2$. Now we substitute $x = 0$ and $y = -3$ into $y = a\sec\left(\frac{x}{2}\right)$ to get

$$-3 = a \sec(0)$$
$$-3 = a$$

Solution: $y = -3\sec\left(\frac{x}{2}\right)$

29. $b = (a \sin B)\csc A$

$$= \left(4.00 \sin \frac{\pi}{4}\right) \csc A$$
$$= 2.83 \csc A$$

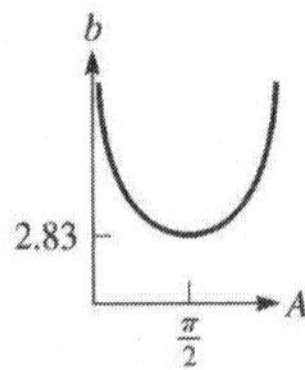

10.5 Applications of the Trigonometric Graphs

1. The displacement of the projection on the y-axis is d and is given by $d = R \cos wt$.

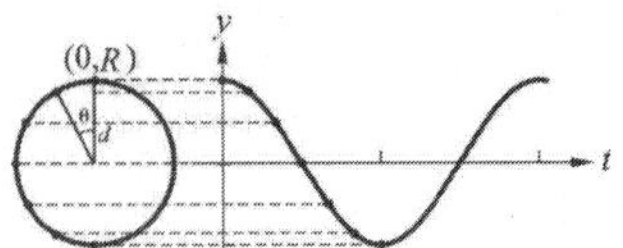

5. $y = R \cos \omega t$

$$= 8.30 \cos\left[(3.20)(2\pi)\right]t$$

Amplitude is 8.30 cm;

period is $\frac{1}{3.20} = 0.3125$ s,

0.625 s for 2 cycles;

displacement is 0 s.

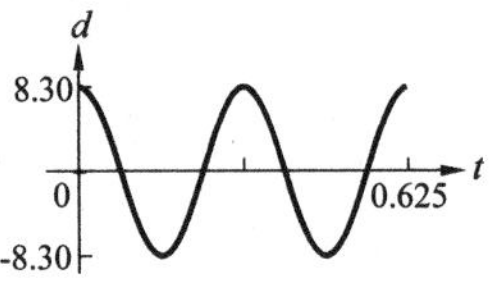

9. $V = E \cos(\omega t + \alpha)$

$$= 170 \cos\left[2\pi(60.0)t - \frac{\pi}{3}\right]$$

Amplitude is 170 V,

period is $\frac{2\pi}{2\pi(60.0)} = 0.016$ s, 0.033 s for 2 cycles; displacement is $\frac{\pi/3}{2\pi(60.0)} = \frac{1}{360}$ s

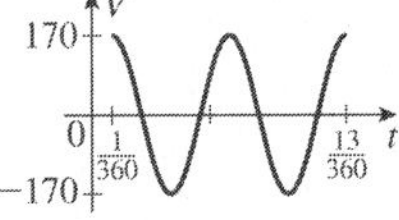

13. $p = p_0 \sin 2\pi ft$

$$= 280 \sin\left[2\pi(2.30)\right]t$$
$$= 280 \sin 14.45t$$

Amplitude is 280 kPa, period is $\frac{2\pi}{14.45} = 0.435$ s for 1 cycle, 0.87 s for 2 cycles; displacement is 0 s

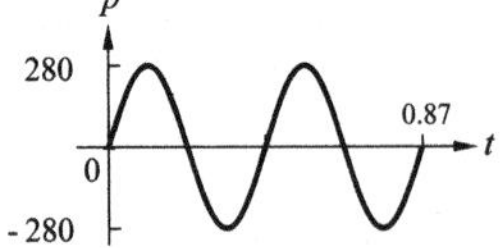

17. $V = 0.014\cos\left(2\pi ft + \frac{\pi}{4}\right)$

$= 0.014\cos\left[2\pi(0.950)t + \frac{\pi}{4}\right]$

Amplitude is 0.014 V,

period is $\frac{2\pi}{2\pi(0.950)} = 1.05$ s, 2.10 s

for 2 cycles; displacement is $\frac{-\pi/4}{2\pi(0.950)} = -0.13$ s

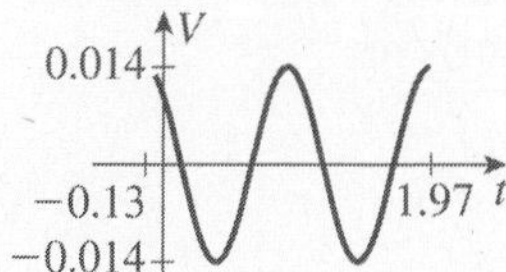

21. amplitude: $a = 1.98$ m;

angular velocity: $\omega = 2\pi f$

$= \frac{2\pi \text{ rad}}{1 \text{ r}} \times \frac{30 \text{ r}}{1 \text{ min}} \times \frac{1 \text{ min}}{60 \text{ s}}$

$\omega = \pi$ rad/s

Solution: $d = 1.98\sin(\pi t)$

Two cycles end at $t = 4$ s

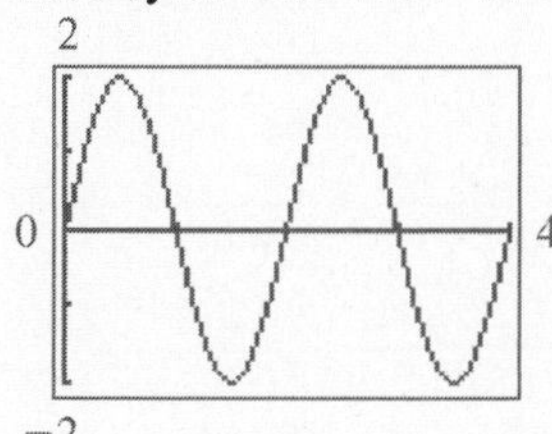

10.6 Composite Trigonometric Curves

1. $y = 1 + \sin x$

x	-2π	$-\frac{3\pi}{2}$	$-\pi$	$-\frac{\pi}{2}$	0	$\frac{\pi}{2}$	π	$\frac{3\pi}{2}$	2π
y	1	2	1	0	1	2	1	0	1

This is a vertical shift of 1 unit of the graph of $y = \sin x$.

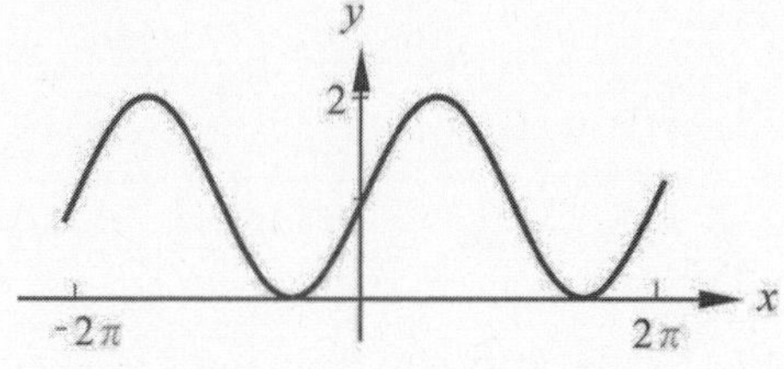

5. $y = \frac{1}{10}x^2 - \sin \pi x$

x	−4	−3.43	−2.55	−1.88	−1.47
y	1.60	0.20	1.64	0	−0.78

x	−1.03	−0.51	0	0.49	0.97	1.53
y	0	1.03	0	−0.98	0	1.23

x	2.15	2.45	2.73	0
y	0	−0.39	4	1.6

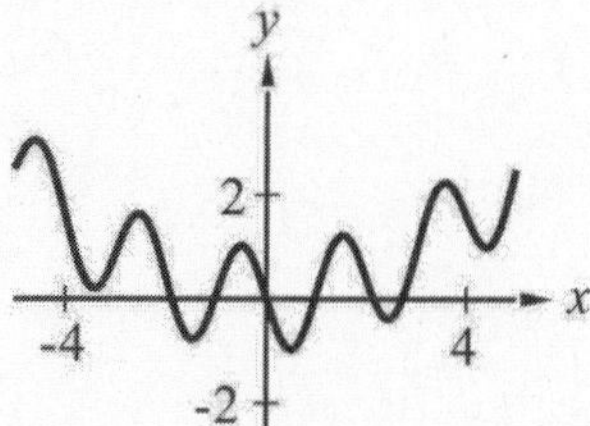

9. Graph $y_1 = x^3 + 10\sin 2x$.

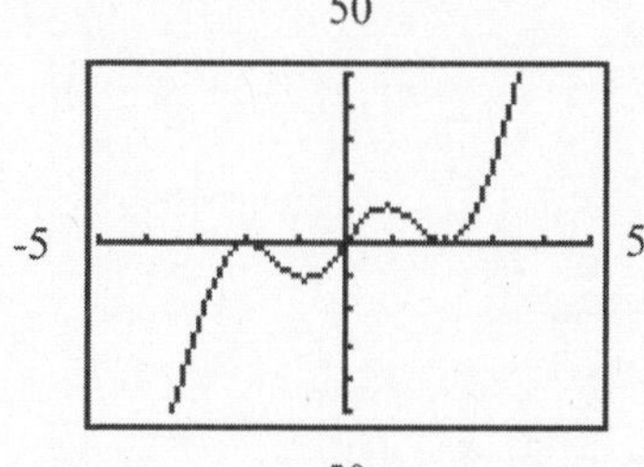

13. Graph $y_1 = 20\cos 2x + 30\sin x$.

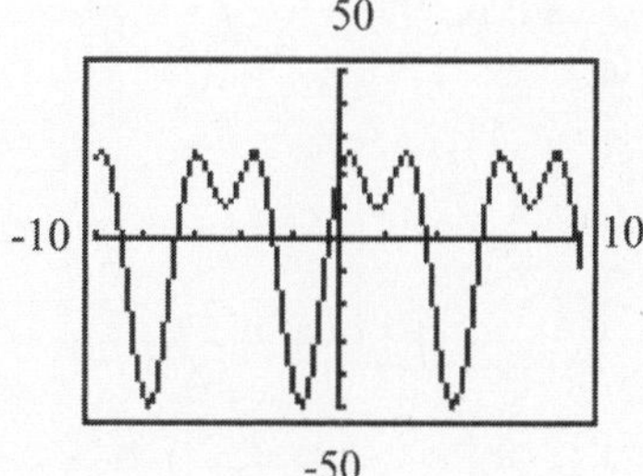

17. Graph $y_1 = \sin \pi x - \cos 2x$.

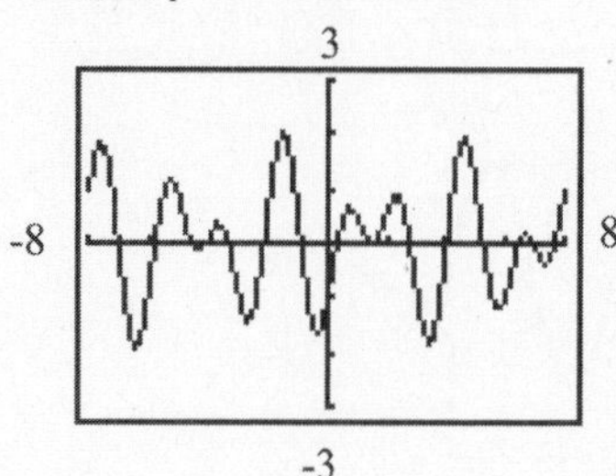

21. In parametric mode graph

$x_{IT} = 3 \sin t,\ y_{IT} = 2 \sin t$

t	$-\frac{\pi}{2}$	$-\frac{\pi}{4}$	0	$\frac{\pi}{4}$	$\frac{\pi}{2}$
x	−3	−2.12	0	2.12	3
y	−2	−1.41	0	1.41	2

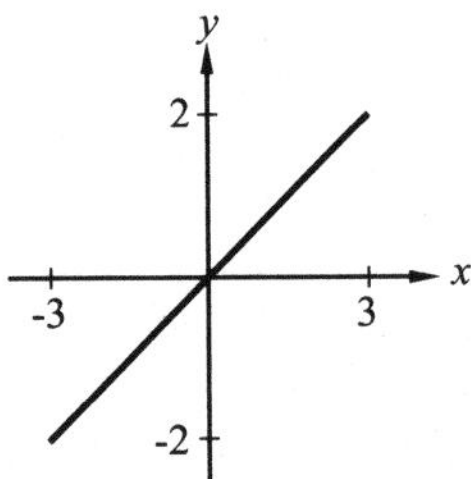

25. In parametric mode, graph

$x_{IT} = \cos \pi\left(t + \frac{1}{6}\right),\ y_{IT} = 2 \sin \pi t$

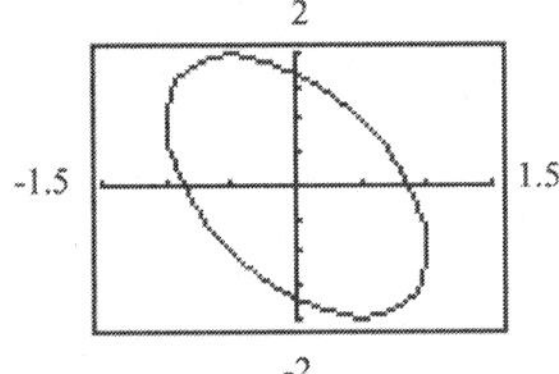

29. In parametric mode, graph

$x_{IT} = \sin\,(t+1),\ y_{IT} = \sin 5t.$

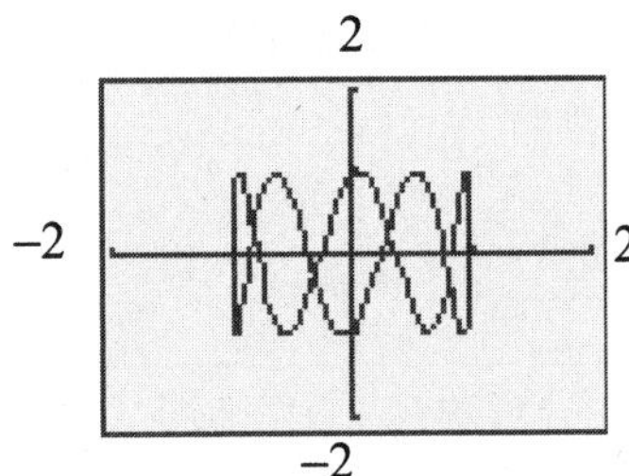

33. $y = 0.4 \sin 4t + 0.3 \cos 4t$

Graph $y_1 = 0.4 \sin 4x + 0.3 \cos 4x$

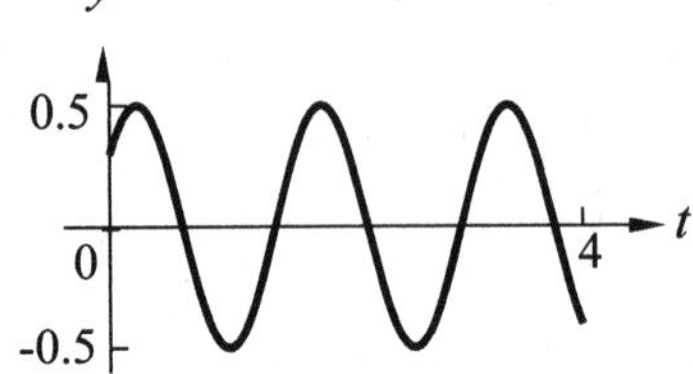

37. Times is in seconds, so we have

$$60 \frac{\text{beats}}{\text{min}} \times \left(\frac{1\,\text{min}}{60\,\text{s}}\right) = 1 \frac{\text{beat}}{\text{s}}$$

This means that the period is 1 s. The pressure varies between a maximum of 120 mmHg and a minimum of 80 mmHg, so the amplitude is half of this difference, or $(120 - 80)/2 = 20$ mmHg.

Pressure starts at 120 mmHg (a maximum), so we can use a cosine function with no displacement. Finally, instead of varying between −20 and 20, the function varies between 80 and 120, so it has been shifted up 100 mmHg. We have

amplitude: $a = 20;$

period: $\frac{2\pi}{b} = 1$

$b = 2\pi;$

displacement: $-\frac{c}{b} = 0$

$c = 0$

shift: $y = 100$

Solution: $y = 100 + 20\cos(2\pi t)$

Graph $y_1 = 100 + 20 \cos(2\pi x)$.

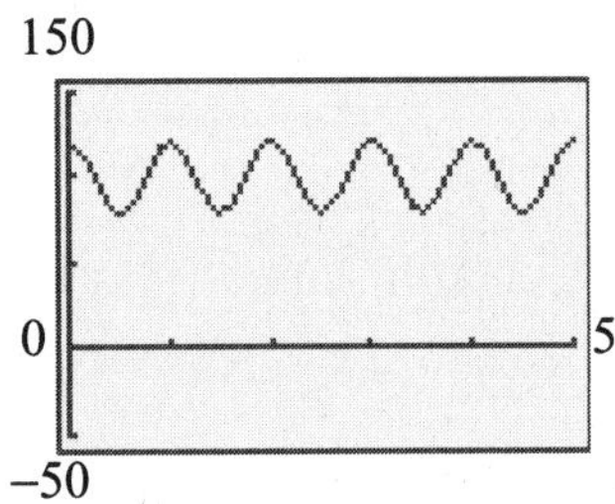

41. $x = 4 \cos \pi t,\ y = 2 \sin 3\pi t$

t	0	$\frac{\pi}{4}$	$\frac{\pi}{2}$	$\frac{3\pi}{4}$	π
x	4	−3.12	0.88	1.75	−3.61
y	0	1.80	1.57	−0.43	−1.94

t	$\frac{5\pi}{4}$	$\frac{3\pi}{2}$	$\frac{7\pi}{4}$	2π
x	3.90	−2.48	−0.028	2.52
y	−1.27	0.84	2.0	0.91

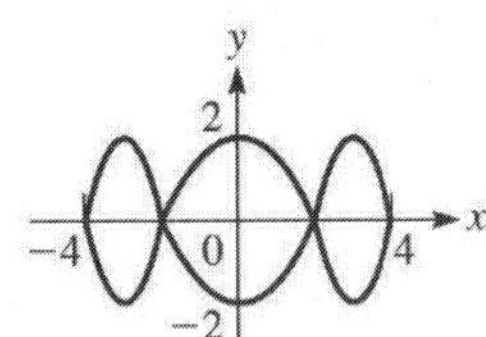

Review Exercises

1. $y=\frac{2}{3}\sin x$ has amplitude 2/3. The table of key values between 0 and 2π is:

x	0	$\pi/2$	π	$3\pi/2$	2π
y	0	2/3	0	-2/3	0

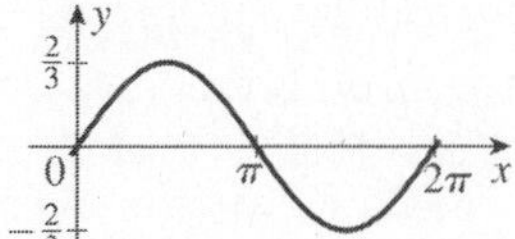

5. $y=2\sin 3x$ has amplitude 2 and period $2\pi/3$, with key values at multiples of $\frac{1}{4}(\frac{2\pi}{3})=\frac{\pi}{6}$. The table of key values between 0 and $2\pi/3$ is:

x	0	$\pi/6$	$\pi/3$	$\pi/2$	$2\pi/3$
y	0	2	0	-2	0

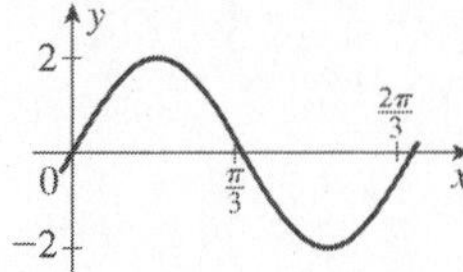

9. $y=3\cos\frac{1}{3}x$ has amplitude 3 and period $2\pi/(1/3)=6\pi$, with key values at multiples of $\frac{1}{4}(6\pi)=\frac{3\pi}{2}$. The table of key values between 0 and 6π is:

x	0	$3\pi/2$	3π	$9\pi/2$	6π
y	3	0	-3	0	3

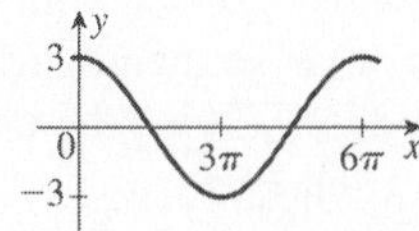

13. $y=5\cos\left(\frac{\pi x}{2}\right)$ has amplitude 5 and period $2\pi/(\pi/2)=4$, with key values at multiples of $\frac{1}{4}(4)=1$. The table of key values between 0 and 4 is:

x	0	1	2	3	4
y	5	0	-5	0	5

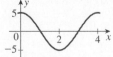

17. $y=2\sin\left(3x-\frac{\pi}{2}\right)$ has amplitude 2, period $2\pi/3$, and displacement $-(-\pi/2)/3=\pi/6$. One-fourth period is $\pi/6$, so key values for one full cycle start at $\pi/6$, end at $5\pi/6$, and are found $\pi/6$ units apart. The table of key values is:

x	$\pi/6$	$\pi/12$	$\pi/4$	$\pi/3$	$5\pi/6$
y	0	2	0	-2	0

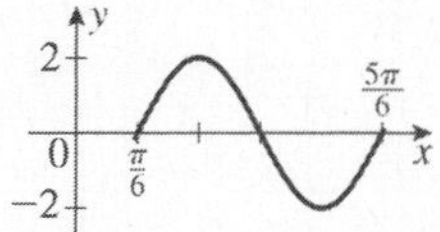

21. $y=-\sin\left(\pi x+\frac{\pi}{6}\right)$ has amplitude 1(inverted), period $2\pi/\pi=2$, and displacement $-(\frac{\pi}{6})/\pi=-\frac{1}{6}$. One-fourth period is 1/2, so key values for one full cycle start at –1/6, end at 11/6, and are found 1/2 units apart. The table of key values is:

x	-1/6	1/3	5/6	4/3	11/6
y	0	-1	0	1	0

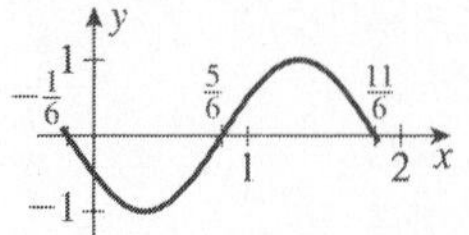

25. $y=0.3\tan 0.5x$ has period 2π instead of period π. Key values occur every half period, with asymptotes at $-\pi$ and π, and a zero at $x=0$.

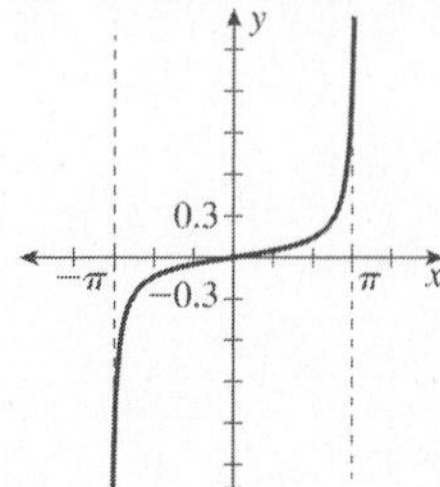

29. $y=2+\frac{1}{2}\sin 2x$ is the result of shifting the function $y=\frac{1}{2}\sin 2x$ vertically 2 units. This function has amplitude ½, period $2\pi/2=\pi$, and no displacement. The table of key values is

x	0	$\pi/4$	$\pi/2$	$3\pi/4$	π
$\frac{1}{2}\sin 2x$	0	1/2	0	−1/2	0
$\frac{1}{2}\sin 2x+2$	2	5/2	2	3/2	0

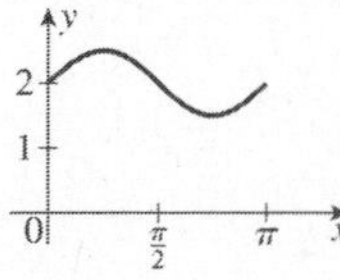

33. Graph $y_1 = 2\sin x - \cos 2x$.

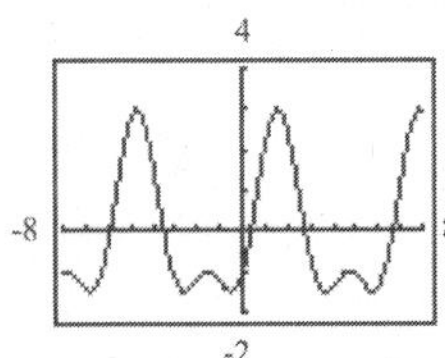

37. Graph $y_1 = \dfrac{\sin x}{x}$.

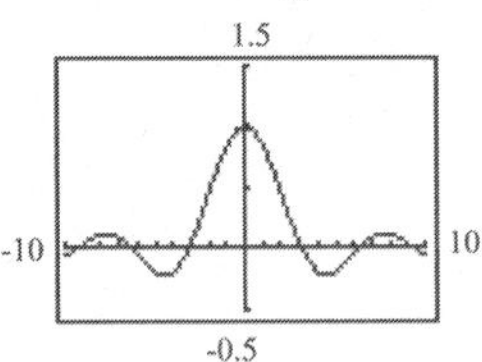

41. amplitude: $a = 2$;

period: $\dfrac{2\pi}{b} = \pi$

$b = 2$;

displacement: $-\dfrac{c}{b} = -\dfrac{\pi}{4}$

$-\dfrac{c}{2} = -\dfrac{\pi}{4}$

$c = \dfrac{\pi}{2}$

Solution: $y = 2\sin\left(2x + \dfrac{\pi}{2}\right)$

45. In parametric mode graph

$x_1 = -\cos 2\pi t,\ y_1 = 2\sin \pi t$

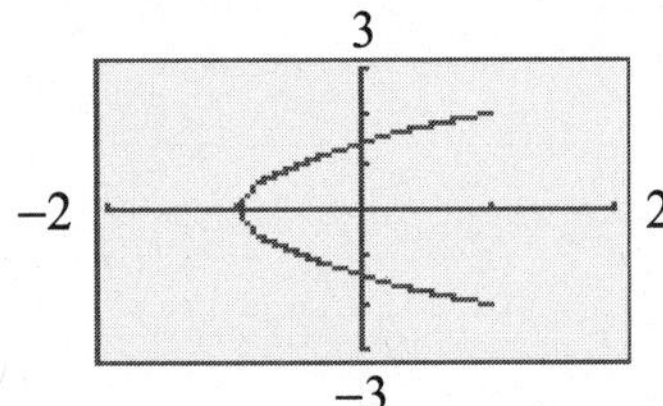

49. Graph $y_1 = 2|2\sin 0.2\pi x| - |\cos 0.4\pi x|$

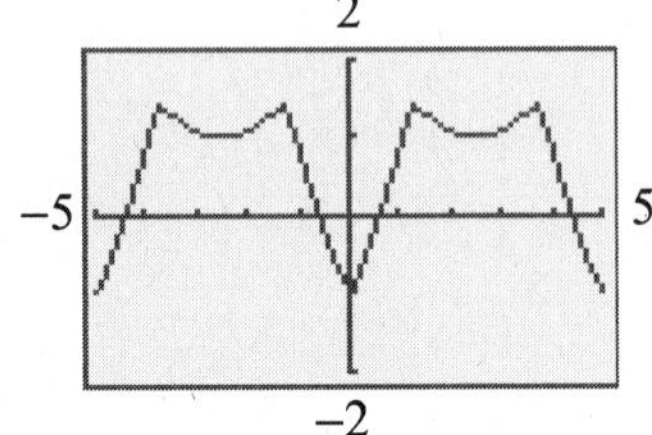

53. The period of $\cos 0.5x$ is $\dfrac{2\pi}{0.5} = 4\pi$.

The period of $\sin 3x$ is $\dfrac{2\pi}{3}$.

The period of $y = 2\cos 0.5x + \sin 3x$ is the least common multiple of 4π and $\dfrac{2\pi}{3}$, which is 4π.

57. Substitute $x = \pi/3$ and $y = -3$ into $y = 3\cos bx$ to get

$$3\cos\frac{\pi}{3}b = -3$$

$$\frac{\pi}{3}b = \pi + 2\pi n$$

from where the smallest positive b is $b = 3$.

Solution: $y = 3\cos 3x$

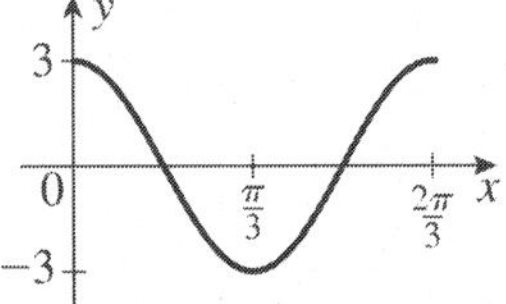

61. Note that θ varies between 0 and π/2.

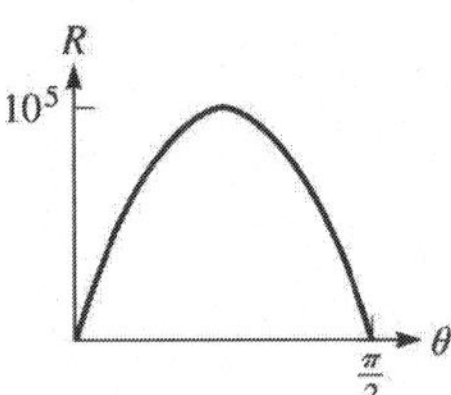

65. The period of $y = \sin 120\pi t$ is $2\pi/120\pi = 1/60$ s. With the absolute value, every negative part of the function is reflected with respect to the x axis to become positive, so the period is reduced by half. The period is 1/120 s. There are six complete cycles between 0 and 0.05 s.

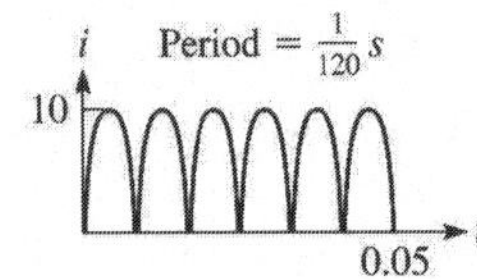

69. Graph $y = 3.0\cos 0.2x + 1.0\sin 0.4x$

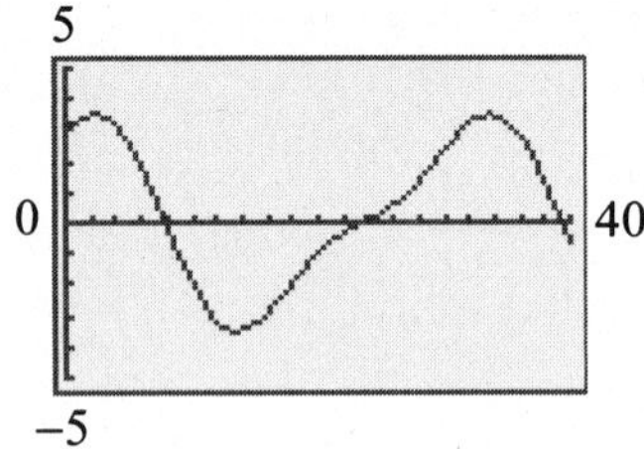

73. We graph the function between 0 and 365.

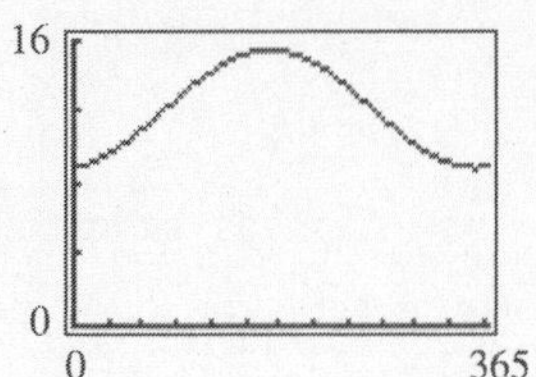

77.

θ	0	$\pi/2$	π	$3\pi/2$	2π
x	0	$\pi/2-1$	π	$3\pi/2+1$	2π
y	0	1	2	1	0

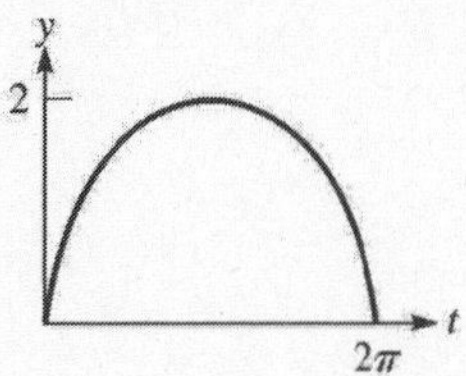

81.

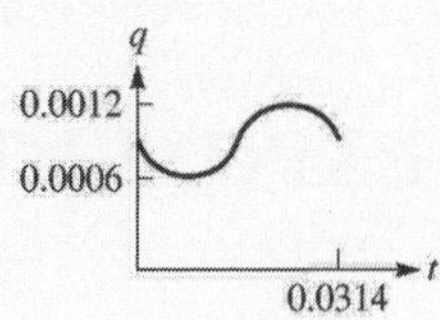

CHAPTER 11

EXPONENTS AND RADICALS

11.1 Simplifying Expressions with Integral Exponents

1. $$\left(x^{-2}y\right)^2\left(\frac{2}{x}\right)^{-2} = \frac{x^{-4}y^2}{\left(\frac{2}{x}\right)^2} = \frac{x^{-4}y^2}{\frac{4}{x^2}} = \frac{x^{-4}y^2}{1}\cdot\frac{x^2}{4} = \frac{x^{-2}y^2}{4} = \frac{y^2}{4x^2}$$

5. $x^7 \cdot x^{-4} = x^{7+(-4)} = x^3$

9. $$5^0 \times 5^{-3} = 5^{0+(-3)} = 5^{-3} = \frac{1}{5^3} = \frac{1}{125}$$

13. $2\left(5an^{-2}\right)^{-1} = 2\times 5^{-1}a^{-1}n^{(-2)(-1)} = \dfrac{2n^2}{5a}$

17. $-7x^0 = -7(1) = -7$

21. $$\left(7a^{-1}x\right)^{-3} = 7^{-3}a^3x^{-3} = \frac{a^3}{7^3x^3} = \frac{a^3}{343x^3}$$

25. $3\left(\dfrac{a}{b^{-2}}\right)^{-3} = \dfrac{3a^{-3}}{b^6} = \dfrac{3}{a^3b^6}$

29. $3x^{-2} + 2y^{-2} = \dfrac{3}{x^2} + \dfrac{2}{y^2} = \dfrac{3y^2 + 2x^2}{x^2y^2}$

33. $$\left(\frac{3a^2}{4b}\right)^{-3}\left(\frac{4}{a}\right)^{-5} = \frac{3^{-3}a^{-6}}{4^{-3}b^{-3}}\times\frac{4^{-5}}{a^{-5}} = \frac{4^3b^3}{3^3a^6}\times\frac{a^5}{4^5} = \frac{b^3}{27a(4^2)} = \frac{b^3}{432a} \quad \text{where } a \neq 0$$

37. $$2a^{-2} + \left(2a^{-2}\right)^4 = 2a^{-2} + 2^4a^{-8} = \frac{2}{a^2} + \frac{16}{a^8} = \frac{2a^6 + 16}{a^8} = \frac{2(a^6 + 8)}{a^8}$$

41. $$\left(R_1^{-1} + R_2^{-1}\right)^{-1} = \frac{1}{\frac{1}{R_1} + \frac{1}{R_2}} = \frac{1}{\frac{R_2 + R_1}{R_1R_2}} = \frac{R_1R_2}{R_1 + R_2}$$

45. $$\frac{6^{-1}}{4^{-2} + 2} = \frac{\frac{1}{6}}{\frac{1}{4^2} + 2} = \frac{\frac{1}{6}}{\frac{1}{16} + 2} = \frac{\frac{1}{6}}{\frac{1+32}{16}} = \frac{1}{6}\times\frac{16}{33} = \frac{8}{99}$$

49. $2t^{-2}+t^{-1}(t+1)=\frac{2}{t^2}+t^0+t^{-1}$

$=\frac{2}{t^2}+1+\frac{1}{t}$

$=\frac{2+t^2+t}{t^2}$

$=\frac{t^2+t+2}{t^2}$

53. If $x<0$, then $x^2>0$ and $x^{-2}>0$

If $x<0$, then $x^1<0$ and $x^{-1}<0$.

$x^{-2}>0>x^{-1}$ so $x^{-2}>x^{-1}$.

No, there are no values for x where $x^{-2}<x^{-1}$

57. (a) $\left(\frac{a}{b}\right)^{-n}=\frac{1}{\left(\frac{a}{b}\right)^n}=\frac{1}{\frac{a^n}{b^n}}=1\times\frac{b^n}{a^n}=\left(\frac{b}{a}\right)^n$

(b) $\left(\frac{a}{b}\right)^{-n}=\left(\frac{b}{a}\right)^n$

$\left(\frac{3.576}{8.091}\right)^{-7}=(0.4419726)^{-7}$

$=303.55182$

$\left(\frac{8.091}{3.576}\right)^{7}=(2.262584)^{7}$

$=303.55182$

Therefore,

$\left(\frac{3.576}{8.091}\right)^{-7}=\left(\frac{8.091}{3.576}\right)^{7}$

61. $2^{5x}=2^7\left(2^{2x}\right)^2$

$2^{5x}=2^7\left(2^{4x}\right)$

$2^{5x}=2^{7+4x}$

$5x=7+4x$

$x=7$

65. $1\ \text{J}=1\ \text{kg}\cdot\text{m}^2\cdot\text{s}^{-2}$, so

$\text{kg}\cdot\text{s}^{-1}\left(\text{m}\cdot\text{s}^{-2}\right)^2=\text{kg}\cdot\text{s}^{-1}\cdot\text{m}^2\cdot\text{s}^{-4}$

$=\text{kg}\cdot\text{m}^2\cdot\text{s}^{-2}\left(\text{s}^{-3}\right)$

$=\text{J/s}^3$

69. $\frac{p(1+i)^{-1}\left[(1+i)^{-n}-1\right]}{(1+i)^{-1}-1}=\frac{p(1+i)^{-1-n}-p(1+i)^{-1}}{(1+i)^{-1}-1}$

$=\frac{p\left[(1+i)^{-(n+1)}-(1+i)^{-1}\right]}{\left(\frac{1}{1+i}-1\right)}$

$=\frac{p\left(\frac{1}{(1+i)^{n+1}}-\frac{1}{(1+i)^1}\right)}{\frac{1-1(1+i)}{1+i}}$

$=\frac{p\left(\frac{1-1(1+i)^n}{(1+i)^{n+1}}\right)}{\frac{-i}{1+i}}$

$=\frac{p\left[1-(1+i)^n\right]}{(1+i)^{n+1}}\cdot\frac{i+1}{-i}$

$=\frac{p\left[1-(1+i)^n\right]}{-i(1+i)^n}$

$=\frac{p\left[(1+i)^n-1\right]}{i(1+i)^n}$

Restriction : $i\neq-1$ (which shouldn't be an issue, since interest rate $i>0$)

11.2 Fractional Exponents

1. $8^{4/3}=\left(8^{1/3}\right)^4=\left(\sqrt[3]{8}\right)^4=2^4=16$

5. $25^{1/2}=\sqrt{25}=5$

9. $100^{25/2}=\left(100^{1/2}\right)^{25}=\left(\sqrt{100}\right)^{25}=10^{25}$

13. $64^{-2/3}=\frac{1}{64^{2/3}}$

$=\frac{1}{\left(\sqrt[3]{64}\right)^2}$

$=\frac{1}{(4)^2}$

$=\frac{1}{16}$

17. $\left(3^6\right)^{2/3}=(3)^{12/3}=3^4=81$

21. $\dfrac{15^{2/3}}{5^2 \cdot 15^{-1/3}} = \dfrac{15^{2/3+1/3}}{5^2} = \dfrac{15^1}{25} = \dfrac{3}{5}$

25. $125^{-2/3} - 100^{-3/2} = \dfrac{1}{125^{2/3}} - \dfrac{1}{100^{3/2}}$

$$= \frac{1}{\left(\sqrt[3]{125}\right)^2} - \frac{1}{\left(\sqrt{100}\right)^3}$$
$$= \frac{1}{5^2} - \frac{1}{10^3}$$
$$= \frac{1}{25} - \frac{1}{1000}$$
$$= \frac{40-1}{1000}$$
$$= \frac{39}{1000}$$

29. $17.98^{1/4} = 2.059$

33. $B^{2/3} \cdot B^{1/2} = B^{2/3+1/2} = B^{4/6+3/6} = B^{7/6}$

37. $\dfrac{x^{3/10}}{x^{-1/5}x^2} = x^{3/10-(-1/5)-2}$

$$= x^{3/10+2/10-20/10}$$
$$= x^{-15/10}$$
$$= x^{-3/2}$$
$$= \frac{1}{x^{3/2}}$$

41. $\left(16a^4b^3\right)^{-3/4} = 16^{-3/4}a^{4(-3/4)}b^{3(-3/4)}$

$$= \left(\sqrt[4]{16}\right)^{-3} a^{-3}b^{-9/4}$$
$$= \frac{1}{2^3a^3b^{9/4}}$$
$$= \frac{1}{8a^3b^{9/4}}$$

45. $\dfrac{1}{2}\left(4x^2+1\right)^{-1/2}(8x) = \dfrac{4x}{\left(4x^2+1\right)^{1/2}}$

49. $\left(T^{-1}+2T^{-2}\right)^{-1/2} = \left(\dfrac{1}{T}+\dfrac{2}{T^2}\right)^{-1/2}$

$$= \left(\frac{T+2}{T^2}\right)^{-1/2}$$
$$= \frac{(T+2)^{-1/2}}{T^{-1}}$$
$$= \frac{T}{(T+2)^{1/2}}$$

53. $\left[\left(a^{1/2}-a^{-1/2}\right)^2+4\right]^{1/2} = \left[\left(a^{1/2}-\dfrac{1}{a^{1/2}}\right)^2+4\right]^{1/2}$

$$= \left[\left(\frac{a-1}{a^{1/2}}\right)^2+4\right]^{1/2}$$
$$= \left[\frac{(a-1)^2}{a}+4\right]^{1/2}$$
$$= \left[\frac{a^2-2a+1+4a}{a}\right]^{1/2}$$
$$= \left[\frac{a^2+2a+1}{a}\right]^{1/2}$$
$$= \left[\frac{(a+1)^2}{a}\right]^{1/2}$$
$$= \frac{a+1}{a^{1/2}}$$

57. $y = f(x) = 3x^{1/2}$

x	0	1	2	4	9
y	0	3	4.24	6	9

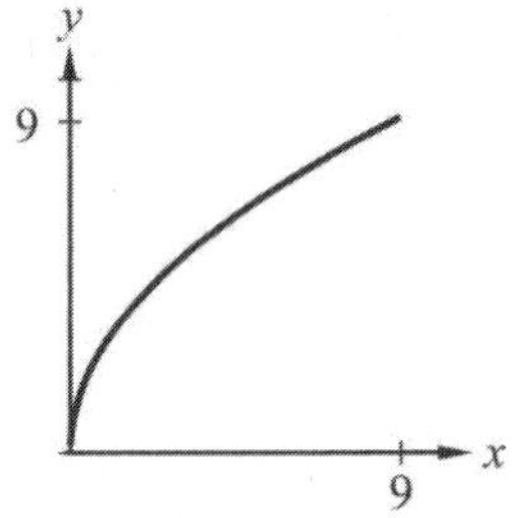

61. $\left(x^{n-1} \div x^{n-3}\right)^{1/3} = \left(x^{n-1-(n-3)}\right)^{1/3}$

$$= \left(x^{n-1-n+3}\right)^{1/3}$$
$$= \sqrt[3]{x^2}$$

65.

$$T^2 = kR^3\left(1+\frac{d}{R}\right)^3$$
$$T^2 = kR^3\left(\frac{R+d}{R}\right)^3$$
$$T^2 = kR^3\frac{(R+d)^3}{R^3}$$
$$T^2 = k(R+d)^3$$
$$(R+d)^3 = \frac{T^2}{k}$$
$$R+d = \left(\frac{T^2}{k}\right)^{1/3}$$
$$R = \frac{T^{2/3}}{k^{1/3}} - d$$

11.3 Simplest Radical Form

1. $\sqrt{a^3b^4} = \sqrt{a^2(b^2)^2 \cdot a} = ab^2\sqrt{a}$

5. $\sqrt{24} = \sqrt{4 \cdot 6} = \sqrt{4} \cdot \sqrt{6} = 2\sqrt{6}$

9. $\sqrt{x^2y^5} = \sqrt{x^2y^4y} = \sqrt{x^2}\sqrt{y^4}\sqrt{y} = xy^2\sqrt{y}$

13. $\sqrt{18R^5TV^4} = \sqrt{9R^4V^4(2RT)}$
$= \sqrt{9R^4V^4} \cdot \sqrt{2RT}$
$= 3R^2V^2\sqrt{2RT}$

17. $\sqrt[5]{96} = \sqrt[5]{32 \cdot 3} = \sqrt[5]{2^5} \cdot \sqrt[5]{3} = 2\sqrt[5]{3}$

21. $\sqrt[4]{64r^3s^4t^5} = \sqrt[4]{16s^4t^4(4r^3t)}$
$= \sqrt[4]{2^4} \cdot \sqrt[4]{s^4} \cdot \sqrt[4]{t^4} \cdot \sqrt[4]{4r^3t}$
$= 2st\sqrt[4]{4r^3t}$

25. $\sqrt[3]{P}\sqrt[3]{P^2V} = \sqrt[3]{P^3V} = P\sqrt[3]{V}$

29. $\sqrt[3]{\frac{3}{4}} = \sqrt[3]{\frac{3}{2^2}} \times \frac{\sqrt[3]{2}}{\sqrt[3]{2}}$
$= \frac{\sqrt[3]{3(2)}}{\sqrt[3]{2^3}}$
$= \frac{\sqrt[3]{6}}{2}$

33. $\sqrt[4]{400} = \sqrt[4]{2^4 \cdot 5^2}$
$= 2^{4/4}5^{2/4}$
$= 2 \cdot 5^{1/2}$
$= 2\sqrt{5}$

37. $\sqrt{4 \times 10^4} = \sqrt{4} \cdot \sqrt{10^4} = 2 \times 10^2 = 200$

41. $\sqrt[4]{4a^2} = (2^2a^2)^{1/4} = 2^{1/2}a^{1/2} = \sqrt{2a}$

45. $\sqrt[4]{\sqrt[3]{16}} = \sqrt[12]{2^4}$
$= 2^{4/12}$
$= 2^{1/3}$
$= \sqrt[3]{2}$

49. $\sqrt{28u^3v^{-5}} = (2^2 \cdot 7u^2uv^{-6}v)^{1/2}$
$= 2^{2/2}7^{1/2}u^{2/2}u^{1/2}v^{-6/2}v^{1/2}$
$= \frac{2u\sqrt{7uv}}{v^3}$

53. $\sqrt{\frac{2x}{3c^4}} = \sqrt{\frac{2x}{3c^4}} \times \frac{\sqrt{3}}{\sqrt{3}}$
$= \frac{\sqrt{6x}}{3^{2/2}c^{4/2}}$
$= \frac{\sqrt{6x}}{3c^2}$

57. $\sqrt{xy^{-1} + x^{-1}y} = \sqrt{\frac{x}{y} + \frac{y}{x}}$
$= \sqrt{\frac{x^2 + y^2}{xy}}$
$= \frac{\sqrt{x^2 + y^2}}{\sqrt{xy}} \times \frac{\sqrt{xy}}{\sqrt{xy}}$
$= \frac{\sqrt{xy(x^2 + y^2)}}{xy}$

61. $\sqrt{a^2 + b^2}$ cannot be simplified any further.

65. $\sqrt{a} = a^{1/2} = a^{3/6} = \sqrt[6]{a^3}$
$\sqrt[3]{b} = b^{1/3} = b^{2/6} = \sqrt[6]{b^2}$
$\sqrt[6]{c} = c^{1/6} = \sqrt[6]{c}$

69. $a\sqrt{\frac{2g}{a}} = a\sqrt{\frac{2g}{a}} \times \frac{\sqrt{a}}{\sqrt{a}}$
$= a\sqrt{\frac{2ag}{a^2}}$
$= \frac{a}{a}\sqrt{2ag}$
$= \sqrt{2ag}$

11.4 Addition and Subtraction of Radicals

1. $3\sqrt{125} - \sqrt{20} + \sqrt{45} = 3\sqrt{25(5)} - \sqrt{4(5)} + \sqrt{9(5)}$
$= 3(5)\sqrt{5} - 2\sqrt{5} + 3\sqrt{5}$
$= 15\sqrt{5} - 2\sqrt{5} + 3\sqrt{5}$
$= 16\sqrt{5}$

5. $\sqrt{28}+\sqrt{5}-3\sqrt{7}=\sqrt{(4)7}+\sqrt{5}-3\sqrt{7}$
$=2\sqrt{7}+\sqrt{5}-3\sqrt{7}$
$=\sqrt{5}-\sqrt{7}$

9. $2\sqrt{3t^2}-3\sqrt{12t^2}=2t\sqrt{3}-3t\sqrt{(4)3}$
$=2t\sqrt{3}-3(2)t\sqrt{3}$
$=2t\sqrt{3}-6t\sqrt{3}$
$=-4t\sqrt{3}$

13. $2\sqrt{28}+3\sqrt{175}=2\sqrt{4(7)}+3\sqrt{25(7)}$
$=2(2)\sqrt{7}+3(5)\sqrt{7}$
$=4\sqrt{7}+15\sqrt{7}$
$=19\sqrt{7}$

17. $3\sqrt{75R}+2\sqrt{48R}-2\sqrt{18R}$
$=3\sqrt{25(3R)}+2\sqrt{16(3R)}-2\sqrt{9(2R)}$
$=3(5)\sqrt{3R}+2(4)\sqrt{3R}-2(3)\sqrt{2R}$
$=15\sqrt{3R}+8\sqrt{3R}-6\sqrt{2R}$
$=23\sqrt{3R}-6\sqrt{2R}$
$=\sqrt{R}\left(23\sqrt{3}-6\sqrt{2}\right)$

21. $\sqrt{\frac{1}{2}}+\sqrt{\frac{25}{2}}-4\sqrt{18}=\frac{1}{\sqrt{2}}\times\frac{\sqrt{2}}{\sqrt{2}}+\frac{5}{\sqrt{2}}\times\frac{\sqrt{2}}{\sqrt{2}}-4\sqrt{(9)2}$
$=\frac{\sqrt{2}}{2}+\frac{5\sqrt{2}}{2}-12\sqrt{2}$
$=\frac{\sqrt{2}+5\sqrt{2}-24\sqrt{2}}{2}$
$=\frac{-18\sqrt{2}}{2}$
$=-9\sqrt{2}$

25. $\sqrt[4]{32}-\sqrt[8]{4}=\sqrt[4]{2^5}-\sqrt[8]{2^2}$
$=2^{5/4}-2^{2/8}$
$=2^{4/4}2^{1/4}-2^{1/4}$
$=\sqrt[4]{2}(2-1)$
$=\sqrt[4]{2}$

29. $\sqrt{6}\sqrt{5}\sqrt{3}-\sqrt{40a^2}=\sqrt{90}-\sqrt{4a^2(10)}$
$=\sqrt{(9)10}-2a\sqrt{10}$
$=3\sqrt{10}-2a\sqrt{10}$
$=(3-2a)\sqrt{10}$

33. $\sqrt{\frac{a}{c^5}}-\sqrt{\frac{c}{a^3}}=\frac{\sqrt{a}}{\sqrt{c^4c}}-\frac{\sqrt{c}}{\sqrt{a^2a}}$
$=\frac{\sqrt{a}}{c^2\sqrt{c}}\times\frac{\sqrt{c}}{\sqrt{c}}-\frac{\sqrt{c}}{a\sqrt{a}}\times\frac{\sqrt{a}}{\sqrt{a}}$
$=\frac{\sqrt{ac}}{c^3}-\frac{\sqrt{ac}}{a^2}$
$=\frac{a^2\sqrt{ac}-c^3\sqrt{ac}}{a^2c^3}$
$=\frac{\left(a^2-c^3\right)\sqrt{ac}}{a^2c^3}$

37. $\sqrt{\frac{T-V}{T+V}}-\sqrt{\frac{T+V}{T-V}}=\frac{\sqrt{T-V}}{\sqrt{T+V}}\times\frac{\sqrt{T+V}}{\sqrt{T+V}}-\frac{\sqrt{T+V}}{\sqrt{T-V}}\times\frac{\sqrt{T-V}}{\sqrt{T-V}}$
$=\frac{\sqrt{T^2-V^2}}{T+V}-\frac{\sqrt{T^2-V^2}}{T-V}$
$=\frac{(T-V)\sqrt{T^2-V^2}-(T+V)\sqrt{T^2-V^2}}{(T+V)(T-V)}$
$=\frac{(T-V-T-V)\sqrt{T^2-V^2}}{T^2-V^2}$
$=\frac{-2V\sqrt{T^2-V^2}}{T^2-V^2}$
$=\frac{2V\sqrt{T^2-V^2}}{V^2-T^2}$

41. $2\sqrt{\frac{2}{3}}+\sqrt{24}-5\sqrt{\frac{3}{2}}=2\frac{\sqrt{2}}{\sqrt{3}}\times\frac{\sqrt{3}}{\sqrt{3}}+\sqrt{4(6)}-5\frac{\sqrt{3}}{\sqrt{2}}\times\frac{\sqrt{2}}{\sqrt{2}}$
$=\frac{2\sqrt{6}}{3}+2\sqrt{6}-\frac{5\sqrt{6}}{2}$
$=\left(\frac{2}{3}+2-\frac{5}{2}\right)\sqrt{6}$
$\left(\frac{4+12-15}{6}\right)\sqrt{6}$
$=\frac{\sqrt{6}}{6}$
$=0.40\,824\,829\ldots$ on a calculator

$2\sqrt{\frac{2}{3}}+\sqrt{24}-5\sqrt{\frac{3}{2}}=0.40\,824\,829\ldots$ on a calculator

45. $x^2-2x-2=0$ has roots
$x=\frac{-b\pm\sqrt{b^2-4ac}}{2a}$
$x=\frac{-(-2)\pm\sqrt{(-2)^2-4(1)(-2)}}{2(1)}$

$$x=\frac{2\pm\sqrt{12}}{2}$$
$$x=\frac{2\pm\sqrt{(4)3}}{2}$$
$$x=\frac{2\pm2\sqrt{3}}{2}$$
$$x=1\pm\sqrt{3}$$

so $x=1+\sqrt{3}$ is the positive root

$x^2+2x-11=0$ has roots

$$x=\frac{-b\pm\sqrt{b^2-4ac}}{2a}$$
$$x=\frac{-2\pm\sqrt{2^2-4(1)(-11)}}{2(1)}$$
$$x=\frac{-2\pm\sqrt{48}}{2}$$
$$x=\frac{-2\pm\sqrt{(16)3}}{2}$$
$$x=\frac{-2\pm4\sqrt{3}}{2}$$
$$x=-1\pm2\sqrt{3}$$

so $x=-1+2\sqrt{3}$ is the positive root

sum of positive roots $=1+\sqrt{3}+(-1+2\sqrt{3})$

sum of positive roots $=3\sqrt{3}$

49.

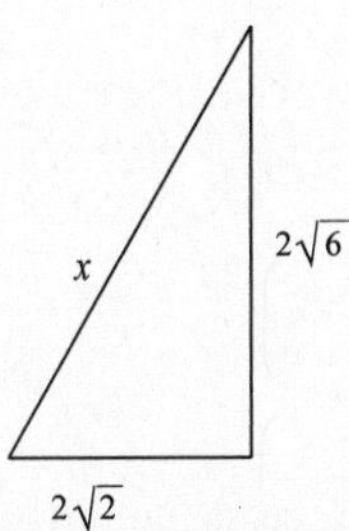

Using Pythagoras' theorem,

$$x^2=\left(2\sqrt{2}\right)^2+\left(2\sqrt{6}\right)^2$$
$$x^2=8+24$$
$$x^2=32$$
$$x=\sqrt{32}$$

$$\begin{aligned}\text{Perimeter}&=\sqrt{32}+2\sqrt{2}+2\sqrt{6}\\&=\sqrt{16(2)}+2\sqrt{2}+2\sqrt{6}\\&=4\sqrt{2}+2\sqrt{2}+2\sqrt{6}\\&=6\sqrt{2}+2\sqrt{6}\text{ units}\end{aligned}$$

11.5 Multiplication and Division of Radicals

1. $$\begin{aligned}\sqrt{2}\left(3\sqrt{5}-4\sqrt{8}\right)&=3\sqrt{10}-4\sqrt{16}\\&=3\sqrt{10}-4(4)\\&=3\sqrt{10}-16\end{aligned}$$

5. $\sqrt{3}\sqrt{10}=\sqrt{3(10)}=\sqrt{30}$

9. $\sqrt[3]{4}\cdot\sqrt[3]{2}=\sqrt[3]{4(2)}=\sqrt[3]{8}=2$

13. $\sqrt{8}\cdot\sqrt{\frac{5}{2}}=\sqrt{\frac{8(5)}{2}}=\sqrt{20}=\sqrt{4(5)}=2\sqrt{5}$

17. $\left(2-\sqrt{5}\right)\left(2+\sqrt{5}\right)=2^2-\sqrt{5}^2=4-5=-1$

21. $$\begin{aligned}&\left(3\sqrt{11}-\sqrt{x}\right)\left(2\sqrt{11}+5\sqrt{x}\right)\\&=6(11)+15\sqrt{11x}-2\sqrt{11x}-5(x)\\&=66+13\sqrt{11x}-5x\end{aligned}$$

25. $$\begin{aligned}\frac{\sqrt{6}-3}{\sqrt{6}}&=\frac{\sqrt{6}-3}{\sqrt{6}}\cdot\frac{\sqrt{6}}{\sqrt{6}}\\&=\frac{6-3\sqrt{6}}{6}\\&=\frac{2-\sqrt{6}}{2}\end{aligned}$$

29. $$\begin{aligned}\sqrt{2}\sqrt[3]{3}&=2^{1/2}3^{1/3}\\&=2^{3/6}3^{2/6}\\&=\left(2^33^2\right)^{1/6}\\&=\sqrt[6]{2^33^2}\\&=\sqrt[6]{72}\end{aligned}$$

33. $$\begin{aligned}\frac{\sqrt{2}-1}{\sqrt{7}-3\sqrt{2}}&=\frac{\sqrt{2}-1}{\sqrt{7}-3\sqrt{2}}\times\frac{\sqrt{7}+3\sqrt{2}}{\sqrt{7}+3\sqrt{2}}\\&=\frac{\sqrt{14}+3(2)-\sqrt{7}-3\sqrt{2}}{7-9(2)}\\&=\frac{\sqrt{14}+6-\sqrt{7}-3\sqrt{2}}{7-18}\\&=-\frac{\sqrt{14}+6-\sqrt{7}-3\sqrt{2}}{11}\end{aligned}$$

37. $\dfrac{2\sqrt{x}}{\sqrt{x}-\sqrt{5}}=\dfrac{2\sqrt{x}}{\sqrt{x}-\sqrt{5}}\times\dfrac{\sqrt{x}+\sqrt{5}}{\sqrt{x}+\sqrt{5}}$

$=\dfrac{2x+2\sqrt{5x}}{x-5}$

$=\dfrac{2\left(x+\sqrt{5x}\right)}{x-5}$

41. $\left(\sqrt{\dfrac{2}{R}}+\sqrt{\dfrac{R}{2}}\right)\left(\sqrt{\dfrac{2}{R}}-2\sqrt{\dfrac{R}{2}}\right)=\dfrac{2}{R}-2\sqrt{\dfrac{2R}{2R}}+\sqrt{\dfrac{2R}{2R}}-2\left(\dfrac{R}{2}\right)$

$=\dfrac{2}{R}-2+1-R$

$=\dfrac{2}{R}-1-R$

$=\dfrac{2-R-R^2}{R}$

$=-\dfrac{R^2+R-2}{R}$

$=-\dfrac{(R+2)(R-1)}{R}$

45. $\dfrac{\sqrt{a}+\sqrt{a-2}}{\sqrt{a}-\sqrt{a-2}}=\dfrac{\sqrt{a}+\sqrt{a-2}}{\sqrt{a}-\sqrt{a-2}}\times\dfrac{\sqrt{a}+\sqrt{a-2}}{\sqrt{a}+\sqrt{a-2}}$

$=\dfrac{a+\sqrt{a}\sqrt{a-2}+\sqrt{a}\sqrt{a-2}+(a-2)}{a-(a-2)}$

$=\dfrac{2a+2\sqrt{a}\sqrt{a-2}-2}{2}$

$=a+\sqrt{a}\sqrt{a-2}-1$

$=a-1+\sqrt{a(a-2)}$

49. $\dfrac{2\sqrt{6}-\sqrt{5}}{3\sqrt{6}-4\sqrt{5}}=\dfrac{2\sqrt{6}-\sqrt{5}}{3\sqrt{6}-4\sqrt{5}}\times\dfrac{3\sqrt{6}+4\sqrt{5}}{3\sqrt{6}+4\sqrt{5}}$

$=\dfrac{6(6)+8\sqrt{30}-3\sqrt{30}-4(5)}{9(6)-16(5)}$

$=\dfrac{16+5\sqrt{30}}{-26}$

$=-\dfrac{16+5\sqrt{30}}{26}$

$=-1.6\,686\,972\ldots$ on a calculator

$\dfrac{2\sqrt{6}-\sqrt{5}}{3\sqrt{6}-4\sqrt{5}}=-1.6\,686\,972\ldots$ on a calculator

53. $\dfrac{x^2}{\sqrt{2x+1}}+2x\sqrt{2x+1}=\dfrac{x^2+2x\sqrt{2x+1}\sqrt{2x+1}}{\sqrt{2x+1}}$

$=\dfrac{x^2+2x(2x+1)}{\sqrt{2x+1}}$

$=\dfrac{x^2+4x^2+2x}{\sqrt{2x+1}}$

$=\dfrac{5x^2+2x}{\sqrt{2x+1}}$

$=\dfrac{x(5x+2)}{\sqrt{2x+1}}$

57. $\dfrac{\sqrt{x+h}-\sqrt{x}}{h}=\dfrac{\sqrt{x+h}-\sqrt{x}}{h}\times\dfrac{\sqrt{x+h}+\sqrt{x}}{\sqrt{x+h}+\sqrt{x}}$

$=\dfrac{(x+h)-(x)}{h\sqrt{x+h}+h\sqrt{x}}$

$=\dfrac{h}{h\left(\sqrt{x+h}+\sqrt{x}\right)}$

$=\dfrac{1}{\sqrt{x+h}+\sqrt{x}}$

61. The quadratic equation

$ax^2+bx+c=0$

has two solutions

$$x=\frac{-b\pm\sqrt{b^2-4ac}}{2a}$$

If the two solutions are reciprocals of each other, then

$$\frac{-b+\sqrt{b^2-4ac}}{2a}=\frac{1}{\dfrac{-b-\sqrt{b^2-4ac}}{2a}}$$

$$\frac{-b+\sqrt{b^2-4ac}}{2a}=\frac{2a}{-b-\sqrt{b^2-4ac}}\quad\text{cross multiply}$$

$$4a^2=\left(-b+\sqrt{b^2-4ac}\right)\left(-b-\sqrt{b^2-4ac}\right)$$

$$4a^2=b^2+b\sqrt{b^2-4ac}-b\sqrt{b^2-4ac}-\left(b^2-4ac\right)$$

$$4a^2=b^2-b^2+4ac$$

$$4a^2=4ac$$

$$\frac{4a^2}{4a}=c$$

$$a=c$$

65. $m^2+bm+k^2=0$

If we substitute $m=\frac{1}{2}\left(\sqrt{b^2-4k^2}-b\right)$ into the equation

$$\left(\frac{1}{2}\left(\sqrt{b^2-4k^2}-b\right)\right)^2+b\cdot\frac{1}{2}\left(\sqrt{b^2-4k^2}-b\right)+k^2=0$$
$$\frac{1}{4}\left[\left(b^2-4k^2\right)-2b\sqrt{b^2-4k^2}+b^2\right]+\frac{b}{2}\sqrt{b^2-4k^2}-\frac{1}{2}b^2+k^2=0$$
$$\frac{1}{4}\left(b^2-4k^2\right)-\frac{b}{2}\sqrt{b^2-4k^2}+\frac{b^2}{4}+\frac{b}{2}\sqrt{b^2-4k^2}-\frac{1}{2}b^2+k^2=0$$
$$\frac{1}{4}b^2-k^2+\frac{1}{4}b^2-\frac{1}{2}b^2+k^2=0$$
$$\frac{1}{2}b^2-\frac{1}{2}b^2=0$$
$$0=0$$

Therefore, it is a solution.

69. $$\frac{2Q}{\sqrt{\sqrt{2}-1}}=\frac{2Q}{\sqrt{\sqrt{2}-1}}\times\frac{\sqrt{\sqrt{2}+1}}{\sqrt{\sqrt{2}+1}}=\frac{2Q\sqrt{\sqrt{2}+1}}{\sqrt{2-1}}=2Q\sqrt{\sqrt{2}+1}$$

Review Exercises

1. $2a^{-2}b^0=2a^{-2}\cdot 1=\frac{2}{a^2}$

5. $$3(25)^{3/2}=3\left[(25)^{1/2}\right]^3=3\left(\sqrt{25}\right)^3=3(5)^3=3\times 125=375$$

9. $$\left(\frac{3}{t^2}\right)^{-2}=\frac{3^{-2}}{t^{-4}}=\frac{t^4}{9}$$

13. $$\left(2a^{1/3}b^{5/6}\right)^6=2^6\cdot\left(a^{1/3}\right)^6\cdot\left(b^{5/6}\right)^6=64\cdot a^{6/3}\cdot b^{6(5/6)}=64a^2b^5$$

17. $$2L^{-2}-4C^{-1}=\frac{2}{L^2}-\frac{4}{C}=\frac{2C-4L^2}{L^2C}=\frac{2(C-2L^2)}{L^2C}$$

21. $$\left(a-3b^{-1}\right)^{-1}=\frac{1}{\left(a-3b^{-1}\right)}=\frac{1}{a-\frac{3}{b}}=\frac{1}{\frac{ab-3}{b}}=\frac{b}{ab-3}$$

25. $$\left(W^2+2WH+H^2\right)^{-1/2}=\frac{1}{\left(W^2+2WH+H^2\right)^{1/2}}=\frac{1}{\left((W+H)^2\right)^{1/2}}=\frac{1}{W+H}$$

29. $\sqrt{68}=\sqrt{4\cdot 17}=\sqrt{4}\cdot\sqrt{17}=2\sqrt{17}$

33. $\sqrt{9a^3b^4}=\sqrt{9a^2b^4a}=3ab^2\sqrt{a}$

37. $\frac{5}{\sqrt{2s}}=\frac{5}{\sqrt{2s}}\times\frac{\sqrt{2s}}{\sqrt{2s}}=\frac{5\sqrt{2s}}{2s}$

41. $$\sqrt[4]{8m^6n^9}=8^{1/4}m^{6/4}n^{9/4}=(2^3)^{1/4}m^{4/4}m^{2/4}n^{8/4}n^{1/4}=2^{3/4}m\cdot m^{1/2}n^2n^{1/4}=mn^2\sqrt{m}\sqrt[4]{8n}$$

45. $$\sqrt{36+4}-2\sqrt{10}=\sqrt{4(10)}-2\sqrt{10}=2\sqrt{10}-2\sqrt{10}=0$$

49. $$a\sqrt{2x^3}+\sqrt{8a^2x^3}=a\sqrt{x^2\cdot 2x}+\sqrt{4\cdot a^2x^2\cdot 2x}=ax\sqrt{2x}+2ax\sqrt{2x}=3ax\sqrt{2x}$$

53. $5\sqrt{5}\left(6\sqrt{5}-\sqrt{35}\right)=30(5)-5\sqrt{175}$
$=150-5\sqrt{25(7)}$
$=150-25\sqrt{7}$
$=25\left(6-\sqrt{7}\right)$

57. $\left(2-3\sqrt{17B}\right)\left(3+\sqrt{17B}\right)=6+2\sqrt{17B}-9\sqrt{17B}-3(17B)$
$=6-51B-7\sqrt{17B}$

61. $\dfrac{\sqrt{3x}}{2\sqrt{3x}-\sqrt{y}}=\dfrac{\sqrt{3x}}{\left(2\sqrt{3x}-\sqrt{y}\right)}\times\dfrac{\left(2\sqrt{3x}+\sqrt{y}\right)}{\left(2\sqrt{3x}+\sqrt{y}\right)}$
$=\dfrac{2(3x)+\sqrt{3x}\cdot\sqrt{y}}{4(3x)-(y)}$
$=\dfrac{6x+\sqrt{3xy}}{12x-y}$

65. $\dfrac{\sqrt{7}-\sqrt{5}}{\sqrt{5}+3\sqrt{7}}=\dfrac{\sqrt{7}-\sqrt{5}}{\sqrt{5}+3\sqrt{7}}\times\dfrac{\sqrt{5}-3\sqrt{7}}{\sqrt{5}-3\sqrt{7}}$
$=\dfrac{\sqrt{35}-3(7)-5+3\sqrt{35}}{5-9(7)}$
$=\dfrac{4\sqrt{35}-21-5}{5-63}$
$=\dfrac{4\sqrt{35}-26}{-58}$
$=\dfrac{2(2\sqrt{35}-13)}{-58}$
$=\dfrac{13-2\sqrt{35}}{29}$

69. $\sqrt{4b^2+1}$ is already in its simplest form

73. $\left(1+6^{1/2}\right)\left(3^{1/2}+2^{1/2}\right)\left(3^{1/2}-2^{1/2}\right)$
$=\left(1+6^{1/2}\right)\left(\left(3^{1/2}\right)^2-\left(2^{1/2}\right)^2\right)$
$=\left(1+6^{1/2}\right)(3-2)$
$=1+\sqrt{6}$

77. $\sqrt{3+n}\left(\sqrt{3+n}-\sqrt{n}\right)^{-1}=\dfrac{\sqrt{3+n}}{\sqrt{3+n}-\sqrt{n}}\times\dfrac{\sqrt{3+n}+\sqrt{n}}{\sqrt{3+n}+\sqrt{n}}$
$=\dfrac{(3+n)+\sqrt{n}\sqrt{3+n}}{(3+n)-(n)}$
$=\dfrac{3+n+\sqrt{n(3+n)}}{3}$

81. $\sqrt{\sqrt{2}-1}\left(\sqrt{2}+1\right)=\sqrt{\sqrt{2}+1}$
$\sqrt{\sqrt{2}-1}\times\dfrac{\sqrt{\sqrt{2}+1}}{\sqrt{\sqrt{2}+1}}\left(\sqrt{2}+1\right)=\sqrt{\sqrt{2}+1}$
$\dfrac{\sqrt{2-1}}{\sqrt{\sqrt{2}+1}}\left(\sqrt{2}+1\right)=\sqrt{\sqrt{2}+1}$
$\dfrac{\sqrt{2}+1}{\sqrt{\sqrt{2}+1}}\left(\dfrac{\sqrt{\sqrt{2}+1}}{\sqrt{\sqrt{2}+1}}\right)=\sqrt{\sqrt{2}+1}$
$\dfrac{\left(\sqrt{2}+1\right)\left(\sqrt{\sqrt{2}+1}\right)}{\sqrt{2}+1}=\sqrt{\sqrt{2}+1}$
$\sqrt{\sqrt{2}+1}=\sqrt{\sqrt{2}+1}$

Check

$\sqrt{\sqrt{2}-1}\left(\sqrt{2}+1\right)=\sqrt{0.41421356}(2.41421356)$
$=1.553774\ldots$ using a calculator
$\sqrt{\sqrt{2}+1}=\sqrt{2.41421356}$
$=1.553774\ldots$ using a calculator

85. $3x^2-2x+5=3\left(\dfrac{1}{2}\left(2-\sqrt{3}\right)\right)^2-2\left(\dfrac{1}{2}\left(2-\sqrt{3}\right)\right)+5$
$=\dfrac{3}{4}\left(4-4\sqrt{3}+3\right)-2+\sqrt{3}+5$
$=3-3\sqrt{3}+\dfrac{9}{4}+3+\sqrt{3}$
$=6+\dfrac{9}{4}-2\sqrt{3}$
$=\dfrac{24+9-8\sqrt{3}}{4}$
$=\dfrac{33-8\sqrt{3}}{4}$
$=\dfrac{c}{(c^2-v^2)^{1/2}}$
$=\dfrac{c}{(c^2-v^2)^{1/2}}$

89. (a) $v = k\sqrt[3]{\frac{P}{w}} = k\left(\frac{P}{w}\right)^{1/3}$

(b) $v = k\sqrt[3]{\frac{P}{w}} \cdot \frac{\sqrt[3]{w^2}}{\sqrt[3]{w^2}}$

$= k\sqrt[3]{\frac{Pw^2}{w^3}}$

$= \frac{k}{w}\sqrt[3]{Pw^2}$

93. $\left(1-\frac{v^2}{c^2}\right)^{-1/2} = \frac{1}{\left(1-\frac{v^2}{c^2}\right)^{1/2}} \times \frac{\left(1-\frac{v^2}{c^2}\right)^{1/2}}{\left(1-\frac{v^2}{c^2}\right)^{1/2}}$

$= \frac{\left(1-\frac{v^2}{c^2}\right)^{1/2}}{1-\frac{v^2}{c^2}}$

$= \frac{\left(1-\frac{v^2}{c^2}\right)^{1/2}}{\frac{c^2-v^2}{c^2}}$

$= \frac{c^2\left(1-\frac{v^2}{c^2}\right)^{1/2}}{c^2-v^2}$

97. $l = \sqrt{3^2+3^2} + \sqrt{2^2+2^2} + \sqrt{1^2+1^2}$

$l = \sqrt{18} + \sqrt{8} + \sqrt{2}$

$l = \sqrt{9(2)} + \sqrt{4(2)} + \sqrt{2}$

$l = 3\sqrt{2} + 2\sqrt{2} + \sqrt{2}$

$l = 6\sqrt{2}$ cm

101. $0.018^{0.13} = 0.018^{13/100} = \sqrt[100]{0.018^{13}}$

$0.018^{0.13}$ is easier to type into calculator than 0.018 to the power of 13 to the 100th root.

CHAPTER 12

COMPLEX NUMBERS

12.1 Basic Definitions

1. $j\sqrt{-6} = j\sqrt{(-1)(6)}$
$= j\left(j\sqrt{6}\right)$
$= j^2\sqrt{6}$
$= -\sqrt{6}$

5. $\sqrt{-81} = \sqrt{81(-1)} = \sqrt{81}\sqrt{-1} = 9j$

9. $\sqrt{-0.36} = \sqrt{0.36(-1)} = \sqrt{0.36}\sqrt{-1} = 0.6j$

13. $\sqrt{-\frac{7}{4}} = \frac{\sqrt{-7}}{\sqrt{4}} = \frac{\sqrt{7(-1)}}{2} = \frac{\sqrt{7}\sqrt{-1}}{2} = j\frac{\sqrt{7}}{2}$

17. **(a)** $\left(\sqrt{-7}\right)^2 = \left(\sqrt{7(-1)}\right)^2 = \left(\sqrt{7}\cdot\sqrt{-1}\right)^2 = \left(\sqrt{7}\cdot j\right)^2$
$= \sqrt{7}^2\cdot j^2 = 7(-1) = -7$

(b) $\sqrt{(-7)^2} = \sqrt{49} = 7$

21. $\sqrt{-\frac{1}{15}}\sqrt{-\frac{27}{5}} = j\sqrt{\frac{1}{15}}j\sqrt{\frac{27}{5}} = j^2\sqrt{\frac{27}{75}}$
$= -\sqrt{\frac{9(3)}{25(3)}} = -\frac{3}{5}$

25. **(a)** $-j^6 = -\left(j^2\right)^3 = -(-1)^3 = -(-1) = 1$

(b) $(-j)^6 = \left((-j)^2\right)^3 = (-1)^3 = -1$

29. $j^{15} - j^{13} = j^{12}\cdot j^3 - j^{12}\cdot j = (1)(-j) - (1)(j)$
$= -j - j = -2j$

33. $2 + \sqrt{-9} = 2 + \sqrt{9(-1)} = 2 + 3j$

37. $\sqrt{-4j^2} + \sqrt{-4} = \sqrt{(-4)(-1)} + 2j = 2 + 2j$

41. $\sqrt{18} - \sqrt{-8} = \sqrt{9\cdot 2} - \sqrt{4\cdot 2}j$
$= 3\sqrt{2} - 2j\sqrt{2}$

45. **(a)** The conjugate of $6 - 7j$ is $6 + 7j$.

(b) The conjugate of $8 + j$ is $8 - j$.

49. Equating real and imaginary parts:
$7x - 2yj = 14 + 4j$
$7x = 14$
$x = 2$, and
$-2y = 4$
$y = -2$

53. We rewrite each side in the form $a + bj$ by first simplifying the expressions involving powers of j. Then we equate real and imaginary parts:
$x - 2j^2 + 7j = yj + 2xj^3$
$x - 2(-1) + 7j = yj + 2x(-j)$
$x + 2 + 7j = 0 + (y - 2x)j$
$x + 2 = 0$ and $7 = y - 2x$
$x = -2$ $\quad 7 = y - 2(-2)$
$7 = y + 4$
$y = 3$

57. $x^2 + 64 = 0$
$8j$ is a solution since
$(8j)^2 + 64 = 0$
$64j^2 + 64 = 0$
$-64 + 64 = 0$
$0 = 0$
$-8j$ is a solution since
$(-8j)^2 + 64 = 0$
$64j^2 + 64 = 0$
$-64 + 64 = 0$
$0 = 0$

61. For a complex number and its conjugate to be equal, it must be a real number. $a + 0j = a - 0j$.

12.2 Basic Operations with Complex Numbers

1. $(7-9j)-(6-4j)=7-9j-6+4j$
$=1-5j$

5. $(3-7j)+(2-j)=(3+2)+(-7-1)j=5-8j$

9. $0.23-(0.46-0.19j)+0.67j$
$=0.23-0.46+0.19j+0.67j$
$=-0.23+0.86j$

13. $(7-j)(7j)=49j-7j^2$
$=49j-7(-1)$
$=7+49j$

17. $(\sqrt{-18}\sqrt{-4})(3j)=(j\sqrt{(9)(2)})(j\sqrt{4})(3j)$
$=(3j\sqrt{2})(2j)(3j)$
$=18j^3\sqrt{2}$
$=18(-j)\sqrt{2}$
$=-18j\sqrt{2}$

21. $j\sqrt{-7}-j^6\sqrt{112}+3j=j\sqrt{7(-1)}-j^6\sqrt{16(7)}+3j$
$=j^2\sqrt{7}-4j^6\sqrt{7}+3j$
$=(-1)\sqrt{7}-4j^4(j^2)\sqrt{7}+3j$
$=-\sqrt{7}-4(1)(-1)\sqrt{7}+3j$
$=-\sqrt{7}+4\sqrt{7}+3j$
$=3\sqrt{7}+3j$

25. $(1-j)^3=(1-j)(1-j)^2=(1-j)(1-2j+j^2)$
$=1-2j+j^2-j+2j^2-j^3$
$=1-3j+3j^2-j^3$
$=1-3j+3(-1)-(-1)j$
$=-2-2j$

29. $\frac{1-j}{3j}=\frac{1-j}{3j}\cdot\frac{-3j}{-3j}$
$=\frac{-3j+3j^2}{-9j^2}$
$=\frac{-3j-3}{9}$
$=\frac{-3(1+j)}{9}$
$=-\frac{1+j}{3}$

33. $\frac{j^2-j}{2j-j^8}=\frac{-1-j}{2j-1}\cdot\frac{-2j-1}{-2j-1}$
$=\frac{1+3j+2j^2}{1^2+2^2}$
$=\frac{1+3j-2}{5}$
$=\frac{-1+3j}{5}$

37. We simplify before squaring:
$(4j^5-5j^4+2j^3-3j^2)^2=(4j-5-2j+3)^2$
$=(-2+2j)^2$
$=4-8j-4$
$=-8j$

41. $\frac{(2-j^3)^4}{(j^8-j^6)^3}+j=\frac{(2+j)^4}{(1+1)^3}+j$
$=\frac{(2+j)^4}{8}+j$
$=\frac{(2+j)^2(2+j)^2}{8}+j$
$=\frac{(3+4j)(3+4j)}{8}+j$
$=\frac{-7+24i}{8}+j$
$=-\frac{7}{8}+4j$

45. Multiply $-3+j$ by its conjugate.
$(-3+j)(-3-j)=9-j^2$
$=9-(-1)$
$=10$

49. $j^{-2}+j^{-3}=\frac{1}{j^2}+\frac{1}{j^3}$
$=\frac{1}{-1}+\frac{1}{-j}\cdot\frac{j}{j}$
$=-1+\frac{j}{-j^2}$
$=-1+j$

53. $f(x) = x + \frac{1}{x}$

$$f(1+3j) = 1+3j+\frac{1}{1+3j}\cdot\frac{1-3j}{1-3j}$$
$$= 1+3j+\frac{1}{10}-\frac{3}{10}j$$
$$= \frac{11}{10}+\frac{27}{10}j$$

57. $V = 85+74j;\ Z = 2500-1200j$

$$I = \frac{V}{Z}$$
$$= \frac{85+74j}{2500-1200j}\cdot\frac{2500+1200j}{2500+1200j}$$
$$= \frac{2\ 12\ 500+88\ 800j^2+(1\ 02\ 000+1\ 85\ 000)j}{62\ 50\ 000-14\ 40\ 000j^2}$$
$$= \frac{2\ 12\ 500+88\ 800(-1)+2\ 87\ 000j}{62\ 50\ 000-14\ 40\ 000(-1)}$$
$$= \frac{1\ 23\ 700+2\ 87\ 000j}{76\ 90\ 000}$$
$= 0.016+0.037j$ amperes

61. $(a+bj)(a-bj) = a^2-b^2j^2 = a^2+b^2$, a positive real number.

12.3 Graphical Representation of Complex Numbers

1. Add $5-2j$ and $-2+j$ graphically.

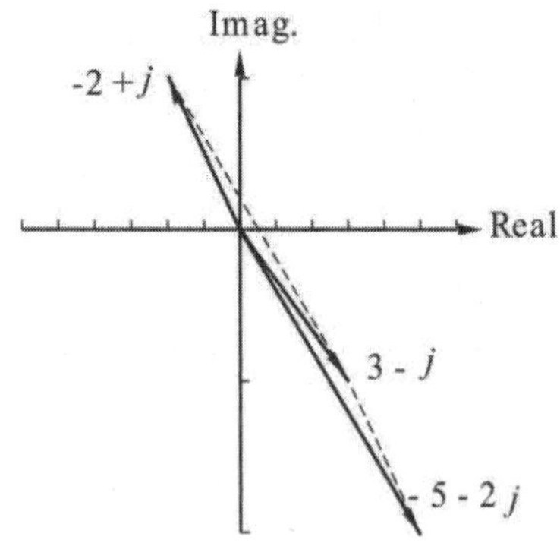

$5-2j+(-2+j) = 3-j$

5.  $-4-3j$

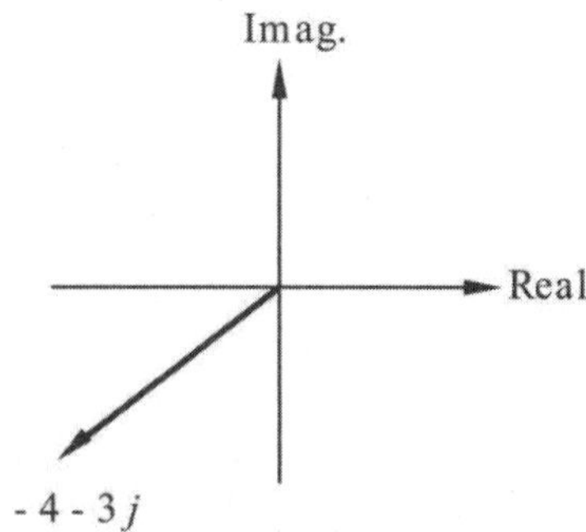

9. $2+3+4j = 5+4j$

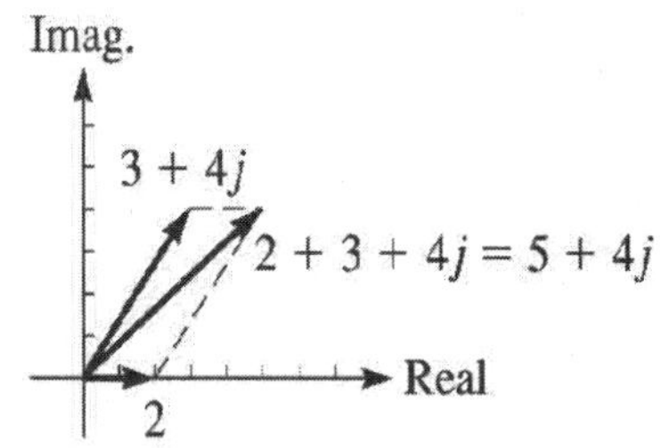

13. $5j-1(1-4j) = 5j-1+4j = -1+9j$

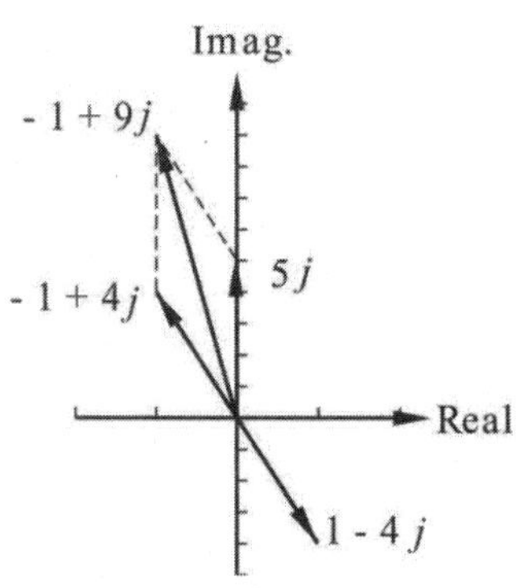

17. $(3-2j)-(4-6j) = 3-2j-4+6j = -1+4j$

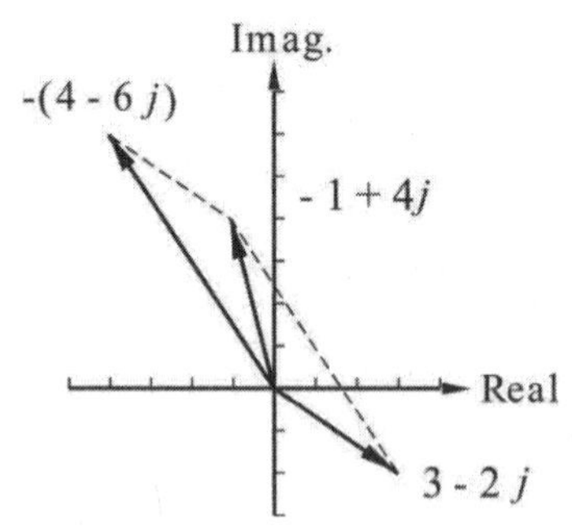

21. $(1.5-0.5j)+(3.0+2.5j)=1.5-0.5j+3.0+2.5j$
$=4.5+2.0j$

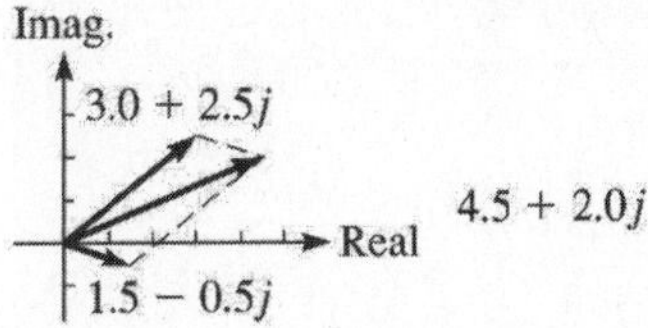

25. $(2j+1)-3j-(j+1)=(2j+1)+(-3j)+(-(j+1))$
$=2j+1-3j-j-1$
$=-2j$

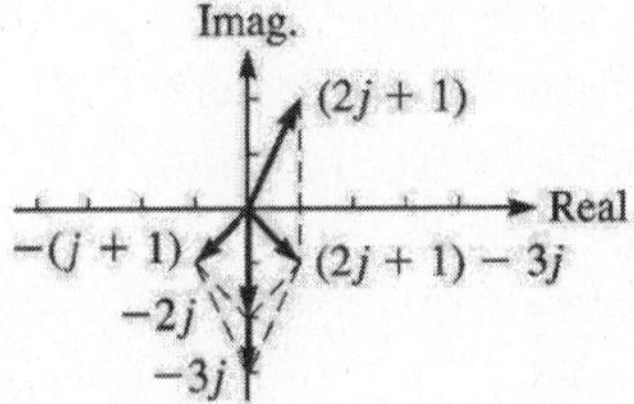

29. $3+2j$

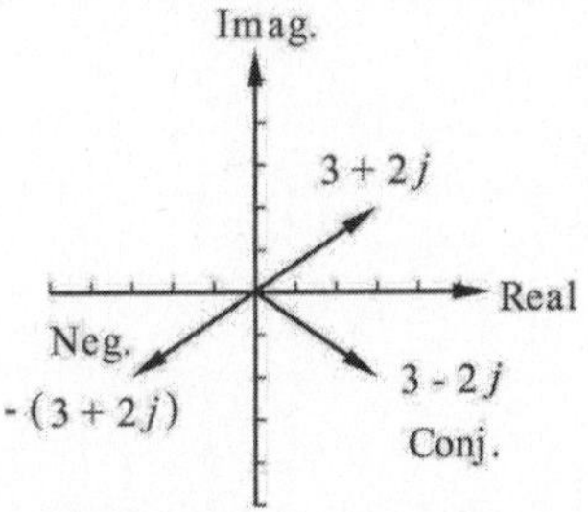

33.
$a+bj=3-j$
$3(a+bj)=9-3j$
$-3(a+bj)=-9+3j$

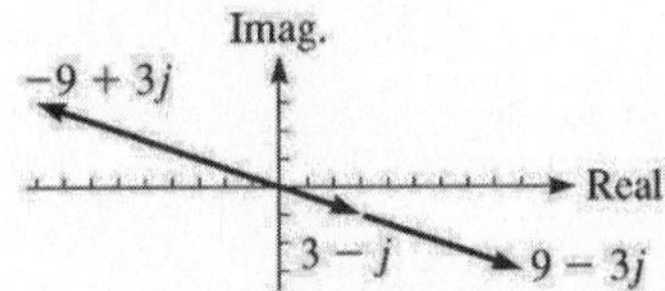

37. $2+4j-(2-4j)=2+4j-2+4j=2(4j)=8j$

The difference of a complex number and its conjugate is twice the imaginary part times the imaginary unit.

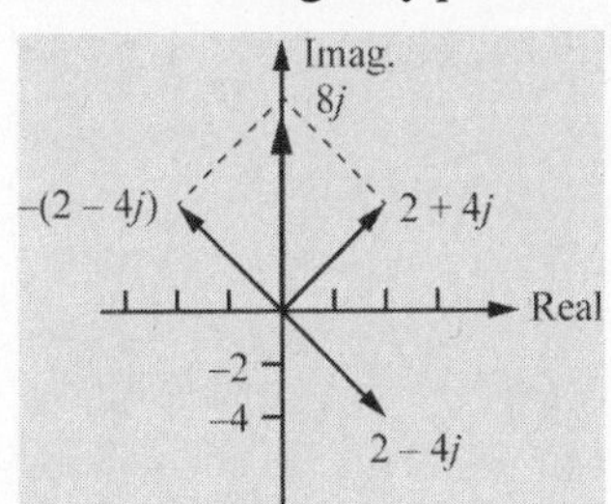

12.4 Polar Form of a Complex Number

1. From the rectangular form we have $x=-3$ and $y=4$, so

$$r=\sqrt{(-3)^2+4^2}=5$$

$$\theta_{\text{ref}}=\tan^{-1}\left(\frac{4}{3}\right)=53.1^\circ.$$

Since θ is in the second quadrant, $\theta=180^\circ-53.1^\circ=126.9^\circ$. Therefore, $-3+4j=5(\cos 126.9^\circ+j\sin 126.9^\circ)$

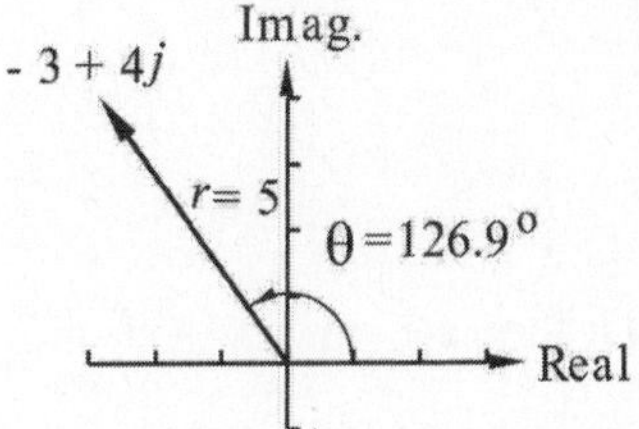

5. From the rectangular form we have $x=30$ and $y=-40$, so

$$r=\sqrt{(30)^2+(-40)^2}=50$$

$$\theta_{\text{ref}}=\tan^{-1}\left(\frac{40}{30}\right)=53.1^\circ.$$

Since θ is in the fourth quadrant, $\theta=360^\circ-53.1^\circ=306.9^\circ$ and $30-40j=50(\cos 306.9^\circ+j\sin 306.9^\circ)$

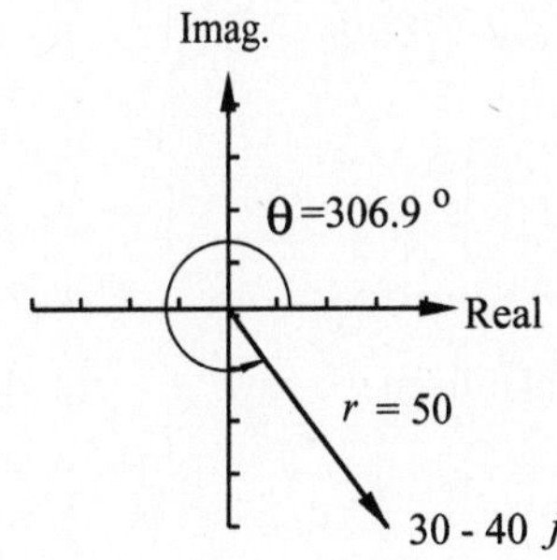

9. From the rectangular form we have $x = -0.55$ and $y = -0.24$, so

$$r = \sqrt{(-0.55)^2 + (-0.24)^2} = 0.60$$

$$\theta_{\text{ref}} = \tan^{-1}\left(\frac{0.24}{0.55}\right) = 24^\circ.$$

Since θ is in the third quadrant, $\theta = 180^\circ + 24^\circ = 204^\circ$ and $-0.55 - 0.24j = 0.60(\cos 204^\circ + j\sin 204^\circ)$

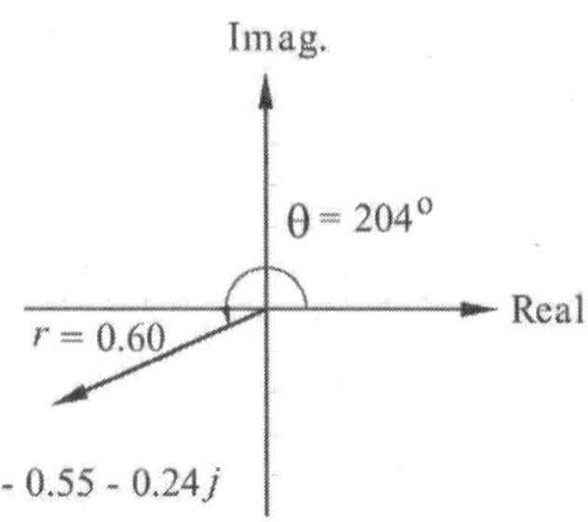

13. From the rectangular form we have $x = 460$ and $y = -460$, so

$$r = \sqrt{(460)^2 + (-460)^2} = 651$$

$$\theta_{\text{ref}} = \tan^{-1}\left(\frac{460}{460}\right) = 45^\circ.$$

Since θ is in the fourth quadrant, $\theta = 360^\circ - 45^\circ = 315^\circ$ and $460 - 460j = 651(\cos 315^\circ + j\sin 315^\circ)$

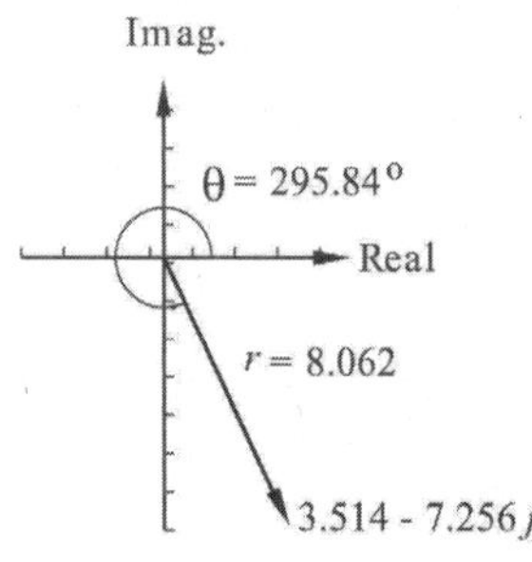

17. From the rectangular form we have $x = 0$ and $y = 9$, so

$$r = \sqrt{0^2 + 9^2} = 9$$

$$\theta = 90^\circ$$

since y is positive, so $9j = 9(\cos 90^\circ + j\sin 90^\circ)$

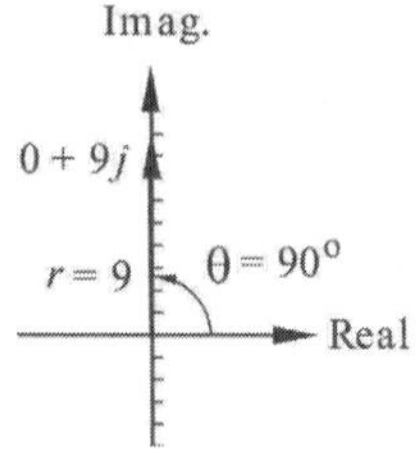

21. $160(\cos 150.0^\circ + j\sin 150.0^\circ)$

$$= 160\cos 150.0^\circ + 160\sin 150.0^\circ \cdot j$$
$$= -140 + 80j$$

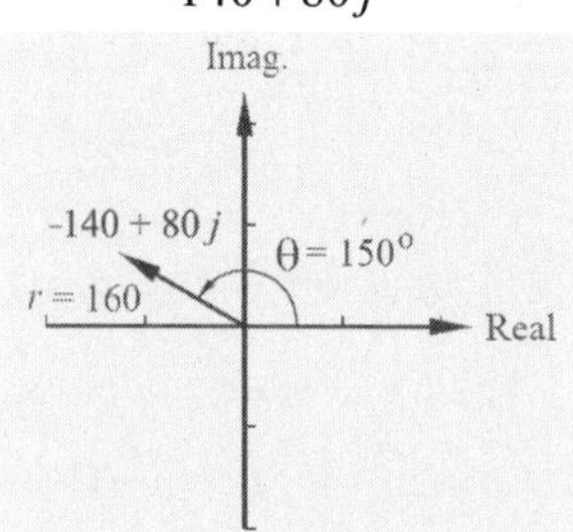

25. $0.08(\cos 360^\circ + j\sin 360^\circ)$

$$= 0.08\cos 360^\circ + 0.08\sin 360^\circ \cdot j$$
$$= 0.08(1) + 0.08(0)j$$
$$= 0.08 + 0j$$
$$= 0.08$$

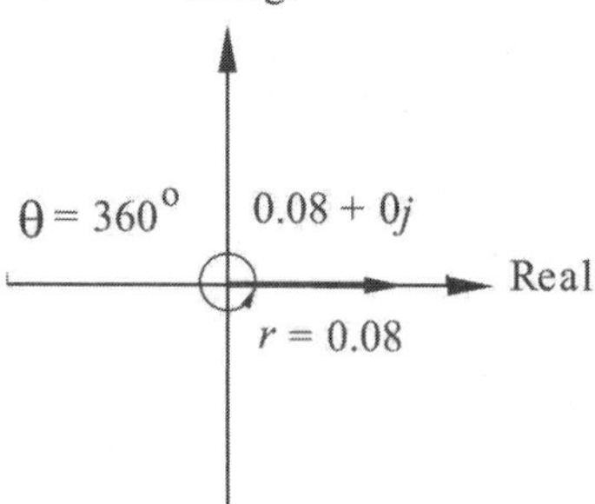

29. $4.75\underline{/172.8^\circ} = 4.75\cos 172.8^\circ + 4.75\sin 172.8^\circ \cdot j$

$$= -4.71 + 0.595j$$

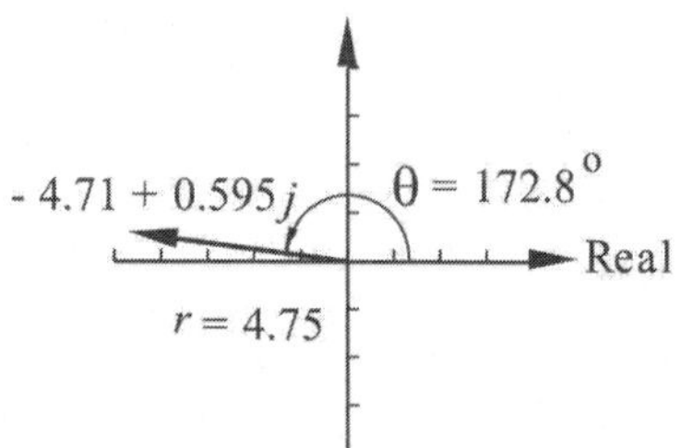

33. $7.32\underline{/-270^\circ} = 7.32\cos(-270^\circ) + 7.32\sin(-270^\circ) \cdot j$

$$= 7.32(0) + 7.32(1)j$$
$$= 7.32j$$

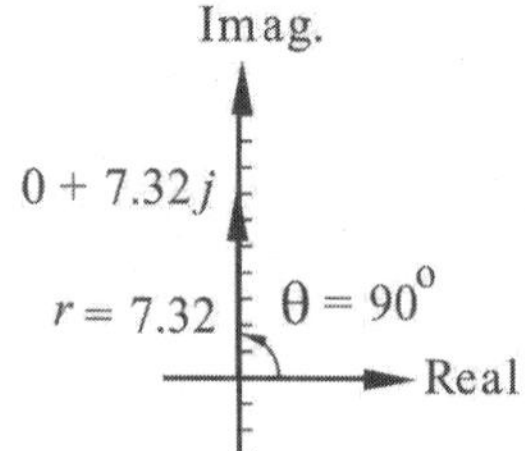

37. The argument for any negative real number is 180°.

41. From the rectangular form we have $x = 2.84$ and $y = -1.04$, so

$$r = \sqrt{2.84^2 + (-1.06)^2} = 3.03$$

$$\theta_{ref} = \tan^{-1}\left(\frac{1.06}{2.84}\right) = 20.5^\circ.$$

Since θ is in the fourth quadrant,
$\theta = 360^\circ - 20.5^\circ = 339.5^\circ$. Therefore,

$$2.84 - 1.06j \text{ kV} = 3.03(\cos 339.5^\circ + j\sin 339.5^\circ)\text{ kV}$$
$$= 3.03\underline{/339.5^\circ}\text{ kV}$$

12.5 Exponential Form of a Complex Number

1. $8.50\angle 226.3^\circ$, $r = 8.50$,

$$\theta = 226.3^\circ\left(\frac{\pi \text{ rad}}{180^\circ}\right) = 3.950$$

The exponential form is

$$8.50\angle 226.3^\circ = 8.50e^{3.95j}$$

5. $3.00\left(\cos 60.0^\circ + j\sin 60.0^\circ\right)$;

$$r = 3.00,\ \theta = 60.0^\circ \cdot \left(\frac{\pi \text{ rad}}{180^\circ}\right) = 1.05 \text{ rad}$$

The exponential form is $3.00e^{1.05j}$

9. $375.5\left(\cos\left(-95.46^\circ\right) + j\sin\left(-95.46^\circ\right)\right)$; $r = 375.5$

$$\theta = -95.46^\circ \cdot \frac{\pi \text{ rad}}{180^\circ} = -1.666 \text{ rad}$$

The exponential form is $375.5e^{-1.666j} = 375.5e^{4.617j}$

13. $4.06\angle -61.4^\circ$; $r = 4.06$;

$$\theta = -61.4^\circ\left(\frac{\pi \text{ rad}}{180^\circ}\right)$$
$$= -1.07 \text{ rad}$$

The exponential form is $4.06e^{-1.07j} = 4.06e^{5.21j}$

17. From the rectangular form we have $x = 3$ and $y = -4$. Using Eq. (12.8), we have

$$r = \sqrt{3^2 + (-4)^2} = 5,$$

$$\theta_{ref} = \tan^{-1}\frac{4}{3} = 0.9273$$

Since θ is in the fourth quadrant,

$$\theta = \theta_{ref} + 2\pi = 5.36 \text{ rad}$$

$$3 - 4j = 5e^{5.36j}$$

21. From the rectangular form we have $x = 5.90$ and $y = 2.40$. Using Eq. (12.8), we have

$$5.90 + 2.40j;\ r = \sqrt{5.90^2 + 2.40^2} = 6.37$$

Since θ is in the first quadrant,

$$\theta = \tan^{-1}\frac{2.40}{5.90} = 0.386 \text{ rad}$$

$$5.90 + 2.40j = 6.37e^{0.386j}$$

25. $3.00e^{0.500j}$; $r = 3$, $\theta = 0.500$ rad$=28.6^\circ$

$$3.00e^{0.500j} = 3\left(\cos 28.6^\circ + j\sin 28.6^\circ\right)$$
$$= 2.63 + 1.44j$$

29. $3.20e^{-5.41j}$

$$r = 3.20$$

$$\theta = -5.41 \text{ rad}\left(\frac{180^\circ}{\pi}\right)$$
$$= -310.0^\circ$$
$$= 50.0^\circ$$

$$3.20e^{-5.41j} = 3.20\left(\cos 50.0^\circ + j\sin 50.0^\circ\right)$$
$$= 2.06 + 2.45j$$

33. $\left(4.55e^{1.32j}\right)^2 = 4.55^2 e^{2.64j}$ from which

$$r = 4.55^2,\ \theta = 2.64 \cdot \frac{180^\circ}{\pi} \text{ and}$$

$$\left(4.55e^{1.32j}\right)^2 = r\left(\cos\theta + j\sin\theta\right)$$
$$= 4.55^2\left(\cos 2.64 \cdot \frac{180^\circ}{\pi} + j\sin 2.64 \cdot \frac{180^\circ}{\pi}\right)$$
$$= -18.2 + 9.95j$$

37. From the rectangular form we have $x = 375$ and $y = 110$, so

$$r = \sqrt{375^2 + 110^2} = 391$$

$$\theta = \tan^{-1}\frac{110}{375}$$
$$= 16.3^\circ\left(\frac{\pi \text{ rad}}{180^\circ}\right)$$
$$= 0.285$$

$$375 + 110j = 391e^{0.285j}$$

The magnitude of the impedance is 391 ohms.

12.6 Products, Quotients, Powers, and Roots of Complex Numbers

1. $2+3j: r_1 = \sqrt{2^2+3^2} = 3.61$

$\tan\theta_1 = \frac{3}{2}$

$\theta_1 = 56.3^\circ$

$1+j: r_2 = \sqrt{1^1+1^2} = 1.41$

$\tan\theta_2 = \frac{1}{1}$

$\theta_2 = 45.0^\circ$

$(2+3j)(1+j)$

$= 3.61(\cos 56.3^\circ + j\sin 56.3^\circ)$

$(1.41)(\cos 45.0^\circ + j\sin 45.0^\circ)$

$= 5.09(\cos 101.3^\circ + j\sin 101.3^\circ)$

5. $[4(\cos 60^\circ + j\sin 60^\circ)][2(\cos 20^\circ + j\sin 20^\circ)]$

$= 4\cdot 2[(\cos(60^\circ+20^\circ)+j\sin(60^\circ+20^\circ)]$

$= 8(\cos 80^\circ + j\sin 80^\circ)$

9. $\frac{8(\cos 100^\circ + j\sin 100^\circ)}{4(\cos 65^\circ + j\sin 65^\circ)}$

$= \frac{8}{4}(\cos(100^\circ - 65^\circ) + j\sin(100^\circ - 65^\circ))$

$= 2(\cos 35^\circ + j\sin 35^\circ)$

13. $\left[0.2\left(\cos 35^\circ + j\sin 35^\circ\right)\right]^3$

$= 0.2^3\left(\cos 3\cdot 35^\circ + j\sin 3\cdot 35^\circ\right)$

$= 0.008\left(\cos 105^\circ + j\sin 105^\circ\right)$

17. $\frac{(50\angle 236^\circ)(2\angle 84^\circ)}{125\angle 47^\circ} = \frac{100\angle 320^\circ}{125\angle 47^\circ}$

$= \frac{4}{5}\angle 273^\circ$

$= \frac{4}{5}\left(\cos 273^\circ + j\sin 273^\circ\right)$

21. $2.78\underline{/56.8^\circ} + 1.37\underline{/207.3^\circ}$

$= 2.78(\cos 56.8^\circ + j\sin 56.8^\circ)$

$+ 1.37(\cos 207.3^\circ + j\sin 207.3^\circ)$

$= 1.52 + 2.33j - 1.22 - 0.628j$

$= 0.30 + 1.702j$

In polar form:

$r = \sqrt{0.30^2 + 1.703^2} = 1.73$

$\theta = \tan^{-1}\left(\frac{1.703}{0.30}\right)$

$= 80.0^\circ$

$2.78\underline{/56.8^\circ} + 1.37\underline{/207.3^\circ} = 1.72\underline{/80.0^\circ}$

25. $3+4j$: $\quad$ $5-12j$:

$r_1 = \sqrt{3^2+4^2}$ $\quad$ $r_2 = \sqrt{5^2+(-12)^2}$

$= 5$ $\quad$ $= 13$

$\theta_{ref} = \tan^{-1}\left(\frac{4}{3}\right)$ $\quad$ $\theta_{ref} = \tan^{-1}\left(\frac{12}{5}\right)$

$= 53.1^\circ$ $\quad$ $= 67.4^\circ$

$\theta_1 = 53.1^\circ$ $\quad$ $\theta_2 = 360^\circ - 67.4^\circ$

$= 292.6$

In polar form:

$(3+4j)(5-12j) = (5\underline{/53.1^\circ})(13\underline{/292.6^\circ})$

$= 65\underline{/345.7^\circ}$

$= 63.0 - 16.1j$

In rectangular form:

$(3+4j)(5-12j) = 15 - 36j + 20j - 48j^2$

$= 15 - 16j + 48$

$= 63 - 16j$

29. 7: $\quad$ $1-3j$:

$r_1 = \sqrt{7^2+0^2}$ $\quad$ $r_2 = \sqrt{1^2+(-3)^2}$

$= 7$ $\quad$ $= 3.16$

$\theta_1 = 0^\circ$ $\quad$ $\theta_{ref} = \tan^{-1}\left(\frac{3}{1}\right)$

$= 71.6^\circ$

$\theta_2 = 360^\circ - 71.6^\circ$

288.4°

In polar form:

$\frac{7}{1-3j} = \frac{7\underline{/0^\circ}}{3.16\underline{/288.4^\circ}}$

$= 2.22\underline{/-288.4^\circ}$

$= 0.70 + 2.1j$

In rectangular form:

$\frac{7}{1-3j}\cdot\frac{1+3j}{1+3j} = \frac{7+21j}{1+9}$

$= 0.7 + 2.1j$

33. $r = \sqrt{3^2 + 4^2} = 5$

$$\theta = \tan^{-1}\left(\frac{4}{3}\right) = 53.1^\circ$$

$$\begin{aligned}(3+4j)^4 &= 5^4(\cos(4(53.1)) + j\sin(4(53.1)))\\ &= 625(\cos 212.5 + j\sin 212.5)\\ &= -527 - 336j\end{aligned}$$

In rectangular form:

$$\begin{aligned}(3+4j)^4 &= [(3+4j)^2]^2\\ &= (9+24j+16j^2)^2\\ &= (-7+24j)^2\\ &= 49 - 336j + 576j^2\\ &= 49 - 336j - 336j\\ &= -527 - 336j\end{aligned}$$

37. The two square roots of $4(\cos 60^\circ + j\sin 60^\circ)$ are

$$r_1 = \sqrt{4}\left[\cos\left(\frac{60^\circ}{2}\right) + j\sin\left(\frac{60^\circ}{2}\right)\right] = 2(\cos 30^\circ + j\sin 30^\circ) = 1.73 + j$$

$$r_2 = \sqrt{4}\left[\cos\left(\frac{60^\circ + 360^\circ}{2}\right) + j\sin\left(\frac{60^\circ + 360^\circ}{2}\right)\right] = 2(\cos 210^\circ + j\sin 210^\circ) = -1.73 - j$$

41. $r = \sqrt{1^2 + 1^2} = 1.41$

$$\theta = \tan^{-1}\left(\frac{1}{1}\right) = 45^\circ$$

So the two square roots of $1 + j$ are:

$$r_1 = \sqrt{1.41}\left[\cos\left(\frac{45^\circ}{2}\right) + j\sin\left(\frac{45^\circ}{2}\right)\right] = 1.19(\cos 22.5^\circ + j\sin 22.5^\circ) = 1.1 + 0.46j$$

$$r_2 = \sqrt{1.41}\left[\cos\left(\frac{45^\circ + 360^\circ}{2}\right) + j\sin\left(\frac{45^\circ + 360^\circ}{2}\right)\right] = 1.19(\cos 202.5^\circ + j\sin 202.5^\circ) = -1.1 - 0.46j$$

45.

$$\begin{aligned}x^3 + 27j &= 0\\ x^3 &= -27j\\ x &= \sqrt[3]{-27j}\end{aligned}$$

To find the three cube roots of $-27j$:

$$r = \sqrt{0^2 + (-27)^2} = 27$$

$$\theta = 270^\circ$$

So the three cube roots of $-27j$ are

$$r_1 = 27^{1/3}\left[\cos\left(\frac{270^\circ}{3}\right) + j\sin\left(\frac{270^\circ}{3}\right)\right] = 3(\cos 90^\circ + j\sin 90^\circ) = 3j$$

$$r_2 = 27^{1/3}\left[\cos\left(\frac{270^\circ + 360^\circ}{3}\right) + j\sin\left(\frac{270^\circ + 360^\circ}{3}\right)\right] = 3(\cos 210^\circ + j\sin 210^\circ) = -\frac{3\sqrt{3}}{2} - \frac{3}{2}j$$

$$r_3 = 27^{1/3}\left[\cos\left(\frac{270^\circ + 2(360)^\circ}{3}\right) + j\sin\left(\frac{270^\circ + 2(360)^\circ}{3}\right)\right] = 3(\cos 330^\circ + j\sin 330^\circ) = \frac{3\sqrt{3}}{2} - \frac{3}{2}j$$

49. $(-125)^{1/3} = (125)^{1/3}(-1)^{1/3} = 5(-1)^{1/3}$

Using the three cube roots of -1 from Example 7, the three cube roots of -125 are

$$-5, \frac{5}{2} + j\frac{5\sqrt{3}}{2}, \frac{5}{2} - j\frac{5\sqrt{3}}{2}.$$

53.

$$\begin{aligned}x^3 + 1 &= 0\\ (x+1)(x^2 - x + 1) &= 0\\ x + 1 &= 0\\ x &= -1\\ x^2 - x + 1 &= 0\\ x &= \frac{-(-1) \pm \sqrt{(-1)^2 - 4(1)(1)}}{2(1)}\\ x &= \frac{1 \pm j\sqrt{3}}{2}\end{aligned}$$

The roots are $-1, \frac{1+j\sqrt{3}}{2}, \frac{1-j\sqrt{3}}{2}$, just as in Example 7.

57. $\dfrac{(8.66\underline{/90.0^\circ})(50.0\underline{/135.0^\circ})}{10.0\underline{/60.0^\circ}}$

$= \dfrac{(8.66)(50.0)}{10.0} 10.0\underline{/(90^\circ + 135.0^\circ - 60.0^\circ)}$

$= 43.3\underline{/165^\circ}$

12.7 An Application to Alternating-current (ac) Circuits

1. $V_R = IR = 2.00(12.0) = 24.0 \text{ V}$

$Z = R + j(X_L - X_C) = 12.0 + j(16.0)$

$|Z| = \sqrt{12.0^2 + 16.0^2} = 20.0\,\Omega$

$V_L = IX_L = 2.00(16.0) = 32.0 \text{ V}$

$V_{RL} = IZ = 2.00(20.0) = 40.0 \text{ V}$

$\theta = \tan^{-1}\dfrac{X_L}{R} = \tan^{-1}\dfrac{16.0}{12.0} = 53.1^\circ$, voltage leads current.

5. **(a)** $|Z| = \sqrt{R^2 + X_L^2} = \sqrt{2250^2 + 1750^2} = 2850\,\Omega$

(b) $\tan\theta = \dfrac{1750}{2250};\ \theta = 37.9^\circ$

(c) $V_{RLC} = IZ = (0.005\ 75)(2850) = 16.4 \text{ V}$

9. **(a)** $X_R = 45.0\,\Omega$

$X_L = 2\pi fL = 2\pi(60)(0.0429) = 16.2\,\Omega$

$Z = 45.0 + 16.2j$

$|Z| = \sqrt{45.0^2 + 16.2^2} = 47.8\,\Omega$

(b) $\tan\theta = \dfrac{16.2}{45.0};\ \theta = 19.8^\circ$

13. $R = 25.3\,\Omega$

$X_C = 1/(2\pi fC)$

$= 1/\left(2\pi\left(1.2\times10^6\right)\left(2.75\times10^{-9}\right)\right) = 48.2\,\Omega$

$= f = 1200 \text{ kHz} = 1.2\times10^6 \text{ Hz}$

$Z = R - X_{Cj} = 25.3 - 48.2j$

$|Z| = \sqrt{25.3^2 + (-48.2)^2} = 54.4\,\Omega$

$\tan\theta = \dfrac{-48.2}{25.3};\ \theta = -62.3^\circ$

17. $L = 12.5\times10^{-6} \text{ H}$

$C = 47.0\times10^{-9} \text{ F}$

$X_L = X_C$

$2\pi fL = \dfrac{1}{2\pi fC}$

$f = \sqrt{\dfrac{1}{4\pi^2 LC}}$

$= \sqrt{\dfrac{1}{4\pi^2\left(12.5\times10^{-6}\right)\left(4.70\times10^{-9}\right)}}$

$= 208 \text{ kHz}$

21. $P = VI\cos\theta$

$V = 225 \text{ mV}$

$\theta = -18.0^\circ = 342^\circ$

$Z = 47.3\,\Omega$

$V = IZ$

$I = \dfrac{225\times10^{-3}}{47.3} = 0.00476 \text{ A}$

$P = \left(225\times10^{-3}\right)(0.00476)\cos 342^\circ$

$= 0.00102 \text{ W} = 1.02 \text{ mW}$

Review Exercises

1. $(6-2j)+(4+j) = 6-2j+4+j = 10-j$

5. $(2+j)(4-j) = 8-2j+4j-j^2 = 8+2j+1 = 9+2j$

9. $\dfrac{3}{7-6j} = \dfrac{3}{(7-6j)}\cdot\dfrac{(7+6j)}{(7+6j)} = \dfrac{21+18j}{7^2+6^2} = \dfrac{21}{85}+\dfrac{18}{85}j$

13. $\dfrac{5j-(3-j)}{4-2j} = \dfrac{(-3+6j)}{(4-2j)}\cdot\dfrac{(4+2j)}{(4+2j)}$

$= \dfrac{-12-6j+24j+12j^2}{4^2+2^2}$

$= \dfrac{-12+18j-12}{16+4}$

$= \dfrac{-24+18j}{20}$

$= -\dfrac{6}{5}+\dfrac{9}{10}j$

17. $3x - 2j = yj - 9$

$3x = -9,\ -2 = y$

$x = -\dfrac{9}{3} \quad y = -2$

$x = -3$

$x = -3,\ y = -2$

21.

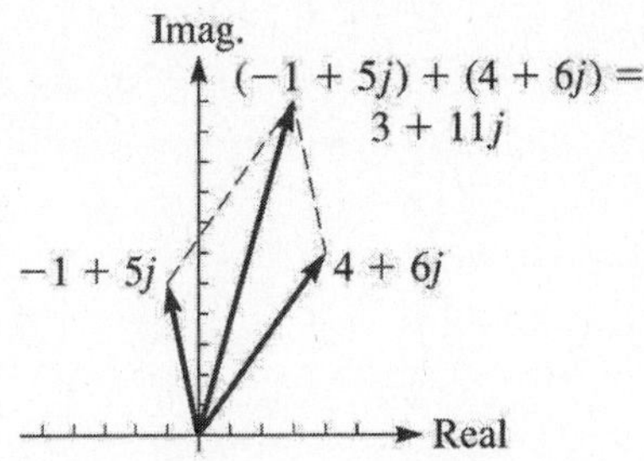

algebraically:

$$(-1+5j)+(4+6j)=-1+5j+4+6j$$
$$=-1+4+5j+6j$$
$$=3+11j$$

25. From the rectangular form we have $x=1$ and $y=-1$, so

$$r=\sqrt{1^2+(-1)^2}$$
$$=\sqrt{2}$$
$$\theta_{\text{ref}}=\tan^{-1}\left(\frac{1}{1}\right)$$
$$=45^\circ$$

Since θ is in the fourth quadrant,

$$\theta=360^\circ-45^\circ$$
$$=315^\circ\left(\frac{\pi\text{ rad}}{180^\circ}\right)$$
$$=5.50\text{ rad}$$
$$1-j=\sqrt{2}(\cos 45^\circ+j\sin 45^\circ)$$
$$=\sqrt{2}\underline{/45^\circ}$$
$$=\sqrt{2}e^{5.50j}$$

29. From the rectangular form we have $x=1.07$ and $y=4.55$, so

$$r=\sqrt{1.07^2+4.55^2}$$
$$=4.67$$
$$\theta=\tan^{-1}\left(\frac{4.55}{1.07}\right)$$
$$=76.8^\circ\left(\frac{\pi\text{ rad}}{180^\circ}\right)$$
$$=1.34\text{ rad}$$

$$1.07+4.55j=4.67(\cos 76.8^\circ+j\sin 76.8^\circ)$$
$$=4.67\underline{/76.8^\circ}$$
$$=4.67\,e^{1.34j}$$

33. $2\left(\cos 225^\circ+j\sin 225^\circ\right)=-\sqrt{2}-\sqrt{2}j$

37. $0.62\angle-72^\circ=0.62\left(\cos\left(-72^\circ\right)+j\sin\left(-72^\circ\right)\right)$

$$=0.19-0.59j$$

41. Using radian mode:

$$2.00e^{0.25j}=2.00(\cos 0.25+j\sin 0.25)$$
$$=1.94+0.495j$$

45. $\left[3\left(\cos 32^\circ+j\sin 32^\circ\right)\right]\cdot\left[5\left(\cos 52^\circ+j\sin 52^\circ\right)\right]$

$$=3\cdot 5\left(\cos\left(32^\circ+52^\circ\right)+j\sin\left(32^\circ+52^\circ\right)\right)$$
$$=15\left(\cos 84^\circ+j\sin 84^\circ\right)$$

49. $\dfrac{24\left(\cos 165^\circ+j\sin 165^\circ\right)}{3\left(\cos 106^\circ+j\sin 106^\circ\right)}$

$$=\frac{24}{3}\cos\left(165^\circ-106^\circ\right)+j\sin\left(165^\circ-106^\circ\right)$$
$$=8\left(\cos 59^\circ+j\sin 59^\circ\right)$$

53. $0.983\underline{/47.2^\circ}+0.366\underline{/95.1^\circ}$

$$=0.983(\cos 47.2^\circ+j\sin 47.2^\circ)$$
$$+0.366(\cos 95.1^\circ+j\sin 95.1^\circ)$$
$$=0.6679+0.7213j-0.03254+0.3646j$$
$$=0.6354+1.0859j$$

In polar form:

$$r=\sqrt{0.6354^2+1.0859^2}=1.26$$
$$\theta=\tan^{-1}\left(\frac{1.0859}{0.6354}\right)$$
$$=59.7^\circ.$$

$$0.983\underline{/47.2^\circ}+0.366\underline{/95.1^\circ}=1.26\underline{/59.7^\circ}$$

57. $\left[2\left(\cos 16^\circ+j\sin 16^\circ\right)\right]^{10}$

$$=2^{10}\left[\cos\left(10\cdot 16^\circ\right)+j\sin\left(10\cdot 16^\circ\right)\right]$$
$$=1024\left(\cos 160^\circ+j\sin 160^\circ\right)$$

61. $1-j=\sqrt{2}\left(\cos 315^\circ + j\sin 315^\circ\right)$ from Problem 25

$$(1-j)^{10}=\left[\sqrt{2}\left(\cos 315^\circ + j\sin 315^\circ\right)\right]^{10}$$
$$=\sqrt{2}^{10}\left(\cos\left(10\cdot 315^\circ\right)+j\sin\left(10\cdot 315^\circ\right)\right)$$
$$=32\left(\cos 3150^\circ + j\sin 3150^\circ\right)$$
$$=32\left(\cos 270^\circ + j\sin 270^\circ\right), \text{ polar form}$$
$$=0-32j, \text{ rectangular form}$$

$$(1-j)^{10}=\left((1-j)^2\right)^5$$
$$=\left(1-2j+j^2\right)^5$$
$$=(-2j)^5$$
$$=-32j^5$$
$$=-32j^4 j$$
$$=-32j$$

65. $x^3+8=0$

$$x^3=-8$$
$$x=\sqrt[3]{-8}$$

To find the three cube roots of –8:

$$r=\sqrt{8^2+0^2}=8$$
$$\theta=180^\circ$$

So the three cube roots of –8 are

$$r_1=8^{1/3}\left[\cos\left(\frac{180^\circ}{3}\right)+j\sin\left(\frac{180^\circ}{3}\right)\right]$$
$$=2(\cos 60^\circ + j\sin 60^\circ)$$
$$=1+j\sqrt{3}$$

$$r_2=8^{1/3}\left[\cos\left(\frac{180^\circ+360^\circ}{3}\right)+j\sin\left(\frac{180^\circ+360^\circ}{3}\right)\right]$$
$$=2(\cos 180^\circ + j\sin 180^\circ)$$
$$=-2+0j$$

$$r_3=8^{1/3}\left[\cos\left(\frac{180^\circ+2(360)^\circ}{3}\right)+j\sin\left(\frac{180^\circ+2(360)^\circ}{3}\right)\right]$$
$$=2(\cos 300^\circ + j\sin 300^\circ)$$
$$=1-j\sqrt{3}$$

69. From the graph, $x=40$, $y=9$. Therefore,

$$r=\sqrt{40^2+9^2}$$
$$=41$$
$$\theta=\tan^{-1}\left(\frac{9}{40}\right)$$
$$=12.7^\circ$$
$$40+9j=41(\cos 12.7^\circ + j\sin 12.7^\circ)$$

73. $x^2-2x+4\Big|_{x=5-2j}=(5-2j)^2-2(5-2j)+4$

$$=25-20j+4j^2-10+4j+4$$
$$=19-16j-4$$
$$=15-16j$$

77. $x=2+j,\ x=2-j$

$$x-(2+j)=0,\ x-(2-j)=0$$
$$\left(x-(2+j)\right)\left(x-(2-j)\right)=0$$
$$x^2-(2+j)x-(2-j)x+(2+j)(2-j)=0$$
$$x^2-2x-jx-2x+jx+4-j^2=0$$
$$x^2-4x+4+1=0$$
$$x^2-4x+5=0$$

81. $(1+jx)^2=1+j-x^2$

$$1+2jx-x^2=1+j-x^2$$
$$2jx=j$$
$$x=\frac{1}{2}$$

85. $V_L=60j,\ V_C=-60j$

$$V=V_R+V_L+V_C$$
$$60=V_R+60j-60j$$
$$V_R=60\text{ V}$$

89. $2\pi fL=\dfrac{1}{2\pi fC}\Rightarrow f=\sqrt{\dfrac{1}{4\pi^2 LC}}$

$$=\sqrt{\frac{1}{4\pi^2(2.65)\left(18.3\times 10^{-6}\right)}}$$
$$f=22.9\text{ Hz}$$

93. $\dfrac{1}{\mu+j\omega n}=\dfrac{1}{(\mu+j\omega n)}\cdot\dfrac{(\mu-j\omega n)}{(\mu-j\omega n)}=\dfrac{\mu-j\omega n}{\mu^2+\omega^2 n^2}$

97. For a positive real number, the argument is always 0°, so the root will be a real number if $(k\cdot 360)/n$ is a multiple of 180, and it will be a pure imaginary number if $(k\cdot 360)/n$ is a multiple of 90, for $0\le k\le n-1$.

For a negative real number, the argument is always 180°, so the root will be a real number if $(180+k360)/n$ is a multiple of 180, and it will be a pure imaginary number if $(180+k\cdot 360)/n$ is a multiple of 90, for $0\le k\le n-1$.

CHAPTER 13

EXPONENTIAL AND LOGARITHMIC FUNCTIONS

13.1 Exponential Functions

1. For $x = -\frac{3}{2}$:

$$y = -2\left(4^x\right)$$
$$= -2\left(4^{-3/2}\right)$$
$$= -2\left(\frac{1}{8}\right)$$
$$= -\frac{1}{4}$$

5. **(a)** $y = -7(-5)^{-x}$, $-5 < 0$; not an exponential function.

(b) $y = -7\left(5^{-x}\right)$ is a real number multiple of an exponential function and therefore an exponential function.

9. $y = 9^x$, $x = -2$:

$$y = 9^{-2}$$
$$= \frac{1}{9^2}$$
$$= \frac{1}{81}$$

13. $y = 4^x$

x	-3	-2	-1	0	1	2	3
y	$\frac{1}{64}$	$\frac{1}{16}$	$\frac{1}{4}$	1	4	16	64

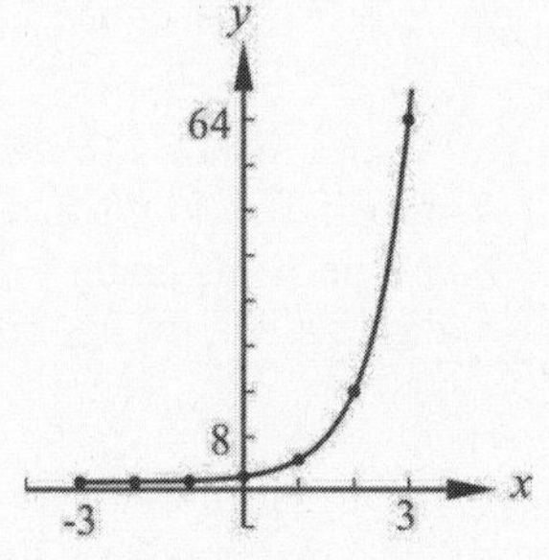

17. $y = 0.5\pi^x$

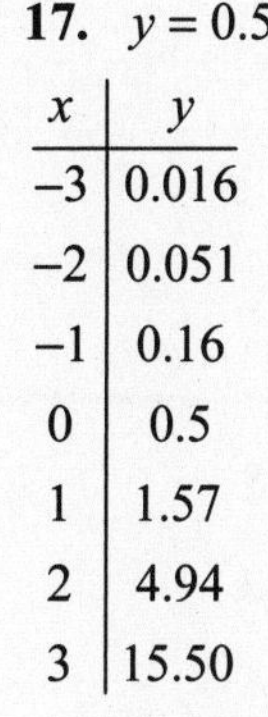

x	y
-3	0.016
-2	0.051
-1	0.16
0	0.5
1	1.57
2	4.94
3	15.50

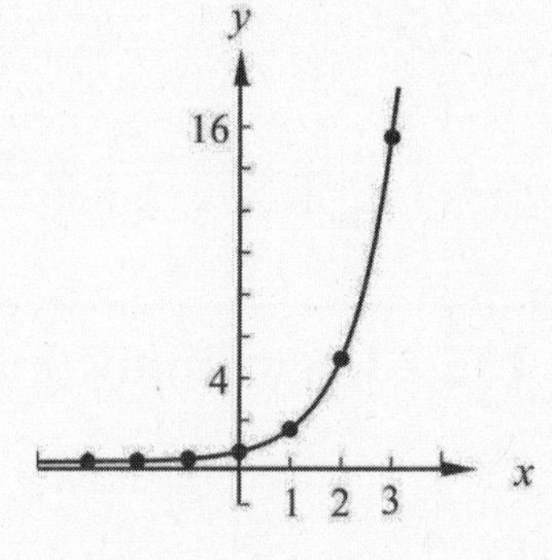

21. $y_1 = 0.1(0.25)^{2x}$

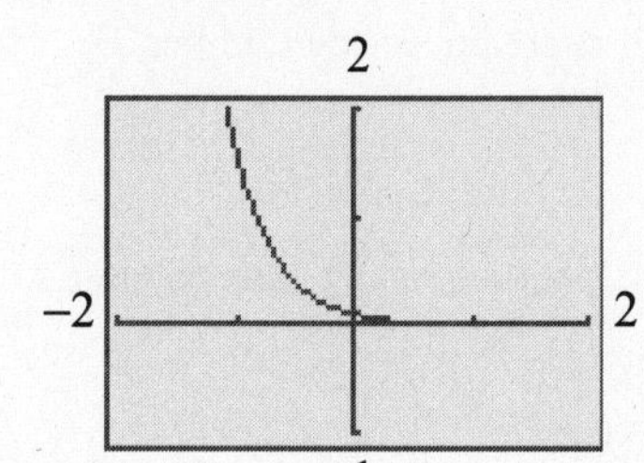
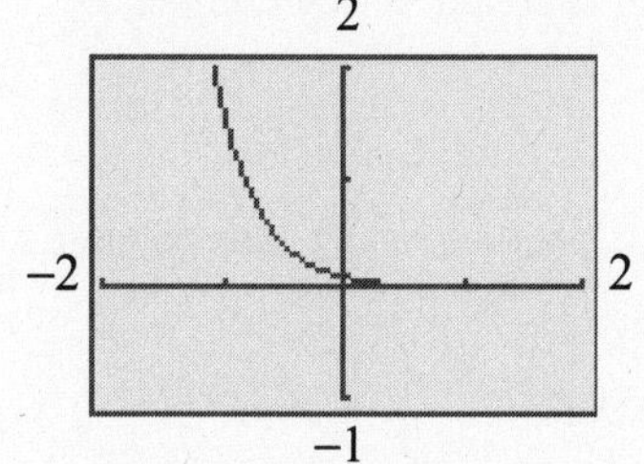

25. Substitute $x = 3$ and $y = 64$ into $y = b^x$:

$$64 = b^3$$
$$4^3 = b^3$$
$$b = 4$$

29. $y = 2^{|x|}$

Graph $y_1 = 2^{\text{abs}(x)}$

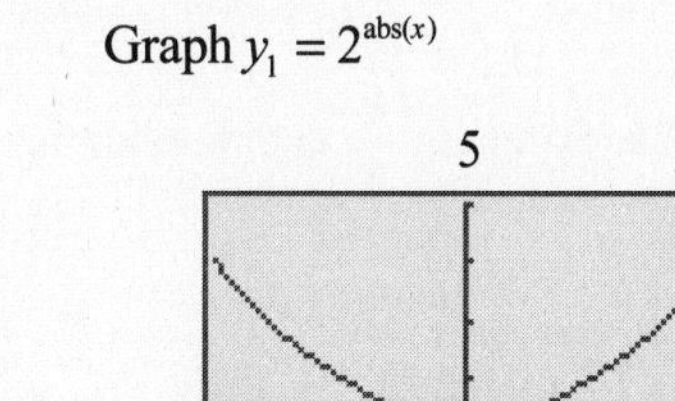
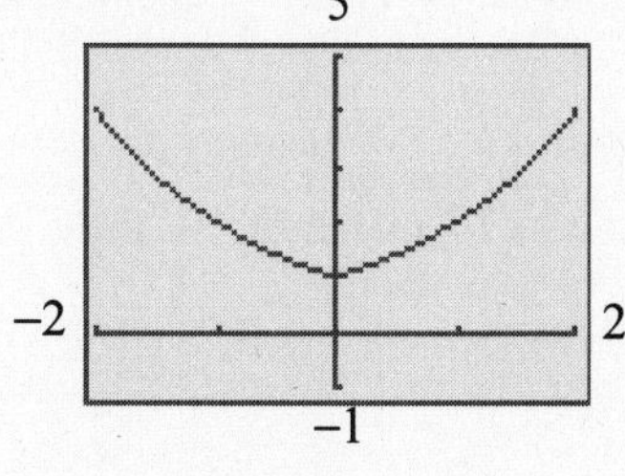

33. $V = 250(1.0500)^t$

$$= 250(1.0500)^4$$
$$= \$304$$

37. $q = 100e^{-10t}$

Graph $y_1 = 100e^{-10x}$.

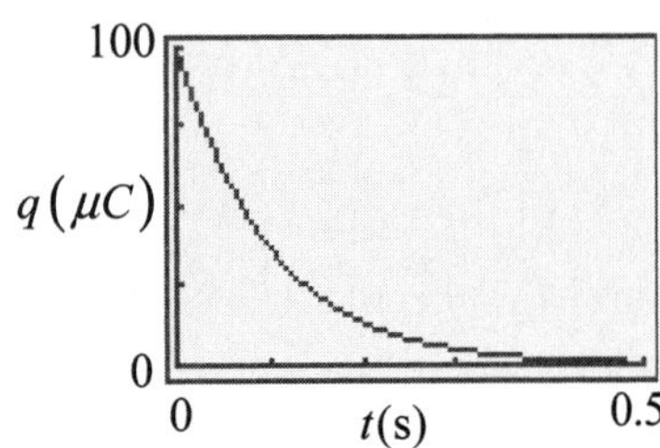

13.2 Logarithmic Functions

1. $32^{4/5} = 16$ in logarithmic form is

$\frac{4}{5} = \log_{32} 16.$

5. $3^3 = 27$ has base 3, exponent 3, and number 27.

$\log_3 27 = 3$

9. $7^{-2} = \frac{1}{49}$ has base 7, exponent -2, and number $\frac{1}{49}$.

$\log_7 \frac{1}{49} = -2$

13. $8^{1/3} = 2$ has base 8, exponent $\frac{1}{3}$, and number 2.

$\log_8 2 = \frac{1}{3}$

17. $\log_3 81 = 4$ has base 3, exponent 4, and number 81.

$3^4 = 81$

21. $\log_{25} 5 = \frac{1}{2}$ has base 25, exponent $\frac{1}{2}$, and number 5.

$25^{1/2} = 5$

25. $\log_{10} 0.1 = -1$ has base 10, exponent -1, and number 0.1.

$0.1 = 10^{-1}$

29. $\log_4 16 = x$ has base 4, exponent x, and number 16.

$4^x = 16$

$4^x = 4^2$

$x = 2$

33. $\log_7 y = 3$ has base 7, exponent 3, and number y.

$7^3 = y$

$y = 343$

37. $\log_b 5 = 2$ has base b, exponent 2, and number 5.

$b^2 = 5,\ b = \sqrt{5}$

41. $\log_{10} 10^{0.2} = x$ has base 10, exponent x, and number $10^{0.2}$

$10^x = 10^{0.2}$

$x = 0.2$

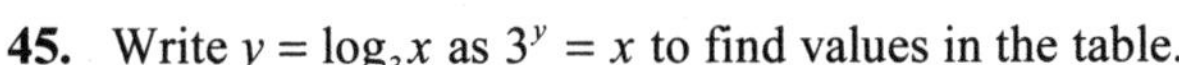

45. Write $y = \log_3 x$ as $3^y = x$ to find values in the table.

x	y
$\frac{1}{27}$	-3
$\frac{1}{9}$	-2
$\frac{1}{3}$	-1
1	0
3	1
9	2
27	3

49. $N = 0.2 \log_4 v$

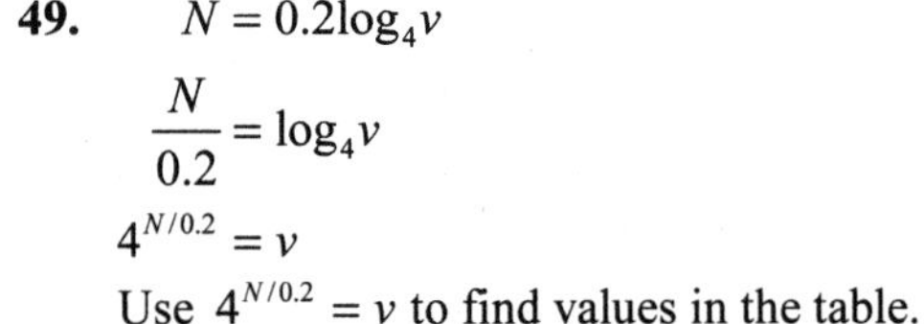

$\frac{N}{0.2} = \log_4 v$

$4^{N/0.2} = v$

Use $4^{N/0.2} = v$ to find values in the table.

v	N
$\frac{1}{16}$	$-.4$
$\frac{1}{4}$	$-.2$
$\frac{1}{2}$	$-.1$
1	0
2	.1
4	.2
16	.4

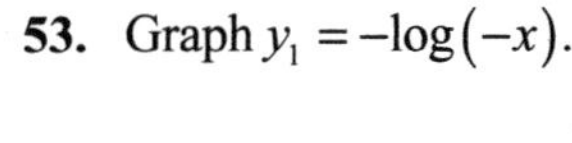

53. Graph $y_1 = -\log(-x)$.

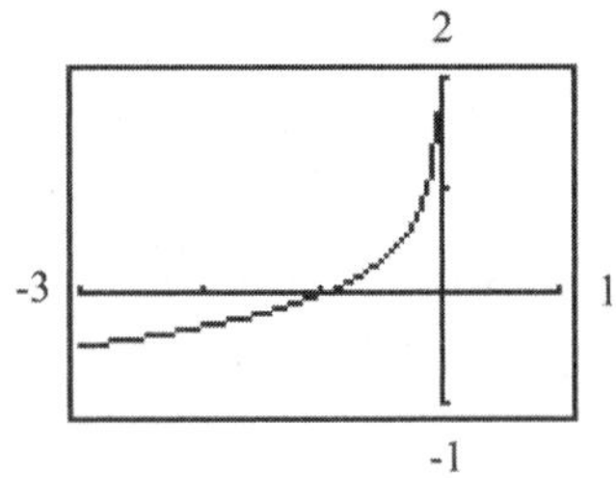

57. **(a)** $f(x) = \log_5 x$

$$f(\sqrt{5}) = \log_5 \sqrt{5}$$
$$= \log_5 5^{1/2}$$
$$N = \log_5 5^{1/2}$$
$$5^N = 5^{1/2}$$
$$N = \frac{1}{2}$$
$$f(\sqrt{5}) = \frac{1}{2}$$

(b) $f(0)$ does not exist.

61. Graph $y_1 = x - 2$ and $y_2 = \log x$ and find their intersection.

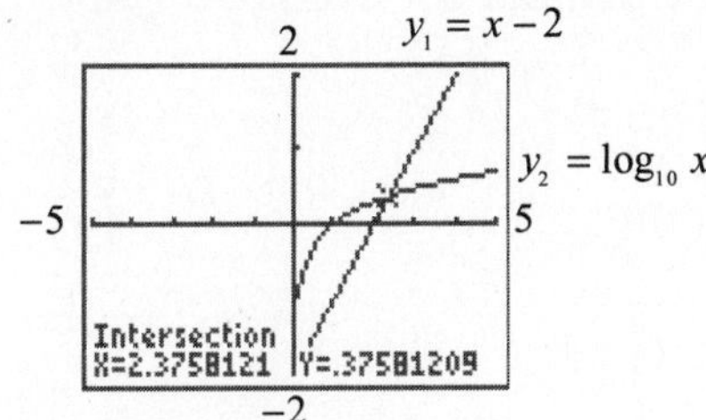

The solution is $x = 2.38$.
The equation can also be solved by finding the *zero* of the function
$y_1 = \log x - x + 2$

65. $\log_e\left(\frac{N}{N_0}\right) = -kt$

$$e^{-kt} = \frac{N}{N_0}$$
$$N = N_0 e^{-kt}$$

69. $t = N + \log_2 N$ where, $N > 0$ and $t > 0$.

N	t
1	1
2	3
4	6
8	11

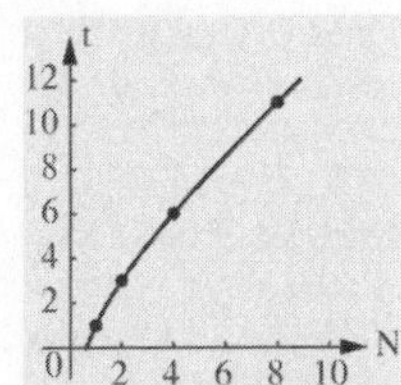

73. Solve $y = 10^{x/2}$ for x by changing into logarithmic form. We get $x = 2\log_{10} y$. Interchange x and y to get $y = 2\log_{10} x$, which is the inverse function.

Graph $y_1 = 10^{x/2}$, $y_2 = 2\log_{10} x$ and $y_3 = x$. For each graph to look like the mirror image of the other across $y = x$, the calculator window must be square. Do this by using the key sequence ZOOM Zsquare.

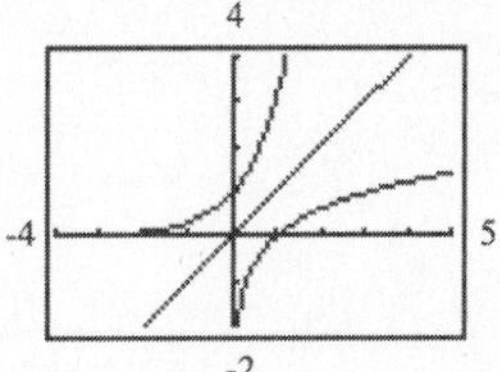

13.3 Properties of Logarithms

1. $\log_4 21 = \log_4(3(7))$
$$= \log_4 3 + \log_4 7$$

5. $\log_3 27 = \log_3 3^3$
$$= 3\log_3 3$$
$$= 3(1)$$
$$= 3$$

9. $\log_5 33 = \log_5(3 \cdot 11)$
$$= \log_5 3 + \log_5 11$$

13. $\log_2(a^3) = 3\log_2 a$

17. $8\log_5 \sqrt[4]{y} = 8\log_5 y^{1/4}$
$$= 2\log_5 y$$

21. $\log_b a + \log_b c = \log_b(ac)$

25. $-\log_b \sqrt{x} + \log_b x^2 = \log_b \frac{x^2}{x^{1/2}}$
$$= \log_b x^{3/2}$$

29. $\log_2\left(\frac{1}{32}\right) = \log_2\left(\frac{1}{2^5}\right)$
$$= \log_2 2^{-5}$$
$$= -5\log_2 2$$
$$= -5$$

33. $6\log_7 \sqrt{7} = 6\log_7 7^{1/2}$
$$= 3\log_7 7$$
$$= 3$$

37. $\log_3 18 = \log_3(9\cdot 2)$
$= \log_3 9 + \log_3 2$
$= \log_3 3^2 + \log_3 2$
$= 2\log_3 3 + \log_3 2$
$= 2 + \log_3 2$

41. $\log_3\sqrt{6} = \log_3(3\cdot 2)^{1/2}$
$= \frac{1}{2}\log_3(3\cdot 2)$
$= \frac{1}{2}\cdot[\log_3 3 + \log_3 2]$
$= \frac{1}{2}\cdot[1 + \log_3 2]$

45. $\log_b y = \log_b 2 + \log_b x$
$\log_b y = \log_b(2x)$
$y = 2x$

49. $\log_{10} y = 2\log_{10} 7 - 3\log_{10} x$
$= \log_{10} 7^2 - \log_{10} x^3$
$= \log_{10} 49 - \log_{10} x^3$
$\log_{10} y = \log_{10}\frac{49}{x^3}$
$y = \frac{49}{x^3}$

53. $\log_2 x + \log_2 y = 1$
$\log_2(xy) = 1$
$2^1 = xy$
$y = \frac{2}{x}$

57. $\log_{10}(x+3) = \log_{10} x + \log_{10} 3$
$= \log_{10}(3x)$
Then $x + 3 = 3x$
$2x = 3$
$x = \frac{3}{2}$

$y_1 = \log x + \log 3$
$y_2 = \log(x+3)$
Intersection X=1.5 Y=.65321251

This means that $x = \frac{3}{2}$ is the only value for which $\log_{10}(x+3) = \log_{10} x + \log_{10} 3$ is true.
For any other x-value, $\log_{10}(x+3) \neq \log_{10} x + \log_{10} 3$ and thus the statement is not true in general.
This can be also be seen from the following graph.

61. $\log_b\sqrt{x^2y^4} = \log_b\left(x^2y^4\right)^{1/2}$
$= \log_b\left(xy^2\right)$
$= \log_b x + 2\log_b y$
$= 2 + 2(3) = 8$

65. $\log_e D = \log_e a - br + cr^2$
$\log_e D - \log_e a = cr^2 - br$
$\log_e\frac{D}{a} = cr^2 - br$
$\frac{D}{a} = e^{cr^2 - br}$
$D = ae^{cr^2 - br}$

13.4 Logarithms to the Base 10

1. $\log 0.3654 = -0.4372$

5. $\log 9.24\times 10^6 = 6.966$

9. $\log\sqrt{274} = 1.219$

13. $10^{-1.3045} = 0.049\ 60$

17. $10^{-2.237\ 46} = 0.005\ 788\ 2$

21. $\log\left[\left(\sqrt[10]{7.32}\right)(2470)^{30}\right] = \log\sqrt[10]{7.32} + \log 2470^{30}$
$= 0.1\log 7.32 + 30\log 2470$
$= 101.867\ 36$

Therefore,
$\left(\sqrt[10]{7.32}\right)(2470)^{30} = 10^{101.867\ 36}$
$= 10^{101}\left(10^{0.867\ 3}\right)$
$= 10^{101}(7.36)$
$= 7.36\times 10^{101}$

25. $\log 81 = 4\log 3$
$\log 81 = 1.908485019$
$4\log 3 = 1.908485019$

29. $1.3\times 10^{-14}\% = 1.3\times 10^{-16}$
$\log 1.3\times 10^{-16} = \log 1.3 + \log 10^{-16}$
$= 0.11 - 16$
$= -15.89$

33. $\log v = 7.423$

$v = 10^{7.423}$

$v = 2.65 \times 10^{7}$ m/s

37. $\log\left(\log 10^{100}\right) = \log(100 \log 10)$

$= \log\left(10^{2}\right)$

$= 2 \log 10$

$= 2$

41. $R = \log\left(\frac{I}{I_0}\right)$; $I = 79\ 000\ 000 I_0$

$R = \log \frac{79\ 000\ 000 I_0}{I_0}$

$= \log 79\ 000\ 000$

$= 7.9$

The 2007 Peruvian earthquake had magnitude 7.9 on the Richter scale.

13.5 Natural Logarithms

1. $\ln 200 = \frac{\log 200}{\log e}$

$= 5.298$

5. $\ln 1.562 = \frac{\log 1.562}{\log e}$

$= \frac{0.1937}{0.4343}$

$= 0.4460$

9. $\log_7 42 = \frac{\log 42}{\log 7}$

$= \frac{1.6232}{0.8451}$

$= 1.92$

13. $\log_{40} 750 = \frac{\log 750}{\log 40}$

$= \frac{2.875}{1.6021}$

$= 1.795$

17. $\ln 1.394 = 0.3322$

21. $\ln 0.012\ 937^4 = -17.390\ 66$

25. $\log 0.685\ 28 = \frac{\ln 0.685\ 28}{\ln 10}$

$= -0.164\ 13$

29. $e^{0.008\ 421\ 0} = 1.0085$

33. $e^{-23.504} = 6.20 \times 10^{-11}$

37. We had shown algebraically in Example 9 of Section 13.2 that the functions are inverses of each other. To verify this using a graphing calculator, we graph $y_1 = 2^x$, $y_2 = \ln x / \ln 2$, and $y_3 = x$. We use the key sequence ZOOM Zsquare and note that $y_1 = 2^x$ and $y_2 = \ln x / \ln 2$ are indeed mirror images of each other—and therefore inverses.

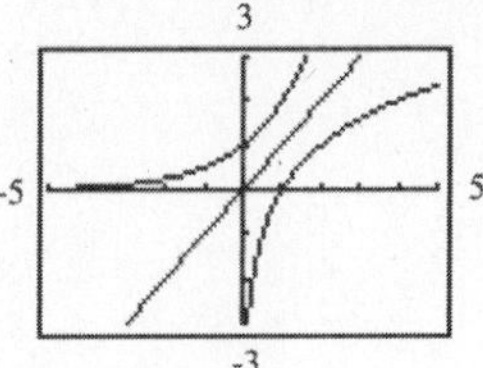

41. $4 \ln 3 = \ln 81$

$4 \ln 3 = 4.394449155$

$\ln 81 = 4.394449155$

45. $\ln(\log x) = 0$

$e^{\ln(\log x)} = e^{0}$

$= 1$

$\log x = 1$

$10^{\log x} = 10^{1}$

$x = 10$

49. $\ln f = 21.619$

$f = e^{21.619} = 2.45 \times 10^{9}$ Hz

53. $t = -\frac{L \cdot \ln\left(\frac{i}{I}\right)}{R}$

$= -\frac{1.25 \ln\left(\frac{0.10 I}{I}\right)}{7.5}$

$= 0.38$ s

13.6 Exponential and Logarithmic Equations

1.
$$\begin{aligned} 3^{x+2} &= 5 \\ \log 3^{x+2} &= \log 5 \\ (x+2)\log 3 &= \log 5 \\ x+2 &= \frac{\log 5}{\log 3} \\ x &= \frac{\log 5}{\log 3} - 2 \\ x &= -0.535 \end{aligned}$$

5.
$$\begin{aligned} 5^x &= 0.3 \\ \log 5^x &= \log 0.3 \\ x \log 5 &= \log 0.3 \\ x &= \frac{\log 0.3}{\log 5} \\ &= \frac{-0.5}{0.69897} \\ &= -0.7 \end{aligned}$$

9.
$$\begin{aligned} 6^{x+1} &= 78 \\ \ln 6^{x+1} &= \ln 78 \\ (x+1)\cdot \ln 6 &= \ln 78 \\ x+1 &= \frac{\ln 78}{\ln 6} \\ x &= \frac{\ln 78}{\ln 6} - 1 \\ x &= 1.432 \end{aligned}$$

13.
$$\begin{aligned} 0.6^x &= 2^{x^2} \\ \ln\left(0.6^x\right) &= \ln 2^{x^2} \\ x\cdot \ln 0.6 &= x^2 \cdot \ln 2 \\ x^2 \cdot \ln 2 - x \ln 0.6 &= 0 \\ x(x\cdot \ln 2 - \ln 0.6) &= 0 \\ x = 0 \quad \text{or} \quad x\cdot \ln 2 - \ln 0.6 &= 0 \\ x &= \frac{\ln 0.6}{\ln 2} \\ &= -0.7 \end{aligned}$$

17.
$$\begin{aligned} \log x^2 &= (\log x)^2 \\ 2\log x &= (\log x)^2 \\ (\log x)^2 - 2\log x &= 0 \\ \log x(\log x - 2) &= 0 \end{aligned}$$

$$\begin{aligned} \log x = 0 \quad &\text{or} \quad \log x = 2 \\ x = 1 \quad &\qquad\quad x = 10^2 \\ &\qquad\quad x = 100 \end{aligned}$$

21.
$$\begin{aligned} 2\log(3-x) &= 1 \\ \log(3-x) &= \frac{1}{2} \\ 3-x &= 10^{1/2} \\ &= 3.162 \\ -x &= 3.162 - 3 \\ &= 0.162 \\ x &= -0.162 \end{aligned}$$

25.
$$\begin{aligned} 3\ln 2 + \ln(x-1) &= \ln 24 \\ \ln 2^3 + \ln(x-1) &= \ln 24 \\ \ln 8 + \ln(x-1) &= \ln 24 \\ \ln\left[8(x-1)\right] &= \ln 24 \\ 8(x-1) &= 24 \\ x-1 &= 3 \\ x &= 4 \end{aligned}$$

29.
$$\begin{aligned} \log(2x-1) + \log(x+4) &= 1 \\ \log\left[(2x-1)(x+4)\right] &= 1 \\ (2x-1)(x+4) &= 10 \\ 2x^2 + 7x - 4 &= 10 \\ 2x^2 + 7x - 14 &= 0 \end{aligned}$$

Use the quadratic formula to solve for x:

$$\begin{aligned} x &= \frac{-7 \pm \sqrt{49 - 4(2)(-14)}}{2(2)} \\ &= \frac{-7 \pm \sqrt{161}}{4} \\ &= \frac{-7 \pm 12.689}{4} \end{aligned}$$

$x = 1.42$ or $x = -4.92$

Since logarithms of negative numbers are undefined, the unique solution is $x = 1.42$.

33. $4\left(3^x\right) = 5$. Graph $y_1 = 4\left(3^x\right) - 5$ and use the zero feature to solve.

$x = 0.203$

2
-1 1
Zero
X=.20311401 Y=0
-2

37. $2\ln 2-\ln x=-1$. Graph $y_1=2\ln 2-\ln x+1$ and use the zero feature to solve.

$x=10.9$

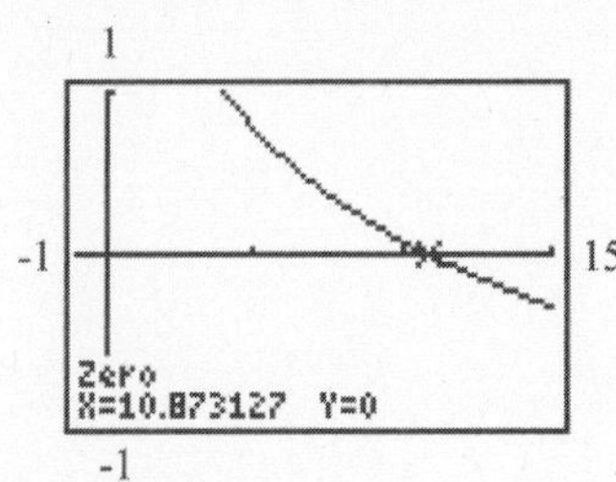

41. $y=1.5e^{-0.90x}$

Substituting $x=7.1$:

$$y=1.5e^{-0.90(7.1)}$$
$$=0.0025$$

45.
$$N=2^x$$
$$2.6\times10^8=2^x$$
$$\log 2^x=\log 2.6\times10^8$$
$$x\cdot\log 2=\log 2.6\times10^8$$
$$x=\frac{\log 2.6\times10^8}{\log 2}$$
$$=27.95393638$$
$$x=28.0$$

49.
$$\text{pH}=-\log\left(\text{H}^+\right)$$
$$4.764=-\log\left(\text{H}^+\right)$$
$$-4.764=\log\left(\text{H}^+\right)$$
$$\text{H}^+=10^{-4.764}$$
$$\text{H}^+=1.72\times10^{-5}\text{ mol/L}$$

53.
$$\ln c=\ln 15-0.20t$$
$$\ln c-\ln 15=-0.20t$$
$$\ln\frac{c}{15}=-0.20t$$
$$\frac{c}{15}=e^{-0.20t}$$
$$c=15e^{-0.20t}$$

57. $2^x+3^x=50$.

Graph $y_1=2^x+3^x-50$ and use the zero feature to solve.

$x=3.35$

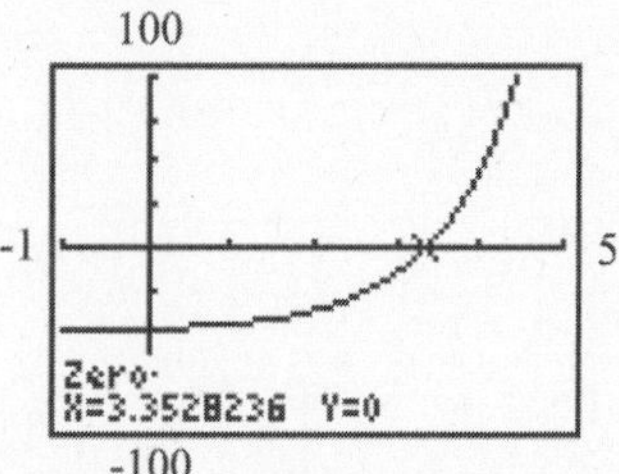

13.7 Graphs on Logarithmic and Semilogarithmic Paper

1. $y=2\left(3^x\right)$

x	−1	0	2	3	4	5
y	0.67	2	18	54	162	486

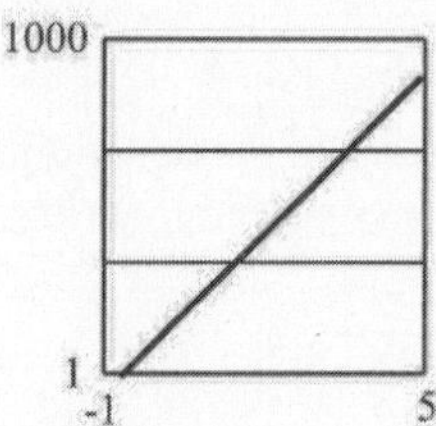

5. $y=5\left(4^x\right)$

x	0	1	2	3	4	5
y	5	20	80	320	1280	5120

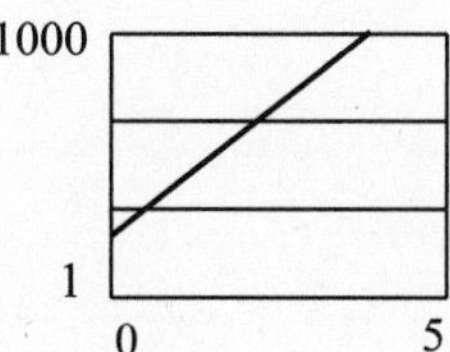

9. $y=2x^3+6x$

x	0	1	2	4	6	8
y	0	8	28	152	468	1072

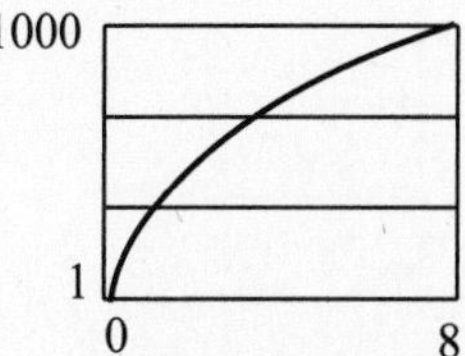

13. $y = x^{2/3}$

x	1	5	10	50	100	500	1000
y	1	2.9	4.6	13.6	21.5	63.0	100

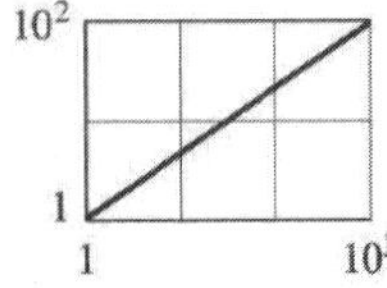

17. $x^2y^2 = 25$

$$y = \sqrt{\frac{25}{x^2}}$$
$$= \frac{5}{x}$$

x	0.1	0.5	1	10	50
y	50	10	5	0.5	0.1

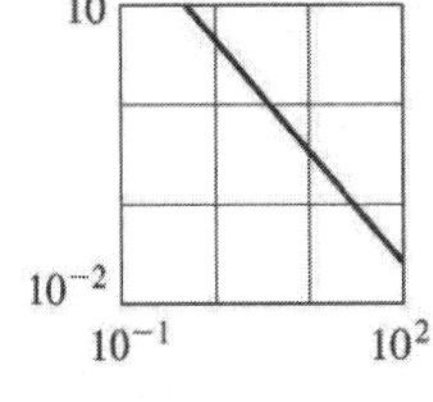

21. $y = 3x^6$, log– log paper

x	1	2	3	4
y	3	192	2187	12288

Taking logarithms on both sides of the equation, we have $\log y = \log 3 + 6\log x$, so we need logarithmic scales along both axes.

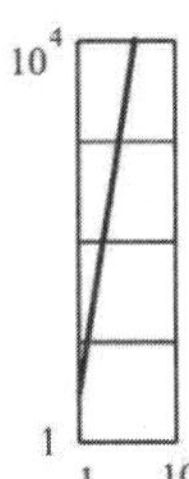

25. $x\sqrt{y} = 4$, $y = \frac{16}{x^2}$, log– log paper

x	1	25	50	75	100
y	16	0.0256	0.0064	$0.0028\overline{4}$	0.0016

Taking logarithms on both sides of the equation, we have $\log y = \log 16 - 2\log x$, so we need logarithmic scales along both axes.

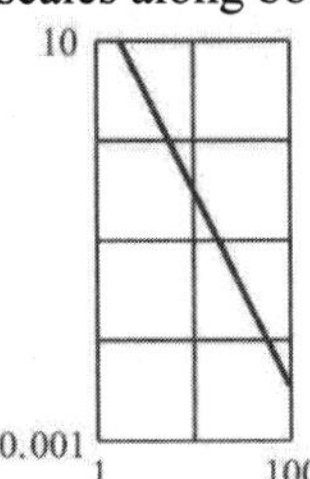

29. $N = N_0 e^{-0.028t}$, $N_0 = 1000$

t	0	25	50	75	100
N	1000	496.6	246.6	122.5	60.81

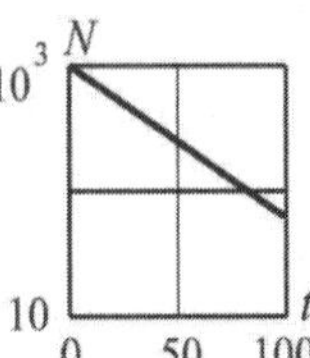

33.

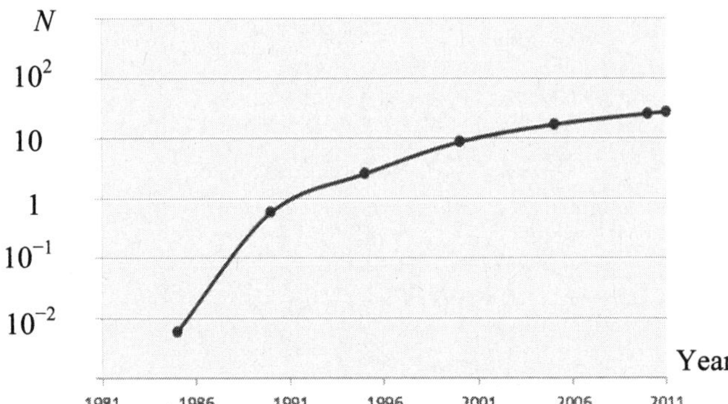

37.

d	0.063	0.13	0.19	0.25	0.38
R	600	190	100	72	46

d	0.50	0.75	1.0	1.5
R	29	17	10	6.0

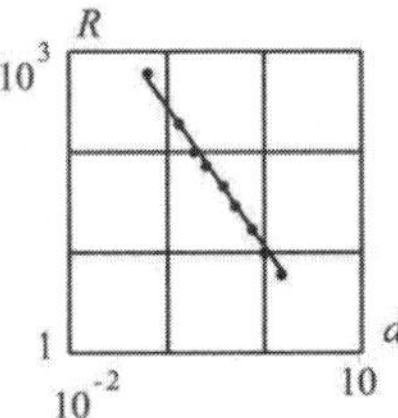

Review Exercises

1. Base is 10, exponent is 4, and number is x:

$\log_{10} x = 4$

$$x = 10^4$$
$$= 10000$$

5. Base is 2, exponent is x, and number is 8^2:

$2\log_2 8 = x$

$$2^x = 64$$
$$= 2^6$$
$$x = 6$$

9. Base is x, exponent is 2, and number is 36:

$$\log_x 36 = 2$$
$$x^2 = 36$$
$$= 6^2$$
$$x = 6$$

13. $\log_3 2x = \log_3 2 + \log_3 x$

17.
$$\log_2 28 = \log_2\left(2^2 \cdot 7\right)$$
$$= \log_2 2^2 + \log_2 7$$
$$= 2\log_2 2 + \log_2 7$$
$$= 2\cdot 1 + \log_2 7$$
$$= 2 + \log_2 7$$

21.
$$\log_3\left(\frac{9}{x}\right) = \log_3 9 - \log_3 x$$
$$= \log_3 3^2 - \log_3 x$$
$$= 2\log_3 3 - \log_3 x$$
$$= 2 - \log_3 x$$

25.
$$\log_6 y = \log_6 4 - \log_6 x$$
$$\log_6 y = \log_6 \frac{4}{x}$$
$$y = \frac{4}{x}$$

29.
$$\log_3 y = \frac{1}{2}\log_3 7 + \frac{1}{2}\log_3 x$$
$$\log_3 y = \log_3 \sqrt{7} + \log_3 \sqrt{x}$$
$$\log_3 y = \log_3 \sqrt{7x}$$
$$y = \sqrt{7x}$$

33.
$$2\left(\log_4 y - 3\log_4 x\right) = 3$$
$$\log_4 y - \log_4 x^3 = \frac{3}{2}$$
$$\log_4 \frac{y}{x^3} = \frac{3}{2}$$
$$\frac{y}{x^3} = 4^{3/2}$$
$$= 8$$
$$y = 8x^3$$

37. $y = 0.5\left(5^x\right)$

41.
$$y = \log_{3.15} x$$

Graph $y_1 = \dfrac{\ln x}{\ln 3.15}$

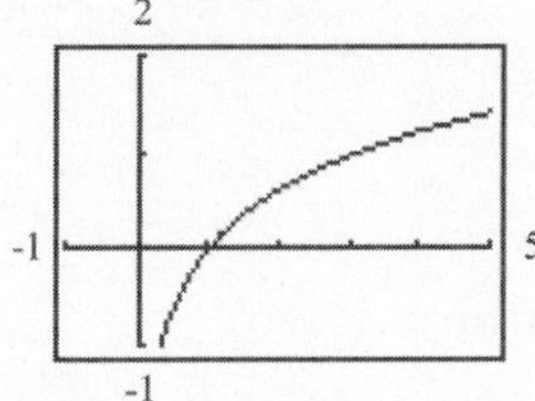

45.
$$\ln 8.86 = \frac{\log_{10} 8.86}{\log_{10} e}$$
$$= 2.182$$

49.
$$\log_{10} 65.89 = \frac{\ln 65.89}{\ln 10}$$
$$= 1.8188$$

53.
$$e^{2x} = 5$$
$$\ln e^{2x} = \ln 5$$
$$2x \cdot \ln e = \ln 5$$
$$2x \cdot 1 = \ln 5$$
$$x = \frac{\ln 5}{2}$$
$$= 0.805$$

57.
$$\log_4 z + \log_4 6 = \log_4 12$$
$$\log_4\left(z \cdot 6\right) = \log_4 12$$
$$6z = 12$$
$$z = 2$$

61. $y = 8^x$

x	1	2	3	4
y	8	64	512	4096

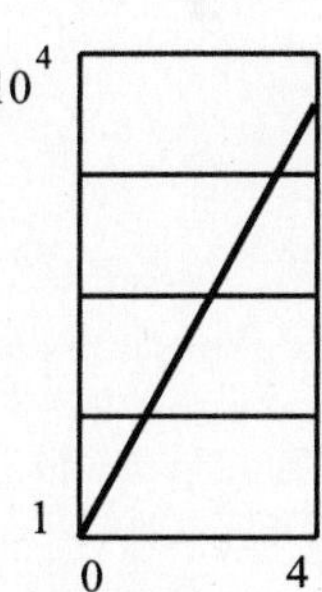

65. $10^{\log 4} = 4$

69.
$$2\log 3 - \log 6 = \log 1.5$$
$$0.1760912591 = 0.1760912591$$

73. If $f(x) = 2\log_b x$ and $f(8) = 3$ we have

$3 = 2\log_b 8$

$3 = \log_b 8^2$

$3 = \log_b 64$

$3 = \log_b 4^3$

Using the exponential form

$b^3 = 4^3$

$b = 4$

Therefore, $f(x) = 2\log_4 x$ and

$f(2) = 2\log_4 2$

$= \log_4 2^2$

$= \log_4 4$

$= 1$

77. Graph $y_1 = \dfrac{\ln x}{\ln 5} - 2x + 7$ and use the zero feature to solve.

$x = 3.92, 1.28\times10^{-5}$

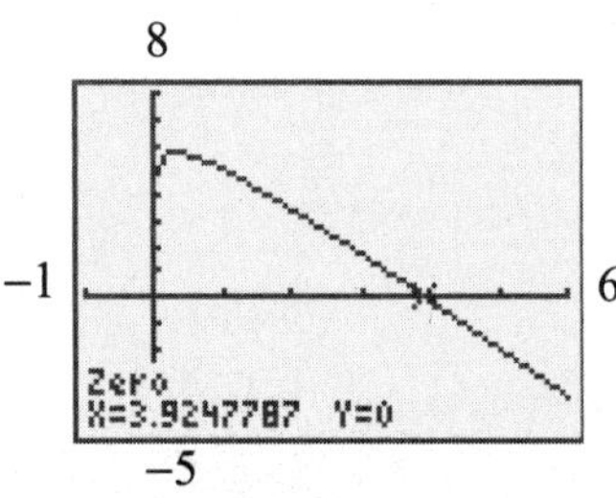

81. $\ln\dfrac{I}{I_0} = -\beta h$

$\dfrac{I}{I_0} = e^{-\beta h}$

$I = I_0 e^{-\beta h}$

85. $P = 937e^{0.0137t}$

Graph $y_1 = 937e^{0.0137x}$

89. $2\ln\omega = \ln 3g + \ln\sin\theta - \ln l$

$\ln\omega^2 = \ln\dfrac{3g\sin\theta}{l}$

$\omega^2 = \dfrac{3g\sin\theta}{l}$

$\sin\theta = \dfrac{\omega^2 l}{3g}$

93. $m_1 - m_2 = 2.5\log\dfrac{b_2}{b_1}$

$-1.4 - 6.0 = 2.5\log\dfrac{b_2}{b_1}$

$-2.96 = \log\dfrac{b_2}{b_1}$

$\dfrac{b_2}{b_1} = 10^{-2.96}$

$\dfrac{b_1}{b_2} = 10^{2.96}$

$= 912$

$b_1 = 912b_2$

Hence Sirius is 912 times brighter than the faintest stars.

97. $x = k\left(\ln I_0 - \ln I\right)$

We substitute $k = 5.00$ and $I = 0.850I_0$ to get

$x = 5.00\left(\ln I_0 - \ln 0.850I_0\right)$

$x = 5.00\ln\dfrac{I_0}{0.850I_0}$

$x = 0.813$ cm

101. $\ln n = -0.04t + \ln 20$

$\ln n - \ln 20 = -0.04t$

$\ln\dfrac{n}{20} = -0.04t$

$\dfrac{n}{20} = e^{-0.04t}$

$n = 20e^{-0.04t}$

105. We manipulate the second equation algebraically to obtain the first one.

$y = (2\ln x)/3 + \ln 4 - \ln(\ln e^2)$ requires $x > 0$

$y = \dfrac{2}{3}\ln x + \ln 4 - \ln(2\ln e)$

$y = \ln x^{2/3} + \ln 4 - \ln 2$

$y = \ln x^{2/3} + \ln\dfrac{4}{2}$

$y = \ln x^{2/3} + \ln 2$

$y = \ln(2x^{2/3})$ which only requires $x \neq 0$ since $2x^{2/3} > 0$ for all $x \neq 0$.

(a) The two equations are equivalent for $x > 0$.

(b) The graph of $y = (2\ln x)/3 + \ln 4 - \ln(\ln e^2)$ contains only the right-hand branch of the graph of $y = \ln(2x^{2/3})$.

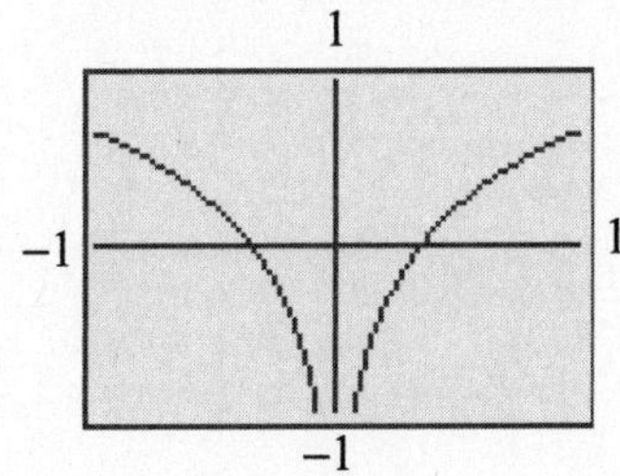

$y = \ln(2x^{2/3})$

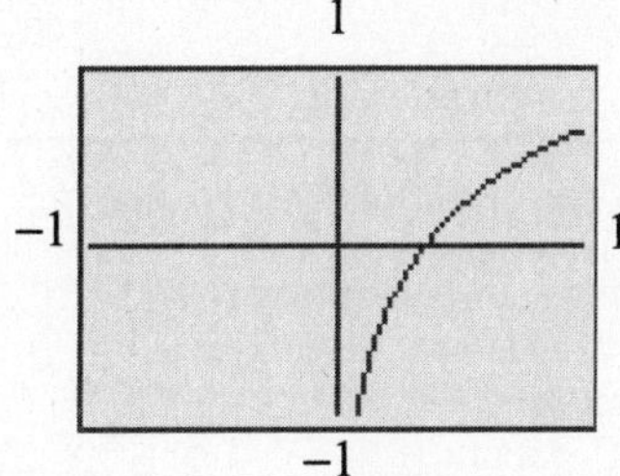

$y = (2\ln x)/3 + \ln 4 - \ln(\ln e^2)$

CHAPTER 14

ADDITIONAL TYPES OF EQUATIONS AND SYSTEMS OF EQUATIONS

14.1 Graphical Solution of Systems of Equations

1. Graph $y_1 = 3x^2 + 6x$.

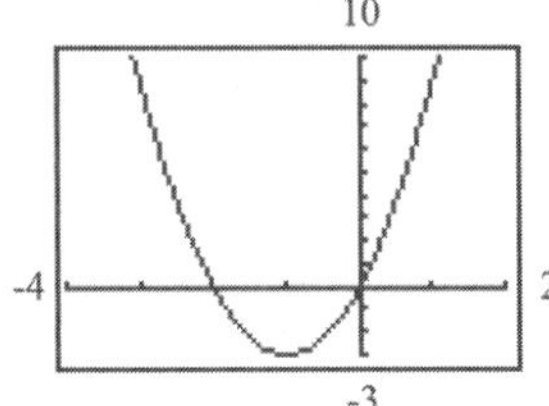

5. $y = 2x$

$x^2 + y^2 = 16 \Rightarrow y = \pm\sqrt{16 - x^2}$.

Graph $y_1 = 2x$, $y_2 = \sqrt{16 - x^2}$, and $y_3 = -\sqrt{16 - x^2}$.

Use the intersect feature to solve.

Solutions:

$x = 1.79,\ y = 3.58$

$x = -1.79,\ y = -3.58$

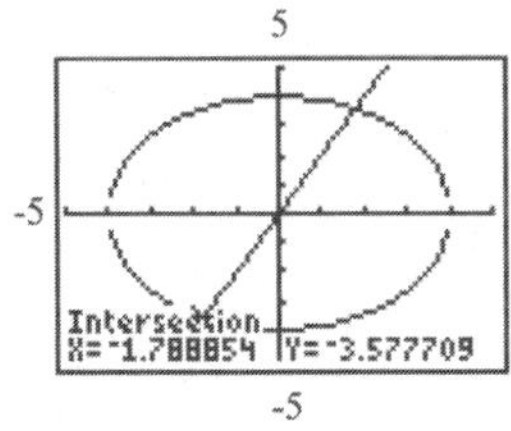

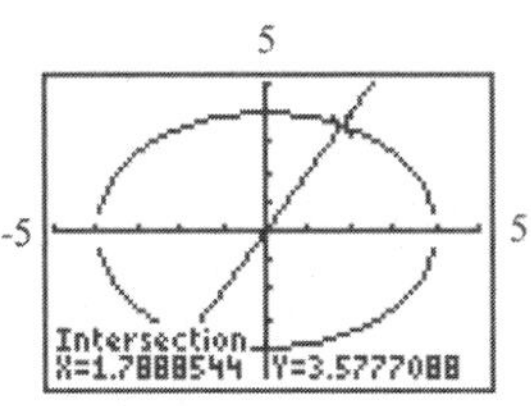

9. $y = x^2 - 2$

$4y = 12x - 7 \Rightarrow y = \dfrac{12x - 7}{4}$

Graph $y_1 = x^2 - 2$ and $y_2 = (12x - 7)/4$.

Use the intersect feature to solve.

Solution:

$x = 1.50,\ y = 0.25$

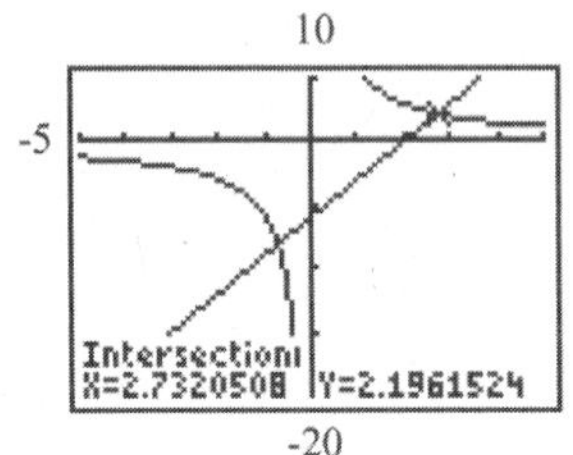

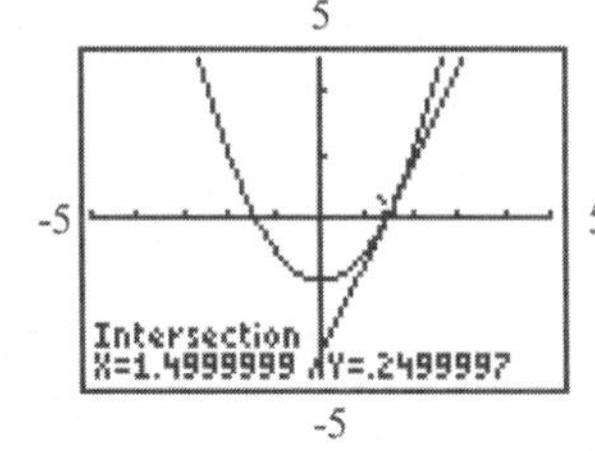

13. $y = -x^2 + 4$

$x^2 + y^2 = 9 \Rightarrow y = \pm\sqrt{9 - x^2}$

Graph $y_1 = \sqrt{9 - x^2}$, $y_2 = -\sqrt{9 - x^2}$, $y_3 = -x^2 + 4$.

Use the intersect feature to solve.

Solutions:

$x = 1.10,\ y = 2.79$

$x = -1.10,\ y = 2.79$

$x = -2.41,\ y = -1.80$

$x = 2.41,\ y = -1.80$

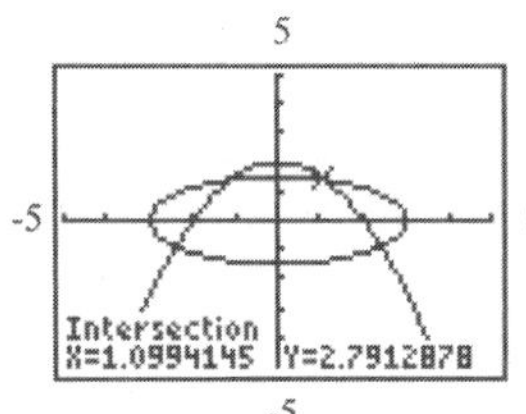

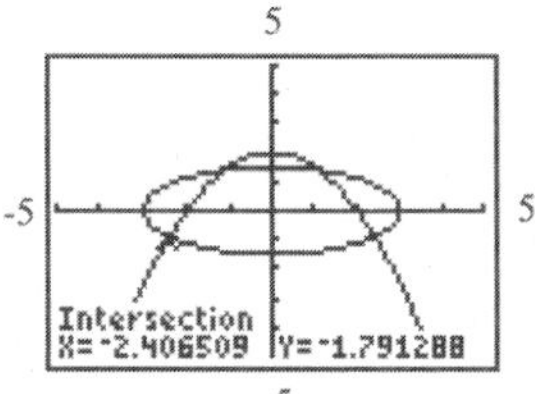

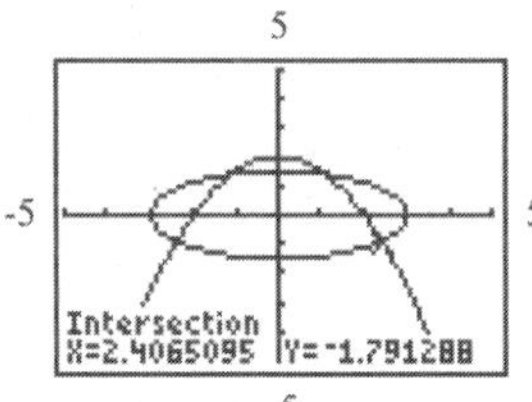

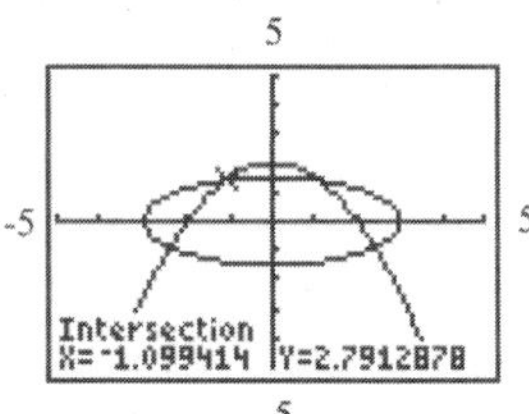

17. $2x^2 + 3y^2 = 19 \Rightarrow y = \pm\sqrt{(19 - 2x^2)/3}$

$x^2 + y^2 = 9 \Rightarrow y = \pm\sqrt{9 - x^2}$.

Graph $y_1 = \sqrt{(19 - 2x^2)/3}$, $y_2 = -\sqrt{(19 - 2x^2)/3}$, $y_3 = \sqrt{9 - x^2}$; $y_4 = -\sqrt{9 - x^2}$ and use the intersect feature to solve.

Solutions:

$x = -2.83,\ y = 1.00$

$x = 2.83,\ y = 1.00$

$x = -2.83,\ y = -1.00$

$x = 2.83,\ y = -1.00$

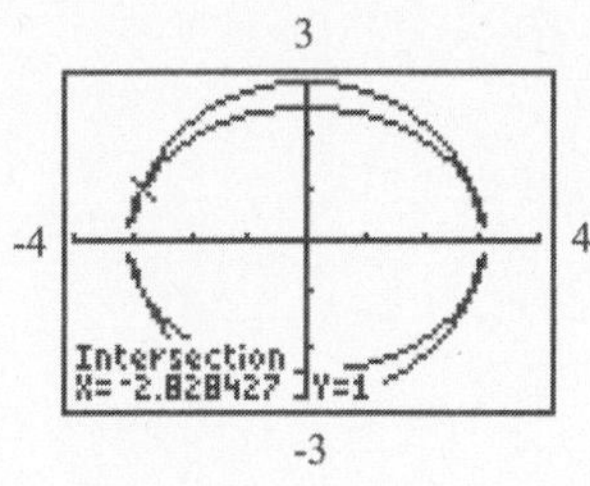

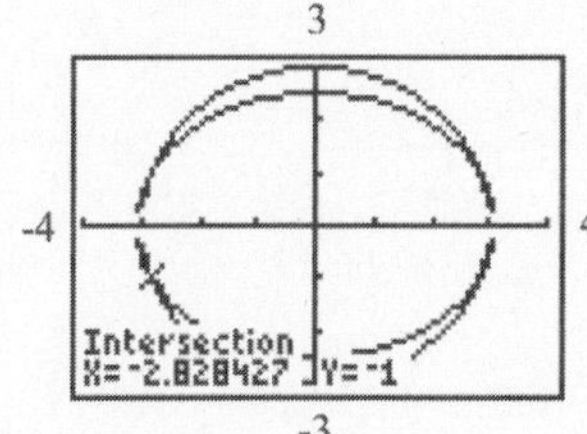

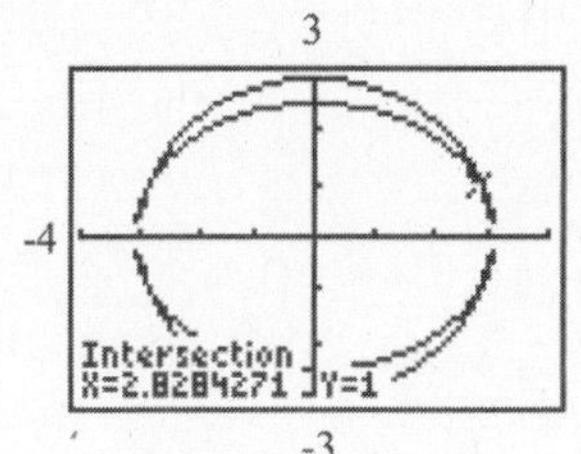

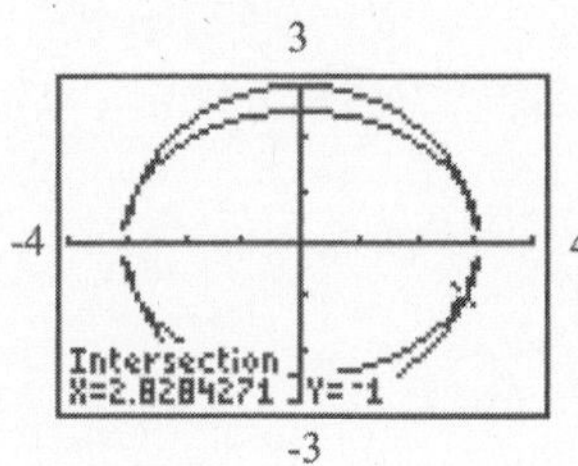

21. $y = x^2$

$y = \sin x$

Graph $y_1 = x^2$ and $y_2 = \sin x$ and use the intersect feature to solve.

Solutions:

$x = 0.00,\ y = 0.00$

$x = 0.88,\ y = 0.77$

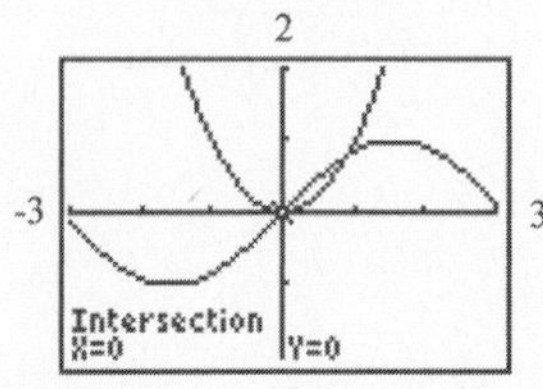

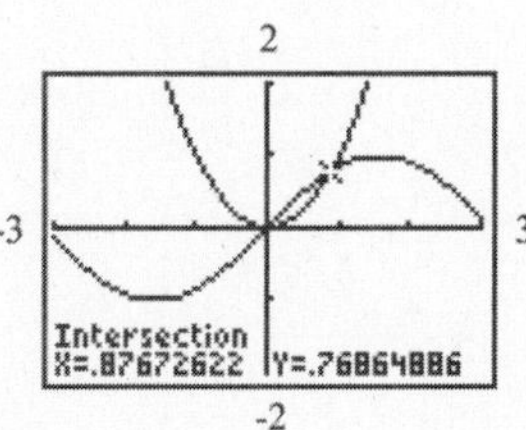

25. $x^2 - y^2 = 7 \Rightarrow y = \pm\sqrt{x^2 - 7},$

$y = 4\log_2 x \Rightarrow y = \dfrac{4\ln x}{\ln 2}.$

Graph $y_1 = \sqrt{x^2 - 7},\ y_2 = -\sqrt{x^2 - 7},$ and $y_3 = \dfrac{4\ln x}{\ln 2}.$

Use the intersect feature to solve.

Solution:

$x = 16.34,\ y = 16.12$

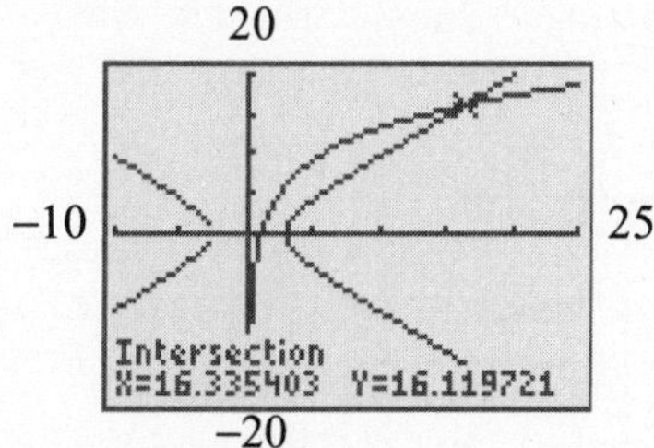

29. $10^{x+y} = 150$

$(x + y)\log 10 = \log 150$

$y = \log 150 - x$

$y = x^2$

Graph $y_1 = \log 150 - x,\ y_2 = x^2$ and use the intersect feature to solve.

Solutions:

$x = -2.06,\ y = 4.23$

$x = 1.06,\ y = 1.12$

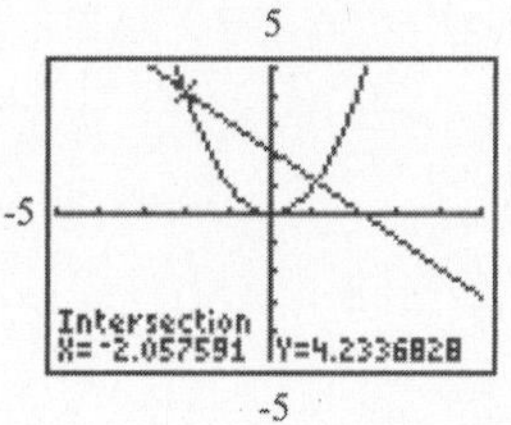

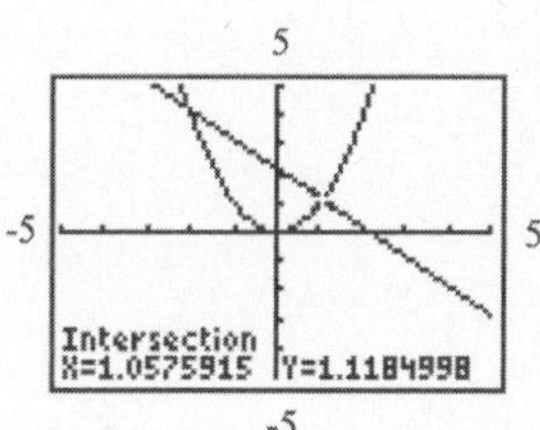

33. Let x = distance east

y = distance north

Then $y = 3x,\ y > 0$

$x^2 + y^2 = 5.2^2$

$= 27.04,\ x > 0,\ y > 0.$

Graph $y_1 = 3x,\ x > 0$ and

$y_2 = \sqrt{27.04 - x^2},\ x > 0$ and use the intersect feature to solve.

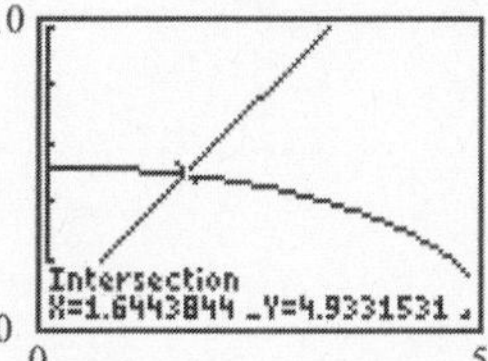

Solution:

$x = 1.64$ Km E, $y = 4.96$ km N.

37. $x^2 + y^2 = 41 \Rightarrow y = \pm\sqrt{41 - x^2}$

$y^2 = 20x + 140 \Rightarrow y = \pm\sqrt{20x + 140}$

Graph $y_1 = \sqrt{41 - x^2}$, $y_2 = -\sqrt{41 - x^2}$,

$y_3 = \sqrt{20x + 140}, y_4 = \sqrt{20x + 140}$

From the graph, there is no intersection. No, the meteorite will not strike the earth.

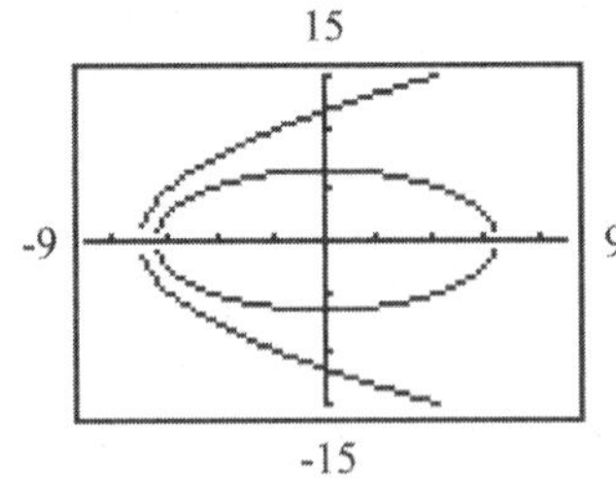

14.2 Algebraic Solution of Systems of Equations

1. $2x + y = 4 \Rightarrow y = 4 - 2x.$

Substitute into second equation: $x^2 - y^2 = 4$

$$x^2 - (4 - 2x)^2 = 4$$
$$x^2 - (16 - 16x + 4x^2) = 4$$
$$x^2 - 16 + 16x - 4x^2 = 4$$
$$3x^2 - 16x + 20 = 0$$
$$(x - 2)(3x - 10) = 0$$

$x - 2 = 0$ or $3x - 10 = 0$

$x = 2$ $\quad x = \dfrac{10}{3}$

$y = 4 - 2(2)$ $\quad y = 4 - 2\left(\dfrac{10}{3}\right) = -\dfrac{8}{3}$

Solutions:

$x = 2,\ y = 0$

$x = \dfrac{10}{3},\ y = -\dfrac{8}{3}$

5. $y = x + 1$

Substitute into: $y = x^2 + 1$

$$x + 1 = x^2 + 1$$
$$x^2 - x = 0$$
$$x(x - 1) = 0$$

$x = 0$ or $x - 1 = 0$

$y = (0) + 1$ $\quad x = 1$

$y = 1$ $\quad y = (1) + 1$

$\quad = 2$

Solutions:

$x = 0,\ y = 1$

$x = 1,\ y = 2$

9. $x + y = 1 \Rightarrow y = 1 - x$

Substitute into: $x^2 - y^2 = 1$

$$x^2 - (1 - x)^2 = 1$$
$$x^2 - 1 + 2x - x^2 = 1$$
$$-1 + 2x = 1$$
$$2x = 2$$
$$x = 1$$
$$y = 1 - (1) = 0$$

Solution:

$x = 1,\ y = 0$

13. $w + h = 2 \Rightarrow h = 2 - w$

Substitute into: $wh = 1$

$$w(2 - w) = 1$$
$$2w - w^2 = 1$$
$$w^2 - 2w + 1 = 0$$
$$(w - 1)^2 = 0$$
$$w - 1 = 0$$
$$w = 1$$
$$(1) + h = 2$$
$$h = 1$$

Solution:

$w = 1,\ h = 1$

17. $y = x^2$

Substitute into: $y = 3x^2 - 50$

$$x^2 = 3x^2 - 50$$
$$2x^2 = 50$$
$$x^2 = 25$$
$$x = \pm 5$$
$$y = (\pm 5)^2$$
$$y = 25$$

Solutions:

$x = 5,\ y = 25$

$x = -5,\ y = 25$

Alternatively, subtracting the second equation from the first also results in $2x^2 = 50$.

21. $D^2 - 1 = R \Rightarrow D^2 = 1 + R$

Substitute into: $D^2 - 2R^2 = 1$

$1 + R - 2R^2 = 1$

$R - 2R^2 = 0$

$R(1 - 2R) = 0$

$R = 0$ or $1 - 2R = 0$

$2R = 1$

$R = \frac{1}{2}$

$D^2 = 1 + (0) \qquad D^2 = 1 + \left(\frac{1}{2}\right)$

$= \frac{6}{4}$

$D = \pm 1 \qquad D = \frac{\pm\sqrt{6}}{2}$

Solutions:

$R = 0, D = 1$

$R = 0, D = -1$

$R = \frac{1}{2}, D = \frac{\sqrt{6}}{2}$

$R = \frac{1}{2}, D = \frac{-\sqrt{6}}{2}$

Alternatively, subtracting the first equation from the second also results in $R - 2R^2 = 0$.

25. $x^2 + 3y^2 = 37$

$2x^2 - 9y^2 = 14$

Multiplying the first equation by 3:

$3x^2 + 9y^2 = 111$

$2x^2 - 9y^2 = 14$

$5x^2 \quad = 125$

$x^2 = 25$

$x = \pm 5$

$(\pm 5)^2 + 3y^2 = 37$

$25 + 3y^2 = 37$

$3y^2 = 12$

$y^2 = 4$

$y = \pm 2$

Solutions:

$x = 5, y = 2$

$x = 5, y = -2$

$x = -5, y = 2$

$x = -5, y = -2$

29. $x - y = a - b \Rightarrow y = x - (a - b)$

Substituting into: $x^2 - y^2 = a^2 - b^2$

$x^2 - (x - (a - b))^2 = a^2 - b^2$

$x^2 - \left(x^2 - 2(a - b)x + (a - b)^2\right) = a^2 - b^2$

$x^2 - x^2 + 2(a - b)x - a^2 + 2ab - b^2 = a^2 - b^2$

$2(a - b)x + 2ab - 2a^2 = 0$

$(a - b)x - a(a - b) = 0$

$x = a$

$y = x - (a - b)$

$= a - a + b$

$y = b$

Solution:

$x = a, y = b$

33.

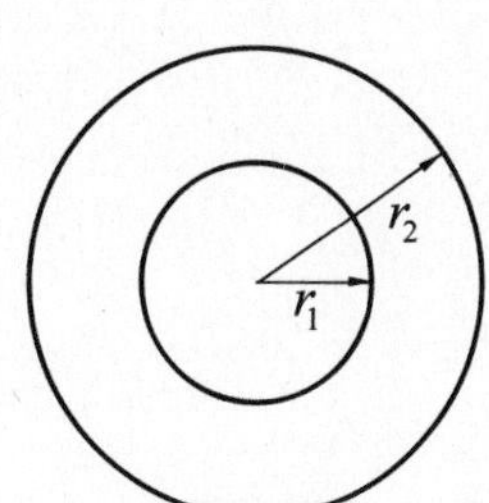

Let r_1 = inner radius, r_2 = outer radius.

$r_1 = r_2 - 2.00$

$\pi r_2^2 - \pi r_1^2 = 37.7$

Substituting $r_1 = r_2 - 2.00$ into $\pi r_2^2 - \pi r_1^2 = 37.7$

$\pi r_2^2 - \pi (r_2 - 2.00)^2 = 37.7$

$\pi r_2^2 - \pi r_2^2 + 4.00\pi r_2 - 4.00\pi = 37.7$

$r_2 = \frac{37.7 + 4.00\pi}{4.00\pi}$

$r_2 = 4.00$

Then $r_1 = r_2 - 2.00$

$= 2.00$

The radii are 2.00 cm and 4.00 cm.

37.

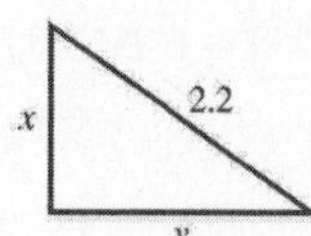

$x + y + 2.2 = 4.6 \Rightarrow y = 2.4 - x$

Substitute into: $x^2 + y^2 = 2.2^2$

$$x^2+(2.4-x)^2=4.84$$
$$x^2+5.76-4.8x+x^2=4.84$$
$$2x^2-4.8x+0.92=0$$
$$x^2-2.4x+0.46=0$$

Using the quadratic formula:

$$x=\frac{-(-2.4)\pm\sqrt{(-2.4)^2-4(1)(0.46)}}{2(1)}$$
$$=\frac{2.4\pm\sqrt{3.92}}{2}$$

$x=2.19$ or $x=0.21$

$(2.19)+y=2.40$ $\quad$ $(0.21)+y=2.40$

$y=0.21$ $\quad$ $y=2.19$

The lengths of the sides of the truss are 2.19 m and 0.21 m.

41.

l, 2.00, w

$$wl=1600\Rightarrow w=\frac{216}{l}$$

Substitute into: $2(w-4.00)(l-4.00)=224$

$$\left(\frac{216}{l}-4.00\right)(l-4.00)=112$$
$$216-\frac{864}{l}-4l+16=112$$
$$120l-4l^2-864=0$$
$$l^2-30l+216=0$$
$$(l-18)(l-12)=0$$

$l-18=0$ or $l-12=0$

$l=18$ $\quad$ $l=12$

$w=\frac{216}{(18)}$ $\quad$ $w=\frac{216}{(12)}$

$=12$ $\quad$ $=18$

The dimensions of the sheet are 18.0 cm. by 12.0 cm.

14.3 Equations in Quadratic Form

1. $2x^4-7x^2=4$

$2x^4-7x^2-4=0$, let $y=x^2$

$$2y^2-7y-4=0$$
$$(y-4)(2y+1)=0$$

$y-4=0$ or $2y+1=0$

$y=4$ $\quad$ $y=-\frac{1}{2}$

$x^2=4$ $\quad$ $x^2=-\frac{1}{2}$

$x=\pm2$ $\quad$ $x=\pm\sqrt{-\frac{1}{2}}$

$$=\pm j\sqrt{\frac{1}{2}\cdot\frac{2}{2}}$$
$$=\pm j\frac{\sqrt{2}}{2}$$

Check:

$$2(\pm2)^4-7(\pm2)^2=32-28=4$$
$$2\left(\pm j\frac{\sqrt{2}}{2}\right)^4-7\left(\pm j\frac{\sqrt{2}}{2}\right)^2=\frac{1}{2}+\frac{7}{2}=4$$

5. $x^{-2}-2x^{-1}-8=0$

Let $y=x^{-1}$, $y^2=x^{-2}$

$$y^2-2y-8=0$$
$$(y-4)(y+2)=0$$

$y-4=0$ or $y+2=0$

$y=4$ $\quad$ $y=-2$

$x^{-1}=4$ $\quad$ $x^{-1}=-2$

$x=\frac{1}{4}$ $\quad$ $x=-\frac{1}{2}$

Check:

$$\left(\frac{1}{4}\right)^{-2}-2\left(\frac{1}{4}\right)^{-1}-8=16-8-8=0$$
$$\left(-\frac{1}{2}\right)^{-2}-2\left(-\frac{1}{2}\right)^{-1}-8=4+4-8=0$$

9. $2x-7\sqrt{x}+5=0$

Let $y=\sqrt{x}$, $y^2=x$

$$2y^2-7y+5=0$$
$$(2y-5)(y-1)=0$$

$2y-5=0$ or $y-1=0$

$2y=5$ $\quad$ $y=1$

$y=\frac{5}{2}$

$\sqrt{x}=\frac{5}{2}$ $\quad$ $\sqrt{x}=1$

$x=\frac{25}{4}$ $\quad$ $x=1$

Check:

$$2\left(\frac{25}{4}\right)-7\sqrt{\frac{25}{4}}+5=\frac{25}{2}-\frac{35}{2}+\frac{10}{2}=0$$

$$2(1)-7\sqrt{1}+5=2-7+5=0$$

13. $x^{2/3}-2x^{1/3}-15=0$

Let $y=x^{1/3}$, $y^2=x^{2/3}$

$y^2-2y-15=0$

$(y-5)(y+3)=0$

$y-5=0$ or $y+3=0$

$y=5$ $\quad y=-3$

$x^{1/3}=5$ $\quad x^{1/3}=-3$

$x=125$ $\quad x=-27$

Check:

$125^{2/3}-2(125)^{1/3}-15=25-10-15=0$

$(-27)^{2/3}-2(-27)^{1/3}-15=9+6-15=0$

17. $(x-1)-\sqrt{x-1}=20$

Let $y=\sqrt{x-1}$, $y^2=x-1$

$y^2-y-20=0$

$(y-5)(y+4)=0$

$y-5=0$ or $y+4=0$

$y=5$ $\quad y=-4$

$\sqrt{x-1}=5$ $\quad \sqrt{x-1}=-4$

$x-1=25$ $\quad$ not possible

$x=26$

Check:

$(26-1)-\sqrt{26-1}=25-\sqrt{25}$

$=25-5$

$=20$

21. $x-3\sqrt{x-2}=6$

Let $y=\sqrt{x-2}$, $y^2=x-2\Rightarrow y^2+2=x$

$y^2+2-3y=6$

$y^2-3y-4=0$

$(y-4)(y+1)=0$

$y-4=0$ or $y+1=0$

$y=4$ $\quad y=-1$

$\sqrt{x-2}=4$ $\quad \sqrt{x-2}=-1$

$x-2=16$ $\quad$ (not possible)

$x=18$

Check:

$18-3\sqrt{18-2}=18-3\sqrt{16}=6$

25. $e^{2x}-e^x=0$

Let $y=e^{2x}$

$y^2-y=0$

$y(y-1)=0$

$y-1=0$ or $y=0$

$y=1$

$e^x=1$ $\quad e^x=0$

$x=0$ $\quad$ (not possible)

Check:

$e^{2(0)}-e^0=1-1=0$

29. $x+2=3\sqrt{x}$

$x-3\sqrt{x}+2=0$

Let $y=\sqrt{x}, y^2=x, y\ge 0, x\ge 0$

$y^2-3y+2=0$

$(y-2)(y-1)=0$

$y-2=0$ or $y-1=0$

$y=2$ $\quad y=1$

$\sqrt{x}=2$ $\quad \sqrt{x}=1$

$x=4$ $\quad x=1$

Check:

$4+2-3\sqrt{4}=6-6=0$

$1+2-3\sqrt{1}=3-3=0$

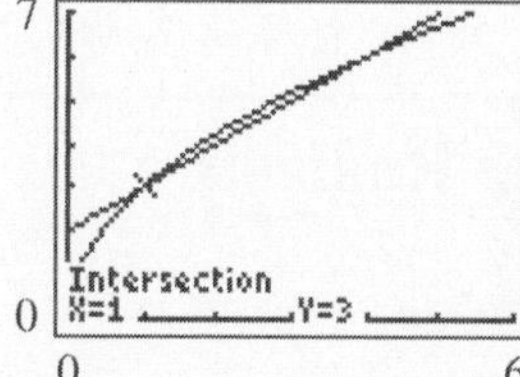

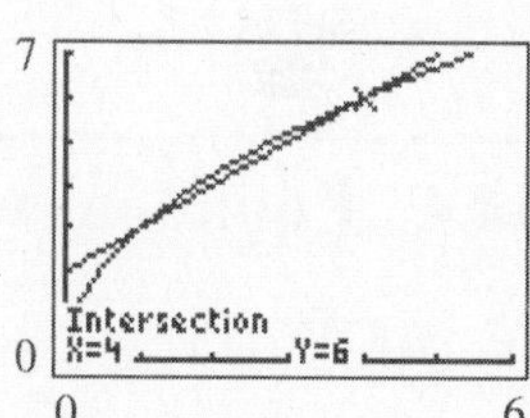

33. $\log\left(x^4+4\right)-\log 5x^2=0$

$$\log\frac{x^4+4}{5x^2}=0$$

$$\frac{x^4+4}{5x^2}=1$$

$x^4+4-5x^2=0$

Let $y=x^2$

$y^2-5y+4=0$

$(y-4)(y-1)=0$

$y-4=0$ or $y-1=0$

$y=4$ $y=1$

$x^2=4$ $x^2=1$

$x=\pm 2$ $x=\pm 1$

Check:

$\log((\pm 2)^4+4)-\log(5(\pm 2)^2)=\log\ 20-\log\ 20=0$

$\log((\pm 1)^4+4)-\log(5(\pm 1)^2)=\log 5-\log 5=0$

37. $\sqrt{F}=\frac{2\sqrt{p}}{(1-p)}$

$\sqrt{16}=\frac{2\sqrt{p}}{1-p}$

$4(1-p)=2\sqrt{p}$

$2(1-p)=\sqrt{p}$

$2-2p=\sqrt{p}$

$2p+\sqrt{p}-2=0$

Let $y=\sqrt{p},\ y\ge 0,\ p\ge 0$

$2y^2+y-2=0$

$y=\frac{-1\pm\sqrt{1^2-4(2)(-2)}}{2(2)}$

$=\frac{-1\pm\sqrt{17}}{4}$

The negative root is discarded.

$\sqrt{p}=\frac{-1+\sqrt{17}}{4}$

$p=0.610$

14.4 Equations with Radicals

1. $2\sqrt{3x-1}=3$; square both sides

$4(3x-1)=9$

$12x-4=9$

$12x=13$

$x=\frac{13}{12}$

Check: $2\sqrt{3\left(\frac{13}{12}\right)-1}\overset{?}{=}3$

$3=3$

$x=\frac{13}{12}$ is the solution.

5. $\sqrt{x-8}=2$; square both sides

$x-8=4$

$x=12$

Check:

$\sqrt{12-8}=\sqrt{4}=2$

$x=12$ is the solution.

9. $\sqrt{3x+2}=3x$; square both sides

$3x+2=9x^2$

$9x^2-3x-2=0$

$(3x+1)(3x-2)=0$

$3x+1=0$ or $3x-2=0$

$3x=-1$ $3x=2$

$x=-\frac{1}{3}$ $x=\frac{2}{3}$

Check:

$\sqrt{3\cdot\frac{1}{3}+2}\overset{?}{=}3\cdot\frac{-1}{3}$

$\sqrt{-1+2}\overset{?}{=}-1$

$1\ne -1$

$x=-\frac{1}{3}$ is not a solution.

Check:

$\sqrt{3\cdot\frac{2}{3}+2}\overset{?}{=}3\cdot\frac{2}{3}$

$\sqrt{2+2}\overset{?}{=}2$

$2=2$

$x=\frac{2}{3}$ is the only solution.

13. $\sqrt[3]{y-5}=3$; cube both sides

$y-5=3^3$

$=27$

$y=32$

Check:

$\sqrt[3]{32-5}=\sqrt[3]{27}=3$

$y=32$ is the solution.

17. $\sqrt{x^2-9}=4$; square both sides

$x^2-9=16$

$x^2=25$

$x=\pm 5$

Check:

$$\sqrt{(\pm5)^2-9}=\sqrt{25-9}$$
$$=\sqrt{16}$$
$$=4$$

The solutions are $x=\pm5$.

21. $\sqrt{5+\sqrt{x}}=\sqrt{x}-1$; square both sides

$$5+\sqrt{x}=x-2\sqrt{x}+1$$
$$x-3\sqrt{x}-4=0$$

Let $y=\sqrt{x}, y^2=x, y\ge0$

$$y^2-3y-4=0$$
$$(y-4)(y+1)=0$$
$$y-4=0 \quad\text{or}\quad y+1=0$$
$$y=4 \qquad y=-1$$
$$x=4^2 \qquad \text{(not possible)}$$
$$=16$$

Check:

$$\sqrt{5+\sqrt{16}}\overset{?}{=}\sqrt{16}-1$$
$$\sqrt{9}\overset{?}{=}4-1$$
$$3=3$$

The only solution is $x=16$.

25. $2\sqrt{x+2}-\sqrt{3x+4}=1$

$$2\sqrt{x+2}=1+\sqrt{3x+4};\text{ square both sides}$$
$$4(x+2)=\left(1+\sqrt{3x+4}\right)^2$$
$$4x+8=1+2\sqrt{3x+4}+3x+4$$
$$x+3=2\sqrt{3x+4};\text{square both sides}$$
$$(x+3)^2=4(3x+4)$$
$$x^2+6x+9=12x+16$$
$$x^2-6x-7=0$$
$$(x-7)(x+1)=0$$
$$x=7 \text{ or } x=-1$$

Check:

$$2\sqrt{-1+2}-\sqrt{3(-1)+4}=2\sqrt{1}-\sqrt{1}=1$$
$$2\sqrt{7+2}-\sqrt{3(7)+4}=2\sqrt{9}-\sqrt{25}=1$$

The solutions are $x=-1$ and $x=7$.

29. $\sqrt{6x-5}-\sqrt{x+4}=2$

$$\sqrt{6x-5}=\sqrt{x+4}+2;\text{ square both sides}$$
$$6x-5=x+4+4\sqrt{x+4}+4$$
$$5x-13=4\sqrt{x+4};\text{ square both sides}$$
$$25x^2-130x+169=16x+64$$
$$25x^2-146x+105=0$$

Using the quadratic formula:

$$x=\frac{-(-146)\pm\sqrt{(-146)^2-4(25)(105)}}{2(25)}$$
$$=\frac{146\pm104}{2(25)}$$
$$x=5 \quad\text{or}\quad x=\frac{21}{25}$$

Check:

$$\sqrt{6(5)-5}-\sqrt{5+4}=\sqrt{25}-\sqrt{9}=2$$
$$\sqrt{6\left(\frac{21}{25}\right)-5}-\sqrt{\frac{21}{25}+4}=\sqrt{\frac{1}{25}}-\sqrt{\frac{121}{25}}=\frac{1}{5}-\frac{11}{5}=-2\ne2$$

The only solution is $x=5$.

33. $\sqrt{x-2}=\sqrt[4]{x-2}+12$

$$\sqrt{x-2}-12=\sqrt[4]{x-2};\text{ square both sides}$$
$$x-2-24\sqrt{x-2}+144=\sqrt{x-2}$$
$$25\sqrt{x-2}=x+142;\text{ square both sides}$$
$$625(x-2)=x^2+284x+20164$$
$$x^2-341x+21414=0$$
$$(x-258)(x-83)=0$$
$$x-258=0 \quad\text{or}\quad x-83=0$$
$$x=258 \qquad x=83$$

Check:

$$\sqrt{258-2}\overset{?}{=}\sqrt[4]{258-2}+12$$
$$\sqrt{256}\overset{?}{=}\sqrt[4]{256}+12$$
$$16=16$$
$$\sqrt{83-2}\overset{?}{=}\sqrt[4]{83-2}+12$$
$$\sqrt{81}\overset{?}{=}\sqrt[4]{81}+12$$
$$16\ne15$$

The only solution is $x=258$.

37. $\sqrt{2x+1}+3\sqrt{x}=9$

$\sqrt{2x+1}=9-3\sqrt{x}$; square both sides

$2x+1=81-54\sqrt{x}+9x$

$54\sqrt{x}=7x+80$; square both sides

$2916x=49x^2+1120x+6400$

$49x^2-1796x+6400=0$

$(x-4)(49x-1600)=0$

$x-4=0$ or $49x=1600$

$x=4$ $\quad x=\dfrac{1600}{49}$

Check:

$\sqrt{2(4)+1}+3\sqrt{4}\overset{?}{=}9$

$3+6\overset{?}{=}9$

$9=9$

$\sqrt{2\left(\dfrac{1600}{49}\right)+1}\overset{?}{=}9$

$\dfrac{57}{7}+\dfrac{120}{7}\overset{?}{=}9$

$\dfrac{177}{9}\neq 9$

$x=4$ is the only solution. We see from the graph that there is only one point of intersection at $x=4$.

41. $\sqrt{x-1}+x=3$

$\sqrt{x-1}=3-x$; square both sides

$x-1=9-6x+x^2$

$x^2-7x+10=0$

$(x-5)(x-2)=0$

$x-5=0$ or $x-2=0$

$x=5$ $\quad x=2$

Check:

$\sqrt{5-1}+5=\sqrt{4}+5=7\neq 3$

$\sqrt{2-1}+2=\sqrt{1}+2=3$

The only solution is $x=2$.

To compare this solution with that of Example 4, we write $\sqrt{x-1}=x-3$ vs. $\sqrt{x-1}=3-x$ as $\sqrt{x-1}=x-3$ vs. $-\sqrt{x-1}=x-3$. We see that one represents the positive square root and the other one the negative root. Squaring both sides gives the same quadratic equation for both problems, but the extraneous root for one is the solution for the other and viceversa.

45. $kC=\sqrt{R_1^2-R_2^2}+\sqrt{r_1^2-r_2^2}-A$

$\sqrt{r_1^2-r_2^2}=kC+A-\sqrt{R_1^2-R_2^2}$; square both sides

$r_1^2-r_2^2=\left(kC+A-\sqrt{R_1^2-R_2^2}\right)^2$

$r_1^2=\left(kC+A-\sqrt{R_1^2-R_2^2}\right)^2+r_2^2$

49.

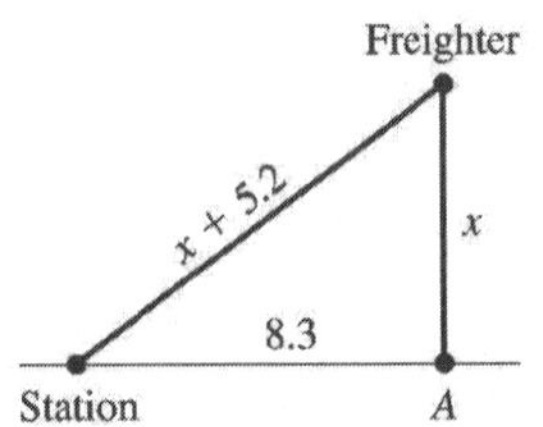

$y=x+5.2$

Substitute into: $y=\sqrt{x^2+8.3^2}$

$x+5.2=\sqrt{x^2+8.3^2}$; square both sides

$x^2+10.4x+27.04=x^2+68.89$

$10.4x=41.85$

$x=4.0$

$y=4.0+5.2$

$=9.2$

Check:

$\sqrt{4.0^2+83.3^2}=9.2$

The station is 9.2 km from the freighter.

Review Exercises

1. Graph $y=\dfrac{6-x}{2}$, $y_2=4x^2$, and use the intersect feature to solve.

Solutions:

$x=-0.93$, $y=3.47$

$x=0.81$, $y=2.60$

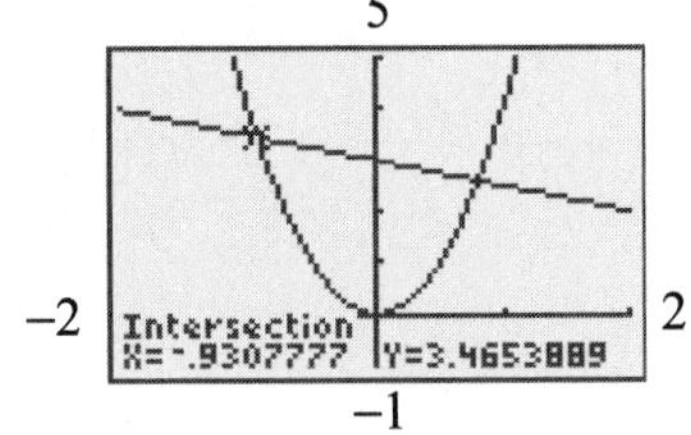

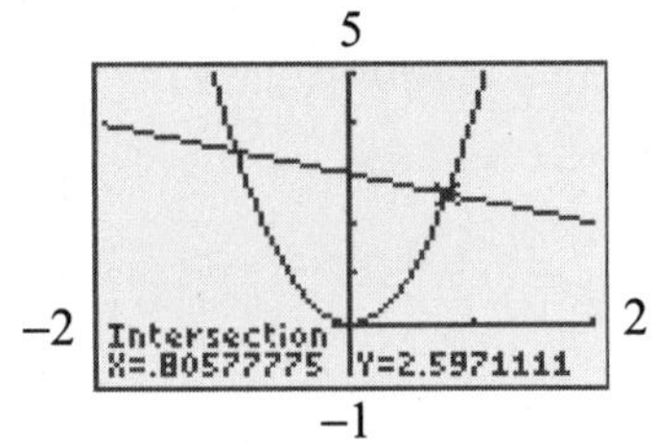

5. Graph $y_1 = x^2 + 1, y_2 = \sqrt{\dfrac{29-4x^2}{16}}, y_3 = -\sqrt{\dfrac{29-4x^2}{16}}$ and use the intersect feature to solve.

Solutions:

$x = -0.56,\ y = 1.32$

$x = 0.56,\ y = 1.32$

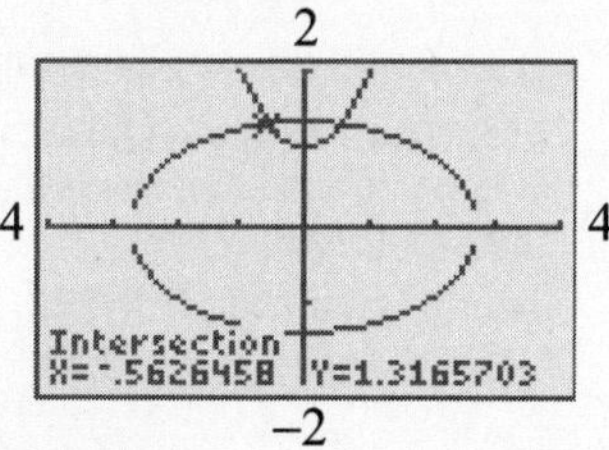

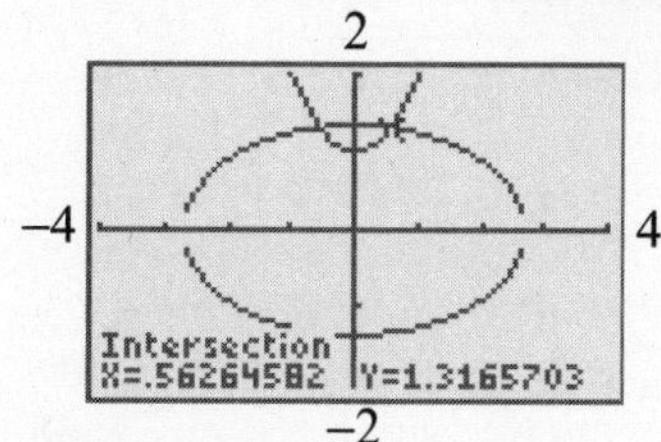

9. Graph $y_1 = x^2 - 2x,\ y_2 = 1 - e^{-x}$, and use the intersect feature to solve.

Solutions:

$x = 0,\ y = 0$

$x = 2.38,\ y = 0.91$

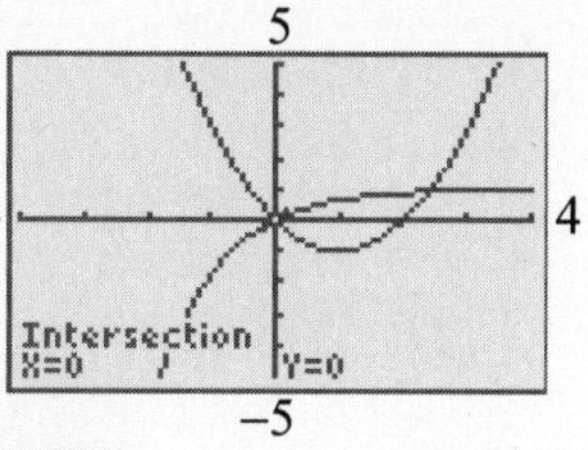

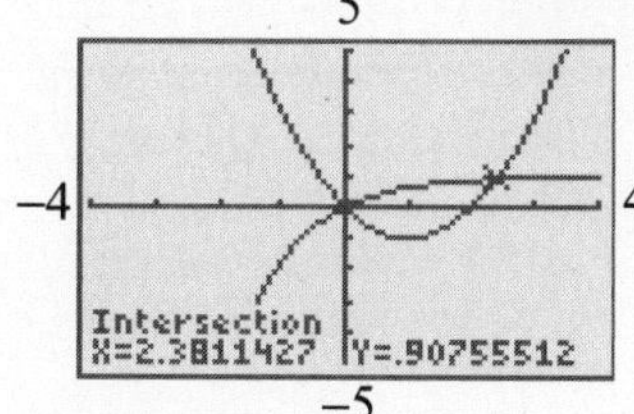

13. Substitute $L^2 = 2R$ into $R^2 + L^2 = 3$

$$R^2 + 2R = 3$$
$$R^2 + 2R - 3 = 0$$
$$(R+3)(R-1) = 0$$
$$R+3 = 0 \quad \text{or} \quad R-1 = 0$$
$$R = -3 \qquad R = 1$$
$$L^2 = 2(-3) \qquad L^2 = 2(1)$$

no solution $\qquad L = \pm\sqrt{2}$

Solutions:

$L = \pm\sqrt{2},\ R = 1$

17. (1) $4x^2 - 7y^2 = 21$

(2) $x^2 + 2y^2 = 99$

Multiply (1) by 4, (2) by 7 and add:

(1) $8x^2 - 14y^2 = 42$

(2) $7x^2 + 14y^2 = 693$

$$15x^2 = 735$$
$$x^2 = 49$$
$$x = \pm 7$$

(2) $(\pm 7)^2 + 2y^2 = 99$

$$49 + 2y^2 = 99$$
$$2y^2 = 50$$
$$y^2 = 25$$
$$y = \pm 5$$

Solutions:

$x = 7,\ y = \pm 5$

$x = -7,\ y = \pm 5$

21. $x^4 - 20x^2 + 64 = 0$

Let $y = x^2,\ y^2 = x^4$

$$y^2 - 20y + 64 = 0$$
$$(y-16)(y-4) = 0$$
$$y - 16 = 0 \quad \text{or} \quad y - 4 = 0$$
$$y = 16 \qquad y = 4$$
$$x^2 = 16 \qquad x^2 = 4$$
$$x = \pm 4 \qquad x = \pm 2$$

Check:

$(\pm 4)^4 - 20(\pm 4)^2 + 64 = 0$

$(\pm 2)^4 - 20(\pm 2)^2 + 64 = 0$

25. $D^{-2} + 4D^{-1} - 21 = 0$

Let $x = D^{-1},\ x^2 = D^{-2}$

$$x^2 + 4x - 21 = 0$$
$$(x+7)(x-3) = 0$$
$$x + 7 = 0 \quad \text{or} \quad x - 3 = 0$$
$$x = -7 \qquad x = 3$$
$$D^{-1} = -7 \qquad D^{-1} = 3$$
$$D = -\frac{1}{7} \qquad D = \frac{1}{3}$$

Check:

$$\left(-\frac{1}{7}\right)^{-2} + 4\left(-\frac{1}{7}\right)^{-1} - 21 = 49 - 28 - 21 = 0$$

$$\left(\frac{1}{3}\right)^{-2} + 4\left(\frac{1}{3}\right)^{-1} - 21 = 9 - 12 - 21 = 0$$

29. $\dfrac{4}{r^2+1}+\dfrac{7}{2r^2+1}=2$

$4(2r^2+1)+7(r^2+1)=2(r^2+1)(2r^2+1)$

$8r^2+4+7r^2+7=2(2r^4+3r^2+1)$

$15r^2+11=4r^4+6r^2+2$

$4r^4-9r^2-9=0$

Let $y=r^2,\ y^2=r^4$

$4y^2-9y-9=0$

$(4y+3)(y-3)=0$

$4y+3=0 \quad \text{or} \quad y-3=0$

$y=-\dfrac{3}{4} \qquad y=3$

$r^2=-\dfrac{3}{4} \qquad r^2=3$

$r=\pm\dfrac{j\sqrt{3}}{2} \qquad r=\pm\sqrt{3}$

Check:

$$\frac{4}{\left(\pm\frac{j\sqrt{3}}{2}\right)^2+1}+\frac{7}{2\left(\pm\frac{j\sqrt{3}}{2}\right)^2+1}$$

$$=\frac{4}{-\frac{3}{4}+1}+\frac{7}{2\left(-\frac{3}{4}\right)+1}=16-14=2$$

$$\frac{4}{(\pm\sqrt{3})^2+1}+\frac{7}{2(\sqrt{3})^2+1}$$

$$=\frac{4}{3+1}+\frac{7}{6+1}=1+1=2$$

33. $\sqrt{5x+9}+1=x$

$\sqrt{5x+9}=x-1$; square both sides

$5x+9=x^2-2x+1$

$x^2-7x-8=0$

$(x-8)(x+1)=0$

$x-8=0 \quad \text{or} \quad x+1=0$

$x=8 \qquad x=-1$

Check:

$\sqrt{5(8)+9}+1\stackrel{?}{=}8$

$\sqrt{49}+1\stackrel{?}{=}8$

$8=8$

$\sqrt{5(-1)+9}+1\stackrel{?}{=}-1$

$\sqrt{4}+1\stackrel{?}{=}-1$

$3\neq-1$

The only solution is $x=8$.

37. $\sqrt{n+4}+2\sqrt{n+2}=3$

$\sqrt{n+4}=3-2\sqrt{n+2}$; square both sides

$n+4=9-12\sqrt{n+2}+4n+8$

$12\sqrt{n+2}=13+3n$; square both sides

$144n+288=169+78n+9n^2$

$9n^2-66n-119=0$

Use the quadratic formula

$$n=\frac{-(-66)\pm\sqrt{(-66)^2-4(9)(-119)}}{2(9)}$$

$$=\frac{66\pm\sqrt{8640}}{18}$$

$$=\frac{11\pm5\sqrt{15}}{3}$$

Check:

$$\sqrt{\frac{11+4\sqrt{15}}{3}+4}+2\sqrt{\frac{11+4\sqrt{15}}{3}+2}=10.2\neq3$$

$$\sqrt{\frac{11-4\sqrt{15}}{3}+4}+2\sqrt{\frac{11-4\sqrt{15}}{3}+2}=3$$

$n=\dfrac{11-4\sqrt{15}}{3}$ is the only solution.

41. $x^3-2x^{3/2}-48=0$

Let $y=x^{3/2}, y^2=x^3, y\ge0$

$y^2-2y-48=0$

$(y-8)(y+6)=0$

$y-8=0 \quad \text{or} \quad y+6=0$

$y=8 \qquad y=-6$

$x^{3/2}=8 \qquad$ not possible

$x=8^{2/3}$

$=4$

150

−1 6

Zero X=4 Y=0

−80

45. $\sqrt[3]{x^3-7}=x-1$; cube both sides

$x^3-7=x^3-3x^2+3x-1$

$3x^2-3x-6=0$

$x^2-x-2=0$

$x-2=0$ or $x+1=0$

$x=2$ $\quad x=-1$

There are two intersections, at $x = 2$ and $x = -1$.

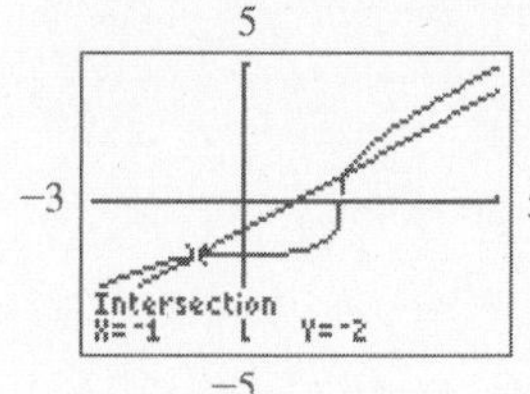

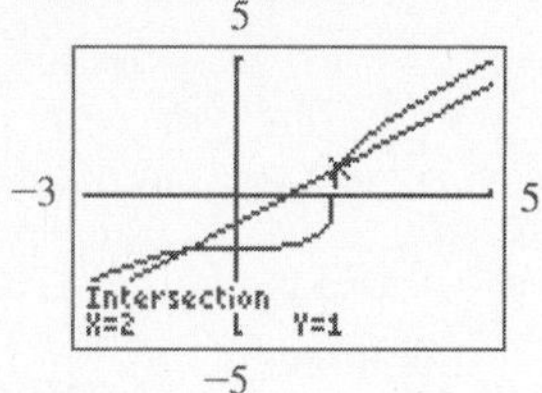

49. $\sqrt{\sqrt{x}-1}=2$; square both sides

$\sqrt{x}-1=4$

$\sqrt{x}=5$; square both sides

$x=25$

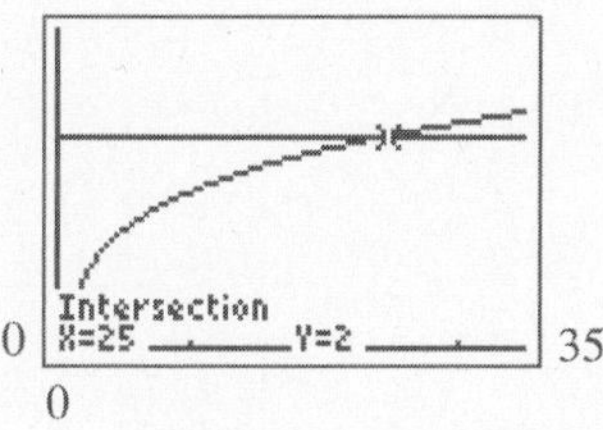

53. $L=\dfrac{h}{2\pi}\sqrt{l(l+1)}$; square both sides

$$L^2=\frac{h^2}{4\pi^2}\left(l^2+l\right)$$

$$\frac{4\pi^2L^2}{h^2}=l^2+l$$

$$l^2+l-\frac{4\pi^2L^2}{h^2}=0$$

Using the quadratic formula,

$$l=\frac{-1\pm\sqrt{1^2-4\cdot1\left(-\frac{4\pi^2L^2}{h^2}\right)}}{2(1)}$$

$$=\frac{-1\pm\sqrt{1+\frac{16\pi^2L^2}{h^2}}}{2}$$

$l=\dfrac{-1+\sqrt{1+\frac{16\pi^2L^2}{h^2}}}{2}$ is the only solution since $l>0$.

57. Substitute $t_2=2t_1$ into $490t_1^2+490t_2^2=392$, where $t_1, t_2>0$

$$490t_1^2+490(2t_1)^2=392$$

$$490t_1^2+1960t_1^2=392$$

$$2450t_1^2=392$$

$$t_1^2=\frac{392}{2450}$$

$$t_1=0.40\text{ s}$$

$$t_2=2\cdot t_1=2(0.40)$$

$$t_2=0.80\text{ s}$$

61.

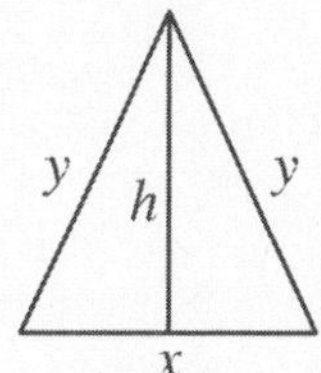

perimeter: $x+2y=72\Rightarrow y=\dfrac{72-x}{2}$

area: $\dfrac{1}{2}x\cdot h=240\Rightarrow h=\dfrac{480}{x}$

$$y^2=\frac{x^2}{4}+h^2\Rightarrow y^2=\frac{x^2}{4}+\frac{480^2}{x^2}$$

Graph $y_1=72-2x$,

$$y_2=\sqrt{\frac{x^2}{4}+\frac{480^2}{x^2}}$$

(only the positive root matters here).

Use the intersect feature to solve.

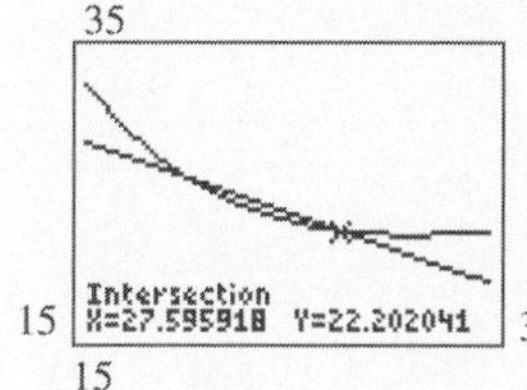

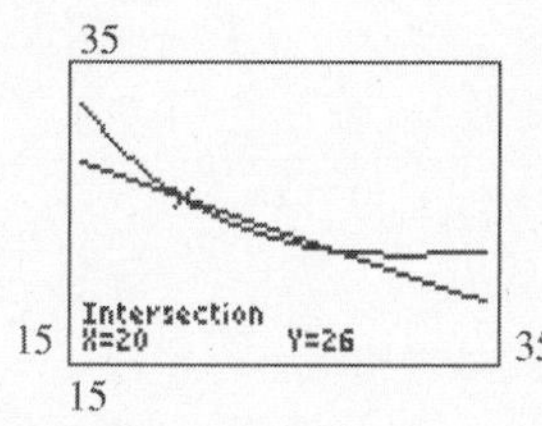

The lengths of the sides the banner can be 27.6 dm, 22.2 dm and 22.2 dm, with height 17.4 dm; or 20 dm, 26 dm and 26 dm, with height 24 dm. Since the height must be longer than the base, the lengths of the sides are 20 dm, 26 dm, and 26 dm, with height 24 dm.

65. Area: $lw=1770\Rightarrow w=\dfrac{1770}{l}$

Substitute into: $l^2+w^2=62^2$

$$l^2+\frac{1770^2}{l^2}=62^2$$

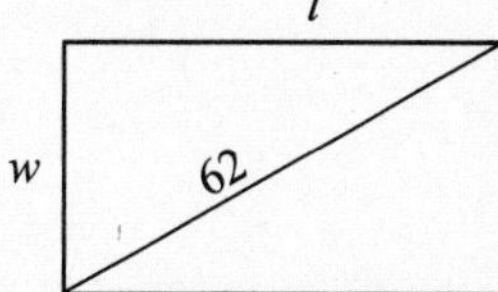

$l^4 - 62^2 l^2 + 1770^2 = 0$

Let $y = l^2$

$y^2 - 62^2 y + 1770^2 = 0$

$$y = \frac{62^2 \pm \sqrt{\left(-62^2\right)^2 - 4\left(1770^2\right)}}{2}$$

$y = 2269$ or $y = 1175$

$l^2 = 2269$ $\quad l^2 = 1175$

$l = 52$ $\quad l = 34$

$w = 34$ $\quad w = 52$

The dimensions of the rectangle are $l = 52$ mm and $w = 34$ mm.

69. Digby to St. John: $72 = v_1 \cdot t_1 \Rightarrow t_1 = \dfrac{72}{v_1}$

St. John to Digby: $72 = (v_1 - 3.20) \cdot t_2 \Rightarrow$

$$t_2 = \frac{72}{v_1 - 3.20}$$

Substitute into: $t_1 + t_2 = 5.7$

$$\frac{72}{v_1} + \frac{72}{v_1 - 3.20} = 5.7$$

$5.7v_1^2 - 162.24v_1 + 230.4 = 0$

Using the quadratic formula

$$v_1 = \frac{162.24 \pm \sqrt{162.24^2 - 4(5.7)(230.4)}}{2(5.7)}$$

$v_1 = 27.0$ or $v_1 = 1.50$

$v_2 = 27.0 - 3.20$ $\quad v_2 = 1.50 - 3.20$

$= 23.8$ $\quad = -1.70$

Solution

$v = 27.0$ km/h from Digby to St. John

$v = 23.8$ km/h from St. John to Digby

CHAPTER 15

EQUATIONS OF HIGHER DEGREE

15.1 The Remainder and Factor Theorems; Synthetic Division

1. Using the remainder theorem find the remainder, for $(3x^3 - x^2 - 20x + 5) \div (x+3)$.

$R = f(-3) = 3(-3)^3 - (-3)^2 - 20(-3) + 5 = -25$

5.
$$\begin{array}{r} x^2 - x + 3 \\ x+1 \overline{)\, x^3 + \quad 2x + 3} \\ \underline{x^3 + x^2} \qquad\qquad \\ -x^2 + 2x \qquad \\ \underline{-x^2 - \ \ x} \qquad \\ 3x + 3 \\ \underline{3x + 3} \\ 0 \end{array}$$

The remainder is $R = 0$.

9.
$$\begin{array}{r} x^3 - x - 9 \\ 2x-3 \overline{)\, 2x^4 - 3x^3 - 2x^2 - 15x - 16} \\ \underline{2x^4 - 3x^3} \qquad\qquad\qquad\qquad \\ -2x^2 - 15x \qquad\quad \\ \underline{-2x^2 + \ 3x} \qquad\quad \\ -18x - 16 \\ \underline{-18x + 27} \\ -43 \end{array}$$

The remainder is $R = -43$.

13. $(2x^4 - 7x^3 - x^2 + 8) \div (x-3);\ r = 3$

$f(3) = 2\cdot 3^4 - 7\cdot 3^3 - 3^2 + 8$

$= -28$

$R = -28$

17. $4x^3 + x^2 - 16x - 4,\ x-2;\ r = 2$

$f(2) = 4(2)^3 + 2^2 - 16(2) - 4$

$= 32 + 4 - 32 - 4 = 0$

$x-2$ is a factor since $f(2) = R = 0$.

21 $x^{61} - 1,\ x+1;\ r = -1$

$f(-1) = (-1)^{61} - 1$

$= -2$

$x+1$ is a not factor since $f(-1) = -2 \neq 0$.

25. $(x^3 + 2x^2 - 3x + 4) \div (x+1)$

$$\begin{array}{rrrr|r} 1 & 2 & -3 & 4 & \underline{-1} \\ & -1 & -1 & 4 & \\ \hline 1 & 1 & -4 & 8 & \end{array}$$

The quotient is $x^2 + x - 4$ and the remainder is 8.

29. $(x^7 - 128) \div (x-2)$

$$\begin{array}{rrrrrrrr|r} 1 & 0 & 0 & 0 & 0 & 0 & 0 & -128 & \underline{2} \\ & 2 & 4 & 8 & 16 & 32 & 64 & 128 & \\ \hline 1 & 2 & 4 & 8 & 16 & 32 & 64 & 0 & \end{array}$$

The quotient is $x^6 + 2x^5 + 4x^4 + 8x^3 + 16x^2 + 32x + 64$ and the remainder is 0.

33. $2x^5 - x^3 + 3x^2 - 4;\ x+1$

$$\begin{array}{rrrrrr|r} 2 & 0 & -1 & 3 & 0 & -4 & \underline{-1} \\ & -2 & 2 & -1 & -2 & 2 & \\ \hline 2 & -2 & 1 & 2 & -2 & -2 & \end{array}$$

$R = -2 \neq 0$, so $x+1$ is not a factor.

37. $2Z^4 - Z^3 - 4Z^2 + 1;\ 2Z - 1$

We write $2Z - 1 = 2\left(Z - \frac{1}{2}\right)$

$$\begin{array}{rrrrr|r} 2 & -1 & -4 & 0 & 1 & \underline{\tfrac{1}{2}} \\ & 1 & 0 & -2 & -1 & \\ \hline 2 & 0 & -4 & -2 & 0 & \end{array}$$

Since $R = 0$, $Z - \frac{1}{2}$ is a factor.

Therefore, $2Z - 1$ is also a factor.

41. $x^4-5x^3-15x^2+5x+14;\ 7$

$$\begin{array}{rrrrr|l} 1 & -5 & -15 & 5 & 14 & \underline{7} \\ & 7 & 14 & -7 & -14 & \\ \hline 1 & 2 & -1 & -2 & 0 & \end{array}$$

$R=0$, 7 is a zero.

45. $f(x)=2x^3+3x^2-19x-4$

$f(x)=(x+4)g(x)$

$g(x)=(2x^3+3x^2-19x-4)\div(x+4)$

$r=-4$

$$\begin{array}{rrrr|l} 2 & 3 & -19 & -4 & \underline{-4} \\ & -8 & 20 & -4 & \\ \hline 2 & -5 & 1 & -8 & \end{array}$$

$g(x)=2x^2-5x+1-\dfrac{8}{x+4}$

49. $f(x)=2x^3+kx^2-x+14;\ x-2$

We want $f(2)=R=0$

$f(2)=2(2)^3+k(2)^2-2+14$

$=16+4k-2+14$

$=28+4k$

$=0$

$4k=-28$

$k=-7$

If $k=-7$ then $x-2$ will be a factor.

53. Suppose r is a zero of $f(x)$, then $f(r)=0$. But $f(r)=-g(r)=0\Rightarrow g(r)=0$, so r is also a zero of $g(x)$. Therefore, if $f(x)=-g(x)$ then $f(x)$ and $g(x)$ have the same zeros.

57. $V^3-6V^2+12V=8$

$V^3-6V^2+12V-8=0$

Let $r=2$

$$\begin{array}{rrrr|l} 1 & -6 & 12 & -8 & \underline{2} \\ & 2 & -8 & 8 & \\ \hline 1 & -4 & 4 & 0 & \end{array}$$

$R=0$, so $V=2\text{ cm}^3$ is indeed a solution.

15.2 The Roots of an Equation

1. $f(x)=(x-1)^3(x^2+2x+1)=0$

$(x-1)^3=0$ or $x^2+2x+1=0$

$x=1$ $\qquad (x+1)^2=0$

A triple root $\qquad x+1=0$

$x=-1$, a double root

the five roots are 1, 1, 1, −1, −1

5. $(x^2+6x+9)(x^2+4)=0$

$(x+3)^2(x^2+4)=0$, by inspection

$x=-3$ double root, $x=\pm 2j$

9. $2x^3+11x^2+20x+12=0\ \left(r_1=-\dfrac{3}{2}\right)$

$$\begin{array}{rrrr|l} 2 & 11 & 20 & 12 & \underline{-\frac{3}{2}} \\ & -3 & -12 & -12 & \\ \hline 2 & 8 & 8 & 0 & \end{array}$$

$2x^3+11x^2+20x+12=\left(x+\dfrac{3}{2}\right)(2x^2+8x+8)$

$=2\left(x+\dfrac{3}{2}\right)(x^2+4x+4)$

$=2\left(x+\dfrac{3}{2}\right)(x+2)(x+2)$

The three roots are $r_1=-\dfrac{3}{2}$, $r_2=-2$, $r_3=-2$

13. $t^4+t^3-2t^2+4t-24=0\ (r_1=2, r_2=-3)$

$$\begin{array}{rrrrr|l} 1 & 1 & -2 & 4 & -24 & \underline{2} \\ & 2 & 6 & 8 & 24 & \\ \hline 1 & 3 & 4 & 12 & 0 & \underline{-3} \\ & -3 & 0 & -10 & & \\ \hline 1 & 0 & 4 & 0 & & \end{array}$$

$t^4+t^3-2t^2+4t-24=(t-2)(t+3)(t^2+4)$

$=(t-2)(t+3)(t-2j)(t+2j)$

The four roots are $r_1=2$, $r_2=-3$, $r_3=-2j$, $r_4=2j$

17. $6x^4+5x^3-15x^2+4=0\left(r_1=-\frac{1}{2}, r_2=\frac{2}{3}\right)$

$$\begin{array}{rrrrr|r} 6 & 5 & -15 & 0 & 4 & -\frac{1}{2} \\ & -3 & -1 & 8 & -4 & \\ \hline 6 & 2 & -16 & 8 & 0 & \frac{2}{3} \\ & 4 & 4 & -8 & 0 & \\ \hline 6 & 6 & -12 & 0 & 0 & \end{array}$$

$$\begin{aligned} &6x^4+5x^3-15x^2+4 \\ &=6\left(x+\frac{1}{2}\right)\left(x-\frac{2}{3}\right)\left(x^2+x-2\right) \\ &=6\left(x+\frac{1}{2}\right)\left(x-\frac{2}{3}\right)(x+2)(x-1) \end{aligned}$$

The four roots are $r_1=-\frac{1}{2}$, $r_2=\frac{2}{3}$, $r_3=-2$, $r_4=1$

21. $x^5-3x^4+4x^3-4x^2+3x-1=0$ (1 is a triple root)

$$\begin{array}{rrrrrr|r} 1 & -3 & 4 & -4 & 3 & -1 & 1 \\ & 1 & -2 & 2 & -2 & 1 & \\ \hline 1 & -2 & 2 & -2 & 1 & 0 & 1 \\ & 1 & -1 & 1 & -1 & & \\ \hline 1 & -1 & 1 & -1 & 0 & & 1 \\ & 1 & 0 & 1 & & & \\ \hline 1 & 0 & 1 & 0 & & & \end{array}$$

$$\begin{aligned} &x^5-3x^4+4x^3-4x^2+3x-1 \\ &=(x-1)^3\left(x^2+1\right) \end{aligned}$$

The five roots are 1, 1, 1, $-j$, j.

25. $x^6+2x^5-4x^4-10x^3-41x^2-72x-36=0$

(-1 is a double root; $2j$ is a root)

The complex conjugate $r_4=-2j$ must also be a root.

$$\begin{array}{rrrrrrr|r} 1 & 2 & -4 & -10 & -41 & -72 & -36 & -1 \\ & -1 & -1 & 5 & 5 & 36 & 36 & \\ \hline 1 & 1 & -5 & -5 & -36 & -36 & 0 & -1 \\ & -1 & 0 & 5 & 0 & 36 & 0 & \\ \hline 1 & 0 & -5 & 0 & -36 & 0 & 0 & 2j \\ & 2j & -4 & -18j & 36 & 0 & 0 & \\ \hline 1 & 2j & -9 & -18j & 0 & 0 & 0 & -2j \\ & -2j & 0 & 18j & 0 & 0 & 0 & \\ \hline 1 & 0 & -9 & 0 & 0 & 0 & 0 & \end{array}$$

$$\begin{aligned} &x^6+2x^5-4x^4-10x^3-41x^2-72x-36 \\ &=(x+1)^2(x-2j)(x+2j)\left(x^2-9\right) \end{aligned}$$

The six roots are $-1, -1, 2j, -2j, -3, 3$.

29. A polynomial of degree 3 has 3 roots, so $f(x)=(x-(1+j))(x-(1-j))(x-r)$ is a polynomial, degree 3 with root $1+j$ (and therefore root $1-j$). To find r, we use the fact that $f(2)=4$.

$$\begin{aligned} 4&=f(2) \\ &=(2-(1+j))(2-(1-j))(2-r) \\ &=(1-j)(1+j)(2-r) \\ &=2(2-r) \\ &=4-2r \end{aligned}$$

Therefore $r=0$ and the required polynomial is

$$\begin{aligned} f(x)&=(x-(1+j))(x-(1-j))x \\ &=x^3-2x^2+2x \end{aligned}$$

15.3 Rational and Irrational Roots

1. $f(x)=4x^5+x^4+4x^3-x^2+5x+6=0$ has two sign changes and thus no more than two positive roots.

$f(-x)=-4x^5+x^4-4x^3-x^2-5x+6=0$ has three sign changes and thus no more than three negative roots.

5. $x^3+2x^2-5x-6=0$; there are 3 roots.

$f(x)=x^3+x^2-5x-6$; there is exactly one positive root.

$f(-x)=-x^3+x^2+5x-6$; there are at most two negative roots.

Possible rational roots are $\pm1, \pm2, \pm3, \pm6$

Trying -1, we have

$$\begin{array}{rrrr|r} 1 & 2 & -5 & -6 & -1 \\ & -1 & -1 & 6 & \\ \hline 1 & 1 & -6 & 0 & \end{array}$$

Hence, -1 is a root and the remaining factor is $x^2+x-6=(x+3)(x-2)$.

The three roots are:

$r_1=-1, r_2=-3, r_3=2$.

9. $3x^3+11x^2+5x-3=0$; there are three roots.
$f(x)=3x^3+11x^2+5x-3$; there is 1 positive root.
$f(-x)=-3x^3+11x^2-5x-3$; there are at most two negative roots.

Possible rational roots are $\pm\frac{1}{3}, \pm1, \pm3$.

Trying $+\frac{1}{3}$ we have:

$$\begin{array}{rrrr|l} 3 & 11 & 5 & -3 & \frac{1}{3} \\ & 1 & 4 & 3 & \\ \hline 3 & 12 & 9 & 0 & \end{array}$$

Remainder is 0 so $\frac{1}{3}$ is a root.

$$3x^3+11x^2+5x-3=\left(x-\frac{1}{3}\right)\left(3x^2+12x+9\right)$$
$$=\left(x-\frac{1}{3}\right)3(x+1)(x+3).$$

The three roots are:

$r_1=\frac{1}{3}, r_2=-1, r_3=-3.$

13. $5n^4-2n^3+40n-16=0$; there are four roots.
$f(n)$ has 3 sign changes, at most 3 positive roots.
$f(-n)=5n^4+2n^3-40n-16$ has one sign change, so exactly one negative root.
Possible rational roots: $\pm16, \pm\frac{16}{5}, \pm2, \pm\frac{2}{5}, \pm4, \pm\frac{4}{5}, \pm8, \pm\frac{8}{5}$

Trying -2:

$$\begin{array}{rrrrr|l} 5 & -2 & 0 & 40 & -16 & -2 \\ & -10 & 24 & -48 & 16 & \\ \hline 5 & -12 & 24 & -8 & 0 & \end{array}$$

Remainder is 0, so -2 is a root

$$5n^4-2n^3+40n-16=(n+2)\left(5n^3-12n^2+24n-8\right)$$

No other roots are negative. Trying $\frac{2}{5}$:

$$\begin{array}{rrrr|l} 5 & -12 & 24 & -8 & \frac{2}{5} \\ & 2 & -4 & 8 & \\ \hline 5 & -10 & 20 & 0 & \end{array}$$

Remainder is 0, so $\frac{2}{5}$ is a root

$$5n^4-2n^3+40n-16$$
$$=(n+2)\left(n-\frac{2}{5}\right)\left(5n^2-10n+20\right)$$
$$=(n+2)\left(n-\frac{2}{5}\right)5\left(n^2-2n+4\right)$$

Using the quadratic formula on the last factor:

$$n=\frac{-(-2)\pm\sqrt{(-2)^2-4(1)(4)}}{2}=1\pm j\sqrt{3}$$

The four roots are: $-2, \frac{2}{5}, 1\pm j\sqrt{3}$

17. $D^5+D^4-9D^3-5D^2+16D+12=0$ has five roots. $f(D)$ has two sign changes and therefore at most two positive roots.
$f(-D)=-D^5+D^4+9D^3-5D^2-16D+12$ has three sign changes and therefore at most three negative roots.
Possible rational roots: $\pm1, \pm2, \pm3, \pm4, \pm6, \pm12$

$$\begin{array}{rrrrrr|l} 1 & 1 & -9 & -5 & 16 & 12 & 2 \\ & 2 & 6 & -6 & -22 & -12 & \\ \hline 1 & 3 & -3 & -11 & -6 & 0 & \end{array}$$

Remainder is 0 so 2 is a root

$$D^5+D^4-9D^3-5D^2+16D+12$$
$$=(D-2)\left(D^4+3D^3-3D^2-11D-6\right)$$

$$\begin{array}{rrrrr|l} 1 & 3 & -3 & -11 & -6 & 2 \\ & 2 & 10 & 14 & 6 & \\ \hline 1 & 5 & 7 & 3 & 0 & \end{array}$$

Remainder is 0 so 2 is a root.

$$D^5+D^4-9D^3-5D^2+16D+12$$
$$=(D-2)(D-2)\left(D^3+5D^2+7D+3\right)$$

$$\begin{array}{rrrr|l} 1 & 5 & 7 & 3 & -1 \\ & -1 & -4 & -3 & \\ \hline 1 & 4 & 3 & 0 & \end{array}$$

Remainder is 0 so -1 is a root.

$$D^5+D^4-9D^3-5D^2+16D+12$$
$$=(D-2)(D-2)(D+1)\left(D^2+4D+3\right)$$
$$=(D-2)(D-2)(D+1)(D+1)(D+3)$$

The five roots are: $2, 2, -1, -1, -3$.

21. $x^3-2x^2-5x+4=0$

Graph $y_1=x^3-2x^2-5x+4$ and use the Zero feature to solve. The Zeros are $-1.86, 0.68, 3.18$.

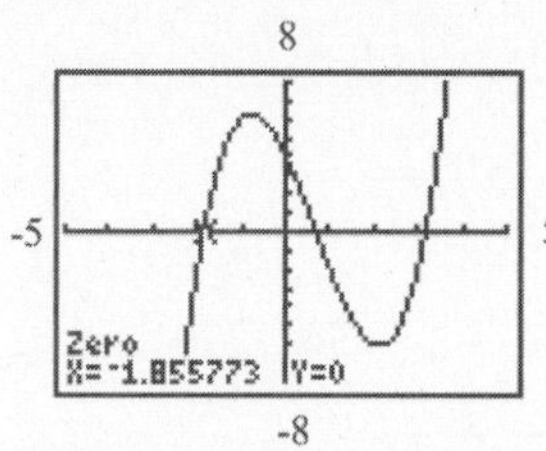

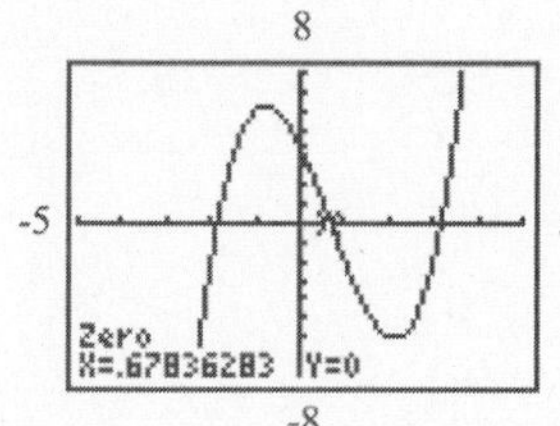

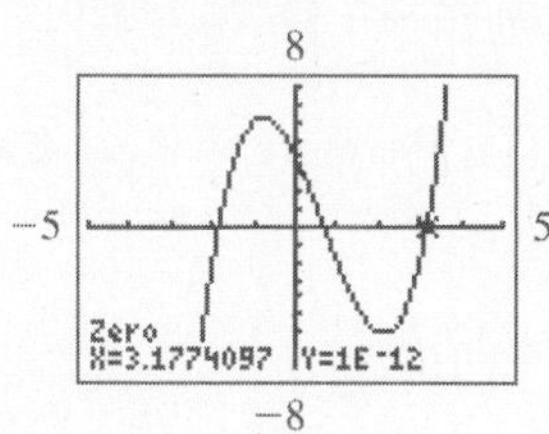

25. $x^3-6x^2+10x-4=0$ (0 and 1)

We evaluate $f(0)=-4$, and $f(1)=1$. This means that $f(x)=0$ between 0 and 1. We evaluate the function at different values, as shown in the table:

Interval	c	$f(c)$	New interval
(0,1)	0.5	−0.375	(0.5,1)
(0.5,1)	0.6	0.056	(0.5,0.6)
(0.5,0.6)	0.58	−0.0233	(0.58,0.6)
(0.58,0.6)	0.59	0.01678	(0.58,0.59)
(0.58,0.59)	0.586	0.00085	(0.58,0.586)

The root is $x=0.59$.

29. Substitute $y=x^4-11x^2$ into $y=12x-4$

$x^4-11x^2=12x-4$

$x^4-11x^2-12x+4=0$; there are 4 roots, with at most 2 positive roots.

$f(-x)=x^4-11x^2+12x+4$, so at most 2 negative roots.

Possible rational roots: $\pm1, \pm2$

Try -2:

$$\begin{array}{rrrrrl} 1 & 0 & -11 & -12 & 4 & \underline{|-2} \\ & -2 & 4 & 14 & -4 & \\ \hline 1 & -2 & -7 & 2 & 0 & \end{array}$$

$$\begin{array}{rrrrl} 1 & -2 & -7 & 2 & \underline{|-2} \\ & -2 & 8 & -2 & \\ \hline 1 & -4 & 1 & 0 & \end{array}$$

$x^4-11x^2-12x+4=(x+2)(x+2)(x^2-4x+1)$

Use quadratic formula:

$$x=\frac{-(-4)\pm\sqrt{(-4)^2-4(1)(1)}}{2(1)}$$

$$=2\pm\sqrt{3}$$

The solutions are:

$x=-2,\ y=-28;$

$x=2+\sqrt{3}, y=20+12\sqrt{3};$

$x=2-\sqrt{3},\ y=20-12\sqrt{3}.$

33. $\alpha=-0.2t^3+t^2, \alpha=2.0 \text{ rad/s}^2$

Solve:

$$-0.2t^3+t^2=2.0$$

$$0.2t^3-t^2+2.0=0$$

Graph $y_1=0.2x^3-x^2+2$. Using the zero feature to solve we find the roots $x=1.76, 4.51$.

Therefore, $\alpha=2.0 \text{ rad/s}^2$ for $t=1.76$s, and 4.51s.

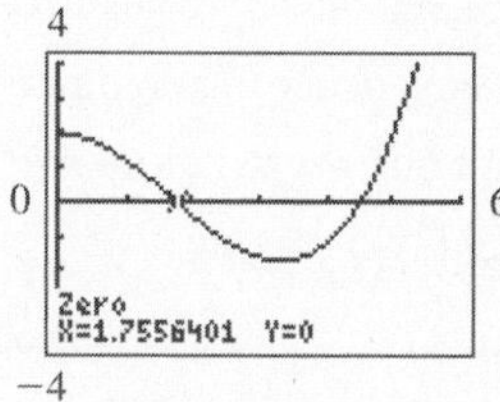

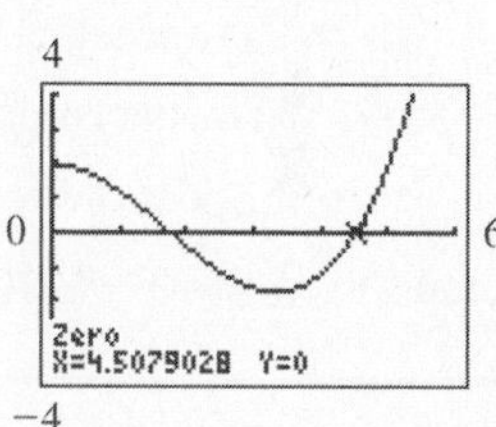

37. $V=0.1t^4-1.0t^3+3.5t^2-5.0t+2.3,\ 0\le t\le 5.0$ s

Graph $y_1=0.1x^4-1.0x^3+3.5x^2-5.0x+2.3$ and use the zero feature to solve.

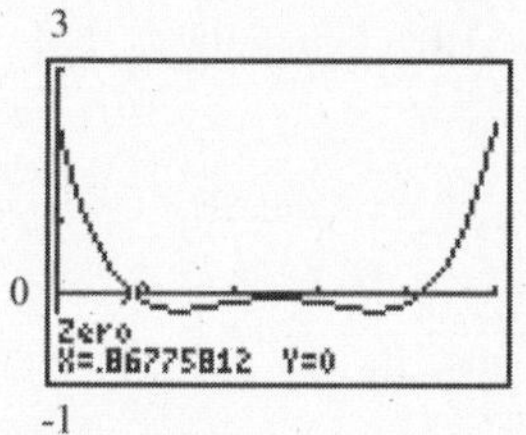

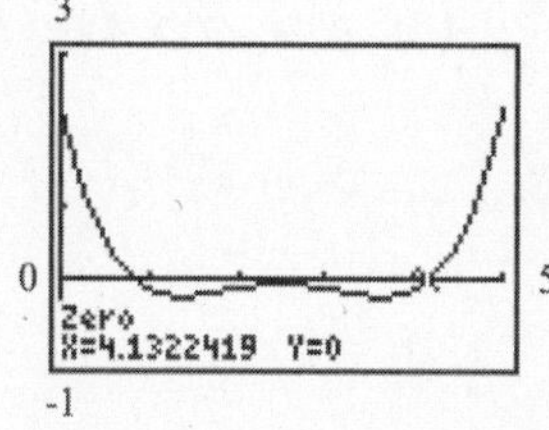

$V=0$ at $t=0.87$ s and $t=4.13$ s.

41. Let r = smallest radius, so that the radii are $r, r+1, r+2,$ and $r+3.$

$$\frac{4}{3}\pi(r+3)^3 = \frac{4}{3}\pi r^3 + \frac{4}{3}\pi(r+1)^3 + \frac{4}{3}\pi(r+2)^3$$

$$r^3+3\cdot r^2\cdot 3+3\cdot r\cdot 3^2+3^3 = r^3+r^3+3r^2+3r+1+r^3+3\cdot r^2\cdot 2+3\cdot r\cdot 2^2+2^3$$

$$r^3+9r^2+27r+27 = 3r^3+9r^2+15r+9$$

$$2r^3-12r-18=0$$

$$r^3-6r-9=0$$

Possible rational roots = $\pm 3, \pm 1$.

Try +3:

$$\begin{array}{rrrr|l} 1 & 0 & -6 & -9 & 3 \\ & 3 & 9 & 9 & \\ \hline 1 & 3 & 3 & 0 & \end{array}$$

Remainder is 0, so 3 is a root.

$$r^3-6r-9=(r-3)(r^2+3r+3)$$

Using the quadratic formula for the last factor:

$$r=\frac{-3\pm\sqrt{3^2-4(1)(3)}}{2}$$

There are no more real solutions.

The radii are 3.0 mm, 4.0 mm, 5.0 mm, 6.0 mm.

45. $f(x)=ax^3-bx^2+c=0.$ There are three roots. Since $f(x)$ has two sign changes, there are at most two positive roots. $f(-x)=-ax^3-bx^2+c=0$ which has one sign change and thus, one negative root. Since the complex roots occur in conjugate pairs, the possible roots are one negative and two positive or one negative and two complex.

Review Exercises

1. $f(x)=2x^3-4x^2-x+4;$ we evaluate $f(1).$

$$2(1)^3-4(1)^2-(1)+4=1$$

$(2x^3-4x^2-x+4)\div(x-1)$ has remainder 1

5. $f(x)=x^4+x^3+x^2-2x-3;$ we evaluate $f(-1).$

$$(-1)^4+(-1)^3+(-1)^2-2(-1)-3=0$$

$f(-1)=0,$ so $x+1$ is a factor of $x^4+x^3+x^2-2x-3$

9.

$$\begin{array}{rrrr|l} 1 & 3 & 6 & 1 & 1 \\ & 1 & 4 & 10 & \\ \hline 1 & 4 & 10 & 11 & \end{array}$$

$$(x^3+3x^2+6x+1)\div(x-1) = (x^2+4x+10)+\frac{11}{x-1}$$

13.

$$\begin{array}{rrrrr|l} 1 & 3 & -20 & -2 & 56 & -6 \\ & -6 & 18 & 12 & -60 & \\ \hline 1 & -3 & -2 & 10 & -4 & \end{array}$$

$$x^4+3x^3-20x^2-2x+56\div x+6 = x^3-3x^2-2x+10+\frac{-4}{x+6}$$

17.

$$\begin{array}{rrrr|l} 1 & 5 & 0 & -6 & -3 \\ & -3 & -6 & 18 & \\ \hline 1 & 2 & -6 & 12 & \end{array}$$

remainder = 12; therefore, -3 is not a root of y^3+5y^2-6

21.

$$\begin{array}{rrrr|l} 1 & -4 & -7 & 10 & 5 \\ & 5 & 5 & -10 & \\ \hline 1 & 1 & -2 & 0 & \end{array}$$

$$x^3-4x^2-7x+10=(x-5)(x^2+x-2) = (x-5)(x+2)(x-1)$$

The three roots are $r_1=5, r_2=-2, r_3=1.$

25.

$$\begin{array}{rrrrr|l} 4 & 0 & -1 & -18 & 9 & 1/2 \\ & 2 & 1 & 0 & -9 & \\ \hline 4 & 2 & 0 & -18 & 0 & 3/2 \\ & 6 & 12 & 18 & & \\ \hline 4 & 8 & 12 & 0 & & \end{array}$$

$$4p^4-p^2-18p+9=\left(p-\frac{1}{2}\right)\left(p-\frac{3}{2}\right)4(p^2+2p+3)$$

Use the quadratic formula for the remaining factor:

$$p=\frac{-2\pm\sqrt{2^2-4(1)(3)}}{2} = -1\pm j\sqrt{2}$$

The four roots are

$$r_1=\frac{1}{2}, r_2=\frac{3}{2}, r_3=-1+j\sqrt{2}, r_4=-1-j\sqrt{2}$$

29.

$$\begin{array}{rrrrrrl} 1 & 3 & -1 & -11 & -12 & -4 & \lfloor -1 \\ & -1 & -2 & 3 & 8 & 4 & \\ \hline 1 & 2 & -3 & -8 & -4 & 0 & \lfloor -1 \\ & -1 & -1 & 4 & 4 & & \\ \hline 1 & 1 & -4 & -4 & 0 & & \lfloor -1 \\ & -1 & 0 & 4 & & & \\ \hline 1 & 0 & -4 & 0 & & & \end{array}$$

$$s^5+3s^4-s^3-11s^2-12s-4$$
$$=(s+1)(s+1)(s+1)(s^2-4)$$
$$=(s+1)(s+1)(s+1)(s-2)(s+2)$$

The five roots are

$r_1=-1, r_2=-1, r_3=-1, r_4=2, r_5=-2$

33. $x^3+x^2-10x+8=0$ has three roots.

$f(x)$ has two sign changes, so at most two positive roots.

$f(-x)=-x^3+x^2+10x+8,$ so exactly one negative root.

Possible rational roots $=\pm1, \pm2, \pm4, \pm8$

$$\begin{array}{rrrrl} 1 & 1 & -10 & 8 & \lfloor 1 \\ & 1 & 2 & -8 & \\ \hline 1 & 2 & -8 & 0 & \end{array}$$

Remainder is 0, so 1 is a root

$$x^3+x^2-10x+8=(x-1)(x^2+2x-8)$$
$$=(x-1)(x+4)(x-2)$$

The three roots are 1, -4, 2

37. $6x^3-x^2-12x-5=0$ has three roots.

$f(x)$ has one sign change, so exactly one positive root. $f(-x)=-6x^3-x^2+12x-5$ has two sign changes, so at most two negative roots.

Possible rational roots $=\dfrac{\pm1, \pm5}{\pm1, \pm2, \pm3, \pm6}$

$$\begin{array}{rrrrl} 6 & -1 & -12 & -5 & \lfloor 5/3 \\ & 10 & 15 & 5 & \\ \hline 6 & 9 & 3 & 0 & \end{array}$$

Remainder is 0, so $\dfrac{5}{3}$ is a root.

$$6x^3-x^2-12x-5=\left(x-\frac{5}{3}\right)(6x^2+9x+3)$$
$$=3\left(x-\frac{5}{3}\right)(2x^2+3x+1)$$
$$=3\left(x-\frac{5}{3}\right)(2x+1)(x+1)$$

The three roots are

$\dfrac{5}{3}, -\dfrac{1}{2}, -1$

41. A polynomial of degree five with real coefficients has five zeros. Since the complex zeros occur in conjugate pairs, the possibilities are 5 real, no complex; 3 real, 2 complex; 1 real, 4 complex.

45. Use synthetic division to divide by -2, and equate the remainder (which is a function of k) to 0. Then solve for k.

$$\begin{array}{rrrrl} 3 & k & -8 & -8 & \lfloor -2 \\ & -6 & 12-2k & -8+4k & \\ \hline 3 & -6+k & 4-2k & -16+4k & \end{array}$$

$$4k-16=0$$
$$4k=16$$
$$k=4$$

If $k=4, x+2$ is a factor of $3x^3+kx^2-8x-8$.

49. $xy=2 \Rightarrow y=\dfrac{2}{x}$

Substitute into $x^2=y+3$

$$x^2=\frac{2}{x}+3$$

$x^3-3x-2=0$

There is one positive root, and it has to be 1 or 2. We try 2:

$$\begin{array}{rrrrl} 1 & 0 & -3 & -2 & \lfloor 2 \\ & 2 & 4 & 2 & \\ \hline 1 & 2 & 1 & 0 & \end{array}$$

$$x^3-3x-2=(x^2+2x+1)(x-2)$$
$$=(x+1)^2(x-2)$$

The solutions are

$x=-1, y=-2;$

$x=2, y=1;$

53. $M=-\dfrac{25}{6}x^3+600x, M=850.$

Solve: $-\dfrac{25}{6}x^3+600x=850$

$$x^3-144x+204=0$$

There are three roots, at most two of which are positive. Graph $y_1 = x^3 - 144x + 204$ and use the zero feature to solve. The zeros are $x = 11.2$ and $x = 1.44$.

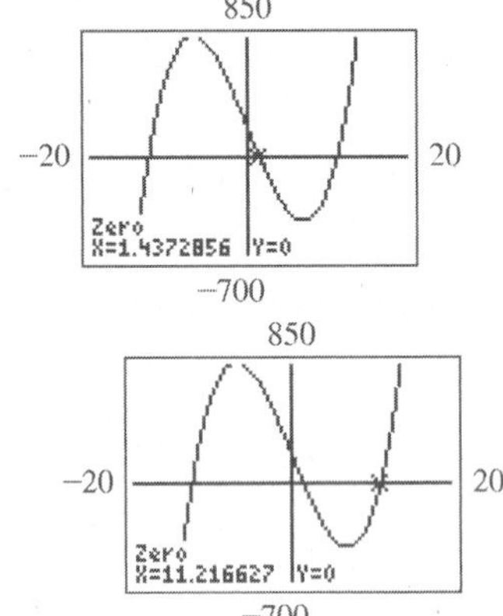

57. $\frac{1}{p} + \frac{1}{q} = \frac{1}{f}$

$f = \frac{p+1}{p}$, so, $\frac{1}{f} = \frac{p}{p+1}$

$$\frac{1}{p} + \frac{1}{p+4} = \frac{p}{p+1}$$

$$p^3 + 4p^2 = 2p^2 + 6p + 4$$

$p^3 + 2p^2 - 6p - 4 = 0$

One sign change so there is only one positive root.

Possible rational roots: $\pm 1, \pm 2, \pm 4$.

```
1  2  -6  -4 |2
   2   8   4
--------------
1  4   2   0
```

The only positive root is thus $p = 2$ cm.

61. $\sqrt{(x+0.300)^2 - x^2}$

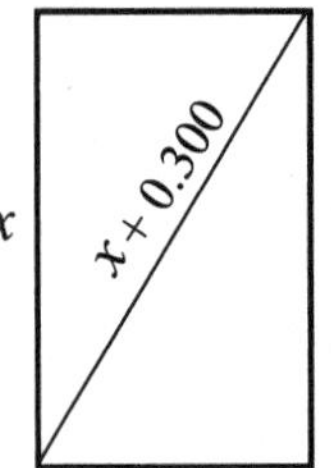

Solve:

$$x\sqrt{(x+0.300)^2 - x^2} = 2.7; \text{square both sides}$$

$$x^2\left(x^2 + 0.600x + 0.090 - x^2\right) = 2.7^2$$

$$0.600x^3 + 0.090x^2 - 2.7^2 = 0$$

Graph $y_1 = 0.600x^3 + 0.090x^2 - 2.7^2$ and use the zero feature to solve.

$x = 2.25$, $\sqrt{(x+0.300)^2 - x^2} = 1.20$

The dimensions are 2.25 m by 1.20 m.

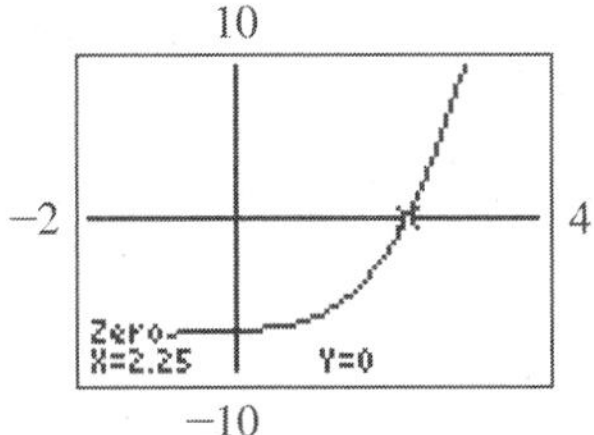

CHAPTER 16

MATRICES; SYSTEMS OF LINEAR EQUATIONS

16.1 Matrices: Definitions and Basic Operations

1. $\begin{bmatrix} 8 & 1 & -5 & 9 \\ 0 & -2 & 3 & 7 \end{bmatrix} + \begin{bmatrix} -3 & 6 & 4 & 0 \\ 6 & 6 & -2 & 5 \end{bmatrix}$

$= \begin{bmatrix} 8+(-3) & 1+6 & -5+4 & 9+0 \\ 0+6 & -2+6 & 3+(-2) & 7+5 \end{bmatrix}$

$= \begin{bmatrix} 5 & 7 & -1 & 9 \\ 6 & 4 & 1 & 12 \end{bmatrix}$

5. $\begin{bmatrix} x & 2y & z \\ r/4 & -s & -5t \end{bmatrix} = \begin{bmatrix} -2 & 10 & -9 \\ 12 & -4 & 5 \end{bmatrix}$

$x = -2, \quad 2y = 10, \quad z = -9$

$y = 5,$

$\frac{r}{4} = 12, \quad -s = -4, \quad -5t = 5$

$r = 48 \quad s = 4 \quad t = -1$

9. $\begin{bmatrix} x-3 & x+y \\ x-z & y+z \\ x+t & y-t \end{bmatrix} = \begin{bmatrix} 5 & 3 \\ 4 & -1 \end{bmatrix}$

This equality is not valid since the matrices have different dimensions.

13. $\begin{bmatrix} 50+(-55) & -82+82 \\ -34+45 & 57+14 \\ -15+26 & 62+(-67) \end{bmatrix} = \begin{bmatrix} -5 & 0 \\ 11 & 71 \\ 11 & -5 \end{bmatrix}$

17. Since A and C do not have the same number of columns, they cannot be added.

21. $A - 2B = \begin{bmatrix} -1 & 4 & -7 \\ 2 & -6 & 11 \end{bmatrix} - 2\begin{bmatrix} 7 & 9 & -6 \\ 4 & -1 & -8 \end{bmatrix}$

$= \begin{bmatrix} -1 & 4 & -7 \\ 2 & -6 & 11 \end{bmatrix} - \begin{bmatrix} 14 & 18 & -12 \\ 8 & -2 & -16 \end{bmatrix}$

$= \begin{bmatrix} -15 & -14 & 5 \\ -6 & -4 & 27 \end{bmatrix}$

25. Since A and C do not have the same number of columns, they cannot be subtracted.

29. $A + 3B = \begin{bmatrix} -1 & 4 & -7 \\ 2 & -6 & 11 \end{bmatrix} + 3\begin{bmatrix} 7 & 9 & -6 \\ 4 & -1 & -8 \end{bmatrix}$

$= \begin{bmatrix} -1 & 4 & -7 \\ 2 & -6 & 11 \end{bmatrix} + \begin{bmatrix} 21 & 27 & -18 \\ 12 & -3 & -24 \end{bmatrix}$

$= \begin{bmatrix} 20 & 31 & -25 \\ 14 & -9 & -13 \end{bmatrix}$

33. $-6B + A = -6\begin{bmatrix} 7 & 9 & -6 \\ 4 & -1 & -8 \end{bmatrix} + \begin{bmatrix} -1 & 4 & -7 \\ 2 & -6 & 11 \end{bmatrix}$

$= \begin{bmatrix} -42 & -54 & 36 \\ -24 & 6 & 48 \end{bmatrix} + \begin{bmatrix} -1 & 4 & -7 \\ 2 & -6 & 11 \end{bmatrix}$

$= \begin{bmatrix} -43 & -50 & 29 \\ -22 & 0 & 59 \end{bmatrix}$

37. $-(A-B) = -\begin{bmatrix} -1-4 & 2+1 & 3+3 & 7-0 \\ 0-5 & -3-0 & -1+1 & 4-1 \\ 9-1 & -1-11 & 0-8 & -2-2 \end{bmatrix}$

$= \begin{bmatrix} 5 & -3 & -6 & -7 \\ 5 & 3 & 0 & -3 \\ -8 & 12 & 8 & 4 \end{bmatrix}$

$B - A = \begin{bmatrix} 4+1 & -1-2 & -3-3 & 0-7 \\ 5-0 & 0+3 & -1+1 & 1-4 \\ 1-9 & 11+1 & 8-0 & 2+2 \end{bmatrix}$

$= \begin{bmatrix} 5 & -3 & -6 & -7 \\ 5 & 3 & 0 & -3 \\ -8 & 12 & 8 & 4 \end{bmatrix}$

41. $\begin{bmatrix} 96 & 75 & 0 & 0 \\ 62 & 44 & 24 & 0 \\ 0 & 35 & 68 & 78 \end{bmatrix} + 2\begin{bmatrix} 96 & 75 & 0 & 0 \\ 62 & 44 & 24 & 0 \\ 0 & 35 & 68 & 78 \end{bmatrix}$

$= \begin{bmatrix} 288 & 225 & 0 & 0 \\ 186 & 132 & 72 & 0 \\ 0 & 105 & 204 & 234 \end{bmatrix}$

16.2 Multiplication of Matrices

1. $A=\begin{bmatrix}1 & 2\\ 0 & -3\\ 2 & 1\end{bmatrix}, B=\begin{bmatrix}-1 & 6 & 5 & -2\\ 3 & 0 & 1 & -4\end{bmatrix}$

$$AB=\begin{bmatrix}1 & 2\\ 0 & -3\\ 2 & 1\end{bmatrix}\begin{bmatrix}-1 & 6 & 5 & -2\\ 3 & 0 & 1 & -4\end{bmatrix}$$

$$AB=\begin{bmatrix}-1+6 & 6+0 & 5+2 & -2-8\\ 0-9 & 0+0 & 0-3 & 0+12\\ -2+3 & 12+0 & 10+1 & -4-4\end{bmatrix}$$

$$AB=\begin{bmatrix}5 & 6 & 7 & -10\\ -9 & 0 & -3 & 12\\ 1 & 12 & 11 & -8\end{bmatrix}$$

5. $\begin{bmatrix}2 & -3 & 1\\ 0 & 7 & -3\end{bmatrix}\begin{bmatrix}90\\ -25\\ 50\end{bmatrix}=\begin{bmatrix}2(90)+(-3)(-25)+1(50)\\ 0(90)+7(-25)+(-3)(50)\end{bmatrix}$

$$=\begin{bmatrix}305\\ -325\end{bmatrix}$$

9. $\begin{bmatrix}-1 & 7\\ 3 & 5\\ 10 & -1\\ -5 & 12\end{bmatrix}\begin{bmatrix}2 & 1\\ 5 & -3\end{bmatrix}$

$$=\begin{bmatrix}-1(2)+7(5) & -1(1)+7(-3)\\ 3(2)+5(5) & 3(1)+5(-3)\\ 10(2)+(-1)(5) & 10(1)+(-1)(-3)\\ -5(2)+(12)(5) & -5(1)+12(-3)\end{bmatrix}$$

$$=\begin{bmatrix}33 & -22\\ 31 & -12\\ 15 & 13\\ 50 & -41\end{bmatrix}$$

13. $\begin{bmatrix}-9.2 & 2.3 & 0.5\\ -3.8 & -2.4 & 9.2\end{bmatrix}\begin{bmatrix}6.5 & -5.2\\ 4.9 & 1.7\\ -1.8 & 6.9\end{bmatrix}=\begin{bmatrix}-49.43 & 55.20\\ -53.02 & 79.16\end{bmatrix}$

17. $AB=\begin{bmatrix}-10 & 25 & 40\\ 42 & -5 & 0\end{bmatrix}\begin{bmatrix}6\\ -15\\ 12\end{bmatrix}$

$$=\begin{bmatrix}(-10)(6)+(25)(-15)+(40)(12)\\ (42)(6)+(-5)(-15)+(0)(12)\end{bmatrix}$$

$$=\begin{bmatrix}45\\ 327\end{bmatrix}$$

BA is not possible because the number of columns in B is not equal to the number of rows in A.

21. $AI=\begin{bmatrix}1 & 3 & -5\\ 2 & 0 & 1\\ 1 & -2 & 4\end{bmatrix}\begin{bmatrix}1 & 0 & 0\\ 0 & 1 & 0\\ 0 & 0 & 1\end{bmatrix}$

$$=\begin{bmatrix}1(1)+3(0)+(-5)(0) & 1(0)+3(1)+(-5)(0)\\ 2(1)+0(0)+1(0) & 2(0)+0(1)+1(0)\\ 1(1)+(-2)(0)+4(0) & 1(0)+(-2)(1)+4(0)\end{bmatrix}$$

$$\begin{bmatrix}1(0)+3(0)+(-5)(1)\\ 2(0)+0(0)+1(1)\\ 1(0)+(-2)(0)+4(1)\end{bmatrix}$$

$$=\begin{bmatrix}1 & 3 & -5\\ 2 & 0 & 1\\ 1 & -2 & 4\end{bmatrix}$$

$$=A$$

$$IA=\begin{bmatrix}1 & 0 & 0\\ 0 & 1 & 0\\ 0 & 0 & 1\end{bmatrix}\begin{bmatrix}1 & 3 & -5\\ 2 & 0 & 1\\ 1 & -2 & 4\end{bmatrix}$$

$$=\begin{bmatrix}1(1)+0(2)+0(1) & 1(3)+0(0)+0(-2)\\ 0(1)+1(2)+0(1) & 0(3)+1(0)+0(-2)\\ 0(1)+0(2)+1(1) & 0(3)+0(0)+1(-2)\end{bmatrix}$$

$$\begin{bmatrix}1(-5)+0(1)+0(4)\\ 0(-5)+1(1)+0(4)\\ 0(-5)+0(1)+1(4)\end{bmatrix}$$

$$=\begin{bmatrix}1 & 3 & -5\\ 2 & 0 & 1\\ 1 & -2 & 4\end{bmatrix}$$

$$=A$$

Therefore, $AI=IA=A$

25. $AB=\begin{bmatrix}1 & -2 & 3\\ 2 & -5 & 7\\ -1 & 3 & -5\end{bmatrix}\begin{bmatrix}4 & -1 & 1\\ 3 & -2 & -1\\ 1 & -1 & -1\end{bmatrix}$

$$=\begin{bmatrix}1(4)+(-2)(3)+3(1) & 1(-1)+(-2)(-2)+3(-1) & 1(1)+(-2)(-1)+3(-1)\\ 2(4)+(-5)(3)+7(1) & 2(-1)+(-5)(-2)+7(-1) & 2(1)+(-5)(-1)+7(-1)\\ -1(4)+3(3)+(-5)(1) & -1(-1)+3(-2)+(-5)(-1) & -1(1)+3(-1)+(-5)(-1)\end{bmatrix}$$

$$=\begin{bmatrix}1 & 0 & 0\\ 0 & 1 & 0\\ 0 & 0 & 1\end{bmatrix}$$

$$BA=\begin{bmatrix}4 & -1 & 1\\ 3 & -2 & -1\\ 1 & -1 & -1\end{bmatrix}\begin{bmatrix}1 & -2 & 3\\ 2 & -5 & 7\\ -1 & 3 & -5\end{bmatrix}$$

$$=\begin{bmatrix}(4)(1)+(-1)(2)+(1)(-1) & (4)(-2)+(-1)(-5)+(1)(3) & (4)(3)+(-1)(7)+(1)(-5)\\ (3)(1)+(-2)(2)+(-1)(-1) & (3)(-2)+(-2)(-5)+(-1)(3) & (3)(3)+(-2)(7)+(-1)(-5)\\ (1)(1)+(-1)(2)+(-1)(-1) & (1)(-2)+(-1)(-5)+(-1)(3) & (1)(3)+(-1)(7)+(-1)(-5)\end{bmatrix}$$

$$=\begin{bmatrix}1 & 0 & 0\\ 0 & 1 & 0\\ 0 & 0 & 1\end{bmatrix}$$

Therefore, $B=A^{-1}$ since $AB=BA=I$.

29. $\begin{bmatrix}3 & 1 & 2\\ 1 & -3 & 4\\ 2 & 2 & 1\end{bmatrix}\begin{bmatrix}-1\\ 2\\ 1\end{bmatrix}=\begin{bmatrix}3(-1)+1(2)+2(1)\\ 1(-1)+(-3)(2)+4(1)\\ 2(-1)+2(2)+1(1)\end{bmatrix}$

$$=\begin{bmatrix}1\\ -3\\ 3\end{bmatrix}$$

$$\neq\begin{bmatrix}1\\ -3\\ 1\end{bmatrix};$$

A is not the proper matrix of solution values.

33. $B^2=\begin{bmatrix}1 & -2 & -6\\ -3 & 2 & 9\\ 2 & 0 & -3\end{bmatrix}\begin{bmatrix}1 & -2 & -6\\ -3 & 2 & 9\\ 2 & 0 & -3\end{bmatrix}$

$$=\begin{bmatrix}-5 & -6 & -6\\ 9 & 10 & 9\\ -4 & -4 & -3\end{bmatrix}$$

$B^3=B^2B$

$$=\begin{bmatrix}-5 & -6 & -6\\ 9 & 10 & 9\\ -4 & -4 & -3\end{bmatrix}\begin{bmatrix}1 & -2 & -6\\ -3 & 2 & 9\\ 2 & 0 & -3\end{bmatrix}$$

$$=\begin{bmatrix}1 & -2 & -6\\ -3 & 2 & 9\\ 2 & 0 & -3\end{bmatrix}$$

$=B$

37. $I=\begin{bmatrix}1 & 0\\ 0 & 1\end{bmatrix}, -I=\begin{bmatrix}-1 & 0\\ 0 & -1\end{bmatrix}$

$$(-I)^2=\begin{bmatrix}-1 & 0\\ 0 & -1\end{bmatrix}\begin{bmatrix}-1 & 0\\ 0 & -1\end{bmatrix}$$

$$=\begin{bmatrix}(-1)(-1)+(0)(0) & (-1)(0)+0(-1)\\ 0(-1)+(-1)(0) & 0(0)+(-1)(-1)\end{bmatrix}$$

$$=\begin{bmatrix}1 & 0\\ 0 & 1\end{bmatrix}$$

$=I$

41. $\sigma_y^2=\begin{bmatrix}0 & -j\\ j & 0\end{bmatrix}\begin{bmatrix}0 & -j\\ j & 0\end{bmatrix}$

$$=\begin{bmatrix}0(0)-j^2 & 0(-j)-j(0)\\ j(0)+0(j) & j(-j)-0(0)\end{bmatrix}$$

$$=\begin{bmatrix}-j^2 & 0\\ 0 & -j^2\end{bmatrix}$$

$$=\begin{bmatrix}-(-1) & 0\\ 0 & -(-1)\end{bmatrix}$$

$$=\begin{bmatrix}1 & 0\\ 0 & 1\end{bmatrix}$$

$\sigma_y^2=I$

45. $$\begin{bmatrix} x & y \end{bmatrix}\begin{bmatrix} 7.10 & -1 \\ 1 & 7.23 \end{bmatrix}\begin{bmatrix} x \\ y \end{bmatrix} = \begin{bmatrix} 5.13\times10^8 \end{bmatrix}$$

$$\begin{bmatrix} 7.10x+y & -x+7.23y \end{bmatrix}\begin{bmatrix} x \\ y \end{bmatrix} = \begin{bmatrix} 5.13\times10^8 \end{bmatrix}$$

$$\begin{bmatrix} 7.10x^2+xy-xy+7.23y^2 \end{bmatrix} = \begin{bmatrix} 5.13\times10^8 \end{bmatrix}$$

$$7.10x^2+7.23y^2 = 5.13\times10^8$$

The equation represents an ellipse

16.3 Finding the Inverse of a Matrix

1. $A = \begin{bmatrix} 2 & -3 \\ 4 & -5 \end{bmatrix}$

$\det A = 2(-5)-(-3)(4) = -10+12 = 2$

$$A^{-1} = \frac{1}{2}\begin{bmatrix} -5 & 3 \\ -4 & 2 \end{bmatrix} = \begin{bmatrix} -\frac{5}{2} & \frac{3}{2} \\ -2 & 1 \end{bmatrix}$$

Check: $AA^{-1} = \begin{bmatrix} 2 & -3 \\ 4 & -5 \end{bmatrix}\begin{bmatrix} -\frac{5}{2} & \frac{3}{2} \\ -2 & 1 \end{bmatrix} = \begin{bmatrix} 1 & 0 \\ 0 & 1 \end{bmatrix}$

5. $\begin{bmatrix} -1 & 5 \\ 4 & 10 \end{bmatrix}$

Find the determinant of the original matrix.

$$\begin{vmatrix} -1 & 5 \\ 4 & 10 \end{vmatrix} = -30 \neq 0$$

Interchange the elements of the principal diagonal and change the signs of the off-diagonal elements.

$$\begin{bmatrix} 10 & -5 \\ -4 & -1 \end{bmatrix}$$

Divide each element of the second matrix by -30.

$$-\frac{1}{30}\begin{bmatrix} 10 & -5 \\ -4 & -1 \end{bmatrix} = \begin{bmatrix} -\frac{1}{3} & \frac{1}{6} \\ \frac{2}{15} & \frac{1}{30} \end{bmatrix}$$

9. $\begin{bmatrix} -50 & -45 \\ 26 & 80 \end{bmatrix}$

Find the determinant of the original matrix.

$$\begin{vmatrix} -50 & -45 \\ 26 & 80 \end{vmatrix} = -2830 \neq 0$$

Interchange the elements of the principal diagonal and change the signs of the off-diagonal elements.

$$\begin{bmatrix} 80 & 45 \\ -26 & -50 \end{bmatrix}$$

Divide each element of the second matrix by -2830.

$$-\frac{1}{2830}\begin{bmatrix} 80 & 45 \\ -26 & -50 \end{bmatrix} = \begin{bmatrix} -\frac{8}{283} & -\frac{9}{566} \\ \frac{13}{1415} & \frac{5}{283} \end{bmatrix}$$

13. $$\left[\begin{array}{cc|cc} 2 & 4 & 1 & 0 \\ -1 & -1 & 0 & 1 \end{array}\right] R1 \to \frac{1}{2}R1 \left[\begin{array}{cc|cc} 1 & 2 & \frac{1}{2} & 0 \\ -1 & -1 & 0 & 1 \end{array}\right]$$

$$R2 \to R2+R1 \left[\begin{array}{cc|cc} 1 & 2 & \frac{1}{2} & 0 \\ 0 & 1 & \frac{1}{2} & 1 \end{array}\right]$$

$$R1 \to R1-2R2 \left[\begin{array}{cc|cc} 1 & 0 & -\frac{1}{2} & -2 \\ 0 & 1 & \frac{1}{2} & 1 \end{array}\right];$$

$$A^{-1} = \begin{bmatrix} -\frac{1}{2} & -2 \\ \frac{1}{2} & 1 \end{bmatrix}$$

17. $$\left[\begin{array}{cc|cc} 25 & 30 & 1 & 0 \\ -10 & -14 & 0 & 1 \end{array}\right] R1 \to \tfrac{1}{25}R1 \left[\begin{array}{cc|cc} 1 & 1.2 & 0.04 & 0 \\ -10 & -14 & 0 & 1 \end{array}\right]$$

$$R2 \to 10R1+R2 \left[\begin{array}{cc|cc} 1 & 1.2 & 0.04 & 0 \\ 0 & -2 & 0.4 & 1 \end{array}\right]$$

$$R2 \to \tfrac{1}{-2}R2 \left[\begin{array}{cc|cc} 1 & 1.2 & 0.04 & 0 \\ 0 & 1 & -0.2 & -0.5 \end{array}\right]$$

$$R1 \to -1.2R2+R1 \left[\begin{array}{cc|cc} 1 & 0 & 0.28 & 0.6 \\ 0 & 1 & -0.2 & -0.5 \end{array}\right];$$

$$A^{-1} = \begin{bmatrix} 0.28 & 0.6 \\ -0.2 & -0.5 \end{bmatrix} = \begin{bmatrix} \frac{7}{25} & \frac{3}{5} \\ -\frac{1}{5} & -\frac{1}{2} \end{bmatrix}$$

21. $$\left[\begin{array}{ccc|ccc} 1 & 3 & 2 & 1 & 0 & 0 \\ -2 & -5 & -1 & 0 & 1 & 0 \\ 2 & 4 & 0 & 0 & 0 & 1 \end{array}\right] \begin{array}{l} R2 \to 2R1+R2 \\ R3 \to -2R1+R3 \end{array}$$

$$\left[\begin{array}{ccc|ccc} 1 & 3 & 2 & 1 & 0 & 0 \\ 0 & 1 & 3 & 2 & 1 & 0 \\ 0 & -2 & -4 & -2 & 0 & 1 \end{array}\right] \begin{array}{l} R1 \to -3R2+R1 \\ R3 \to 2R2+R3 \end{array}$$

$$\left[\begin{array}{ccc|ccc} 1 & 0 & -7 & -5 & -3 & 0 \\ 0 & 1 & 3 & 2 & 1 & 0 \\ 0 & 0 & 2 & 2 & 2 & 1 \end{array}\right] R3 \to \tfrac{1}{2}R3$$

$$\left[\begin{array}{ccc|ccc} 1 & 0 & -7 & -5 & -3 & 0 \\ 0 & 1 & 3 & 2 & 1 & 0 \\ 0 & 0 & 1 & 1 & 1 & \frac{1}{2} \end{array}\right] \begin{array}{l} R2 \to -3R3+R2 \\ R1 \to 7R3+R1 \end{array}$$

$$\left[\begin{array}{ccc|ccc} 1 & 0 & 0 & 2 & 4 & \frac{7}{2} \\ 0 & 1 & 0 & -1 & -2 & \frac{-3}{2} \\ 0 & 0 & 1 & 1 & 1 & \frac{1}{2} \end{array}\right]; A^{-1} = \begin{bmatrix} 2 & 4 & \frac{7}{2} \\ -1 & -2 & \frac{-3}{2} \\ 1 & 1 & \frac{1}{2} \end{bmatrix}$$

25.

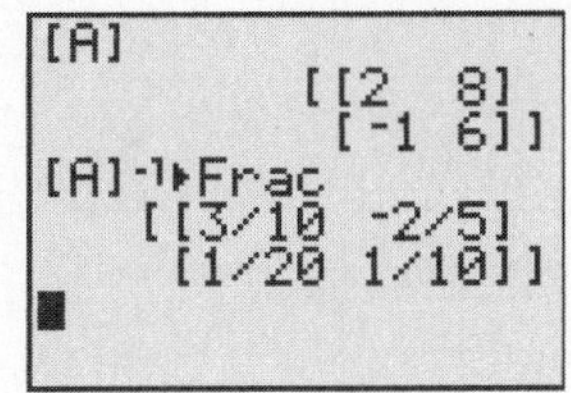

29.

```
      [[2  4  0 ]
       [3  4 -2]
       [-1 1  2 ]]
[A]-1
[[2.5  -2   -2]
 [-1    1    1 ]
 [1.75 -1.5 -1]]
```

33.

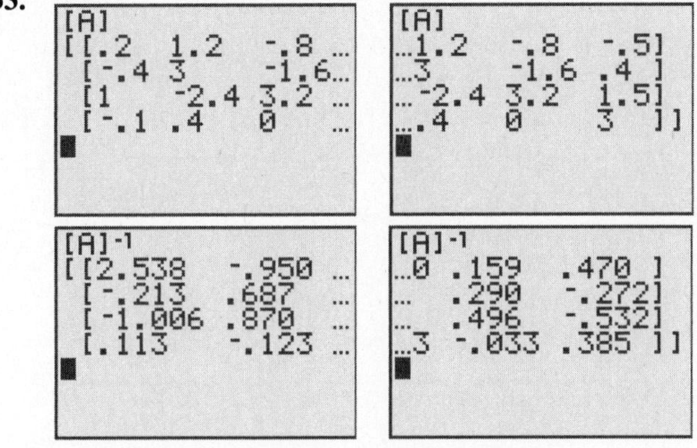

$$A^{-1}=\begin{bmatrix} 2.538 & -0.950 & 0.159 & 0.470 \\ -0.213 & 0.687 & 0.290 & -0.272 \\ -1.006 & 0.870 & 0.496 & -0.532 \\ 0.113 & -0.123 & -0.033 & 0.385 \end{bmatrix}$$

37. $\frac{1}{ad-bc}\cdot\begin{bmatrix} a & b \\ c & d \end{bmatrix}\cdot\begin{bmatrix} d & -b \\ -c & a \end{bmatrix}$

$$=\frac{1}{ad-bc}\cdot\begin{bmatrix} ad-bc & -ab+ab \\ cd-cd & -bc+ad \end{bmatrix}$$

$$=\frac{1}{ad-bc}\cdot\begin{bmatrix} ad-bc & 0 \\ 0 & ad-bc \end{bmatrix}$$

$$=\begin{bmatrix} \frac{ad-bc}{ad-bc} & 0 \\ 0 & \frac{ad-bc}{ad-bc} \end{bmatrix}$$

$$=\begin{bmatrix} 1 & 0 \\ 0 & 1 \end{bmatrix}$$

41. $\left[\begin{array}{ccc|ccc} 0.8 & 0.0 & -0.6 & 1 & 0 & 0 \\ 0.0 & 1.0 & 0.0 & 0 & 1 & 0 \\ 0.6 & 0.0 & 0.8 & 0 & 0 & 1 \end{array}\right] R1 \to \frac{1}{0.8}R1$

$$\left[\begin{array}{ccc|ccc} 1 & 0.0 & -0.75 & 1.25 & 0 & 0 \\ 0.0 & 1.0 & 0.0 & 0 & 1 & 0 \\ 0.6 & 0.0 & 0.8 & 0 & 0 & 1 \end{array}\right] R3 \to R3-0.6R1$$

$$\left[\begin{array}{ccc|ccc} 1 & 0 & -0.75 & 1.25 & 0 & 0 \\ 0 & 1 & 0 & 0 & 1 & 0 \\ 0 & 0 & 1.25 & -0.75 & 0 & 1 \end{array}\right] R3 \to 0.8R3$$

$$\left[\begin{array}{ccc|ccc} 1 & 0 & -0.75 & 1.25 & 0 & 0 \\ 0 & 1 & 0 & 0 & 1 & 0 \\ 0 & 0 & 1 & -0.6 & 0 & 1 \end{array}\right] R1 \to R1+0.75R3$$

$$\left[\begin{array}{ccc|ccc} 1 & 0 & 0 & 0.8 & 0 & 0.6 \\ 0 & 1 & 0 & 0 & 1.0 & 0 \\ 0 & 0 & 1 & -0.6 & 0 & 0.8 \end{array}\right]:$$

$$R^{-1}=\begin{bmatrix} 0.8 & 0.0 & 0.6 \\ 0.0 & 1.0 & 0.0 \\ -0.6 & 0.0 & 0.8 \end{bmatrix}$$

16.4 Matrices and Linear Equations

1. $2x-y=7$

$5x-3y=19$

$$A=\begin{bmatrix} 2 & -1 \\ 5 & -3 \end{bmatrix},\ C=\begin{bmatrix} 7 \\ 19 \end{bmatrix},\ A^{-1}=\begin{bmatrix} 3 & -1 \\ 5 & -2 \end{bmatrix}$$

$$A^{-1}C=\begin{bmatrix} 3 & -1 \\ 5 & -2 \end{bmatrix}\begin{bmatrix} 7 \\ 19 \end{bmatrix}$$

$$=\begin{bmatrix} 2 \\ -3 \end{bmatrix}$$

The solution is $x=2,\ y=-3$.

5. $x+2y=7$

$2x+3y=11$

$$A=\begin{bmatrix} 1 & 2 \\ 2 & 3 \end{bmatrix};\ C=\begin{bmatrix} 7 \\ 11 \end{bmatrix};\ A^{-1}=\begin{bmatrix} -3 & 2 \\ 2 & -1 \end{bmatrix}$$

$$A^{-1}C=\begin{bmatrix} -3 & 2 \\ 2 & -1 \end{bmatrix}\begin{bmatrix} 7 \\ 11 \end{bmatrix}$$

$$=\begin{bmatrix} 1 \\ 3 \end{bmatrix}$$

The solution is $x=1,\ y=3$.

9. $A=\begin{bmatrix}1&3&2\\-2&-5&-1\\2&4&0\end{bmatrix};\ C=\begin{bmatrix}5\\-1\\-2\end{bmatrix};\ A^{-1}=\begin{bmatrix}2&4&\frac{7}{2}\\-1&-2&-\frac{3}{2}\\1&1&\frac{1}{2}\end{bmatrix}$

$$A^{-1}C=\begin{bmatrix}2&4&\frac{7}{2}\\-1&-2&-\frac{3}{2}\\1&1&\frac{1}{2}\end{bmatrix}\begin{bmatrix}5\\-1\\-2\end{bmatrix}$$

$$=\begin{bmatrix}-1\\0\\3\end{bmatrix}$$

The solution is $x=-1,\ y=0,\ z=3$.

13. $A=\begin{bmatrix}2.5&2.8\\3.5&-1.6\end{bmatrix};\ C=\begin{bmatrix}-3.0\\9.6\end{bmatrix};\ \begin{vmatrix}2.5&2.8\\3.5&-1.6\end{vmatrix}=-13.8;$

$$A^{-1}=-\frac{1}{13.8}\begin{bmatrix}-1.6&-2.8\\-3.5&2.5\end{bmatrix}$$

$$A^{-1}C=-\frac{1}{13.8}\begin{bmatrix}-1.6&-2.8\\-3.5&2.5\end{bmatrix}\begin{bmatrix}-3.0\\9.6\end{bmatrix}$$

$$=\begin{bmatrix}1.6\\-2.5\end{bmatrix}$$

The solution is $x=1.6,\ y=-2.5$.

17. $A=\begin{bmatrix}2&4&1\\-2&-2&-1\\-1&2&1\end{bmatrix};\ C=\begin{bmatrix}5\\-6\\0\end{bmatrix}.$

$$\left[\begin{array}{ccc|ccc}2&4&1&1&0&0\\-2&-2&-1&0&1&0\\-1&2&1&0&0&1\end{array}\right]\ R1\to\tfrac{1}{2}R1$$

$$\left[\begin{array}{ccc|ccc}1&2&\frac{1}{2}&\frac{1}{2}&0&0\\-2&-2&-1&0&1&0\\-1&2&1&0&0&1\end{array}\right]\begin{array}{l}\\R2\to R2+2R1\\R3\to R3+R1\end{array}$$

$$\left[\begin{array}{ccc|ccc}1&2&\frac{1}{2}&\frac{1}{2}&0&0\\0&2&0&1&1&0\\0&4&\frac{3}{2}&\frac{1}{2}&0&1\end{array}\right]\ R2\to\tfrac{1}{2}R2$$

$$\left[\begin{array}{ccc|ccc}1&2&\frac{1}{2}&\frac{1}{2}&0&0\\0&1&0&\frac{1}{2}&\frac{1}{2}&0\\0&4&\frac{3}{2}&\frac{1}{2}&0&1\end{array}\right]\begin{array}{l}R1\to R1-2R2\\ \\R3\to R3-4R2\end{array}$$

$$\left[\begin{array}{ccc|ccc}1&0&\frac{1}{2}&-\frac{1}{2}&-1&0\\0&1&0&\frac{1}{2}&\frac{1}{2}&0\\0&0&\frac{3}{2}&-\frac{3}{2}&-2&1\end{array}\right]\to\tfrac{2}{3}R3$$

$$\left[\begin{array}{ccc|ccc}1&0&\frac{1}{2}&-\frac{1}{2}&-1&0\\0&1&0&\frac{1}{2}&\frac{1}{2}&0\\0&0&1&-1&-\frac{4}{3}&\frac{2}{3}\end{array}\right]\ R1\to R1-\tfrac{1}{2}R3$$

$$\left[\begin{array}{ccc|ccc}1&0&0&0&-\frac{1}{3}&-\frac{1}{3}\\0&1&0&0.5&0.5&0\\0&0&1&-1&-\frac{4}{3}&\frac{2}{3}\end{array}\right]$$

$$A^{-1}C=\begin{bmatrix}0&-\frac{1}{3}&-\frac{1}{3}\\\frac{1}{2}&\frac{1}{2}&0\\-1&-\frac{4}{3}&\frac{2}{3}\end{bmatrix}\begin{bmatrix}5\\-6\\0\end{bmatrix}$$

$$=\begin{bmatrix}2\\-\frac{1}{2}\\3\end{bmatrix};$$

The solution is $x=2,\ y=-\dfrac{1}{2},\ z=3$.

21. $A=\begin{bmatrix}1&-3&-2\\3&2&6\\4&-1&3\end{bmatrix};C=\begin{bmatrix}9\\20\\25\end{bmatrix};A^{-1}=\begin{bmatrix}-\frac{12}{11}&-1&\frac{14}{11}\\-\frac{15}{11}&-1&\frac{12}{11}\\1&1&-1\end{bmatrix}$

$$A^{-1}C=\begin{bmatrix}-\frac{12}{11}&-1&\frac{14}{11}\\-\frac{15}{11}&-1&\frac{12}{11}\\1&1&-1\end{bmatrix}\begin{bmatrix}9\\20\\25\end{bmatrix}$$

$$=\begin{bmatrix}2\\-5\\4\end{bmatrix}$$

The solution is $u=2, v=-5, w=4$.

25. $A=\begin{bmatrix}2&3&1&-1&-2\\6&-2&-1&3&-1\\1&3&-4&2&3\\3&-1&-1&7&4\\1&6&6&-4&-1\end{bmatrix};C=\begin{bmatrix}6\\21\\-9\\5\\-4\end{bmatrix}$

$$A^{-1}=\begin{bmatrix}-\frac{97}{427}&\frac{107}{427}&\frac{43}{427}&-\frac{40}{427}&\frac{8}{61}\\\frac{358}{1281}&-\frac{166}{1281}&\frac{19}{427}&\frac{86}{1281}&-\frac{5}{183}\\-\frac{46}{1281}&-\frac{79}{2562}&-\frac{117}{854}&\frac{257}{2562}&\frac{17}{183}\\\frac{571}{1281}&-\frac{551}{2562}&-\frac{135}{854}&\frac{625}{2562}&-\frac{32}{183}\\-\frac{703}{1281}&\frac{190}{1281}&\frac{76}{427}&-\frac{83}{1281}&\frac{41}{183}\end{bmatrix}$$

$$A^{-1}C=\begin{bmatrix}-\frac{97}{427} & \frac{107}{427} & \frac{43}{427} & -\frac{40}{427} & \frac{8}{61}\\ \frac{358}{1281} & -\frac{166}{1281} & \frac{19}{427} & \frac{86}{1281} & -\frac{5}{183}\\ -\frac{46}{1281} & -\frac{79}{2562} & -\frac{117}{854} & \frac{257}{2562} & \frac{17}{183}\\ \frac{571}{1281} & -\frac{551}{2562} & -\frac{135}{854} & \frac{625}{2562} & -\frac{32}{183}\\ -\frac{703}{1281} & \frac{190}{1281} & \frac{76}{427} & -\frac{83}{1281} & \frac{41}{183}\end{bmatrix}\begin{bmatrix}6\\21\\-9\\5\\-4\end{bmatrix}$$

$$=\begin{bmatrix}2\\-1\\\frac{1}{2}\\\frac{3}{2}\\-3\end{bmatrix}$$

The solution is $v=2, w=-1, x=\frac{1}{2}, y=\frac{3}{2}, z=-3$.

29. $2x-y=4$

$3x+y=1$

$$A=\begin{bmatrix}2 & -1\\3 & 1\end{bmatrix}, C=\begin{bmatrix}4\\1\end{bmatrix}, \begin{vmatrix}2 & -1\\3 & 1\end{vmatrix}=5$$

$$A^{-1}=\begin{bmatrix}\frac{1}{5} & \frac{1}{5}\\-\frac{3}{5} & \frac{2}{5}\end{bmatrix}$$

$$A^{-1}C=\begin{bmatrix}\frac{1}{5} & \frac{1}{5}\\-\frac{3}{5} & \frac{2}{5}\end{bmatrix}\begin{bmatrix}4\\1\end{bmatrix}$$

$$=\begin{bmatrix}1\\-2\end{bmatrix}$$

The solution is $x=1,\ y=-2$. The solution is valid for the other two pairs:

$$\begin{bmatrix}2 & -1\\1 & -2\end{bmatrix}\begin{bmatrix}1\\-2\end{bmatrix}=\begin{bmatrix}4\\5\end{bmatrix} \text{ and } \begin{bmatrix}3 & 1\\1 & -2\end{bmatrix}\begin{bmatrix}1\\-2\end{bmatrix}=\begin{bmatrix}1\\5\end{bmatrix}$$

Every pair of equations indeed have the same solution. Graphically, this means that the graphs of all three equations intersect at the same point, namely $(1,-2)$.

33. $x+y=18$

$x-2y=-6$

$$A=\begin{bmatrix}1 & 1\\1 & -2\end{bmatrix}, C=\begin{bmatrix}18\\-6\end{bmatrix}, \begin{vmatrix}1 & 1\\1 & -2\end{vmatrix}=-3$$

$$A^{-1}=\begin{bmatrix}\frac{2}{3} & \frac{1}{3}\\\frac{1}{3} & -\frac{1}{3}\end{bmatrix}$$

$$A^{-1}C=\begin{bmatrix}\frac{2}{3} & \frac{1}{3}\\\frac{1}{3} & -\frac{1}{3}\end{bmatrix}\begin{bmatrix}18\\-6\end{bmatrix}$$

$$=\begin{bmatrix}10\\8\end{bmatrix}$$

The voltages of the batteries are 10 V and 8 V.

37. $x+y+z=10$

$4y=x$

$0.05y+0.06z=10(0.02)$

$$A=\begin{bmatrix}1 & 1 & 1\\-1 & 4 & 0\\0 & 0.05 & 0.06\end{bmatrix}; C=\begin{bmatrix}10\\0\\0.2\end{bmatrix}$$

To find A^{-1}:

$$\left[\begin{array}{ccc|ccc}1 & 1 & 1 & 1 & 0 & 0\\-1 & 4 & 0 & 0 & 1 & 0\\0 & 0.05 & 0.06 & 0 & 0 & 1\end{array}\right] R2\to R2+R1$$

$$\left[\begin{array}{ccc|ccc}1 & 1 & 1 & 1 & 0 & 0\\0 & 5 & 1 & 1 & 1 & 0\\0 & 0.05 & 0.06 & 0 & 0 & 1\end{array}\right] R2\to\tfrac{1}{5}R2$$

$$\left[\begin{array}{ccc|ccc}1 & 1 & 1 & 1 & 0 & 0\\0 & 1 & \frac{1}{5} & \frac{1}{5} & \frac{1}{5} & 0\\0 & 0.05 & 0.06 & 0 & 0 & 1\end{array}\right] \begin{matrix}R1\to R1-R2\\ \\ R3\to R3-0.05R2\end{matrix}$$

$$\left[\begin{array}{ccc|ccc}1 & 0 & \frac{4}{5} & \frac{4}{5} & -\frac{1}{5} & 0\\0 & 1 & \frac{1}{5} & \frac{1}{5} & \frac{1}{5} & 0\\0 & 0 & 0.05 & -0.01 & -0.01 & 1\end{array}\right] R3\to 20R3$$

$$\left[\begin{array}{ccc|ccc}1 & 0 & \frac{4}{5} & \frac{4}{5} & -\frac{1}{5} & 0\\0 & 1 & \frac{1}{5} & \frac{1}{5} & \frac{1}{5} & 0\\0 & 0 & 1 & -0.2 & -0.2 & 20\end{array}\right] \begin{matrix}R1\to R1-\frac{4}{5}R3\\ R2\to R2-\frac{1}{5}R3\\ \end{matrix}$$

$$\left[\begin{array}{ccc|ccc}1 & 0 & 0 & 0.96 & -0.04 & -16\\0 & 1 & 0 & 0.24 & 0.24 & -4\\0 & 0 & 1 & -0.2 & -0.2 & 20\end{array}\right];$$

$$A^{-1}C=\begin{bmatrix}0.96 & -0.04 & -16\\0.24 & 0.24 & -4\\-0.2 & -0.2 & 20\end{bmatrix}\begin{bmatrix}10\\0\\0.2\end{bmatrix}$$

$$=\begin{bmatrix}6.4\\1.6\\2.0\end{bmatrix}$$

The mixture should have 6.4 L of gasoline without additive,1.6 L of gasoline with 5.0% additive and 2.0 L of gasoline with 6.0% additive.

16.5 Gaussian Elimination

1. $-2x+y=4 \quad R1 \to -\frac{1}{2}R1$

$-3x-2y=3$

$x-\frac{1}{2}y=-2 \quad R2 \to 3R1+R2$

$-3x-2y=3$

$x-\frac{1}{2}y=-2$

$-\frac{7}{2}y=-3 \quad R \to -\frac{2}{7}R2$

$x-\frac{1}{2}y=-2$

$y=\frac{6}{7}$

$x-\frac{1}{2}\left(\frac{6}{7}\right)=-2$

$x \quad =-\frac{11}{7}$

The solution is $x=-\frac{11}{7},\ y=\frac{6}{7}$.

5. $5x-3y=2 \quad R1 \to \frac{1}{5}R1$

$-2x+4y=3$

$x-\frac{3}{5}y=\frac{2}{5} \quad R2 \to 2R1+R2$

$-2x+4y=3$

$x-\frac{3}{5}y=\frac{2}{5}$

$\frac{14}{5}y=\frac{19}{5} \quad R2 \to \frac{5}{14}R2$

$x-\frac{3}{5}y=\frac{2}{5}$

$y=\frac{19}{14}$

$x-\frac{3}{5}\left(\frac{19}{14}\right)=\frac{2}{5}$

$x \quad =\frac{17}{14}$

The solution is $x=\frac{17}{14},\ y=\frac{19}{14}$.

9. $x+3y+3z=-3 \quad R2 \to -2R1+R2$

$2x+2y+z=-5 \quad R3 \to 2R1+R3$

$-2x-y+4z=6$

$x+3y+3z=-3$

$-4y-5z=1 \quad R2 \to -\frac{1}{4}R2$

$5y+10z=0$

$x+3y+3z=-3$

$y+\frac{5}{4}z=-\frac{1}{4} \quad R3 \to -5R2+R3$

$5y+10z=0$

$x+3y+3z=-3$

$y+\frac{5}{4}z=-\frac{1}{4}$

$\frac{15}{4}z=\frac{5}{4}$

$z=\frac{1}{3}$

$y+\frac{5}{4}\left(\frac{1}{3}\right)=-\frac{1}{4}$

$y=-\frac{2}{3}$

$x+3\left(-\frac{2}{3}\right)+3\left(\frac{1}{3}\right)=-3$

$x=-2$

The solution is $x=-2,\ y=-\frac{2}{3},\ z=\frac{1}{3}$

13. $x-4y+z=2 \quad R2 \to -3R1+R2$

$3x-y+4z=-4$

$x-4y+z=2$

$11y+z=-10$

$y=-\frac{10}{11}-\frac{1}{11}z$

The value of z can be chosen arbitrarily, so there is an unlimited number of solutions. For example, if $z=1$, then $y=-1$ and $x=-3$. If $z=-10$, then $y=0$ and $x=12$.

17. $x+3y+z=4$

$2x-6y-3z=10 \quad R2 \to -2R1+R2$

$4x-9y+3z=4 \quad R3 \to -4R1+R3$

$x+3y+z=4$

$-12y-5z=2 \quad R2 \to -\frac{1}{12}R2$

$-21y-z=-12$

$x+3y+z=4$

$y+\frac{5}{12}z=-\frac{1}{6} \quad R3 \to R3+21R2$

$-21y-z=-12$

$x+3y+z=4$

$y+\frac{5}{12}z=-\frac{1}{6}$

$\frac{31}{4}z=-\frac{31}{2}$

$z=-2$

$-12y-5(-2)=2$

$y=\frac{2}{3}$

$x+3\left(\frac{2}{3}\right)+(-2)=4$

$x=4$

The solution is $x=4,\ y=\frac{2}{3},\ z=-2$.

21. $$\begin{aligned} 3x+5y&=-2 \\ 24x-18y&=13 \quad R2\to -8R1+R2 \\ 15x-33y&=19 \quad R3\to -5R1+R3 \\ 6x+68y&=-33 \quad R4\to -2R1+R4 \end{aligned}$$

$$\begin{aligned} 3x+5y&=-2 \\ -58y&=29 \quad R2\to -\tfrac{1}{58}R2 \\ -58y&=29 \quad R3\to R3-R2 \\ 58y&=-29 \quad R4\to R4+R2 \end{aligned}$$

$$\begin{aligned} 3x+5y&=-2 \\ y&=-\tfrac{1}{2} \\ 0&=0 \\ 0&=0 \end{aligned}$$

$$3x+5\left(-\tfrac{1}{2}\right)=-2$$
$$x=\tfrac{1}{6}$$

The solution is $x=\frac{1}{6}$, $y=-\frac{1}{2}$.

25. $$\begin{aligned} s+2t-3u&=2 \\ 3s+6t-9u&=6 \quad R2\to -3R1+R2 \\ 7s+14t-21u&=13 \quad R3\to -7R1+R3 \end{aligned}$$

$$\begin{aligned} s+2t-3u&=2 \\ 0&=0 \\ 0&=-1 \end{aligned}$$

The system is inconsistent.

29. $$\begin{aligned} a_1x+b_1y&=c_1 \\ a_2x+b_2y&=c_2 \quad R2\to -\frac{a_2}{a_1}R1+R2 \end{aligned}$$

$$a_1x+b_1y=c_1$$
$$\left(\frac{-a_2b_1}{a_1}+b_2\right)y=\frac{-a_2c_1}{a_1}+c_2$$

$$y=\frac{-\frac{a_2c_1}{a_1}+c_2}{-\frac{a_2b_1}{a_1}+b_2}=\frac{a_1c_2-a_2c_1}{a_1b_2-a_2b_1}=\frac{\begin{vmatrix} a_1 & c_1 \\ a_2 & c_2 \end{vmatrix}}{\begin{vmatrix} a_1 & b_1 \\ a_2 & b_2 \end{vmatrix}}$$

$$a_1x+b_1\cdot\frac{a_1c_2-a_2c_1}{a_1b_2-a_2b_1}=c_1$$

$$x=\frac{-b_1\left(a_1c_2-a_2c_1\right)}{a_1b_2-a_2b_1}+\frac{c_1\left(a_1b_2-a_2b_1\right)}{a_1\left(a_1b_2-a_2b_1\right)}$$

$$x=\frac{-a_1b_1c_2+a_2b_1c_1+a_1b_2c_1-a_2b_1c_1}{a_1\left(a_1b_2-a_2b_1\right)}$$

$$x=\frac{a_1\left(b_2c_1-b_1c_2\right)}{a_1\left(a_1b_2-a_2b_1\right)}=\frac{\begin{vmatrix} c_1 & b_1 \\ c_2 & b_2 \end{vmatrix}}{\begin{vmatrix} a_1 & b_1 \\ a_2 & b_2 \end{vmatrix}}$$

33. $$\begin{aligned} x+y+z&=650 \\ -x+2y-z&=10 \quad R2\to R1+R2 \\ 3x+2y+2z&=1550 \quad R3\to -3R1+R3 \end{aligned}$$

$$\begin{aligned} x+y+z&=650 \\ 3y&=660 \quad R2\to \tfrac{1}{3}R2 \\ -y-z&=-400 \quad R3\to R3+\tfrac{1}{3}R2 \end{aligned}$$

$$\begin{aligned} x+y+z&=650 \\ y&=220 \\ -z&=-180 \end{aligned}$$

$$x+220+180=650$$
$$x=250$$

The production rates are 250 parts/h, 220 parts/h, and 180 parts/h.

16.6 Higher-order Determinants

1. $\begin{vmatrix} 3 & 0 & 0 \\ 1 & 1 & 0 \\ 2 & 1 & 3 \end{vmatrix}=9$, switch first and third column

$$\begin{vmatrix} 0 & 0 & 3 \\ 0 & 1 & 1 \\ 3 & 1 & 2 \end{vmatrix}=-9$$

5. $\begin{vmatrix} 3 & -2 & 4 & 2 \\ 5 & -1 & 2 & -1 \\ 3 & -2 & 4 & 2 \\ 0 & 3 & -6 & 0 \end{vmatrix}=0$

Row 1 and Row 3 are identical, so the determinant is zero.

9. $\begin{vmatrix} 2 & -3 & -1 \\ -4 & 1 & -3 \\ 1 & -3 & 2 \end{vmatrix} = -40,$

Column 3 has been multiplied by -1, so the value of the determinant was multiplied by -1.

13. Expand by first row,

$$\begin{vmatrix} 3 & 1 & 0 \\ -2 & 3 & -1 \\ 4 & 2 & 5 \end{vmatrix} = 3\begin{vmatrix} 3 & -1 \\ 2 & 5 \end{vmatrix} - (1)\begin{vmatrix} -2 & -1 \\ 4 & 5 \end{vmatrix} + 0\begin{vmatrix} -2 & 3 \\ 4 & 2 \end{vmatrix}$$

$$= 3(3(5)-2(-1)) - (-2(5)-4(-1))$$

$$= 57$$

17. Expand by first column,

$$\begin{vmatrix} 1 & 3 & -3 & 5 \\ 4 & 2 & 1 & 2 \\ 3 & 2 & -2 & 2 \\ 0 & 1 & 2 & -1 \end{vmatrix} = 1\begin{vmatrix} 2 & 1 & 2 \\ 2 & -2 & 2 \\ 1 & 2 & -1 \end{vmatrix} - 4\begin{vmatrix} 3 & -3 & 5 \\ 2 & -2 & 2 \\ 1 & 2 & -1 \end{vmatrix}$$

$$+ 3\begin{vmatrix} 3 & -3 & 5 \\ 2 & 1 & 2 \\ 1 & 2 & -1 \end{vmatrix} - 0\begin{vmatrix} 3 & -3 & 5 \\ 2 & 1 & 2 \\ 2 & -2 & 2 \end{vmatrix}$$

$$= 1(12) - 4(12) + 3(-12)$$

$$= -72$$

21. $\begin{vmatrix} 3 & 0 & 0 \\ -2 & 1 & 4 \\ 4 & -2 & 5 \end{vmatrix} \; R2 \to -\dfrac{4}{5}R3 + R2$

$$\begin{vmatrix} 3 & 0 & 0 \\ -2 & 1 & 4 \\ 4 & -2 & 5 \end{vmatrix} = \begin{vmatrix} 3 & 0 & 0 \\ -\frac{26}{5} & \frac{13}{5} & 0 \\ 5 & -2 & 5 \end{vmatrix}$$

$$= 3\left(\frac{13}{5}\right)(5)$$

$$= 39$$

25. $\begin{vmatrix} 4 & 3 & 6 & 0 \\ 3 & 0 & 0 & 4 \\ 5 & 0 & 1 & 2 \\ 2 & 1 & 1 & 7 \end{vmatrix} \; \begin{matrix} \\ R2 \to -\frac{3}{4}R1 + R2 \\ R3 \to -\frac{5}{4}R1 + R3 \\ R4 \to -\frac{1}{2}R1 + R4 \end{matrix}$

$$= \begin{vmatrix} 4 & 3 & 6 & 0 \\ 0 & -\frac{9}{4} & -\frac{9}{2} & 4 \\ 0 & -\frac{15}{4} & -\frac{13}{2} & 2 \\ 0 & -\frac{1}{2} & -2 & 7 \end{vmatrix} \; \begin{matrix} \\ \\ R3 \to -\frac{15}{9}R2 + R3 \\ R4 \to -\frac{2}{9}R2 + R4 \end{matrix}$$

$$= \begin{vmatrix} 4 & 3 & 6 & 0 \\ 0 & -\frac{9}{4} & -\frac{9}{2} & 4 \\ 0 & 0 & 1 & -\frac{14}{3} \\ 0 & 0 & -1 & \frac{55}{9} \end{vmatrix} \; R4 \to R3 + R4$$

$$= \begin{vmatrix} 4 & 3 & 6 & 0 \\ 0 & -\frac{9}{4} & -\frac{9}{2} & 4 \\ 0 & 0 & 1 & -\frac{14}{3} \\ 0 & 0 & 0 & \frac{13}{9} \end{vmatrix}$$

$$= 4\left(-\tfrac{9}{4}\right)(1)\left(\tfrac{13}{9}\right)$$

$$= -13$$

29. $\begin{vmatrix} 1 & 2 & 0 & 1 & 0 \\ 0 & 2 & 1 & 0 & 1 \\ 1 & 0 & -1 & 1 & -1 \\ -2 & 0 & -1 & 2 & 1 \\ 1 & 0 & 2 & -1 & -2 \end{vmatrix} \; \begin{matrix} \\ \\ R3 \to -R1 + R3 \\ R4 \to 2R1 + R4 \\ R5 \to -R1 + R5 \end{matrix}$

$$= \begin{vmatrix} 1 & 2 & 0 & 1 & 0 \\ 0 & 2 & 1 & 0 & 1 \\ 0 & -2 & -1 & 0 & -1 \\ 0 & 4 & -1 & 4 & 1 \\ 0 & -2 & 2 & 2 & -2 \end{vmatrix} \; \begin{matrix} \\ \\ R3 \to R2 + R3 \\ R4 \to -2R2 + R4 \\ R5 \to R2 + R5 \end{matrix}$$

$$= \begin{vmatrix} 1 & 2 & 0 & 1 & 0 \\ 0 & 2 & 1 & 0 & 1 \\ 0 & 0 & 0 & 0 & 0 \\ 0 & 4 & -1 & 4 & 1 \\ 0 & -2 & 2 & -2 & -2 \end{vmatrix}$$

$$= 0$$

Expanding by minors using the third row results in the determinant being 0 (without having to calculate the minors).

33.
$$\begin{aligned} x+2y-z \quad &= 6 \\ y-2z-3t &= -5 \\ 3x-2y \quad +t &= 2 \\ 2x+y+z-t &= 0 \end{aligned}$$

$$\begin{vmatrix} 1 & 2 & -1 & 0 \\ 0 & 1 & -2 & -3 \\ 3 & -2 & 0 & 1 \\ 2 & 1 & 1 & -1 \end{vmatrix} = \begin{vmatrix} 1 & 2 & -1 & 0 \\ 0 & 1 & -2 & -3 \\ 0 & -8 & 3 & 1 \\ 0 & -3 & 3 & -1 \end{vmatrix}$$

$$= 1\begin{vmatrix} 1 & -2 & -3 \\ -8 & 3 & 1 \\ -3 & 3 & -1 \end{vmatrix}$$

$$= \begin{vmatrix} 1 & -2 & -3 \\ 0 & -13 & -23 \\ 0 & -3 & -10 \end{vmatrix}$$

$$= 1\begin{vmatrix} -13 & -23 \\ -3 & -10 \end{vmatrix}$$

$$= 61$$

$$x = \frac{\begin{vmatrix} 6 & 2 & -1 & 0 \\ -5 & 1 & -2 & -3 \\ 2 & -2 & 0 & 1 \\ 0 & 1 & 1 & -1 \end{vmatrix}}{61} = \frac{\begin{vmatrix} 6 & 2 & -1 & 0 \\ 0 & \frac{8}{3} & -\frac{17}{6} & -3 \\ 0 & -\frac{8}{3} & \frac{1}{3} & 1 \\ 0 & 1 & 1 & -1 \end{vmatrix}}{61}$$

$$= \frac{6}{61}\begin{vmatrix} \frac{8}{3} & -\frac{17}{6} & -3 \\ -\frac{8}{3} & \frac{1}{3} & 1 \\ 1 & 1 & -1 \end{vmatrix}$$

$$= \frac{6}{61}\begin{vmatrix} \frac{8}{3} & -\frac{17}{6} & -3 \\ 0 & -\frac{5}{2} & -2 \\ 0 & \frac{33}{16} & \frac{1}{8} \end{vmatrix}$$

$$= \frac{6}{61}\cdot\frac{8}{3}\cdot\left(-\frac{5}{2}\cdot\frac{1}{8} - \frac{-2(33)}{16}\right)$$

$$= 1$$

$$y = \frac{\begin{vmatrix} 1 & 6 & -1 & 0 \\ 0 & -5 & -2 & -3 \\ 3 & 2 & 0 & 1 \\ 2 & 0 & 1 & -1 \end{vmatrix}}{61} = \frac{\begin{vmatrix} 1 & 6 & -1 & 0 \\ 0 & -5 & -2 & -3 \\ 0 & -16 & 3 & 1 \\ 0 & -12 & 3 & -1 \end{vmatrix}}{61}$$

$$= \frac{1}{61}\begin{vmatrix} -5 & -2 & -3 \\ -16 & 3 & 1 \\ -12 & 3 & -1 \end{vmatrix}$$

$$= \frac{1}{61}\begin{vmatrix} 31 & -11 & 0 \\ -28 & 6 & 0 \\ -12 & 3 & -1 \end{vmatrix}$$

$$= \frac{-(-122)}{61}$$

$$= 2$$

$$z = \frac{\begin{vmatrix} 1 & 2 & 6 & 0 \\ 0 & 1 & -5 & -3 \\ 3 & -2 & 2 & 1 \\ 2 & 1 & 0 & -1 \end{vmatrix}}{61} = \frac{\begin{vmatrix} 1 & 2 & 6 & 0 \\ 0 & 1 & -5 & -3 \\ 0 & -8 & -16 & 1 \\ 0 & -3 & -12 & -1 \end{vmatrix}}{61}$$

$$= \frac{1}{61}\begin{vmatrix} 1 & -5 & -3 \\ -8 & -16 & 1 \\ -3 & -12 & -1 \end{vmatrix}$$

$$= \frac{1}{61}\begin{vmatrix} 1 & -5 & -3 \\ 0 & -56 & -23 \\ 0 & -27 & -10 \end{vmatrix}$$

$$= \frac{1(-61)}{61}$$

$$= -1$$

$$t = \frac{\begin{vmatrix} 1 & 2 & -1 & 6 \\ 0 & 1 & -2 & -5 \\ 3 & -2 & 0 & 2 \\ 2 & 1 & 1 & 0 \end{vmatrix}}{61} = \frac{\begin{vmatrix} 1 & 2 & -1 & 6 \\ 0 & 1 & -2 & -5 \\ 0 & -8 & 3 & -16 \\ 0 & -3 & 3 & -12 \end{vmatrix}}{61}$$

$$= \frac{1}{61}\begin{vmatrix} 1 & -2 & -5 \\ -8 & 3 & -16 \\ -3 & 3 & -12 \end{vmatrix}$$

$$= \frac{1}{61}\begin{vmatrix} 1 & -2 & -5 \\ 0 & -13 & -56 \\ 0 & -3 & -27 \end{vmatrix}$$

$$= \frac{1(183)}{61}$$

$$= 3$$

The solution is $x = 1$, $y = 2$, $z = -1$, $t = 3$.

37.
$$\begin{aligned} D+E+2F \quad &= 1 \\ 2D-E+ \quad G &= -2 \\ D-E-F-2G &= 4 \\ 2D-E+2F-G &= 0 \end{aligned}$$

$$\begin{vmatrix} 1 & 1 & 2 & 0 \\ 2 & -1 & 0 & 1 \\ 1 & -1 & -1 & -2 \\ 2 & -1 & 2 & -1 \end{vmatrix} = \begin{vmatrix} 1 & 1 & 2 & 0 \\ 0 & -3 & -4 & 1 \\ 0 & -2 & -3 & -2 \\ 0 & -3 & -2 & -1 \end{vmatrix}$$

$$= \begin{vmatrix} 1 & 1 & 2 & 0 \\ 0 & -3 & -4 & 1 \\ 0 & 0 & -\frac{1}{3} & -\frac{8}{3} \\ 0 & 0 & 2 & -2 \end{vmatrix}$$

$$= \begin{vmatrix} 1 & 1 & 2 & 0 \\ 0 & -3 & -4 & 1 \\ 0 & 0 & -\frac{1}{3} & -\frac{8}{3} \\ 0 & 0 & 0 & -18 \end{vmatrix}$$

$$= (1)(-3)(-\tfrac{1}{3})(-18)$$

$$= -18$$

$$D=\frac{\begin{vmatrix}1&1&2&0\\-2&-1&0&1\\4&-1&-1&-2\\0&-1&2&-1\end{vmatrix}}{-18}=\frac{\begin{vmatrix}1&1&2&0\\0&1&4&1\\0&-5&-9&-2\\0&-1&2&-1\end{vmatrix}}{-18}=\frac{\begin{vmatrix}1&1&2&0\\0&1&4&1\\0&0&11&3\\0&0&6&0\end{vmatrix}}{-18}=\frac{\begin{vmatrix}1&1&2&0\\0&1&4&1\\0&0&11&3\\0&0&0&-\frac{18}{11}\end{vmatrix}}{-18}=\frac{(1)(1)(11)(-\frac{18}{11})}{-18}=1$$

$$E=\frac{\begin{vmatrix}1&1&2&0\\2&-2&0&1\\1&4&-1&-2\\2&0&2&-1\end{vmatrix}}{-18}=\frac{\begin{vmatrix}1&1&2&0\\0&-4&-4&1\\0&3&-3&-2\\0&-2&-2&-1\end{vmatrix}}{-18}=\frac{\begin{vmatrix}1&1&2&0\\0&-4&-4&1\\0&0&-6&-\frac{5}{4}\\0&0&0&-\frac{3}{2}\end{vmatrix}}{-18}=\frac{(1)(-4)(-6)(-\frac{3}{2})}{-18}=2$$

$$F=\frac{\begin{vmatrix}1&1&1&0\\2&-1&-2&1\\1&-1&4&-2\\2&-1&0&-1\end{vmatrix}}{-18}=\frac{\begin{vmatrix}1&1&1&0\\0&-3&-4&1\\0&-2&3&-2\\0&-3&-2&-1\end{vmatrix}}{-18}=\frac{\begin{vmatrix}1&1&1&0\\0&-3&-4&1\\0&0&\frac{17}{3}&-\frac{8}{3}\\0&0&2&-2\end{vmatrix}}{-18}=\frac{\begin{vmatrix}1&1&1&0\\0&-3&-4&1\\0&0&\frac{17}{3}&-\frac{8}{3}\\0&0&0&-\frac{18}{17}\end{vmatrix}}{-18}=\frac{(1)(-3)(\frac{17}{3})(-\frac{18}{17})}{-18}=-1$$

$$G=\frac{\begin{vmatrix}1&1&2&1\\2&-1&0&-2\\1&-1&-1&4\\2&-1&2&0\end{vmatrix}}{-18}=\frac{\begin{vmatrix}1&1&2&1\\0&-3&-4&-4\\0&-2&-3&3\\0&-3&-2&-2\end{vmatrix}}{-18}=\frac{\begin{vmatrix}1&1&2&1\\0&-3&-4&-4\\0&0&-\frac{1}{3}&\frac{17}{3}\\0&0&2&2\end{vmatrix}}{-18}=\frac{\begin{vmatrix}1&1&2&1\\0&-3&-4&-4\\0&0&-\frac{1}{3}&\frac{17}{3}\\0&0&0&36\end{vmatrix}}{-18}=\frac{(1)(-3)(-\frac{1}{3})(36)}{-18}=-2$$

The solution is $D=1$, $E=2$, $F=-1$, $G=-2$.

41. $$\begin{vmatrix}2a&2b&2c\\2d&2e&2f\\2g&2h&2i\end{vmatrix}=2\begin{vmatrix}a&b&c\\2d&2e&2f\\2g&2h&2i\end{vmatrix}=2(2)\begin{vmatrix}a&b&c\\d&e&f\\2g&2h&2i\end{vmatrix}=2(2)(2)\begin{vmatrix}a&b&c\\d&e&f\\g&h&i\end{vmatrix}$$

The value of the determinant is changed by a factor of 8.

45. $$\begin{vmatrix}C&-1&0&0\\-1&C&-1&0\\0&-1&C&-1\\0&0&-1&C\end{vmatrix}=C\begin{vmatrix}C&-1&0\\-1&C&-1\\0&-1&C\end{vmatrix}+\begin{vmatrix}-1&0&0\\-1&C&-1\\0&-1&C\end{vmatrix}$$

$$=C(C^3-2C)+(-C^2+1)$$
$$=C^4-3C^2+1$$
$$=0$$

There are four roots, at most two of which are positive. ± 1 are not roots, so the roots are irrational. We solve graphically to get $C=0.618$ or $C=1.618$.

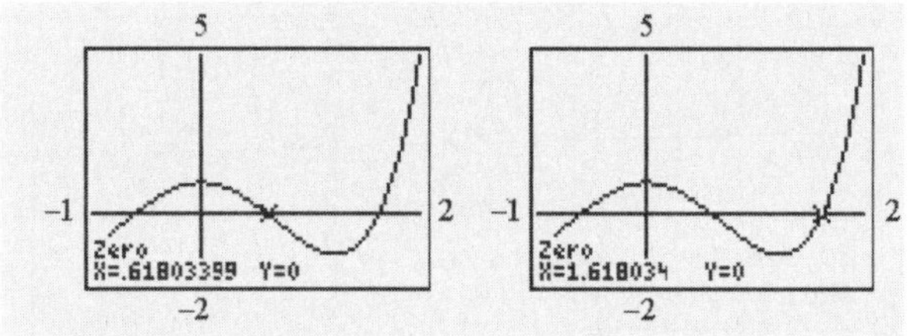

Review Exercises

1. $\begin{bmatrix}2a\\a-b\end{bmatrix}=\begin{bmatrix}8\\5\end{bmatrix}$;

$$2a=8$$
$$a=4;$$
$$a-b=5$$
$$4-b=5$$
$$b=-1$$

The solution is $a=4$, $b=-1$.

5. $\begin{bmatrix} \cos\pi & \sin\frac{\pi}{6} \\ x+y & x-y \end{bmatrix} = \begin{bmatrix} x & y \\ a & b \end{bmatrix}$

$x = \cos\pi$

$x = -1$

$y = \sin\dfrac{\pi}{6}$

$y = \dfrac{1}{2}$

9. $B - A = \begin{bmatrix} -1 & 0 \\ 4 & -6 \\ -3 & -2 \\ 1 & -7 \end{bmatrix} - \begin{bmatrix} 2 & -3 \\ 4 & 1 \\ -5 & 0 \\ 2 & -3 \end{bmatrix}$

$= \begin{bmatrix} -3 & 3 \\ 0 & -7 \\ 2 & -2 \\ -1 & -4 \end{bmatrix}$

13. $\begin{bmatrix} 2 & -1 \\ -2 & 1 \end{bmatrix}\begin{bmatrix} 1 & -1 \\ 2 & -2 \end{bmatrix} = \begin{bmatrix} 2(1)-1(2) & 2(-1)-1(-2) \\ -2(1)+1(2) & -2(-1)+1(-2) \end{bmatrix}$

$= \begin{bmatrix} 0 & 0 \\ 0 & 0 \end{bmatrix}$

17. Find the determinant of the original matrix.

$\begin{vmatrix} 2 & -5 \\ 2 & -4 \end{vmatrix} = 2$

Interchange elements of principal diagonal and change signs of off-diagonal elements.

$\begin{bmatrix} -4 & 5 \\ -2 & 2 \end{bmatrix}$

Divide each element of second matrix by 2.

$\dfrac{1}{2}\begin{bmatrix} -4 & 5 \\ -2 & 2 \end{bmatrix} = \begin{bmatrix} -2 & \frac{5}{2} \\ -1 & 1 \end{bmatrix}$

21. $\left[\begin{array}{ccc|ccc} 1 & 1 & -2 & 1 & 0 & 0 \\ -1 & -2 & 1 & 0 & 1 & 0 \\ 0 & 3 & 4 & 0 & 0 & 1 \end{array}\right]$

$R1+R2 \to R2 \left[\begin{array}{ccc|ccc} 1 & 1 & -2 & 1 & 0 & 0 \\ 0 & -1 & -1 & 1 & 1 & 0 \\ 0 & 3 & 4 & 0 & 0 & 1 \end{array}\right]$

$3R2+R3 \to R3 \left[\begin{array}{ccc|ccc} 1 & 1 & -2 & 1 & 0 & 0 \\ 0 & -1 & -1 & 1 & 1 & 0 \\ 0 & 0 & 1 & 3 & 3 & 1 \end{array}\right]$

$R2+R1 \to R1 \left[\begin{array}{ccc|ccc} 1 & 0 & -3 & 2 & 1 & 0 \\ 0 & -1 & -1 & 1 & 1 & 0 \\ 0 & 0 & 1 & 3 & 3 & 1 \end{array}\right]$

$R3+R2 \to R2 \left[\begin{array}{ccc|ccc} 1 & 0 & -3 & 2 & 1 & 0 \\ 0 & -1 & 0 & 4 & 4 & 1 \\ 0 & 0 & 1 & 3 & 3 & 1 \end{array}\right]$

$3R3+R1 \to R1 \left[\begin{array}{ccc|ccc} 1 & 0 & 0 & 11 & 10 & 3 \\ 0 & -1 & 0 & 4 & 4 & 1 \\ 0 & 0 & 1 & 3 & 3 & 1 \end{array}\right]$

$-R2 \to R2 \left[\begin{array}{ccc|ccc} 1 & 0 & 0 & 11 & 10 & 3 \\ 0 & 1 & 0 & -4 & -4 & -1 \\ 0 & 0 & 1 & 3 & 3 & 1 \end{array}\right]$

$A^{-1} = \begin{bmatrix} 11 & 10 & 3 \\ -4 & -4 & -1 \\ 3 & 3 & 1 \end{bmatrix}$

25. $A = \begin{bmatrix} 2 & -3 \\ 4 & -1 \end{bmatrix}; C = \begin{bmatrix} -9 \\ -13 \end{bmatrix}; \begin{vmatrix} 2 & -3 \\ 4 & -1 \end{vmatrix} = 10;$

$A^{-1} = \dfrac{1}{10}\begin{bmatrix} -1 & 3 \\ -4 & 2 \end{bmatrix}$

$= \begin{bmatrix} -\frac{1}{10} & \frac{3}{10} \\ -\frac{4}{10} & \frac{2}{10} \end{bmatrix}$

$A^{-1}C = \begin{bmatrix} -\frac{1}{10} & \frac{3}{10} \\ -\frac{4}{10} & \frac{2}{10} \end{bmatrix}\begin{bmatrix} -9 \\ -13 \end{bmatrix}$

$= \begin{bmatrix} -3 \\ 1 \end{bmatrix}$

The solution is $x = -3,\ y = 1$.

29. $A = \begin{bmatrix} 2 & -3 & 2 \\ 3 & 1 & -3 \\ 1 & 4 & 1 \end{bmatrix}; C = \begin{bmatrix} 7 \\ -6 \\ -13 \end{bmatrix}$

$\left[\begin{array}{ccc|ccc} 2 & -3 & 2 & 1 & 0 & 0 \\ 3 & 1 & -3 & 0 & 1 & 0 \\ 1 & 4 & 1 & 0 & 0 & 1 \end{array}\right]$

$\frac{1}{2}R1 \to R1 \quad \left[\begin{array}{ccc|ccc} 1 & -\frac{3}{2} & 1 & \frac{1}{2} & 0 & 0 \\ 3 & 1 & -3 & 0 & 1 & 0 \\ 1 & 4 & 1 & 0 & 0 & 1 \end{array}\right]$

$\begin{array}{r} -3R1+R2 \to R2 \\ -R1+R2 \end{array} \left[\begin{array}{ccc|ccc} 1 & -\frac{3}{2} & 1 & \frac{1}{2} & 0 & 0 \\ 0 & \frac{11}{2} & -6 & -\frac{3}{2} & 1 & 0 \\ 0 & \frac{11}{2} & 0 & -\frac{1}{2} & 0 & 1 \end{array}\right]$

$$\frac{2}{11}R2 \to R1 \quad \left[\begin{array}{ccc|ccc} 1 & -\frac{3}{2} & 1 & \frac{1}{2} & 0 & 0 \\ 0 & 1 & -\frac{12}{11} & -\frac{3}{11} & \frac{2}{11} & 0 \\ 0 & \frac{11}{2} & 0 & -\frac{1}{2} & 0 & 1 \end{array}\right]$$

$$\begin{array}{l} \frac{3}{2}R2 + R1 \to R1 \\ -\frac{11}{2}R2 + R3 \to R3 \end{array} \left[\begin{array}{ccc|ccc} 1 & 0 & -\frac{7}{11} & \frac{1}{11} & \frac{3}{11} & 0 \\ 0 & 1 & -\frac{12}{11} & -\frac{3}{11} & \frac{2}{11} & 0 \\ 0 & 0 & 6 & 1 & -1 & 1 \end{array}\right]$$

$$\frac{1}{6}R3 \to R3 \quad \left[\begin{array}{ccc|ccc} 1 & 0 & -\frac{7}{11} & \frac{1}{11} & \frac{3}{11} & 0 \\ 0 & 1 & -\frac{12}{11} & -\frac{3}{11} & \frac{2}{11} & 0 \\ 0 & 0 & 1 & \frac{1}{6} & -\frac{1}{6} & \frac{1}{6} \end{array}\right]$$

$$\begin{array}{l} \frac{7}{11}R3 + R1 \to R1 \\ \frac{12}{11}R3 + R2 \to R2 \end{array} \left[\begin{array}{ccc|ccc} 1 & 0 & 0 & \frac{13}{66} & \frac{11}{66} & \frac{7}{66} \\ 0 & 1 & 0 & -\frac{1}{11} & 0 & \frac{2}{11} \\ 0 & 0 & 1 & \frac{1}{6} & -\frac{1}{6} & \frac{1}{6} \end{array}\right]$$

$$A^{-1}C = \begin{bmatrix} \frac{13}{66} & \frac{11}{66} & \frac{7}{66} \\ -\frac{1}{11} & 0 & \frac{2}{11} \\ \frac{1}{6} & -\frac{1}{6} & \frac{1}{6} \end{bmatrix} \begin{bmatrix} 7 \\ -6 \\ -13 \end{bmatrix} = \begin{bmatrix} -1 \\ -3 \\ 0 \end{bmatrix}$$

The solution is $u = -1, v = -3, w = 0$.

33. $2x - 3y = -9$
$4x - y = -13 \quad R2 \to -2R1 + R2$

$2x - 3y = -9$
$5y = 5$

$y = 1$
$2x - 3(1) = -9$
$x = -3$
The solution is $x = -3,\ y = 1$.

37. $x + 2y + 3z = 1$
$3x - 4y - 3z = 2 \quad R2 \to -3R1 + R2$
$7x - 6y + 6z = 2 \quad R3 \to -7R1 + R3$

$x + 2y + 3z = 1$
$-10y - 12z = -1$
$-20y - 15z = -5 \quad R3 \to -2R2 + R3$

$x + 2y + 3z = 1$
$-10y - 12z = -1$
$9z = -3$

$z = -\frac{1}{3}$
$-10y - 12\left(-\frac{1}{3}\right) = -1$
$y = \frac{1}{2}$
$x + 2\left(\frac{1}{2}\right) + 3\left(-\frac{1}{3}\right) = 1$
$x = 1$
The solution is $x = 1,\ y = \frac{1}{2},\ z = -\frac{1}{3}$.

41. $2u - 3v + 2w = 7$
$3u + v - 3w = -6$
$u + 4v + w = -13$

$$\begin{vmatrix} 2 & -3 & 2 \\ 3 & 1 & -3 \\ 1 & 4 & 1 \end{vmatrix} = 66$$

$$u = \frac{\begin{vmatrix} 7 & -3 & 2 \\ -6 & 1 & -3 \\ -13 & 4 & 1 \end{vmatrix}}{66} = \frac{-66}{66} = -1$$

$$v = \frac{\begin{vmatrix} 2 & 7 & 2 \\ 3 & -6 & -3 \\ 1 & -13 & 1 \end{vmatrix}}{66} = \frac{-198}{66} = -3$$

$$w = \frac{\begin{vmatrix} 2 & -3 & 7 \\ 3 & 1 & -6 \\ 1 & 4 & -13 \end{vmatrix}}{66} = \frac{0}{66} = 0$$

The solution is u $= -1,\ v = -3,\ w = 0$.

45. $3x - 2y + z = 6$
$2x + 0y + 3z = 3$
$4x - y + 5z = 6$

$$A = \begin{bmatrix} 3 & -2 & 1 \\ 2 & 0 & 3 \\ 4 & -1 & 5 \end{bmatrix};\ C = \begin{bmatrix} 6 \\ 3 \\ 6 \end{bmatrix};\ A^{-1} = \begin{bmatrix} 1 & 3 & -2 \\ \frac{2}{3} & \frac{11}{3} & -\frac{7}{3} \\ -\frac{2}{3} & -\frac{5}{3} & \frac{4}{3} \end{bmatrix}$$

$$A^{-1}C = \begin{bmatrix} 1 & 3 & -2 \\ \frac{2}{3} & \frac{11}{3} & -\frac{7}{3} \\ -\frac{2}{3} & -\frac{5}{3} & \frac{4}{3} \end{bmatrix} \begin{bmatrix} 6 \\ 3 \\ 6 \end{bmatrix} = \begin{bmatrix} 3 \\ 1 \\ -1 \end{bmatrix}$$

The solution is $x = 3, y = 1, z = -1$.

49. $3x - y + 6z - 2t = 8$
$2x + 5y + z + 2t = 7$
$4x - 3y + 8z + 3t = -17$
$3x + 5y - 3z + t = 8$

$$A = \begin{bmatrix} 3 & -1 & 6 & -2 \\ 2 & 5 & 1 & 2 \\ 4 & -3 & 8 & 3 \\ 3 & 5 & -3 & 1 \end{bmatrix},\ C = \begin{bmatrix} 8 \\ 7 \\ -17 \\ 8 \end{bmatrix};$$

$$A^{-1} = \begin{bmatrix} \frac{41}{666} & -\frac{17}{74} & \frac{7}{74} & \frac{199}{666} \\ \frac{11}{222} & \frac{17}{74} & -\frac{7}{74} & -\frac{17}{222} \\ \frac{47}{666} & \frac{13}{74} & -\frac{1}{74} & -\frac{113}{666} \\ -\frac{49}{222} & \frac{5}{74} & \frac{11}{74} & -\frac{5}{222} \end{bmatrix}$$

$$A^{-1}C=\begin{bmatrix} \frac{41}{666} & -\frac{17}{74} & \frac{7}{74} & \frac{199}{666} \\ \frac{11}{222} & \frac{17}{74} & -\frac{7}{74} & -\frac{17}{222} \\ \frac{47}{666} & \frac{13}{74} & -\frac{1}{74} & -\frac{113}{666} \\ -\frac{49}{222} & \frac{5}{74} & \frac{11}{74} & -\frac{5}{222} \end{bmatrix}\begin{bmatrix} 8 \\ 7 \\ -17 \\ 8 \end{bmatrix}=\begin{bmatrix} -\frac{1}{3} \\ 3 \\ \frac{2}{3} \\ -4 \end{bmatrix}$$

The solution is $x=-\frac{1}{3},\ y=3, z=\frac{2}{3},\ t=-4.$

53. $$A^2=\begin{bmatrix} 1 & 0 \\ 3 & 4 \end{bmatrix}\begin{bmatrix} 1 & 0 \\ 3 & 4 \end{bmatrix}=\begin{bmatrix} 1 & 0 \\ 15 & 16 \end{bmatrix}$$

$$A^3=A^2A=\begin{bmatrix} 1 & 0 \\ 15 & 16 \end{bmatrix}\begin{bmatrix} 1 & 0 \\ 3 & 4 \end{bmatrix}=\begin{bmatrix} 1 & 0 \\ 63 & 64 \end{bmatrix}$$

$$A^4=A^3A=\begin{bmatrix} 1 & 0 \\ 63 & 64 \end{bmatrix}\begin{bmatrix} 1 & 0 \\ 3 & 4 \end{bmatrix}=\begin{bmatrix} 1 & 0 \\ 255 & 256 \end{bmatrix}$$

57. Expanding by the first row:

$$\begin{vmatrix} 4 & 2 & 3 \\ 1 & -5 & -2 \\ -3 & 4 & -3 \end{vmatrix}$$
$$=4(-5(-3)-4(-2))-2(1(-3)-(-3)(-2))$$
$$+3(1(4)-(-3)(-5))$$
$$=77$$

61. $$\begin{vmatrix} 4 & 2 & 3 \\ 1 & -5 & -2 \\ -3 & 4 & -1 \end{vmatrix} \begin{matrix} -\frac{1}{4}R1+R2\to R2 \\ \frac{3}{4}R1+R3\to R3 \end{matrix}$$

$$=\begin{vmatrix} 4 & 2 & 3 \\ 0 & -\frac{11}{2} & -\frac{11}{4} \\ 0 & \frac{11}{2} & -\frac{3}{4} \end{vmatrix} R2+R3\to R3$$

$$=\begin{vmatrix} 4 & 2 & 3 \\ 0 & -\frac{11}{2} & -\frac{11}{4} \\ 0 & 0 & -\frac{7}{2} \end{vmatrix}$$

$$=4\left(-\tfrac{11}{2}\right)\left(-\tfrac{7}{2}\right)$$
$$=77$$

65. $$N=\begin{bmatrix} 0 & -1 \\ 1 & 0 \end{bmatrix};$$

$$N^{-1}=\frac{1}{0-(-1)}\begin{bmatrix} 0 & 1 \\ -1 & 0 \end{bmatrix}=1\begin{bmatrix} 0 & 1 \\ -1 & 0 \end{bmatrix}=-N$$

69. $$A=\begin{bmatrix} 1 & -2 \\ 0 & 3 \end{bmatrix}, B=\begin{bmatrix} -3 & 1 \\ 2 & -1 \end{bmatrix}$$

$$(A+B)(A-B)$$
$$=\left(\begin{bmatrix} 1 & -2 \\ 0 & 3 \end{bmatrix}+\begin{bmatrix} -3 & 1 \\ 2 & -1 \end{bmatrix}\right)\left(\begin{bmatrix} 1 & -2 \\ 0 & 3 \end{bmatrix}-\begin{bmatrix} -3 & 1 \\ 2 & -1 \end{bmatrix}\right)$$
$$=\begin{bmatrix} -2 & -1 \\ 2 & 2 \end{bmatrix}\begin{bmatrix} 4 & -3 \\ -2 & 4 \end{bmatrix}$$
$$=\begin{bmatrix} -6 & 2 \\ 4 & 2 \end{bmatrix}$$

$$A^2-B^2=\begin{bmatrix} 1 & -2 \\ 0 & 3 \end{bmatrix}^2-\begin{bmatrix} -3 & 1 \\ 2 & -1 \end{bmatrix}^2$$
$$=\begin{bmatrix} 1 & -8 \\ 0 & 9 \end{bmatrix}-\begin{bmatrix} 11 & -4 \\ -8 & 3 \end{bmatrix}$$
$$=\begin{bmatrix} -10 & -4 \\ 8 & 6 \end{bmatrix}$$

$$(A+B)(A-B)\neq A^2-B^2$$

73. $$A=\begin{bmatrix} 2 & 3 \\ 3 & 2 \end{bmatrix}, C=\begin{bmatrix} 26 \\ 24 \end{bmatrix}, \begin{vmatrix} 2 & 3 \\ 3 & 2 \end{vmatrix}=-5;$$

$$A^{-1}=\begin{bmatrix} -\frac{2}{5} & \frac{3}{5} \\ \frac{3}{5} & -\frac{2}{5} \end{bmatrix}$$

$$A^{-1}C=\begin{bmatrix} -\frac{2}{5} & \frac{3}{5} \\ \frac{3}{5} & -\frac{2}{5} \end{bmatrix}\begin{bmatrix} 26 \\ 24 \end{bmatrix}=\begin{bmatrix} 4 \\ 6 \end{bmatrix}$$

The solution is $R_1=4\Omega,\ R_2=6\Omega.$

77. $$2R_1+3R_2=26$$
$$3R_1+2R_2=24 \quad -\tfrac{3}{2}R1+R2\to R2$$

$$2R_1+3R_2=26$$
$$-\tfrac{5}{2}R_2=-15$$
$$R_2=6$$
$$2R_1+3(6)=26$$
$$R_1=4$$

The solution is $R_1=4\Omega,\ R_2=6\Omega.$

81. $180t - d = 0$, suspect

$$225t - d = \frac{225(3.0)}{60}, \text{ police}$$

$$A = \begin{bmatrix} 180 & -1 \\ 225 & -1 \end{bmatrix}, C = \begin{bmatrix} 0 \\ \frac{225(3.0)}{60} \end{bmatrix}, A^{-1} = \begin{bmatrix} -\frac{1}{45} & \frac{1}{45} \\ -5 & 4 \end{bmatrix}$$

$$A^{-1}C = \begin{bmatrix} -\frac{1}{45} & \frac{1}{45} \\ -5 & 4 \end{bmatrix}\begin{bmatrix} 0 \\ 11.25 \end{bmatrix}$$

$$= \begin{bmatrix} 0.25 \\ 45 \end{bmatrix} = \begin{bmatrix} t \\ d \end{bmatrix}$$

$$t = 0.25, \quad t - \tfrac{3.0}{60} = 0.20$$

The police overtake the suspect 0.20 h, or 12 minutes, after passing the intersection.

85.

$$\begin{aligned} -0.3x + 0.5y + 0.32z &= 0 \quad R1 \leftrightarrow R4 \\ 0.2x - 0.8y + 0.3z &= 0 \\ 0.1x + 0.3y - 0.62z &= 0 \\ x + y + z &= 1 \end{aligned}$$

$$\begin{aligned} x + y + z &= 1 \quad -0.2R1 + R2 \rightarrow R2 \\ 0.2x - 0.8y + 0.3z &= 0 \quad -0.1R1 + R3 \rightarrow R3 \\ 0.1x + 0.3y - 0.62z &= 0 \quad 0.3R1 + R4 \rightarrow R4 \\ -0.3x + 0.5y + 0.32z &= 0 \end{aligned}$$

$$\begin{aligned} x + y + z &= 1 \quad 0.2R2 + R3 \rightarrow R3 \\ -y + 0.1z &= -0.2 \quad 0.8R2 + R4 \rightarrow R4 \\ 0.2y - 0.72z &= -0.1 \\ 0.8y + 0.62z &= 0.3 \end{aligned}$$

$$\begin{aligned} x + y + z &= 1 \\ -y + 0.1z &= -0.2 \\ -0.7z &= -0.14 \quad \tfrac{1}{-0.7}R3 \rightarrow R3 \\ 0.7z &= 0.14 \quad R3 + R4 \rightarrow R4 \end{aligned}$$

$$\begin{aligned} x + y + z &= 1 \\ -y + 0.1z &= -0.2 \\ z &= 0.2 \\ 0 &= 0 \end{aligned}$$

$$-y + 0.1(0.2) = -0.2$$
$$y = 0.22$$
$$x + 0.22 + 0.2 = 1$$
$$x = 0.58$$

The solution is $x = 0.58, y = 0.22, z = 0.2$. That is, in the long run the first brand will have 58% of the market, the second brand will have 22%, and the third one will have 20%.

89. $$\begin{bmatrix} 25 & 21 & 22 \\ 3 & 10 & 10 \\ 7 & 25 & 6 \end{bmatrix}\begin{bmatrix} 17 \\ 16 \\ 37 \end{bmatrix} = \begin{bmatrix} 1575 \\ 581 \\ 741 \end{bmatrix}$$

beef stew 1575 kJ
coleslaw 581 kJ
ice cream 741 kJ

CHAPTER 17

INEQUALITIES

17.1 Properties of Inequalities

1. The inequality $x+1<0$ is true for all values of x less than -1. Therefore, the values of x that satisfy this inequality are written as $x<-1$, or as the interval $(-\infty,-1)$.

5.
$$4<9$$
$$4+3<9+3$$
$$7<12$$
(property 1)

9.
$$4<9$$
$$\frac{4}{-1}>\frac{9}{-1}$$
$$-4>-9$$
(property 3)

13. $x>-2$

17. $1<x<7$

21. $x<1$ or $3<x\le 5$

25. x is greater than 0 and less than or equal to 2.

29. $x<3$

$(-\infty,3)$

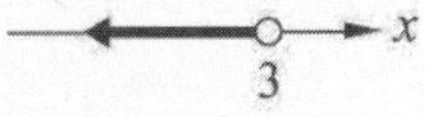

33. $0\le x<5$

$[0,5)$

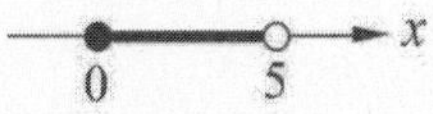

37. $x<-1$ or $1\le x<4$

$(-\infty,-1)$ or $[1,4)$

41. $t<-0.3$ or $t>-0.3$

$(-\infty,-0.3)$ or $(-0.3,\infty)$

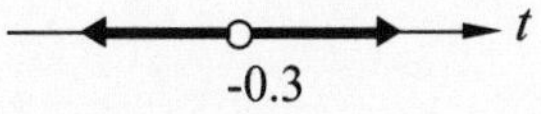

45. Suppose $0<a<b$.
Since a and b are both positive, and the power 2 is a positive integer, then by property 4,
$$a^2<b^2$$
Hence this is an absolute inequality since it is always true for a,b satisfying the original inequality.

49. For $x>0,\ y<0$ we have
$$xy<0<|x||y|$$
Multiplying both members by 2 does not change the sense of the inequality, so
$$2xy<|x||y|$$
We add x^2+y^2 (which is the same as $|x|^2+|y|^2$) to both members, and the sense of the inequality does not change
$$x^2+2xy+y^2<|x|^2+2|x||y|+|y|^2$$
Factoring and taking the square root of two positive numbers does not change the sense of the inequality. Hence we conclude
$$(x+y)^2<(|x|+|y|)^2$$
$$|x+y|<|x|+|y|$$

53. $2000\le M\le 1\,000\,000$

57. $0<n\le 2565$ steps

17.2 Solving Linear Inequalities

1.
$$21-2x\ge 15$$
$$-2x\ge -6$$
$$x\le 3,$$
or $(-\infty,3]$

5. $x-3>-4$

$x>-4+3$

$x>-1,$

or $(-1,\infty)$

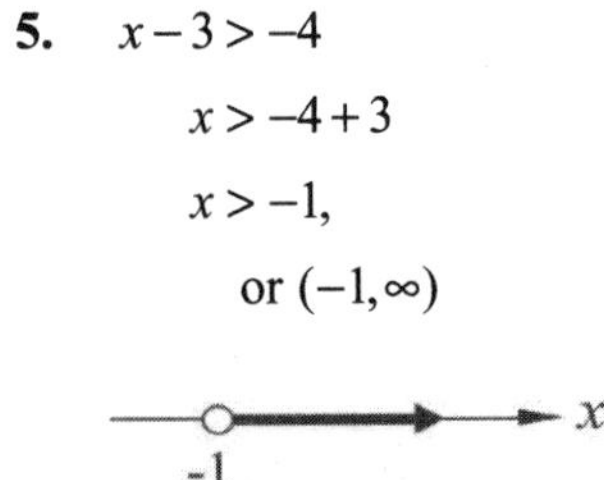

9. $3x-5\le -11$

$3x\le -11+5$

$3x\le -6$

$x\le -2,$

or $(-\infty,-2]$

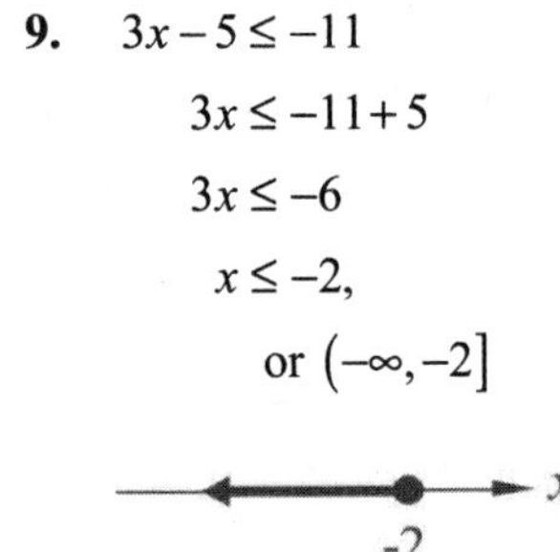

13. $\dfrac{4x-5}{2}\le x$

$4x-5\le 2x$

$2x\le 5$

$x\le \dfrac{5}{2},$

or $\left(-\infty,\frac{5}{2}\right]$

17. $2.50(1.50-3.40x)<3.84-8.45x$

$3.75-8.50x<3.84-8.45x$

$-0.05x<0.09$

$x>-1.80,$

or $(-1.80,\infty)$

21. $-1<2x+1<3$

$-2<2x<2$

$-1<x<1,$

or $(-1,1)$

25. $2x<x-1\le 3x+5$

$2x<x-1$ and $x-1\le 3x+5$

$x<-1$ and $-2x\le 6$

$x<-1$ and $x\ge -3$

$-3\le x<-1,$

or $[-3,-1)$

29. $f(x)=\sqrt{2x-10}$

$2x-10\ge 0$

$2x\ge 10$

$x\ge 5,$

or $[5,\infty)$

33. $x^2-kx+9=0$ has roots $\dfrac{k\pm\sqrt{k^2-36}}{2}$ which will be imaginary for $k^2-36<0$, or $(k+6)(k-6)<0$.

This requires:

$(k+6>0 \text{ and } k-6<0)$ or $(k+6<0 \text{ and } k-6>0)$

$(k>-6 \text{ and } k<6)$ or $(k<-6 \text{ and } k>6)$

$(-6<k<6)$ or (no values in common)

Therefore, the roots are imaginary for $-6<k<6$, or $(-6,6)$.

37. $9\,C=5(F-32)$

$C=\dfrac{5(F-32)}{9}$

$10<\dfrac{5(F-32)}{9}<20$

$90<5\,F-160<180$

$250<5\,F<340$

$50<F<68$

The temperatures that correspond to temperatures between 10°C and 20°C are between 50°F and 68°F, or (50°F, 68°F).

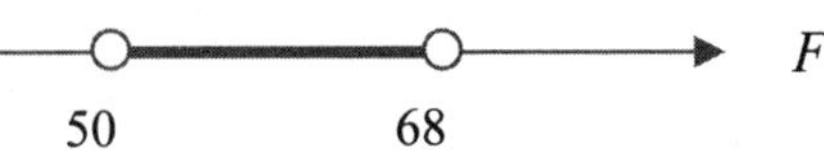

41.
$$m = 125 + 15.0t$$
$$131 < 125 + 15.0t < 164$$
$$6 < 15.0t < 39$$
$$0.400 < t < 2.60 \text{ h},$$
$$\text{or } (0.400 \text{ h}, 2.60 \text{ h})$$

45. Let t = length of stops. Since 1 h = 60 min,

$$30\ \frac{\text{km}}{\text{h}} = 30\frac{\text{km}}{\text{h}}\cdot\frac{1\text{ h}}{60\text{ min}} = \frac{1}{2}\frac{\text{km}}{\text{min}}.$$

The average speed is given by $\frac{0(5t)+40}{5t+60}$.

Since it must be at least $\frac{1}{2}$, we write

$$\frac{40}{5t+60} \ge \frac{1}{2}$$
$$80 \ge 5t + 60$$
$$-5t \ge -20$$
$$t \le 4$$

Stops must also be at least 2 min, so the length of stops satisfies 2 min $\le t \le$ 4 min, or is within $[2 \text{ min}, 4 \text{ min}]$.

17.3 Solving Nonlinear Inequalities

1.
$$x^2 + 3 > 4x$$
$$x^2 - 4x + 3 > 0$$
$$(x-3)(x-1) > 0$$

The critical values are 1, 3.

Interval	$(x-3)$	$(x-1)$	Sign
$x < 1$ $(-\infty, 1)$	−	−	+
$1 < x < 3$ $(1, 3)$	−	+	−
$x > 3$ $(3, \infty)$	+	+	+

$x^2 - 4x + 3 > 0$ when $x < 1$ or $x > 3$, or in $(-\infty, 1)$ or $(3, \infty)$.

5.
$$x^2 - 16 < 0$$
$$(x+4)(x-4) < 0$$

The critical values are $x = -4$ and $x = 4$.

Interval	$(x+4)$	$(x-4)$	Sign
$x < -4$ $(-\infty, -4)$	−	−	+
$-4 < x < 4$ $(-4, 4)$	+	−	−
$x > 4$ $(4, \infty)$	+	+	+

$x^2 - 16 < 0$ for $-4 < x < 4$, or in $(-4, 4)$.

9.
$$2x^2 - 12 \le -5x$$
$$2x^2 + 5x - 12 \le 0$$
$$(2x-3)(x+4) \le 0$$

The critical values are $x = \frac{3}{2}$, $x = -4$.

Interval	$(2x-3)$	$(x+4)$	Sign
$x < -4$ $(-\infty, -4)$	−	−	+
$-4 < x < \frac{3}{2}$ $(-4, \frac{3}{2})$	−	+	−
$x > \frac{3}{2}$ $(\frac{3}{2}, \infty)$	+	+	+

$(2x-3)(x+4) \le 0$ for $-4 \le x \le \frac{3}{2}$, or in $[-4, \frac{3}{2}]$.

13. $R^2 + 4 > 0$

This is an absolute inequality, so all values of R are solutions. The solution is $(-\infty, \infty)$.

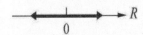

17.
$$s^3+2s^2-s\ge 2$$
$$s^3+2s^2-s-2\ge 0$$
$$s^2(s+2)-1(s+2)\ge 0$$
$$(s^2-1)(s+2)\ge 0$$
$$(s+1)(s-1)(s+2)\ge 0$$

The critical values are $s=-1$, $s=1$, $s=-2$.

Interval	$(s-1)(s+1)(s+2)$	Sign
$s<-2$ $(-\infty,-2)$	$-\ -\ -$	$-$
$-2<s<-1$ $(-2,-1)$	$-\ -\ +$	$+$
$-1<s<1$ $(-1,1)$	$-\ +\ +$	$-$
$s>1$ $(1,\infty)$	$+\ +\ +$	$+$

$(s+1)(s-1)(s+2)\ge 0$ for $-2\le s\le -1$ or $s\ge 1$, or in $[-2,-1]$ or $[1,\infty)$.

21. $\frac{2x-3}{x+6}\le 0 \quad (x\ne -6)$

The critical values are $x=\frac{3}{2}$, $x=-6$.

Interval	$(2x-3)\div(x+6)$	Sign
$x<-6$ $(-\infty,-6)$	$-\ -$	$+$
$-6<x<\frac{3}{2}$ $(-6,\frac{3}{2})$	$-\ +$	$-$
$x>\frac{3}{2}$ $(\frac{3}{2},\infty)$	$+\ +$	$+$

$\frac{2x-3}{x+6}\le 0$ for $-6<x\le\frac{3}{2}$, or in $(-6,\frac{3}{2}]$.

25.
$$\frac{x}{x+1}>1$$
$$\frac{x}{x+1}-1>0$$
$$\frac{x-(x+1)}{(x+1)}>0$$
$$\frac{-1}{(x+1)}>0$$

The critical value is -1.

Interval	$x+1$	$\frac{-1}{(x+1)}$
$x<-1$ $(-\infty,-1)$	$-$	$+$
$x>-1$ $(-1,\infty)$	$+$	$-$

$\frac{-1}{(x+1)}>0$ for $x<-1$, or in $(-\infty,-1)$.

29.
$$\frac{6-x}{3-x-4x^2}\ge 0$$
$$\frac{6-x}{(1+x)(3-4x)}\ge 0;\ \left(x\ne -1,\ x\ne\frac{3}{4}\right)$$

The critical values are $x=6$, $x=-1$, and $x=\frac{3}{4}$.

Interval	$(6-x)\div(1+x)(3-4x)$	Sign
$x<-1$ $(-\infty,-1)$	$+\ -\ +$	$-$
$-1<x<\frac{3}{4}$ $(-1,\frac{3}{4})$	$+\ +\ +$	$+$
$\frac{3}{4}<x<6$ $(\frac{3}{4},6)$	$+\ +\ -$	$-$
$x>6$ $(6,\infty)$	$-\ +\ -$	$+$

$\frac{6-x}{(1+x)(3-4x)}\ge 0$ for $-1<x<\frac{3}{4}$ or $x\ge 6$, or in $(-1,\frac{3}{4})$ or $[6,\infty)$.

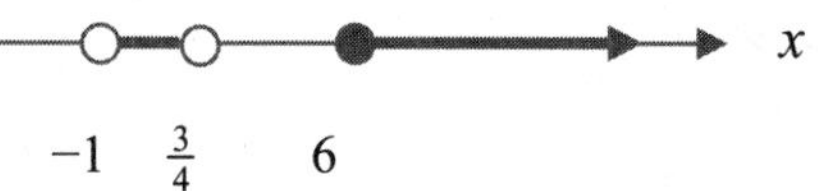

33. $\sqrt{(x-1)(x+2)}$ is real if $(x-1)(x+2)\ge 0$

The critical values are $x=1$ and $x=-2$.

Interval	$(x-1)(x+2)$	Sign
$x<-2$ $(-\infty,-2)$	$-\quad -$	$+$
$-2<x<1$ $(-2,1)$	$-\quad +$	$-$
$x>1$ $(1,\infty)$	$+\quad +$	$+$

$(x-1)(x+2)>0$ for $x\le -2$ or $x\ge 1$, or in $(-\infty,-2)$ or $(1,\infty)$.

37. To solve $x^3-x>2$ using a graphing calculator, let $y_1=x^3-x-2$.
$y>0$ for $x>1.52$, or in $(1.52,\infty)$.

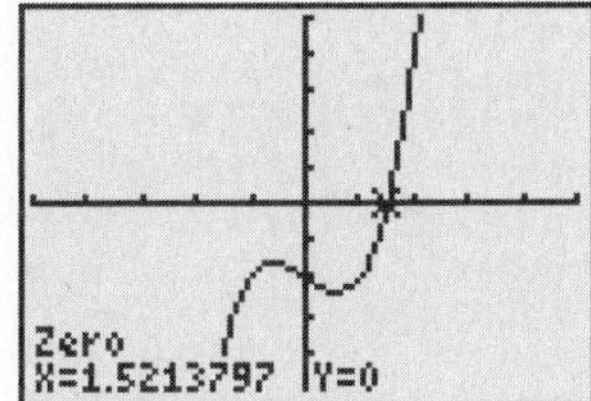

41. To solve $2^x>x+2$ using a graphing calculator, let $y_1=2^x-x-2$.
$y>0$ for $x<-1.69$, $x>2.00$, or in $(-\infty,-1.69)$ or $(2.00,\infty)$.

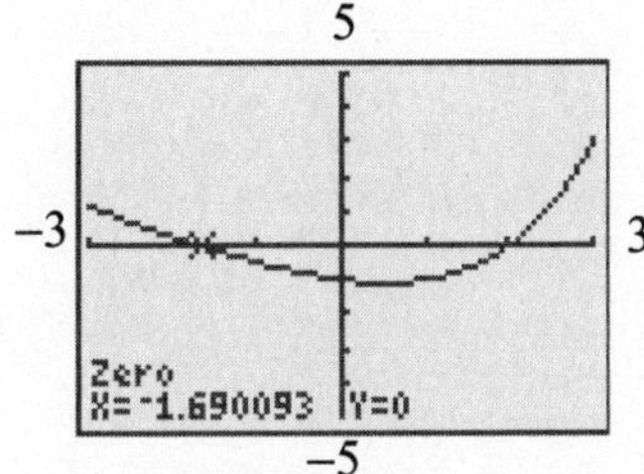

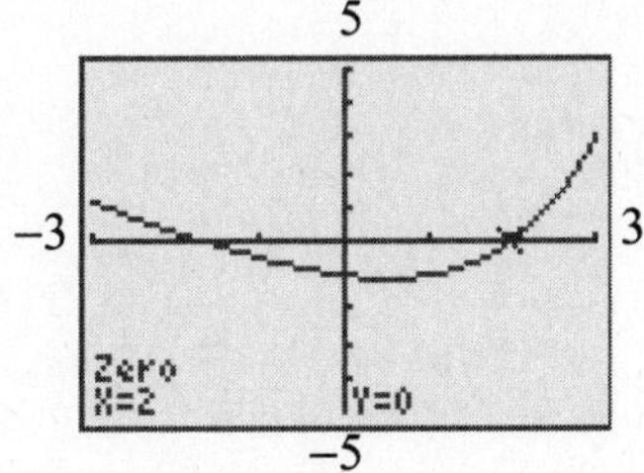

45.
$$x^2>x$$
$$x^2-x>0$$
$$x(x-1)>0$$

The critical values are $x=0$, $x=1$.

Interval	$x(x-1)$	Sign
$x<0$ $(-\infty,0)$	$-\quad -$	$+$
$0<x<1$ $(0,1)$	$+\quad -$	$-$
$x>1$ $(1,\infty)$	$+\quad +$	$+$

$x(x-1)>0$ for $x<0$ or $x>1$, or in $(-\infty,0)$ or $(1,\infty)$.
Therefore, $x^2>x$ is only true for $x<0$ or $x>1$.
$x^2>x$ is not true for $0\le x\le 1$.

49.
$$2^{x+2}>3^{2x-3}$$
$$2^x2^2>3^{2x}3^{-3}$$
$$\log 2^x+\log 2^2>\log 3^{2x}+\log 3^{-3}$$
$$x\log 2+\log 4>2x\log 3-\log 27$$
$$x\log 2-2x\log 3>-\log 4-\log 27$$
$$x(\log 2-2\log 3)>-(\log 4+\log 27)$$
$$x(\log 2-\log 9)>-\log 108$$
$$x<-\frac{\log 108}{\log 2-\log 9}$$
$$x<3.11, \text{ or } (-\infty,3.11)$$

53. $P=6i-4i^2$,
$$6i-4i^2>2 \text{ and } i>0.$$
$$4i^2-6i+2<0$$
$$2i^2-3i+1<0$$
$$(2i-1)(i-1)<0$$

Interval	$(2i-1)(i-1)$	Sign
$0<i<\frac{1}{2}$ $(0,\frac{1}{2})$	$-\quad -$	$+$
$\frac{1}{2}<i<1$ $(\frac{1}{2},1)$	$+\quad -$	$-$
$i>1$ $(1,\infty)$	$+\quad +$	$+$

$(2i-1)(i-1)<0$ for $\frac{1}{2}<i<1$ A, or in $(\frac{1}{2}\text{A}, 1\text{A})$.

57.

$$C > 1.00$$
$$C^{-1} < 1$$
$$C_1^{-1} + C_2^{-1} = C^{-1}$$
$$C_1^{-1} + C_2^{-1} < 1$$
$$C_1^{-1} + 4.00^{-1} < 1$$
$$C_1^{-1} < 0.750$$
$$C_1 > 1.33\ \mu\text{F}$$

61. $l = w + 2.0;\ w(w+2.0) < 35;\ w \geq 3.0$ mm

$$w^2 + 2.0w - 35 < 0$$
$$(w+7.0)(w-5.0) < 0$$

The critical values are $w = -7.0$ and $w = 5.0$. Width must be at least 3.0 mm, so the table of signs is

Interval	$(w+7)(w-5)$	Sign
$3 \leq w < 5$ $[3,5)$	$+\ \ -$	$-$
$w > 5$ $(5,\infty)$	$+\ \ +$	$+$

$(w+7.0)(w-5.0) < 0$ for $3 \leq w < 5.0$ mm, or in $[3\text{ mm}, 5\text{ mm})$.

17.4 Inequalities Involving Absolute Values

1. $|2x-1| < 5$

$$-5 < 2x-1 < 5$$
$$-4 < 2x < 6$$
$$-2 < x < 3$$
$$(-2,3)$$

-2 3 x

5. $|5x+4| > 6$

$$5x+4 < -6 \text{ or } 5x+4 > 6$$
$$5x < -10 \text{ or } 5x > 2$$
$$x < -2 \text{ or } x > \frac{2}{5}$$
$$(-\infty,-2) \text{ or } \left(\tfrac{2}{5},\infty\right)$$

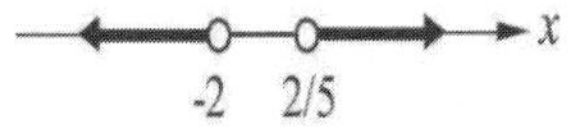

-2 2/5 x

9. $|3-4x| > 3$

$$3-4x < -3 \text{ or } 3-4x > 3$$
$$-4x < -6 \text{ or } -4x > 0$$
$$x > \frac{3}{2} \text{ or } x < 0$$
$$(-\infty,0) \text{ or } \left(\tfrac{3}{2},\infty\right)$$

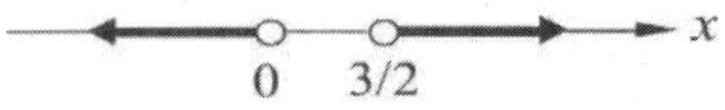

0 3/2 x

13.

$$|20x+85| \leq 43$$
$$-43 \leq 20x+85 \leq 43$$
$$-128 \leq 20x \leq -42$$
$$-6.4 \leq x \leq -2.1$$
$$[-6.4,-2.1]$$

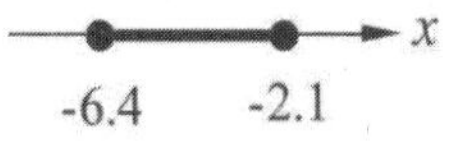

-6.4 -2.1 x

17. $8+3|3-2x| < 11$

$$3|3-2x| < 3$$
$$|3-2x| < 1$$
$$-1 < 3-2x < 1$$
$$-4 < -2x < -2$$
$$2 > x > 1$$
$$1 < x < 2$$
$$(1,2)$$

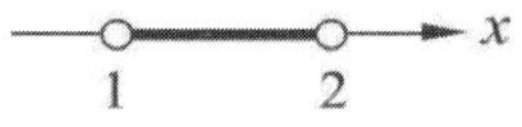

1 2 x

21.

$$\left|\frac{3R}{5}+1\right| < 8$$
$$-8 < \frac{3R}{5}+1 < 8$$
$$-9 < \frac{3R}{5} < 7$$
$$-45 < 3R < 35$$
$$-15 < R < \frac{35}{3}$$
$$\left(-15,\tfrac{35}{3}\right)$$

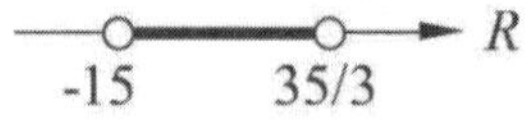

-15 35/3 R

25. $\left|x^2+x-4\right|>2$

$x^2+x-4>2$ or $x^2+x-4<-2$

$x^2+x-6>0$ or $x^2+x-2<0$

(A) $(x+3)(x-2)>0$

(B) $(x-1)(x+2)<0$

(A) Critical values are $x=-3$, $x=2$.

Interval	$(x+3)(x-2)$	Sign
$x<-3$ $(-\infty,-3)$	$-\quad -$	$+$
$-3<x<2$ $(-3,2)$	$+\quad -$	$-$
$x>2$ $(2,\infty)$	$+\quad +$	$+$

$(x+3)(x-2)>0$ for $x<-3$ or $x>2$, or $(-\infty,-3)$ or $(2,\infty)$.

(B) Critical values are $x=1$, $x=-2$.

Interval	$(x-1)(x+2)$	Sign
$(-\infty,-2)$	$-\quad -$	$+$
$(-2,1)$	$-\quad +$	$-$
$(1,\infty)$	$+\quad +$	$+$

$(x-1)(x+2)<0$ for $-2<x<1$, or $(-2,1)$.

The solution consists of values of x that are in (A) or (B): $x<-3$, $-2<x<1$, $x>2$, that is, $(-\infty,-3)$ or $(-2,1)$ or $(2,\infty)$.

29. Solve for x if $|x|<a$ and $a\le 0$.

$|x|<a\le 0$

$|x|<0$, no solutions since $|x|\ge 0$.

33. $|t-27|\le 23$

$-23\le t-27\le 23$

$4\le t\le 50$

$[4 \text{ km}, 50 \text{ km}]$

4 km, minimum thickness of earth's crust.

50 km, maximum thickness of earth's crust.

37. $3.675-0.002\le d\le 3.675+0.002$

$-0.002\le d-3.675\le 0.002$

$|d-3.675|\le 0.002$ cm

41. $h=190t-4.9t^2$

We need to solve $\left|190t-4.9t^2\right|>1300$

$190t-4.9t^2>1300$ or $190t-4.9t^2<-1300$

$4.9t^2-190t+1300<0$ or $4.9t^2-190t-1300>0$

Use the quadratic formula to find critical values:

$4.9t^2-190t+1300=0$:

$$t=\frac{190\pm\sqrt{190^2-4(4.9)(1300)}}{2(4.9)}=8.87 \text{ s}, 29.9 \text{ s}$$

Interval	$(t-8.87)(t-29.9)$	Sign
$t<8.87$ $(-\infty,8.87)$	$-\quad -$	$+$
$8.87<t<29.9$ $(8.87,29.9)$	$+\quad -$	$-$
$t>29.9$ $(29.9,\infty)$	$+\quad +$	$+$

$4.9t^2-190t+1300=0>0$ for $8.87 \text{ s}<t<29.9$ s, or in $(8.87 \text{ s}, 29.9 \text{ s})$.

$4.9t^2-190t-1300>0$

Use quadratic formula to find critical values:

$$t=\frac{190\pm\sqrt{190^2-4(4.9)(-1300)}}{2(4.9)}=-5.93, 44.7$$

Time must be non-negative, so we discard the negative root and start at $t=0$.

Also, height cannot go below -3000, so we find the upper limit through:

$$190t-4.9t^2=-3000$$

$4.9t^2-190t-3000=0$

$$t=\frac{190\pm\sqrt{190^2-4(4.9)(-3000)}}{2(4.9)}=-12.0,\ 50.8 \text{ s}$$

We now check the sign of the function.

Interval	$(t-44.7)(t+5.87)$	Sign
$0\le t<44.7$ $[0,44.7)$	$-\quad +$	$-$
$44.7<t\le 50.8$ $(44.7,50.8]$	$+\quad +$	$+$

$4.9t^2-190t-1300>0$ for $44.7 \text{ s}<t<50.8$ s, or in $(44.7, 50.8 \text{ s}]$.

The complete solution is $(8.9 \text{ s}, 30 \text{ s})$ or $(45 \text{ s}, 51 \text{ s}]$.

17.5 Graphical Solution of Inequalities with Two Variables

1. $y < 3 - x$
Graph $y = 3 - x$. Used a dashed line to indicate that points on the line do not satisfy the inequality. Shade in the region below the line.

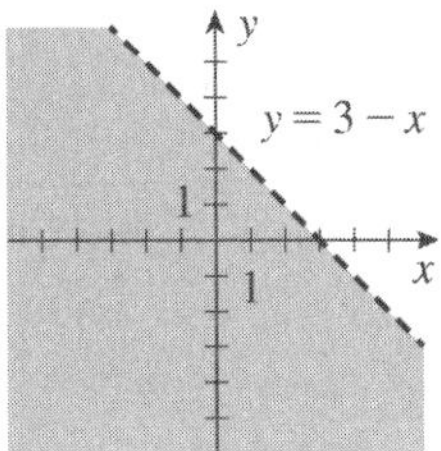

5. $y \geq 2x + 5$
Graph $y = 2x + 5$. Use a solid line to indicate that points on it satisfy the inequality. Shade in the region above the line.

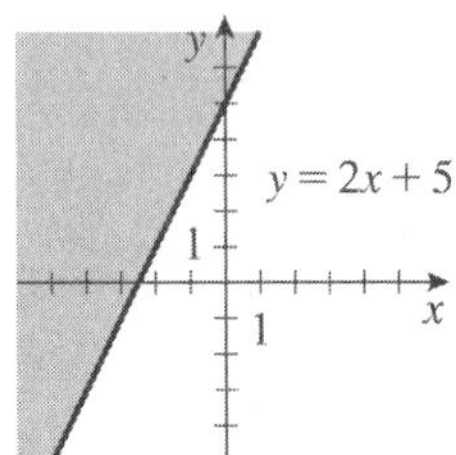

9. $y < x^2$
Graph $y = x^2$. Use a dashed curve to indicate that points on it do not satisfy the inequality. Shade in the region below the curve.

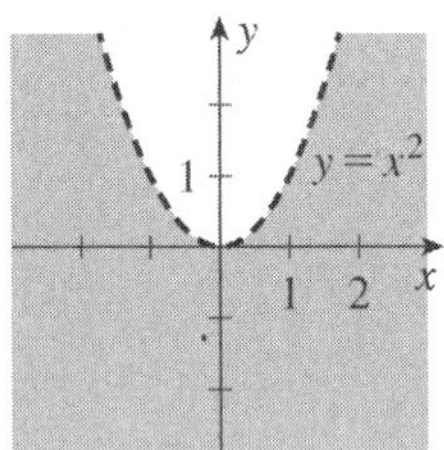

13. $y < 32x - x^4$
Graph $y = 32x - x^4$. Use a dashed curve to indicate that points on it do not satisfy the inequality. Shade in the region below the curve.

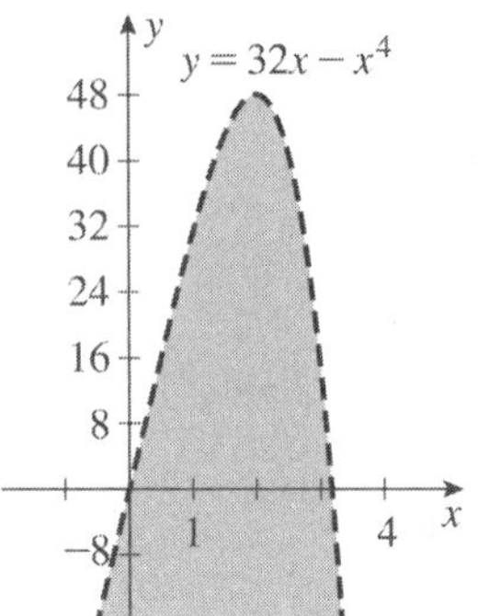

17. $y > 1 + \sin 2x$
Graph $y = 1 + \sin 2x$. Use a dashed curve to indicate that the points on it do not satisfy the inequality. Shade in the region above the curve.

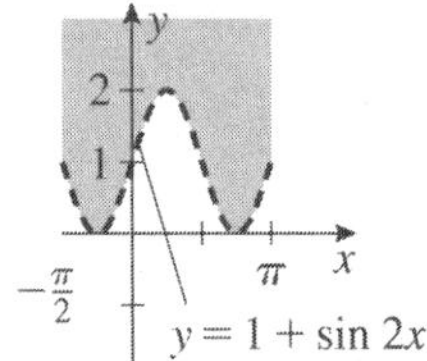

21. $|y| > |x|$.
For $y > 0$, $|y| > |x|$ becomes $y > |x|$.
Graph $y = |x|$ with dashed line and shade in the region above the graph.
For $y < 0$, $|y| > |x|$ becomes $-y > |x|$, so that $y < -|x|$. Graph $y = -|x|$ with dashed line and shade in the region below the graph.
The solution consists of both regions.

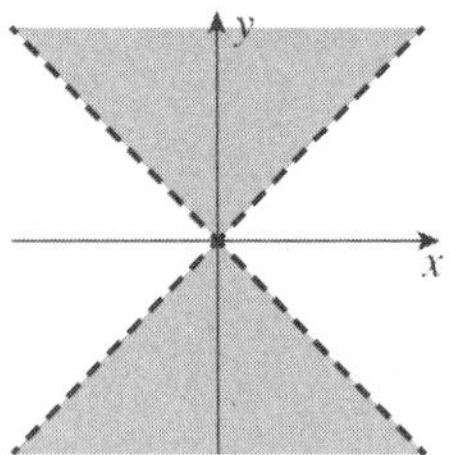

25. $y \le 2x^2$ and $y > x-2$. Graph $y = 2x^2$ using a solid curve. Shade the region below the curve. Graph $y = x-2$ using a dashed line. Shade the region above the line. The region where the shadings overlap satisfies both inequalities.

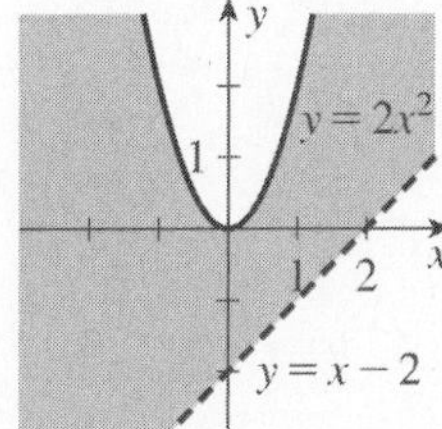

29. $y \ge 0$ and $y \le \sin x$; $0 \le x \le 3\pi$. Graph $y = \sin x$ using a solid curve. Shade the region below the curve and above the x-axis for $0 \le x \le 3\pi$.

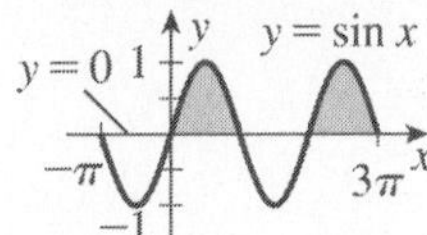

33. $2x + y < 5$

$y < -2x + 5$

Graph $y_1 = -2x + 5$.

The boundary line is dashed and the region below the graph is shaded in.

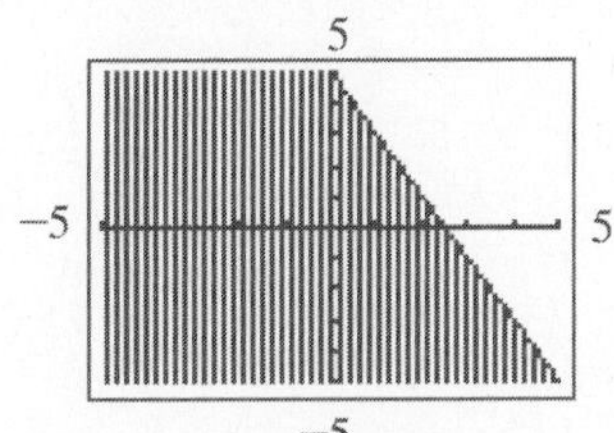

37. $y > 2x - 1$. Graph $y_1 = 2x - 1$.

$y < x^4 - 8$. Graph $y_2 = x^4 - 8$.

The boundary lines are dashed. The solution is the intersection of the region above the line and below the curve.

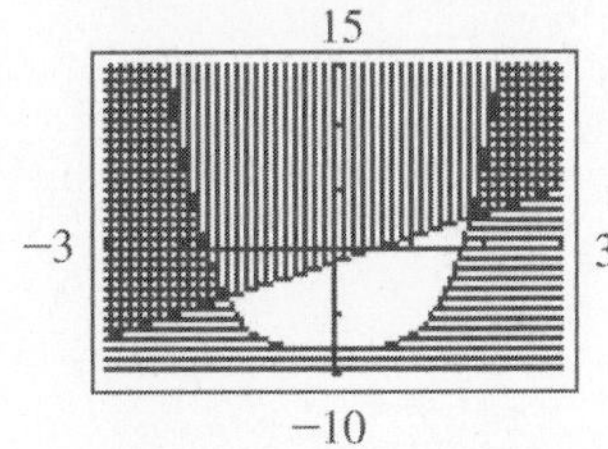

41. $y \le |2x-3|$. Graph $y_1 = |2x-3|$. The boundary line is solid.

$y > 1 - 2x^2$. Graph $y_2 = 1 - 2x^2$. The boundary line is dashed.

The solution is the intersection of the region below the absolute value and the region above the parabola.

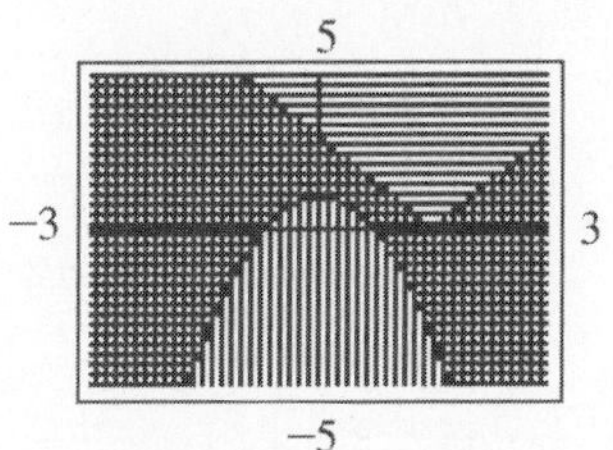

45. $Ax + By > C$

$By > -Ax + C$

Since $B < 0$ division of both sides by B gives

$$y < -\frac{A}{B}x + \frac{C}{B}$$

The solution is to shade in below the dashed line

49. $0 \le A \le 300$ m; $200 \text{ m} \le B \le 400$ m

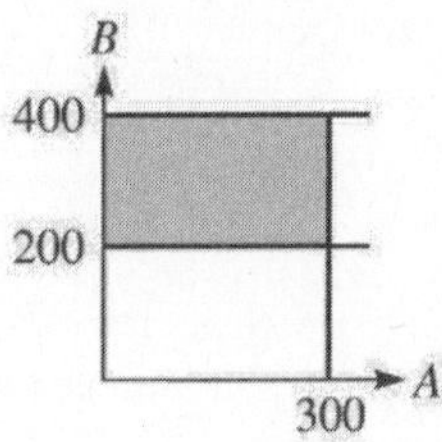

53. Let x and y = pumping rates.

$250x + 150y > 15\,000$

$x \ge 0,\ y \ge 0$

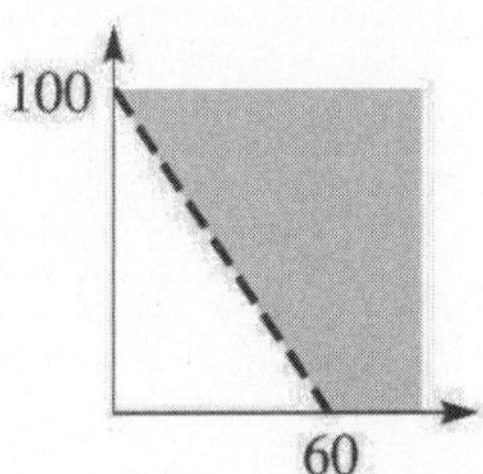

17.6 Linear Programming

1. Maximize $F = 2x + 3y$, subject to $x \ge 0$, $y \ge 0$, $x + y \le 6$, $2x + y \le 8$.

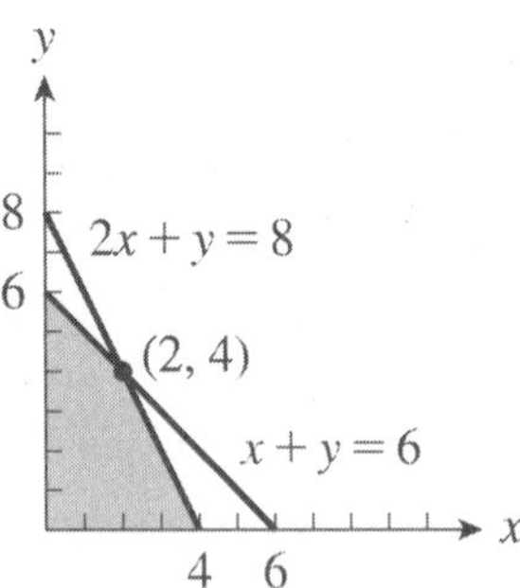

point	value of F
$(0, 0)$	0
$(0, 6)$	18
$(2, 4)$	16
$(4, 0)$	8

Maximum value of F is 18 at $(0, 6)$

5. Maximum P: $P = 3x + 5y$, subject to
$x \ge 0,\ y \ge 0$
$2x + y \le 6$

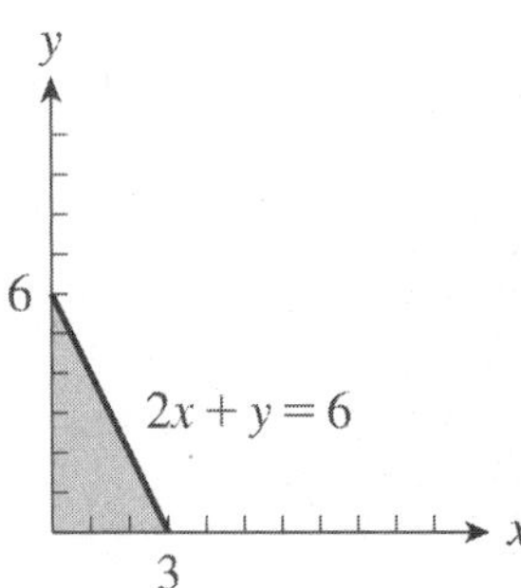

Vertex	$P = 3x + 5y$
$(0, 0)$	0
$(0, 6)$	30
$(3, 0)$	9

max $P = 30$ at $(0, 6)$

9. Minimum C: $C = 4x + 6y$, subject to
$x \ge 0,\ y \ge 0$
$x + y \ge 5$
$x + 2y \ge 7$

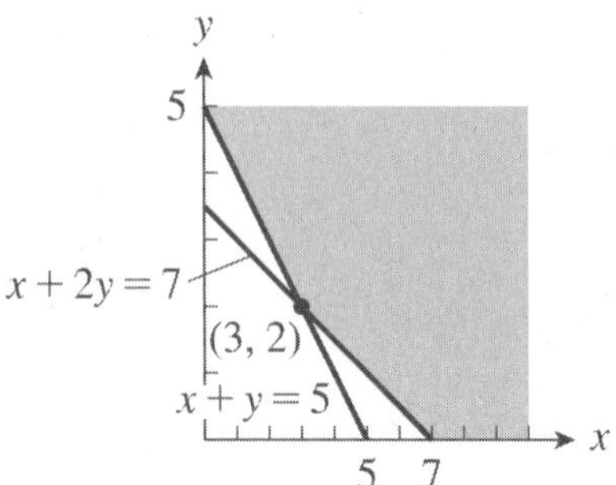

Vertex	$C = 4x + 6y$
$(0, 5)$	30
$(3, 2)$	24
$(7, 0)$	28

min $C = 24$ at $(3, 2)$

13. Maximum P: $P = 9x + 2y$, subject to
$x \ge 0,\ y \ge 0$
$2x + 5y \le 10$
$4x + 3y \le 12$

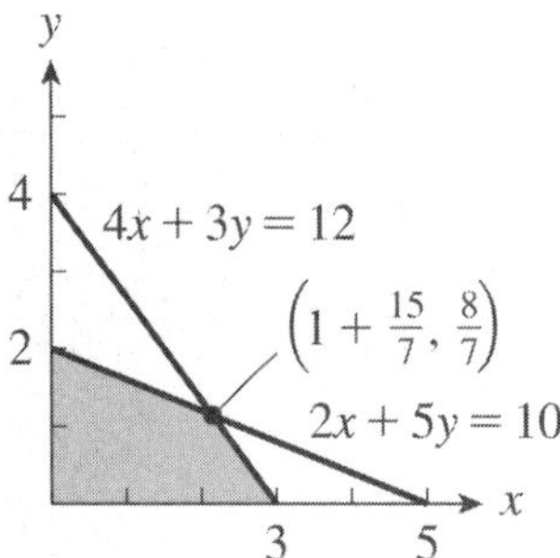

Vertex	$P = 9x + 2y$
$(0, 0)$	0
$(0, 2)$	4
$\left(\frac{15}{7}, \frac{8}{7}\right)$	$\frac{151}{7}$
$(3, 0)$	27

max $P = 27$ at $(3, 0)$

17. $x =$ amount invested at 6%

$y =$ amount invested at 5%

Maximum $I = 0.06x + 0.05y$, subject to

$x \geq 0,\ y \geq 0$

$x + y \leq 9000$

$x \leq 2y$

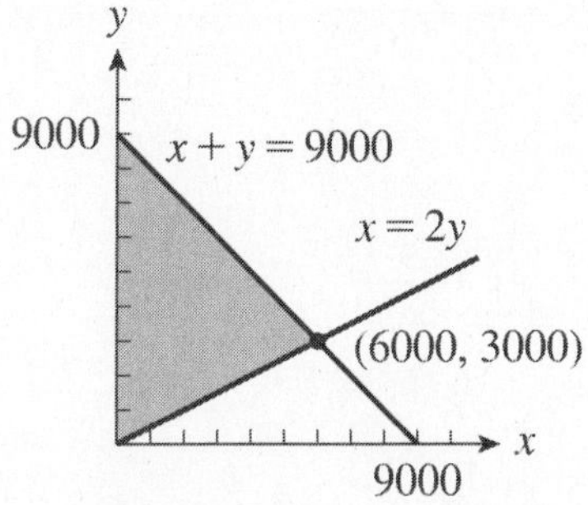

Vertex	$I = 0.06x + 0.05y$
(0, 0)	0
(0, 9000)	450
(6000, 3000)	510

max $I =$ \$510 with \$6000 at 6% and \$3000 at 5%

21. $x =$ servings of cereal A, $y =$ servings of cereal B

Minimum $C = 12x + 18y$, subject to

$x \geq 0,\ y \geq 0$

$x + 2y \geq 10$

$5x + 3y \geq 30$

Vertex	$C = 12x + 18y$
(0,10)	180
(10,0)	120
$\left(\frac{30}{7}, \frac{20}{7}\right)$	$\frac{720}{7}$

Minimum $C = \frac{720}{7}$, when $x = \dfrac{30}{7},\ y = \dfrac{20}{7}$.

Since each serving is 30 g, the optimal solution is $\dfrac{900}{7}$ g of cereal A, and $\dfrac{600}{7}$ g of cereal B.

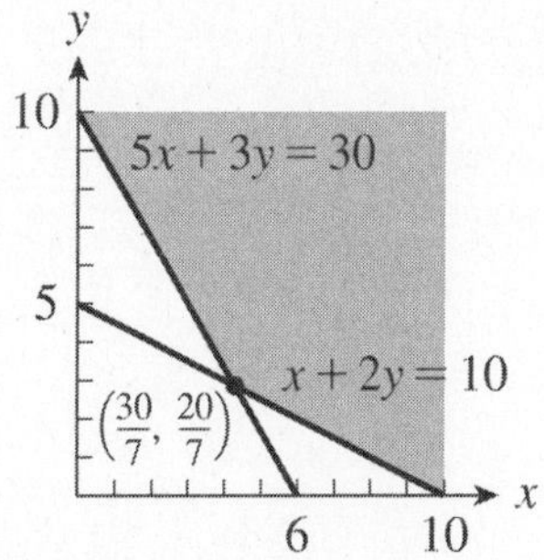

Review Exercises

1. $2x - 12 > 0$

$2x > 12$

$x > 6$

$(6, \infty)$

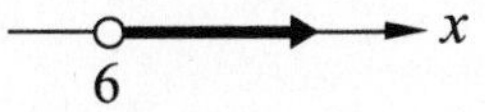

5. $4 < 2x - 1 < 11$

$5 < 2x < 12$

$\dfrac{5}{2} < x < 6$

$\left(\frac{5}{2}, 6\right)$

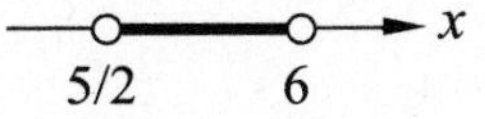

9. $5x^2 + 9x < 2$

$(x+2)(5x-1) < 0$

Interval	$(x+2)$	$(5x-1)$	Sign
$x < -2$ $(-\infty, -2)$	−	−	+
$-2 < x < \frac{1}{5}$ $\left(-2, \frac{1}{5}\right)$	+	−	−
$x > \frac{1}{5}$ $\left(\frac{1}{5}, \infty\right)$	+	+	+

$(x+2)(5x-1)<0$ for $-2<x<\frac{1}{5}$, or in $\left(-2,\frac{1}{5}\right)$.

-2 1/5 x

13. $\dfrac{(2x-1)(3-x)}{(x+4)}>0$

Critical values are $-4, \dfrac{1}{2}, 3$.

Interval	$2x-1$	$3-x$	$x+4$	$\frac{(2x-1)(3-x)}{(x+4)}$
$x<-4$ $(-\infty,-4)$	$-$	$+$	$-$	$+$
$-4<x<\frac{1}{2}$ $\left(-4,\frac{1}{2}\right)$	$-$	$+$	$+$	$-$
$\frac{1}{2}<x<3$ $\left(\frac{1}{2},3\right)$	$+$	$+$	$+$	$+$
$x>3$ $(3,\infty)$	$+$	$-$	$+$	$-$

$\frac{(2x-1)(3-x)}{(x+4)}>0$ for $x<-4$ or $\frac{1}{2}<x<3$, or in $(-\infty,-4)$ or $\left(\frac{1}{2},3\right)$.

-4 1/2 3 x

17. $\dfrac{8-R}{2R+1}\le 0$

Critical values are $-\dfrac{1}{2}, 8$.

Interval	$8-R$	$2R+1$	$\frac{8-R}{2R+1}$
$R<-\frac{1}{2}$ $\left(-\infty,-\frac{1}{2}\right)$	$+$	$+$	$-$
$-\frac{1}{2}<R<8$ $\left(-\frac{1}{2},8\right)$	$-$	$+$	$+$
$R>8$ $(8,\infty)$	$-$	$+$	$-$

$\dfrac{8-R}{2R+1}\le 0$ for $R<-\frac{1}{2}$ or $R\ge 8$, or in $\left(-\infty,-\frac{1}{2}\right)$ or $[8,\infty)$.

$-\frac{1}{2}$ 8 R

21. $|3-5x|>7$

$3-5x<-7$ or $3-5x>7$

$-5x<-10$ $\quad$ $-5x>4$

$x>2$ $\quad$ $x<-\dfrac{4}{5}$

$(2,\infty)$ or $\left(-\infty,-\frac{4}{5}\right)$

-4/5 2 x

25. $x^3+x+1<0$

Graph $y_1=x^3+x+1$ and use the zero feature to find the root.

$x^3+x+1<0$ for $x<-0.68$, or $(-\infty,-0.68)$

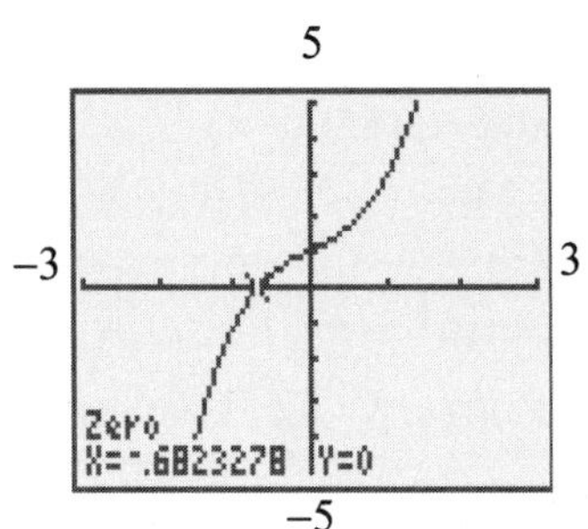

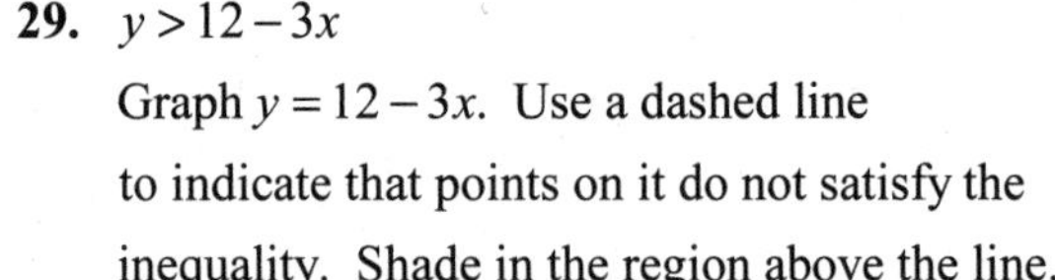

29. $y>12-3x$

Graph $y=12-3x$. Use a dashed line to indicate that points on it do not satisfy the inequality. Shade in the region above the line.

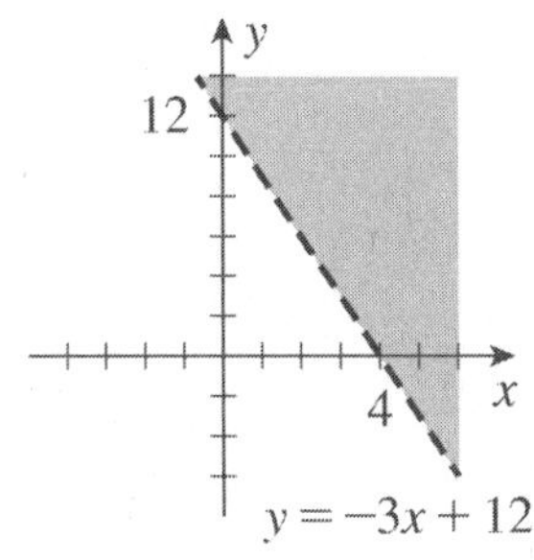

33. $y > 6 - 2x^2$

Graph $y = 6 - 2x^2$. Use a dashed curve to indicate points on the boundary curve are not part of the solution. Shade in the region above the curve.

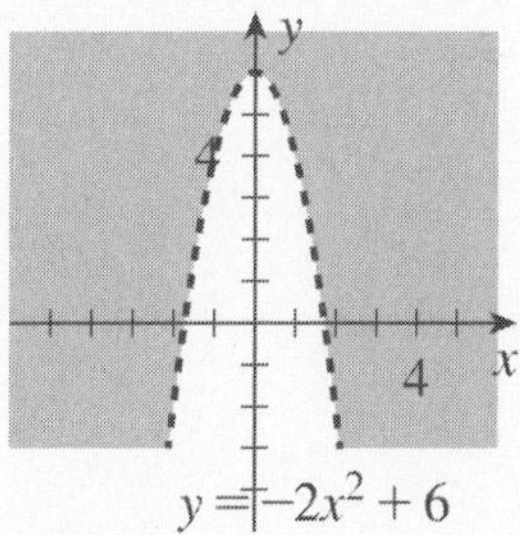

37. $y > x + 1,\ y < 4 - x^2$

Graph $y = x + 1$ and $y = 4 - x^2$. Use dashed lines to indicate that the boundary lines are not part of the solution. The solution is the intersection of the region below the parabola with the region above the line.

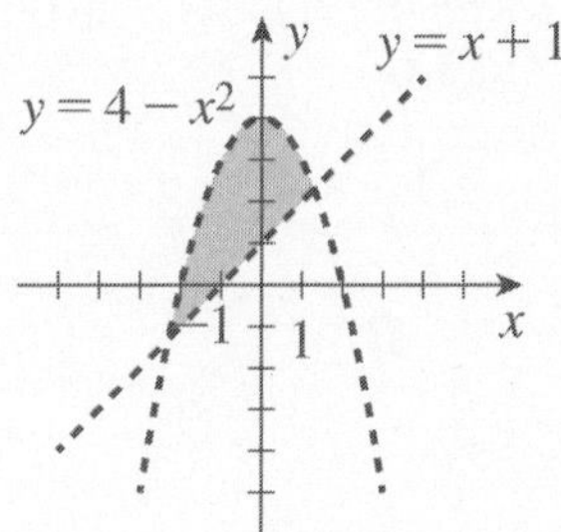

41. $y < 3x + 5$. Graph $y_1 = 3x + 5$ and shade below the line. Boundary line is not part of solution.

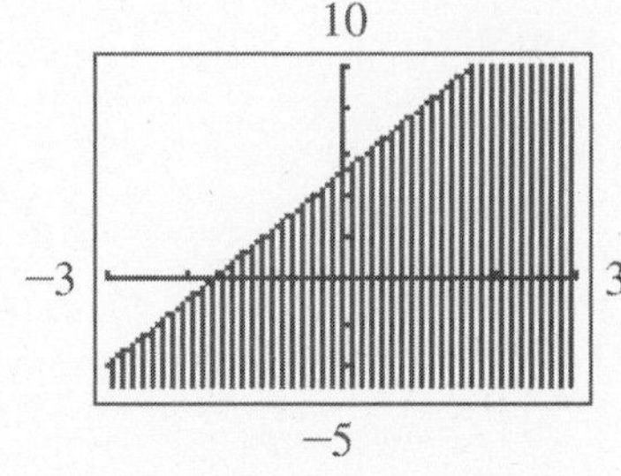

45. $y < 32x - x^4$. Graph $y_1 = 32x - x^4$ and shade below curve. Boundary line is not part of solution.

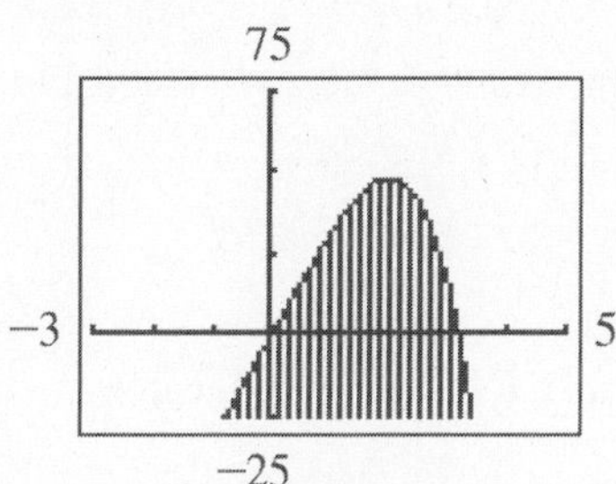

49. $\sqrt{3-x}$ is a real number for

$3 - x \geq 0$

$3 \geq x$

$x \leq 3$

$(-\infty, 3]$

53. Maximize P: $P = 2x + 9y$, subject to

$x \geq 0,\ y \geq 0$

$x + 4y \leq 13$

$3y - x \leq 8$

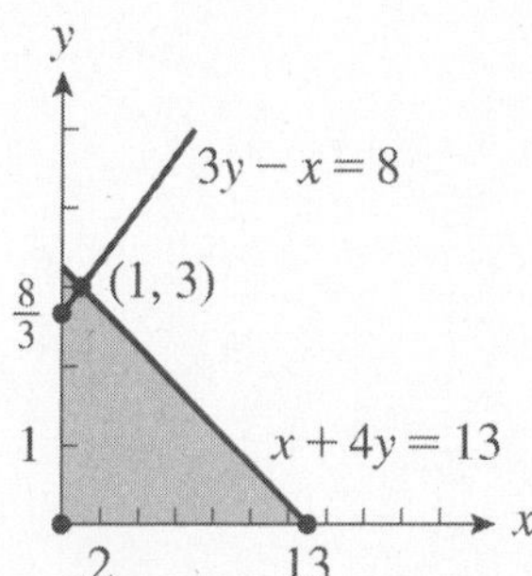

Vertex	$P = 2x + 9y$
(0, 0)	0
$(0, \frac{8}{3})$	24
(1, 3)	29
(13, 0)	26

max $P = 29$ at $(1, 3)$

57. When is $|a+b|<|a|+|b|$? There are 4 cases.

(1) $a \ge 0, b \ge 0 \Rightarrow a+b \ge 0,$ the given inequality is

$$a+b<a+b$$
$$0<0, F$$

(2) $a<0, b<0 \Rightarrow a+b<0,$ the given inequality is

$$-(a+b)<-a+(-b)$$
$$-(a+b)<-(a+b)$$
$$0<0, F$$

(3) $a<0, b>0, |a|>b \Rightarrow a+b<0,$ the given inequality is

$$-(a+b)<-a+b$$
$$-a-b<-a+b$$
$$-b<b, T$$

(4) $a<0, b>0, |a|<b \Rightarrow a+b>0,$ the given inequality is

$$a+b<-a+b$$
$$a<-a, T$$

Note: in cases (3) and (4) a and b can be reversed without loss of generality.

$|a+b|<|a|+|b|$ when a and b have opposite signs.

61. $-2<x<5$

$-2<x$ and $x<5$

$x+2>0$ and $x-5<0$

$(x-5)(x+2)<0$

$x^2-3x-10<0$

65. $f(x)=(x-2)(x-3)$ has critical values 2, 3

Interval	$(x-2)(x-3)$		sign
$x<2$ $(-\infty,2)$	−	−	+
$2<x<3$ $(2,3)$	+	−	−
$x>3$ $(3,\infty)$	+	+	+

$(x-2)(x-3)<0$ in $(2,3)$

$(x-2)(x-3)>0$ in $(-\infty,2)$ or $(3,\infty)$

Moreover, f is a parabola with $f(0)=6$.

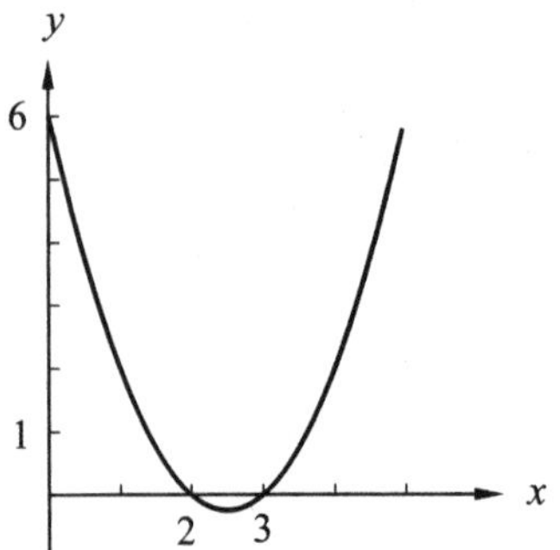

69. $2x+5y=50 \Rightarrow y=\dfrac{50-2x}{5}$

$$5<y<8$$
$$5<\frac{50-2x}{5}<8$$
$$25<50-2x<40$$
$$-25<-2x<-10$$
$$12.5>x>5$$
$$5<x<12.5$$
$$(\$5,\ \$12.50)$$

The cost of production for the first type is between \$5 and \$12.50.

73. $0.8x+0.9y=360, \quad 0 \le x \le 261$

$$x=\frac{360-0.9y}{0.8}$$
$$0 \le \frac{360-0.9y}{0.8} \le 261$$
$$0 \le 360-0.9y \le 208.8$$
$$-360 \le -0.9y \le -151.2$$
$$400 \ge y \ge 168$$
$$168 \le y \le 400 \text{ MJ}$$
$$[168 \text{ MJ}, 400 \text{ MJ}]$$

77. $E=100\left(1-r^{-0.4}\right)$

$$100\left(1-r^{-0.4}\right)>50$$
$$100-100r^{-0.4}>50$$
$$r^{-0.4}<\frac{1}{2}$$
$$r^{0.4}>2$$
$$r>2^{\frac{1}{0.4}}$$
$$r>5.7$$

81. x = the number of regular models
y = the number of deluxe models
$x + y \leq 450$
$P = 8x + 15y$ is the profit
t = time spent on one regular model
$2t$ = time spent on one deluxe model
xt = time spent on regular models
$2yt$ = time spent on deluxe models
$xt + 2yt$ = total time = $600t \Rightarrow$
$x + 2y \leq 600$
Maximize $P = 8x + 15y$, subject to
$x \geq 0,\ y \geq 0$
$x + y \leq 450,\ x + 2y \leq 600$

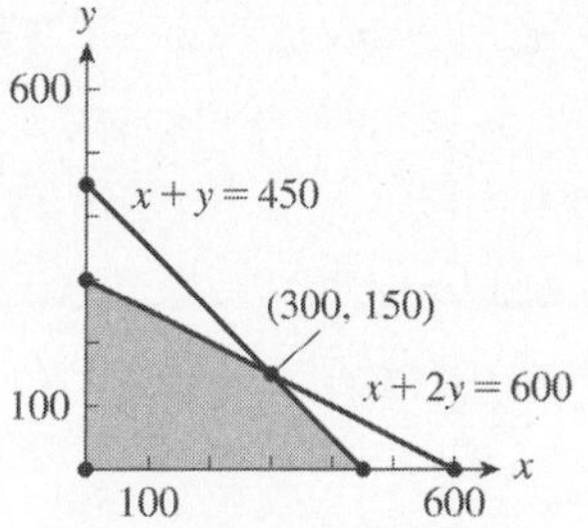

vertex	$P = 8x + 15y$
$(0, 0)$	0
$(0, 300)$	4500
$(300, 150)$	4650
$(450, 0)$	3600

Produce 300 regular, 150 deluxe for maximum profit

CHAPTER 18

VARIATION

18.1 Ratio and Proportion

1. $\frac{3.5\text{ cm}}{x}=\frac{1\text{ cm}}{16\text{ km}}$

$x=(3.5)(16)$

$x=56\text{ km}$

5. $\frac{96\text{ h}}{3\text{ days}}=\frac{96\text{ h}}{72\text{ h}}$

$=\frac{4}{3}$

9. $\frac{0.14\text{ kg}}{3500\text{ mg}}=\frac{0.14\text{ kg}}{3500\text{ mg}}\cdot\left(\frac{10^6\text{mg}}{1\text{ kg}}\right)$

$=40$

13. $\frac{450\text{ N}}{1100\text{ N}}=0.41$

17. $R=\frac{s^2}{A_w}$

$=\frac{10.0^2}{18.0}$

$=5.56$

21. $\frac{2540-2450}{2540}=\frac{90}{2540}$

$=0.035$

$=3.5\%$

25. $\frac{36.6\text{ kPa}}{84.4\text{ kPa}}=\frac{0.0447\text{ m}^3}{V_1\text{ m}^3}$

$36.6V_1=3.77$

$V_1=0.103\text{ m}^3$

29. $\frac{1\text{ha}}{10\ 000\text{m}^2}=\frac{x}{4500\text{m}^2}$

$x=4500\cdot\frac{1}{10\ 000}$

$=0.45\text{ ha}$

33. $\frac{10^4\text{ cm}^2}{10^6\text{ mm}^2}=\frac{x}{2.50\times10^5\text{ mm}^2}$

$10^6x=2.50\times10^9$

$x=2.50\times10^3\text{ cm}^2$

37. $\frac{62\ 500\text{ cm}^3}{2.00\text{ min}}=\frac{x\text{ cm}^3}{0.750\text{ min}}$

$x=\frac{62\ 500(0.750)}{2.00}$

$=23\ 400\text{ cm}^3$

41. Let x be one of the lengths, so that $7.5-x$ is the other length.

$\frac{2}{3}=\frac{x}{7.5-x}$

$2(7.5-x)=3x$

$15-2x=3x$

$5x=15$

$x=3$

The lengths are 3 cm and 4.5 cm.

45. $\frac{17\text{ defectives}}{500\text{ total chips}}=\frac{595}{x}$

$17x=297\ 500$

$x=17\ 500\text{ chips}$

18.2 Variation

1. $c=kd$

$c=\pi d$

$k=\pi$

5. $v=kr$

9. $p=\frac{k}{\sqrt{A}}$

13. The area varies directly as the square of the radius.

17. $V = kH^2$

$2 = k \cdot 64^2$

$k = \frac{2}{64^2};$

$V = \frac{2H^2}{64^2}$

$V = \frac{H^2}{2048}$

21. $y = kx$

$20 = k(8)$

$k = 2.5$

$y = 2.5x$

$y = 2.5(10)$

$y = 25$

25. $y = \frac{kx}{z}$

$60 = \frac{k(4)}{10}$

$k = 150$

$y = \frac{150x}{z}$

$y = \frac{150(6)}{5}$

$y = 180$

29. $A = k_1x,\ B = k_2x$

$A + B = k_1x + k_2x$

$= (k_1 + k_2)x,$

so that $A + B$ varies directly as x.

33. $H = km$

$2.93 \times 10^5 = k(875)$

$k = 335$ J/g

$H = 335m$

$H = 335(625)$

$= 2.09 \times 10^5$ J

37. $E = kp$

$1200 = k(0.75)$

$k = \frac{1200}{0.75}$

$= 1600$

$E = 1600p$

$E = 1600(0.35)$

$= 560$ kJ

41. **(a)** a varies inversely with mass: as mass increases, a decreases

(b) $a = \frac{k}{m}$

$30 = \frac{k}{2.0}$

$k = 60\ \text{g} \cdot \text{cm/s}^2$

$a = \frac{60}{m}$

45. $F = kAv^2$

$76.5 = k(0.372)(9.42)^2$

$k = 2.32$

$F = 2.32\ Av^2$

49. $s = k\sqrt{T}$

$460 = k\sqrt{273}$

$k = \frac{460}{\sqrt{273}}$

$= 27.8$

$s = 27.8\sqrt{T}$

$s = 27.8\sqrt{300}$

$= 480$ m/s

53. $P = kRI^2$

$10.0 = k(40.0)(0.500)^2$

$k = \frac{10.0}{10.0} = 1.00$

$P = RI^2$

$P = 20.0(2.00^2)$

$= 80.0$ W

57. *Note*: Make sure calculator is in rad mode.

$x = k\omega^2(\cos \omega t)$

$-11.4 = k(0.524^2)\cos[(1.00)(0.524)]$

$k = \frac{-11.4}{0.524^2 \cos(0.524)}$

$= -48.0$

$x = -48.0\omega^2(\cos \omega t)$

$x = (-48.0)(0.524^2)\cos(0.524(2.00))$

$= -6.58\ \text{cm/s}^2$

18 Review Exercises

1. $\frac{4\text{ Mg}}{20\text{ kg}}\cdot\frac{1000\text{kg}}{1\text{Mg}}=\frac{4000}{20}$

$=200$

5. $\pi=\frac{c}{d}$

$\frac{4.2736}{1.3603}=3.1417$

9. $p=\frac{F}{A}$

$=\frac{37.4\text{ N}}{(2.25\text{ cm})^2}$

$=7.39\text{ N/cm}^2$

13. $\text{Commission rate}=\frac{\text{Commission}}{\text{Selling price}}$

$=\frac{20\ 900}{380\ 000}$

$=0.055$

$=5.5\%$

17. $\frac{37\text{ mm}}{300\text{ km}}=\frac{78\text{ mm}}{x}$

$x=630\text{ km}$

21. $2\text{ min }5\text{ s}=125\text{ s}$

$10.0\text{ min}=600\text{ s}$

$\frac{50\text{ pages}}{125\text{ s}}=\frac{x}{600\text{ s}}$

$x=\frac{(600)(50)}{125}$

$x=240\text{ pages}$

25. $\frac{25.0\text{ m}}{2.00\text{ mm}}=\frac{x}{5.75\text{ mm}};$

$x=\frac{(25.0)(5.75)}{2.00}$

$x=71.9\text{ m}$

29. $\frac{80.0}{98.0}=\frac{x}{37.0}$

$x=\frac{(80.0)(37.0)}{98.0}$

$x=30.2\text{ kg}$

33. $y=kx^2$

$27=k(3^2)$

$k=3$

$y=3x^2$

37. $\frac{F_1}{F_2}=\frac{L_2}{L_1}$

$\frac{4.50}{6.75}=\frac{L_2}{17.5}$

$(6.75)L_2=(4.50)(17.5)$

$L_2=11.7\text{ cm}$

41. $R=kA$

$850=k(100)$

$k=8.5$

$R=8.5\text{ A}$

45. F varies inversely as L, so

$F=\frac{k}{L}$

$250=\frac{k}{22}$

$k=5500$

$F=\frac{5500}{L}$

Check:

$F=\frac{5500}{10}$

$=550\text{ N}$

49. $C=kV$

$6.3=k(220)$

$k=\frac{6.3}{220}$

$C=\frac{6.3}{220}V$

$C=\frac{6.3}{220}(150)$

$=4.3\ \mu\text{C}$

53.

$$d = kt^2$$
$$19.62 = k(2.00)^2$$
$$k = \frac{19.62}{(2.00)^2}$$
$$= 4.90$$
$$d = 4.90t^2$$
$$d = 4.90(10.43)^2$$
$$= 533 \text{ m}$$

57.

$$f = \frac{k}{\sqrt{C}}$$
$$25.0 = \frac{k}{\sqrt{95.0}}$$
$$k = 25.0\sqrt{95.0}$$
$$f = \frac{25.0\sqrt{95.0}}{\sqrt{C}}$$
$$f = \frac{25.0\sqrt{95.0}}{\sqrt{25.0}}$$
$$= 48.7 \text{ Hz}$$

61.

$$w = k \cdot L^3$$
$$15\ 400 = k \cdot 15^3$$
$$k = \frac{15\ 400}{15^3}$$
$$= 4.56$$
$$w = 4.56L^3$$
$$w = 4.56(5.5)^3$$
$$= 759 \text{ N}$$

65.

$$R = kv_0^2 \sin 2\theta$$
$$5.12\times10^4 = k(850)^2 \sin\left(2\left(22.0^\circ\right)\right)$$
$$k = \frac{5.12\times10^4}{850^2 \sin\left(44.0^\circ\right)}$$
$$= 0.102$$
$$R = 0.102v_0 \ \sin 2\theta$$
$$R = (0.102)(750^2)\sin\left(2\left(43.2^\circ\right)\right)$$
$$R = 5.73\times10^4 \text{ m}$$

69.

$$R = \frac{kr^4}{d}$$
$$R_1 = \frac{k(r+0.25r)^4}{d-.02d}$$
$$= \frac{k(1.25)^4 r^4}{.98d}$$
$$= 2.49\frac{kr^4}{d}$$
$$R_1 = 2.49R$$

The increase is: $2.49R - R = 1.49R$, or 150%

73.

$$L = \frac{kt}{d}$$
$$1.20 = \frac{k \cdot 30}{20.0};$$
$$k = 0.800$$
$$L = \frac{0.800t}{d}$$
$$L = \frac{0.800(90)}{15.0}$$
$$= 4.80 \text{ MJ}$$

77.

$$\frac{12.5 \text{ mm}}{200 \text{ km}} = \frac{18 \text{ mm}}{x}$$
$$x = \frac{(18)(200)}{12.5}$$
$$x = 290 \text{ km}$$

CHAPTER 19

SEQUENCES AND THE BINOMIAL THEOREM

19.1 Arithmetic Sequences

1. $a_1 = 5,\ a_{32} = -88,\ n = 32$

$-88 = 5 + (32-1)d$

$31d = -93$

$d = -3$

5. 4, 6, 8, 10, 12

9. $d = 4 - 1$

$= 3$

$a_8 = 1 + 3(8-1)$

$= 22$

13. $a_{80} = -0.7 + (80-1)0.4$

$= 30.9$

17. $S_{20} = \frac{20}{2}(4+40)$

$= 440$

21. $45 = 5 + (n-1)8$

$45 = 8n - 3$

$n = 6$

$S_6 = \frac{6}{2}(5+45)$

$= 150$

25. $a_{30} = a_1 + (29)(3)$

$= a_1 + 87$

$1875 = \frac{30}{2}(a_1 + a_1 + 87)$

$125 = 2a_1 + 87$

$a_1 = 19$

$a_{30} = 106$

29. $a_n = 5k + (n-1)(0.5k)$

$S_n = \frac{n}{2}\left[5k + (5k + (n-1)(0.5k))\right]$

$104k = \frac{n}{2}(5k + 5k + 0.5kn - 0.5k)$

$208k = n(9.5k + 0.5kn)$

$n^2 + 19n - 416 = 0$

$(n+32)(n-13) = 0$

$n = -32$ (not valid)

$n = 13$

$a_{13} = 5k + (12)(0.5k)$

$= 5k + 6k$

$= 11k$

33. $d = \frac{720-560}{10-6} = 40$

$560 = a_1 + (5)(40)$

$a_1 = 360$

$S_{10} = 5(360 + 720)$

$= 5400$

37. $a_{n+1} = a_n + 2 \Rightarrow d = 2$

$a_n = a_1 + (n-1)d$

$= 3 + (n-1)2$

$= 2n + 1$

41. $a_1,\ b,\ c,\ a_4,\ a_5$

$b + d = c \Rightarrow d = c - b$

$a_1 = b - d$

$= b - (c - b)$

$= 2b - c$

$a_4 = c + d$

$= c + (c - b)$

$= 2c - b$

$a_5 = c + 2d$

$= c + 2(c - b)$

$= 3c - 2b$

45. $3 - x,\ -x,\ \sqrt{9-2x}$

$d = -x - (3 - x)$

$= -3$

$-x - 3 = \sqrt{9-2x}$; square both sides

$x^2+6x+9=9-2x$

$x^2+8x+9=0$

$x(x+8)=0$

$x=-8$ or $x=0$

Check: $x=-8$

$-(-8)-3=\sqrt{9-2(-8)}$

$5=5$

Check: $x=0$

$0-3=\sqrt{9-2(0)}$

$-3\neq 3$

The only solution is $x=-8$, which gives the sequence 11, 8, 5,... with $d=-3$.

49. $a_1=20,\ d=-1,\ n=15$

$a_{15}=20+(15-1)(-1)=6$

$S_{15}=\frac{15}{2}(20+6)$

$S_{15}=195$ logs in pile

53. $a_1=1800,\ d=-150,$

$a_n=0$ when

$0=1800+(n-1)(-150)$

$=1800-150n+150$

$150n=1950$

$n=13$

$S_{13}=\frac{13}{2}(1800+0)=\$11\,700$, the sum of all depreciations, which is the cost of the car.

57. $S_n=\frac{n}{2}(a_1+a_n)$

$=\frac{n}{2}\left[a_1+(a_1+(n-1)d)\right]$

$=\frac{n}{2}\left[2a_1+(n-1)d\right]$

19.2 Geometric Sequences

1. Find a_{10} for $a_2=3,\ a_4=9$

Use $n=3$ for a new sequence starting at the second term, so $a_1=3, a_3=9$ and

$9=3r^2$

$r=\sqrt{3}$

Find the first term of the original sequence:

$3=a_1\left(\sqrt{3}\right)^{2-1}$

$a_1=\sqrt{3}$

Find the 10th term:

$a_{10}=\sqrt{3}\left(\sqrt{3}\right)^{10-1}=\sqrt{3}\left(\sqrt{3}\right)^{9}=243$

5. $\frac{1}{6},\frac{1}{6}\cdot 3,\frac{1}{6}\cdot 3^2,\frac{1}{6}\cdot 3^3,\frac{1}{6}\cdot 3^4$

$\frac{1}{6},\frac{1}{2},\frac{3}{2},\frac{9}{2},\frac{27}{2}$

9. $r=-25\div 125=-0.2,\ a_1=125,\ n=7$

$a_7=125(-0.2)^{7-1}=\frac{1}{125}$

13. $10^{100}, -10^{98}, 10^{96},...; n=51$

$r=\frac{-10^{98}}{10^{100}}=-10^{-2}$

$a_{51}=10^{100}\left(-10^{-2}\right)^{50}$

$=10^{100}(-1)^{100}(10^{-100})$

$=1$

17. 384, 192, 96, $\cdots$

$r=\frac{192}{384}=\frac{1}{2}$

$S_7=\frac{384\left(1-\left(\frac{1}{2}\right)^7\right)}{1-\frac{1}{2}}=762$

21. $a_6=\left(\frac{1}{16}\right)(4)^{6-1}=\left(\frac{1}{16}\right)(4)^5=64$

$S_6=\frac{\frac{1}{16}\left(1-4^6\right)}{1-4}=\frac{\frac{1}{16}(1-4096)}{-3}=\frac{4095}{48}=\frac{1365}{16}$

25. $27=a_1r^{4-1}$

$a_1=\frac{27}{r^3}$

$40=a_1\frac{\left(1-r^4\right)}{1-r}$

$=a_1\frac{\left(1+r^2\right)(1+r)(1-r)}{1-r}$

$=a_1\left(1+r^2\right)(1+r)$

Substitute a from first equation in second equation:

$40=\frac{27}{r^3}(1+r^2)(1+r)$

$40r^3=27+27r+27r^2+27r^3$

$13r^3-27r^2-27r-27=0$

There is one change of sign so there is exactly one positive root. There are at most two negative roots. We first test the possible integer roots: $\pm1,\pm3,\pm9,\pm27$

Using synthetic division, 3 gives a remainder of zero (see Section 15.3):

13	−27	−27	−27	⌊3
	39	36	27	
13	12	9	0	

The remaining roots are solutions to the equation $13r^2+12r+9=0$

Using the quadratic formula:

$r=\frac{-12\pm\sqrt{-288}}{26}$

Since the other solutions are complex, the only solution is $r=3$. Now

$27=a_1(3^{4-1})$

$27a_1=27$

$a_1=1$

29. $\frac{3^{x+1}}{3}=3^x$

$\frac{3^{2x+1}}{3^{x+1}}=3^x$

The sequence $3, 3^{x+1}, 3^{2x+1},\cdots$ is a geometric sequence since every term in the sequence can be obtained from the preceding one by multiplying it by 3^x.

Therefore

$a_1=3,\ r=3^x$

$a_{20}=3\cdot(3^x)^{20-1}$

$=3^{19x+1}$

33. G.S: $2,\ 6,\ 2x+8,\cdots$

$r=\frac{6}{2}$

$=3$

$6\cdot r=2x+8$

$6\cdot3=2x+8$

$x=5$

37. $7a_1=\frac{a_1(1-r^3)}{1-r}, r\neq1$

$7-7r=1-r^3$

$r^3-7r+6=0$

$(r+3)(r-2)(r-1)=0$

$r=-3,\ r=2,\ r=1$ (reject since $r\neq1$)

If $r=1$, then the sum of the first three terms is $3a_1$, so we would have the equality $3a_1=7a_1$, which is not valid. Hence the only solutions are $r=-3$ and $r=2$.

41. $r=1-0.125=0.875,\ a_1=3.27\text{ mA},\ n=9.2$

$a_{9.2}=3.27(0.875)^{8.2}=1.09\text{ mA}$

45. $a_1=9800, r=1-0.1=0.9$

$n=\frac{4000}{800}=5$ periods of 800 years

$a_6=(9800)0.9^5$

$=5800^\circ\text{C}$

49. $D_1=100-20=80$

$r=1-.035=0.65$

$D_{11}=80(0.65)^{10}=1.1$

$20.0^\circ+1.1^\circ=21.1^\circ\text{C}$

53. $a_n=a_1r^{n-1}$

$=a_1r^nr^{-1}$

$=\frac{a_1r^n}{r}$

$a_1r^n=a_nr$

$S_n=\frac{a_1(1-r^n)}{1-r}$

$=\frac{a_1-a_1r^n}{1-r}$

$=\frac{a_1-ra_n}{1-r}$

57. A.S.: $8,x,y,\ldots$

$x-8=y-x$

$y=2x-8$

G.S.: $x,y,36,\ldots$

$\frac{y}{x}=\frac{36}{y}$

$y^2=36x$

Substitute $y = 2x - 8$ into $y^2 = 36x$

$$(2x-8)^2 = 36x$$
$$4x^2 - 32x + 64 = 36x$$
$$4x^2 - 68x + 64 = 0$$
$$x^2 - 17x + 16 = 0$$
$$(x-16)(x-1) = 0$$
$x = 16$ or $x = 1$

$y = 2(16) - 8 \quad y = 2(1) - 8$

$= 24 \quad = -6$

A.S.: 8, 16, 24, …

G.S.: 16, 24, 36, …

or

A.S.: 8, 1, −6,…

G.S.: 1, −6, 36,…

19.3 Infinite Geometric Series

1. Given the G.S. $4 + \frac{1}{2} + \frac{1}{16} + \frac{1}{128} + \cdots$ find the sum.

$a_1 = 4,\ r = \frac{1}{8}$

$$S = \frac{a}{1-r} = \frac{4}{1-\frac{1}{8}}$$
$$S = \frac{32}{7}$$

5. $a_1 = 0.5, S = 0.625$

$$0.625 = \frac{0.5}{1-r}$$
$$1 - r = 0.8$$
$$r = \frac{1}{5}$$

9. $a_1 = 1,\ r = \frac{7}{8}$

$$S = \frac{1}{1-\frac{7}{8}}$$
$$= 8$$

13. $a_1 = 2+\sqrt{3},\ r = \frac{1}{2+\sqrt{3}}$

$$S = \frac{2+\sqrt{3}}{1-\frac{1}{2+\sqrt{3}}}$$
$$= \frac{2+\sqrt{3}}{\frac{2+\sqrt{3}-1}{2+\sqrt{3}}}$$
$$= \frac{\left(2+\sqrt{3}\right)\left(2+\sqrt{3}\right)}{1+\sqrt{3}}$$
$$= \frac{7+4\sqrt{3}}{1+\sqrt{3}} \times \frac{1-\sqrt{3}}{1-\sqrt{3}}$$
$$= \frac{7-7\sqrt{3}+4\sqrt{3}-12}{1-3}$$
$$= \frac{-5-3\sqrt{3}}{-2}$$
$$= \frac{1}{2}\left(5+3\sqrt{3}\right)$$

17. $0.499\,99\ldots = 0.4 + 0.09 + 0.009 + 0.0009 + \cdots$

$a_1 = 0.09, r = 0.01$

$$S = 0.4 + \frac{0.09}{1-\frac{1}{10}}$$
$$= 0.5$$

21. $0.181\,818\ldots = 0.18 + 0.0018 + 0.000\,018 + \ldots$

$a_1 = 0.18,\ r = 0.01$

$$S = \frac{0.18}{1-0.01}$$
$$= \frac{2}{11}$$

25. $0.366\,66\ldots = 0.3 + 0.06 + 0.006 + 0.0006 + \ldots$

For the G.S., $a_1 = 0.06,\ r = 0.1$

$$S = \frac{0.06}{1-0.1}$$
$$= \frac{0.06}{0.9}$$
$$= \frac{1}{15}$$

Therefore,

$$0.366\,66\ldots = \frac{3}{10} + \frac{1}{15}$$
$$= \frac{11}{30}$$

29. $50,\ a_2,\ 2, \cdots$

$$\frac{a_2}{50} = \frac{2}{a_2}$$
$$a_2^2 = 100$$
$$a_2 = \pm 10$$

(1) 50, 10, 2,... has $r=\frac{1}{5}$

$$S=\frac{50}{1-\frac{1}{5}}=\frac{125}{2}$$

(2) 50, −10, 2,...has $r=-\frac{1}{5}$

$$S=\frac{50}{1+\frac{1}{5}}=\frac{125}{3}$$

33. $a_1=5.882$ g

The sequence is approximately geometric, since the ratios of successive terms are constant (to four significant digits).

$$r=\frac{5.782}{5.882}=\frac{5.684}{5.782}=\frac{5.587}{5.684}=0.9830$$

$$\begin{aligned}S&=\frac{5.882}{1-0.9830}\\&=\frac{5.882}{0.0170}\\&=346\text{ g}\end{aligned}$$

37. $1+2x+4x^2+\cdots=\frac{2}{3}$

$$\begin{aligned}&=\frac{1}{1-2x}\\2-4x&=3\\4x&=-1\\x&=-\frac{1}{4}\end{aligned}$$

19.4 The Binomial Theorem

1. $$\begin{aligned}(2x+3)^5&=(2x)^5+5(2x)^4(3)+\frac{5(4)}{2!}(2x)^3(3)^2\\&+\frac{5(4)(3)}{3!}(2x)^2(3)^3\\&+\frac{5(4)(3)(2)}{4!}(2x)^1(3)^4+(3)^5\\(2x+3)^5&=32x^5+240x^4+720x^3+1080x^2\\&+810x+243\end{aligned}$$

5. $$\begin{aligned}(2x-1)^4&=(2x)^4+4(2x)^3(-1)+\frac{4(3)}{2}(2x)^2(-1)^2\\&+\frac{4(3)(2)}{6}(2x)(-1)^3+\frac{4(3)(2)(1)}{24}(-1)^4\\&=16x^4-32x^3+24x^2-8x+1\end{aligned}$$

9. $$\begin{aligned}(n+2\pi)^5&=n^5+5n^4(2\pi)+\frac{5(4)}{2!}n^3(2\pi)^2\\&+\frac{5(4)(3)}{3!}n^2(2\pi)^3\\&+\frac{5(4)(3)(2)}{4!}n(2\pi)^4+(2\pi)^5\\(n+2\pi)^5&=n^5+10\pi n^4+40\pi^2n^3+80\pi^3n^2\\&+80\pi^4n+32\pi^5\end{aligned}$$

13. From Pascal's triangle, the coefficients for $n=4$ are 1, 4, 6, 4, 1.

$$\begin{aligned}(5x-3)^4&=[5x+(-3)]^4\\&=1(5x)^4+4(5x)^3(-3)+6(5x)^2(-3)^2\\&+4(5x)(-3)^3+(-3)^4\\&=625x^4-1500x^3+1350x^2-540x+81\end{aligned}$$

17. $$\begin{aligned}(x+2)^{10}&=x^{10}+10x^9(2)+\frac{(10)(9)}{2}x^8(2)^2\\&+\frac{(10)(9)(8)}{6}x^7(2)^3+\cdots\\&=x^{10}+20x^9+180x^8+960x^7+\cdots\end{aligned}$$

21. $$\begin{aligned}(x^{1/2}-4y)^{12}&=(x^{1/2})^{12}+12(x^{1/2})^{11}(-4y)\\&+\frac{12\cdot11}{2!}(x^{1/2})^{10}(-4y)^2\\&+\frac{12\cdot11\cdot10}{3!}(x^{1/2})^9(-4y)^3\\&=x^6-48x^{11/2}y+1056x^5y^2\\&-14\ 080x^{9/2}y^3+\cdots\end{aligned}$$

25. $$\begin{aligned}(1.05)^6&=(1+0.05)^6\\&=1^6+6(1)^5(0.05)+\frac{6(5)}{2!}(1)^4(0.05)^2\\&=1.3375\\&=1.338\text{ to 3 decimal places using three terms}\end{aligned}$$

From a calculator, $(1.05)^6=1.340$ to 3 decimal places.

29. $$\begin{aligned}(1+x)^8&=1+8x+\frac{8(7)}{2}x^2+\frac{8(7)(6)}{6}x^3+\cdots\\&=1+8x+28x^2+56x^3+\cdots\end{aligned}$$

33. $\sqrt{1+x}=(1+x)^{1/2}=1+\frac{1}{2}x+\frac{\frac{1}{2}\left(-\frac{1}{2}\right)}{2}x^2+\frac{\frac{1}{2}\left(-\frac{1}{2}\right)\left(-\frac{3}{2}\right)}{6}x^3+\cdots$

$$=1+\frac{1}{2}x-\frac{1}{8}x^2+\frac{1}{16}x^3\cdots$$

37. (a) $17!+4!=3.557\times10^{14}$

(b) $21!=5.109\times10^{19}$

(c) $17!\times4!=8.536\times10^{15}$

(d) $68!=2.480\times10^{96}$

41. The term involving b^5 will be the sixth term.

$r=5,\ n=8$

The sixth term is $\frac{8(7)(6)(5)(4)}{5(4)(3)(2)}a^3b^5=56a^3b^5$.

45. If $n>4$, $n!=n(n-1)(n-2)\cdots5\cdot4\cdot3\cdot2\cdot1$

$$=n(n-1)(n-2)\cdots4\cdot3\cdot(5\cdot2)$$
$$=n(n-1)(n-2)\cdots4\cdot3\cdot(10)$$

which will always end in a zero

49. $\sqrt{6}=\sqrt{4(1.5)}=2\sqrt{1+0.5}$

$$=2(1+0.5)^{1/2}$$
$$=2\left[1+\frac{1}{2}(0.5)+\frac{\frac{1}{2}\left(\frac{1}{2}-1\right)}{2!}(0.5)^2+\frac{\frac{1}{2}\left(\frac{1}{2}-1\right)\left(\frac{1}{2}-2\right)}{3!}(0.5)^3\right]$$
$$=2.453125$$

$\sqrt{6}=2.45$ to hundredths using four terms

53. $\left(a^2+x^2\right)^{-1/2}=\frac{1}{a}\left[1+\left(\frac{x}{a}\right)^2\right]^{-1/2}$

$$=\frac{1}{a}\left[1-\frac{1}{2}\left(\frac{x^2}{a^2}\right)+\frac{3}{8}\left(\frac{x^4}{a^4}\right)+\cdots\right]$$

Therefore

$$1-\frac{x}{\sqrt{a^2+x^2}}=1-x\left(\frac{1}{a}\right)\left[1-\frac{x^2}{2a^2}+\frac{3x^4}{8a^4}+\cdots\right]$$
$$=1-\frac{x}{a}+\frac{x^3}{2a^3}-\cdots$$

Review Exercises

1. This is an A.S. with $d=6-1=5, a_1=1$, so

$$a_{17}=1+(17-1)5$$
$$=1+80$$
$$=81$$

5. This is an A.S. with $d=8-3.5=4.5, a_1=-1$, so

$$a_{16}=-1+(16-1)(4.5)$$
$$=66.5$$

9. $S_{15}=\frac{15}{2}(-4+17)$

$$=\frac{15}{2}(13)$$
$$=\frac{195}{2}$$

13. $a_n=17+(9-1)(-2)$

$$=17-16$$
$$=1$$
$$S_n=\frac{9}{2}(1+17)=81$$

17. $a_1=80,\ a_n=-25,\ S_n=220$

$$220=\frac{n}{2}(80-25)$$
$$440=55n$$
$$n=8$$
$$-25=80+(8-1)d$$
$$=80+7d$$
$$d=\frac{-105}{7}$$
$$=-15$$

21. $S_{12}=\frac{12}{2}(-1+32)$

$$=186$$

25. $r=\frac{6}{9}=\frac{2}{3}$

$$S=\frac{a_1}{1-r}$$
$$=\frac{0.9}{1-\frac{2}{3}}$$
$$=2.7$$

29. $0.030\ 303... = 0.03 + 0.000\ 3 + 0.000\ 003...$

$a_1 = 0.03;\ r = 0.01$

$$S = \frac{0.03}{1-0.01} = \frac{0.03}{0.99} = \frac{1}{33}$$

33. $$(x-2)^4 = [x+(-2)]^4 = x^4 + 4x^3(-2) + \frac{4(3)}{2}x^2(-2)^2 + \frac{4(3)(2)}{3(2)}x(-2)^3 + (-2)^4 = x^4 - 8x^3 + 24x^2 - 32x + 16$$

37. $$(a+2e)^{10} = a^{10} + 10a^{10-1}(2e) + \frac{10(10-1)}{2!}a^{10-2}(2e)^2 + \frac{10(10-1)(10-2)}{3!}a^{10-3}(2e)^3 + \cdots$$
$$= a^{10} + 10a^9(2e) + 45a^8(4e^2) + 120a^7(8e^3) + \cdots$$
$$= a^{10} + 20a^9e + 180a^8e^2 + 960a^7e^3 + \cdots$$

41. $$(1+x)^{12} = 1 + 12x + \frac{12(12-1)}{2!}x^2 + \frac{12(12-1)(12-2)}{3!}x^3 + \cdots = 1 + 12x + 66x^2 + 220x^3 + \cdots$$

45. $$\left[1+\left(-a^2\right)\right]^{1/2} = 1 + \frac{1}{2}\left(-a^2\right) + \frac{\frac{1}{2}\left(-\frac{1}{2}\right)}{2}\left(-a^2\right)^2 + \frac{\frac{1}{2}\left(-\frac{1}{2}\right)\left(-\frac{3}{2}\right)}{3(2)}\left(-a^2\right)^3 = 1 - \frac{1}{2}a^2 - \frac{1}{8}a^4 - \frac{1}{16}a^6 - \cdots$$

49. $a = 2,\ d = 2,\ n = 1000$

$$a_n = 2 + (1000-1)2 = 2 + 999(2) = 2000$$

$$S = \frac{1000}{2}(2 + 2000) = 1\ 001\ 000$$

53. Let a_3, a_4, a_5, a_6, a_7 be the GS 6, a_4, 9, a_6, 12

Suppose r = common ratio, the GS is 6, $6r$, $6r^2$, $6r^3$, $6r^4$ which gives

$9 = 6r^2$ and $6r^4 = 12$

$r^2 = \frac{9}{6}$ $\quad r^4 = 2$

$r^2 = \sqrt{2}$

Since $\frac{9}{6} \neq \sqrt{2}$, 6, a_4, 9, a_6, 12 cannot be a GS.

57. $$(1.06)^{-6} = (1+0.06)^{-6} = 1^{-6} + \frac{-6}{1}1^{-7}(0.06)^1 + \frac{-6(-7)}{1\cdot 2}1^{-8}(0.06)^2 = 0.7156$$

From a calculator,

$(1.06)^{-6} = 0.705$

61. 24, 22, 20,$\cdots$ is an AS with $a_1 = 24$, $d = -2$

$4 = 24 + (n-1)(-2)$

$n = 11$

65. The length of the vertical pieces is an A.S. with $a_1 = 254$ mm. The common difference corresponds to the adjacent side of a right triangle with opposite side equal to $\frac{5690}{14}$, and angle 84.8°, so that

$$d = \frac{5690}{14\tan 84.8^\circ}.$$

The total length of the vertical pieces is S_{15}:

$$S_{15} = \frac{15}{2}\left(254 + \left(254 + (15-1)\frac{5690}{14\tan 84.8^\circ}\right)\right) = 7690 \text{ mm}$$

69. $0.015,\ 0.015(2),\ 0.015(2)^2, \cdots, 0.015(2)^{40}$

This is a G.S. with $a_1 = 0.15$ and $r = 2$.

We are interested in the 41st term.

$$a_{41} = 0.015(2)^{40} = 1.65 \times 10^{10} \text{ cm} = 165\ 000 \text{ km}$$

73. This is a G.S. with $a_1 = 10, r = 0.9$

$$S = \frac{10}{1-0.9} = 100 \text{ cm}$$

77. This is a G.S. with $a_1 = 250$ and $r = 0.40$

$$a_5 = 250(0.4)^4 = \$6.40$$

81. Let $x=\frac{a-1}{2}m^2$ and $y=\frac{a}{a-1}$

$(1+x)^y=1+yx+\frac{y(y-1)}{2}x^2 \ldots$ (3 terms)

$=1+\left(\frac{a}{a-1}\right)\left(\frac{a-1}{2}m^2\right)+\frac{\left(\frac{a}{a-1}\right)\left(\frac{a}{a-1}-1\right)}{2}\left(\frac{a-1}{2}m^2\right)^2$

$=1+\frac{a}{2}m^2+\frac{\left(\frac{a}{a-1}\right)\left(\frac{1}{a-1}\right)}{2}\left(\frac{(a-1)^2}{2^2}m^4\right)$

$=1+\frac{a}{2}m^2+\frac{a}{2^3}m^4$

$=1+\frac{1}{2}am^2+\frac{1}{8}am^4$

85. The percent of insects remaining is a G.S. with $a_1=100$ and

$r=1-0.75$

$=0.25$

99.9% of insects will be destroyed when the percent remaining is 0.1.

$0.1=100(0.25)^n$

$0.001=0.25^n$

$\log 0.001=\log 0.25^n$

$\log 0.001=n\log 0.25$

$n=\frac{\log 0.001}{\log 0.25}$

$=5$ applications

89. For odd n, the middle position of a sequence is term $\frac{n+1}{2}$. Therefore, the middle term of an arithmetic sequence is

$a_{\frac{n+1}{2}}=a+\left(\frac{n+1}{2}-1\right)d$

$=a+\frac{n-1}{2}d$

$\frac{S_n}{n}=\frac{1}{n}\left[\frac{n}{2}(a+(a+(n-1)d))\right]$

$=\frac{1}{2}[2a+(n-1)d]$

$=a+\frac{n-1}{2}d$

$=a_{\frac{n+1}{2}}$

93. Let A = initial deposit, and let n be the number of compounding periods in one year. Then the interest in one compounding period is $\frac{r}{n}$.

$V=A+A\frac{r}{n}=A\left(1+\frac{r}{n}\right)$, after one compounding period

$V=\left[A\left(1+\frac{r}{n}\right)\right]\left(1+\frac{r}{n}\right)=A\left(1+\frac{r}{n}\right)^2$, after two compounding periods

In general,

$V=A\left(1+\frac{r}{n}\right)^n$, at the end of one year with n compounding periods per year.

For $A=\$1000$ and $r=0.1=10\%$

$V=1000\left(1+\frac{0.1}{n}\right)^n$.

Let us tabulate some values of n and V.

n	$V=1000(1+\frac{0.1}{n})^n$
1	1100
5	1104.08
10	1104.62
50	1105.06
100	1105.12
1000	1105.17
10000	1105.17
100000	1105.17

As n increases, even though the interest per period is smaller, the number of times the investment is compounded increases, so the value of the investment increases. However, the value does not increase indefinitely. We can see that for large n, the value of the investment is the same (up to hundredths), and the maximum value is \$1105.17.

(As n increases, it can be shown that $\lim_{n\to\infty}1000\left(1+0.1\cdot\frac{1}{n}\right)^n=e^{0.1}$, where e is the base of the natural logarithms. See Section 27.5.)

CHAPTER 20

ADDITIONAL TOPICS IN TRIGONOMETRY

20.1 Fundamental Trigonometric Identities

1. We simplify the right side:

$$\frac{\tan x}{\sec x} = \frac{\frac{\sin x}{\cos x}}{\frac{1}{\cos x}}$$
$$= \frac{\sin x}{\cos x} \cdot \frac{\cos x}{1}$$
$$= \sin x$$

5. Verify $\sin^2 \theta + \cos^2 \theta = 1$ for $\theta = \frac{4\pi}{3}$

$$\left(\sin\frac{4\pi}{3}\right)^2 = \left(-\frac{1}{2}\sqrt{3}\right)^2 = \frac{3}{4}$$
$$\left(\cos\frac{4\pi}{3}\right)^2 = \left(-\frac{1}{2}\right)^2 = \frac{1}{4}$$
$$\frac{3}{4} + \frac{1}{4} = 1$$

9. $\cos\theta \cot\theta(\sec\theta - 2\tan\theta)$

$$= \cos\theta\cot\theta\left(\frac{1}{\cos\theta} - 2\frac{1}{\cot\theta}\right)$$
$$= \cot\theta - 2\cos\theta$$

13. $\sin x + \sin x \tan^2 x = \sin x\left(1 + \tan^2 x\right)$

$$= \sin x \sec^2 x$$
$$= \sin x \cdot \frac{1}{\cos x} \cdot \sec x$$
$$= \tan x \sec x$$

17. $\csc^4 y - 1 = \left(\csc^2 y + 1\right)\left(\csc^2 y - 1\right)$

$$= \left(\csc^2 y + 1\right)\left(\cot^2 y\right)$$

21. $\sin x \sec x = \sin x \frac{1}{\cos x}$

$$= \frac{\sin x}{\cos x}$$
$$= \tan x$$

25. $\sin x\left(1 + \cot^2 x\right) = \sin x\left(\csc^2 x\right)$

$$= \sin x\left(\frac{1}{\sin^2 x}\right)$$
$$= \frac{1}{\sin x}$$
$$= \csc x$$

29. $\cot\theta\sec^2\theta - \cot\theta = \frac{\cos\theta}{\sin\theta}\frac{1}{\cos^2\theta} - \frac{\cos\theta}{\sin\theta}$

$$= \frac{1}{\sin\theta\cos\theta} - \frac{\cos\theta}{\sin\theta}$$
$$= \frac{1 - \cos^2\theta}{\sin\theta\cos\theta}$$
$$= \frac{\sin^2\theta}{\sin\theta\cos\theta}$$
$$= \frac{\sin\theta}{\cos\theta}$$
$$= \tan\theta$$

33. $\cos^2 x - \sin^2 x = 1 - \sin^2 x - \sin^2 x$

$$= 1 - 2\sin^2 x$$

37. $2\sin^4 x - 3\sin^2 x + 1 = \left(2\sin^2 x - 1\right)\left(\sin^2 x - 1\right)$

$$= \left(2\sin^2 x - 1\right)\left(-\cos^2 x\right)$$
$$= \cos^2 x\left(1 - 2\sin^2 x\right)$$

41. $1 + \sin^2 x + \sin^4 x \cdots +$

This is a geometric series with $a_1 = 1$, $r = \sin^2 x$

$$S = \frac{1}{1 - \sin^2 x}$$
$$= \frac{1}{\cos^2 x}$$
$$= \sec^2 x$$

The series is finite because $\sin^2 x < 1$ for $-\frac{\pi}{2} < x < \frac{\pi}{2}$.

45. $\cot x(\sec x - \cos x) = \dfrac{\cos x}{\sin x}\cdot\dfrac{1}{\cos x} - \dfrac{\cos x}{\sin x}\cdot\cos x$

$$= \frac{1}{\sin x} - \frac{\cos^2 x}{\sin x}$$

$$= \frac{1-\cos^2 x}{\sin x}$$

$$= \frac{\sin^2 x}{\sin x}$$

$$= \sin x$$

49. $\dfrac{\cos x + \sin x}{1+\tan x} = \dfrac{\cos x + \sin x}{1+\frac{\sin x}{\cos x}}$

$$= \frac{(\cos x + \sin x)}{\frac{\cos x + \sin x}{\cos x}}$$

$$= \cos x$$

53.

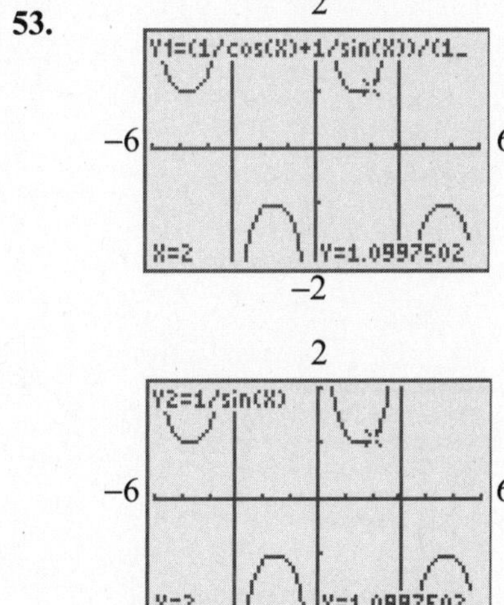

57. No. $\dfrac{2\cos^2 x - 1}{\sin x\cos x} \neq \tan x - \cot x$

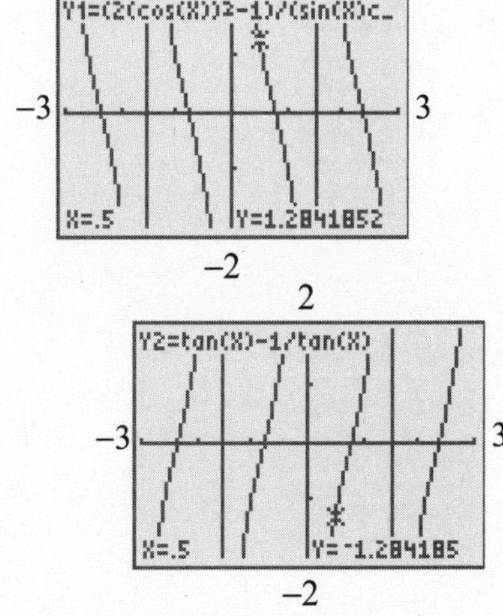

61. $l = a\csc\theta + a\sec\theta$

$$= a(\csc\theta + \sec\theta)$$

$$= a\left(\frac{1}{\sin\theta} + \frac{1}{\cos\theta}\right)$$

$$= a\left(\frac{1}{\sin\theta} + \frac{\sin\theta}{\cos\theta}\cdot\frac{1}{\sin\theta}\right)$$

$$= a\left(\frac{1}{\sin\theta} + \frac{\tan\theta}{\sin\theta}\right)$$

$$= a\frac{(1+\tan\theta)}{\sin\theta}$$

65. $\sin^2 x(1-\sec^2 x) + \cos^2 x(1+\sec^4 x)$

$$= \sin^2 x - \sin^2 x\sec^2 x + \cos^2 x + \cos^2 x\sec^4 x$$

$$= \sin^2 x - \frac{\sin^2 x}{\cos^2 x} + \cos^2 x + \frac{\cos^2 x}{\cos^4 x}$$

$$= \sin^2 x - \tan^2 x + \cos^2 x + \sec^2 x$$

$$= 1 - \tan^2 x + \sec^2 x$$

$$= 1 - (\sec^2 x - 1) + \sec^2 x$$

$$= 1 - \sec^2 x + 1 + \sec^2 x$$

$$= 2$$

69. $x = \cos\theta;$

$$\sqrt{1-x^2} = \sqrt{1-\cos^2\theta}$$

$$= \sqrt{\sin^2\theta}$$

$$= \sin\theta$$

20.2 The Sum and Difference Formulas

1. $\sin\alpha = \dfrac{12}{13}$ (α in first quadrant) and $\sin\beta = -\dfrac{3}{5}$ for β in third quadrant.

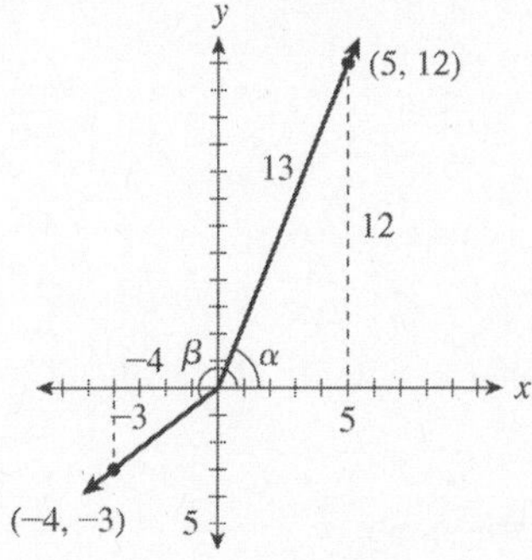

$$\cos(\alpha+\beta) = \cos\alpha\cos\beta - \sin\alpha\sin\beta$$

$$= \frac{5}{13}\cdot\frac{-4}{5} - \frac{12}{13}\cdot\frac{-3}{5} = \frac{16}{65}$$

5. Given $15^\circ = 60^\circ - 45^\circ$

$$\cos(\alpha-\beta) = \cos\alpha\cos\beta + \sin\alpha\sin\beta$$
$$\cos 15^\circ = \cos(60^\circ - 45^\circ)$$
$$= \cos 60^\circ \cos 45^\circ + \sin 60^\circ \sin 45^\circ$$
$$= \frac{1}{2}\times\frac{\sqrt{2}}{2} + \frac{\sqrt{3}}{2}\times\frac{\sqrt{2}}{2}$$
$$= \frac{\sqrt{2}}{4} + \frac{\sqrt{6}}{4}$$
$$= 0.966$$

9. Using the results of Exercise 7:

$$\cos\alpha = \frac{3}{5}$$
$$\cos\beta = -\frac{12}{13}$$
$$\sin\alpha = \frac{4}{5}$$
$$\sin\beta = \frac{5}{13}$$
$$\cos(\alpha+\beta) = \cos\alpha\cos\beta - \sin\alpha\sin\beta$$
$$= \frac{3}{5}\left(-\frac{12}{13}\right) - \frac{4}{5}\left(\frac{5}{13}\right)$$
$$= \frac{-36-20}{65} = -\frac{56}{65}$$

13. $\cos\pi\cos x + \sin\pi\sin x = \cos(\pi - x)$

Also,

$$\cos\pi\cos x + \sin\pi\sin x = (-1)\cos x + (0)\sin x$$
$$= -\cos x$$

Therefore

$$\cos(\pi - x) = -\cos x$$

17.
$$\tan(x-\pi) = \frac{\tan x + \tan\pi}{1 - \tan x\tan\pi}$$
$$= \frac{\tan x + 0}{1 - \tan x(0)}$$
$$= \tan x$$

21. $\sin 122^\circ \cos 32^\circ - \cos 122^\circ \sin 32^\circ$ is of the form $\sin\alpha\cos\beta - \cos\alpha\sin\beta = \sin(\alpha-\beta)$, with $\alpha = 122^\circ$ and $\beta = 32^\circ$ so

$$\sin 122^\circ \cos 32^\circ - \cos 122^\circ \sin 32^\circ$$
$$= \sin(122^\circ - 32^\circ)$$
$$= \sin 90^\circ$$
$$= 1$$

25. $\sin(x+y)\sin(x-y)$

$$= (\sin x\cos y + \cos x\sin y)(\sin x\cos y - \cos x\sin y)$$
$$= \sin^2 x\cos^2 y - \cos^2 x\sin^2 y$$
$$= \sin^2 x(1-\sin^2 y) - (1-\sin^2 x)(\sin^2 y)$$
$$= \sin^2 x - \sin^2 x\sin^2 y - \sin^2 y + \sin^2 x\sin^2 y$$
$$= \sin^2 x - \sin^2 y$$

29.

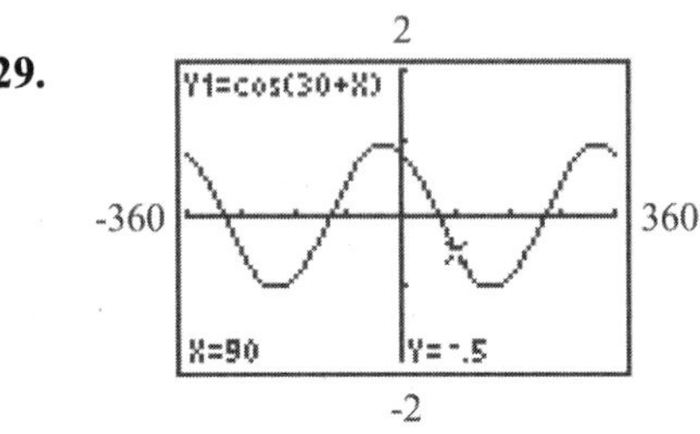

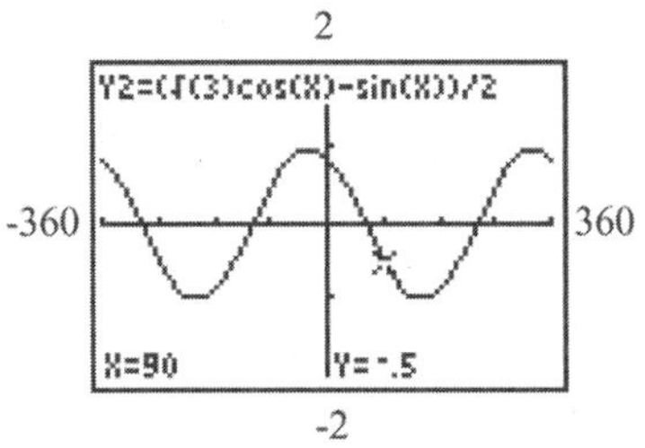

33. $\tan(\alpha\pm\beta)$

$$= \frac{\sin(\alpha\pm\beta)}{\cos(\alpha\pm\beta)} = \frac{\sin\alpha\cos\beta \pm \cos\alpha\sin\beta}{\cos\alpha\cos\beta \mp \sin\alpha\sin\beta}$$

(divide numerator and denominator by $\cos\alpha\cos\beta$)

$$= \frac{\dfrac{\sin\alpha\cos\beta}{\cos\alpha\cos\beta} \pm \dfrac{\cos\alpha\sin\beta}{\cos\alpha\cos\beta}}{\dfrac{\cos\alpha\cos\beta}{\cos\alpha\cos\beta} \mp \dfrac{\sin\alpha\sin\beta}{\cos\alpha\cos\beta}}$$
$$= \frac{\dfrac{\sin\alpha}{\cos\alpha} \pm \dfrac{\sin\beta}{\cos\beta}}{1 \mp \dfrac{\sin\alpha}{\cos\alpha}\times\dfrac{\sin\beta}{\cos\beta}}$$
$$= \frac{\tan\alpha \pm \tan\beta}{1 \mp \tan\alpha\tan\beta}$$

37. $\alpha+\beta=x;\ \alpha-\beta=y;\ \alpha=\frac{1}{2}(x+y);\ \beta=\frac{1}{2}(x-y)$

We will use Eq. (20.14) in its equivalent form

$\sin(\alpha+\beta)+\sin(\alpha-\beta)=2\sin\alpha\cos\beta.$

$$\begin{aligned}\sin x+\sin y &= \sin(\alpha+\beta)+\sin(\alpha-\beta)\\ &= 2\sin\alpha\cos\beta\\ &= 2\sin\frac{1}{2}(x+y)\cos\frac{1}{2}(x-y)\end{aligned}$$

41. $$\begin{aligned}\frac{\sin 2x}{\sin x} &= \frac{\sin(x+x)}{\sin x}\\ &= \frac{\sin x\cos x+\sin x\cos x}{\sin x}\\ &= 2\cos x\end{aligned}$$

45. $$\begin{aligned}V &= 20\sqrt{2}\sin(120\pi t+\pi/4)\\ &= 20\sqrt{2}(\sin 120\pi t\cos\pi/4+\sin\pi/4\cos 120\pi t)\\ &= 20\sqrt{2}\left(\tfrac{1}{\sqrt{2}}\right)(\sin 120\pi t+\cos 120\pi t)\\ &= 20\sin 120\pi t+20\cos 120\pi t\\ &= V_1+V_2\end{aligned}$$

49. $$\begin{aligned}\tan\alpha(R+\cos\beta) &= \sin\beta\\ R\tan\alpha+\tan\alpha\cos\beta &= \sin\beta\\ R\tan\alpha &= \sin\beta-\tan\alpha\cos\beta\\ R &= \frac{\sin\beta-\tan\alpha\cos\beta}{\tan\alpha}\\ &= \frac{\sin\beta-\frac{\sin\alpha}{\cos\alpha}\cos\beta}{\frac{\sin\alpha}{\cos\alpha}}\\ &= \frac{\sin\beta\cos\alpha-\sin\alpha\cos\beta}{\sin\alpha}\\ &= \frac{\sin(\beta-\alpha)}{\sin\alpha}\end{aligned}$$

20.3 Double-angle Formulas

1. If $\alpha=\frac{\pi}{3}$,

$$\begin{aligned}\tan\frac{2\pi}{3} &= \tan\left(2\cdot\frac{\pi}{3}\right)=\frac{2\tan\frac{\pi}{3}}{1-\tan^2\frac{\pi}{3}}\\ &= \frac{2(\sqrt{3})}{1-(\sqrt{3})^2}\\ &= -\sqrt{3}\end{aligned}$$

5. $60^\circ=2(30^\circ);\ \sin 2\alpha=2\sin\alpha\cos\alpha$

$$\begin{aligned}\sin 2(30^\circ) &= 2\sin 30^\circ\cos 30^\circ\\ &= 2\left(\frac{1}{2}\right)\left(\frac{\sqrt{3}}{2}\right)\\ &= \frac{\sqrt{3}}{2}\end{aligned}$$

9. $\sin 258^\circ=-0.978$

$$\begin{aligned}\sin 258^\circ &= \sin 2(129^\circ)\\ &= 2\sin 129^\circ\cos 129^\circ\\ &= -0.978\end{aligned}$$

13. $\tan\frac{2\pi}{5}=3.08$ (calculator in radian mode)

$\tan\frac{2\pi}{5}=\frac{2\tan\frac{\pi}{5}}{1-\tan^2\frac{\pi}{5}}=3.08$

17. $\sin x=0.5$ (QII)

$$\begin{aligned}\cos^2 x &= 1-\sin^2 x\\ &= 1-\frac{1}{4}\\ &= \frac{3}{4}\\ \cos x &= -\frac{\sqrt{3}}{2}\\ \tan x &= \frac{1}{2}\div\frac{-\sqrt{3}}{2}=-\frac{1}{\sqrt{3}}\\ \tan 2x &= \frac{2\tan x}{1-\tan^2 x}\\ &= \frac{2\cdot\frac{-1}{\sqrt{3}}}{1-\frac{1}{3}}\\ &= -\sqrt{3}\end{aligned}$$

21. $$\begin{aligned}1-2\sin^2 4x &= \cos 2(4x)\\ &= \cos 8x\end{aligned}$$

25. $$\begin{aligned}4\sin^2 2x-2 &= -2(1-2\sin^2 2x)\\ &= -2\cos 2(2x)\\ &= -2\cos 4x\end{aligned}$$

29. $\frac{\sin 3x}{\sin x} - \frac{\cos 3x}{\cos x} = \frac{\sin 3x \cos x - \sin x \cos 3x}{\sin x \cos x}$

$= \frac{\sin(3x - x)}{\sin x \cos x}$

$= \frac{\sin 2x}{\sin x \cos x}$

$= \frac{2 \sin x \cos x}{\sin x \cos x}$

$= 2$

33. $\frac{\cos x - \tan x \sin x}{\sec x} = \frac{\cos x - \frac{\sin x \sin x}{\cos x}}{\frac{1}{\cos x}}$

$= \frac{\cos^2 x - \sin^2 x}{\cos x} \times \frac{\cos x}{1}$

$= \cos^2 x - \sin^2 x$

$= \cos 2x$

37. $1 - \cos 2\theta = 1 - (1 - 2\sin^2 \theta)$

$= 2\sin^2 \theta$

$= \frac{2}{\csc^2 \theta}$

$= \frac{2}{1 + \cot^2 \theta}$

41. Both graphs are the same.

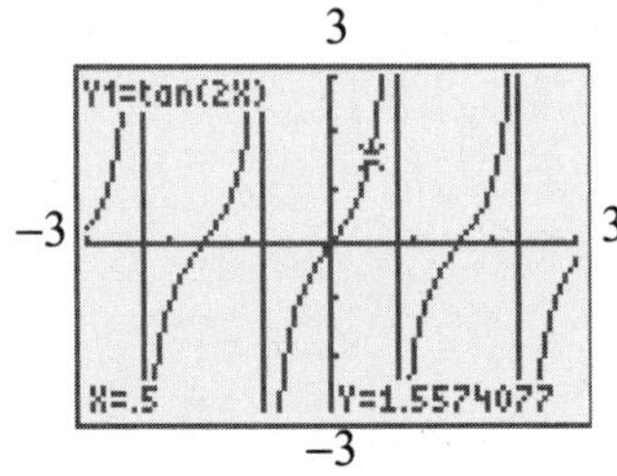

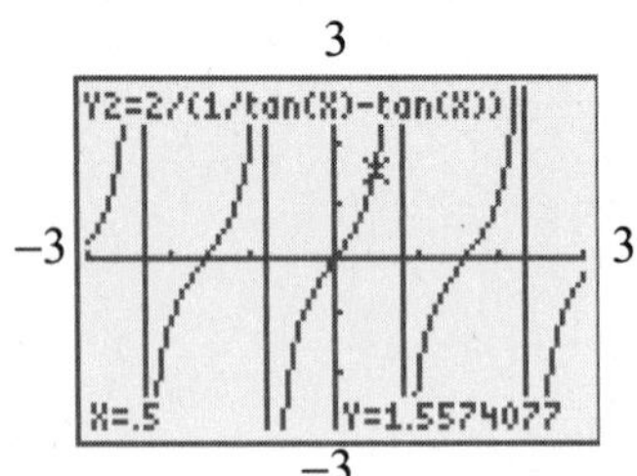

45. $\sin 3x$

$= \sin(2x + x)$

$= \sin 2x \cos x + \cos 2x \sin x$

$= (2 \sin x \cos x)(\cos x) + (\cos^2 x - \sin^2 x)(\sin x)$

$= 2 \sin x \cos^2 x + \sin x \cos^2 x - \sin^3 x$

$= 3 \sin x \cos^2 x - \sin^3 x$

$= 3 \sin x (1 - \sin^2 x) - \sin^3 x$

$= 3 \sin x - 4 \sin^3 x$

49. $\cos 2x + \sin 2x \tan x = \cos^2 x - \sin^2 x + 2 \sin x \cos x \cdot \frac{\sin x}{\cos x}$

$= \cos^2 x - \sin^2 x + 2\sin^2 x$

$= \cos^2 x + \sin^2 x$

$= 1$

53. $y = \sqrt{(\sin x + \cos x)^2}$

$= \sqrt{\sin^2 x + 2 \sin x \cos x + \cos^2 x}$

$y = \sqrt{1 + 2 \sin x \cos x}$ and since $\sqrt{x^2} = |x|$

$y = \sqrt{(\sin x + \cos x)^2} = |\sin x + \cos x|$

Graph $y_1 = \sqrt{(\sin x + \cos x)^2}$, $y_2 = \sqrt{1 + 2 \sin x \cos x}$, $y_3 = |\sin x + \cos x|$

All three graphs are the same.

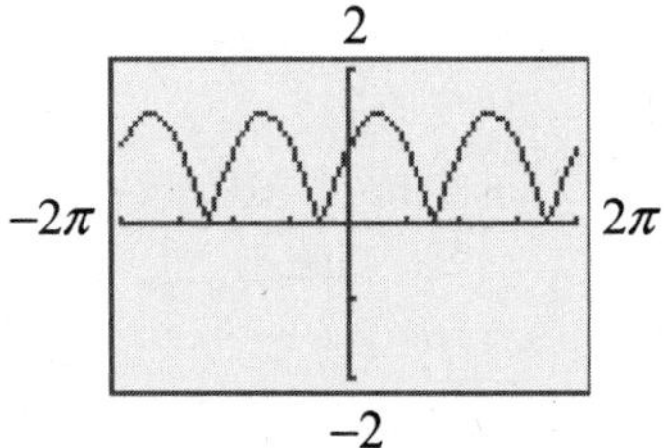

57. $R = vt \cos \alpha$; $t = \frac{(2v \sin \alpha)}{g}$

$R = v\left(\frac{2v \sin \alpha}{g}\right) \cos \alpha$

$= \frac{v^2 (2 \sin \alpha \cos \alpha)}{g}$

$= \frac{v^2 \sin 2\alpha}{g}$

20.4 Half-angle Formulas

1. $\sqrt{\dfrac{1+\cos 114^\circ}{2}} = \cos\dfrac{1}{2}\left(114^\circ\right) = \cos 57^\circ$

$\sqrt{\dfrac{1+\cos 114^\circ}{2}} = 0.544\ 639\ 035$

$\cos 57^\circ = 0.544\ 639\ 035$

5. $\sin 105^\circ = \sin\dfrac{210^\circ}{2}$

$= \sqrt{\dfrac{1-\cos 210^\circ}{2}}$

$= \sqrt{\dfrac{1-\left(-\frac{\sqrt{3}}{2}\right)}{2}}$

$= \dfrac{\sqrt{2+\sqrt{3}}}{2}$

$= 0.966$

9. $\sqrt{\dfrac{1-\cos 236^\circ}{2}} = \sin\dfrac{1}{2}\left(236^\circ\right)$

$= \sin 118^\circ$

$= 0.883$

$\sqrt{\dfrac{1-\cos 236^\circ}{2}} = 0.883$

13. $\sin\dfrac{\alpha}{2} = \sqrt{\dfrac{1-\cos\alpha}{2}}$

$\sqrt{\dfrac{1-\cos 6x}{2}} = \sin\dfrac{6x}{2}$

$= \sin 3x$

17. $\sqrt{4-4\cos 10\theta} = \sqrt{\dfrac{(2)(4)(1-\cos 10\theta)}{2}}$

$= 2\sqrt{2}\sin\dfrac{10\theta}{2}$

$= 2\sqrt{2}\sin 5\theta$

21. $0 < \dfrac{\alpha}{2} < 45^\circ$, so $\sin\dfrac{\alpha}{2} > 0$

$\sin\dfrac{\alpha}{2} = \sqrt{\dfrac{1-\cos\alpha}{2}}$

$= \sqrt{\dfrac{1-\frac{12}{13}}{2}}$

$= \sqrt{\dfrac{1}{26}}$

$= \sqrt{\dfrac{1}{26}\dfrac{26}{26}}$

$= \dfrac{1}{26}\sqrt{26}$

25. $\csc\dfrac{\alpha}{2} = \dfrac{1}{\sin\frac{\alpha}{2}}$

$= \dfrac{1}{\pm\sqrt{\frac{1-\cos\alpha}{2}}}$

$= \pm\sqrt{\dfrac{2}{1-\cos\alpha}}$

$= \pm\sqrt{\dfrac{2}{1-\frac{1}{\sec\alpha}}}$

$= \pm\sqrt{\dfrac{2\sec\alpha}{\sec\alpha-1}}$

29. $\dfrac{1-\cos\alpha}{2\sin\frac{\alpha}{2}} = \dfrac{1-\cos\alpha}{2\sqrt{\frac{1-\cos\alpha}{2}}} \times \dfrac{\sqrt{\frac{1-\cos\alpha}{2}}}{\sqrt{\frac{1-\cos\alpha}{2}}}$

$= \dfrac{(1-\cos\alpha)\sqrt{\frac{1-\cos\alpha}{2}}}{2\left(\frac{1-\cos\alpha}{2}\right)}$

$= \sqrt{\dfrac{1-\cos\alpha}{2}}$

$= \sin\dfrac{\alpha}{2}$

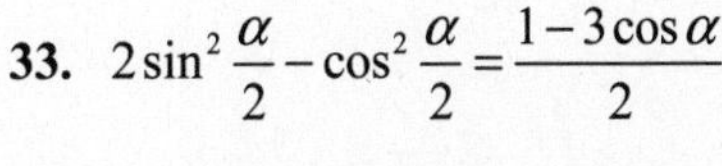

33. $2\sin^2\dfrac{\alpha}{2} - \cos^2\dfrac{\alpha}{2} = \dfrac{1-3\cos\alpha}{2}$

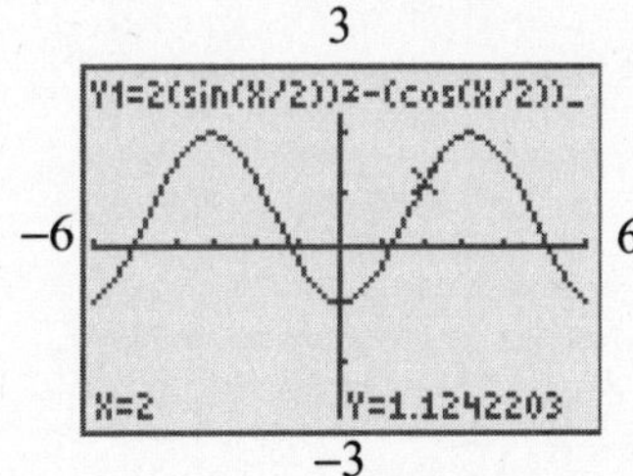

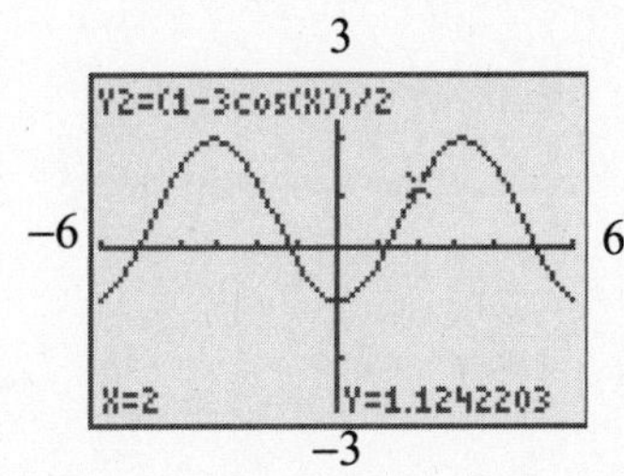

37. $\sin\frac{\theta}{2}=\sqrt{\frac{1-\cos\theta}{2}}$

$=\frac{3}{5}$

$1-\cos\theta=2\cdot\frac{9}{25}$

$\cos\theta=1-\frac{18}{25}$

$=\frac{7}{25}$

$\sin\theta=\pm\sqrt{1-\frac{49}{625}}$

$=\pm\frac{24}{25}$

$\tan\theta=\pm\frac{24}{7}$

41. $\sin^2\omega t=\left(\sqrt{\frac{1-\cos 2\omega t}{2}}\right)^2$

$=\frac{1-\cos 2\omega t}{2}$

20.5 Solving Trigonometric Equations

1. $\tan\theta-1=0,\ 0\le\theta<2\pi$

$\tan\theta=1$

$\theta=\tan^{-1}1$

$\theta=\frac{\pi}{4},\frac{5\pi}{4}$

5. $\sin x-1=0,\ 0\le x<2\pi;$

$\sin x=1$

$x=\frac{\pi}{2}$

9. $4\cos^2 x-1=0;\ 0\le x<2\pi$

$4\cos^2 x=1$

$\cos^2 x=\frac{1}{4}$

$\cos x=\pm\frac{1}{2}$

$x=\frac{\pi}{3},\frac{2\pi}{3},\frac{4\pi}{3},\frac{5\pi}{3}$

13. $\sin 2x\sin x+\cos x=0,\ 0\le x<2\pi$

$(2\sin x\cos x)(\sin x)+\cos x=0$

$2\sin^2 x\cos x+\cos x=0$

$\cos x(2\sin^2 x+1)=0$

$\cos x=0$ or $2\sin^2 x+1=0$

$x=\frac{\pi}{2},\frac{3\pi}{2}$ $\quad$ $2\sin^2 x=-1$

no real solution

17. $4\tan x-\sec^2 x=0$

$4\tan x-(1+\tan^2 x)=0$

$4\tan x-1-\tan^2 x=0$

$\tan^2 x-4\tan x+1=0$

Let $y=\tan x$

$y^2-4y+1=0$

$y=\frac{4\pm\sqrt{16-4(1)(1)}}{2}$

$=2\pm\sqrt{3}$

$\tan x=2+\sqrt{3}$ or $\tan x=2-\sqrt{3}$

$x=1.31, 4.45$ $\quad$ $x=0.260, 3.40$

21. $\tan x+1=0$

$\tan x=-1$

$x_{\text{ref}}=\frac{\pi}{4}$

(tan negative QII, QIV)

$x=\pi-\frac{\pi}{4}=\frac{3\pi}{4}$

$x=2\pi-\frac{\pi}{4}=\frac{7\pi}{4}$

25. $4\sin^2 x-3=0$

$4\sin^2 x=3$

$\sin^2 x=\frac{3}{4};$

$\sin x=\pm\frac{\sqrt{3}}{2}$

$x_{\text{ref}}=\frac{\pi}{3}$

(sin positive or negative—all quadrants)

$x=\frac{\pi}{3},\frac{2\pi}{3},\frac{4\pi}{3},\frac{5\pi}{3}$

29. $2\sin x - \tan x = 0$

$$2\sin x - \frac{\sin x}{\cos x} = 0$$

$$\sin x\left(2 - \frac{1}{\cos x}\right) = 0$$

$$\sin x = 0 \quad \text{or} \quad 2 - \frac{1}{\cos x} = 0$$

$$x = 0, \pi \qquad \cos x = \frac{1}{2}$$

$$x = \frac{\pi}{3}, \frac{5\pi}{3}$$

33. $\tan x + 3\cot x = 4$

$$\tan x + \frac{3}{\tan x} = 4$$

$$\tan^2 x + 3 = 4\tan x$$

$$\tan^2 x - 4\tan x + 3 = 0;$$

$$(\tan x - 1)(\tan x - 3) = 0$$

$$\tan x = 1 \qquad \tan x = 3$$

$$x = \frac{\pi}{4}, \frac{5\pi}{4} \qquad x = 1.25, \pi + 1.25$$

37. $2\sin 2x - \cos x \sin^3 x = 0$

$$2(2\sin x \cos x) - \cos x \sin^3 x = 0$$

$$\sin x \cos x\left(4 - \sin^2 x\right) = 0$$

$$\sin x \cos x = 0 \qquad 4 - \sin^2 x = 0$$

$$\sin x = 0 \quad \cos x = 0 \qquad \sin x = \pm 2$$

$$x = 0, \pi \qquad x = \frac{\pi}{2}, \frac{3\pi}{2} \qquad \text{no solution}$$

41. $\sin\theta + \cos\theta + \tan\theta + \cot\theta + \sec\theta + \csc\theta = f(\theta) = 1$

θ is a positive acute angle $\Rightarrow 0 < \theta < \frac{\pi}{2}$ for which $0 < \sin\theta < 1$, $0 < \cos\theta < 1$, $\tan\theta > 0$, $\cot\theta > 0$, $\sec\theta > 1$, $\csc\theta > 1$ which implies $f(\theta) = \sin\theta + \cos\theta + \tan\theta + \cot\theta + \sec\theta + \csc\theta > 1 \neq 1$.

$f(\theta) = 1$ has no solution.

45. $g = 9.8000 = 9.7805\left(1 + 0.0053\sin^2\theta\right)$

$$\sin^2\theta = \frac{\frac{9.8000}{9.7805} - 1}{0.0053}$$

$$\sin\theta = \sqrt{\frac{\frac{9.8000}{9.7805} - 1}{0.0053}}$$

$$\theta = 37.8^\circ$$

49. $\dfrac{p^2\tan\theta}{0.0063 + p\tan\theta} = 1.6;\ p = 4.8; 0 < \theta < \dfrac{\pi}{2}$

$$\frac{4.8^2\tan\theta}{0.0063 + 4.8\tan\theta} = 1.6$$

$$23.04\tan\theta = 1.6(0.0063 + 4.8\tan\theta)$$

$$23.04\tan\theta = 0.01008 + 7.68\tan\theta$$

$$15.36\tan\theta = 0.01008$$

$$\tan\theta = 6.5625 \times 10^{-4}$$

$$\theta = 6.6 \times 10^{-4} \text{ rad or}$$

$$\theta = 0.038^\circ$$

53. $3\sin x - x = 0$. Graph $y_1 = 3\sin x - x$.

Use the Zero feature to solve.

$x = -2.28, 0.00, 2.28.$

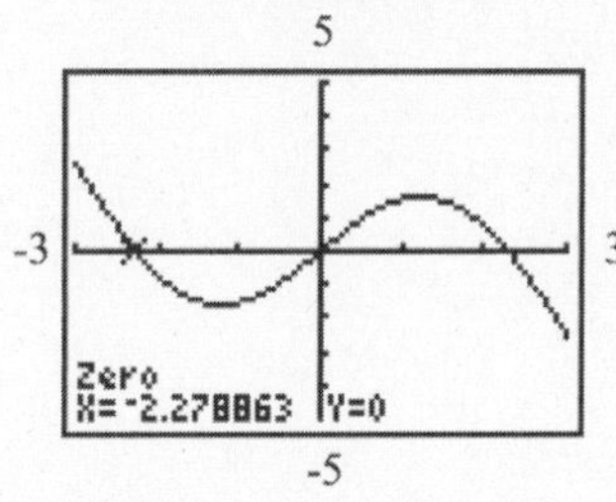

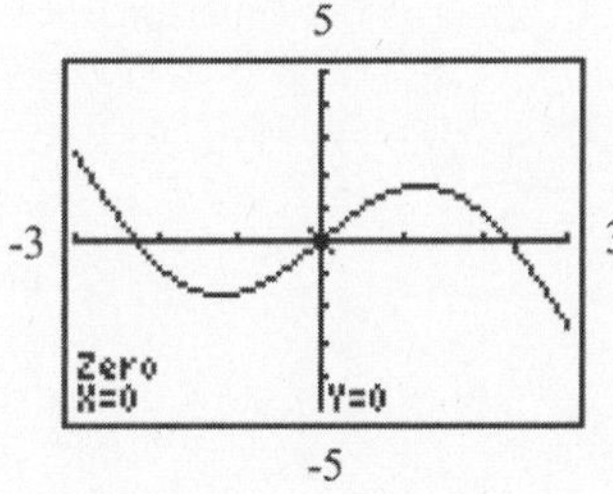

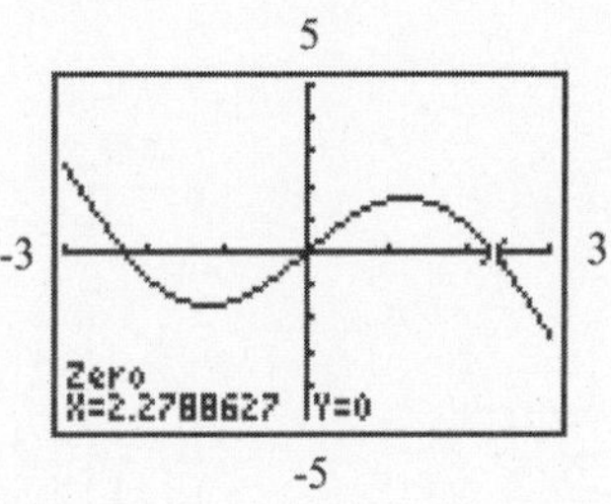

57. $2\ln x = 1 - \cos 2x;\ 2\ln x - 1 + \cos 2x = 0$

Graph $y_1 = 2\ln x - 1 + \cos 2x$. Use the Zero feature to solve. $x = 2.10$.

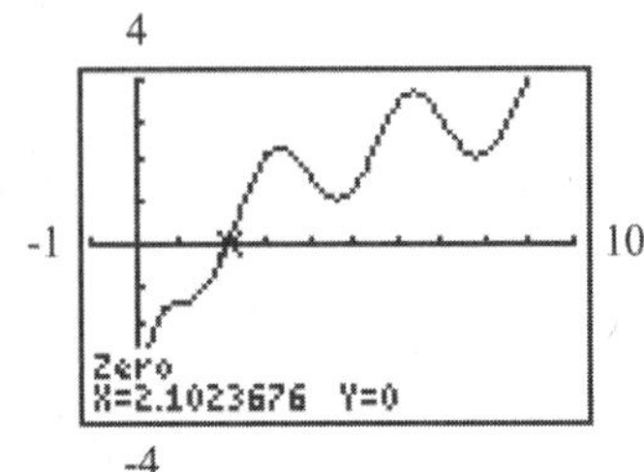

20.6 The Inverse Trigonometric Functions

1. $y = \tan^{-1} 3A$ is read as "y is the angle whose tangent is $3A$." In this case, $3A = \tan y$.

5. y is the angle whose cotangent is $3x$.

9. y is five times the angle whose cosine is $2x - 1$.

13. $\tan^{-1} 1 = \dfrac{\pi}{4}$ since $\tan \dfrac{\pi}{4} = 1$ and $-\dfrac{\pi}{2} < \dfrac{\pi}{4} < \dfrac{\pi}{2}$

17. Let $\sec^{-1} 0.5 = x$, then

$$\sec x = 0.5$$
$$\frac{1}{\cos x} = 0.5$$
$$\cos x = 2$$

and since $-1 \le \cos x \le 1$ there is no value for x.

21. $\sin\left(\tan^{-1}\sqrt{3}\right) = \sin\dfrac{\pi}{3} = \dfrac{1}{2}\sqrt{3}$

2 $\sqrt{3}$ 1

25. $\cos\left[\tan^{-1}(-5)\right] = \dfrac{1}{\sqrt{26}}$

$\tan^{-1}(-5)$ 1 $\sqrt{26}$ -5

29. $\tan^{-1} x = \sin^{-1}\dfrac{2}{5}$

$$x = \tan\left(\sin^{-1}\frac{2}{5}\right)$$
$$x = \frac{2}{\sqrt{21}}$$

5 2 $\sqrt{21}$

33. $\tan^{-1}(-3.7321) = -1.3090$

Note $-\dfrac{\pi}{2} < -1.3090 < \dfrac{\pi}{2}$

37. $\tan\left[\cos^{-1}(-0.6281)\right] = \tan 2.250 = -1.239$

41.
$$y = \sin 3x$$
$$3x = \sin^{-1} y$$
$$x = \frac{1}{3}\sin^{-1} y$$

45.
$$1 - y = \cos^{-1}(1 - x)$$
$$\cos(1 - y) = 1 - x$$
$$x = 1 - \cos(1 - y)$$

49. Let $\alpha - \sin^{-1} x$, $\beta = \cos^{-1} y$

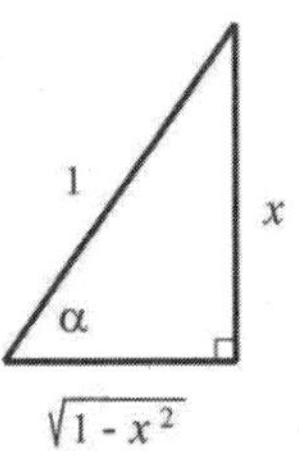

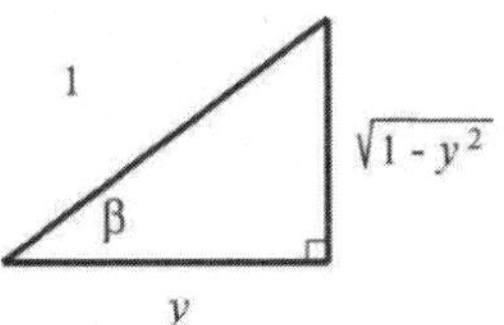

$$\begin{aligned}\sin\left(\sin^{-1} x + \cos^{-1} y\right) &= \sin(\alpha + \beta) \\ &= \sin\alpha\cos\beta + \sin\beta\cos\alpha \\ &= xy + \sqrt{1 - y^2}\sqrt{1 - x^2} \\ &= xy + \sqrt{1 - y^2 - x^2 + x^2y^2}\end{aligned}$$

53. $\sin\left(2\sin^{-1} x\right) = \sin 2\theta = 2\sin\theta\cos\theta$

$$= 2\left(\frac{x}{1}\right)\left(\frac{\sqrt{1-x^2}}{1}\right)$$

$$= 2x\sqrt{1-x^2}$$

In a triangle, θ is set up such that its sine is x. This gives an opposite side of x, hypotenuse 1, and adjacent side $\sqrt{1-x^2}$.

57. $y = A\cos 2(\omega t + \phi)$

$$\frac{y}{A} = \cos 2(\omega t + \phi)$$

$$\cos^{-1}\frac{y}{A} = 2(\omega t + \phi) = 2\omega t + 2\phi$$

$$\cos^{-1}\frac{y}{A} - 2\phi = 2\omega t$$

$$\frac{\cos^{-1}\frac{y}{A} - 2\phi}{2\omega} = t$$

$$t = \frac{1}{2\omega}\cos^{-1}\frac{y}{A} - \frac{\phi}{\omega}$$

61. Let $\alpha = \sin^{-1}\frac{3}{5}$ and $\beta = \sin^{-1}\frac{5}{13}$;

$$\sin\alpha = \frac{3}{5}$$

$$\cos\alpha = \sqrt{1-\tfrac{9}{25}} = \frac{4}{5}$$

$$\sin\beta = \frac{5}{13}$$

$$\cos\beta = \sqrt{1-\tfrac{25}{169}} = \frac{12}{13}$$

$$\sin^{-1}\frac{3}{5} + \sin^{-1}\frac{5}{13} = \alpha + \beta$$

$$\sin(\alpha+\beta) = \sin\alpha\cos\beta + \cos\alpha\sin\beta$$

$$= \frac{3}{5}\left(\frac{12}{13}\right) + \frac{4}{5}\left(\frac{5}{13}\right)$$

$$= \frac{36}{65} + \frac{20}{65} = \frac{56}{65}$$

Since $\sin(\alpha+\beta) = \frac{56}{65}$,

$$\alpha + \beta = \sin^{-1}\frac{56}{65}.$$

65. $y = \sin^{-1} x + \sin^{-1}(-x)$

$$\sin y = \sin\left(\sin^{-1} x + \sin^{-1}(-x)\right)$$

$$\sin y = \sin\left(\sin^{-1} x\right)\cos\left(\sin^{-1}(-x)\right) + \cos\left(\sin^{-1} x\right)\sin\left(\sin^{-1}(-x)\right)$$

$$= x\cos\left(\sin^{-1} x\right) - x\cos\left(\sin^{-1} x\right)$$

$$= 0$$

Here we have used the fact that if $-\frac{\pi}{2} \le x \le \frac{\pi}{2}$, $\cos\left(\sin^{-1}(-x)\right) = \cos\left(\sin^{-1} x\right)$.

69. $\tan B = \frac{y}{b} \Rightarrow y = b\tan B$

$$\tan A = \frac{y}{a} = \frac{b\tan B}{a}$$

$$A = \tan^{-1}\frac{b\tan B}{a}$$

73. $\tan\alpha = \frac{y}{x} \Rightarrow a = \tan^{-1}\left(\frac{y}{x}\right)$

$$\tan(\alpha+\theta) = \frac{y+50}{x} \Rightarrow \alpha+\theta = \tan^{-1}\left(\frac{y+50}{x}\right)$$

$$\tan^{-1}\left(\frac{y}{x}\right) + \theta = \tan^{-1}\left(\frac{y+50}{x}\right)$$

$$\theta = \tan^{-1}\left(\frac{y+50}{x}\right) - \tan^{-1}\left(\frac{y}{x}\right)$$

Review Exercises

1. $\sin 120^\circ = \sin\left(90^\circ + 30^\circ\right)$

$$= \sin 90^\circ \cos 30^\circ + \cos 90^\circ \sin 30^\circ$$

$$= 1\left(\frac{\sqrt{3}}{2}\right) + 0\left(\frac{1}{2}\right)$$

$$= \frac{\sqrt{3}}{2}$$

5. $\cos\pi = \cos 2\left(\frac{\pi}{2}\right)$

$= \cos^2\frac{\pi}{2} - \sin^2\frac{\pi}{2}$

$= 0 - (1)^2$

$= -1$

9. $\sin 14^\circ \cos 38^\circ + \cos 14^\circ \sin 38^\circ$

$= \sin\left(14^\circ + 38^\circ\right)$

$= \sin 52^\circ$

$0.788 = 0.788$

13. $\cos 73^\circ \cos\left(-142^\circ\right) + \sin 73^\circ \sin\left(-142^\circ\right)$

$= \cos 73^\circ \cos 142^\circ - \sin 73^\circ \sin 142^\circ$

$= \cos\left(73^\circ + 142^\circ\right)$

$= \cos 215^\circ$

$-0.819 = -0.819$

17. $\sin\alpha\cos\beta + \cos\alpha\sin\beta$

$= \sin(\alpha+\beta)$ where $\alpha = 2x,\ \beta = 3x$

$\sin 2x\cos 3x + \cos 2x\sin 3x$

$= \sin(2x+3x)$

$= \sin 5x$

21. $2 - 4\sin^2 6x = 2\left(1 - 2\sin^2 6x\right)$

$= 2\cos 12x$

25. $\sin^{-1}(-1) = -\frac{\pi}{2}$ since $\sin\left(-\frac{\pi}{2}\right) = -1$ and

$-\frac{\pi}{2} \le -\frac{\pi}{2} \le \frac{\pi}{2}$

29. $\tan\left[\sin^{-1}(-0.5)\right] = \tan\left(-\frac{\pi}{6}\right)$

$= -\frac{\sqrt{3}}{3}$

33. $\frac{\sec y}{\cos y} - \frac{\tan y}{\cot y} = \sec^2 y - \tan^2 y$

$= 1 + \tan^2 y - \tan^2 y$

$= 1$

37. $\frac{\sec^4 x - 1}{\tan^2 x} = \frac{\left(\sec^2 x - 1\right)\left(\sec^2 x + 1\right)}{\tan^2 x}$

$= \frac{\left(\sec^2 x + 1\right)\left(\tan^2 x\right)}{\tan^2 x}$

$= \sec^2 x + 1$

$= 1 + \tan^2 x + 1$

$= 2 + \tan^2 x$

41. $\frac{1-\sin^2\theta}{1-\cos^2\theta} = \frac{\cos^2\theta}{\sin^2\theta}$ since $\sin^2\theta + \cos^2\theta = 1$

$= \left(\frac{\cos\theta}{\sin\theta}\right)^2$

$= (\cot\theta)^2$

$= \cot^2\theta$

45. $\frac{\sec x}{\sin x} - \sec x\sin x = \frac{1}{\cos x\sin x} - \frac{\sin x}{\cos x}\cdot\frac{\sin x}{\sin x}$

$= \frac{1-\sin^2 x}{\sin x\cos x}$

$= \frac{\cos^2 x}{\sin x\cos x}$

$= \frac{\cos x}{\sin x}$

$= \cot x$

49. $\frac{\sin x\cot x + \cos x}{2\cot x} = \frac{\sin x}{2} + \frac{\cos x}{2\frac{\cos x}{\sin x}}$

$= \frac{\sin x}{2} + \frac{\sin x}{2}$

$= \sin x$

53.

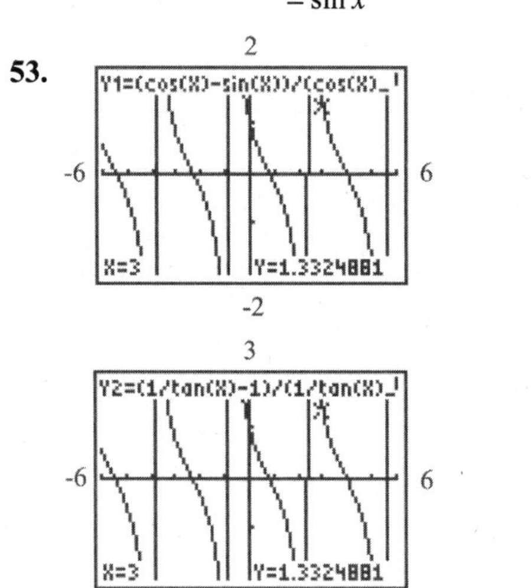

57.

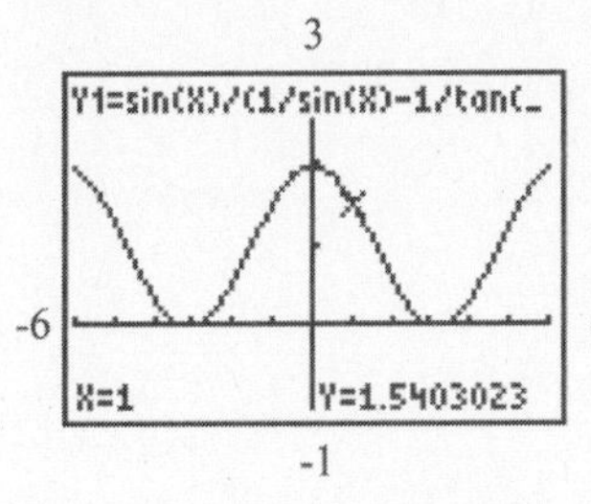

3
Y2=1+cos(X)
-6
6
X=1
Y=1.5403023
-1

61. $y = 2\cos 2x$

$\frac{y}{2} = \cos 2x$

$\cos^{-1}\frac{y}{2} = 2x$

$x = \frac{1}{2}\cos^{-1}\frac{1}{2}y$

65. $3(\tan x - 2) = 1 + \tan x$

$3\tan x - 6 = 1 + \tan x$

$2\tan x = 7$

$\tan x = \frac{7}{2}$

$x = \tan^{-1}\frac{7}{2}$

$= 1.29$

Since $\tan x$ is positive in QIII also, $\pi + 1.29 = 4.43$ is also a value for x that is within the specified range of values for x.

69. $2\sin^2\theta + 3\cos\theta - 3 = 0$

$2(1 - \cos^2\theta) + 3\cos\theta - 3 = 0$

$2 - 2\cos^2\theta + 3\cos\theta - 3 = 0$

$2\cos^2\theta - 3\cos\theta + 1 = 0$

$(\cos\theta - 1)(2\cos\theta - 1) = 0$

$\cos\theta - 1 = 0$ or $2\cos\theta - 1 = 0$

$\theta = 0$ $\quad\cos\theta = \frac{1}{2}$

$\theta = \frac{\pi}{3}, \frac{5\pi}{3}$

73. $\sin 2x = \cos 3x$

$\sin 2x = \cos(2x + x)$

$2\sin x\cos x = \cos 2x\cos x - \sin 2x\sin x$

$2\sin x\cos x = \cos 2x\cos x - 2\sin x\cos x\sin x$

$2\sin x\cos x - \cos 2x\cos x + 2\sin^2 x\cos x = 0$

$\cos x(2\sin x - \cos 2x + \sin^2 x) = 0$

$\cos x = 0 \Rightarrow x = \frac{\pi}{2}, \frac{3\pi}{2}$ or

$2\sin x - (1 - 2\sin^2 x) + 2\sin^2 x = 0$

$4\sin^2 x + 2\sin x - 1 = 0$

$\sin x = \frac{-2 \pm \sqrt{2^2 - 4(4)(-1)}}{2(4)} = \frac{-1 \pm \sqrt{5}}{4}$

$\sin x = \frac{-1 + \sqrt{5}}{4}$ or $\sin x = \frac{-1 - \sqrt{5}}{4}$

$x = \frac{\pi}{10}, \frac{9\pi}{10}$ $\quad x = \frac{13\pi}{10}, \frac{17\pi}{10}$

$\sin 2x = \cos 3x$ has solutions

$\left\{\frac{\pi}{10}, \frac{\pi}{2}, \frac{9\pi}{10}, \frac{13\pi}{10}, \frac{3\pi}{2}, \frac{17\pi}{10}\right\}$

77. $\tan x + \cot x = \frac{\sin x}{\cos x} + \frac{\cos x}{\sin x}$

$= \frac{\sin^2 x + \cos^2 x}{\sin x\cos x}$

$= \frac{1}{\sin x}\frac{1}{\cos x}$

$= \csc x\sec x$, identity

81. $x + \ln x - 3\cos^2 x = 2$

$x + \ln x - 3\cos^2 x - 2 = 0$

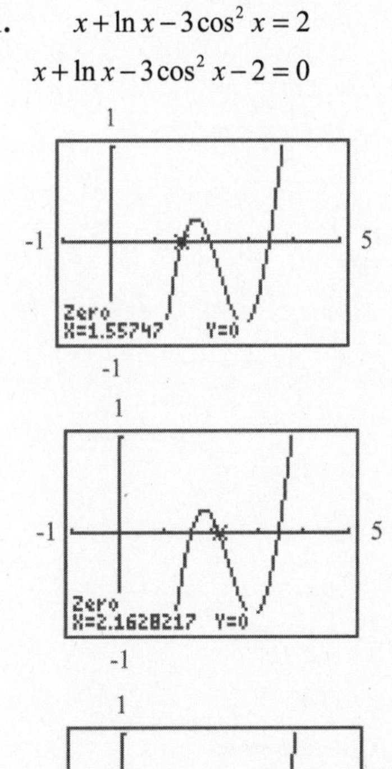

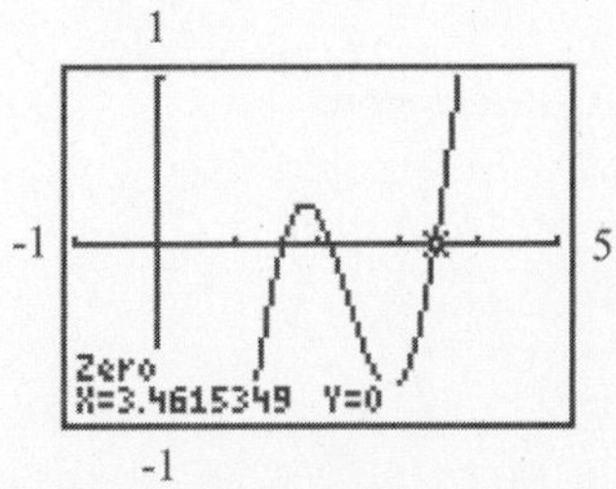

85. $\tan\left(\cot^{-1}x\right) = \tan\theta = \frac{1}{x}$

$$\theta = \cot^{-1}x$$

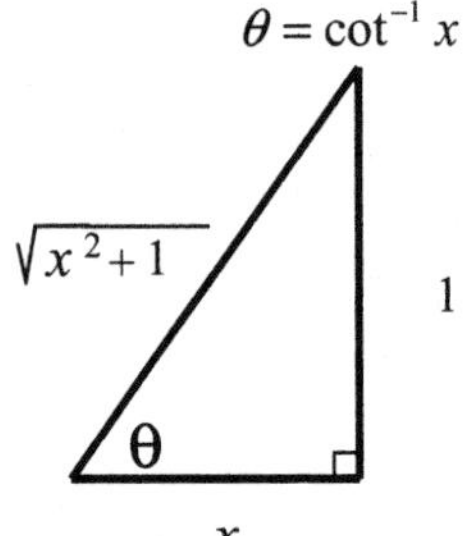

89.

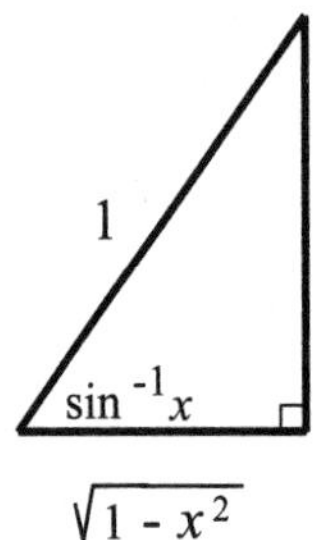

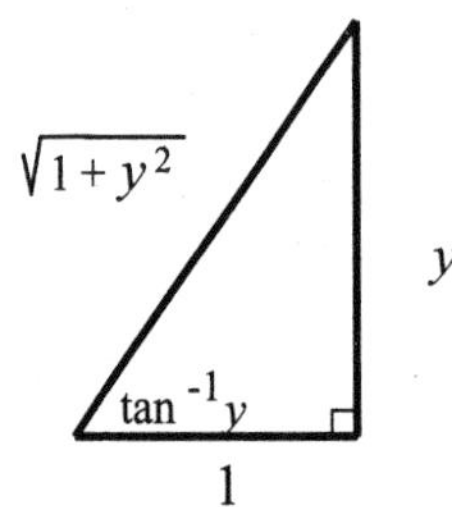

$$\begin{aligned}\cos\left(\sin^{-1}x + \tan^{-1}y\right) &= \cos\left(\sin^{-1}x\right)\cos\left(\tan^{-1}y\right) \\ &\quad - \sin\left(\sin^{-1}x\right)\sin\left(\tan^{-1}y\right) \\ &= \sqrt{1-x^2}\frac{1}{\sqrt{1+y^2}} - x\frac{y}{\sqrt{1+y^2}} \\ &= \frac{\sqrt{1-x^2} - xy}{\sqrt{1+y^2}}\end{aligned}$$

93. $\dfrac{x}{\sqrt{1+x^2}} = \dfrac{\tan\theta}{\sqrt{1+\tan^2\theta}}$

$$\begin{aligned}&= \frac{\tan\theta}{\sqrt{\sec^2\theta}} \\ &= \frac{\tan\theta}{\sec\theta} \\ &= \frac{\sin\theta}{\cos\theta}\cdot\cos\theta \\ &= \sin\theta\end{aligned}$$

97. $\sin 2x > 2\cos x, \quad 0 \le x < 2\pi$

$2\sin x\cos x - 2\cos x > 0$

$\cos x(\sin x - 1) > 0$, there are two cases

I $\quad \cos x > 0$ and $\sin x - 1 > 0$

$0 < x < \frac{\pi}{2} \qquad \sin x > 1$, no solution

$\frac{3\pi}{2} < x < 2\pi$

II $\quad \cos x < 0$ and $\sin x - 1 < 0$

$\frac{\pi}{2} < x < \frac{3\pi}{2} \qquad \sin x < 1$

$0 < x < \frac{\pi}{2}$ or $\frac{\pi}{2} < x < 2\pi$

The solution is $\frac{\pi}{2} < x < \frac{3\pi}{2}$

101. $R = \sqrt{Rx^2 + Ry^2}$

$$\begin{aligned}&= \sqrt{\left(A\cos\theta - B\sin\theta\right)^2 + \left(A\sin\theta + B\cos\theta\right)^2} \\ &= \sqrt{A^2\cos^2\theta + A^2\sin^2\theta + B^2\cos^2\theta + B^2\sin^2\theta} \\ &= \sqrt{A^2\left(\cos^2\theta + \sin^2\theta\right) + B^2\left(\cos^2\theta + \sin^2\theta\right)} \\ &= \sqrt{\left(\cos^2\theta + \sin^2\theta\right)\left(A^2 + B^2\right)} \\ &= \sqrt{\left(A^2 + B^2\right)}\end{aligned}$$

105.

$$\begin{aligned}\omega t &= \sin^{-1}\frac{\theta - \alpha}{R} \\ \sin\left(\omega t\right) &= \frac{\theta - \alpha}{R} \\ R\sin\left(\omega t\right) &= \theta - \alpha \\ \theta &= R\sin\left(\omega t\right) + \alpha\end{aligned}$$

109. $P = VI\cos\phi\cos^2\omega t - VI\sin\phi\cos\omega t\sin\omega t$

$P = VI\cos\omega t\left(\cos\phi\cos\omega t - \sin\phi\sin\omega t\right)$

$P = VI\cos\omega t\left[\cos\left(\omega t + \phi\right)\right]$

113.

$$\begin{aligned}2\sin\theta\cos^2\theta - \sin^3\theta &= 0, \quad 0 < \theta < 90^\circ \\ \sin\theta\left(2\cos^2\theta - \sin^2\theta\right) &= 0 \\ \sin\theta\left(2\cos^2\theta - \left(1 - \cos^2\theta\right)\right) &= 0 \\ \sin\theta\left(3\cos^2\theta - 1\right) &= 0\end{aligned}$$

$\sin\theta = 0 \quad$ or $\quad \cos^2\theta = \frac{1}{3}$

no solution $\qquad \cos\theta = \sqrt{\frac{1}{3}}$

$\theta = 54.7^\circ$

CHAPTER 21

PLANE ANALYTIC GEOMETRY

21.1 Basic Definitions

1. The distance between $(3, -1)$ and $(-2, 5)$ is

$$d=\sqrt{(3-(-2))^2+(-1-5)^2}$$
$$=\sqrt{61}$$

5. Given $(x_1, y_1)=(3, 8); (x_2, y_2)=(-1, -2)$

$$d=\sqrt{(x_2-x_1)^2+(y_2-y_1)^2}$$
$$=\sqrt{(-1-3)^2+(-2-8)^2}$$
$$=\sqrt{(-4)^2+(-10)^2}$$
$$=\sqrt{16+100}$$
$$=\sqrt{116}$$
$$=\sqrt{4\times 29}$$
$$=2\sqrt{29}$$

9. Given $(x_1, y_1)=(-12, 20); (x_2, y_2)=(32, -13)$

$$d=\sqrt{(x_2-x_1)^2+(y_2-y_1)^2}$$
$$=\sqrt{(32+12)^2+(-13-20)^2}$$
$$=\sqrt{(44)^2+(-33)^2}$$
$$=\sqrt{1936+1089}$$
$$=\sqrt{3025}$$
$$=55$$

13. Given $(x_1, y_1)=(1.22, -3.45);$

$(x_2, y_2)=(-1.07, -5.16)$

$$d=\sqrt{(x_2-x_1)^2+(y_2-y_1)^2}$$
$$=\sqrt{(-1.07-1.22)^2+(-5.16-(-3.45))^2}$$
$$=\sqrt{(-2.29)^2+(-5.16+3.45)^2}$$
$$=\sqrt{(-2.29)^2+(-1.71)^2}$$
$$=\sqrt{8.1682}$$
$$=2.86$$

17. Given $(x_1, y_1)=(4, -5); (x_2, y_2)=(4, -8)$

$$m=\frac{y_2-y_1}{x_2-x_1}=\frac{-8-(-5)}{4-4}$$

Since $x_2-x_1=4-4=0$, the slope is undefined.

21. Given $(x_1, y_1)=\left(\sqrt{32}, -\sqrt{18}\right);$

$(x_2, y_2)=\left(-\sqrt{50}, \sqrt{8}\right)$

$$m=\frac{y_2-y_1}{x_2-x_1}=\frac{\sqrt{8}-\left(-\sqrt{18}\right)}{-\sqrt{50}-\sqrt{32}}$$
$$=\frac{2\sqrt{2}+3\sqrt{2}}{-5\sqrt{2}-4\sqrt{2}}$$
$$=\frac{5\sqrt{2}}{-9\sqrt{2}}$$
$$=-\frac{5}{9}$$

25. Given $\alpha=30^\circ$; $m=\tan\alpha$, $0^\circ<\alpha<180^\circ$

$$\tan 30^\circ=\frac{\sqrt{3}}{3}$$

29. Given $m=0.364$; $m=\tan\alpha$; $0.364=\tan\alpha$;

$\alpha=20.0^\circ$

33. Given $(x_1, y_1)=(6, -1); (x_2, y_2)=(4, 3)$

$(x_3, y_3)=(-5, 2); (x_4, y_4)=(-7, 6)$

$$m_1=\frac{y_2-y_1}{x_2-x_1}=\frac{3-(-1)}{4-6}=\frac{4}{-2}=-2$$
$$m_2=\frac{y_4-y_3}{x_4-x_3}=\frac{6-2}{-7-(-5)}=\frac{4}{-2}=-2$$

$m_1=m_2$ so the lines are parallel.

37. Given distance between $(-1, 3)$ and $(11, k)$ is 13.

$$d=\sqrt{(x_1-x_2)^2+(y_1-y_2)^2}$$
$$13=\sqrt{(-1-11)^2+(3-k)^2}$$
$$=\sqrt{(-12)^2+(3-k)^2}$$

$$169 = 144 + (3-k)^2$$
$$(3-k)^2 = 25$$
$$3-k = \pm 5$$
$$-k = -3 \pm 5$$
$$k = -2, 8$$

41. $d_1 = \sqrt{(9-7)^2 + [4-(-2)]^2}$
$= \sqrt{2^2 + 6^2} = \sqrt{40} = 2\sqrt{10}$
$d_2 = \sqrt{(9-3)^2 + (4-2)^2} = \sqrt{6^2 + 2^2}$
$= \sqrt{40} = 2\sqrt{10}$
$d_1 = d_2$ so the triangle is isosceles.

45. $d_1 = \sqrt{(3-5)^2 + (-1-3)^2} = \sqrt{(-2)^2 + (-4)^2}$
$= \sqrt{4+16} = \sqrt{20}$
$m_1 = \frac{y-y_1}{x-x_1} = \frac{5-3}{3-(-1)} = \frac{5-3}{3+1} = \frac{2}{4} = \frac{1}{2}$
$d_2 = \sqrt{(5-1)^2 + (3-5)^2} = \sqrt{(4)^2 + (-2)^2}$
$= \sqrt{16+4} = \sqrt{20}$
$m_2 = \frac{y-y_1}{x-x_1} = \frac{5-1}{3-5} = \frac{4}{-2} = -2$
$m_1 = \frac{-1}{m_2},\ m_1 \perp m_2$
$A = \frac{1}{2}d_1 d_2 = \frac{1}{2}\sqrt{20}\sqrt{20} = \frac{1}{2}(20) = 10$

49. $\left(\frac{-4+6}{2}, \frac{9+1}{2}\right) = \left(\frac{2}{2}, \frac{10}{2}\right) = (1, 5)$

53. The distance between (x, y) and $(0, 0) = 3$.
$\sqrt{(x-0)^2 + (y-0)^2} = 3$
$x^2 + y^2 = 9$

57. $m = \frac{y_2 - y_1}{x_2 - x_1} = \frac{5-0}{-2-x} = 3$
$x = -\frac{11}{3}$

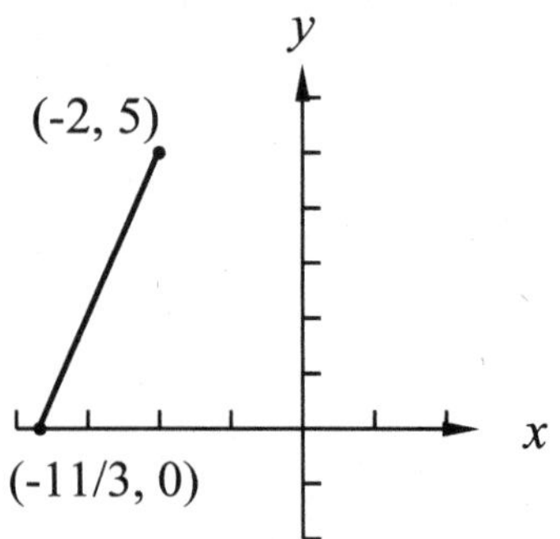

61. The three sides are equal, so
$a^2 + b^2 = x^2 = (a-x)^2 + b^2$
$a^2 + b^2 = a^2 - 2ax + x^2 + b^2$
$x - 2a = 0 \quad \text{or} \quad x = 0$
$x = 2a$
$a^2 + b^2 = 4a^2$
$a = \frac{b}{\sqrt{3}}$
$m_1 = \frac{b-0}{\frac{b}{\sqrt{3}} - 0} = \sqrt{3}$
$m_2 = \frac{b-0}{\frac{b}{\sqrt{3}} - \frac{2b}{\sqrt{3}}} = -\sqrt{3}$

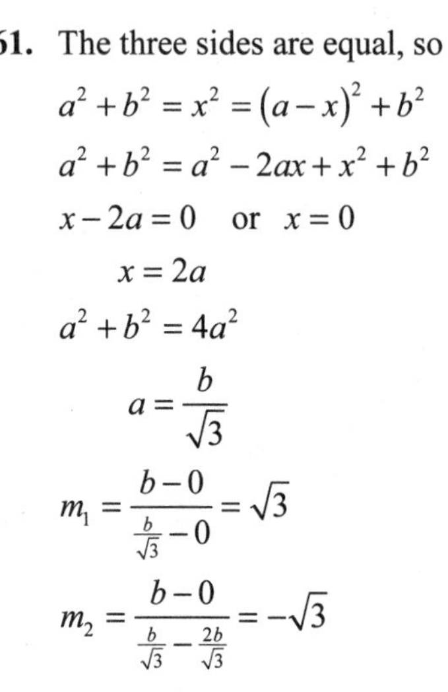

21.2 The Straight Line

1. $m = -1/2, (x_1, y_1) = (4, -1)$
$y - y_1 = m(x - x_1)$
$y - (-1) = -\frac{1}{2}(x-4)$
$y + 1 = -\frac{1}{2}x + 2$
$2y + 2 = -x + 4$
$2y - x - 2 = 0$

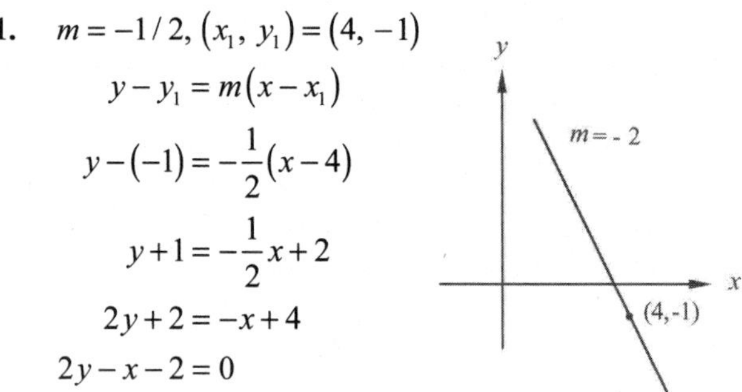

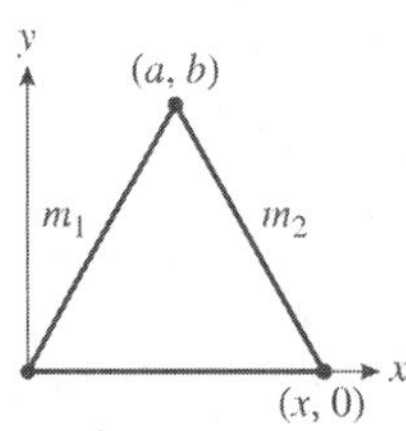

5. Given $m = 4; (x_1, y_1) = (-3, 8)$

$$y - y_1 = m(x - x_1)$$
$$y - 8 = 4[x - (-3)]$$
$$= 4(x+3)$$
$$= 4x + 12$$
$$y = 4x + 20 \text{ or } 4x - y + 20 = 0$$

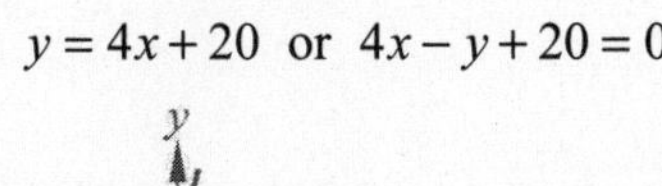

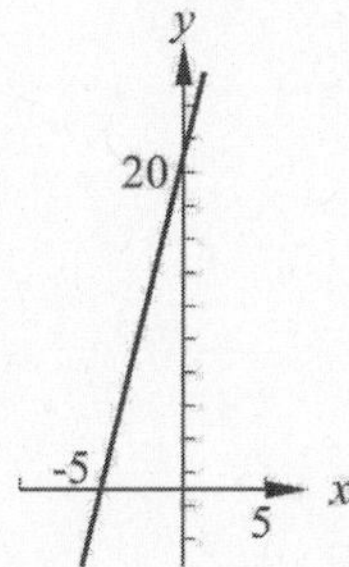

9. Given $(x_1, y_1) = (-7, 12)\ \alpha = 45^\circ$

$$m = \tan\alpha = \tan 45^\circ = 1$$
$$y - y_1 = m(x - x_1)$$
$$y - 12 = 1(x+7)$$
$$= x + 7$$
$$y = x + 19 \text{ or } x - y + 19 = 0$$

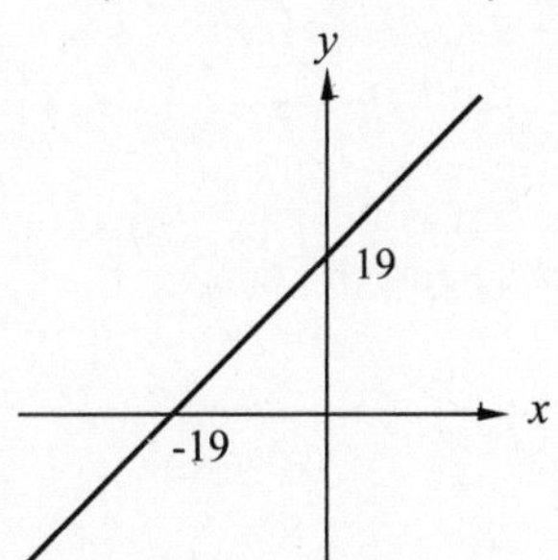

13. Parallel to y-axis and 3 units left of y-axis.

$$x = -3$$

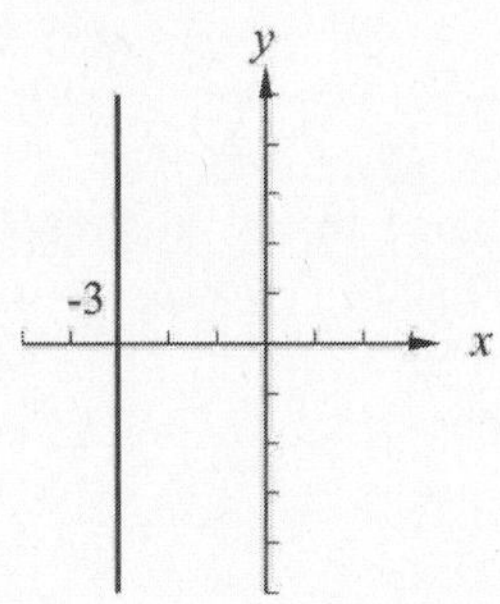

17. $\dfrac{\alpha - 2}{0 - 5} = \dfrac{2 - 0}{5 - \alpha}$

$$\alpha^2 - 7\alpha = 0$$
$$\alpha(\alpha - 7) = 0$$
$$\alpha = 7$$
$$y - y_1 = m(x - x_1)$$
$$y - 2 = \frac{7-2}{-5}(x - 5)$$
$$y = -x + 7 \text{ or } x + y - 7 = 0$$

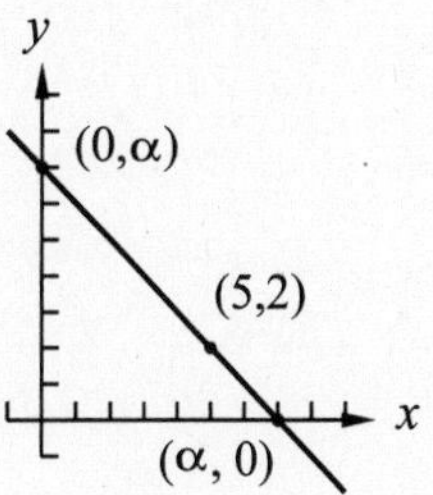

21. Given $4x - y = 8$,

$y = 4x - 8$, so $m = 4$, $b = -8$

When $x = 0$, $y = -8$

$y = 0$, $x = 2$

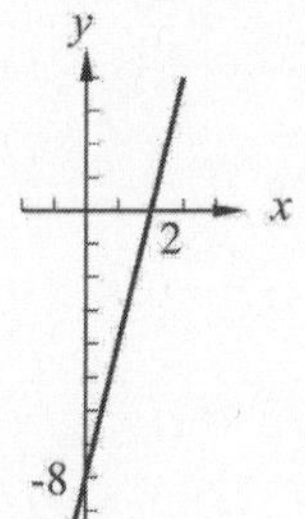

25. Given $3x - 2y - 1 = 0$

$$y = \frac{-3}{-2}x + \frac{1}{-2};$$
$$y = \frac{3}{2}x - \frac{1}{2}, \text{ so } m = \frac{3}{2}, b = -\frac{1}{2}$$

When $x = 0, y = -\dfrac{1}{2}$

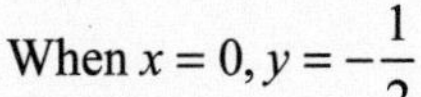

$y = 0, x = \dfrac{1}{3}$

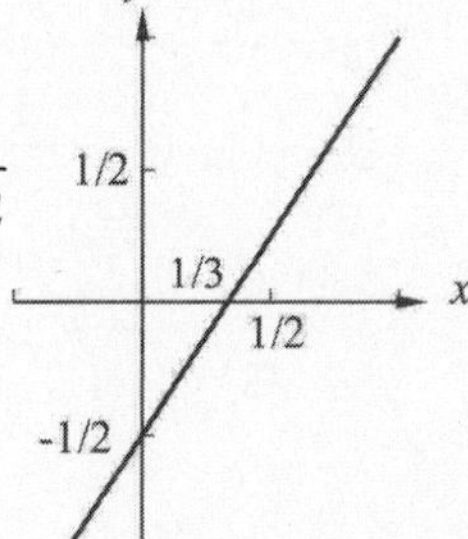

29. $3x-2y+5=0;$

$-2y=-3x-5$

$y=\frac{-3}{-2}x+\frac{-5}{-2}$

$y=\frac{3}{2}x+\frac{5}{2}$;slope $m_1=\frac{3}{2}$

$4y=6x-1$

$y=\frac{6}{4}x-\frac{1}{4}$; slope $m_2=\frac{3}{2}$

$m_1=m_2$ so the lines are parallel.

33. $5x+2y=3$

$y=\frac{-5}{2}x+\frac{3}{2}, m_1=-\frac{5}{2}$

$10y=7-4x$

$y=\frac{-4}{10}x+\frac{7}{10}, m_2=-\frac{2}{5}$

$m_1\cdot m_2=\frac{-5}{2}\cdot\frac{-2}{5}=1\neq -1$

$m_1\neq m_2$

The lines are neither perpendicular nor parallel.

37. Given: $4x-ky=6\|6x+3y+2=0$

$6x+3y+2=0$

$3y=-6x-2$

$y=\frac{-6}{3}x-\frac{2}{3}$

$y=-2x-\frac{2}{3}$; slope is -2

$4x-ky=6$

$-ky=-4x+6$

$y=\frac{-4}{-k}x+\frac{6}{-k}$

$y=\frac{4}{k}x-\frac{6}{k}$; slope is $\frac{4}{k}$

Since the lines are parallel, the slopes are equal.

$\frac{4}{k}=-2$

$k=-2$

41.

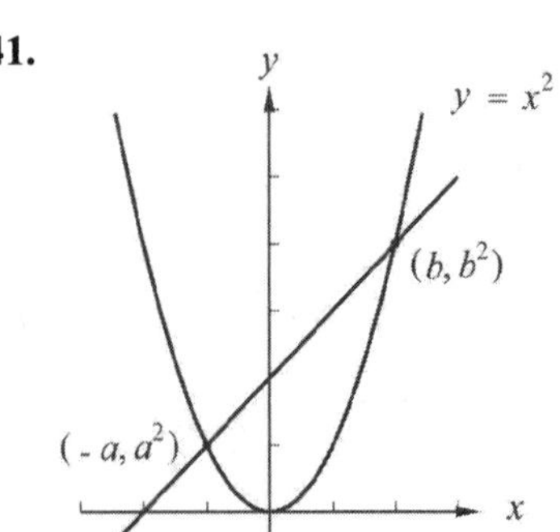

$m=\frac{b^2-a^2}{b-(-a)}$

$m=\frac{(b+a)(b-a)}{b+a}$

$m=b-a$

45. $8x+10y=3\Rightarrow y_1=-\frac{4}{5}x+\frac{3}{10}$

$2x-3y=5\Rightarrow y_2=\frac{2}{3}x-\frac{5}{3}$

$4x-6y=-3\Rightarrow y_3=\frac{2}{3}x+\frac{1}{2}$

$5y+4x=1\Rightarrow y_4=-\frac{4}{5}x+\frac{1}{5}$

$m_1=m_4=-\frac{4}{5}$ and $m_2=m_3=\frac{2}{3}$

showing the lines form a parallelogram.

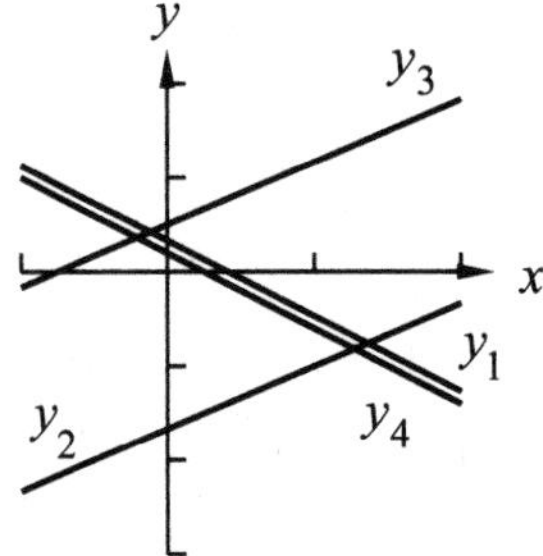

49. If we let Celsius temperature be a function of Réaumur temperature, the function will go through the two points (0,0) and (80,100).

Therefore, $b = 0$ and

$$m = \frac{100-0}{80-0} = \frac{5}{4}$$

Therefore, from the slope-intercept form,

$$C = \frac{5}{4}R$$

53. The line goes through two points:
$(x_1, T_1) = (0,3); (x_2, T_2) = (15,23)$
The T-intercept is $b = 3$. The slope is

$$m = \frac{23-3}{15-0} = \frac{4}{3}$$

The equation of the line is

$$T = \frac{4}{3}x + 3.$$

The slope is the increase in temperature (in °C) as the distance from the outside increases in 1 cm. For every cm further inside the wall, the temperature is increased by $\left(\frac{4}{3}\right)^{\circ}$ C.

57. $m = \tan(180^{\circ} - 0.0032^{\circ})$

$m = -5.6 \times 10^{-5}$

$$b = \frac{48}{2}\mu\text{m} = 24 \times 10^{-6}\text{m} = 2.4 \times 10^{-5}\text{ m}$$

$$y = mx + b = -5.6 \times 10^{-5}x + 2.4 \times 10^{-5}$$

$$y = (-5.6x + 2.4)10^{-5}$$

61. $n = 1200\sqrt{t} + 0$

$m = 1200$

$b = 0$

t	$\sqrt{t}$	h
0	0	0
1	1	1200
4	2	2400

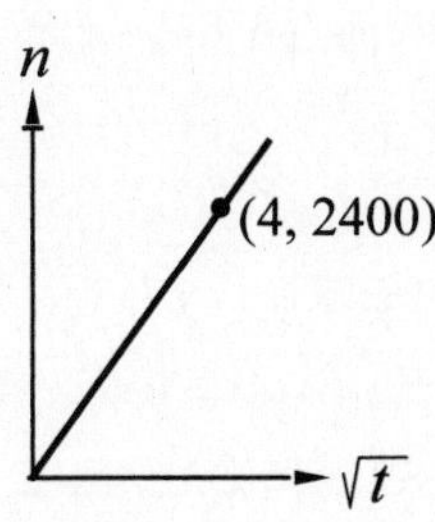

65. $y = ax^n$; $y = 3x^4$

$a = 3$, $n = 4$

x	$\log x$	y	$\log y$
1	0	3	$\log 3 = 0.48$
2	0.30	48	1.7
3	0.48	243	2.4
4	0.60	768	2.9

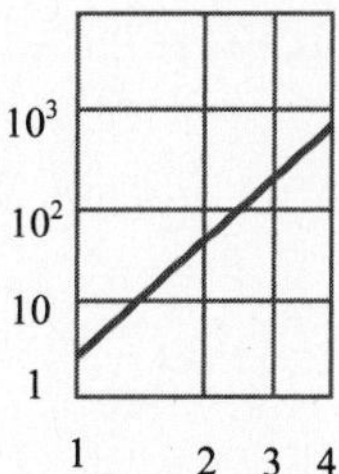

From the table above we see that the slope is

$$m = \frac{2.9 - 0.48}{0.60} = 4.0$$

The $\log y$-intercept is found where $x = 1$, so that $\log x = 0$ and

$$b = \log 3 = 0.48$$

The line is

$\log y = 4.0 \log x + 0.48$

This verifies the fact that the equation is

$$\log y = \log a + n \log x$$

$$\log y = \log 3 + 4 \log x = 0.48 + 4 \log x$$

21.3 The Circle

1. $(x-1)^2 + (y+1)^2 = 16$

$(x-1)^2 + (y-(-1))^2 = 16$

has centre at $(1, -1)$ and $r = 4$

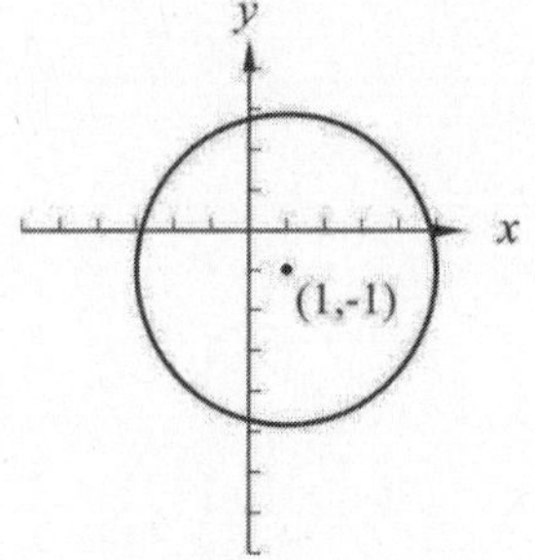

5. $(x-2)^2+(y-1)^2=25$
$C(2, 1)$, radius is 5.

9. $(x-h)^2+(y-k)^2=r^2$; $C(0, 0)$, $r=3$
$(x-0)^2+(y-0)^2=3^2$
$x^2+y^2=9$
$x^2+y^2-9=0$

13. $(x-h)^2+(y-k)^2=r^2; C(12, -15), r=18$
$(x-12)^2+(y-(-15))^2=18^2$
$(x-12)^2+(y+15)^2=324$
$x^2+y^2-24x+30y+45=0$

17. Concentric with $(x-2)^2+(y-1)^2=4$ gives centre at $(2, 1)$. The standard equation is
$(x-2)^2+(y-1)^2=r^2$
Since it is satisfied by $(4,-1)$ we get
$r=\sqrt{(4-2)^2+(-1-1)^2}$
$=2\sqrt{2}$
The equations are
$(x-2)^2+(y-1)^2=8$
$x^2+y^2-4x-2y-3=0$

21. The centre has coordinates
$\left(\frac{-4+0}{2}, \frac{4+0}{2}\right)=(-2,2)$
The radius is 2. The equations are
$(x+2)^2+(y-2)^2=2^2$
$x^2+y^2+4x-4y+4=0$

25. $x^2+(y-3)^2=4$ is the same as
$(x-0)^2+(y-3)^2=2^2$, so
$C(0, 3), r=2$

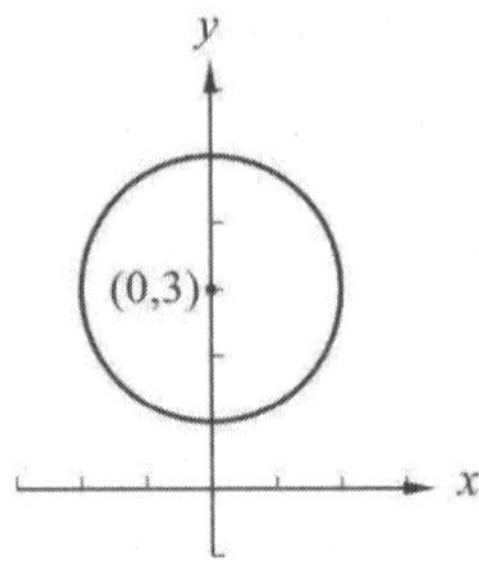

29. $x^2+y^2-2x-8=0$
$x^2-2x+1+y^2=8+1$
$(x-1)^2+(y-0)^2=3^2$
$C(1, 0)$, $r=3$

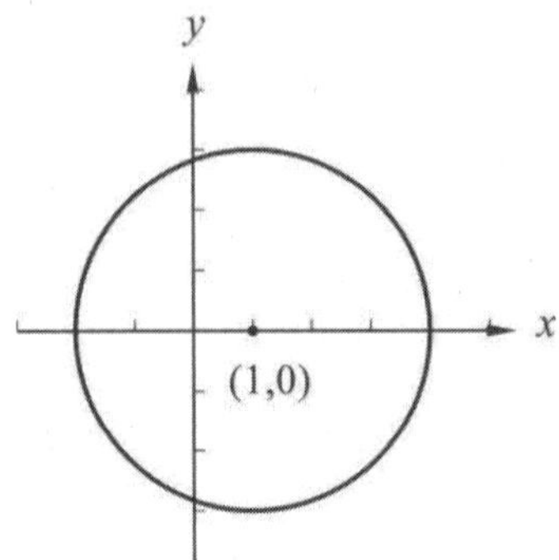

33. $4x^2+4y^2-16y=9$
$x^2+y^2-4y=\frac{9}{4}$
$x^2+y^2-4y+4=\frac{9}{4}+\frac{16}{4}$
$(x-0)^2+(y-2)=\left(\frac{5}{2}\right)^2$
$C(0, 2)$, $r=\frac{5}{2}$

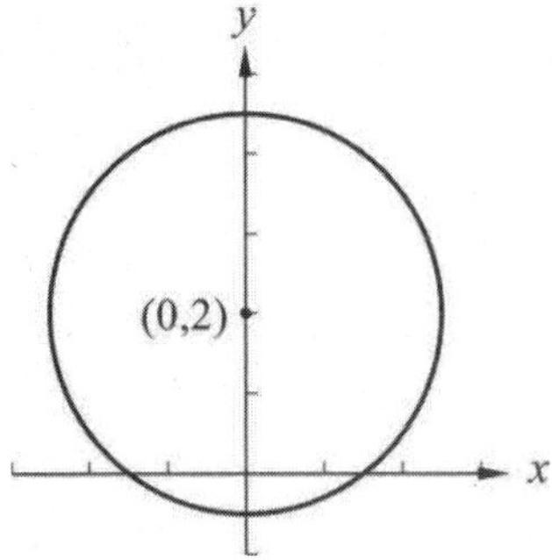

37. Replace x by $-x$:
$(-x)^2+y^2=100$; $x^2+y^2=100$
Symmetric to y-axis.
Replace x by $-x$ and y by $-y$:
$(-x)^2+(-y)^2=100$; $x^2+y^2=100$
Symmetric to origin.
Replace y by $-y$:
$x^2+(-y)^2=100$; $x^2+y^2=100$
Symmetric to x-axis.

41. Find all points for which $y = 0$.

$$x^2 - 6x + (0)^2 - 7 = 0$$
$$x^2 - 6x - 7 = 0$$
$$(x+1)(x-7) = 0$$
$$x = -1 \text{ or } x = 7$$

Yes, the circle crosses the x-axis at $(-1, 0)$ and at $(7, 0)$

45. $x^2 + y^2 + 5y - 4 = 0$

$y^2 + 5y + (x^2 - 4) = 0$; solve for y

$$y = \frac{-5 \pm \sqrt{5^2 - 4(x^2 - 4)}}{2}$$
$$= \frac{-5 \pm \sqrt{41 - 4x^2}}{2}$$
$$= -2.5 \pm \sqrt{10.25 - x^2}$$

To graph:

$x_{\min} = -6,\ x_{\max} = 6,\ y_{\min} = -6,\ y_{\max} = 2$

$y_1 = -2.5 + \sqrt{10.25 - x^2}$

$y_2 = -2.5 - \sqrt{10.25 - x^2}$

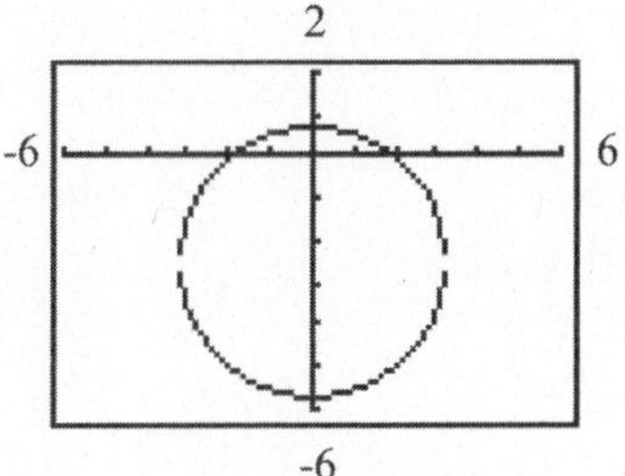

49. $x^2 + y^2 - 2x - 4y + 3 = 0$

$x^2 - 2x + 1 + y^2 - 4y + 4 = -3 + 1 + 4 = 2$

$(x-1)^2 + (y-2)^2 = 2$

The equation is a circle with $C(1, 2)$, $r = \sqrt{2}$

Substituting $(0.1, 3.1)$ into the left hand side of the equation,

$(0.1-1)^2 + (3.1-2)^2 = 2.02 > 2$

Therefore, the point is outside the circle.

53. The thickness is the difference in the radii of the circles.

$$2.00x^2 + 2.00y^2 = 5.73$$
$$x^2 + y^2 = \frac{5.73}{2.00}$$
$$C(0,0), r_1 = \sqrt{\frac{5.73}{2.00}}$$
$$2.80x^2 + 2.80y^2 = 8.91$$
$$x^2 + y^2 = \frac{8.91}{2.80}$$
$$C(0,0), r_2 = \sqrt{\frac{8.91}{2.80}}$$
$$\text{thickness} = r_2 - r_1$$
$$= \sqrt{\frac{8.91}{2.80}} - \sqrt{\frac{5.73}{2.00}}$$
$$= 0.0912 \text{ cm}$$

57. $60 \text{ Hz} = 60 \text{ cycles/s} = 37.7 \text{ m/s};\ (h, k) = (0, 0)$

$60 \text{ cycles} = 37.7 \text{ m}$

$1 \text{ cycle} = 0.628 \text{ m}$

$r = 0.628 \text{ m} \div 2\pi$

$r = 0.10 \text{ m}$

The equation of the path is

$x^2 + y^2 = (0.10)^2$

$x^2 + y^2 = 0.0100$

61.

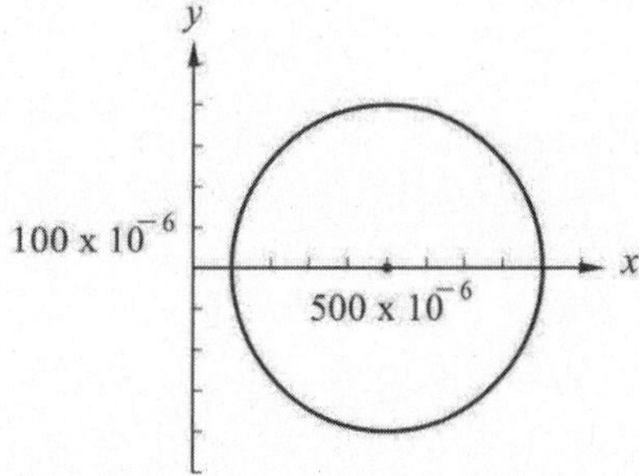

The centre of circle is at the midpoint between $(100\times10^{-6}, 0)$ and $(900\times10^{-6}, 0)$, or at

$$\left(\frac{100\times10^{-6} + 900\times10^{-6}}{2}, 0\right) = (500\times10^{-6}, 0)$$

The radius of the circle is half the distance between $(100\times10^{-6}, 0)$ and $(900\times10^{-6}, 0)$, or

$$r = \frac{900\times10^{-6} - 100\times10^{-6}}{2}$$
$$r = 400\times10^{-6}$$

The equation is

$$(x - 500\times10^{-6})^2 + (y-0)^2 = (400\times10^{-6})^2$$

21.4 The Parabola

1. $y^2 = 20x$

$4p = 20$

$p = 5$

$F(5, 0)$; directrix $x = -5$

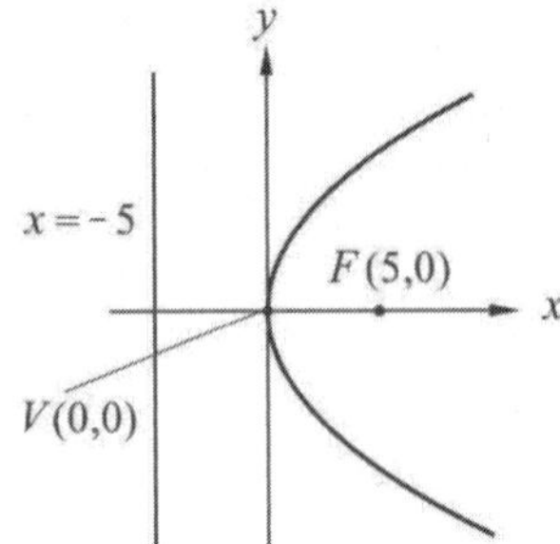

5. $y^2 = 4x$

$4p = 4$

$p = 1$

$F(1, 0)$; directrix $x = -1$

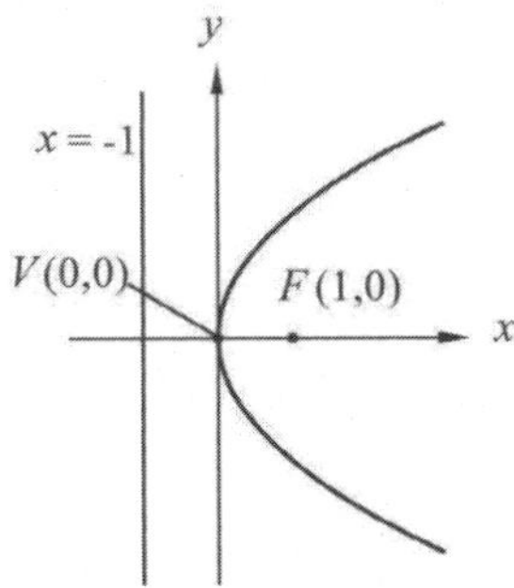

9. $x^2 = 72y$

$4p = 72$

$p = 18$

$F(0, 18)$; directrix $y = -18$

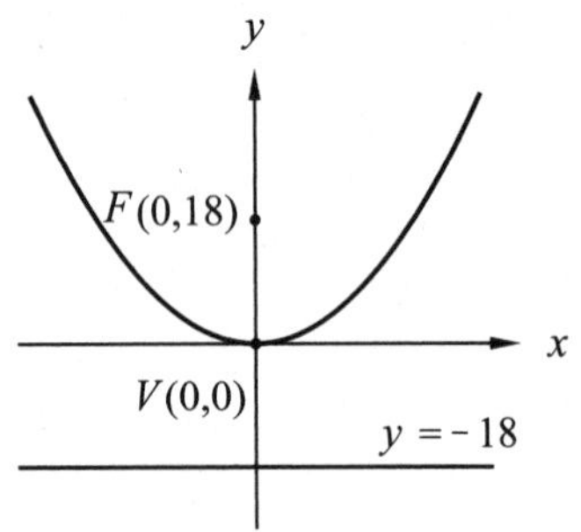

13. $2y^2 - 5x = 0$

$2y^2 = 5x$

$4p = 5$

$p = \frac{5}{8}$

$F\left(\frac{5}{8}, 0\right)$; directrix $x = -\frac{5}{8}$

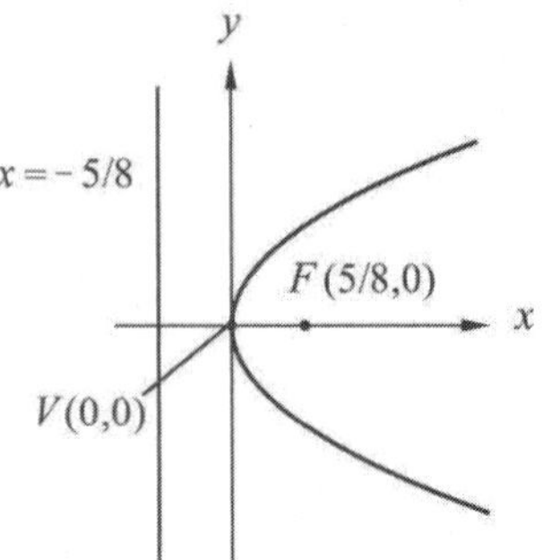

17. $F(3, 0); V(0,0)$

directrix $x = -3,\ p = 3$

$y^2 = 4px$

$y^2 = 4(3)x$

$y^2 = 12x$

21. $V(0, 0)$, directrix $y = -0.16$

$F(0, 0.16),\ p = 0.16$

$x^2 = 4(0.16)y$

$x^2 = 0.64y$

25. $V(0, 0)$, axis $x = 0$

Substitute $(-1,8)$ into $x^2 = 4py$:

$(-1)^2 = 4p(8)$

$1 = 32p$

$p = \frac{1}{32}$

Therefore,

$x^2 = 4 \cdot \frac{1}{32} y$

$x^2 = \frac{1}{8} y.$

29. Passes through $(3,3), (12,6)$

Axis is the x-axis.

Substitute $(3,3)$ into $y^2 = 4px$:

$3^2 = 4p(3)$

$4p = 3$

$y^2 = 3x$

33. $y^2 = 2x$

$x = \frac{y^2}{2},$

Substitute into $x^2 = -16y$:

$\left(\frac{y^2}{2}\right)^2 = -16y$

$y^4 + 64y = 0$

$y(y^3 + 64) = 0$

$y = 0 \quad \text{or} \quad y^3 + 64 = 0$

$y^3 = -64$

$y = -4$

$x = \frac{0^2}{2} \qquad x = \frac{(-4)^2}{2}$

$x = 0 \qquad x = 8$

The points of intersection are $(0, 0)$ and $(8, -4)$.

37. $y^2 + 2x + 8y + 13 = 0$; solve for y

$y^2 + 8y + (2x + 13) = 0$

$y = \frac{-8 \pm \sqrt{8^2 - 4(2x+13)}}{2}$

$y = \frac{-8 \pm \sqrt{12 - 8x}}{2}$

$y_1 = -4 + \sqrt{3 - 2x},\ y_2 = -4 - \sqrt{3 - 2x}$

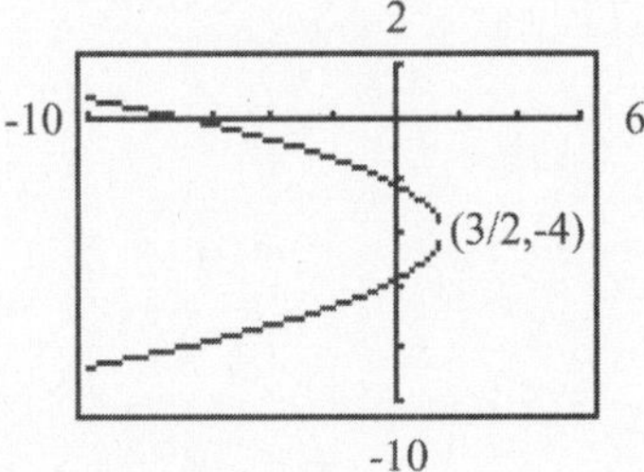

41. $y^2 = 4px, F(p, 0)$

When $x = p$,

$y^2 = 4p^2$

$y = \pm 2p$

Therefore, the latus rectum intersects the parabola at $(p, 2p)$ and at $(p, -2p)$.

The length of latus rectum is

$|2p - (-2p)| = 4|p|$.

45. Let the vertex of the parabola be at the origin. $x^2 = 4py$. A point on the parabola will be (236.5, 55.0). Substitute this into the equation and solve for p.

$236.5^2 = 4p(55.0)$

$4p = 1017$

$x^2 = 1020y$

49. Place the vertex at the origin. Then the equation of the parabola is

$y^2 = 4px$

Substitute $(0.00625, 1.20)$ into the equation:

$(1.20)^2 = 4p(0.00625)$

$p = 57.6$

The focal length is 57.6 m.

53. $y^2 = 4px$

Substitute $(6.5, 7.5)$:

$7.5^2 = 4p(6.5)$

$p = 2.16$ cm

The filament should be placed 2.16 cm from the vertex.

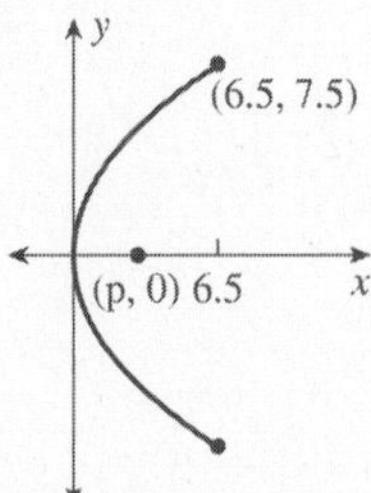

57. The path of the ship channel is a parabola with focus at $(0, 2)$ and vertex $(0, 0)$ and directrix $y = -2$. Therefore, $p = 2$.

The parabola is of the type

$x^2 = 4py$;

therefore, $x^2 = 8y$.

21.5 The Ellipse

1. $\frac{x^2}{25}+\frac{y^2}{36}=1,$

$a^2=36,\ a=6,$

$b^2=25,\ b=5$

$V(0,\pm 6)$, minor axis: $(\pm 5,\ 0)$

$a^2=b^2+c^2$

$36=25+c^2$

$c=\sqrt{11}$

$F(0,\pm\sqrt{11})$

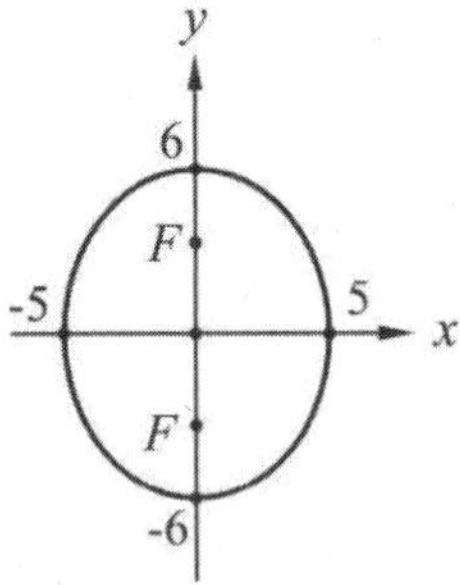

5. $\frac{x^2}{25}+\frac{y^2}{144}=1$

$a^2=144,\ a=12,$

$b^2=25,\ b=5$

$V(0,\pm 12)$, minor axis: $(\pm 5,\ 0)$

$a^2=b^2+c^2$

$144=25+c^2$

$c=\sqrt{119}=10.9$

$F(0,\pm\sqrt{119})$

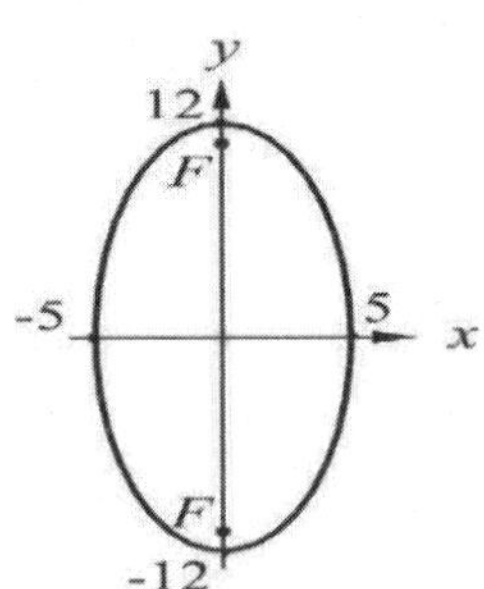

$V(0,\pm 12)$, $F(0,\pm\sqrt{119})$, x-intercepts $(\pm 5,\ 0)$

9. $4x^2+9y^2=324$

$\frac{x^2}{81}+\frac{y^2}{36}=1$

$a^2=81,\ a=9$

$b^2=36, b=6$

$c^2=81-36$

$=45,\ c=3\sqrt{5}$

$V(\pm 9,\ 0)$, $F(\pm 3\sqrt{5},\ 0)$,

y-intercepts $(0,\pm 6)$

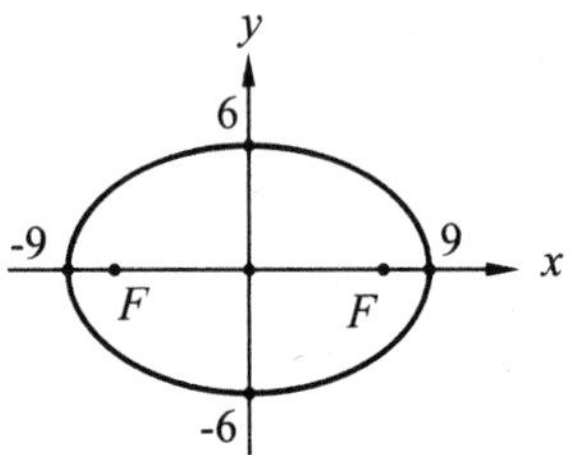

13. $y^2=8(2-x^2)$

$8x^2+y^2=16$

$\frac{8x^2}{16}+\frac{y^2}{16}=1$

$\frac{x^2}{2}+\frac{y^2}{16}=1$

$\frac{y^2}{16}+\frac{x^2}{2}=1$

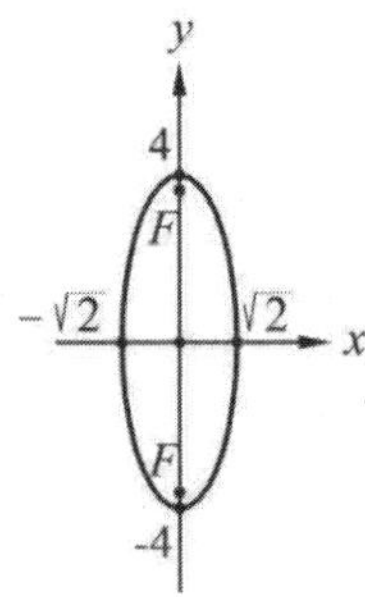

$a^2=16, a=4$

$b^2=2, b=\sqrt{2}$

$c^2=16-2=14, c=\sqrt{14}$

$V(0,\pm 4)$, $F(0,\pm\sqrt{14})$, x-intercepts $(\pm\sqrt{2},\ 0)$

17. $V(15, 0);\ F(9, 0)$

$a = 15,\ a^2 = 225;$

$c = 9,\ c^2 = 81;$

$a^2 - c^2 = b^2$

$b^2 = 144;$

$\frac{x^2}{a^2} + \frac{y^2}{b^2} = 1;$

$\frac{x^2}{225} + \frac{y^2}{144} = 1$

$144x^2 + 225y^2 = 32\ 400$

21. $F(8, 0) \Rightarrow c = 8$ and the major axis is the x – axis.

end of minor axis: $(0, 12) \Rightarrow b = 12$

$a^2 = b^2 + c^2$

$a^2 = 12^2 + 8^2$

$a = \sqrt{208}$

$\frac{x^2}{\sqrt{208}^2} + \frac{y^2}{12^2} = 1$

$\frac{x^2}{208} + \frac{y^2}{144} = 1$

25. $(x_1, y_1) = (2, 2), (x_2, y_2) = (1, 4)$

$\frac{x^2}{b^2} + \frac{y^2}{a^2} = 1$

Substitute: $\frac{4}{b^2} + \frac{4}{a^2} = 1$

Therefore, $4a^2 + 4b^2 = a^2b^2$

$\frac{1}{b^2} + \frac{16}{a^2} = 1$

Therefore, $a^2 + 16b^2 = a^2b^2$

$16b^2 = a^2b^2 - a^2 = a^2(b^2 - 1)$

$a^2 = \frac{16b^2}{b^2 - 1}$

Substitute:

$4\frac{16b^2}{b^2 - 1} + 4b^2 = \frac{16b^2}{b^2 - 1}b^2$

$64b^2 + 4b^4 - 4b^2 = 16b^4$

$-12b^4 + 60b^2 = 0$

$12b^2(-b^2 + 5) = 0$

$b^2 = 5$

$a^2 = \frac{16(5)}{4} = 20$

The equation is $\frac{y^2}{20} + \frac{x^2}{5} = 1$

or: $5y^2 + 20x^2 = 100$

$4x^2 + y^2 = 20$

29. $4x^2 + 9y^2 = 40,\quad y^2 = 4x$

$4x^2 + 9(4x) = 40$

$x^2 + 9x - 10 = 0$

$(x + 10)(x - 1) = 0$

$x = -10$ or $x = 1$

$y^2 = 4(-10)$ $\qquad y^2 = 4(1)$

$y^2 = -40$, no solution $\qquad y = \pm 2$

Graphs intersect at $(1, 2), (1, -2)$

33. $4x^2 + 3y^2 + 16x - 18y + 31 = 0$; solve for y

$3y^2 - 18y + (4x^2 + 16x + 31) = 0$

$y = \frac{18 \pm \sqrt{(-18)^2 - 4(3)(4x^2 + 16x + 31)}}{2(3)}$

$= \frac{18 + \sqrt{-48x^2 - 192x - 48}}{2}$

$y_1 = 3 + \frac{\sqrt{-12x^2 - 48x - 12}}{3}$

$y_{2(3)} = 3 - \frac{\sqrt{-12x^2 - 48x - 12}}{3}$

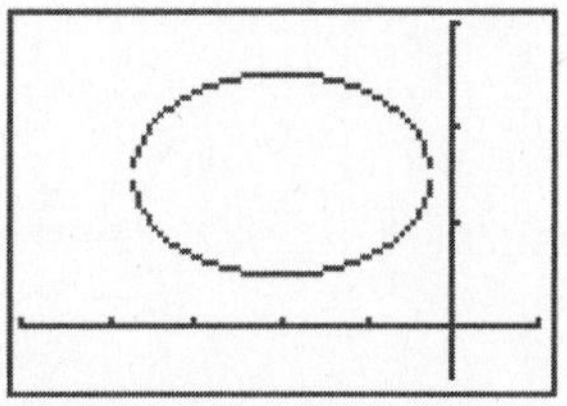

6

-5 1

-1

37. $x^2 + k^2y^2 = 1$

Therefore, $\frac{x^2}{1} + \frac{y^2}{\frac{1}{k^2}} = 1$

The vertices will be on the y-axis if the denominator of y^2 is greater than that of x^2, so

$\frac{1}{k^2} > 1$

$k^2 < 1$

Therefore, $|k| < 1$

41. $100x^2+49y^2 \le 4900$

$100x^2+49y^2 = 4900$

$\frac{x^2}{49}+\frac{y^2}{100}=1$ is the elliptical boundary

It is an ellipse with centre at the origin, vertices at $(0,10)$ and $(0,-10)$, and minor axis ending at $(7,0)$ and $(-7,0)$.

Test point $(0, 0)$

$100(0)^2+49(0)^2 < 4900$

$0 < 4900$

So $(0,0)$ satisfies the inequality and we shade inside the ellipse

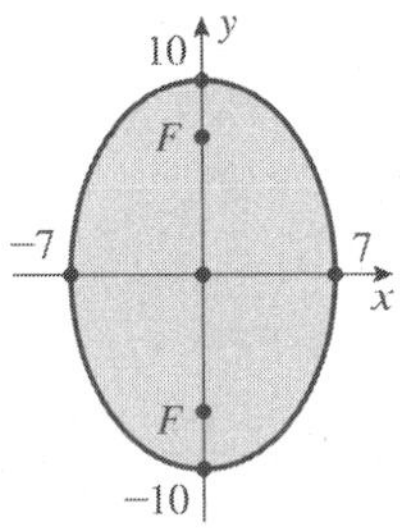

45. $P = Ri^2$

$P_T = R_1 i_1^2 + R_2 i_2^2$

$64 = 2i_1^2 + 8i_2^2$

$\frac{2}{64}i_1^2+\frac{8}{64}i_2^2=1$

$\frac{i_1^2}{32}+\frac{i_2^2}{8}=1$

$a^2=32;\ a=\sqrt{32}=5.7$

$b^2=8;\ b=\sqrt{8}=2.8$

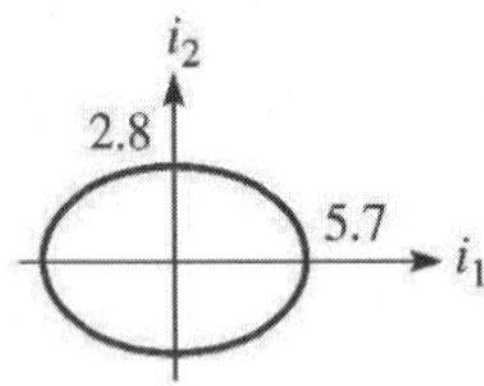

49. If the two vertices of each base are fixed at $(-3, 0)$ and $(3, 0)$, and the sum of the two leg lengths is also fixed, the third vertex lies on an ellipse. The base is 6 cm, so

$d_1+d_2 = 14 \text{ cm} - 6 \text{ cm} = 8 \text{ cm}$

$(-3, 0)$ and $(3, 0)$ are foci $(-c, 0)$ and $(c, 0)$

$d_1+d_2=2a=8;\ a=4$

$a^2-c^2=b^2$

$4^2-3^2=b^2$

$b^2=7,\ a^2=16$

The equation is $\frac{x^2}{16}+\frac{y^2}{7}=1,\ 7x^2+16y^2=112$

53. $2a=19.6, a=9.8,$

$b=5.5$

Let the centre be at the origin. The equation of the ellipse is

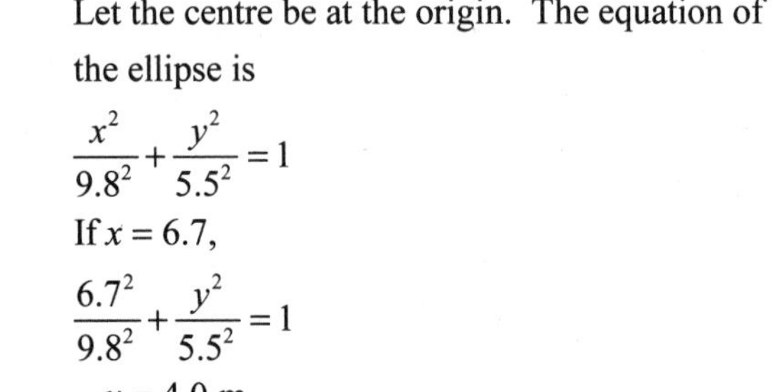

$\frac{x^2}{9.8^2}+\frac{y^2}{5.5^2}=1$

If $x=6.7$,

$\frac{6.7^2}{9.8^2}+\frac{y^2}{5.5^2}=1$

$y=4.0 \text{ m}$

21.6 The Hyperbola

1. $\frac{y^2}{16}-\frac{x^2}{4}=1$

$a^2=16,\ a=4$

$b^2=4,\ b=2$

$c^2=a^2+b^2=20$

$c=2\sqrt{5}$

$V(0, \pm 4)$

conjugate axis: $(\pm 2, 0)$

$F(0, \pm 2\sqrt{5})$

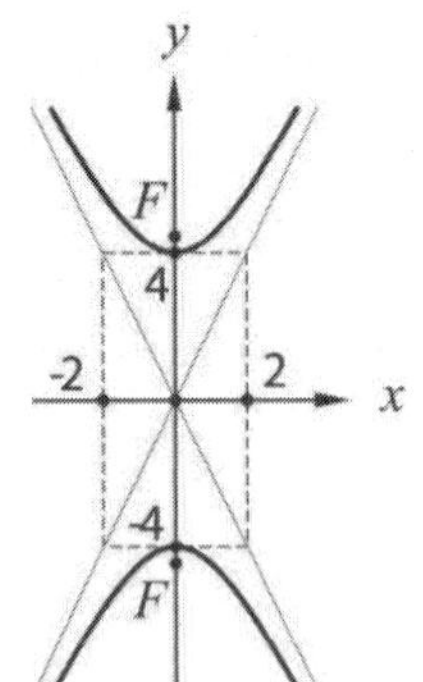

5. $\dfrac{y^2}{9}-\dfrac{x^2}{1}=1$

$a^2=9,\ a=3$

$b^2=1,\ b=1$

$c^2=a^2+b^2$

$c^2=10$

$c=\sqrt{10}$

$V(0,\pm 3),\ F\left(0,\pm\sqrt{10}\right)$

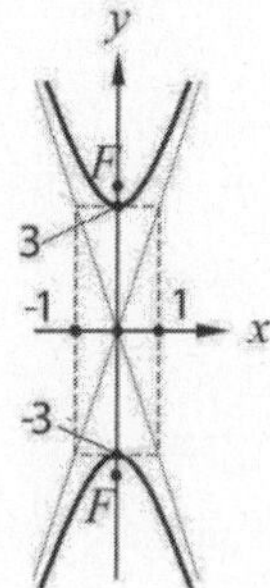

9. $4x^2-y^2=4$

$\dfrac{4x^2}{4}-\dfrac{y^2}{4}=1;$

$\dfrac{x^2}{1}-\dfrac{y^2}{4}=1$

$a^2=1, a=1$

$b^2=4, b=2$

$c^2=4+1=5$

$c=\sqrt{5}$

$V(\pm 1,\ 0),\ F\left(\pm\sqrt{5},\ 0\right)$

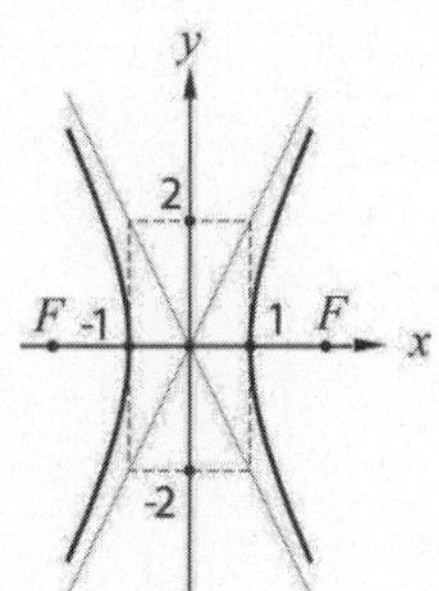

13. $y^2=4\left(x^2+1\right)$

$y^2-4x^2=4$

$\dfrac{y^2}{4}-\dfrac{x^2}{1}=1$

$a^2=4, a=2$

$b^2=1, b=1$

$c^2=5, c=\sqrt{5}$

$V(0,\pm 2),\ F\left(0,\pm\sqrt{5}\right)$

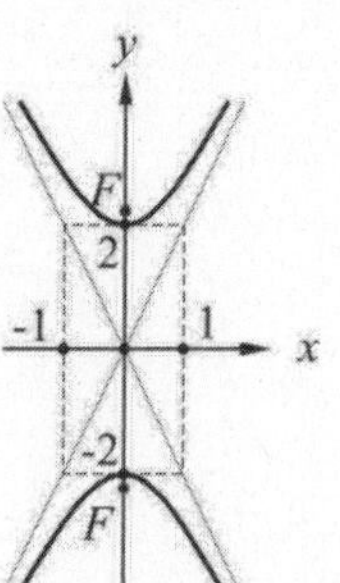

17. $V(3,\ 0);\ F(5,\ 0)$

$a=3;\ c=5;\ a^2=9;\ c^2=25$

$b^2=c^2-a^2=25-9=16$

$\dfrac{x^2}{a^2}-\dfrac{y^2}{b^2}=1$

$\dfrac{x^2}{9}-\dfrac{y^2}{16}=1;$

$16x^2-9y^2=144$

21. $(x,\ y)$ is $(2,\ 3)$; $F(2,\ 0),\ (-2,\ 0);\ c=\pm 2,\ c^2=4$

$d_1=\sqrt{(2-(-2))^2+(3-0)^2}$

$=\sqrt{4^2+3^2}=\sqrt{16+9}$

$=\sqrt{25}=5$

$d_2=\sqrt{(2-2)^2+(3-0)^2}$

$=\sqrt{0+9}=\sqrt{9}=3$

$d_1-d_2=2a;\ 5-3=2a;\ 2=2a;\ 1=a;\ a^2=1$

$c^2=4;\ b^2=c^2-a^2=3$

$\dfrac{x^2}{a^2}-\dfrac{y^2}{b^2}=1;\ \dfrac{x^2}{1}-\dfrac{y^2}{3}=1$

$3x^2-y^2=3$

Alternatively, solve the system

$a^2+b^2=4$

$\dfrac{4}{a^2}-\dfrac{9}{b^2}=1$

Substitute $a^2=4-b^2$:

$\dfrac{4}{4-b^2}-\dfrac{9}{b^2}=1$

$4b^2 - 36 + 9b^2 = 4b^2 - b^4$
$b^4 + 9b^2 - 36 = 0$
$(b^2 + 12)(b^2 - 3) = 0$
$b^2 = 3$ is the only solution, and we obtain the same equation as before.

25. $V(1, 0)$

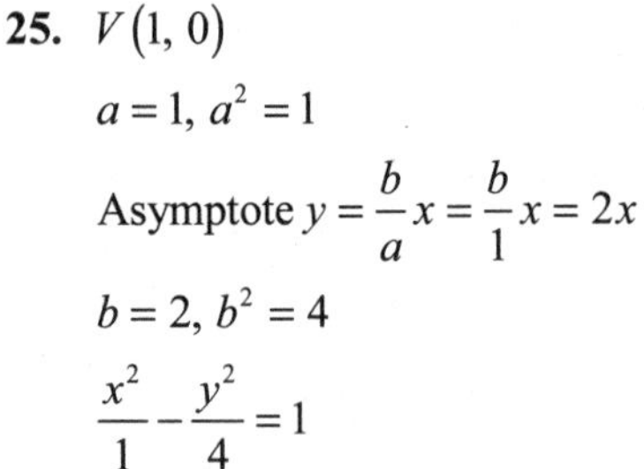

$a = 1,\ a^2 = 1$

Asymptote $y = \frac{b}{a}x = \frac{b}{1}x = 2x$

$b = 2,\ b^2 = 4$

$\frac{x^2}{1} - \frac{y^2}{4} = 1$

29. $xy = 2;\ y = \frac{2}{x}$

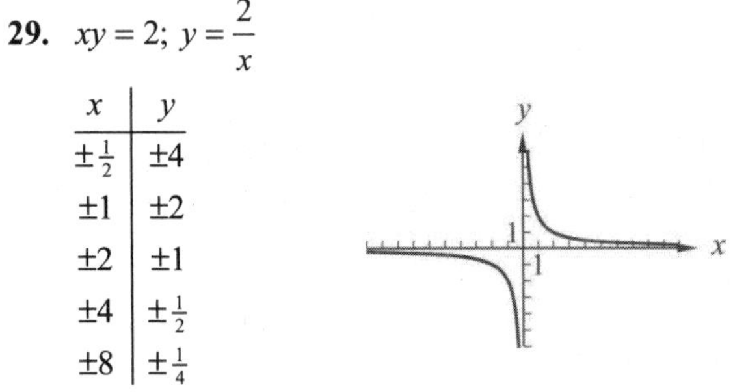

x	y
$\pm\frac{1}{2}$	± 4
± 1	± 2
± 2	± 1
± 4	$\pm\frac{1}{2}$
± 8	$\pm\frac{1}{4}$

33. $y^2 - x^2 = 5$

Substitute $y^2 = x^2 + 5$ into $2x^2 + y^2 = 17$:

$2x^2 + x^2 + 5 = 17$

$3x^2 = 12$

$x^2 = 4$

$x = \pm 2$

$y^2 = x^2 + 5$

$= 4 + 5$

$= 9$

$y = \pm 3$

points of intersection $(2, \pm 3), (-2, \pm 3)$

37. $x^2 - 4y^2 + 4x + 32y - 64 = 0$; solve for y

$4y^2 - 34y + (-x^2 - 4x + 64) = 0$

$$y = \frac{32 \pm \sqrt{(-32)^2 - 4(4)(-x^2 - 4x + 64)}}{2(4)}$$

$$= \frac{32 \pm \sqrt{16x^2 + 64x}}{8}$$

$y_1 = 4 + 0.5\sqrt{x^2 + 4x},\ y_2 = 4 - 0.5\sqrt{x^2 + 4x}$

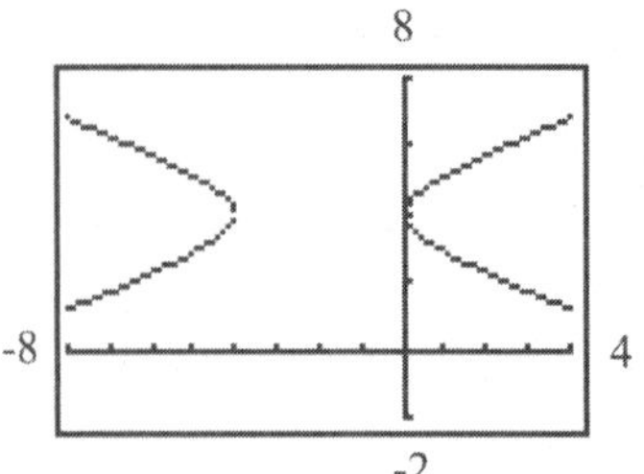

41. $V(0, 1), F(0, \sqrt{3});\ c^2 = a^2 + b^2$ where $c = \sqrt{3}$ and

$a = 1;\ b^2 = \sqrt{3}^2 - 1^2 = 2$

$\frac{y^2}{1^2} - \frac{x^2}{\sqrt{2}^2} = 1$

The transverse axis of the first equation is length $2a = 2\sqrt{1}$ along the y-axis. Its conjugate axis is length $2b = 2\sqrt{2}$ along the x-axis.

The transverse axis of the conjugate hyperbola is length $2\sqrt{2}$ along the x-axis, and its conjugate axis is length $2\sqrt{1}$ along the y-axis.

The equation, then, is $\frac{x^2}{\sqrt{2}^2} - \frac{y^2}{\sqrt{1}^2} = 1$

$\frac{x^2}{2} - \frac{y^2}{1} = 1$ or $x^2 - 2y^2 = 2$

45. $\frac{x^2}{169} + \frac{y^2}{144} = 1$

$\frac{x^2}{13^3} + \frac{y^2}{12^2} = 1$

$13^3 = 12^2 + c^2$

$c = 5$

$\frac{x^2}{a^2} - \frac{y^2}{b^2} = 1,$

$c^2 = a^2 + b^2 \Rightarrow 5^2 = a^2 + b^2,\ 0 < a < 5$

$b^2 = 25 - a^2$

Substitute the point $(4\sqrt{2}, 3)$:

$\frac{(4\sqrt{2})^2}{a^2} - \frac{9}{25 - a^2} = 1$

$800 - 32a^2 - 9a^2 = 25a^2 - a^4$

$a^4 - 66a^2 + 800 = 0$

$(a^2 - 16)(a^2 - 50) = 0$

$a^4-16=0 \quad$ or $\quad a^2-50=0$

$a=4 \qquad a=\sqrt{50}>5$, reject

$b^2=25-4^2=9$

$b=3$

$$\frac{x^2}{16}-\frac{y^2}{9}=1$$

49. $600=vt$

$$v=\frac{600}{t}$$

53.

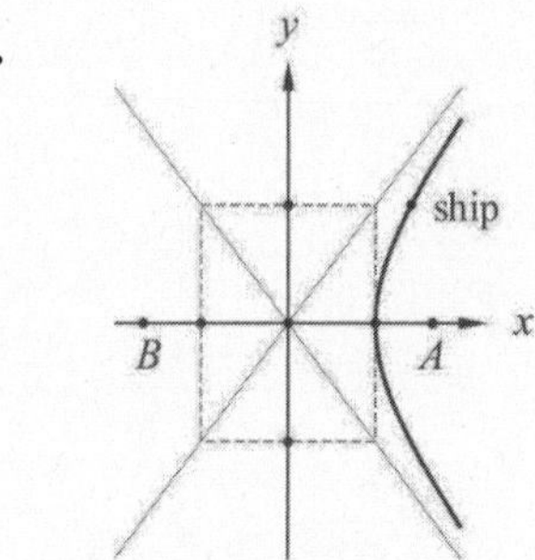

$d_1-d_2=$ constant

Let t_1 be time for signal to go from B to ship.

Let t_2 be time for signal to go from A to ship.

Then $t_2=t_1-1.20$ ms

$v=\frac{s}{t}$, therefore, $s=vt$

Therefore, $d_1=300t_1$ and $d_2=300(t_1-1.20)$

$d_1-d_2=2a$

$300t_1-300(t_1-1.20)=\text{constant}=360\text{ km}=2a$

Therefore, the ship could lie anywhere on the hyperbolic arc sketched.

Foci at $(\pm300, 0)$; therefore, $c=300$

Vertices at $(\pm180, 0)$; therefore, $a=180$; therefore, $b=240$

21.7 Translation of Axes

1. $\frac{(x-3)^2}{25}-\frac{(y-2)^2}{9}=1$, hyperbola: $a=5$, $b=3$

Centre: $(3, 2)$. Transverse axis parallel to x-axis.

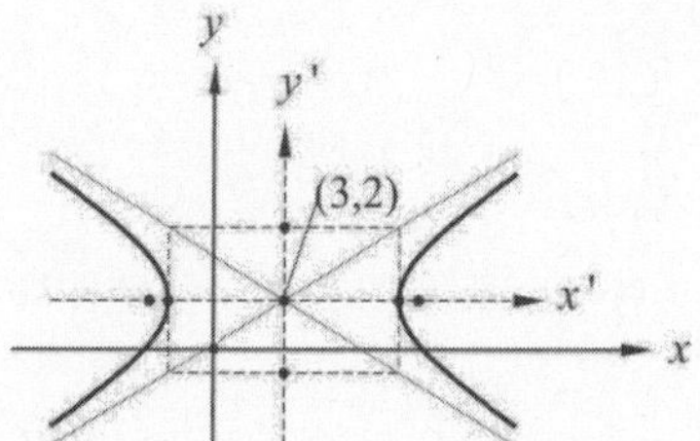

5. $\frac{(x-1)^2}{4}-\frac{(y-2)^2}{9}=1$; hyperbola with transverse axis parallel to x-axis

Centre: $(1, 2)$;

$a^2=4, a=2$;

$b^2=9$, $b=3$

Vertices are at $(1\pm2, 2)$ or at $(-1, 2)$ and at $(3, 2)$.

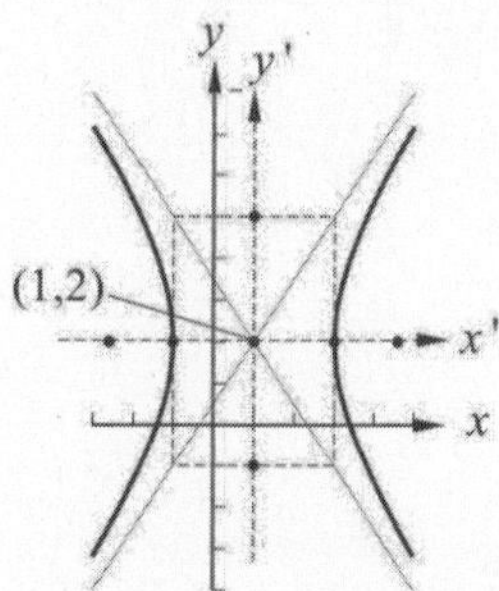

9. $(x+3)^2=-12(y-1)$; parabola with axis parallel to the y-axis

$x'=x+3$; $y'=y-1$

$(x')^2=-12y'$

Origin O' at $(h, k)=(-3, 1)$

$4p=-12$

$p=-3$

Vertex $(-3, 1)$

Focus $(-3,1+p)=(-3,-2)$

Directrix $y=1-p=4$

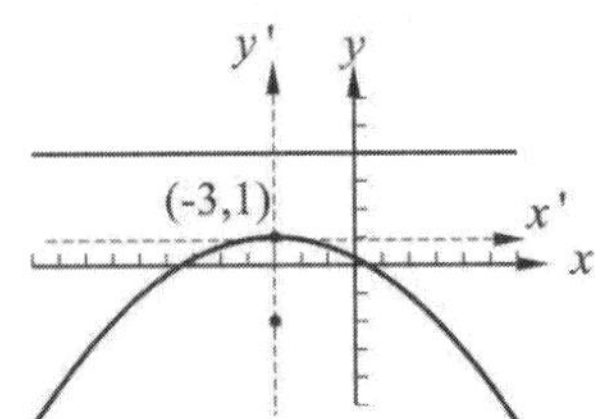

13. $F(12,0)$, axis, directrix are coordinate axis

If the directrix is a coordinate axis, it can only be the y-axis, so the axis of the parabola is the x-axis

$V\left(\frac{12+0}{2},0\right)=(6,0)$

$p=\frac{12-0}{2}=6$

$(y-k)^2=4p(x-h)$

$(y-0)^2=4\cdot 6(x-6)$

$y^2=24(x-6)$

17. Ellipse: centre $(-2,1)$, vertex $(-2,5)$, passes through $(0,1)$.

$a=5-1=4$

Major axis parallel to the x-axis

$\frac{(y-1)^2}{4^2}+\frac{(x+2)^2}{b^2}=1$

Substitute $(0,1)$:

$\frac{(1-1)^2}{4^2}+\frac{(0+2)^2}{b^2}=1$

$b^2=4$

$\frac{(y-1)^2}{16}+\frac{(x+2)^2}{4}=1$

21. Hyperbola: $V(2,1)$, $V(-4,1)$, $F(-6,1)$

Centre: $\left(\frac{-4+2}{2},1\right)=(-1,1)$

$a=\frac{2-(-4)}{2}=3$

$c=-1-(-6)=5$

$b^2=c^2-a^2=25-9=16$

Transverse axis parallel to x-axis.

Therefore, $\frac{(x-h)^2}{a^2}-\frac{(y-k)^2}{b^2}=1$

$\frac{(x+1)^2}{9}-\frac{(y-1)^2}{16}=1$

or $16x^2-9y^2+32x+18y-137=0$

25. $x^2+4y=24$

$x^2=-4y+24$

$x^2=-4(y-6)$

Parabola with vertex at $V(0,6)$ and the y-axis as its axis

$4p=-4$

$p=-1$

Focus is at $(0,6-1)=(0,5)$

Directrix is $y=6+1=7$

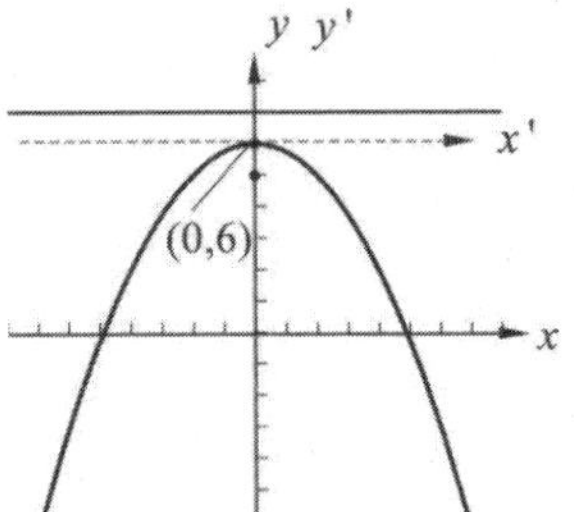

29. $9x^2-y^2+8y=7$

$9x^2-(y^2-8y+16)=7-16$

$9x^2-(y-4)^2=-9$

$\frac{9x^2}{-9}-\frac{(y-4)^2}{-9}=1$

$-x^2+\frac{(y-4)^2}{9}=1$

$\frac{(y-4)^2}{9}-\frac{x^2}{1}=1$

Hyperbola whose transverse axis is the y-axis, with centre at $(0,4)$

Vertices are at $(0,4\pm 3)$ or at $(0,7)$ and $(0,1)$.

The ends of the conjugate axis are at $(0\pm 1,4)$ or at $(1,4)$ and $(-1,4)$.

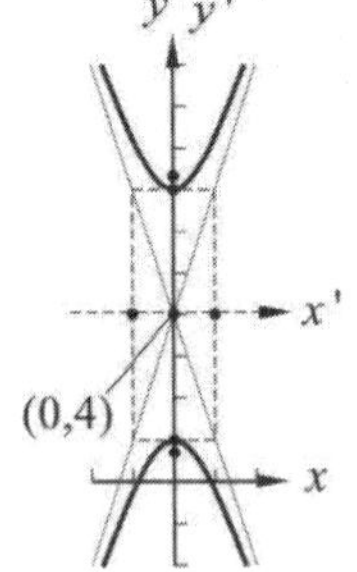

33. $4x^2 - y^2 + 32x + 10y + 35 = 0$

$4(x^2 + 8x) - (y^2 - 10y) = -35$

$4(x^2 + 8x + 16) - (y^2 - 10y + 25) = -35 + 64 - 25$

$\frac{(x+4)^2}{1^2} - \frac{(y-5)^2}{2^2} = 1$

Hyperbola with centre at $(-4, 5)$ and transverse axis parallel to the x-axis.

Vertices are at $(-4 \pm 1, 5)$ or at $(-3, 5)$ and at $(-5, 5)$.

The conjugate axis ends at $(-4, 5 \pm 2)$ or at $(-4, 7)$ and $(-4, 3)$.

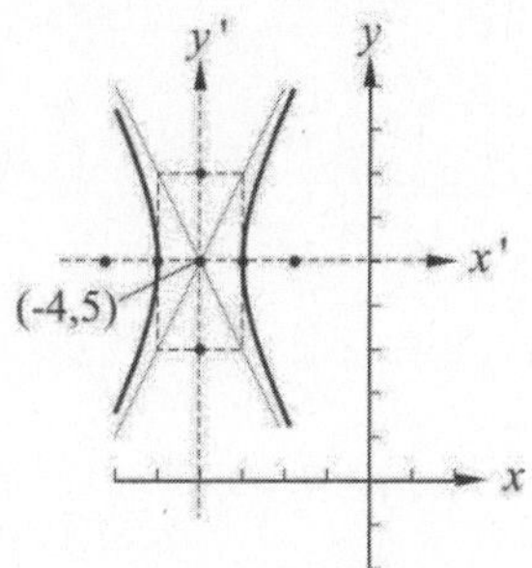

37. $5x^2 - 3y^2 + 95 = 40x$

$5(x^2 - 8x + 16) - 3y^2 = -95 + 80 = -15$

$\frac{y^2}{5} - \frac{(x-4)^2}{3} = 1$

Hyperbola with centre at $(4, 0)$ and transverse axis parallel to the y-axis.

Vertices are at $(4, 0 \pm \sqrt{5}) = (4, \pm 2.24)$

Conjugate axis ends at $(4 \pm \sqrt{3}, 0)$ or at $(5.73, 0)$ and $(2.27, 0)$.

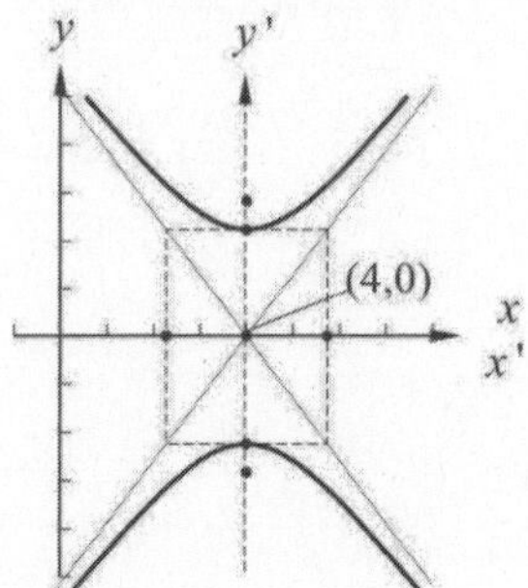

41. Hyperbola: asymptotes: $x - y = -1$ or $y = x + 1$, and $x + y = -3$ or $y = -x - 3$; vertix $(3, -1)$.

The centre is at the point of intersection of the asymptotes. The equations for the asymptotes are solved simultaneously:

$y = x + 1$

$\underline{y = -x - 3}$

$2y = -2$

$y = -1$

$x = -1 - 1$

$x = -2.$

Therefore, the coordinates of the centre are $(-2, -1)$.

Since the slopes of the asymptotes are 1 and -1, $a = b$, where a is the distance from the centre $(-2, -1)$ to the vertex $(3, -1)$;

$a = 3 - (-2) = 5,\ b = 5.$

$\frac{(x-h)^2}{a^2} - \frac{(y-k)^2}{b^2} = 1;$

$\frac{[x-(-2)]^2}{25} - \frac{[y-(-1)]^2}{25} = 1$

$\frac{(x+2)^2}{25} - \frac{(y+1)^2}{25} = 1$

$x^2 + 4x + 4 - (y^2 + 2y + 1) = 25;$

$x^2 + 4x + 4 - 2y - 1 = 25$

$x^2 - y^2 + 4x - 2y - 22 = 0$

45. Parabola: vertex and focus on x-axis. Vertex will be $(h, 0)$ and axis will be the x-axis.

$y^2 = 4p(x - h)$

49. $(x-h)^2 = 4p(y-k)$

$(x-28)^2 = 4p(y-18)$

Solve for $4p$ using $(x, y) = (0, 0)$

$(-28)^2 = 4p(-18)$

$4p = \frac{-28^2}{18}$

$(x-28)^2 = \frac{-28^2}{18}(y-18)$

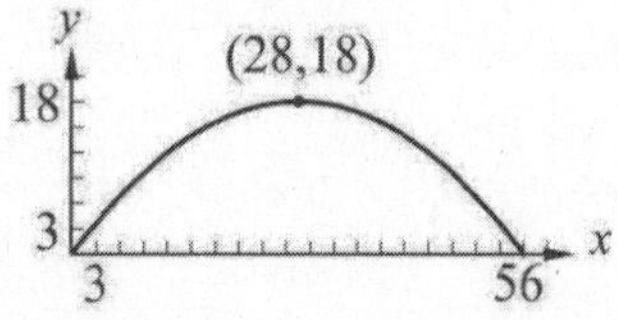

21.8 The Second-degree Equation

1. $2x^2 = 3 + 2y^2$

$2x^2 - 2y^2 - 3 = 0$, A, C have different signs, $B = 0$, hyperbola

5. $2x^2 - y^2 - 1 = 0$

A and C have different signs, $B = 0$; hyperbola

9. $2.2x^2 - x - y = 1.6$

$A \neq 0$; $C = 0$; $B = 0$; parabola

13. $36x^2 = 12y(1 - 3y) + 1$

$36x^2 = 12y - 36y^2 + 1$

$36x^2 + 36y^2 - 12x - 1 = 0$,

$A = C$, $B = 0$, circle

17. $2xy + x - 3y = 6$

$2xy + x - 3y - 6 = 0$

$A = 0$; $B \neq 0$; $C = 0$; hyperbola

21. $(x+1)^2 + (y+1)^2 = 2(x + y + 1)$

$x^2 + 2x + 1 + y^2 + 2y + 1 = 2x + 2y + 2$

$x^2 + y^2 = 0$, point $(0, 0)$ is the only solution

25. $x^2 = 8(y - x - 2)$

$x^2 = 8y - 8x - 16$

$x^2 + 8x - 8y + 16 = 0$

$A \neq 0$; $B = 0$; $C = 0$; parabola

$x^2 + 8x - 8y + 16 = 0$

$x^2 + 8x + 16 = 8y$

$(x+4)^2 = 4(2)y$

$p = 2$

Vertex $(-4, 0)$, focus $(-4, 2)$, directrix $y = -2$.

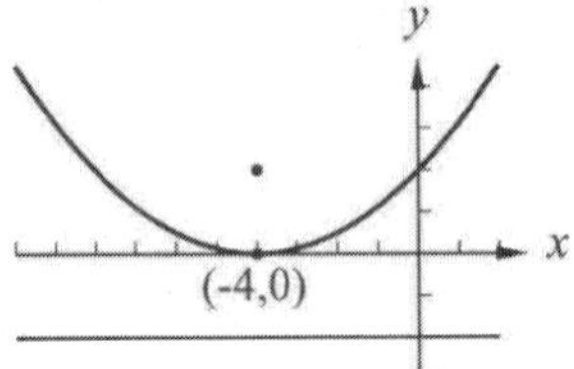

29. $y^2 + 42 = 2x(10 - x)$

$y^2 + 42 = 20x - 2x^2$

$y^2 + 2x^2 - 20x + 42 = 0$;

$A \neq C$, same sign, $B = 0$, ellipse

$\frac{y^2}{2} + x^2 - 10x = -21$

$\frac{y^2}{2} + x^2 - 10x + 25 = -21 + 25$

$\frac{y^2}{2} + (x-5)^2 = 4$

$\frac{y^2}{8} + \frac{(x-5)^2}{4} = 1$

(h, k) at $(5, 0)$, $V(5, \pm 2\sqrt{2})$

$a = \sqrt{8} = 2\sqrt{2}$; $b = 2$

Minor axis ends at $(5 \pm 2, 0)$ or at $(7, 0)$ and $(3, 0)$.

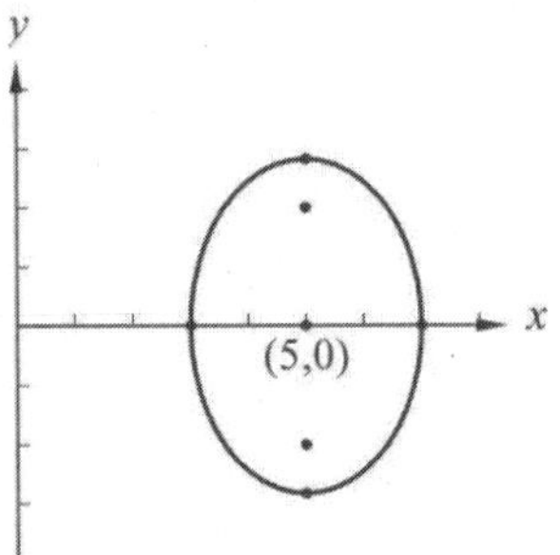

33. $4(y^2 - 4x - 2) = 5(4y - 5)$

$4y^2 - 16x - 8 = 20y - 25$

$4y^2 - 20y - 16x + 17 = 0$

$A = 0$; $C = 4$; $B = 0$; parabola

$y^2 - 5y - 4x + \frac{17}{4} = 0$

To graph, solve for y.

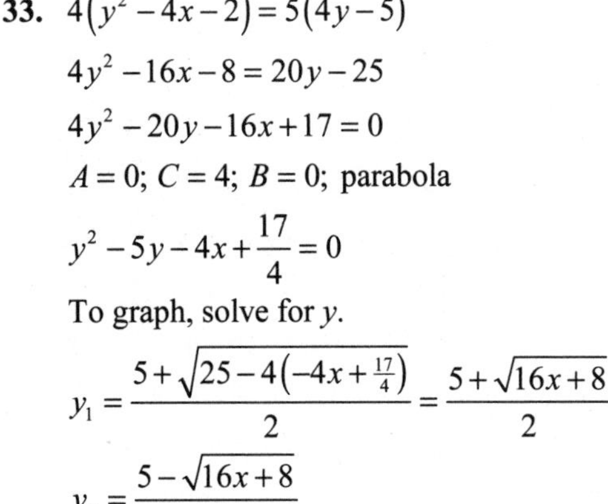

$y_1 = \frac{5 + \sqrt{25 - 4\left(-4x + \frac{17}{4}\right)}}{2} = \frac{5 + \sqrt{16x + 8}}{2}$

$y_2 = \frac{5 - \sqrt{16x + 8}}{2}$

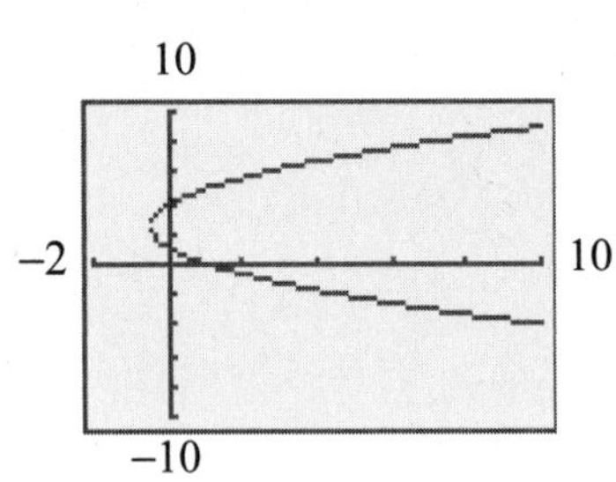

37. $x^2+ky^2=a^2$

(a) If $k=1, x^2+(1)y^2=a^2$

$x^2+y^2=a^2$ $(A=C, B=0, \text{circle})$

(b) If $k<0, x^2-|k|y^2=a^2$

$\frac{x^2}{a^2}-\frac{y^2}{a^2/|k|}=1$ $(A\neq C$, different sign, $B=0$, hyperbola)

(c) If $k>0(k\neq 1)$

$\frac{x^2}{a^2}+\frac{y^2}{a^2/k}=1$ $(A\neq C$, same sign, $B=0$, ellipse)

41. In $Ax^2+Bxy+Cy^2+Dx+Ey+F=0$, if $A=B=C=0$, $D\neq 0$, $E\neq 0$, $F\neq 0$, then the equation is $Dx+Ey+F=0$ whose locus is a straight line.

45. **(a)** Beam is perpendicular to floor. We have a circle.

(b) Beam is not perpendicular to floor. We have an ellipse.

*See conic section diagrams, Fig. 21.89 in the text.

21.9 Rotation of Axes

1. $x^2-y^2=25, \theta=45^\circ;$

$x=x'\cos 45^\circ-y'\sin 45^\circ$

$=\frac{x'}{\sqrt{2}}-\frac{y'}{\sqrt{2}}$

$y=x'\sin 45^\circ+y'\cos 45^\circ$

$=\frac{x'}{\sqrt{2}}+\frac{y'}{\sqrt{2}}$

$$x^2-y^2=\left(\frac{x'}{\sqrt{2}}-\frac{y'}{\sqrt{2}}\right)^2-\left(\frac{x'}{\sqrt{2}}+\frac{y'}{\sqrt{2}}\right)^2=25;$$

$$\frac{x'^2}{2}-\frac{2x'y'}{2}+\frac{y'^2}{2}-\frac{x'^2}{2}-\frac{2x'y'}{2}-\frac{y'^2}{2}=25$$

$2x'y'+25=0$, hyperbola

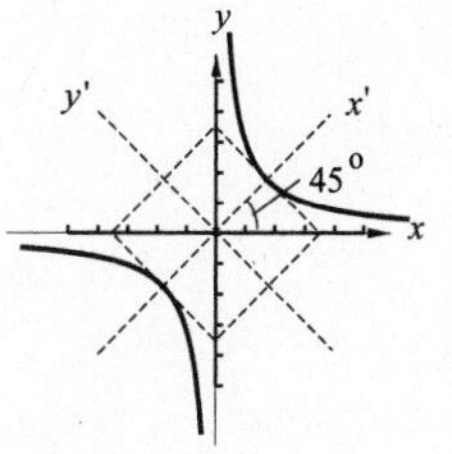

5. $x^2+2xy+x-y-3=0$

$B^2-4AC=2^2-4(1)(0)=4>0$, hyperbola

9. $13x^2+10xy+13y^2+6x-42y-27=0$

$B^2-4AC=10^2-4(13)(13)=-576<0$, ellipse

13. $3x^2+4xy=4$

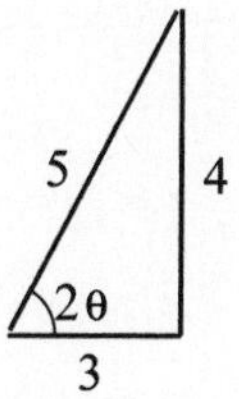

$\tan 2\theta=\frac{B}{A-C}=\frac{4}{3-0}=\frac{4}{3}$

$\cos 2\theta=\frac{3}{5}$

$\sin\theta=\sqrt{\frac{1-\cos 2\theta}{2}}=\sqrt{\frac{1-\frac{3}{5}}{2}}=\frac{1}{\sqrt{5}};$

$\cos\theta=\sqrt{\frac{1+\cos 2\theta}{2}}=\sqrt{\frac{1+\frac{3}{5}}{2}}=\frac{2}{\sqrt{5}};$

$x=\frac{2x'-y'}{\sqrt{5}}, y=\frac{x'+2y'}{\sqrt{5}}$

$$3\left(\frac{2x'-y'}{\sqrt{5}}\right)^2+4\left(\frac{2x'-y'}{\sqrt{5}}\right)\left(\frac{x'+2y'}{\sqrt{5}}\right)=4;$$

$$3(4x'^2-4x'y'+y'^2)+4(2x'^2+3x'y'-2y'^2)=20$$

$20x'^2-5y'^2=20$

$4x'^2-y'^2=4$, hyperbola

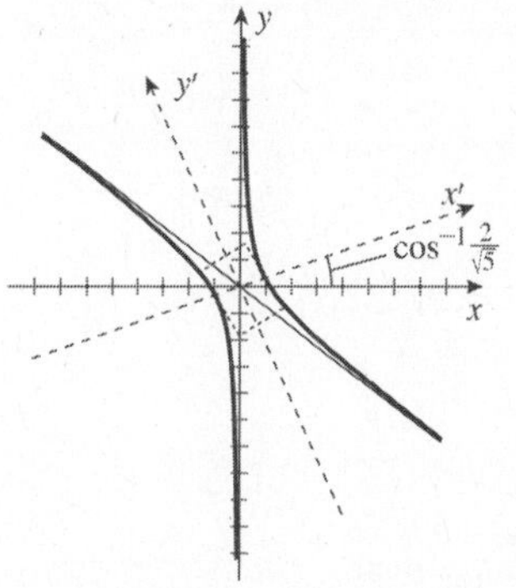

17. $16x^2-24xy+9y^2-60x-80y+400=0$

$\tan 2\theta=\frac{B}{A-C}=\frac{-24}{16-9}=\frac{-24}{7}$

$\cos 2\theta=-\frac{7}{25}$

$\sin\theta=\sqrt{\frac{1-\cos 2\theta}{2}}=\sqrt{\frac{1-\frac{-7}{25}}{2}}=\frac{4}{5}$

$\cos\theta=\sqrt{\frac{1+\cos 2\theta}{2}}=\sqrt{\frac{1+\frac{-7}{25}}{2}}=\frac{3}{5}$

$$x=\frac{3x'-4y'}{5},\ y=\frac{4x'+3y'}{5}$$

$$16\left(\frac{3x'-4y'}{5}\right)^2-24\left(\frac{3x'-4y'}{5}\right)\left(\frac{4x'+3y'}{5}\right)$$

$$+9\left(\frac{4x'+3y'}{5}\right)^2-60\left(\frac{3x'-4y'}{5}\right)-80\left(\frac{4x'+3y'}{5}\right)$$

$$+400=0$$

$$16\left(9x'^2-24x'y'+16y'^2\right)$$

$$-24\left(12x'^2+9x'y'-16x'y'-12y'^2\right)$$

$$+9\left(16x'^2+24x'y'+9y'^2\right)-300(3x'-4y')$$

$$-400(4x'+3y')+10\ 000=0$$

$$625y'^2-2500x'+10\ 000=0$$

$$y'^2-4x'+16=0$$

$$(y'-0)^2-4(x'-4)=0$$

$$y''^2-4x''=0$$

$$y''^2=4x'',\text{ parabola}$$

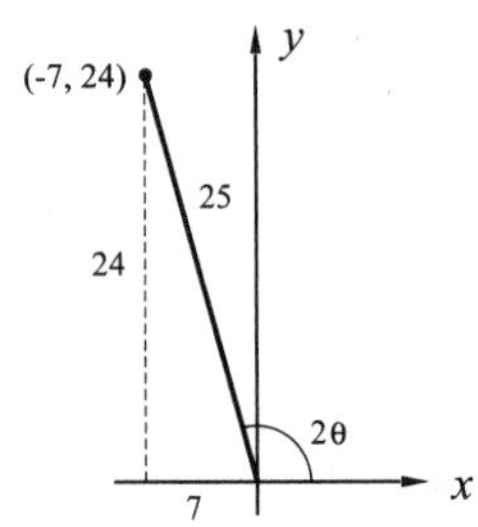

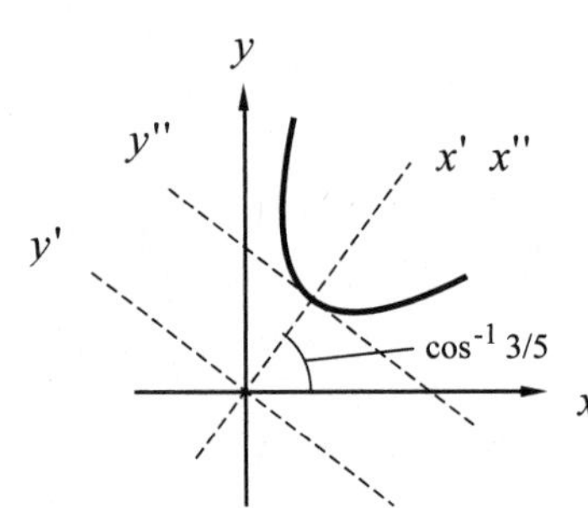

21.10 Polar Coordinates

1. $\left(3,\frac{\pi}{3}\right)$ and $\left(3,-\frac{5\pi}{3}\right)$ represent the same point.

$\left(-3,\frac{\pi}{3}\right)$ and $\left(3,-\frac{2\pi}{3}\right)$ represent the same point

on the opposite side of the pole.

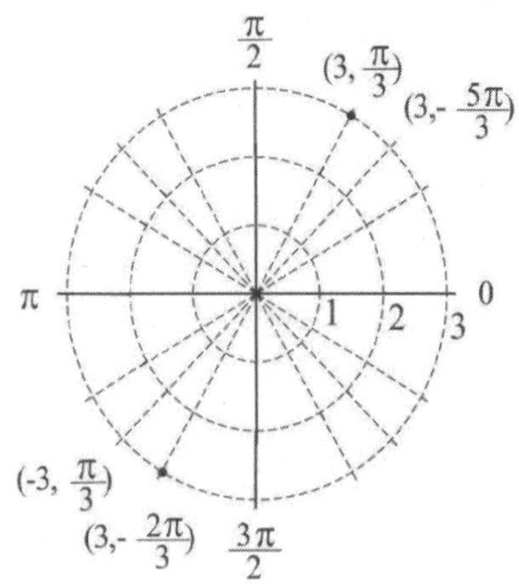

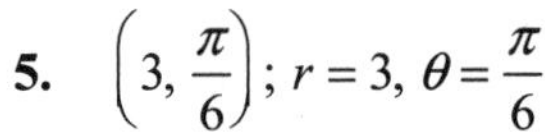

5. $\left(3,\frac{\pi}{6}\right);\ r=3,\ \theta=\frac{\pi}{6}$

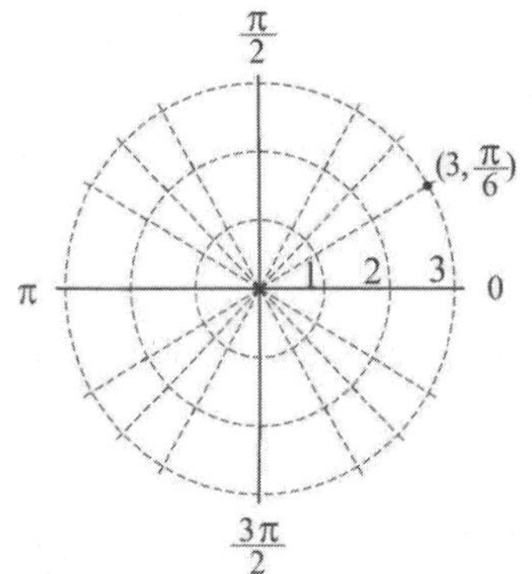

9. $\left(-8,\frac{7\pi}{6}\right)$; negative r is reversed in direction from positive r.

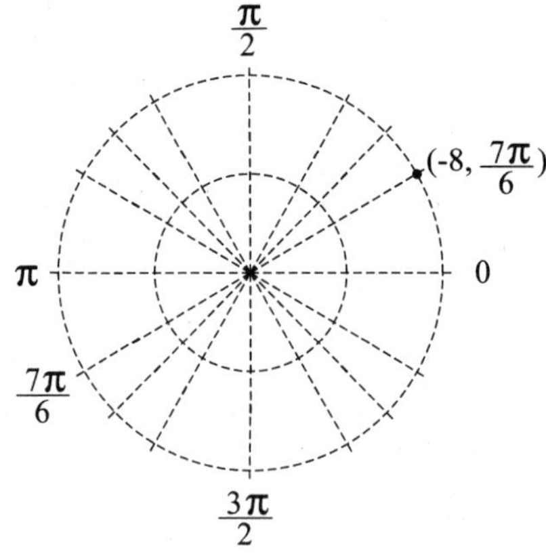

13. $(2, 2)$

$\frac{2}{\pi} = 0.64$, so $2 = 0.64\pi$

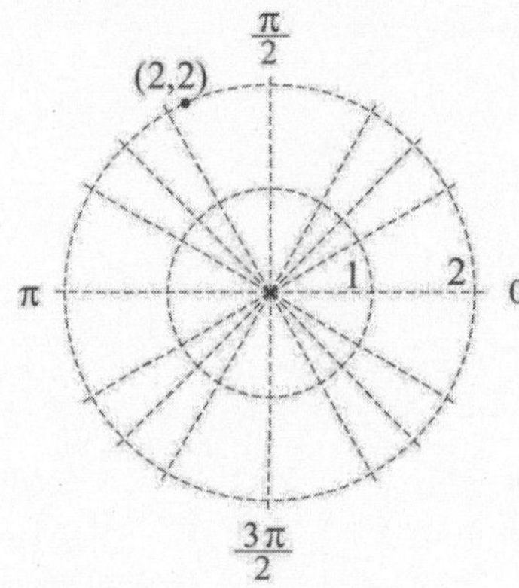

17. $(\sqrt{3}, 1)$ is (x, y), QI

$$\tan\theta = \frac{y}{x}$$

$$\theta = \tan^{-1}\frac{y}{x} = \tan^{-1}\frac{1}{\sqrt{3}} = \tan^{-1}\frac{\sqrt{3}}{3};$$

$$\theta = 30^\circ = \frac{\pi}{6}$$

$$r = \sqrt{x^2+y^2} = \sqrt{(\sqrt{3})^2 + 1^2}$$

$$= \sqrt{3+1} = \sqrt{4} = 2$$

(r, θ) is $\left(2, \frac{\pi}{6}\right)$

21. (r, θ) is $\left(8, \frac{4\pi}{3}\right)$, QIII

$$x = r\cos\theta = 8\cos\frac{4\pi}{3} = 8\left(-\frac{1}{2}\right) = -4$$

$$y = r\sin\theta = 8\left(-\frac{\sqrt{3}}{2}\right) = -4\sqrt{3}$$

(x, y) is $\left(-4, -4\sqrt{3}\right)$

25. $x = 3$

$$r\cos\theta = x = 3$$

$$r = \frac{3}{\cos\theta} = 3\sec\theta$$

29. $x^2 + (y-2)^2 = 4$

$$x^2 + y^2 - 4y + 4 = 4$$

$$r^2 - 4\cdot r\sin\theta = 0$$

$$r = 4\sin\theta$$

33. $x^2 + y^2 = 6y$

$$r^2 = 6\cdot r\sin\theta$$

$$r = 6\sin\theta$$

37. $r = \sin\theta$

$$r^2 = r\sin\theta$$

Substitute $r^2 = x^2 + y^2$ and $y = r\sin\theta$:

$$x^2 + y^2 = y$$

$$x^2 + y^2 - y = 0,$$

$A = C$, $B = 0$, circle

41. $r = \frac{2}{\cos\theta - 3\sin\theta}$

$$r\cos\theta - 3r\sin\theta = 2$$

$x - 3y = 2$, line

45. $r = 2(1+\cos\theta)$

$$r^2 = 2r + 2r\cos\theta$$

$$x^2 + y^2 = 2\sqrt{x^2+y^2} + 2x$$

$$x^2 + y^2 - 2x = 2\sqrt{x^2+y^2}$$

$$\left(x^2+y^2-2x\right)^2 = 4\left(x^2+y^2\right)$$

$$x^4 + y^4 - 4x^3 + 2x^2y^2 - 4xy^2 + 4x^2 = 4x^2 + 4y^2$$

$$x^4 + y^4 - 4x^3 + 2x^2y^2 - 4xy^2 + 4x^2 - 4x^2 - 4y^2 = 0$$

$$x^4 + y^4 - 4x^3 + 2x^2y^2 - 4xy^2 - 4y^2 = 0$$

49. As the graph shows, the point $(2, 3\pi/4)$ is on the curve $r = 2\sin 2\theta$ even though $(2, 3\pi/4)$ is not a solution to $r = 2\sin 2\theta$. $(2, 3\pi/4)$ and $(-2, 7\pi/4)$ are the same point and $(-2, 7\pi/4)$ is a solution to $r = 2\sin 2\theta$.

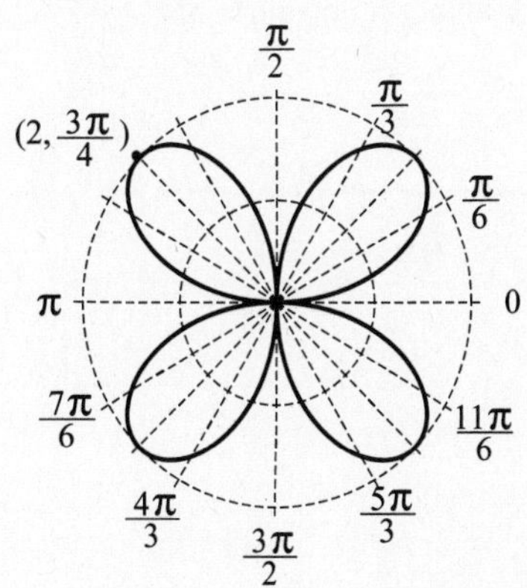

53. Each pair of vertices subtends a central angle of $\frac{\pi}{3}$ at the pole. The coordinates of the other vertices are $(2, 0), \left(2, \frac{\pi}{3}\right), \left(2, \frac{2\pi}{3}\right), \left(2, \frac{4\pi}{3}\right), \left(\ , \frac{5\pi}{3}\right)$

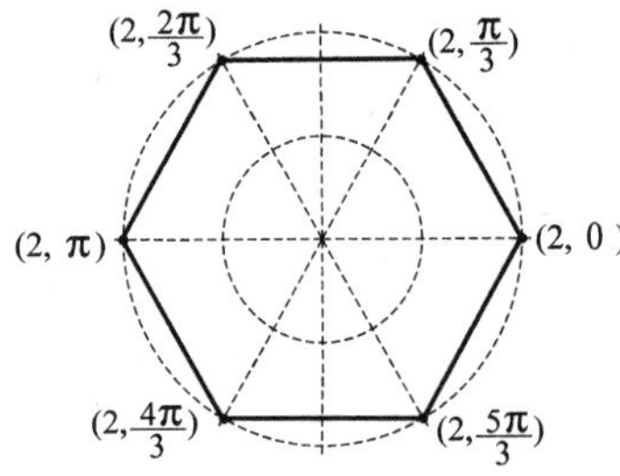

57. $r = 3 - \sin\theta$; $y = r\sin\theta$

$$\sqrt{x^2+y^2} = 3 - \frac{y}{r} = 3 - \frac{y}{\sqrt{x^2+y^2}}$$

$$\sqrt{x^2+y^2} = \frac{3\sqrt{x^2+y^2} - y}{\sqrt{x^2+y^2}}$$

$$x^2+y^2 = 3\sqrt{x^2+y^2} - y$$

$$x^2+y^2+y = 3\sqrt{x^2+y^2}$$

Square both sides.

$$x^4 + 2x^2y^2 + 2x^2y + y^4 + 2y^3 + y^2 = 9\left(x^2+y^2\right)$$
$$= 9x^2 + 9y^2$$

Therefore,

$$x^4 + 2x^2y^2 + 2x^2y + y^4 + 2y^3 - 9x^2 - 8y^2 = 0$$

21.11 Curves in Polar Coordinates

1. The graph of $\theta = \frac{5\pi}{6}$ is a straight line through the pole. $\theta = \frac{5\pi}{6}$ for all possible values of r.

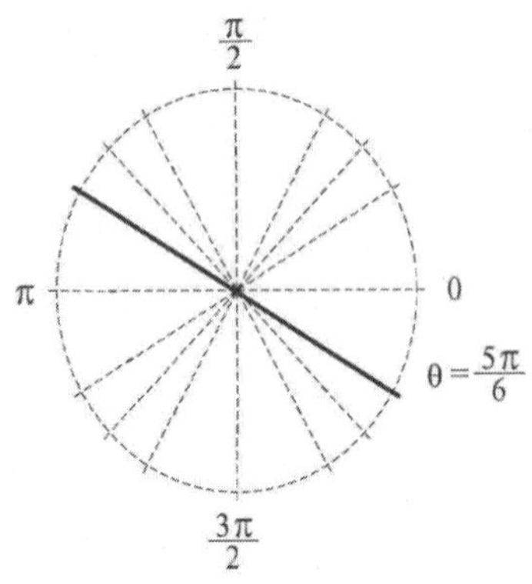

5. $r = 5$ for all θ. Graph is a circle with radius 5.

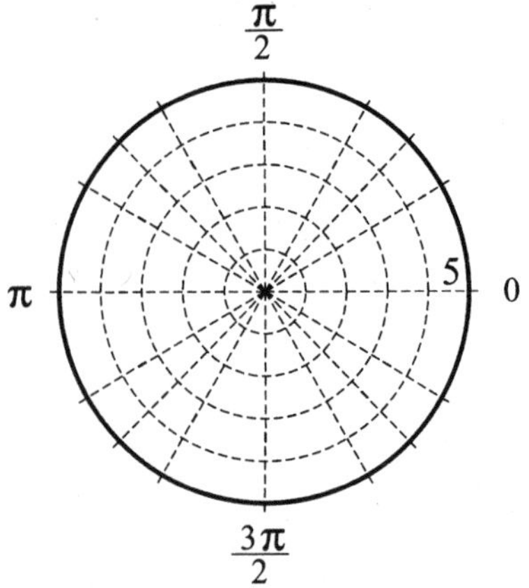

9. $r = 4\sec\theta = \frac{4}{\cos\theta}$

$r\cos\theta = 4$

$x = 4$

This is a vertical line at $x = 4$

θ	r
0	4
$\frac{\pi}{6}$	4.6
$\frac{\pi}{4}$	5.7
$\frac{\pi}{3}$	8
$\frac{\pi}{2}$	*
$\frac{2\pi}{3}$	−8
$\frac{3\pi}{4}$	−5.7
$\frac{5\pi}{6}$	4.6
π	−4
$\frac{5\pi}{4}$	−5.7
$\frac{3\pi}{2}$	*
$\frac{7\pi}{4}$	5.7
2π	4

* denotes undefined

13. $1 - r = \cos\theta$

$r = 1 - \cos\theta$; cardioid

θ	r
0	0
$\frac{\pi}{4}$	0.3
$\frac{\pi}{2}$	1
$\frac{3\pi}{4}$	1.7
π	2
$\frac{5\pi}{4}$	1.7
$\frac{3\pi}{2}$	1
$\frac{7\pi}{4}$	0.3

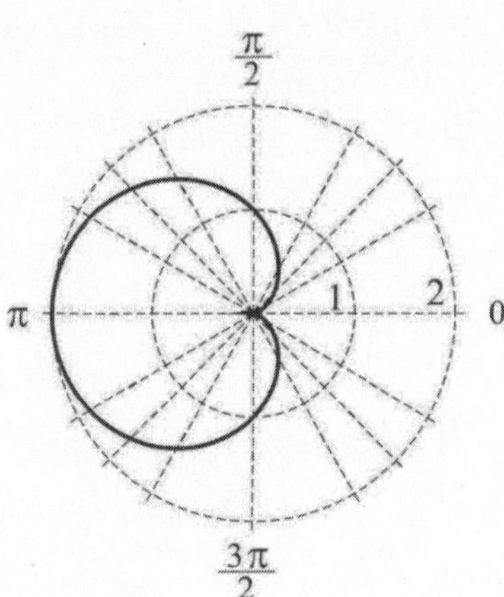

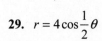

17. $r = 4\sin 2\theta$; rose (4 petals)

θ	r
0	0
$\frac{\pi}{8}$	2.8
$\frac{\pi}{4}$	4
$\frac{3\pi}{8}$	−2.8
$\frac{\pi}{2}$	0
$\frac{5\pi}{8}$	2.8
$\frac{3\pi}{4}$	−4
$\frac{7\pi}{8}$	−2.8
π	0
$\frac{9\pi}{8}$	2.8
$\frac{5\pi}{4}$	4
$\frac{11\pi}{8}$	2.8
$\frac{3\pi}{2}$	0
$\frac{13\pi}{8}$	−2.8
$\frac{7\pi}{4}$	−4
2π	−2.8

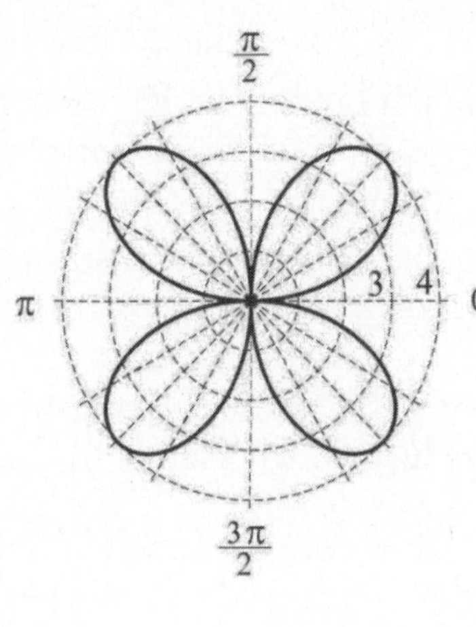

21. $r = 2^{\theta}$; spiral

θ	r
0	1
$\frac{\pi}{4}$	1.7
$\frac{\pi}{2}$	3.0
$\frac{3\pi}{4}$	5.1
π	8.8
$\frac{5\pi}{4}$	15.2
$\frac{3\pi}{2}$	26.2
$\frac{7\pi}{4}$	45.2
2π	77.9

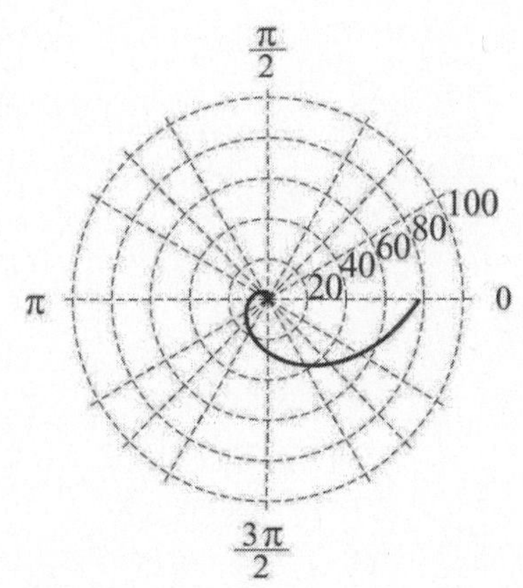

25. $r = \dfrac{3}{2-\cos\theta}$; ellipse

θ	r
0	3
$\frac{\pi}{4}$	2.32
$\frac{\pi}{2}$	1.5
$\frac{3\pi}{4}$	1.11
π	1
$\frac{5\pi}{4}$	1.11
$\frac{3\pi}{2}$	1.5
$\frac{7\pi}{4}$	2.32
2π	3

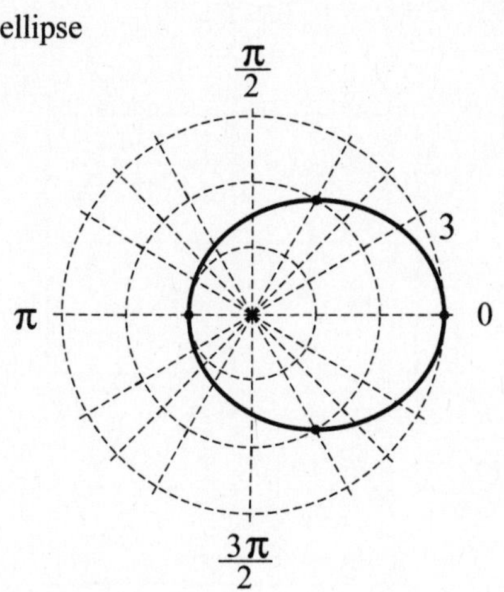

29. $r = 4\cos\dfrac{1}{2}\theta$

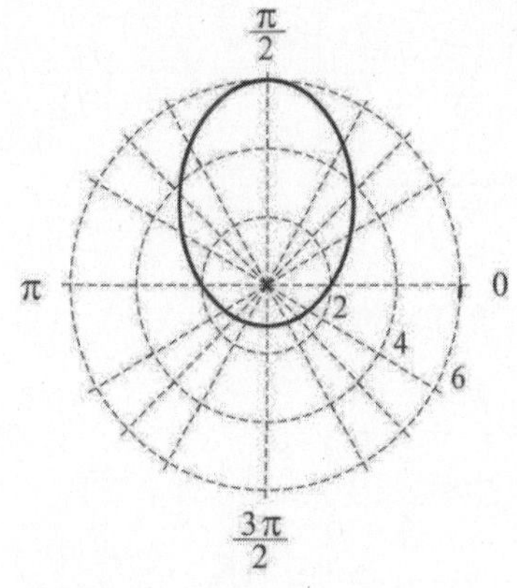

θ	r	θ	r
0	4.0	$\frac{13\pi}{6}$	−3.9
$\frac{\pi}{6}$	3.9	$\frac{9\pi}{4}$	−3.7
$\frac{\pi}{4}$	3.7	$\frac{7\pi}{3}$	−3.5
$\frac{\pi}{3}$	3.5	$\frac{5\pi}{2}$	−2.8
$\frac{\pi}{2}$	2.8	$\frac{8\pi}{3}$	−2.0
$\frac{2\pi}{3}$	2.0	$\frac{11\pi}{4}$	−1.5
$\frac{3\pi}{4}$	1.5	$\frac{17\pi}{6}$	−1.0
$\frac{5\pi}{6}$	1.0	3π	0
π	0	$\frac{19\pi}{6}$	1.0
$\frac{7\pi}{6}$	−1.0	$\frac{13\pi}{4}$	1.5
$\frac{5\pi}{4}$	−1.5	$\frac{10\pi}{3}$	2.0
$\frac{4\pi}{3}$	−2.0	$\frac{7\pi}{2}$	2.8
$\frac{3\pi}{2}$	−2.8	$\frac{11\pi}{3}$	3.5
$\frac{5\pi}{3}$	−3.5	$\frac{15\pi}{4}$	3.7
$\frac{7\pi}{4}$	−3.7	$\frac{23\pi}{6}$	3.9
$\frac{11\pi}{6}$	−3.9	4π	4.0
2π	−4.0		

33. $r = \theta \quad (-20 \le \theta \le 20)$

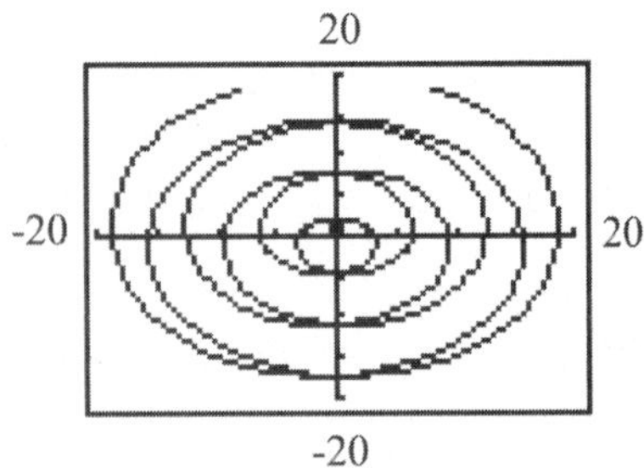

37. $r = 3\cos 4\theta$

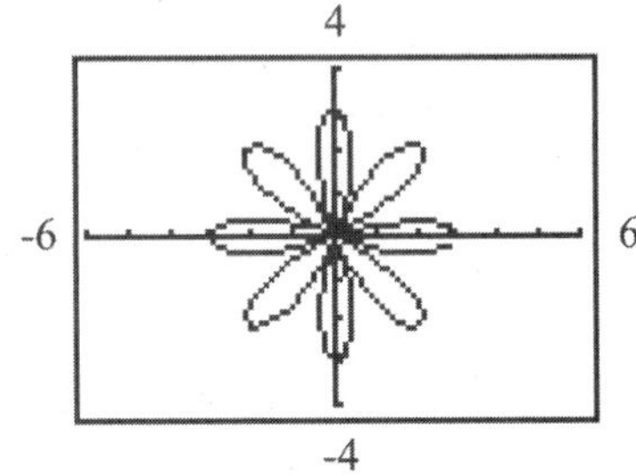

41. $r + 2 = \cos 2\theta \Rightarrow r = \cos 2\theta - 2$

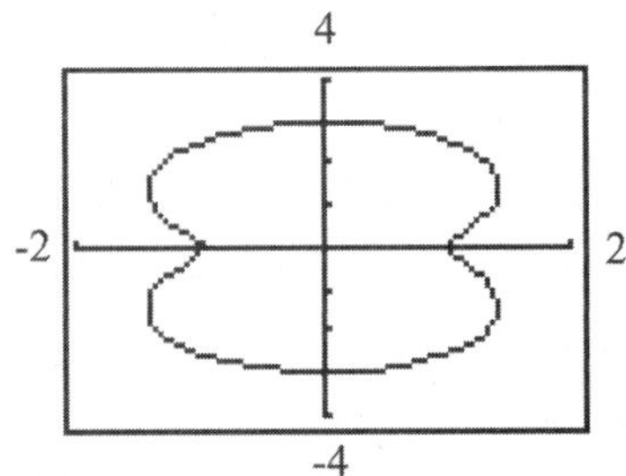

45. From the calculator screen the curves intersect at $(0, 0)$ and $(1, 1)$, where the tangent lines are horizontal and vertical, showing the curves intersect at right angles.

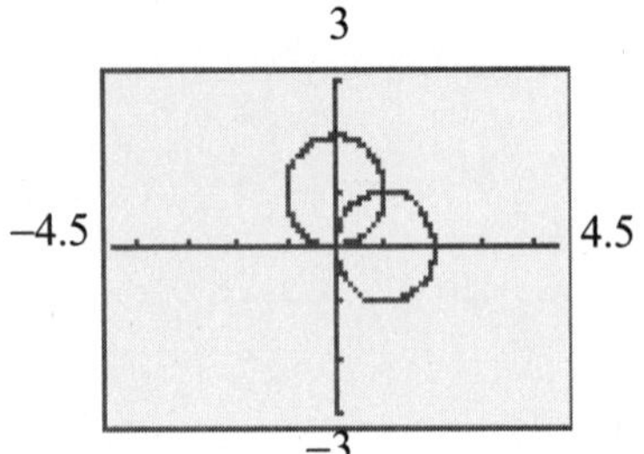

49. $r = \dfrac{25}{10 + 4\cos\theta}$

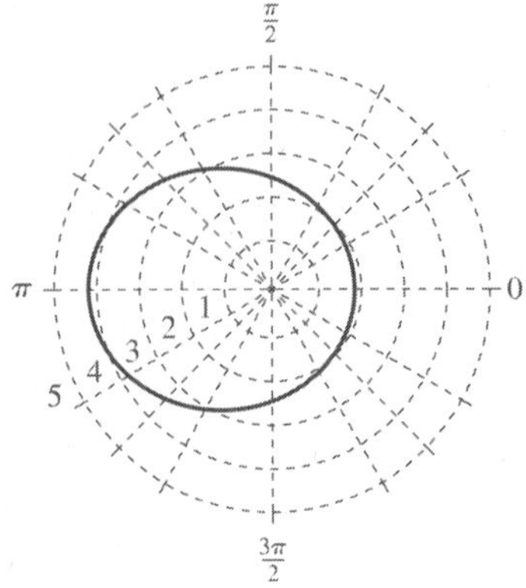

53.

exercise	equation	graph
17	$r = 4\sin 2\theta$	rose with 4 petals
18	$r = 2\sin 3\theta$	rose with 3 petals
37	$r = 3\cos 4\theta$	rose with 8 petals
38	$r = 3\sin 5\theta$	rose with 5 petals

The graph of $r = a\sin n\theta$ or $r = b\cos n\theta$ is a rose with n loops for n odd and a rose with $2n$ loops for n even.

Review Exercises

1. Given straight line; (x_1, y_1) is $(1, -7)$; $m = 4$

$$y - y_1 = m(x - x_1)$$
$$y - (-7) = 4(x - 1)$$
$$y + 7 = 4x - 4$$
$$y = 4x - 4 - 7$$
$$y = 4x - 11 \text{ or } 4x - y - 11 = 0$$

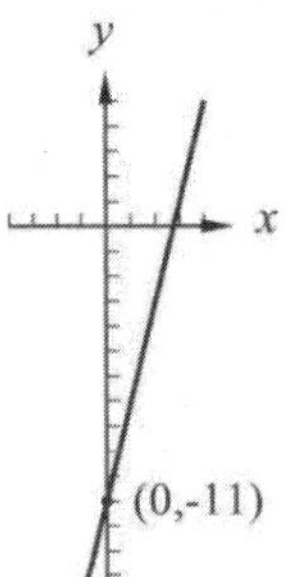

5. $x^2 + y^2 = 6x$

$x^2 - 6x + 9 + y^2 = 9$

$(x-3)^2 + y^2 = 3^2$; circle, centre $(3, 0)$, $r = 3$

The concentric circle has equation

$(x-3)^2 + y^2 = r^2$ and passes through $(4, -3)$

Substitute $(4, -3)$:

$$(4-3)^2 + (-3)^2 = r^2$$
$$r^2 = 10$$
$$(x-3)^2 + y^2 = 10$$
$$x^2 - 6x + 9 + y^2 = 10$$
$$x^2 - 6x + y^2 - 1 = 0$$

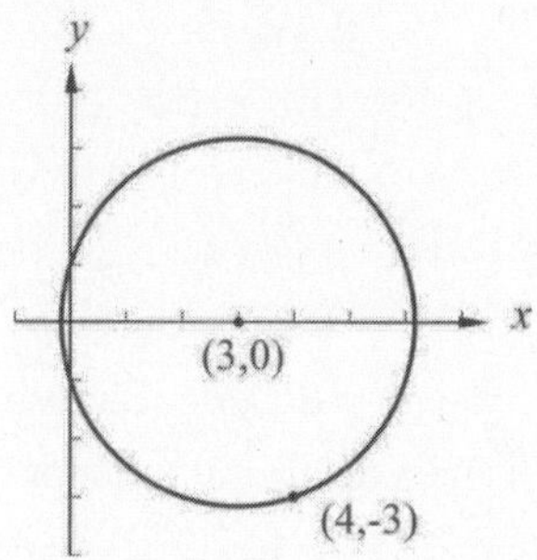

9. $V(10,0), F(8,0)$, tangent to $x = -10$

The major axis is the x-axis. Since the ellipse has vertex $V(10,0)$ and is tangent to $x = -10$, the other vertex is $(-10,0)$, $a = 10$ and the centre is the origin. Therefore, $c = 8$

$$a^2 = b^2 + c^2$$
$$100 = b^2 + 8^2$$
$$b^2 = 36$$
$$\frac{x^2}{100} + \frac{y^2}{36} = 1 \text{ or } 9x^2 + 25y^2 = 900$$

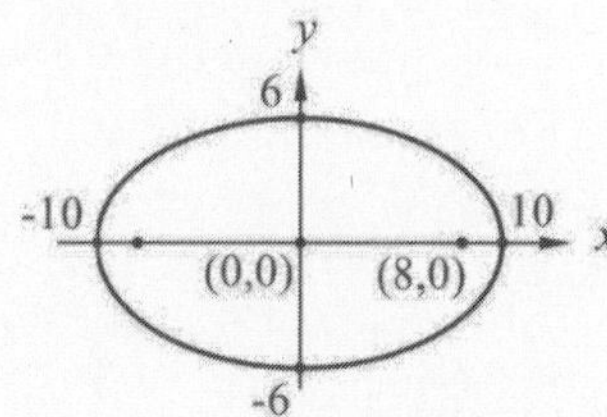

13. Given $x^2 + y^2 + 6x - 7 = 0$

$$(x^2 + 6x) + (y^2) = 7$$
$$(x^2 + 6x + 9) + y^2 = 7 + 9$$
$$(x+3)^2 + (y+0)^2 = 16$$
$$[x-(-3)]^2 + (y-0)^2 = 4^2$$
$$C(h, k) = (-3, 0);\ r = 4$$

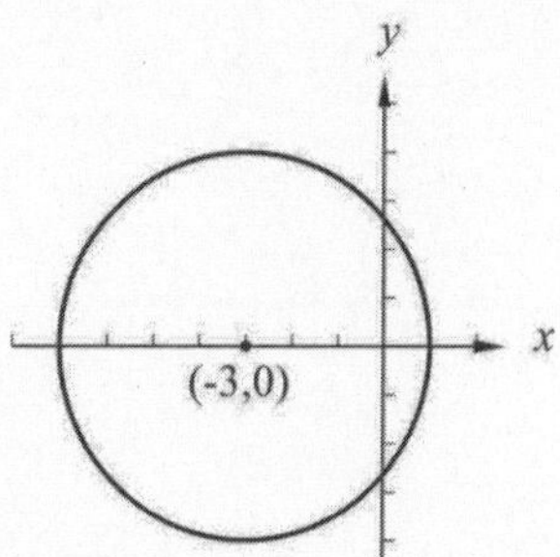

17. Given $4x^2 + y^2 = 1$

$$\frac{x^2}{\frac{1}{4}} + \frac{y^2}{1} = 1$$

$a^2 = 1,\ b^2 = \frac{1}{4},\ c = \sqrt{1 - \frac{1}{4}} = \frac{\sqrt{3}}{2}$

vertices $(0, 1), (0, -1)$

foci: $\left(0, \frac{\sqrt{3}}{2}\right), \left(0, \frac{-\sqrt{3}}{2}\right)$

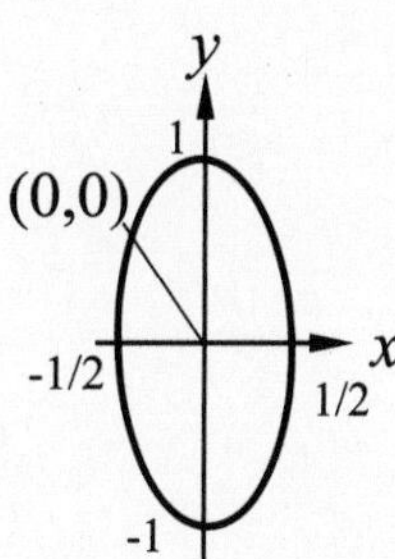

21. Given $x^2 - 8x - 4y - 16 = 0$

$x^2 - 8x = 4y + 16; x^2 - 8x + 16 = 4y + 16 + 16$

$(x-4)^2 = 4y + 32; (x-4)^2 = 4(y+8)$

$4p = 4; p = 1$

vertex (h, k) is $(4, -8)$; focus is $(4, -7)$

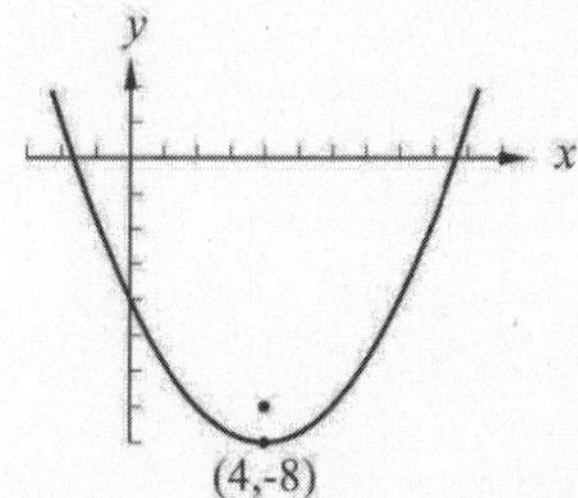

25. $x^2 - 2xy + y^2 + 4x + 4y = 0$

$B^2 - 4AC = (-2)^2 - 4(1)(1) = 0$, parabola

$A = C$, so $\theta = 45^\circ$

$$\left(x'\cdot\frac{1}{\sqrt{2}} - y'\cdot\frac{1}{\sqrt{2}}\right)^2 - 2\left(x'\cdot\frac{1}{\sqrt{2}} - y'\cdot\frac{1}{\sqrt{2}}\right)$$
$$\left(x'\cdot\frac{1}{\sqrt{2}} + y'\cdot\frac{1}{\sqrt{2}}\right) + \left(x'\cdot\frac{1}{\sqrt{2}} + y'\cdot\frac{1}{\sqrt{2}}\right)^2$$
$$+4\left(x'\cdot\frac{1}{\sqrt{2}} - y'\cdot\frac{1}{\sqrt{2}}\right) + 4\left(x'\cdot\frac{1}{\sqrt{2}} + y'\cdot\frac{1}{\sqrt{2}}\right) = 0$$

$y'^2 = -2\sqrt{2}x'$

$V(0, 0)$

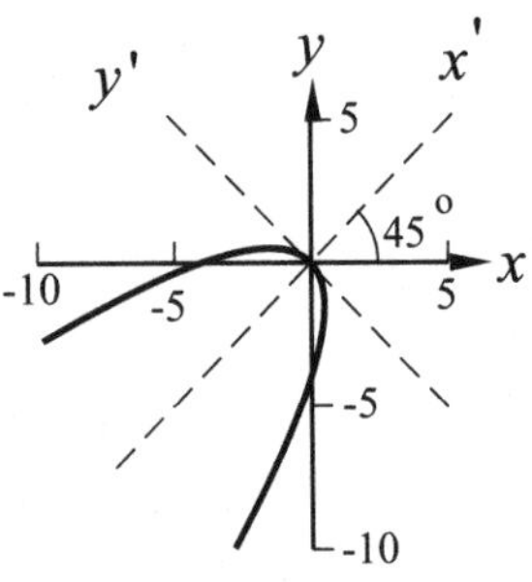

29. $r = 4\cos 3\theta$

Let $\theta = 0$ to π in steps of $\dfrac{\pi}{48}$.

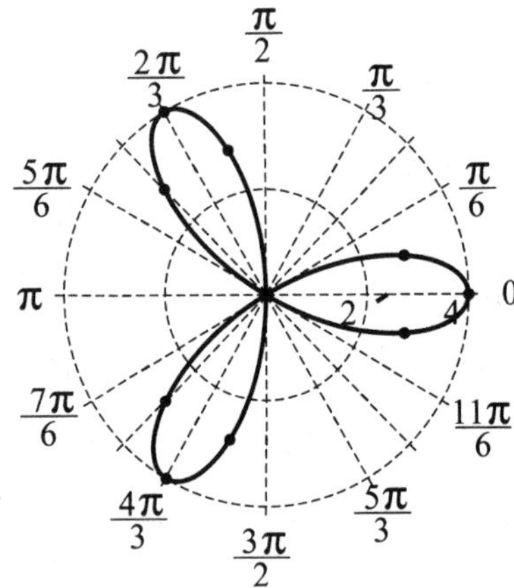

33. $r = 2\sin\dfrac{\theta}{2}$

Let $\theta = 0$ to 4π in steps of $\dfrac{\pi}{24}$.

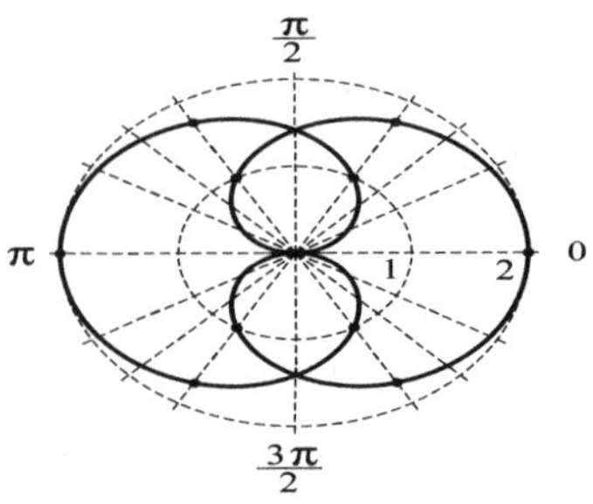

37. $x^2 + xy + y^2 = 2$

$x^2 + y^2 + xy = 2$

$r^2 + (r\cos\theta)(r\sin\theta) = 2$

$r^2 = \dfrac{2}{1 + \sin\theta\cos\theta}$

41. $r = \dfrac{4}{2 - \cos\theta}$

$2r - r\cos\theta = 4$

$2r = 4 + r\cos\theta = 4 + x$

$4r^2 = 16 + 8x + x^2$

$4(x^2 + y^2) = x^2 + 8x + 16$

$4x^2 + 4y^2 = x^2 + 8x + 16$

$3x^2 + 4y^2 - 8x - 16 = 0$

45. $x^2 + y^2 - 4y - 5 = 0$

$y^2 - 4x^2 - 4 = 0$

From the graph, two real solutions.

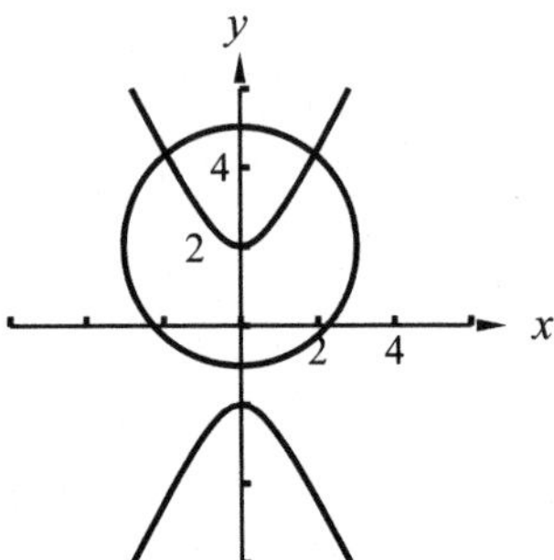

49. $x^2 + 3y + 2 - (1 + x)^2 = 0$

$3y = (1 + x)^2 - x^2 - 2$

$y = \dfrac{(1 + x)^2 - x^2 - 2}{3}$

$= \dfrac{1 + 2x + x^2 - x^2 - 2}{3}$

$$= \frac{2x-1}{3}$$
$$= \frac{2}{3}x - \frac{1}{3}$$

Graph $y_1 = \frac{2}{3}x - \frac{1}{3}$

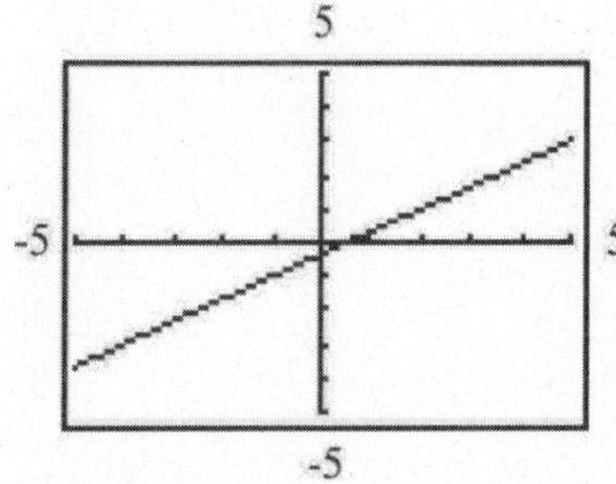

53. $x^2 - 4y^2 + 4x + 24y - 48 = 0.$ Solve for y by completing the square.

$$y^2 - 6y = 0.25x^2 + x - 12$$
$$y^2 - 6y + 9 = 0.25x^2 + x - 3$$
$$(y-3)^2 = 0.25x^2 + x - 3$$
$$y = \pm\sqrt{0.25x^2 + x - 3} + 3$$

Graph $y_1 = \sqrt{0.25x^2 + x - 3} + 3$

$y_2 = -\sqrt{0.25x^2 + x - 3} + 3$

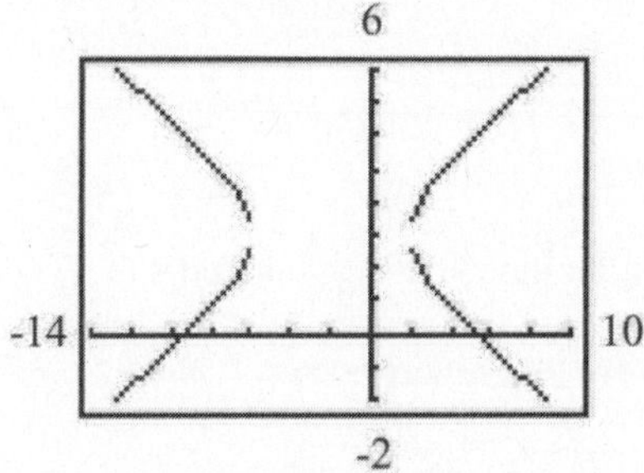

57. $r = 2 - 3\csc\theta = 2 - \frac{3}{\sin\theta}$

Graph $r_1 = 2 - \frac{3}{\sin\theta}$

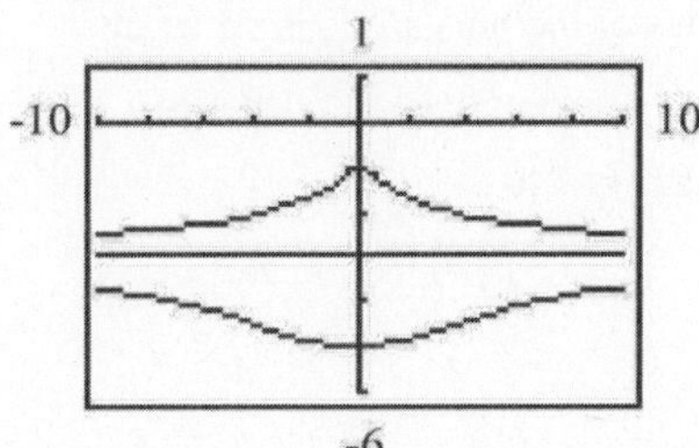

61.

y

$P(x, y)$

x

d_1 d_2

(1,-3) (7,-3)

$d_1 + d_2 = 8$ describes an ellipse with centre $(4, -3)$,

$$2a = 8$$
$$a = 4, a^2 = 16$$
$$c = 3, c^2 = 9.$$
$$a^2 = b^2 + c^2$$
$$16 = b^2 + 9$$
$$b^2 = 7$$
$$\frac{(x-4)^2}{16} + \frac{(y+3)^2}{7} = 1$$

65. The slopes between every pair of points must be the same:

$$\frac{x-(-3)}{13-3} = \frac{-3-(-5)}{3-(-2)}$$
$$x = 1$$

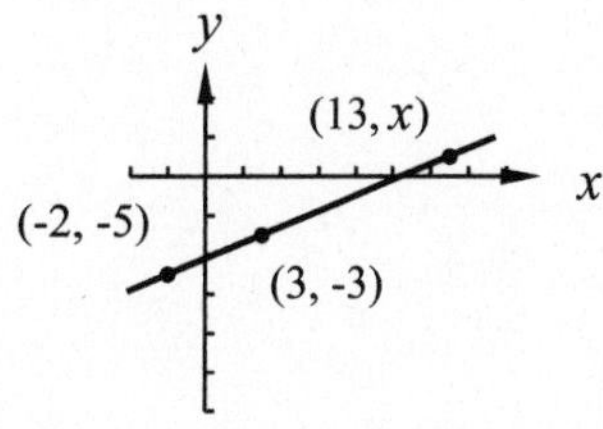

69. $a^2 + b^2 = (-3-2)^2 + (11+1)^2 + (14-2)^2 + (4+1)^2$

$$a^2 + b^2 = 338$$
$$c^2 = (14+3)^2 + (4-11)^2 = 338,$$

points form a right triangle

$$m_a = \frac{11+1}{-3-2} = \frac{12}{-5},\ m_b = \frac{4+1}{14-2} = \frac{5}{12}$$
$$m_a \cdot m_b = \frac{12}{-5} \cdot \frac{5}{12} = -1 \Rightarrow a \perp b,$$

points form a right triangle

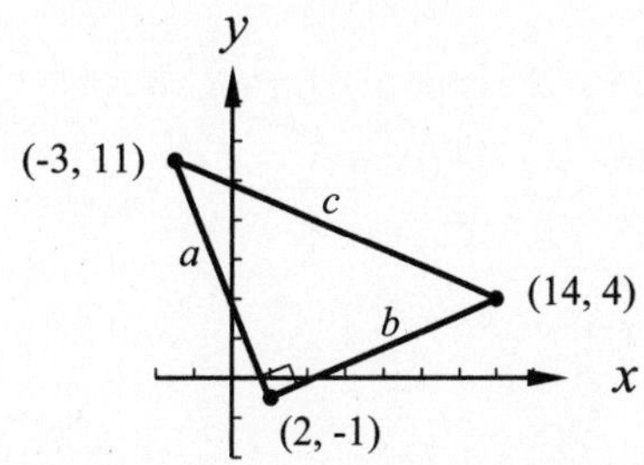

73. $(x+jy)^2+(x-jy)^2=2$

$x^2+2jxy-y^2+x^2-2jxy-y^2=2$

$x^2-y^2-1=0$, hyperbola

77. From definition,

$(x-3)^2+(y-1)^2=(y+3)^2$

$x^2-6x+9+y^2-2y+1=y^2+6y+9$

$x^2-6x-8y+1=0$

From translation of axes,

$(h, k)=(3, -1),\ p=2 \Rightarrow 4p=8$

$(x-h)^2=4p(y-k)$

$(x-3)^2=8(y+1)$

$x^2-6x+9=8y+8$

$x^2-6x-8y+1=0$

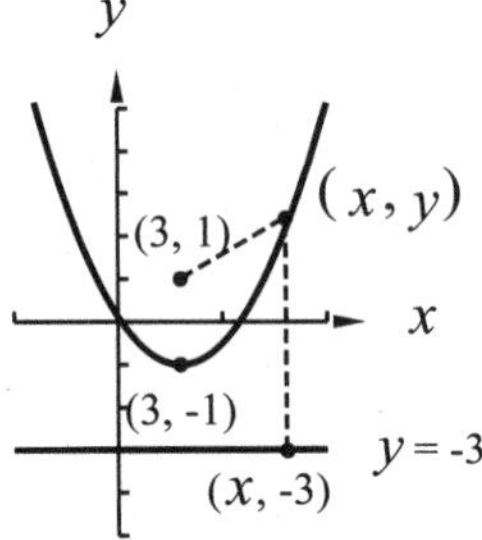

81. $v=v_0+at$

$6.20=1.92+a(5.50)$

$a=0.778$

$v=1.92+0.778t$

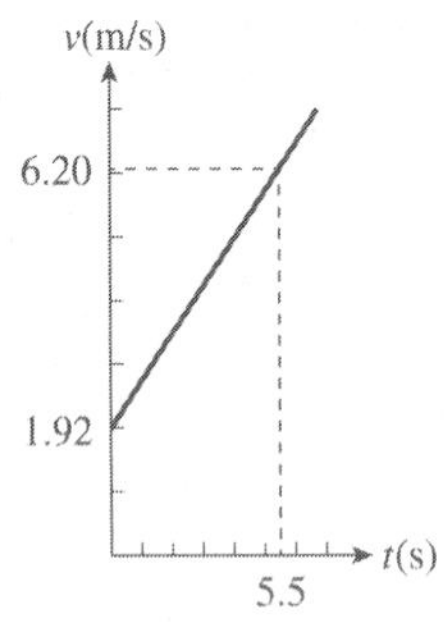

85. The intersection of the cone and the highway is a circle. Its radius is $r=170\tan 7^\circ$

$A=\pi r^2=\pi(170\tan 7^\circ)^2$

$A=1400\text{ m}^2$

89. $y^2=4px$

$40^2=4p(50)$

$4p=32$

$y^2=32x$

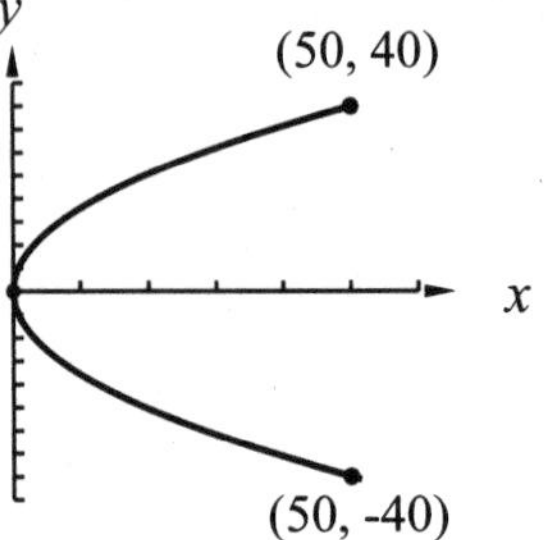

93. $P=12.0i-0.500i^2$

$i^2-24i=-2P$

$i^2-24i+144=-2(P-72)$

$(i-12)^2=-2(P-72)$

This is a parabola with vertex at $(12, 72)$, with axis parallel to the P-axis and opening downward.

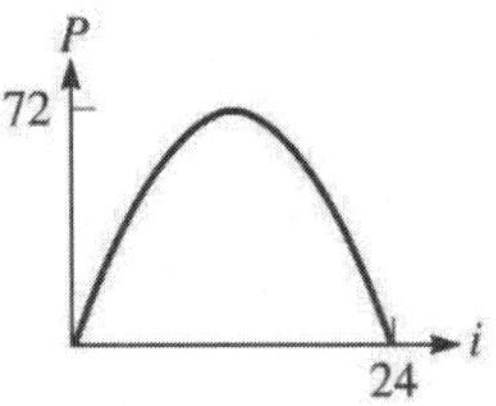

97.

If the pins are on the x-axis and centred, the equation of the ellipse has $a=\frac{10}{2}=5$ and

$b=\frac{6}{2}=3$, so $\frac{x^2}{5^2}+\frac{y^2}{3^2}=1$

$a^2=b^2+c^2$

$5^2=3^2+c^2$

$c=4$

The coordinates of the pins are $(\pm 4, 0)$ so they are 8 cm apart. The length of the string is

$l=2a+2c=2(3)+2(4)$

$l=18\text{ cm}$

101. Let $P(x, y)$ be the coordinates of the recorder in a coordinate system with origin at the target. Let the rifle be at $(0, r)$. Using the distance from the recorder to the rifle, we have

$\sqrt{x^2+(y-r)^2}=v_s(t_0+t_1)$ where t_0 is the time for the bullet to reach the target and t_1 is the time for sound to reach the detector from the target.

Using the distance from the recorder to the target,

$\sqrt{x^2+y^2}=v_s t_1$

Subtracting

$\sqrt{x^2+(y-r)^2}-\sqrt{x^2+y^2}=v_s t_0=\text{constant}=2a$

By definition, if the difference of distances from two points is constant, the locus is a hyperbola. We now obtain its second-degree equation.

$\sqrt{x^2+(y-r)^2}=2a+\sqrt{x^2+y^2}$; square both sides

$x^2+y^2-2ry+r^2=4a^2+4a\sqrt{x^2+y^2}+x^2+y^2$

$-2ry+(r^2-4a^2)=4a\sqrt{x^2+y^2}$; square both sides

$4r^2y^2-4r(r^2-4a^2)y+(r^2-4a^2)^2=16a^2(x^2+y^2)$

$-16a^2x^2+4(r^2-4a^2)y^2-4r(r^2-4a^2)y+(r^2-4a^2)^2=0$

which has the form $Ax^2+Cy^2+Ey+F=0$

Since the bullet travels faster than the speed of sound (at speed v_B),

$r=v_B t_0>v_s t_0=2a$, and $r^2-4a^2>0$.

Therefore, A and C differ in sign, with $B=0$, so the equation represents a hyperbola.

Summary:

The difference of distances from P to the rifle and from P to the target is a constant, so the locus is a hyperbola. Moreover, in the second-degree equation obtained A and C differ in sign with $B=0$.

105. $r=200(\sec\theta+\tan\theta)^{-5}/\cos\theta,\quad 0<\theta<\pi/2$

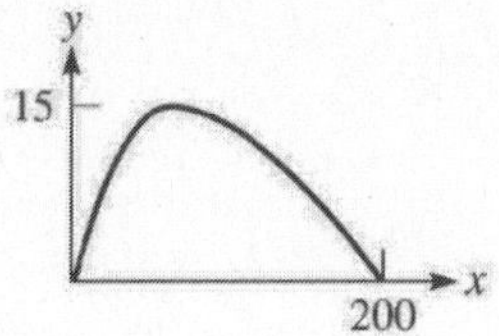

CHAPTER 22

INTRODUCTION TO STATISTICS

22.1 Tabular and Graphical Representation of Data

1.

Est. (hrs)	0 – 5	6 – 11	12 – 17	18 – 23	24 – 29
Freq.	5	12	19	9	5

5. 5.3, 5.4, 5.4, 5.4, 5.5, 5.6, 5.6, 5.6, 5.7, 5.7, 5.8, 5.8, 5.8, 5.8, 5.9, 5.9, 5.9, 6.0, 6.1, 6.3

Litres of gasoline	Frequency
5.3	1
5.4	3
5.5	1
5.6	3
5.7	2
5.8	4
5.9	3
6.0	1
6.1	1
6.2	0
6.3	1

9.

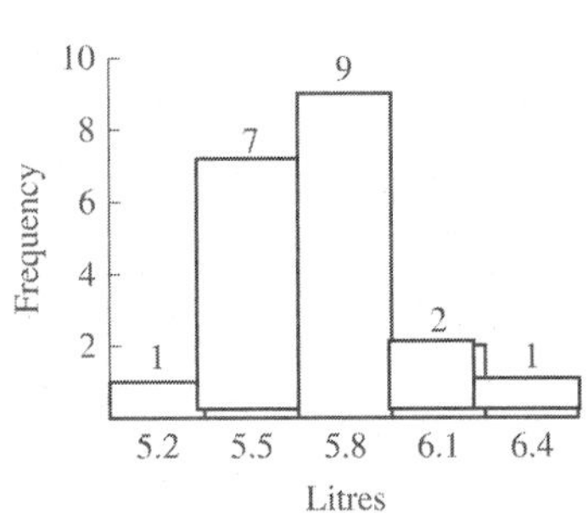

13. Since there are only eight different values for the 15 tests, we use the number of instructions to build the frequency table.

Number	18	19	20	21	22	23	24	25
Frequency	1	3	2	4	3	1	0	1

17.

Time (s)	Frequency
2.21	2
2.22	7
2.23	18
2.24	41
2.25	56
2.26	32
2.27	8
2.28	3
2.29	3

21.

Dist. (m)	Rel. freq. (%)
47 – 49	1.7
50 – 52	12.5
53 – 55	26.7
56 – 58	30.0
59 – 61	20.0
62 – 64	8.3
65 – 67	0.8

25. The class width has to be larger than (0.451 − 0.383) / 6 = 0.011, so we pick a class width of 0.015. If the first class mark is 0.380, the first lower class limit is 0.380 − (0.015 / 2) = 0.373.

Dosage (mSv)	Frequency
0.373 – 0.387	1
0.388 – 0.402	2
0.403 – 0.417	2
0.418 – 0.432	7
0.433 – 0.447	7
0.448 – 0.462	1

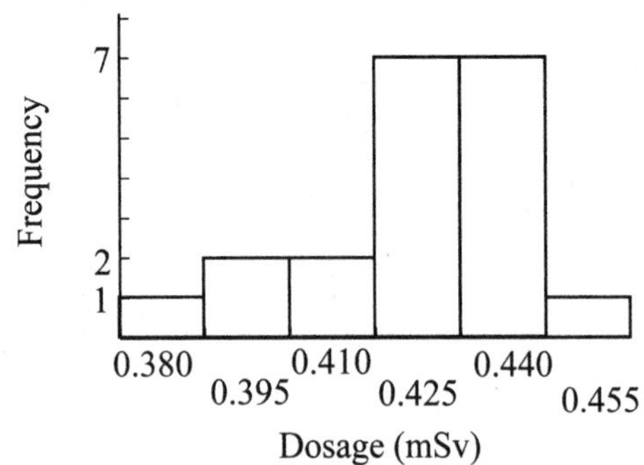

29.

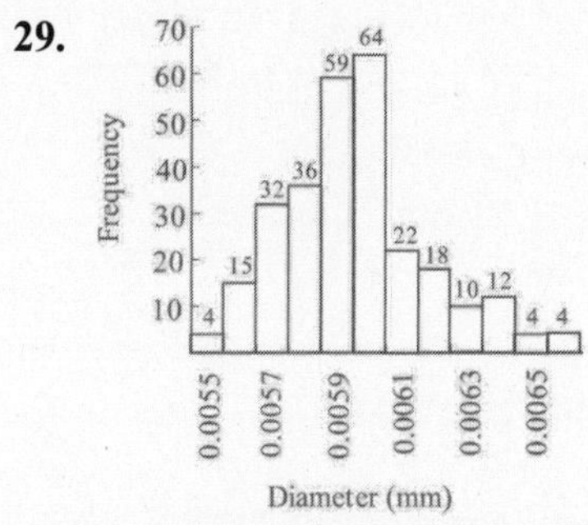

22.2 Summarizing Data

1. 1, 2, 2, 3, 4, 4, 4, 6, 7, 7, 8, 9, 9, 11

There are 14 numbers. The median is halfway between the seventh and eighth numbers.

$$\text{median} = \frac{4+6}{2} = 5$$

5.

x	$x-\bar{x}$	$(x-\bar{x})^2$
6	2	4
5	1	1
4	0	0
7	3	9
6	2	4
2	−2	4
1	−3	9
1	−3	9
5	1	1
3	−1	1
40		42

$\bar{x} = 40/10 = 4$

$$\frac{\sum(x-\bar{x})^2}{n-1} = \frac{42}{10-1} = \frac{42}{9}$$

$$s = \sqrt{\frac{42}{9}} = 2.2$$

9. Arrange in ascending order: 23, 23, 24, 25, 25, 25, 26, 26, 27, 28, 28;

$n = 11$;

(a) The mean is $\bar{x} = \frac{\sum x}{n} = \frac{280}{11} = 25.5$

(b) The median is the sixth observation, so the median is 25.

(c) The mode is the most common observation, so it is 25.

13.

x	$x-\bar{x}$	$(x-\bar{x})^2$	x^2
23	−2.455	6.0248	529
23	−2.455	6.0248	529
24	−1.455	2.1157	576
25	−0.4545	0.206 61	625
25	−0.4545	0.206 61	625
25	−0.4545	0.206 61	625
26	0.545 45	0.297 52	676
26	0.545 45	0.297 52	676
27	1.5455	2.3884	729
28	2.5455	6.479 73	784
28	2.5455	6.479 73	784
280		30.72803	7158

$$\bar{x} = \frac{\sum x}{n} = \frac{280}{11} = 25.455$$

(a) Using Eq. (22.2):

$$s = \sqrt{\frac{\sum(x-\bar{x})^2}{n-1}} = \sqrt{\frac{30.072803}{11-1}} = 1.8$$

(b) Using Equation (22.3):

$$s = \sqrt{\frac{\sum x^2 - (\sum x)^2}{n(n-1)}}$$

$$= \sqrt{\frac{11(7158) - (280)^2}{11(10)}}$$

$$= 1.8$$

17. $s = \sqrt{\frac{\sum(x-\bar{x})^2}{n-1}}$

$$= \sqrt{\frac{1.2575}{19}}$$

$$= 0.257 \text{ L/100 km}$$

21. Mean: $\bar{t} = (2(2.21) + 7(2.22) + 18(2.23) + 41(2.24) + 56(2.25) + 32(2.26) + 8(2.27) + 3(2.28) + 3(2.29))/170$

$$= \frac{382.13}{170} = 2.248 \text{ s}$$

Median: There are 170 observations, so the median is the average of observations 85 and 86. Since both of these have class mark of 2.25, the median is 2.25 s.

25. Mean:

$$\bar{d} = (0.383+0.390+0.396+0.409+0.415+0.418 +0.421+0.423+0.425+0.426+0.427+0.429 +0.433+0.434+2(0.436)+0.437+0.441 +0.444+0.451)/20 = 0.4237 \text{ mSv}$$

Median: There are 20 observations, so the median is the average between observation 10 and observation 11, or $\frac{0.426+0.427}{2} = 0.4265$ mSv

Mode: The most repeated value is 0.436, so the mode is 0.436.

29. Mean: $\bar{x} = \frac{\sum xf}{\sum f} = \frac{1.6673}{280} = 0.005\,95$ mm

Median: There are 280 observations, so the median is the average between observation 140 and observation 141. Since both of these are 0.0059, the median is 0.0059 mm.

33. Range $= 800 - 525 = 275$

$$s = \sqrt{\frac{n\sum x^2 - \left(\sum x\right)^2}{n(n-1)}} = \sqrt{\frac{14(5\,975\,625)-(9075)^2}{14(13)}} = \$84.62$$

37. Arrange in order: 0.14, 0.15, 0.15, 0.17, 0.17, 0.18, 0.18, 0.18, 0.19, 0.20, 0.22, 0.22, 0.23, 0.23, 0.24, 0.26, 0.27, 0.32

There are 18 observations, so the median is between observation 9 and observation 10, or $(0.19+0.20)\div 2 = 0.195$ ppm

Mode: The most common value is 0.18, so the mode is 0.18 ppm

41. Lowest = 525; highest = 800;

midrange $= \frac{1325}{2} = \$662.5$

45. $\bar{x} = \frac{\sum xf}{\sum f} = \frac{12\,375}{14} = \$883.9$

The mean increased by \$235.9, from \$648 to \$883.9. The mean no longer represents the centre of the data, since it is bigger than all the other observations.

49. We can write

$55.7 = 62 - 3(2.1)$, and $68.3 = 62 + 3(2.1)$, so we use Chebychev's theorem with $k = 3$. Therefore, at least 89% of observations are within 55.7 and 68.3.

22.3 Normal Distributions

1. $\mu = 10$, $\alpha = 5$ and $\mu = 20$, $\alpha = 5$ result in the same curve, with the first centred at $x = 10$ and the second centred at $x = 20$.

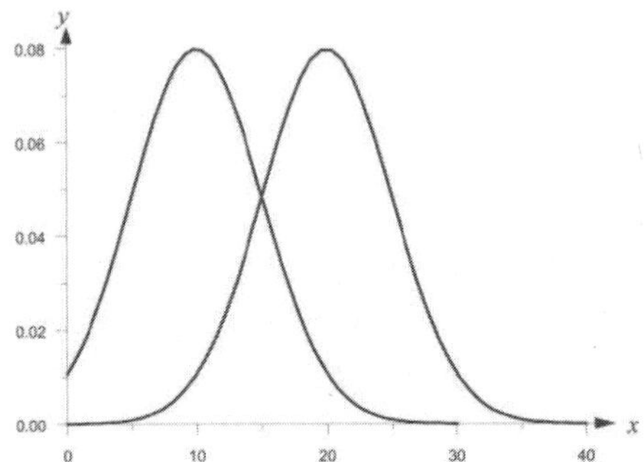

5. $\mu = 10$, $\sigma = 5$ gives the same graph as the left curve in Fig. 22.7.

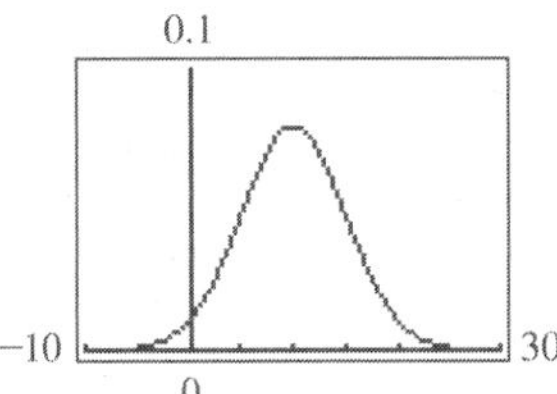

9. 68% of bags have a weight that is one standard deviation from the mean. From a sample of 200 bags, it is expected that 68% of them will have a weight that is one standard deviation away from the mean. This is the same as $0.6826 * 200 = 136.52 = 137$ bags

13. $\mu = 1.50$, $\sigma = 0.05$. x between 1.45 and 1.55

$$z = \frac{1.45-1.50}{0.05} = -1,\ z = \frac{1.55-1.50}{0.05} = 1$$

The area under the curve between -1 and 1 is $0.3413 + 0.3413 = 0.6826$, so $(0.6826)(500) = 341.3$, and 341 batteries are expected to have a voltage between 1.45 V and 1.55 V.

17. $\mu = 100\ 000$, $\sigma = 10\ 000$. x between 85 000 and 100 000

$$z = \frac{85\ 000 - 100\ 000}{10\ 000} = -1.5$$

$$z = \frac{100\ 000 - 100\ 000}{10\ 000} = 0$$

The area under the curve between -1.5 and 0 is 0.4332, so the number of tires that are expected to last between 85 000 km and 100 000 km is $(0.4332)(5000) = 2166$ tires.

21. $\sigma_{\bar{x}} = \dfrac{\sigma}{\sqrt{n}} = \dfrac{10\ 000}{\sqrt{5000}} = 141.$

About 68% of all samples of size 5000 have a mean lifetime from 99 859 km to 100 141 km.

25. Since 50% of area is to right of 0, 25.8% of area will be between z and 0 (to the left of 0). This gives a z − value of 0.7, from the table, but since we know that z is to the left of 0, the required value is -0.7.

29. From the given values for the mean and the standard deviation, we have:

$\bar{x} + s = 2.262$ and $\bar{x} - s = 2.234$.

The readings within these bounds are 2.24, 2.25, and 2.26, with frequencies of $32 + 56 + 41 = 129$. Thus 129 of the 170 readings, or 76%, fall within these bounds. The normal distribution percentage would be 68%.

22.4 Confidence Intervals

1. We substitute the given values of $\bar{x}$, σ, and n into Eq. (22.10). The resulting 95% confidence interval is

$$\bar{x} \pm 1.96 \cdot \frac{\sigma}{\sqrt{n}} = 32.8 \pm 1.96 \cdot \frac{2.3}{\sqrt{140}}$$
$$= 32.8 \pm 0.381$$
$$= (32.4, 33.2)$$

5. We use Eq. (22.16) with $E = 0.025$ and $z_{\alpha/2} = 1.96$. The required sample size is

$$n = \frac{1}{4}\left[\frac{z_{\alpha/2}}{E}\right]^2$$
$$= \frac{1}{4}\left[\frac{1.96}{0.025}\right]^2$$
$$= 1536.6,$$

so that 1537 parts must be sampled for 95% confidence.

9. We have $\hat{p} = \frac{38}{215}, n\hat{p} = 38 \geq 5$, and $n(1 - \hat{p}) = 177 \geq 5$. Therefore, we can apply Eq. (22.14) with $z_{\alpha/2} = 1.96$. The desired 95% confidence interval is

$$\hat{p} \pm z_{\alpha/2}\sqrt{\frac{\hat{p}(1-\hat{p})}{n}} = \frac{38}{215} \pm 1.96\sqrt{\frac{\frac{38}{215}\cdot\frac{177}{215}}{215}}$$
$$= 0.177 \pm 0.051$$
$$= (0.126, 0.228)$$

13. A 90% confidence interval when $n = 50$ is given by

$$\bar{x} \pm z_{\alpha/2} \cdot \frac{\sigma}{\sqrt{n}} = 83.7 \pm 1.645 \cdot \frac{3.14}{\sqrt{50}}$$
$$= 83.7 \pm 0.73$$
$$= (83.0, 84.4)$$

17. A 90% confidence interval is given by

$$\bar{x} \pm z_{\alpha/2} \cdot \frac{s}{\sqrt{n}} = 107.2 \pm 1.645 \cdot \frac{9.2}{\sqrt{45}}$$
$$= 107.2 \pm 2.26$$
$$= (104.9, 109.5)$$

21. The desired margin of error, written as a proportion, is $E = 0.04$. Since no prior information is used, Eq. (22.16) gives

$$n = \frac{1}{4}\left[\frac{z_{\alpha/2}}{E}\right]^2$$
$$= \frac{1}{4}\left[\frac{1.96}{0.04}\right]^2$$
$$= 600.25$$

Therefore, 601 observations must be taken for a margin of error of 0.04 with 95% confidence.

22.5 Statistical Process Control

1. With the new observations, the mean of Subgroup 1 changes to 497.8, and the range R of the subgroup stays the same ($R = 9$). Knowing the sum of the means from before, we can recalculate the new sum without constructing a complete table by subtracting 502.2 (the previous mean for Subgroup 1) and adding 497.8 (the new mean for Subgroup 1). Therefore,

$$\bar{\bar{x}} = \frac{9998 - 502.2 + 497.8}{20}$$
$$= 499.7$$

Hence,

$$\text{UCL}(\bar{x}) = \bar{\bar{x}} + A_2\bar{R} = 499.7 + 0.577(8.7) = 504.7 \text{ mg}$$
$$\text{LCL}(\bar{x}) = \bar{\bar{x}} - A_2\bar{R} = 499.7 - 0.577(8.7) = 494.7 \text{ mg}$$

5.

Hour	Torques	(N·m)	of Five	Engines	
1	366	352	354	360	362
2	370	374	362	366	356
3	358	357	365	372	361
4	360	368	367	359	363
5	352	356	354	348	350
6	366	361	372	370	363
7	365	366	361	370	362
8	354	363	360	361	364
9	361	358	356	364	364
10	368	366	368	358	360
11	355	360	359	362	353
12	365	364	357	367	370
13	360	364	372	358	365
14	348	360	352	360	354
15	358	364	362	372	361
16	360	361	371	366	346
17	354	359	358	366	366
18	362	366	367	361	357
19	363	373	364	360	358
20	372	362	360	365	367

Subgroup	Mean $\bar{x}$	Range R
1	358.8	14
2	365.6	18
3	362.6	15
4	363.4	9
5	352.0	8
6	366.4	11
7	364.8	9
8	360.4	10
9	360.6	8
10	364.0	10
11	357.8	9
12	364.6	13
13	363.8	14
14	354.8	12
15	363.4	14
16	360.8	25
17	360.6	12
18	362.6	10
19	363.6	15
20	365.2	12
Sum	7235.8	248
Mean	361.79	12.4

$$\text{CL}: \bar{\bar{x}} = 361.79 \text{ N}\cdot\text{m}$$
$$\text{UCL}(\bar{x}) = \bar{\bar{x}} + A_2\bar{R} = 361.79 + 0.577(12.4)$$
$$= 368.9 \text{ N}\cdot\text{m}$$
$$\text{LCL}(\bar{x}) = \bar{\bar{x}} - A_2\bar{R} = 361.79 - 0.577(12.4)$$
$$= 354.6 \text{ N}\cdot\text{m}$$

9.

Subgroup	Output	Voltages	Five	Adaptors	
1	9.03	9.08	8.85	8.92	8.90
2	9.05	8.98	9.20	9.04	9.12
3	8.93	8.96	9.14	9.06	9.00
4	9.16	9.08	9.04	9.07	8.97
5	9.03	9.08	8.93	8.88	8.95
6	8.92	9.07	8.86	8.96	9.04
7	9.00	9.05	8.90	8.94	8.93
8	8.87	8.99	8.96	9.02	9.03
9	8.89	8.92	9.05	9.10	8.93
10	9.01	9.00	9.09	8.96	8.98
11	8.90	8.97	8.92	8.98	9.03
12	9.04	9.06	8.94	8.93	8.92
13	8.94	8.99	8.93	9.05	9.10
14	9.07	9.01	9.05	8.96	9.02
15	9.01	8.82	8.95	8.99	9.04
16	8.93	8.91	9.04	9.05	8.90
17	9.08	9.03	8.91	8.92	8.96
18	8.94	8.90	9.05	8.93	9.01
19	8.88	8.82	8.89	8.94	8.88
20	9.04	9.00	8.98	8.93	9.05
21	9.00	9.03	8.94	8.92	9.05
22	8.95	8.95	8.91	8.90	9.03
23	9.12	9.04	9.01	8.94	9.02
24	8.94	8.99	8.93	9.05	9.07

Subgroup	Mean $\overline{\overline{x}}$	Range R
1	8.956	0.23
2	9.078	0.22
3	9.018	0.21
4	9.064	0.19
5	8.974	0.20
6	8.970	0.21
7	8.964	0.15
8	8.974	0.16
9	8.978	0.21
10	9.008	0.13
11	8.960	0.13
12	8.978	0.14
13	9.002	0.17
14	9.022	0.11
15	8.962	0.22
16	8.966	0.15
17	8.980	0.17
18	8.966	0.15
19	8.882	0.12
20	9.000	0.12
21	8.988	0.13
22	8.948	0.13
23	9.026	0.18
24	8.996	0.14
Sum	215.66	3.97
Mean	8.986	0.1654

$$\text{CL}: \overline{\overline{x}} = 8.986 \text{ V}$$

$$\text{UCL}(\overline{x}) = \overline{\overline{x}} + A_2\overline{R} = 8.986 + 0.577(0.1654) = 9.081 \text{ V}$$

$$\text{LCL}(\overline{x}) = \overline{\overline{x}} - A_2\overline{R} = 8.986 - 0.577(0.1654) = 8.891 \text{ V}$$

13. $\text{CL}: \ \mu = 2.725 \text{ cm}$

$$\text{UCL}(\overline{x}) = \mu + A\sigma = 2.725 + 1.342(0.0032) = 2.729 \text{ cm}$$

$$\text{LCL}(\overline{x}) = \mu - A\sigma = 2.725 - 1.342(0.0032) = 2.721 \text{ cm}$$

17.

Week	Accounts with Errors	Proportion with Errors
1	52	0.052
2	36	0.036
3	27	0.027
4	58	0.058
5	44	0.044
6	21	0.021
7	48	0.048
8	63	0.063
9	32	0.032
10	38	0.038
11	27	0.027
12	43	0.043
13	22	0.022
14	35	0.035
15	41	0.041
16	20	0.020
17	28	0.028
18	37	0.037
19	24	0.024
20	42	0.042
Total	738	

$$\text{CL}: \ \bar{p} = \frac{738}{1000(20)} = 0.0369$$

$$\sigma_{\bar{p}} = \sqrt{\frac{\bar{p}(1-\bar{p})}{n}}$$

$$= \sqrt{\frac{0.0369(1-0.0369)}{1000}} = 0.005\ 96$$

$$\text{UCL}(p) = 0.0369 + 3(0.005\ 96) = 0.0548$$

$$\text{LCL}(p) = 0.0369 - 3(0.005\ 96) = 0.0190$$

22.6 Linear Regression

1.

x	y	xy	x^2
1	3	3	1
2	7	14	4
3	9	27	9
4	9	36	16
5	12	60	25
15	40	140	55

$n = 5$

$$m = \frac{n\sum xy - \sum x \sum y}{n\sum x^2 - \left(\sum x\right)^2}$$

$$= \frac{5(140) - 15(40)}{5(55) - 15^2} = 2$$

$$b = \frac{\sum x^2 \sum y - \sum xy \sum x}{n\sum x^2 - \left(\sum x\right)^2}$$

$$= \frac{55(40) - 140(15)}{5(55) - 15^2} = 2$$

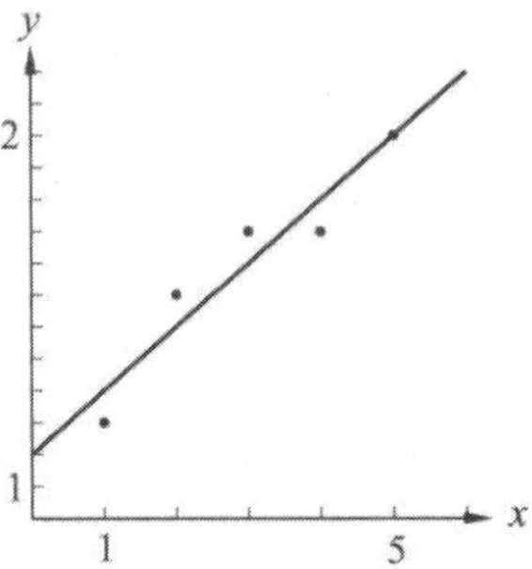

The equation of the least-squares line is $y = 2x + 2$.

5.

t(h)	1.0	2.0	4.0	8.0	10.0	12.0
y(mg/dL)	8.7	8.4	7.7	7.3	5.7	5.2

t	y	ty	t^2
1.0	8.7	8.7	1.0
2.0	8.4	16.8	4.0
4.0	7.7	30.8	16.0
8.0	7.3	58.4	64.0
10.0	5.7	57	100
12.0	5.2	62.4	144
37.0	43	234.1	329

$n = 6$

$$m = \frac{n\sum ty - \sum t \sum y}{n\sum t^2 - \left(\sum t\right)^2}$$

$$= \frac{6(234.1) - 37.0(43)}{6(329) - 37^2} = -0.308$$

$$b = \frac{\sum t^2 \sum y - \sum ty \sum t}{n\sum t^2 - \left(\sum t\right)^2}$$

$$= \frac{329(43) - 234.1(37)}{6(329) - 37^2} = 9.07$$

The least-squares line is

$y = -0.308t + 9.07$

To plot the line we use two points:

t	y
0	9.07
5	7.53

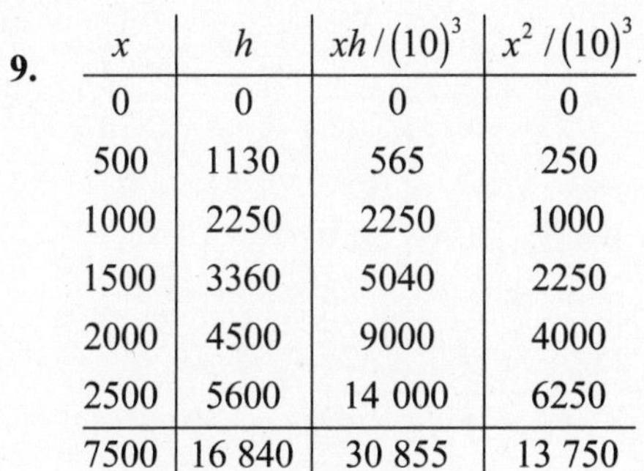

9.

x	h	$xh/(10)^3$	$x^2/(10)^3$
0	0	0	0
500	1130	565	250
1000	2250	2250	1000
1500	3360	5040	2250
2000	4500	9000	4000
2500	5600	14 000	6250
7500	16 840	30 855	13 750

$n = 6$

$$m = \frac{n\sum xh - \sum x\sum h}{n\sum x^2 - \left(\sum x\right)^2}$$

$$= \frac{6\left(30\ 855\times10^3\right)-(7500)(16\ 840)}{6\left(13,750\times10^3\right)-(7500)^2} = 2.24$$

$$b = \frac{\sum x^2\sum y - \sum xy\sum x}{n\sum x^2 - \left(\sum x\right)^2}$$

$$= \frac{\left(13\ 750\times10^3\right)(16\ 840)-\left(30\ 855\times10^3\right)(7500)}{6\left(13\ 750\times10^3\right)-(7500)^2}$$

$= 5.24$

The least-squares line is

$h = 2.24x + 5.24$

To plot the line we use two points:

x	h
750	1690
2250	5045

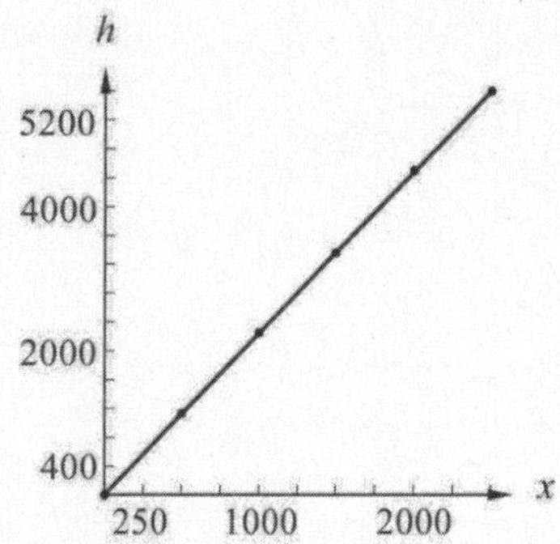

13.

f	V	fV	f^2
0.550	0.350	0.192 50	0.302 500
0.605	0.600	0.363 00	0.366 025
0.660	0.850	0.561 00	0.435 600
0.735	1.10	0.808 50	0.540 225
0.805	1.45	1.167 25	0.648 025
0.880	1.80	1.584 00	0.774 400
4.235	6.15	4.676 25	3.066 775

$n = 6$

$$m = \frac{n\sum fV - \sum f\sum V}{n\sum f^2 - \left(\sum f\right)^2}$$

$$= \frac{6(4.676\ 25)-(4.235)(6.15)}{6(3.066\ 775)-(4.235)^2} = 4.32$$

$$b = \frac{\sum f^2\sum V - \sum fV\sum f}{n\sum f^2 - \left(\sum f\right)^2}$$

$$= \frac{(3.066\ 775)(6.15)-(4.676\ 25)(4.235)}{6(3.066\ 775)-(4.235)^2} = -2.03$$

The least-squares line is

$V = 4.32f - 2.03$

To find the threshold frequency, we set V to 0 and solve for f,

$$f_0 = \frac{2.03}{4.32} = 0.470 \text{ PHz}$$

To plot the line we use two points:

f (PHz)	V
0.600	0.562
0.800	1.426

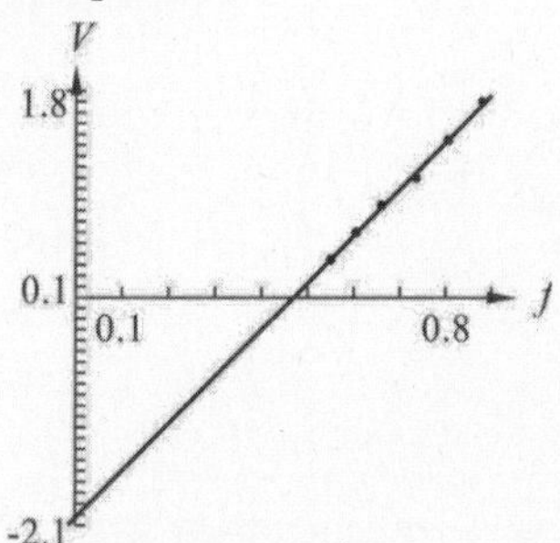

17.

x	y	x^2	y^2
1	15	1	225
3	12	9	144
6	10	36	100
5	8	25	625
8	9	64	81
10	2	100	4
4	11	16	121
7	9	49	81
3	11	9	121
8	7	64	49
55	94	373	990

$n = 10$

$$s_x = \sqrt{\frac{n\left(\sum x^2\right) - \left(\sum x\right)^2}{n(n-1)}}$$
$$= \sqrt{\frac{10(373) - (55)^2}{10(9)}}$$
$$= 2.80$$

$$s_y = \sqrt{\frac{n\left(\sum y^2\right) - \left(\sum y\right)^2}{n(n-1)}}$$
$$= \sqrt{\frac{10(990) - (94)^2}{10(9)}}$$
$$= 3.44$$

$m = -1.1064$

$$r = -1.1064\left(\frac{2.80}{3.44}\right) = -0.901$$

22.7 Nonlinear Regression

1.

x	$f(x) = x^2$	y	x^2y	$(x^2)^2$
0	0	2	0	0
1	1	3	3	1
2	4	10	40	16
3	9	25	225	81
4	16	44	704	256
5	25	65	1625	625
	55	149	2597	979

$n = 6$

$$m = \frac{6(2597) - 55(149)}{6(979) - 55^2} = 2.59$$

$$b = \frac{979(149) - 2597(55)}{6(979) - 55^2} = 1.07$$

$y = 2.59x^2 + 1.07$

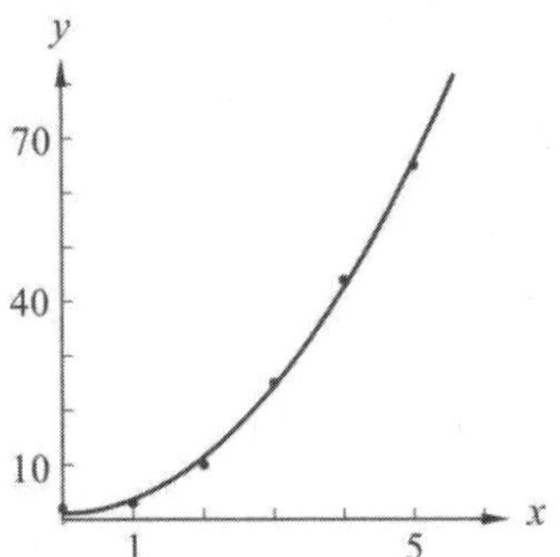

5. $y = mt^2 + b$

t	y	t^2	yt^2	$(t^2)^2$
1.0	6.0	1.0	6.0	1.0
2.0	23	4.0	92	16.0
3.0	55	9.0	495	81.0
4.0	98	16.0	1568	256
5.0	148	25.0	3700	625
	330	55.0	5861.0	979.0

$n = 5$

$$m = \frac{5(5861) - (55.0)(330)}{5(979.0) - (55.0)^2} = 5.97$$

$$b = \frac{(979.0)(330) - (5861)(55.0)}{5(979.0) - (55.0)^2} = 0.38$$

$y = 5.97t^2 + 0.38$

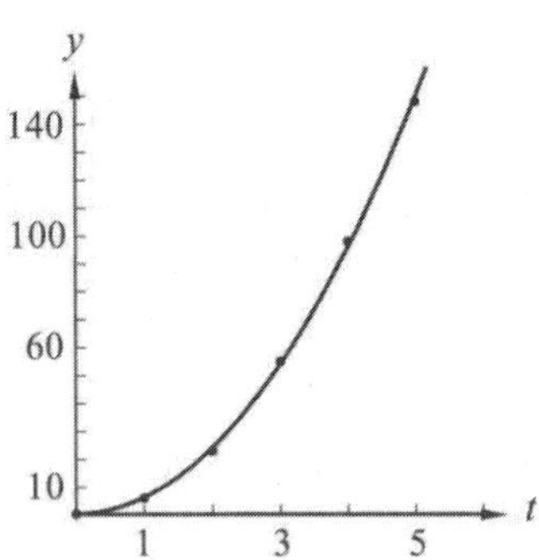

9.

S	$\frac{1}{S}$	P	$\left(\frac{1}{S}\right)P$	$\left(\frac{1}{S}\right)^2$
240	0.004 166	5.60	0.023 333	1.7361×10^{-5}
305	0.003 278	4.40	0.014 426	1.0749×10^{-5}
420	0.002 380	3.20	0.007 619	5.6689×10^{-5}
480	0.002 083	2.80	0.005 833	4.3402×10^{-5}
560	0.001 785	2.40	0.004 285	3.1887×10^{-5}
	0.013 695	18.40	0.055 497	4.1308×10^{-5}

$n=5$

$$m=\frac{5(0.554\,976)-(0.013\,695)(18.40)}{5(4.130\,89\times10^{-5})-(0.013\,695)^2}=1343$$

$$b=\frac{(4.130\,89\times10^{-5})(18.40)-(0.055\,49)(0.013\,69)}{5(4.130\,89\times10^{-5})-(0.013\,695)^2}$$

$=1.226\,612\times10^{-3}$ or 0 (since P is in dollars)

$$P=1343\left(\frac{1}{S}\right)+0=\frac{1343}{S}$$

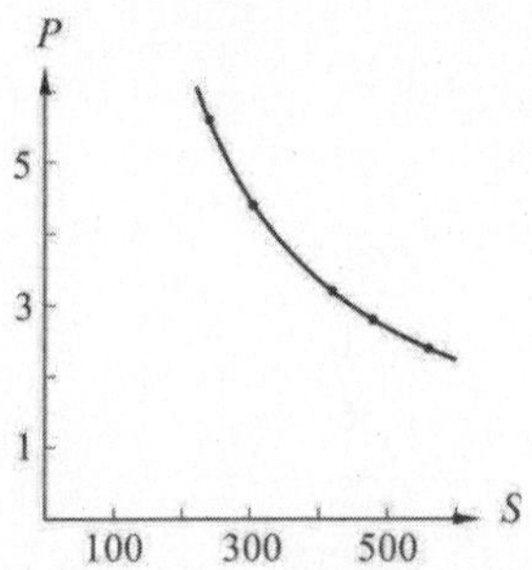

Review Exercises

1. Arrange the data in ascending order:
67, 69, 70, 72, 72, 73, 74,75, 75, 76
77, 77, 77, 78, 78, 80, 80, 82, 85, 86
There are 20 observations, so the median is the average between observation 10 and observation 11, or $\frac{76+77}{2}=76.5$ %

5.

Percent of On-time Flights	Frequency
67 − 70	3
71 − 74	4
75 − 78	8
79 − 82	3
83 − 86	2
Total	20

9.

Percent of On-time Flights	Cumulative Frequency
less than 71	3
less than 75	7
less than 79	15
less than 83	18
less than 87	20

13. $s=\sqrt{\dfrac{n\sum x^2-\left(\sum x\right)^2}{n(n-1)}}$

$=\sqrt{\dfrac{12(0.8397)-(3.17)^2}{12(11)}}$

$=0.0144$ Pa · s

17. There are 121 observations, so the median is observation 61, which falls in the 700 W class. Therefore, the median is 700 W.

21. $s=\sqrt{\dfrac{n\sum x^2 f-\left(\sum xf\right)^2}{n(n-1)}}$

$=\sqrt{\dfrac{121(58\,864\,900)-(84\,370)^2}{121(120)}}$

$=17.3$ W

25. A 95% confidence interval is given by

$\bar{x}\pm z_{\alpha/2}\cdot\dfrac{s}{\sqrt{n}}=696\pm1.96\cdot\dfrac{17.7}{\sqrt{121}}$

$=696\pm3.15$

$=(692.8, 699.2)$

The true mean power output is between 692.8 W and 699.2 W with 95% confidence.

29.

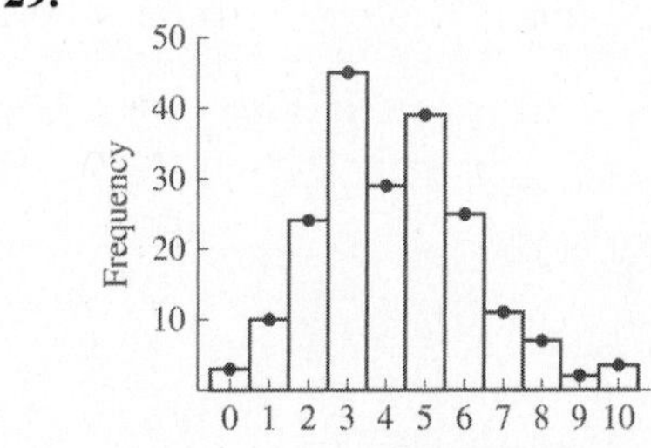

33. $s = \sqrt{\dfrac{n\sum x^2 - \left(\sum x\right)^2}{n(n-1)}}$

$= \sqrt{\dfrac{110(491\,275) - (7285)^2}{110(109)}}$

$= 9.0$ km/h

37. We can write
$27.8 = 32 - 2(2.1)$, and $36.2 = 32 + 2(2.1)$,
so we use Chebychev's theorem with $k = 2$.
Therefore, at least 75% of observations are within 27.8 and 36.2.

41. If $\hat{p} = \frac{36}{185}$ is used, Eq. (22.15) gives

$n = \hat{p}(1-\hat{p})\left[\dfrac{z_{\alpha/2}}{E}\right]^2$

$= \dfrac{36}{185} \cdot \dfrac{149}{185}\left[\dfrac{1.645}{0.032}\right]^2$

$= 414.2$

Therefore, 415 samples in total must be taken for a margin of error of 0.032 with 90% confidence.

45. CL: $\overline{\overline{x}} = 4.9885$

$\overline{R} = 0.1144$

$\text{UCL}(\overline{x}) = \overline{\overline{x}} + A_2\overline{R}$

$= 4.9885 + 0.577(0.1144)$

$= 5.055$

$\text{LCL}(\overline{x}) = \overline{\overline{x}} - A_2\overline{R}$

$= 4.9885 - 0.577(0.1144)$

$= 4.922$

Sample Mean (mm)
5.100
4.900
UCL = 5.055
$\overline{\overline{x}} = 4.989$
LCL = 4.922
1
16
Subgroup

49. $\mu = 2.20$, $\sigma = 0.50$. x above 1.00

$z = \dfrac{1.00 - 2.20}{0.50} = -2.4$

The area under the curve to the right of -2.6 is $0.5 + 0.4918 = 0.9918$, so $(0.9918)(500) = 495.9$, and 495 readings are expected to be above 1.00 $\mu\text{g/m}^3$.

53.

t (min)	s (Mach number)	ts	t^2
0.00	0.88	0.00	0
0.60	0.97	0.582	0.36
1.20	1.03	1.236	1.44
1.80	1.11	1.998	3.24
2.40	1.19	2.856	5.76
3.00	1.25	3.750	9.00
9.00	6.43	10.422	19.8

$m = \dfrac{n\sum ts - \sum t\sum s}{n\sum t^2 - \left(\sum t\right)^2}$

$= \dfrac{6(10.422) - 9.00(6.43)}{6(19.8) - (9.00)^2}$

$= 0.123$

$b = \dfrac{\left(\sum t^2\right)\sum s - \sum ts\sum t}{n\sum t^2 - \left(\sum t\right)^2}$

$= \dfrac{19.8(6.43) - 10.422(9.00)}{6(19.8) - (9.00)^2} = 0.887$

The least-squares line is

$s = 0.123t + 0.887$

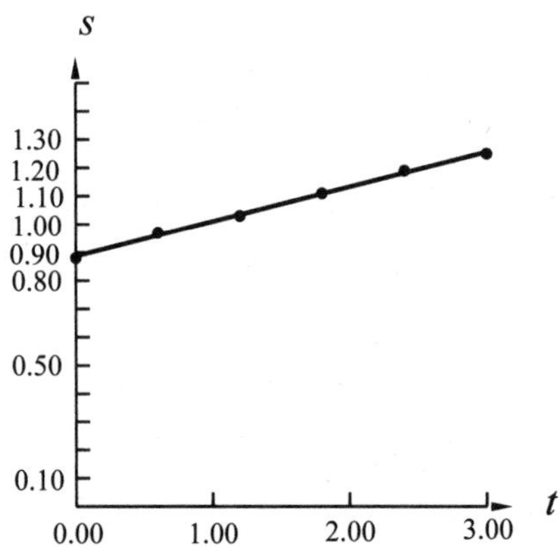

57.

	x	y	xy	x^2
$t(\text{s})$	$e^{-0.500t}$	$i(\text{A})$		
0.00	1.00	0.00	0.00	1.00
2.00	0.367 88	2.52	0.927 06	0.135 34
4.00	0.135 34	3.45	0.466 91	0.018 32
6.00	0.049 79	3.80	0.189 19	0.002 48
8.00	0.018 32	3.92	0.0718	3.4×10^{-4}
	1.571 317 432	13.69	1.654 951 083	1.156 465137

$$m=\frac{n\sum xy-\sum x\sum y}{n\sum x^2-\left(\sum x\right)^2}$$
$$=\frac{5(1.654\,951\,083)-1.571\,317\,432(13.69)}{5(1.156\,465\,137)-(1.571\,318\,432)^2}$$
$$=-3.994\,999\,34$$
$$b=\frac{\sum x^2\sum y-\sum xy\sum x}{n\sum x^2-\left(\sum x\right)^2}$$
$$=\frac{1.156\,465\,137(13.69)-1.654\,951\,08(1.571\,317\,43)}{5(1.156\,465\,137)-(1.571\,317\,432)^2}$$
$$=3.993\,482\,242$$
$$i=3.99\left(1-e^{-0.500t}\right)$$

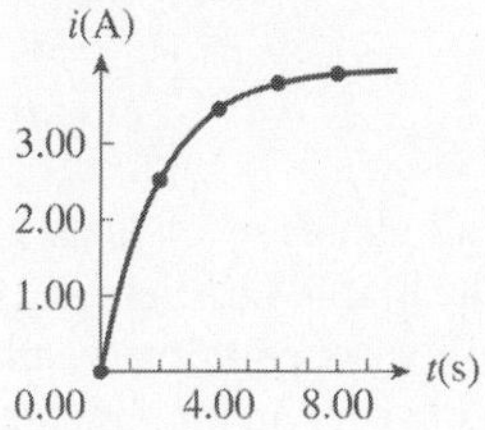

$i \to 3.99A$ as $t \to \infty$

61. Using a calculator, software, or a spreadsheet, the linear regression equation is found to be
$y=0.997x+0.581,$
with $r=0.998$

65. The area between z_1=0.5 and z_2 is (area between 0 and z_2) $-0.1915=0.305$. Therefore, the area between 0 and z_2 is 0.305+0.1915=0.4965, which gives $z_2=2.7$.

69.
$$m\bar{x}+b=\frac{n\sum xy-\sum x\sum y}{n\sum x^2-\left(\sum x\right)^2}\cdot\frac{\sum x}{n}+\frac{\sum x^2\sum y-\sum xy\sum x}{n\sum x^2-\left(\sum x\right)^2}$$
$$=\frac{\sum xy\sum x-\left(\sum x\right)^2\cdot\frac{\sum y}{n}+\sum x^2\sum y-\sum xy\sum x}{n\sum x^2-\left(\sum x\right)^2}$$
$$=\frac{\sum x^2\sum y\cdot\frac{n}{n}-\left(\sum x\right)^2\frac{\sum y}{n}}{n\sum x^2-\left(\sum x\right)^2}$$
$$=\frac{\frac{\sum y}{n}\left(n\sum x^2-\left(\sum x\right)^2\right)}{n\sum x^2-\left(\sum x\right)^2}=\bar{y}$$

CHAPTER 23

THE DERIVATIVE

23.1 Limits

1. $f(x) = \dfrac{1}{x+2}$

The function is not continuous at $x = -2$, due to a division by zero error at that point, making the function undefined at $x = -2$. The condition for continuity that the function must exist at that point is not satisfied.

5. $f(x) = 3x - 2$

The function is continuous over all values of x. The graph is linear, no points are undefined, and there are no jumps of gaps in the value of the function, as any small change in x produces only a small change in the value of the function. Continuous on $x \in (-\infty, \infty)$ or $x \in \mathbb{R}$.

9. $f(x) = \sqrt{\dfrac{x}{x-2}}$

The function is not continuous at $x = 2$ because the function is undefined due to a division by zero error. The function will also be discontinuous anywhere the argument of the square root is negative (if the sign of the numerator differs from the denominator). Therefore, if $0 < x < 2$ the numerator is positive and denominator negative, yielding a non-real solution. When $x > 2$ both numerator and denominator are positive, and if $x < 0$ both numerator and denominator are negative, yielding a positive argument for the square root. At $x = 0$, the function is also defined. So the function is continuous on intervals $x \le 0$ and $x > 2$. {Continuous on $x \in (-\infty, 0]$, $x \in (2, \infty)$ }

13. Function is continuous for all values of x except $x = 2$, since there is a "jump" in the graph, i.e. there is a large change in the value of the function for a small change in the value of x near $x = 2$. So the function is continuous on intervals $x < 2$ and $x > 2$. {Continuous on $x \in (-\infty, 2)$, $x \in (2, \infty)$ }

17. **(a)** $f(2) = -1$

(b) $\lim\limits_{x\to 2} f(x)$ does not exist since

$\lim\limits_{x\to 2^-} f(x) = -1$ and $\lim\limits_{x\to 2^+} f(x) = 2$ do not agree.

21. $f(x) = \begin{cases} x^2 & \text{for } x < 2 \\ 2 & \text{for } x \ge 2 \end{cases}$

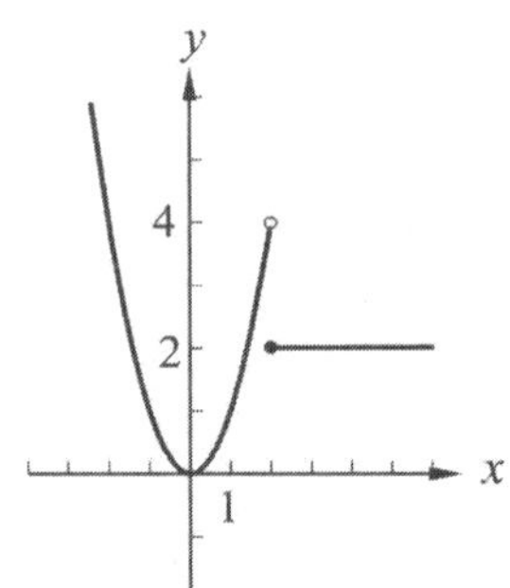

Function is continuous for all values of x except $x = 2$ since there is a "jump" in the graph, i.e. there is a large change in the value of the function for a small change in the value of x near $x = 2$. So the function is continuous on intervals $x < 2$ and $x \ge 2$. {Continuous on $x \in (-\infty, 2)$, $x \in [2, \infty)$ }

25. Find $\lim\limits_{x\to 1} \dfrac{x^3 - x}{x - 1}$

x	0.900	0.990	0.999	1.001	1.010	1.100
$f(x)$	1.710	1.970	1.997	2.003	2.030	2.310

$\lim\limits_{x\to 1^-} \dfrac{x^3 - x}{x - 1} = 2$ and $\lim\limits_{x\to 1^+} \dfrac{x^3 - x}{x - 1} = 2$

Therefore, $\lim\limits_{x\to 1} \dfrac{x^3 - x}{x - 1} = 2$

29. Find $\lim\limits_{x\to \infty} \dfrac{2x+1}{5x-3}$

x	10	100	1000	10 000
$f(x)$	0.446 81	0.404 43	0.400 44	0.400 04

Therefore, $\lim\limits_{x\to \infty} \dfrac{2x+1}{5x-3} = 0.4 = \dfrac{2}{5}$

33. $\lim\limits_{x\to 0} \dfrac{x^2 + x}{x} = \dfrac{0^2 + 0}{0} = \dfrac{0}{0} =$ indeterminate

Try factoring.

$\lim\limits_{x\to 0} \dfrac{x^2 + x}{x} = \lim\limits_{x\to 0} \dfrac{x(x+1)}{x}$

$= \lim_{x\to 0}(x+1)$

$= 0+1$

$\lim_{x\to 0}\frac{x^2+x}{x} = 1$

37. $\lim_{h\to 3}\frac{h^3-27}{h-3} = \frac{(3)^3-27}{3-3} = \frac{0}{0} =$ indeterminate

Try factoring.

$\lim_{h\to 3}\frac{h^3-27}{h-3} = \lim_{h\to 3}\frac{(h-3)(h^2+3h+9)}{(h-3)}$

$= \lim_{h\to 3}(h^2+3h+9)$

$= (3)^2+3(3)+9$

$\lim_{h\to 3}\frac{h^3-27}{h-3} = 27$

41. $\lim_{p\to -1}\sqrt{p}(p+1.3) = \sqrt{-1}(-1+1.3)$ has no real solution.

The root is not real for domain $p<0$, so $\lim_{p\to -1}\sqrt{p}(p+1.3)$ does not exist.

45. $\lim_{x\to\infty}\frac{3x^2+4.5}{x^2-1.5} = \frac{\infty}{\infty} =$ indeterminate

Try dividing numerator and denominator by the highest power of x.

$\lim_{x\to\infty}\frac{3x^2+4.5}{x^2-1.5} = \lim_{x\to\infty}\frac{\frac{3x^2}{x^2}+\frac{4.5}{x^2}}{\frac{x^2}{x^2}-\frac{1.5}{x^2}}$

$= \lim_{x\to\infty}\frac{3+\frac{4.5}{x^2}}{1-\frac{1.5}{x^2}}$

$= \frac{3+0}{1-0}$

$\lim_{x\to\infty}\frac{3x^2+4.5}{x^2-1.5} = 3$

49. **Table Method**

x	−0.100	−0.010	−0.001	0.001	0.010	0.100
$f(x)$	−3.100	−3.010	−3.001	−2.999	−2.990	−2.900

$\lim_{x\to 0^-}\frac{x^2-3x}{x} = -3$ and $\lim_{x\to 0^+}\frac{x^2-3x}{x} = -3$

$\therefore \lim_{x\to 0}\frac{x^2-3x}{x} = -3$

Analytical Method

$\lim_{x\to 0}\frac{x^2-3x}{x} = \frac{(0)^2-3(0)}{0} = \frac{0}{0} =$ indeterminate

Try factoring.

$\lim_{x\to 0}\frac{x^2-3x}{x} = \lim_{x\to 0}\frac{x(x-3)}{x}$

$= \lim_{x\to 0}(x-3)$

$= (0-3)$

$\lim_{x\to 0}\frac{x^2-3x}{x} = -3$

53. The object's temperature T (in °C) decreases 10% every minute. This means that 90% of the temperature remains after each minute. This produces a geometric sequence with first term 100°C and ratio 0.900.

$T = 100(0.9)^t$ where t is time (in min)

We can find the limits of the temperature as $t\to 10$ and as $t\to\infty$

$\lim_{t\to 10} T = \lim_{t\to 10} 100(0.9)^t$

$= 100(0.9)^{10}$

$= 100(0.9)^{10}$

$\lim_{t\to 10} T = 34.9$ °C

$\lim_{t\to\infty} T = \lim_{t\to\infty} 100(0.9)^t$

$= 100\lim_{t\to\infty}(0.9)^t$

$= 100(0)$

$\lim_{t\to\infty} T = 0$ °C

These limits can be confirmed with a table:

t (min)	T (°C)
0	100.0
1	90.0
2	81.0
3	72.9
4	65.6
5	59.0
6	53.1
7	47.8
8	43.0
9	38.7
10	34.9
50	0.515
100	0.003

57.

x	$f(x)$
1.9000	2.6787
1.9900	2.7630
1.9990	2.7716
1.9999	2.7725
2.0001	2.7727
2.0010	2.7735
2.0100	2.7822
2.1000	2.8709

$$\therefore \lim_{x\to 2}\frac{2^x-4}{x-2}\approx 2.77$$

61. **(a)** $\lim_{x\to 2^-} f(x) = -1$

(b) $\lim_{x\to 2^+} f(x) = 2$

(c) $\lim_{x\to 2} f(x)$ does not exist. The limits from above and below must agree for the limit to exist.

65. Remember that

$$|a| = \begin{cases} a & \text{for } a \ge 0 \\ -a & \text{for } a < 0 \end{cases}$$

$$\lim_{x\to 0^-}\frac{x}{|x|} = \lim_{x\to 0^-}\frac{x}{-x} = \lim_{x\to 0^-} -1 = -1$$

$$\lim_{x\to 0^+}\frac{x}{|x|} = \lim_{x\to 0^+}\frac{x}{x} = \lim_{x\to 0^+} 1 = 1$$

Since the limits from above and from below do not agree, and since the function is not defined at $x = 0$, the function is not continuous at $x = 0$.

23.2 The Slope of a Tangent to a Curve

1. Point P has the coordinates $(3,18)$. The coordinates of any point Q can be expressed as $(3+\Delta x, f(3+\Delta x))$, so in this case, $x = 3$.

$$f(x) = x^2 + 3x$$
$$y + \Delta y = f(x+\Delta x)$$
$$y + \Delta y = (x+\Delta x)^2 + 3(x+\Delta x)$$
$$y + \Delta y = x^2 + 2x\Delta x + \Delta x^2 + 3x + 3\Delta x$$
$$\Delta y = x^2 + 2x\Delta x + \Delta x^2 + 3x + 3\Delta x - x^2 - 3x$$
$$\Delta y = 2x\Delta x + \Delta x^2 + 3\Delta x$$
$$\frac{\Delta y}{\Delta x} = \frac{2x\Delta x + \Delta x^2 + 3\Delta x}{\Delta x}$$
$$\frac{\Delta y}{\Delta x} = 2x + \Delta x + 3$$
$$m_{PQ} = 2(3) + \Delta x + 3$$
$$m_{PQ} = 9 + \Delta x$$
$$m_{\tan} = \lim_{\Delta x\to 0}(9+\Delta x)$$
$$m_{\tan} = 9$$

5. $y = 2x^2 + 5x$ $\qquad P(-2, -2)$

Point	*Q*1	*Q*2	*Q*3	*Q*4	*P*
x_2	−1.5	−1.9	−1.99	−1.999	−2
y_2	−3	−2.28	−2.0298	−2.002 998	−2
$y_2 - y$	−1	−0.28	−0.0298	−0.002 998	
$x_2 - x$	0.5	0.1	0.01	0.001	
$m = \Delta y/\Delta x$	−2	−2.8	−2.98	−2.998	

The slope of PQ approaches −3 as Q approaches P.

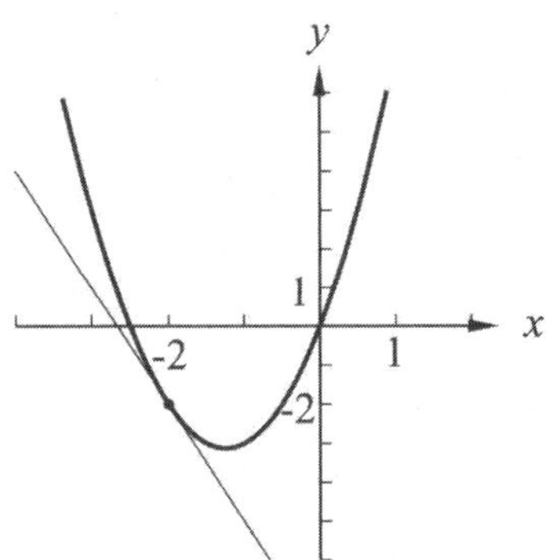

9.

$$f(x) = 2x^2 + 5x$$
$$y + \Delta y = f(x+\Delta x)$$
$$y + \Delta y = 2(x+\Delta x)^2 + 5(x+\Delta x)$$
$$y + \Delta y = 2x^2 + 4x\Delta x + 2\Delta x^2 + 5x + 5\Delta x$$
$$\Delta y = 2x^2 + 4x\Delta x + 2\Delta x^2 + 5x + 5\Delta x - 2x^2 - 5x$$
$$\Delta y = 4x\Delta x + 2\Delta x^2 + 5\Delta x$$
$$\frac{\Delta y}{\Delta x} = \frac{4x\Delta x + 2\Delta x^2 + 5\Delta x}{\Delta x}$$
$$\frac{\Delta y}{\Delta x} = 4x + 2\Delta x + 5$$
$$m_{\tan} = \lim_{\Delta x\to 0}(4x + 2\Delta x + 5)$$
$$m_{\tan} = 4x + 5$$
$$m_{\tan}\big|_{x=-2} = 4(-2) + 5 = -3$$
$$m_{\tan}\big|_{x=0.5} = 4(0.5) + 5 = 7$$

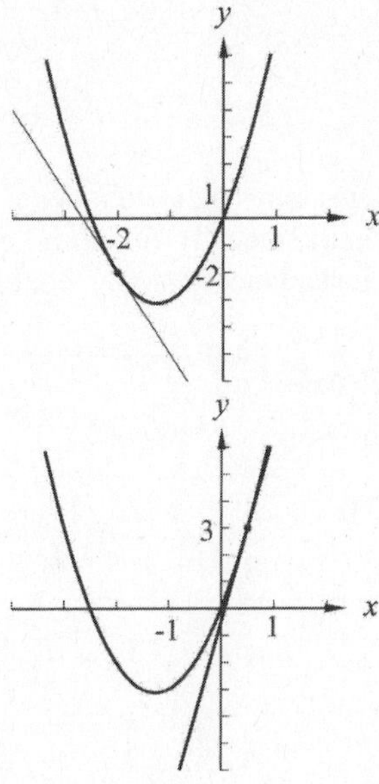

13. $f(x)=6x-x^2$

$y+\Delta y=f(x+\Delta x)$

$y+\Delta y=6(x+\Delta x)-(x+\Delta x)^2$

$y+\Delta y=6x+6\Delta x-x^2-2x\Delta x-\Delta x^2$

$\Delta y=6x+6\Delta x-x^2-2x\Delta x-\Delta x^2-(6x-x^2)$

$\Delta y=6\Delta x-2x\Delta x-\Delta x^2$

$\frac{\Delta y}{\Delta x}=\frac{6\Delta x-2x\Delta x-\Delta x^2}{\Delta x}$

$\frac{\Delta y}{\Delta x}=6-2x-\Delta x$

$m_{\tan}=\lim_{\Delta x\to 0}(6-2x-\Delta x)$

$m_{\tan}=6-2x$

$m_{\tan}\big|_{x=-2}=6-2(-2)=10$

$m_{\tan}\big|_{x=3}=6-2(3)=0$

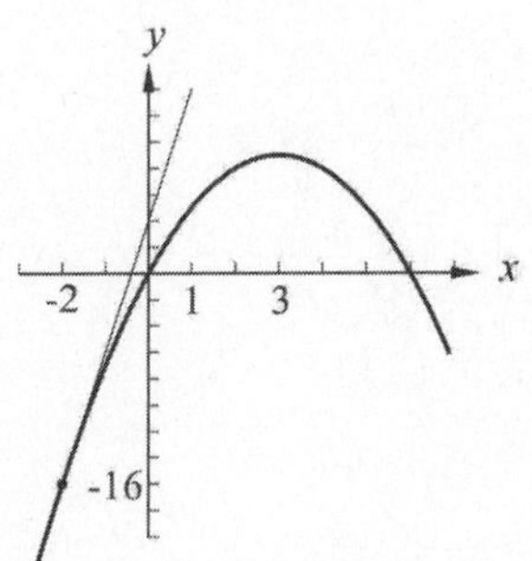

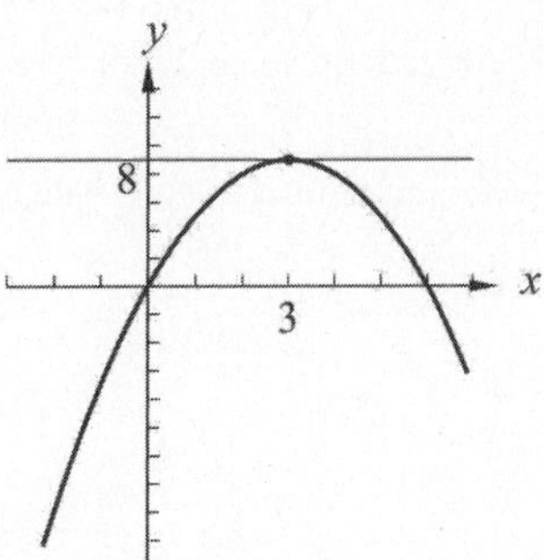

17. $f(x)=x^5$

$y+\Delta y=f(x+\Delta x)$

$y+\Delta y=(x+\Delta x)^5=(x+\Delta x)^2(x+\Delta x)^2(x+\Delta x)$

From Question 16 above

$y+\Delta y=(x^4+4x^3\Delta x+6x^2\Delta x^2$
$+4x\Delta x^3+\Delta x^4)(x+\Delta x)$

$y+\Delta y=x^5+4x^4\Delta x+6x^3\Delta x^2+4x^2\Delta x^3+x\Delta x^4$
$+x^4\Delta x+4x^3\Delta x^2+6x^2\Delta x^3+4x\Delta x^4+\Delta x^5$

$y+\Delta y=x^5+5x^4\Delta x+10x^3\Delta x^2+10x^2\Delta x^3$
$+5x\Delta x^4+\Delta x^5$

$\Delta y=x^5+5x^4\Delta x+10x^3\Delta x^2+10x^2\Delta x^3$
$+5x\Delta x^4+\Delta x^5-x^5$

$\frac{\Delta y}{\Delta x}=\frac{5x^4\Delta x+10x^3\Delta x^2+10x^2\Delta x^3+5x\Delta x^4+\Delta x^5}{\Delta x}$

$\frac{\Delta y}{\Delta x}=5x^4+10x^3\Delta x+10x^2\Delta x^2+5x\Delta x^3+\Delta x^4$

$m_{\tan}=\lim_{\Delta x\to 0}(5x^4+10x^3\Delta x+10x^2\Delta x^2+5x\Delta x^3+\Delta x^4)$

$m_{\tan}=5x^4$

$m_{\tan}\big|_{x=0}=5(0)^4=0$

$m_{\tan}\big|_{x=0.5}=5(0.5)^4=\frac{5}{16}$

$m_{\tan}\big|_{x=1}=5(1)^4=5$

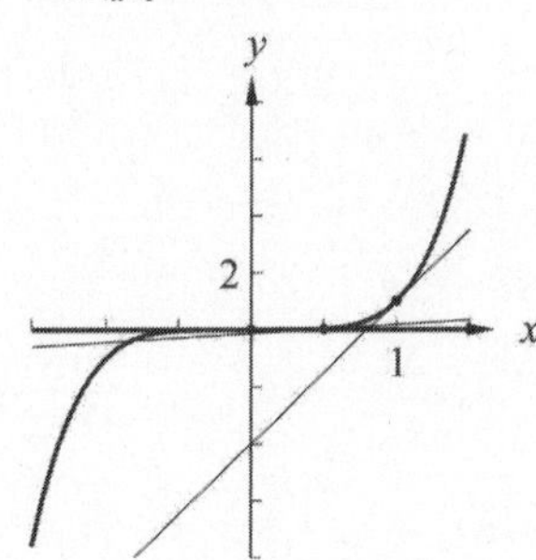

21. Graph $y = \frac{1}{3}x^6$. Use the *draw* feature to sketch a line tangent to the curve and the $\frac{dy}{dx}$ function to calculate a numerical value for the slope.

$x = 0,\ m_{\tan} = 0$

$x = 0.5,\ m_{\tan} = 0.0625$

$x = 1,\ m_{\tan} = 2$

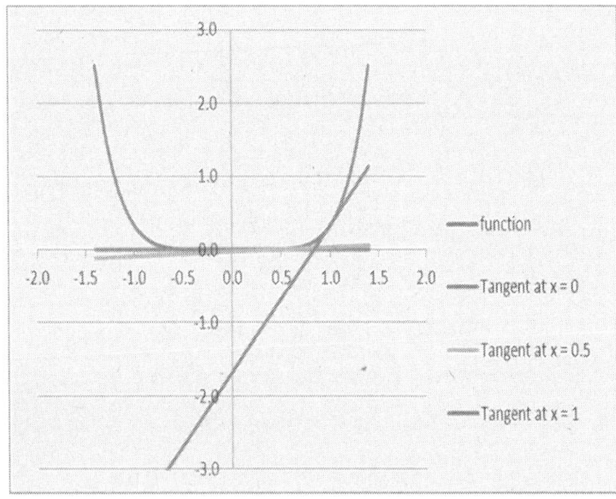

25. $y = 9 - x^3,\ P(2,\ 1),\ Q(2.1,\ -0.261)$

Average slope can be calculated from the fundamental definition:

$$m = \frac{\text{rise}}{\text{run}} = \frac{\Delta y}{\Delta x} = \frac{y_2 - y_1}{x_2 - x_1}$$

$$m = \frac{-0.261 - 1}{2.1 - 2}$$

$$m = \frac{-1.261}{0.1}$$

$$m = -12.6$$

Average slope and instantaneous slope can be found from the delta process:

$$f(x) = 9 - x^3$$

$$y + \Delta y = f(x + \Delta x)$$

$$y + \Delta y = 9 - (x + \Delta x)^3$$

$$y + \Delta y = 9 - (x^3 + 3x^2\Delta x + 3x\Delta x^2 + \Delta x^3)$$

$$\Delta y = 9 - x^3 - 3x^2\Delta x - 3x\Delta x^2 - \Delta x^3 - (9 - x^3)$$

$$\Delta y = -3x^2\Delta x - 3x\Delta x^2 - \Delta x^3$$

$$\frac{\Delta y}{\Delta x} = \frac{-3x^2\Delta x - 3x\Delta x^2 - \Delta x^3}{\Delta x}$$

$$\frac{\Delta y}{\Delta x} = -3x^2 - 3x\Delta x - \Delta x^2$$

In this case, the interval begins at $x = 2$ for point P and changes by $\Delta x = 0.1$ to arrive at point Q.

$$\frac{\Delta y}{\Delta x} = -3(2)^2 - 3(2)(0.1) - (0.1)^2$$

$$\frac{\Delta y}{\Delta x} = -12.6$$

This confirms the average slope calculation done above with the definition of slope. To find the slope of the tangent line (the instantaneous slope), continue with the delta process:

$$m_{\tan} = \lim_{\Delta x \to 0}(-3x^2 - 3x\Delta x - \Delta x^2)$$

$$m_{\tan} = -3x^2$$

$$m_{\tan}\big|_{x=2} = -3(2)^2 = -12.0$$

29. $f(x) = 8 - 3x^2$

$$y + \Delta y = f(x + \Delta x)$$

$$y + \Delta y = 8 - 3(x + \Delta x)^2$$

$$y + \Delta y = 8 - 3x^2 - 6x\Delta x - 3\Delta x^2$$

$$\Delta y = 8 - 3x^2 - 6x\Delta x - 3\Delta x^2 - (8 - 3x^2)$$

$$\Delta y = -6x\Delta x - 3\Delta x^2$$

$$\frac{\Delta y}{\Delta x} = \frac{-6x\Delta x - 3\Delta x^2}{\Delta x}$$

$$\frac{\Delta y}{\Delta x} = -6x - 3\Delta x$$

$$m_{\tan} = \lim_{\Delta x \to 0}(-6x - 3\Delta x)$$

$$m_{\tan} = -6x$$

$$m_{\tan}\big|_{x=-1} = -6(-1) = 6$$

Perpendicular lines have negative reciprocal slope, so

$$m_{\perp} = -\frac{1}{6}$$

33. $f(x) = x^3 + 1$

$$y + \Delta y = f(x + \Delta x)$$

$$y + \Delta y = (x + \Delta x)^3 + 1$$

$$y + \Delta y = x^3 + 3x^2\Delta x + 3x\Delta x^2 + \Delta x^3 + 1$$

$$\Delta y = x^3 + 3x^2\Delta x + 3x\Delta x^2 + \Delta x^3 + 1 - x^3 - 1$$

$$\Delta y = 3x^2\Delta x + 3x\Delta x^2 + \Delta x^3$$

$$\frac{\Delta y}{\Delta x} = \frac{3x^2\Delta x + 3x\Delta x^2 + \Delta x^3}{\Delta x}$$

$$\frac{\Delta y}{\Delta x} = 3x^2 + 3x\Delta x + \Delta x^2$$

$$m_{\tan} = \lim_{\Delta x \to 0}(3x^2 + 3x\Delta x + \Delta x^2)$$

$$m_{\tan} = 3x^2$$

$$m_{\tan}\big|_{x=-1} = 3(-1)^2 = 3$$

23.3 The Derivative

1. $y = 4x^2 + 3x$

$$f(x+h) = 4(x+h)^2 + 3(x+h)$$
$$f(x+h) = 4(x^2 + 2xh + h^2) + 3x + 3h$$
$$f(x+h) = 4x^2 + 8xh + 4h^2 + 3x + 3h$$
$$f(x+h) - f(x) = 4x^2 + 8xh + 4h^2 + 3x + 3h - 4x^2 - 3x$$
$$f(x+h) - f(x) = 8xh + 4h^2 + 3h$$
$$\frac{f(x+h) - f(x)}{h} = \frac{8xh + 4h^2 + 3h}{h}$$
$$\frac{f(x+h) - f(x)}{h} = 8x + 4h + 3$$
$$\lim_{h\to 0}\frac{f(x+h) - f(x)}{h} = \lim_{h\to 0}(8x + 4h + 3)$$
$$f'(x) = 8x + 3$$

5. $y = 1 - 2x$

$$f'(x) = \lim_{h\to 0}\frac{f(x+h) - f(x)}{h}$$
$$f'(x) = \lim_{h\to 0}\frac{1 - 2(x+h) - (1 - 2x)}{h}$$
$$f'(x) = \lim_{h\to 0}\frac{1 - 2x - 2h - 1 + 2x}{h}$$
$$f'(x) = \lim_{h\to 0}\frac{-2h}{h}$$
$$f'(x) = \lim_{h\to 0} -2$$
$$f'(x) = -2$$

9. $y = \pi x^2$

$$f'(x) = \lim_{h\to 0}\frac{f(x+h) - f(x)}{h}$$
$$f'(x) = \lim_{h\to 0}\frac{\pi(x+h)^2 - \pi x^2}{h}$$
$$f'(x) = \lim_{h\to 0}\frac{\pi x^2 + 2\pi xh + \pi h^2 - \pi x^2}{h}$$
$$f'(x) = \lim_{h\to 0}\frac{2\pi xh + \pi h^2}{h}$$
$$f'(x) = \lim_{h\to 0}(2\pi x + \pi h)$$
$$f'(x) = 2\pi x$$

13. $y = 8x - 2x^2$

$$f'(x) = \lim_{h\to 0}\frac{f(x+h) - f(x)}{h}$$
$$f'(x) = \lim_{h\to 0}\frac{8(x+h) - 2(x+h)^2 - (8x - 2x^2)}{h}$$
$$f'(x) = \lim_{h\to 0}\frac{8x + 8h - 2x^2 - 4xh - 2h^2 - 8x + 2x^2}{h}$$
$$f'(x) = \lim_{h\to 0}\frac{8h - 4xh - 2h^2}{h}$$
$$f'(x) = \lim_{h\to 0}(8 - 4x - 2h)$$
$$f'(x) = 8 - 4x$$
$$f'(x) = 4(2 - x)$$

17. $y = \frac{\sqrt{3}}{x+2}$

$$f'(x) = \lim_{h\to 0}\frac{f(x+h) - f(x)}{h}$$
$$f'(x) = \lim_{h\to 0}\frac{\frac{\sqrt{3}}{x+h+2} - \left(\frac{\sqrt{3}}{x+2}\right)}{h}$$
$$f'(x) = \lim_{h\to 0}\frac{\frac{\sqrt{3}(x+2) - \sqrt{3}(x+h+2)}{(x+h+2)(x+2)}}{h}$$
$$f'(x) = \lim_{h\to 0}\frac{\frac{\sqrt{3}x + 2\sqrt{3} - \sqrt{3}x - \sqrt{3}h - 2\sqrt{3}}{(x+h+2)(x+2)}}{h}$$
$$f'(x) = \lim_{h\to 0}\frac{\frac{-\sqrt{3}h}{(x+h+2)(x+2)}}{h}$$
$$f'(x) = \lim_{h\to 0}\frac{-\sqrt{3}}{(x+h+2)(x+2)}$$
$$f'(x) = -\frac{\sqrt{3}}{(x+2)^2}$$

21. $y = \frac{2}{x^2}$

$$f'(x) = \lim_{h\to 0}\frac{f(x+h) - f(x)}{h}$$
$$f'(x) = \lim_{h\to 0}\frac{\frac{2}{(x+h)^2} - \frac{2}{x^2}}{h}$$
$$f'(x) = \lim_{h\to 0}\frac{\frac{2x^2 - 2(x+h)^2}{x^2(x+h)^2}}{h}$$
$$f'(x) = \lim_{h\to 0}\frac{2x^2 - 2(x^2 + 2xh + h^2)}{hx^2(x+h)^2}$$

$$f'(x)=\lim_{h\to 0}\frac{2x^2-2x^2-4xh-2h^2}{hx^2(x+h)^2}$$
$$f'(x)=\lim_{h\to 0}\frac{-4xh-2h^2}{hx^2(x+h)^2}$$
$$f'(x)=\lim_{h\to 0}\frac{-4x-2h}{x^2(x+h)^2}$$
$$f'(x)=\frac{-4x}{x^4}$$
$$f'(x)=-\frac{4}{x^3}$$

25. $$y=x^4-\frac{2}{x}$$
$$f(x+h)=(x+h)^4-\frac{2}{x+h}$$
$$f(x+h)=x^4+4x^3h+6x^2h^2+4xh^3+h^4-\frac{2}{x+h}$$
$$f(x+h)-f(x)=x^4+4x^3h+6x^2h^2+4xh^3+h^4-\frac{2}{x+h}-x^4+\frac{2}{x}$$
$$f(x+h)-f(x)=4x^3h+6x^2h^2+4xh^3+h^4+\frac{-2(x)+2(x+h)}{x(x+h)}$$
$$f(x+h)-f(x)=4x^3h+6x^2h^2+4xh^3+h^4+\frac{2h}{x(x+h)}$$
$$\frac{f(x+h)-f(x)}{h}=4x^3+6x^2h+4xh^2+h^3+\frac{2}{x(x+h)}$$
$$f'(x)=\lim_{h\to 0}\left(4x^3+6x^2h+4xh^2+h^3+\frac{2}{x(x+h)}\right)$$
$$f'(x)=4x^3+\frac{2}{x^2}$$
$$f'(x)=2\left(2x^3+\frac{1}{x^2}\right)$$

29. $$y=\frac{11}{3x+2}\quad \text{Point}(3,\ 1)$$
$$f(x+h)-f(x)=\frac{11}{3(x+h)+2}-\frac{11}{3x+2}$$
$$f(x+h)-f(x)=\frac{11(3x+2)-11(3x+3h+2)}{(3x+3h+2)(3x+2)}$$
$$f(x+h)-f(x)=\frac{33x+22-33x-33h-22}{(3x+3h+2)(3x+2)}$$
$$f(x+h)-f(x)=\frac{-33h}{(3x+3h+2)(3x+2)}$$
$$\frac{f(x+h)-f(x)}{h}=\frac{-33}{(3x+3h+2)(3x+2)}$$
$$f'(x)=\lim_{h\to 0}\frac{-33}{(3x+3h+2)(3x+2)}$$
$$f'(x)=\frac{-33}{(3x+2)(3x+2)}$$
$$f'(x)=-\frac{33}{(3x+2)^2}$$
$$f'(x)\big|_{x=3}=-\frac{33}{(3(3)+2)^2}=-\frac{33}{121}$$
$$f'(x)\big|_{x=3}=-\frac{3}{11}$$

33. $$y=\frac{3}{x^2-1}$$
$$\frac{f(x+h)-f(x)}{h}=\frac{\frac{3}{(x+h)^2-1}-\frac{3}{x^2-1}}{h}$$
$$\frac{f(x+h)-f(x)}{h}=\frac{\frac{3}{(x^2+2xh+h^2-1)}-\frac{3}{x^2-1}}{h}$$
$$\frac{f(x+h)-f(x)}{h}=\frac{3(x^2-1)-3(x^2+2xh+h^2-1)}{h(x^2+2xh+h^2-1)(x^2-1)}$$
$$\frac{f(x+h)-f(x)}{h}=\frac{3x^2-3-3x^2-6xh-3h^2+3}{h(x^2+2xh+h^2-1)(x^2-1)}$$
$$\frac{f(x+h)-f(x)}{h}=\frac{-6xh-3h^2}{h(x^2+2xh+h^2-1)(x^2-1)}$$
$$\frac{f(x+h)-f(x)}{h}=\frac{-6x-3h}{(x^2+2xh+h^2-1)(x^2-1)}$$
$$f'(x)=\lim_{h\to 0}\frac{-6x-3h}{(x^2+2xh+h^2-1)(x^2-1)}$$
$$f'(x)=\frac{-6x}{(x^2-1)(x^2-1)}$$
$$f'(x)=-\frac{6x}{(x^2-1)^2}$$

This function is differentiable for all $x^2-1\neq 0$ or $x^2\neq 1$ which is $x\neq \pm 1$.

37. $y = 2x^2 - 16x$

$f(x+h) - f(x) = 2(x+h)^2 - 16(x+h) - (2x^2 - 16x)$

$f(x+h) - f(x) = 2x^2 + 4xh + 2h^2 - 16x$
$\quad -16h - 2x^2 + 16x$

$f(x+h) - f(x) = 4xh + 2h^2 - 16h$

$\dfrac{f(x+h) - f(x)}{h} = 4x + 2h - 16$

$f'(x) = \lim\limits_{h\to 0}(4x + 2h - 16)$

$f'(x) = 4x - 16$

$f'(x) = 4(x-4)$

If the slope of the tangent line is horizontal, i.e. a slope of 0, then

$0 = 4(x-4)$

$0 = x - 4$

$x = 4$

$y = 2(4^2) - 16(4) = -32$

Point $(4, -32)$

41. For the function $y = x^4 + x^3 + x^2 + x$, the derivative was $\dfrac{dy}{dx} = 4x^3 + 3x^2 + 2x + 1$. It seems like there is a pattern that has developed, where in the derivative, the power over the x in each term is multiplied by that term, and then the power is decreased by one. So that if $y = x^n$, when $n > 0$ then $\dfrac{dy}{dx} = nx^{n-1}$.

23.4 The Derivative as an Instantaneous Rate of Change

1. $s(t) = 14.0t - 4.90t^2$

$s(t+h) - s(t) = 14.0(t+h) - 4.90(t+h)^2$
$\quad - (14.0t - 4.90t^2)$

$s(t+h) - s(t) = 14.0t + 14.0h - 4.90t^2$
$\quad -9.80ht - 4.90h^2 - 14.0t + 4.90t^2$

$s(t+h) - s(t) = 14.0h - 9.80ht - 4.90h^2$

$\dfrac{s(t+h) - s(t)}{h} = 14.0 - 9.80t - 4.90h$

$\dfrac{ds}{dt} = \lim\limits_{h\to 0}(14.0 - 9.80t - 4.90h)$

$\dfrac{ds}{dt} = 14.0 - 9.80t$

$\left.\dfrac{ds}{dt}\right|_{t=2} = 14.0 - 9.80(2) = -5.60 \text{ m/s}$

$\left.\dfrac{ds}{dt}\right|_{t=4} = 14.0 - 9.80(4) = -25.2 \text{ m/s}$

5. $y = \dfrac{16}{3x+1}$ Point $(-3, -2)$

$\dfrac{dy}{dx} = \lim\limits_{h\to 0}\dfrac{f(x+h) - f(x)}{h}$

$\dfrac{dy}{dx} = \lim\limits_{h\to 0}\dfrac{\frac{16}{3(x+h)+1} - \frac{16}{3x+1}}{h}$

$\dfrac{dy}{dx} = \lim\limits_{h\to 0}\dfrac{\frac{16(3x+1) - 16(3x+3h+1)}{(3x+3h+1)(3x+1)}}{h}$

$\dfrac{dy}{dx} = \lim\limits_{h\to 0}\dfrac{48x + 16 - 48x - 48h - 16}{h(3x+3h+1)(3x+1)}$

$\dfrac{dy}{dx} = \lim\limits_{h\to 0}\dfrac{-48h}{h(3x+3h+1)(3x+1)}$

$\dfrac{dy}{dx} = \lim\limits_{h\to 0}\dfrac{-48}{(3x+3h+1)(3x+1)}$

$\dfrac{dy}{dx} = -\dfrac{48}{(3x+1)^2}$

$\left.\dfrac{dy}{dx}\right|_{x=-3} = -\dfrac{48}{(3(-3)+1)^2}$

$\left.\dfrac{dy}{dx}\right|_{x=-3} = -\dfrac{48}{64} = -\dfrac{3}{4}$

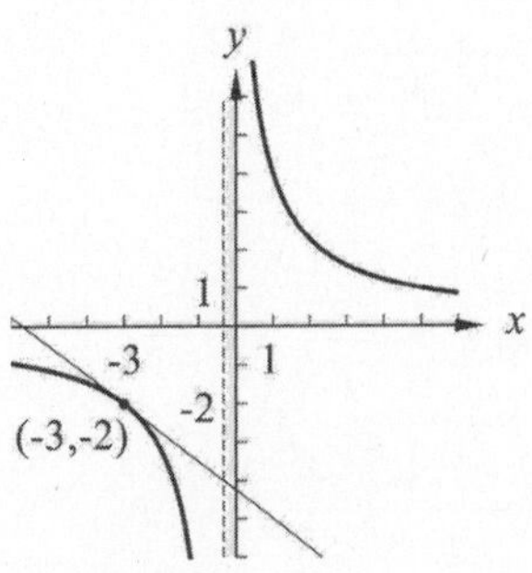

9. $s = 3t^2 - 4t;\ t = 2$

$s = 3(2)^2 - 4(2) = 4$

t (s)	1.0	1.5	1.9	1.99	1.999
s (m)	−1.0	0.75	3.23	3.9203	3.9092003
$4 - s$ (m)	5.0	3.25	0.77	0.0797	0.007997
$h = 2 - t$ (s)	1.0	0.5	0.1	0.01	0.001
$v = \frac{4-s}{h}$ (m/s)	5.00	6.50	7.70	7.97	7.997

$v = 8.00$ m/s when $t = 2$ s

13. $s = 3t^2 - 4t$

$$v = \lim_{h\to 0} \frac{s(t+h) - s(t)}{h}$$

$$v = \lim_{h\to 0} \frac{3(t+h)^2 - 4(t+h) - 3t^2 + 4t}{h}$$

$$v = \lim_{h\to 0} \frac{3t^2 + 6th + 3h^2 - 4t - 4h - 3t^2 + 4t}{h}$$

$$v = \lim_{h\to 0} \frac{6th + 3h^2 - 4h}{h}$$

$$v = \lim_{h\to 0} (6t + 3h - 4)$$

$$v = 6t - 4$$

$$v = 2(3t - 2)$$

$$v\big|_{t=2} = 2(3(2) - 2) = 8.00 \text{ m/s}$$

17. $s = 12t^2 - t^3$

$$v = \lim_{h\to 0} \frac{s(t+h) - s(t)}{h}$$

$$v = \lim_{h\to 0} \frac{12(t+h)^2 - (t+h)^3 - 12t^2 + t^3}{h}$$

$$v = \lim_{h\to 0} \frac{12t^2 + 24th + 12h^2 - (t^3 + 3t^2h + 3th^2 + h^3) - 12t^2 + t^3}{h}$$

$$v = \lim_{h\to 0} \frac{24th + 12h^2 - 3t^2h - 3th^2 - h^3}{h}$$

$$v = \lim_{h\to 0} (24t + 12h - 3t^2 - 3th - h^2)$$

$$v = 24t - 3t^2$$

$$v = 3t(8 - t)$$

21. $v = 6t^2 - 4t + 2$

$$a = \frac{dv}{dt} = \lim_{h\to 0} \frac{v(t+h) - v(t)}{h}$$

$$a = \lim_{h\to 0} \frac{6(t+h)^2 - 4(t+h) + 2 - 6t^2 + 4t - 2}{h}$$

$$a = \lim_{h\to 0} \frac{6t^2 + 12th + 6h^2 - 4t - 4h + 2 - 6t^2 + 4t - 2}{h}$$

$$a = \lim_{h\to 0} \frac{12th + 6h^2 - 4h}{h}$$

$$a = \lim_{h\to 0} (12t + 6h - 4)$$

$$a = 12t - 4$$

25. $c = 2\pi r$

$$\frac{dc}{dr} = \lim_{h\to 0} \frac{c(r+h) - c(r)}{h}$$

$$\frac{dc}{dr} = \lim_{h\to 0} \frac{2\pi(r+h) - 2\pi r}{h}$$

$$\frac{dc}{dr} = \lim_{h\to 0} \frac{2\pi h}{h}$$

$$\frac{dc}{dr} = \lim_{h\to 0} 2\pi$$

$$\frac{dc}{dr} = 2\pi = 6.28 \text{ cm/cm}$$

29. $q = 30 - 2t$

$$i = \frac{dq}{dt} = \lim_{h\to 0} \frac{q(t+h) - q(t)}{h}$$

$$i = \lim_{h\to 0} \frac{30 - 2(t+h) - 30 + 2t}{h}$$

$$i = \lim_{h\to 0} \frac{-2h}{h}$$

$$i = \lim_{h\to 0} (-2)$$

$$i = -2$$

33. $P = 500 + 250m^2$

$$\frac{dP}{dm} = \lim_{h\to 0} \frac{P(m+h) - P(m)}{h}$$

$$\frac{dP}{dm} = \lim_{h\to 0} \frac{500 + 250(m+h)^2 - 500 - 250m^2}{h}$$

$$\frac{dP}{dm} = \lim_{h\to 0} \frac{250m^2 + 500mh + 250h^2 - 250m^2}{h}$$

$$\frac{dP}{dm} = \lim_{h\to 0} \frac{500mh + 250h^2}{h}$$

$$\frac{dP}{dm} = \lim_{h\to 0} (500m + 250h)$$

$$\frac{dP}{dm} = 500m$$

$$\left.\frac{dP}{dm}\right|_{m=0.920} = 500(0.920)$$

$$\left.\frac{dP}{dm}\right|_{m=0.920} = 460 \text{ W} = 0.460 \text{ kW}$$

37. $V = \dfrac{48}{t+3}$

$$\frac{dV}{dt} = \lim_{h\to 0} \frac{V(t+h) - V(t)}{h}$$

$$\frac{dV}{dt} = \lim_{h\to 0} \frac{\frac{48}{t+h+3} - \frac{48}{t+3}}{h}$$

$$\frac{dV}{dt} = \lim_{h\to 0} \frac{\frac{48(t+3) - 48(t+h+3)}{(t+h+3)(t+3)}}{h}$$

$$\frac{dV}{dt}=\lim_{h\to 0}\frac{48t+144-48t-48h-144}{h(t+h+3)(t+3)}$$

$$\frac{dV}{dt}=\lim_{h\to 0}\frac{-48h}{h(t+h+3)(t+3)}$$

$$\frac{dV}{dt}=\lim_{h\to 0}\frac{-48}{(t+h+3)(t+3)}$$

$$\frac{dV}{dt}=-\frac{48}{(t+3)^2}$$

$$\left.\frac{dV}{dt}\right|_{t=3}=-\frac{48}{(3+3)^2}=-1.33 \text{ (thousands of \$/year)}$$

$$\left.\frac{dV}{dt}\right|_{t=3}=-\$1330\text{/year}$$

41. $r=k\sqrt{\lambda}$

If $r=3.72\times 10^{-2}$ m when $\lambda=592\times 10^{-9}$ m then

$$k=\frac{r}{\sqrt{\lambda}}=\frac{3.72\times 10^{-2}\text{ m}}{\sqrt{592\times 10^{-9}\text{ m}}}=48.3484\ \sqrt{\text{m}}$$

$$\frac{dr}{d\lambda}=\lim_{h\to 0}\frac{r(\lambda+h)-r(\lambda)}{h}$$

$$\frac{dr}{d\lambda}=\lim_{h\to 0}\frac{k\sqrt{\lambda+h}-k\sqrt{\lambda}}{h}\cdot\frac{k\sqrt{\lambda+h}+k\sqrt{\lambda}}{k\sqrt{\lambda+h}+k\sqrt{\lambda}}$$

$$\frac{dr}{d\lambda}=\lim_{h\to 0}\frac{k^2(\lambda+h)-k^2(\lambda)}{h\left(k\sqrt{\lambda+h}+k\sqrt{\lambda}\right)}$$

$$\frac{dr}{d\lambda}=\lim_{h\to 0}\frac{k^2h}{h\left(k\sqrt{\lambda+h}+k\sqrt{\lambda}\right)}$$

$$\frac{dr}{d\lambda}=\lim_{h\to 0}\frac{k^2}{\left(k\sqrt{\lambda+h}+k\sqrt{\lambda}\right)}$$

$$\frac{dr}{d\lambda}=\frac{k^2}{2k\sqrt{\lambda}}=\frac{k}{2\sqrt{\lambda}}$$

$$\frac{dr}{d\lambda}=\frac{48.3484}{2\sqrt{\lambda}}$$

$$\frac{dr}{d\lambda}=\frac{24.2}{\sqrt{\lambda}}\text{ m/m}$$

23.5 Derivatives of Polynomials

1. $v=r^9$

$$\frac{dv}{dr}=\frac{d\left(r^9\right)}{dr}$$

$$\frac{dv}{dr}=9r^{9-1}$$

$$\frac{dv}{dr}=9r^8$$

5. $y=x^5$

$$\frac{dy}{dx}=\frac{d\left(x^5\right)}{dx}$$

$$\frac{dy}{dx}=5x^4$$

9. $y=5x^4-3\pi$

$$\frac{dy}{dx}=5\frac{d\left(x^4\right)}{dx}-\frac{d(3\pi)}{dx}$$

$$\frac{dy}{dx}=5\left(4x^3\right)-0$$

$$\frac{dy}{dx}=20x^3$$

13. $p=5r^3-2r+1$

$$\frac{dp}{dr}=5\frac{d\left(r^3\right)}{dr}-2\frac{d(r)}{dr}+\frac{d(1)}{dr}$$

$$\frac{dp}{dr}=5\left(3r^2\right)-2(1)+0$$

$$\frac{dp}{dr}=15r^2-2$$

17. $f(x)=-6x^7+5x^3+\pi^2$

$$f'(x)=-6\frac{d\left(x^7\right)}{dx}+5\frac{d\left(x^3\right)}{dx}+\frac{d\left(\pi^2\right)}{dx}$$

$$f'(x)=-6\left(7x^6\right)+5\left(3x^2\right)+0$$

$$f'(x)=-42x^6+15x^2$$

$$f'(x)=-3x^2(14x^4-5)$$

21. $y=6x^2-8x+1$

$$\frac{dy}{dx}=6\left(2x^1\right)-8(1)+0$$

$$\frac{dy}{dx}=12x-8$$

$$\left.\frac{dy}{dx}\right|_{x=2}=12(2)-8$$

$$\left.\frac{dy}{dx}\right|_{x=2}=16$$

25. $y = 2x^6 - 4x^2$

$$\frac{dy}{dx} = 2(6x^5) - 4(2x)$$

$$\frac{dy}{dx} = 12x^5 - 8x$$

$$\left.\frac{dy}{dx}\right|_{x=-1} = 12(-1)^5 - 8(-1)$$

$$\left.\frac{dy}{dx}\right|_{x=-1} = -4$$

29. $s = 6t^5 - 5t + 2$

$$v = \frac{ds}{dt} = 6(5t^4) - 5(1) + 0$$

$$v = 30t^4 - 5$$

$$v = 5(6t^4 - 1)$$

33. $s = 2t^3 - 4t^2$

$$v = \frac{ds}{dt} = 2(3t^2) - 4(2t)$$

$$v = 6t^2 - 8t$$

$$v|_{t=4} = 6(4)^2 - 8(4)$$

$$v|_{t=4} = 64.0 \text{ m/ s}$$

37. $y = 3x^2 - 6x$

$$\frac{dy}{dx} = 3(2x) - 6(1)$$

$$\frac{dy}{dx} = 6x - 6$$

$\frac{dy}{dx} = 6(x-1)$ if slope is parallel to x-axis, slope is 0, factor the derivative

$$0 = 6(x-1)$$

$$x = 1$$

41. Parabola

$$y = 2x^2 - 7x$$

$$\frac{dy}{dx} = 2(2x) - 7(1)$$

$$\frac{dy}{dx} = 4x - 7$$

Line

$$x - 3y = 16$$

$$-3y = -x + 16$$

$$y = \frac{1}{3}x - \frac{16}{3}$$

This line has slope 1/3. Perpendicular lines have negative reciprocal slopes, so any curve perpendicular to this one at this location will have instantaneous slope –3.

So, for the parabola to be perpendicular to the given line:

$$\frac{dy}{dx} = -3$$

$$-3 = 4x - 7$$

$$4 = 4x$$

$$x = 1$$

45. In general, the volume of a cylinder is $V = \pi r^2 h$, where r is the radius and h is the height of the cylinder. In the particular case where the radius and height are equivalent, $h = r$, so

$$V = \pi r^2 r = \pi r^3$$

Taking the derivative with respect to a change in variable r yields

$$\frac{dV}{dr} = \frac{d}{dr}(\pi r^3)$$

$$\frac{dV}{dr} = \pi(3r^2)$$

$$\frac{dV}{dr} = 3\pi r^2$$

r

h

49. $R = 16.0 + 0.450T + 0.0125T^2$

$$\frac{dR}{dT} = 0 + 0.450(1) + 0.0125(2T)$$

$$\frac{dR}{dT} = 0.450 + 0.0250T$$

$$\left.\frac{dR}{dT}\right|_{T=115\,^\circ\text{C}} = 0.450 + 0.0250(115)$$

$$\left.\frac{dR}{dT}\right|_{T=115\,^\circ\text{C}} = 3.325\ \frac{\Omega}{^\circ\text{C}}$$

$$\left.\frac{dR}{dT}\right|_{T=115\,^\circ\text{C}} = 3.32\ \frac{\Omega}{^\circ\text{C}}$$

53. $h = 0.000104x^4 - 0.0417x^3 + 4.21x^2 - 8.33x$

$$\frac{dh}{dx} = 0.000104(4x^3) - 0.0417(3x^2) + 4.21(2x) - 8.33(1)$$

$$\frac{dh}{dx} = 0.000416x^3 - 0.1251x^2 + 8.42x - 8.33$$

$$\left.\frac{dh}{dx}\right|_{x=125\text{ km}} = 0.000416(125)^3 - 0.1251(125)^2 + 8.42(125) - 8.33$$

$$\left.\frac{dh}{dx}\right|_{x=125\text{ km}} = -98.0\ \frac{\text{m}}{\text{km}}$$

57. Given $g = 9.81\text{ m/s}^2$ and $K = 0.500$,

$$h_L = \frac{K}{2g}\cdot v^2$$
$$\frac{dh_L}{dv} = \frac{K}{2g}(2v)$$
$$\frac{dh_L}{dv} = \frac{Kv}{g}$$
$$\left.\frac{dh_L}{dv}\right|_{v=3.25\text{ m/s}} = \frac{(0.500)(3.25)}{9.81}\ \frac{\text{m/s}}{\text{m/s}^2}$$
$$\left.\frac{dh_L}{dv}\right|_{v=3.25\text{ m/s}} = 0.166\ \frac{\text{m}}{\text{m/s}}$$

23.6 Derivatives of Products and Quotients of Functions

1. $p(x) = (5-3x^2)(3-2x)$

Identify this as a product of two functions u and v, where

$$u = 5-3x^2 \qquad v = 3-2x$$
$$\frac{du}{dx} = -3(2x) = -6x \qquad \frac{dv}{dx} = -2$$

The product rule derivative is

$$\frac{d(u\cdot v)}{dx} = u \cdot \frac{dv}{dx} + v \cdot \frac{du}{dx}$$
$$\downarrow \quad \downarrow \quad \downarrow \quad \downarrow \quad \downarrow$$
$$\frac{d(p(x))}{dx} = (5-3x^2)\cdot(-2) + (3-2x)\cdot(-6x)$$
$$p' = -10+6x^2-18x+12x^2$$
$$p' = 18x^2-18x-10$$
$$p' = 2(9x^2-9x-5)$$

5. $s = (3t+2)(2t-5)$

Identify this as a product of two functions u and v, where

$$u = 3t+2 \quad \text{and} \quad v = 2t-5$$
$$\frac{du}{dt} = 3 \qquad \frac{dv}{dt} = 2$$
$$\frac{d(u\cdot v)}{dt} = u\cdot\frac{dv}{dt} + v\cdot\frac{du}{dt}$$
$$\frac{ds}{dt} = (3t+2)(2) + (2t-5)(3)$$
$$\frac{ds}{dt} = 6t+4+6t-15$$
$$\frac{ds}{dt} = 12t-11$$

9. $y = (2x-7)(5-2x)$

$$u = (2x-7) \quad \text{and} \quad v = (5-2x)$$
$$\frac{du}{dx} = 2 \qquad \frac{dv}{dx} = -2$$
$$\frac{d(u\cdot v)}{dx} = u\cdot\frac{dv}{dx} + v\cdot\frac{du}{dx}$$
$$\frac{dy}{dx} = (2x-7)\cdot(-2) + (5-2x)\cdot(2)$$
$$\frac{dy}{dx} = -4x+14+10-4x$$
$$\frac{dy}{dx} = -8x+24$$
$$\frac{dy}{dx} = -8(x-3)$$

$$y = (2x-7)(5-2x)$$
$$y = 10x-4x^2-35+14x$$
$$y = -4x^2+24x-35$$
$$\frac{dy}{dx} = -8x+24$$
$$\frac{dy}{dx} = -8(x-3)$$

13. $y = \dfrac{x}{2x+3}$

$$u = x \quad \text{and} \quad v = 2x+3$$
$$\frac{du}{dx} = 1 \qquad \frac{dv}{dx} = 2$$
$$\frac{d\left(\frac{u}{v}\right)}{dx} = \frac{v\cdot\frac{du}{dx} - u\cdot\frac{dv}{dx}}{v^2}$$
$$\frac{dy}{dx} = \frac{(2x+3)\cdot(1) - x\cdot(2)}{(2x+3)^2}$$
$$\frac{dy}{dx} = \frac{2x+3-2x}{(2x+3)^2}$$
$$\frac{dy}{dx} = \frac{3}{(2x+3)^2}$$

17. $$y = \frac{6x^2}{3-2x}$$

$u = 6x^2$ and $v = 3-2x$

$\frac{du}{dx} = 12x$ $\qquad$ $\frac{dv}{dx} = -2$

$$\frac{d\left(\frac{u}{v}\right)}{dx} = \frac{v \cdot \frac{du}{dx} - u \cdot \frac{dv}{dx}}{v^2}$$

$$\frac{dy}{dx} = \frac{(3-2x)\cdot(12x) - (6x^2)\cdot(-2)}{(3-2x)^2}$$

$$\frac{dy}{dx} = \frac{36x - 24x^2 + 12x^2}{(3-2x)^2}$$

$$\frac{dy}{dx} = \frac{36x - 12x^2}{(3-2x)^2}$$

$$\frac{dy}{dx} = \frac{12x(3-x)}{(3-2x)^2}$$

21. $$f(x) = \frac{3x+8}{x^2+4x+2}$$

$u = 3x+8$ and $v = x^2+4x+2$

$\frac{du}{dx} = 3$ $\qquad$ $\frac{dv}{dx} = 2x+4$

$$\frac{d\left(\frac{u}{v}\right)}{dx} = \frac{v \cdot \frac{du}{dx} - u \cdot \frac{dv}{dx}}{v^2}$$

$$\frac{df(x)}{dx} = \frac{(x^2+4x+2)\cdot(3) - (3x+8)\cdot(2x+4)}{(x^2+4x+2)^2}$$

$$\frac{df(x)}{dx} = \frac{3x^2+12x+6-6x^2-12x-16x-32}{(x^2+4x+2)^2}$$

$$\frac{df(x)}{dx} = \frac{-3x^2-16x-26}{(x^2+4x+2)^2}$$

25. $$y = (3x-1)(4-7x)$$

$u = 3x-1$ and $v = 4-7x$

$\frac{du}{dx} = 3$ $\qquad$ $\frac{dv}{dx} = -7$

$$\frac{d(u\cdot v)}{dx} = u \cdot \frac{dv}{dx} + v \cdot \frac{du}{dx}$$

$$\frac{dy}{dx} = (3x-1)\cdot(-7) + (4-7x)\cdot(3)$$

$$\frac{dy}{dx} = -21x + 7 + 12 - 21x$$

$$\frac{dy}{dx} = -42x + 19$$

$$\left.\frac{dy}{dx}\right|_{x=3} = -42(3) + 19$$

$$\left.\frac{dy}{dx}\right|_{x=3} = -107$$

29. $$y = \frac{3x-5}{2x+3}$$

$u = 3x-5$ and $v = 2x+3$

$\frac{du}{dx} = 3$ $\qquad$ $\frac{dv}{dx} = 2$

$$\frac{d\left(\frac{u}{v}\right)}{dx} = \frac{v \cdot \frac{du}{dx} - u \cdot \frac{dv}{dx}}{v^2}$$

$$\frac{dy}{dx} = \frac{(2x+3)\cdot(3) - (3x-5)\cdot(2)}{(2x+3)^2}$$

$$\frac{dy}{dx} = \frac{6x+9-6x+10}{(2x+3)^2}$$

$$\frac{dy}{dx} = \frac{19}{(2x+3)^2}$$

$$\left.\frac{dy}{dx}\right|_{x=-2} = \frac{19}{(2(-2)+3)^2}$$

$$\left.\frac{dy}{dx}\right|_{x=-2} = 19$$

33. If $v = c$, the product rule

$$\frac{d(u\cdot v)}{dx} = u \cdot \frac{dv}{dx} + v \cdot \frac{du}{dx}$$

becomes

$$\frac{d(u\cdot c)}{dx} = u \cdot \frac{dc}{dx} + c \cdot \frac{du}{dx}$$

$$\frac{d(c\cdot u)}{dx} = u \cdot (0) + c \cdot \frac{du}{dx}$$

$$\frac{d(c\cdot u)}{dx} = c \cdot \frac{du}{dx}$$

This is Eq. (23.10), the derivative of a constant multiplied by a function.

37. $$y = x^2 \cdot f(x)$$

$u = x^2$ and $v = f(x)$

$\frac{du}{dx} = 2x$ $\qquad$ $\frac{dv}{dx} = \frac{df(x)}{dx}$

$$\frac{d(u\cdot v)}{dx} = u \cdot \frac{dv}{dx} + v \cdot \frac{du}{dx}$$

$$\frac{dy}{dx} = x^2 \cdot \frac{df(x)}{dx} + f(x) \cdot 2x$$

$$\frac{dy}{dx} = x^2 \cdot f'(x) + 2x \cdot f(x)$$

41. $$y = (4x+1)(x^4-1)$$

$$\frac{dy}{dx} = (4x+1)\cdot(4x^3) + (x^4-1)\cdot 4$$

$$\frac{dy}{dx} = 16x^4 + 4x^3 + 4x^4 - 4$$

$$\frac{dy}{dx} = 20x^4 + 4x^3 - 4$$

$$\frac{dy}{dx} = 4(5x^4 + x^3 - 1)$$

$$\left.\frac{dy}{dx}\right|_{x=-1} = 4(5-1-1) = 12.0$$

45. $$P = VI$$

$$P = (0.048 - 1.20t^2)(2.00 - 0.800t)$$

$$\frac{dP}{dt} = (0.048 - 1.20t^2)\cdot(-0.800) + (2.00 - 0.800t)\cdot(-2.40t)$$

$$\frac{dP}{dt} = -0.0384 + 0.960t^2 - 4.80t + 1.92t^2$$

$$\frac{dP}{dt} = -0.0384 + 0.960t^2 - 4.80t + 1.92t^2$$

$$\frac{dP}{dt} = 2.88t^2 - 4.80t - 0.0384$$

$$\left.\frac{dP}{dt}\right|_{t=0.150\text{ s}} = 2.88(0.150)^2 - 4.80(0.150) - 0.0384$$

$$\left.\frac{dP}{dt}\right|_{t=0.150\text{ s}} = -0.694\ \frac{\text{W}}{\text{s}}$$

49. $$s = (t^2 - 8t)(2t^2 + t + 1)$$

$$v = \frac{ds}{dt} = (t^2 - 8t)\cdot(4t+1) + (2t^2 + t + 1)\cdot(2t - 8)$$

$$v = 4t^3 + t^2 - 32t^2 - 8t + 4t^3 - 16t^2 + 2t^2 - 8t + 2t - 8$$

$$v = 8t^3 - 45t^2 - 14t - 8$$

53. $$r_f = \frac{2(R^2 + Rr + r^2)}{3(R+r)}$$

$$\frac{dr_f}{dR} = \frac{3(R+r)\cdot 2(2R+r) - 2(R^2 + Rr + r^2)\cdot(3)}{9(R+r)^2}$$

$$\frac{dr_f}{dR} = \frac{6(R+r)(2R+r) - 6(R^2 + Rr + r^2)}{9(R+r)^2}$$

$$\frac{dr_f}{dR} = \frac{12R^2 + 6Rr + 12Rr + 6r^2 - 6R^2 - 6Rr - 6r^2}{9(R+r)^2}$$

$$\frac{dr_f}{dR} = \frac{6R^2 + 12Rr}{9(R+r)^2}$$

$$\frac{dr_f}{dR} = \frac{6R(R+2r)}{9(R+r)^2}$$

$$\frac{dr_f}{dR} = \frac{2R(R+2r)}{3(R+r)^2}$$

57. $$p = \frac{9800h^2}{h+1}$$

$$\frac{dp}{dh} = \frac{(h+1)\cdot(19600h) - 9800h^2\cdot(1)}{(h+1)^2}$$

$$\frac{dp}{dh} = \frac{19600h^2 + 19600h - 9800h^2}{(h+1)^2}$$

$$\frac{dp}{dh} = \frac{9800h^2 + 19600h}{(h+1)^2}$$

$$\frac{dp}{dh} = \frac{9800h(h+2)}{(h+1)^2}$$

$$\left.\frac{dp}{dh}\right|_{h=48.0\text{ m}} = \frac{9800(48)(50)}{(49)^2}$$

$$\left.\frac{dp}{dh}\right|_{h=48.0\text{ m}} = 9795.9\text{ Pa/m} = 9.80\times 10^3\text{ Pa/m}$$

23.7 The Derivative of a Power of a Function

1. $p(x) = (2 + 3x^3)^4$

Identify this as a power of function u u and v, where

$u = 2 + 3x^3$ and $n = 4$

$$\frac{du}{dx} = 9x^2$$

$$\frac{d(u^n)}{dx} = nu^{n-1}\cdot\frac{du}{dx}$$

$$\frac{dp(x)}{dx} = 4(2+3x^3)^3(9x^2)$$

$$\frac{dp(x)}{dx} = 36x^2(2+3x^3)^3$$

5. $y = \sqrt{x}$

$y = x^{1/2}$

Recognize form $y = u^n$ with $n = 1/2$

$$\frac{dy}{dx} = \frac{1}{2}x^{1/2-1}$$
$$\frac{dy}{dx} = \frac{1}{2}x^{-1/2}$$
$$\frac{dy}{dx} = \frac{1}{2}\left(\frac{1}{x^{1/2}}\right)$$
$$\frac{dy}{dx} = \frac{1}{2\sqrt{x}}$$

9. $y = \dfrac{3}{\sqrt[3]{x}}$

$$y = \frac{3}{x^{1/3}}$$
$$y = 3x^{-1/3}$$
$$\frac{dy}{dx} = 3\left(-\frac{1}{3}x^{-1/3-1}\right)$$
$$\frac{dy}{dx} = -1x^{-4/3}$$
$$\frac{dy}{dx} = -\frac{1}{x^{4/3}}$$

13. $y = (x^2+1)^5$

$$\frac{dy}{dx} = 5\left(x^2+1\right)^4(2x)$$
$$\frac{dy}{dx} = 10x\left(x^2+1\right)^4$$

17. $y = \left(2x^3-3\right)^{1/3}$

$$\frac{dy}{dx} = \frac{1}{3}\left(2x^3-3\right)^{-2/3}\left(6x^2\right)$$
$$\frac{dy}{dx} = \frac{2x^2}{\left(2x^3-3\right)^{2/3}}$$

21. $y = 4\left(2x^4-5\right)^{0.75}$

$$\frac{dy}{dx} = 4(0.75)\left(2x^4-5\right)^{-0.25}\left(8x^3\right)$$
$$\frac{dy}{dx} = \frac{24x^3}{\left(2x^4-5\right)^{0.25}}$$

25. $u = v\sqrt{8v+5}$

$$u = v(8v+5)^{1/2}$$
$$\frac{du}{dv} = v\left(\frac{1}{2}\right)(8v+5)^{-1/2}(8) + (8v+5)^{1/2}(1)$$
$$\frac{du}{dv} = 4v(8v+5)^{-1/2} + (8v+5)^{1/2}$$
$$\frac{du}{dv} = (8v+5)^{-1/2}\left[4v + (8v+5)\right]$$
$$\frac{du}{dv} = \frac{12v+5}{(8v+5)^{1/2}}$$

29. $y = \dfrac{2x\sqrt{x+2}}{x+4}$

$$y = \frac{2x(x+2)^{1/2}}{x+4}$$

The numerator contains a product rule,
to a whole function is a quotient rule.

$$\frac{dy}{dx} = \frac{(x+4)\left[2x\cdot\frac{1}{2}(x+2)^{-1/2} + (x+2)^{1/2}(2)\right] - 2x(x+2)^{1/2}(1)}{(x+4)^2}$$
$$\frac{dy}{dx} = \frac{x(x+4)(x+2)^{-1/2} + 2(x+4)(x+2)^{1/2} - 2x(x+2)^{1/2}}{(x+4)^2}$$
$$\frac{dy}{dx} = \frac{(x+2)^{-1/2}\left[x(x+4) + 2(x+4)(x+2) - 2x(x+2)\right]}{(x+4)^2}$$
$$\frac{dy}{dx} = \frac{x^2+4x+2x^2+4x+8x+16-2x^2-4x}{(x+4)^2\sqrt{x+2}}$$
$$\frac{dy}{dx} = \frac{x^2+12x+16}{(x+4)^2\sqrt{x+2}}$$

33. $y = \sqrt{3x+4}$

$$y = (3x+4)^{1/2}$$

$u = 3x+4$ and $n = \dfrac{1}{2}$

$$\frac{du}{dx} = 3$$
$$\frac{dy}{dx} = \frac{1}{2}(3x+4)^{-1/2}(3)$$

$$\frac{dy}{dx}=\frac{3}{2}(3x+4)^{-1/2}$$
$$\frac{dy}{dx}=\frac{3}{2\sqrt{3x+4}}$$
$$\left.\frac{dy}{dx}\right|_{x=7}=\frac{3}{2\sqrt{3(7)+4}}$$
$$\left.\frac{dy}{dx}\right|_{x=7}=\frac{3}{2\sqrt{25}}=\frac{3}{10}$$

37. $y=x^{3/2}$
$$y=x\left(x^{1/2}\right)$$
$$\frac{dy}{dx}=x\left(\frac{1}{2}x^{-1/2}\right)+x^{1/2}(1)$$
$$\frac{dy}{dx}=\frac{1}{2}x^{1/2}+x^{1/2}$$
$$\frac{dy}{dx}=\frac{3}{2}x^{1/2}$$

41. $y=\frac{x^2}{\sqrt{x^2+1}}$
$$y=x^2\left(x^2+1\right)^{-1/2}$$
$$\frac{dy}{dx}=x^2\tfrac{-1}{2}\left(x^2+1\right)^{-3/2}(2x)+\left(x^2+1\right)^{-1/2}2x$$
$$\frac{dy}{dx}=-x^3\left(x^2+1\right)^{-3/2}+2x\left(x^2+1\right)^{-1/2}$$
$$\frac{dy}{dx}=x\left(x^2+1\right)^{-3/2}\left[-x^2+2(x^2+1)\right]$$
$$\frac{dy}{dx}=\frac{x(x^2+2)}{\left(x^2+1\right)^{3/2}}$$

If the derivative is zero,
$$0=\frac{x(x^2+2)}{\left(x^2+1\right)^{3/2}}$$
$$0=x(x^2+2)$$
$x=0$ or $x=\pm\sqrt{-2}$ (imaginary). Therefore,
$\frac{dy}{dx}=0$ for $x=0$.

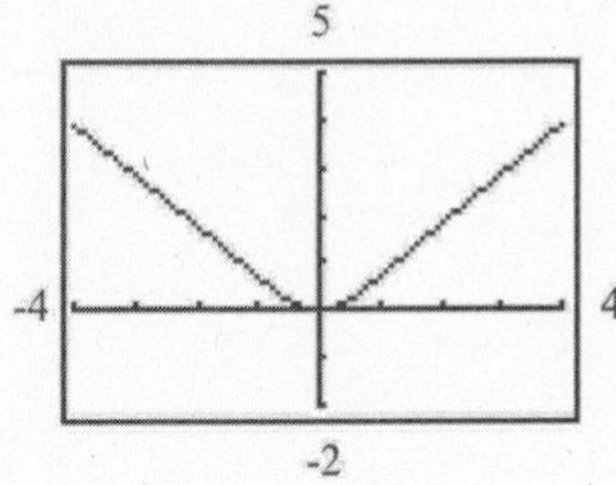

45. $y^2=4x$
$$y=\sqrt{4x}$$
$$y=2\sqrt{x}$$
$$y=2x^{1/2}$$
$$\frac{dy}{dx}=2\left(\frac{1}{2}\right)x^{-1/2}$$
$$\frac{dy}{dx}=\frac{1}{\sqrt{x}}$$
$$\left.\frac{dy}{dx}\right|_{x=1}=\frac{1}{1}=1$$

49. $s=\left(8t-t^2\right)^{2/3}$
$$v=\frac{ds}{dt}=\frac{2}{3}\left(8t-t^2\right)^{-1/3}(8-2t)$$
$$v=\frac{2(8-2t)}{3\sqrt[3]{8t-t^2}}$$
$$v|_{t=6.25\text{ s}}=\frac{2(8-2(6.25))}{3\sqrt[3]{8(6.25)-(6.25)^2}}$$
$$v|_{t=6.25\text{ s}}=-1.35\text{ cm/s}$$

53. $v=k\sqrt{\frac{l}{a}+\frac{a}{l}}$
$$v=k\left(\frac{1}{a}l+al^{-1}\right)^{1/2}$$
$$\frac{dv}{dl}=k\frac{1}{2}\left(\frac{l}{a}+\frac{a}{l}\right)^{-1/2}\left(\frac{1}{a}(1)-al^{-2}\right)$$
$$\frac{dv}{dl}=\frac{k\left(\frac{1}{a}-\frac{a}{l^2}\right)}{2\sqrt{\frac{l}{a}+\frac{a}{l}}}$$

If the derivative is set to zero,
$$0=\frac{k\left(\frac{1}{a}-\frac{a}{l^2}\right)}{2\sqrt{\frac{l}{a}+\frac{a}{l}}}$$
$$0=\frac{1}{a}-\frac{a}{l^2}$$
$$\frac{1}{a}=\frac{a}{l^2}$$
$$l^2=a^2$$
$$l=a$$

57. $\lambda_r = \dfrac{2a\lambda}{\sqrt{4a^2-\lambda^2}} = 2a\lambda\left(4a^2-\lambda^2\right)^{-1/2}$

$$\lambda_r = 2a\lambda\left(4a^2-\lambda^2\right)^{-1/2}$$

$$\frac{d\lambda_r}{d\lambda} = 2a\lambda\left(-\tfrac{1}{2}\right)\left(4a^2-\lambda^2\right)^{-3/2}(-2\lambda) + \left(4a^2-\lambda^2\right)^{-1/2}(2a)$$

$$\frac{d\lambda_r}{d\lambda} = 2a\lambda^2\left(4a^2-\lambda^2\right)^{-3/2} + 2a\left(4a^2-\lambda^2\right)^{-1/2}$$

$$\frac{d\lambda_r}{d\lambda} = 2a\left(4a^2-\lambda^2\right)^{-3/2}\left[\lambda^2+\left(4a^2-\lambda^2\right)\right]$$

$$\frac{d\lambda_r}{d\lambda} = \frac{8a^3}{\left(4a^2-\lambda^2\right)^{3/2}}$$

23.8 Differentiation of Implicit Functions

1. $y^3 + 2x^2 = 5$

$$\frac{d}{dx}\left(y^3+2x^2\right) = \frac{d}{dx}(5)$$

$$3y^2\cdot\frac{dy}{dx} + 4x = 0$$

$$3y^2\cdot\frac{dy}{dx} = -4x$$

$$\frac{dy}{dx} = -\frac{4x}{3y^2}$$

5. $\dfrac{d}{dx}\left(\dfrac{1}{xy}\right) = \dfrac{xy(0) - 1\left(x\frac{dy}{dx} + y(1)\right)}{x^2y^2}$

$$\frac{d}{dx}\left(\frac{1}{xy}\right) = \frac{-x\dfrac{dy}{dx} - y}{x^2y^2}$$

9. $4y - 3x^2 = x$

$$\frac{d}{dx}\left(4y-3x^2\right) = \frac{d}{dx}(x)$$

$$4\frac{dy}{dx} - 6x = 1$$

$$4\frac{dy}{dx} = 1 + 6x$$

$$\frac{dy}{dx} = \frac{1+6x}{4}$$

13. $y^5 = x^2 - 1$

$$\frac{d}{dx}\left(y^5\right) = \frac{d}{dx}\left(x^2-1\right)$$

$$5y^4\frac{dy}{dx} = 2x + 0$$

$$\frac{dy}{dx} = \frac{2x}{5y^4}$$

17. $y + 3xy - 4 = 0$

$$\frac{d}{dx}(y+3xy-4) = \frac{d}{dx}(0)$$

$$1\frac{dy}{dx} + 3\left(x\frac{dy}{dx} + y(1)\right) - 0 = 0$$

$$\frac{dy}{dx} + 3x\frac{dy}{dx} + 3y = 0$$

$$\frac{dy}{dx}(1+3x) = -3y$$

$$\frac{dy}{dx} = \frac{-3y}{1+3x}$$

21. $\dfrac{3x^2}{y^2+1} + y = 3x + 1$

$$\frac{d}{dx}\left(\frac{3x^2}{y^2+1} + y\right) = \frac{d}{dx}(3x+1)$$

$$\frac{\left(y^2+1\right)6x - 3x^2\left(2y\frac{dy}{dx}\right)}{\left(y^2+1\right)^2} + 1\cdot\frac{dy}{dx} = 3 + 0$$

$$\frac{6x}{y^2+1} - \frac{6x^2y}{\left(y^2+1\right)^2}\frac{dy}{dx} + 1\cdot\frac{dy}{dx} = 3$$

$$\frac{dy}{dx}\left(-\frac{6x^2y}{\left(y^2+1\right)^2} + 1\right) = 3 - \frac{6x}{y^2+1}$$

$$\frac{dy}{dx} = \frac{3 - \dfrac{6x}{y^2+1}}{-\dfrac{6x^2y}{\left(y^2+1\right)^2} + 1}$$

$$\frac{dy}{dx} = \frac{\dfrac{3(y^2+1) - 6x}{y^2+1}}{\dfrac{-6x^2y + \left(y^2+1\right)^2}{\left(y^2+1\right)^2}}$$

$$\frac{dy}{dx}=\frac{3(y^2+1)-6x}{y^2+1}\cdot\frac{(y^2+1)^2}{-6x^2y+(y^2+1)^2}$$

$$\frac{dy}{dx}=\frac{3(y^2+1-2x)(y^2+1)}{(y^2+1)^2-6x^2y}$$

25. $2(x^2+1)^3+(y^2+1)^2=17$

$$\frac{d}{dx}\left[2(x^2+1)^3+(y^2+1)^2\right]=\frac{d}{dx}(17)$$

$$6(x^2+1)^2(2x)+2(y^2+1)^1 2y\frac{dy}{dx}=0$$

$$4y(y^2+1)\frac{dy}{dx}=-12x(x^2+1)^2$$

$$\frac{dy}{dx}=\frac{-12(x^2+1)^2}{4y(y^2+1)}$$

$$\frac{dy}{dx}=\frac{-3x(x^2+1)^2}{y(y^2+1)}$$

29. $5y^4+7=x^4-3y$

$$\frac{d}{dx}(5y^4+7)=\frac{d}{dx}(x^4-3y)$$

$$20y^3\frac{dy}{dx}+0=4x^3-3\frac{dy}{dx}$$

$$\frac{dy}{dx}(20y^3+3)=4x^3$$

$$\frac{dy}{dx}=\frac{4x^3}{20y^3+3}$$

$$\left.\frac{dy}{dx}\right|_{(3,-2)}=\frac{4(3)^3}{20(-2)^3+3}=-\frac{108}{157}$$

33. $x^2+y^2=4x$

$$\frac{d}{dx}(x^2+y^2)=\frac{d}{dx}(4x)$$

$$2x+2y\frac{dy}{dx}=4$$

$$\frac{dy}{dx}=\frac{4-2x}{2y}$$

$$\frac{dy}{dx}=\frac{2-x}{y}$$

If tangent is horizontal, slope is zero

$$0=\frac{2-x}{y}$$

$$0=2-x$$

$$x=2$$

$$2^2+y^2=4(2)$$

$$y^2=4$$

$$y=\pm 2$$

The function will have a horizontal tangent at (2, 2) and (2, –2).

37. $xy+y^2+2=0$

$$\frac{d}{dx}(xy+y^2+2)=\frac{d}{dx}(0)$$

$$x\frac{dy}{dx}+y(1)+2y\frac{dy}{dx}+0=0$$

$$\frac{dy}{dx}(x+2y)=-y$$

$$\frac{dy}{dx}=\frac{-y}{x+2y}$$

$$\left.\frac{dy}{dx}\right|_{(-3,1)}=\frac{-1}{-3+2}=1$$

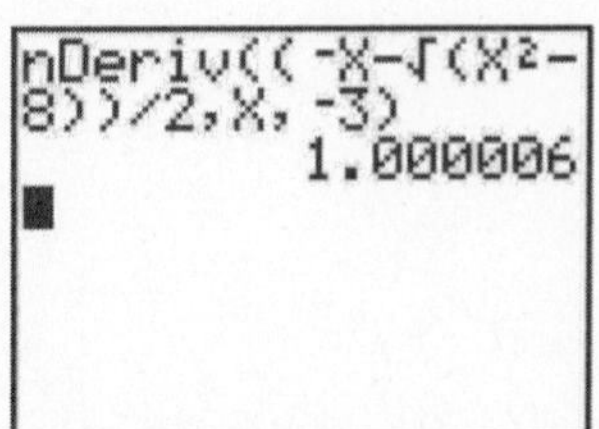

41. $PV=n\left(RT+aP-\frac{bP}{T}\right)$ where a,b,n,R,V are constant

$$PV=nRT+naP-nbPT^{-1}$$

$$\frac{d}{dT}(PV)=\frac{d}{dT}(nRT+naP-nbPT^{-1})$$

$$V\frac{dP}{dT}=nR+na\frac{dP}{dT}-nbP(-T^{-2})+T^{-1}\left(-nb\frac{dP}{dT}\right)$$

$$\frac{dP}{dT}\left(V-na+\frac{nb}{T}\right)=nR+\frac{nbP}{T^2}$$

$$\frac{dP}{dT}\left(\frac{VT-naT+nb}{T}\right)=\frac{nRT^2+nbP}{T^2}$$

$$\frac{dP}{dT}=\frac{nRT^2+nbP}{T^2}\cdot\frac{T}{VT-naT+nb}$$

$$\frac{dP}{dT}=\frac{nRT^2+nbP}{T(VT-naT+nb)}$$

45.
$$r^2 = 2rR + 2R - 2r$$
$$\frac{d}{dr}\left(r^2\right) = \frac{d}{dr}\left(2rR + 2R - 2r\right)$$
$$2r = (2r)\frac{dR}{dr} + R(2) + 2\frac{dR}{dr} - 2$$
$$2r - 2R + 2 = \frac{dR}{dr}(2r+2)$$
$$\frac{dR}{dr} = \frac{2(r-R+1)}{2(r+1)}$$
$$\frac{dR}{dr} = \frac{r-R+1}{r+1}$$

23.9 Higher Derivatives

1.
$$y = 5x^3 - 2x^2$$
$$y' = 15x^2 - 4x$$
$$y'' = 30x - 4$$
$$y''' = 30$$
$$y^{(4)} = 0$$
$$y^{(n)} = 0 \text{ for } n \ge 4$$

5.
$$f(x) = x^3 - 6x^4$$
$$f'(x) = 3x^2 - 24x^3$$
$$f''(x) = 6x - 72x^2$$
$$f'''(x) = 6 - 144x$$
$$f^{(4)}(x) = -144$$
$$f^{(n)}(x) = 0 \text{ for } n \ge 5$$

9.
$$f(r) = r(4r+9)^3$$
$$f'(r) = r \cdot 3(4r+9)^2(4) + (4r+9)^3$$
$$f'(r) = 12r(4r+9)^2 + (4r+9)^3$$
$$f'(r) = (4r+9)^2(12r + (4r+9))$$
$$f'(r) = (4r+9)^2(16r+9)$$
$$f''(r) = (4r+9)^2(16) + (16r+9)(2)(4r+9)(4)$$
$$f''(r) = 16(4r+9)^2 + 8(16r+9)(4r+9)$$
$$f''(r) = 8(4r+9)(2(4r+9) + 16r+9)$$
$$f''(r) = 8(4r+9)(24r+27)$$
$$f''(r) = 24(4r+9)(8r+9)$$
$$f'''(r) = 24(4r+9)(8) + (8r+9)(24)(4)$$
$$f'''(r) = 768r + 1728 + 768r + 864$$
$$f'''(r) = 1536r + 2592$$
$$f'''(r) = 96(16r+27)$$
$$f^{(4)}(r) = 1536$$
$$f^{(n)}(r) = 0 \text{ for } n \ge 5$$

13.
$$y = 2x + \sqrt{x}$$
$$y = 2x + x^{1/2}$$
$$y' = 2 + \frac{1}{2}x^{-1/2}$$
$$y'' = \frac{1}{2}\left(-\frac{1}{2}\right)x^{-3/2}$$
$$y'' = -\frac{1}{4x^{3/2}}$$

17.
$$f(p) = \frac{4.8\pi}{\sqrt{1+2p}}$$
$$f(p) = 4.8\pi(1+2p)^{-1/2}$$
$$f'(p) = -2.4\pi(1+2p)^{-3/2}(2)$$
$$f'(p) = -4.8\pi(1+2p)^{-3/2}$$
$$f''(p) = -4.8\pi\left(\frac{-3}{2}\right)(1+2p)^{-5/2}(2)$$
$$f''(p) = \frac{14.4\pi}{(1+2p)^{5/2}}$$

21.
$$y = \left(3x^2-1\right)^5$$
$$y' = 5\left(3x^2-1\right)^4(6x)$$
$$y' = 30x\left(3x^2-1\right)^4$$
$$y'' = 30x(4)\left(3x^2-1\right)^3(6x) + \left(3x^2-1\right)^4(30)$$
$$y'' = 720x^2\left(3x^2-1\right)^3 + 30\left(3x^2-1\right)^4$$
$$y'' = 30\left(3x^2-1\right)^3\left(24x^2 + (3x^2-1)\right)$$
$$y'' = 30\left(3x^2-1\right)^3\left(27x^2-1\right)$$

25.
$$u = \frac{v^2}{v+15}$$
$$u' = \frac{(v+15)2v - v^2(1)}{(v+15)^2}$$
$$u' = \frac{2v^2 + 30v - v^2}{(v+15)^2}$$

$$u' = \frac{v^2+30v}{(v+15)^2}$$
$$u' = (v^2+30v)(v+15)^{-2}$$
$$u'' = (v^2+30v)(-2)(v+15)^{-3}(1)+(v+15)^{-2}(2v+30)$$
$$u'' = (v+15)^{-3}\left[(-2)(v^2+30v)+(v+15)(2v+30)\right]$$
$$u'' = \frac{(-2v^2-60v+2v^2+30v+30v+450)}{(v+15)^3}$$
$$u'' = \frac{450}{(v+15)^3}$$

29. $x^2 - xy = 1 - y^2$

$$\frac{d}{dx}(x^2-xy) = \frac{d}{dx}(1-y^2)$$
$$2x-(x\cdot y'+y(1)) = -2y\cdot y'$$
$$2x-xy'-y=-2yy'$$
$$2yy'-xy'=y-2x$$
$$y' = \frac{y-2x}{2y-x}$$
$$y'' = \frac{(2y-x)(y'-2)-(y-2x)(2y'-1)}{(2y-x)^2}$$
$$y'' = \frac{\left[(2y-x)\left(\frac{y-2x}{2y-x}-2\right)-(y-2x)\left(2\frac{y-2x}{2y-x}-1\right)\right]}{(2y-x)^2}$$
$$y'' = \frac{\left[y-2x-2(2y-x)-\frac{2(y-2x)^2}{2y-x}+(y-2x)\right]}{(2y-x)^2}$$
$$y'' = \frac{\left[y-2x-4y+2x+y-2x-\frac{2(y-2x)^2}{2y-x}\right]}{(2y-x)^2}$$
$$y'' = \frac{\left[-2y-2x-\frac{2(y-2x)^2}{(2y-x)}\right]}{(2y-x)^2}$$
$$y'' = \frac{\left((-2y-2x)(2y-x)-2(y-2x)^2\right)}{(2y-x)^3}$$
$$y'' = \frac{\left(-4y^2+2xy-4xy+2x^2-2y^2+8xy-8x^2\right)}{(2y-x)^3}$$
$$y'' = \frac{\left(-6y^2+6xy-6x^2\right)}{(2y-x)^3}$$
$$y'' = \frac{-6\left(x^2-xy+y^2\right)}{(2y-x)^3}$$

Look at original function

$$x^2-xy+y^2=1$$
$$y'' = \frac{-6}{(2y-x)^3}$$

33. $y = 3x^{2/3} - \frac{2}{x}$

$$y = 3x^{2/3}-2x^{-1}$$
$$y' = 2x^{-1/3}+2x^{-2}$$
$$y'' = -\frac{2}{3}x^{-4/3}-4x^{-3}$$
$$y'' = -\frac{2}{3x^{4/3}}-\frac{4}{x^3}$$
$$y''\big|_{x=-8} = -\frac{2}{3(-8)^{4/3}}-\frac{4}{(-8)^3}$$
$$y''\big|_{x=-8} = -\frac{2}{48}+\frac{4}{512} = -\frac{1}{24}+\frac{1}{128}$$
$$y''\big|_{x=-8} = \frac{-16+3}{384} = -\frac{13}{384}$$

37. $s = 26.0t - 4.90t^2$

$$v = \frac{ds}{dt} = 26.0-9.80t$$
$$a = \frac{dv}{dt} = -9.80$$
$$a\big|_{t=3.00\text{ s}} = -9.80\text{ m/s}^2$$

41. $\frac{d(u\cdot v)}{dx} = u\cdot\frac{dv}{dx}+v\cdot\frac{du}{dx}$

$$\frac{d^2(u\cdot v)}{dx^2} = u\cdot\frac{d^2v}{dx^2}+\frac{dv}{dx}\cdot\frac{du}{dx}+v\cdot\frac{d^2u}{dx^2}+\frac{du}{dx}\cdot\frac{dv}{dx}$$
$$\frac{d^2(u\cdot v)}{dx^2} = u\cdot\frac{d^2v}{dx^2}+2\frac{du}{dx}\cdot\frac{dv}{dx}+v\cdot\frac{d^2u}{dx^2}$$

45. $P(t) = 8000(1+0.02t+0.005t^2)$

$$P'(t) = 8000(0.02+0.01t)$$
$$P''(t) = 8000(0.01) = 80.0\text{ people/year}^2$$

49. $V = L\left(\frac{d^2q}{dt^2}\right)$

$q = \sqrt{2t+1} - 1$

$q = (2t+1)^{1/2} - 1$

$\frac{dq}{dt} = \frac{1}{2}(2t+1)^{-1/2}(2)$

$\frac{dq}{dt} = (2t+1)^{-1/2}$

$\frac{d^2q}{dt^2} = -\frac{1}{2}(2t+1)^{-3/2}(2)$

$\frac{d^2q}{dt^2} = \frac{-1}{(2t+1)^{3/2}}$

$V = 1.60\left(\frac{-1}{(2t+1)^{3/2}}\right)$

$V = -\frac{1.60}{(2t+1)^{3/2}}$

53. $s = \frac{1}{6}t^4 - \frac{7}{6}t^3 - 2t^2 + 3$

$v = \frac{ds}{dt} = \frac{1}{6}(4t^3) - \frac{7}{6}(3t^2) - 2(2t)$

$v = \frac{2}{3}t^3 - \frac{7}{2}t^2 - 4t$

$a = \frac{dv}{dt} = \frac{2}{3}(3t^2) - \frac{7}{2}(2t) - 4$

$a = 2t^2 - 7t - 4$

$a = 2t^2 - 8t + t - 4$

$a = 2t(t-4) + (t-4)$

$a = (t-4)(2t+1)$

If acceleration is zero,

$0 = (t-4)(2t+1)$

$t = 4.00$ s or $t = -0.500$ s

But since $t > 0$

$t = 4.00$ s is the solution

Review Exercises

1. $\lim_{x\to 4}(8-3x) = 8 - 3(4) = -4$

5. $\lim_{x\to 2}\frac{4x-8}{x^2-4} = \lim_{x\to 2}\frac{4(x-2)}{(x-2)(x+2)}$

$\lim_{x\to 2}\frac{4x-8}{x^2-4} = \lim_{x\to 2}\frac{4}{x+2}$

$\lim_{x\to 2}\frac{4x-8}{x^2-4} = \frac{4}{2+2} = 1$

9. $\lim_{x\to\infty}\frac{2+\frac{1}{x+4}}{3-\frac{1}{x^2}} = \frac{2+0}{3-0} = \frac{2}{3}$

13. $y = 7 + 5x$

$\frac{dy}{dx} = \lim_{h\to 0}\frac{f(x+h)-f(x)}{h}$

$\frac{dy}{dx} = \lim_{h\to 0}\frac{7+5(x+h)-7-5x}{h}$

$\frac{dy}{dx} = \lim_{h\to 0}\frac{5h}{h}$

$\frac{dy}{dx} = \lim_{h\to 0}5$

$\frac{dy}{dx} = 5$

17. $y = \frac{2}{x^2}$

$\frac{dy}{dx} = \lim_{h\to 0}\frac{f(x+h)-f(x)}{h}$

$\frac{dy}{dx} = \lim_{h\to 0}\frac{\frac{2}{(x+h)^2}-\frac{2}{x^2}}{h}$

$\frac{dy}{dx} = \lim_{h\to 0}\frac{2x^2-2(x+h)^2}{hx^2(x+h)^2}$

$\frac{dy}{dx} = \lim_{h\to 0}\frac{2x^2-2x^2-4xh-2h^2}{hx^2(x+h)^2}$

$\frac{dy}{dx} = \lim_{h\to 0}\frac{-4xh-2h^2}{hx^2(x+h)^2}$

$\frac{dy}{dx} = \lim_{h\to 0}\frac{-4x-2h}{x^2(x+h)^2}$

$\frac{dy}{dx} = \frac{-4x}{x^4} = -\frac{4}{x^3}$

21. $y = 2x^7 - 3x^2 + 5$

$\frac{dy}{dx} = 2(7x^6) - 3(2x) + 0$

$\frac{dy}{dx} = 14x^6 - 6x$

$\frac{dy}{dx} = 2x(7x^5 - 3)$

25. $f(y)=\dfrac{3y}{1-5y}$

$$\frac{df(y)}{dy}=\frac{(1-5y)(3)-3y(-5)}{(1-5y)^2}$$

$$\frac{df(y)}{dy}=\frac{3-15y+15y}{(1-5y)^2}$$

$$\frac{df(y)}{dy}=\frac{3}{(1-5y)^2}$$

29. $y=\dfrac{3\pi}{\left(5-2x^2\right)^{3/4}}$

$$y=3\pi\left(5-2x^2\right)^{-3/4}$$

$$\frac{dy}{dx}=3\pi\left(\frac{-3}{4}\right)\left(5-2x^2\right)^{-7/4}(-4x)$$

$$\frac{dy}{dx}=\frac{9\pi x}{\left(5-2x^2\right)^{7/4}}$$

33. $y=\dfrac{\sqrt{4x+3}}{2x}=\dfrac{(4x+3)^{1/2}}{2x}$

$$\frac{dy}{dx}=\frac{2x\cdot\left(\frac{1}{2}\right)(4x+3)^{-1/2}(4)-(4x+3)^{1/2}(2)}{(2x)^2}$$

$$\frac{dy}{dx}=\frac{4x(4x+3)^{-1/2}-2(4x+3)^{1/2}}{4x^2}$$

$$\frac{dy}{dx}=\frac{2(4x+3)^{-1/2}\left[2x-(4x+3)\right]}{4x^2}$$

$$\frac{dy}{dx}=\frac{(-2x-3)}{2x^2(4x+3)^{1/2}}$$

$$\frac{dy}{dx}=-\frac{2x+3}{2x^2\sqrt{4x+3}}$$

37. $y=\dfrac{4}{x}+2\sqrt[3]{x}$

$$y=4x^{-1}+2x^{1/3}$$

$$\frac{dy}{dx}=4(-1)x^{-2}+2\left(\frac{1}{3}x^{-2/3}\right)$$

$$\frac{dy}{dx}=\frac{-4}{x^2}+\frac{2}{3x^{2/3}}$$

$$\left.\frac{dy}{dx}\right|_{x=8}=\frac{-4}{8^2}+\frac{2}{3(8)^{2/3}}$$

$$\left.\frac{dy}{dx}\right|_{x=8}=\frac{-4}{64}+\frac{2}{3(4)}$$

$$\left.\frac{dy}{dx}\right|_{x=8}=\frac{-1}{16}+\frac{1}{6}=\frac{-3}{48}+\frac{8}{48}=\frac{5}{48}$$

41. $y=3x^4-\dfrac{1}{x}=3x^4-x^{-1}$

$$y'=12x^3+x^{-2}$$

$$y''=36x^2-2x^{-3}$$

$$y''=2x^{-3}\left(18x^5-1\right)$$

$$y''=\frac{2\left(18x^5-1\right)}{x^3}$$

45. The slopes of the three functions are

$$\frac{d}{dx}\left(\frac{1}{x}\right)=-\frac{1}{x^2}$$

$$\frac{d}{dx}\left(\frac{1}{x^2}\right)=-\frac{2}{x^3}$$

$$\frac{d}{dx}\left(\frac{1}{\sqrt{x}}\right)=\frac{d}{dx}\left(x^{-1/2}\right)=-\frac{1}{2x^{3/2}}$$

The derivative that is the most negative as x approaches 0 is the derivative of $1/\sqrt{x}$, so it is the function that increases most rapidly.

49. $y=\dfrac{2\left(x^2-4\right)}{x-2}$

$$y=\frac{2(x+2)(x-2)}{(x-2)}$$

$$y=2(x+2),x\neq 2$$

$$\lim_{x\to 2}2(x+2)=2(2+2)=8.00$$

Just to the left of $x=2$, the trace feature gives $x=1.9787234$, $y=7.9574468$ and just to the right of $x=2$, the trace feature gives $x=2.0319149$, $y=8.0638298$ which would appear to give $y=8$ for $x=2$. However, using the value feature shows there is no y-value for $x=2$.

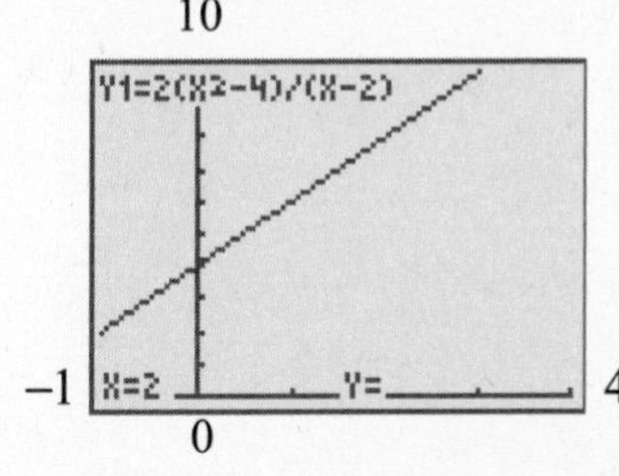

53. $y = 7x^4 - x^3$

$$\frac{dy}{dx} = 28x^3 - 3x^2$$

$$\left.\frac{dy}{dx}\right|_{(-1,8)} = 28(-1)^3 - 3(-1)^2$$

$$\left.\frac{dy}{dx}\right|_{(-1,8)} = -31.0$$

57. $A = 5000(1+0.250i)^8$

$$\frac{dA}{di} = 5000(8)(1+0.250i)^7(0.250)$$

$$\frac{dA}{di} = 10\,000(1+0.250i)^7$$

61. $y = 0.0015x^2 + C$

$$\frac{dy}{dx} = 2(0.0015)x$$

$$\frac{dy}{dx} = 0.003x$$

If this slope is along a line

$y = 0.3x - 10$, it must have slope 0.3

$0.3 = 0.003x$

$x = 100$

To be tangent to

$y = 0.3x - 10$, it must intersect that curve

$y = 0.3(100) - 10$

$y = 20$

The curve must pass through point (100, 20)

$y = 0.0015x^2 + C$

$20 = 0.0015(100)^2 + C$

$20 = 15 + C$

$C = 5$

65. $E = \dfrac{k}{r^2} = kr^{-2}$

$$\frac{dE}{dr} = -2kr^{-3} = \frac{-2k}{r^3}$$

69. $r_f = \dfrac{2(R^3 - r^3)}{3(R^2 - r^2)}$

$$\frac{dr_f}{dR} = \frac{3(R^2 - r^2)(6R^2) - 2(R^3 - r^3)(6R)}{9(R^2 - r^2)^2}$$

$$\frac{dr_f}{dR} = \frac{18R^4 - 18R^2r^2 - 12R^4 + 12Rr^3}{9(R^2 - r^2)^2}$$

$$\frac{dr_f}{dR} = \frac{6R^4 - 18R^2r^2 + 12Rr^3}{9(R^2 - r^2)^2}$$

$$\frac{dr_f}{dR} = \frac{6R(R^3 - 3Rr^2 + 2r^3)}{9((R+r)(R-r))^2}$$

$$\frac{dr_f}{dR} = \frac{2R(R^3 - 3Rr^2 + 2r^3)}{3(R+r)^2(R-r)^2}$$

Numerator can be factored

$$\frac{dr_f}{dR} = \frac{2R(R-r)(R^2 + Rr - 2r^2)}{3(R+r)^2(R-r)^2}$$

$$\frac{dr_f}{dR} = \frac{2R(R^2 + Rr - 2r^2)}{3(R+r)^2(R-r)}$$

$$\frac{dr_f}{dR} = \frac{2R(R+2r)(R-r)}{3(R+r)^2(R-r)}$$

$$\frac{dr_f}{dR} = \frac{2R(R+2r)}{3(R+r)^2}$$

73. $y = kx(x^4 + 450x^2 - 950)$

$y = kx^5 + 450kx^3 - 950kx$

$$\frac{dy}{dx} = 5kx^4 + 1350kx^2 - 950k$$

$$\frac{dy}{dx} = 5k(x^4 + 270x^2 - 190)$$

77. $T = \dfrac{10(1-t)}{0.5t+1}$

$$\frac{dT}{dt} = \frac{(0.5t+1)(-10) - 10(1-t)(0.5)}{(0.5t+1)^2}$$

$$\frac{dT}{dt} = \frac{-5t - 10 - 5 + 5t}{(0.5t+1)^2}$$

$$\frac{dT}{dt} = \frac{-15}{(0.5t+1)^2}$$

81. $A = lw = 75.0$

$l = \dfrac{75.0}{w}$

$p = 2l + 2w$

$p = \dfrac{150}{w} + 2w$

$$\frac{dp}{dw} = \frac{-150}{w^2} + 2$$

$$\frac{dp}{dw} = 2\left(1 - \frac{75.0}{w^2}\right)$$

85. Using Pythagoras' Theorem

$$r^2 = x^2 + y^2$$

Where $y = 0.500$ km constant and

$$\frac{dx}{dt} = 400 \text{ km/h}$$

$$\frac{d}{dt}\left(r^2\right) = \frac{d}{dt}\left(x^2 + y^2\right)$$

$$2r\frac{dr}{dt} = 2x\cdot\frac{dx}{dt} + 0$$

$$\frac{dr}{dt} = \frac{x}{r}\cdot\frac{dx}{dt}$$

After $t = 0.600$ min the plane has moved

$$x = (400 \text{ km/h})(0.600 \text{ min})\left(\frac{1 \text{ h}}{60 \text{ min}}\right)$$

$$= 4.00 \text{ km}$$

$$r^2 = 4^2 + 0.5^2 = 4.0311 \text{ km}$$

$$\frac{dr}{dt} = \frac{4.00 \text{ km}}{4.0311 \text{ km}}\cdot 400 \text{ km/h} = 397 \text{ km/h}$$

CHAPTER 24
APPLICATIONS OF THE DERIVATIVE

24.1 Tangents and Normals

1. $x^2+4y^2=17, (1, 2)$

$2x+8yy'=0$

$y'=\left.\frac{-x}{4y}\right|_{(1,\,2)}$

$=\frac{-1}{4(2)}=-\frac{1}{8}$

$y-2=-\frac{1}{8}(x-1)$

$8y-16=-x+1$

$x+8y-17=0$

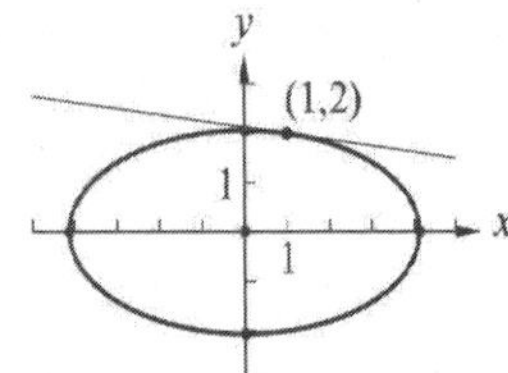

5. $y=\frac{1}{x^2+1}$ at $\left(1, \frac{1}{2}\right)$

$y=\left(x^2+1\right)^{-1}$

$\frac{dy}{dx}=-\left(x^2+1\right)^{-2}(2x)$

$m_{\tan}=\left.\frac{-2}{\left(x^2+1\right)^2}\right|_{x=1}$

$=-\frac{1}{2}$

Eq. T.L.:

$y-\frac{1}{2}=-\frac{1}{2}(x-1)$

$2y-1=-x+1$

$y=-\frac{1}{2}x+1$

$x+2y-2=0$

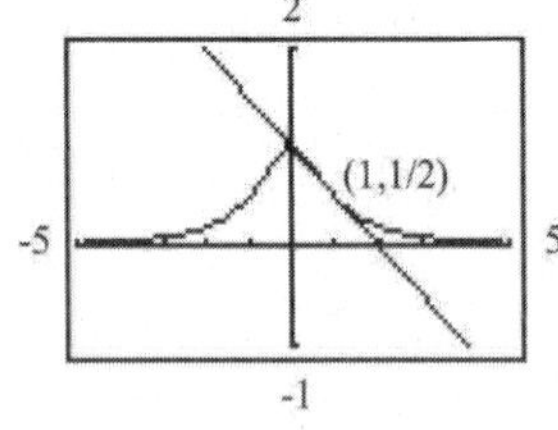

9. $y=\frac{6}{\left(x^2+1\right)^2}$ at $\left(1, \frac{3}{2}\right)$

$y=6\left(x^2+1\right)^{-2}$

$\frac{dy}{dx}=-12\left(x^2+1\right)^{-3}(2x)$

$m_{\tan}=\left.\frac{-24x}{\left(x^2+1\right)^3}\right|_{x=1}=-3;$

$m_{\text{normal}}=\frac{1}{3}$

Eq. of normal:

$y-\frac{3}{2}=\frac{1}{3}(x-1)$

$2x-6y+7=0$

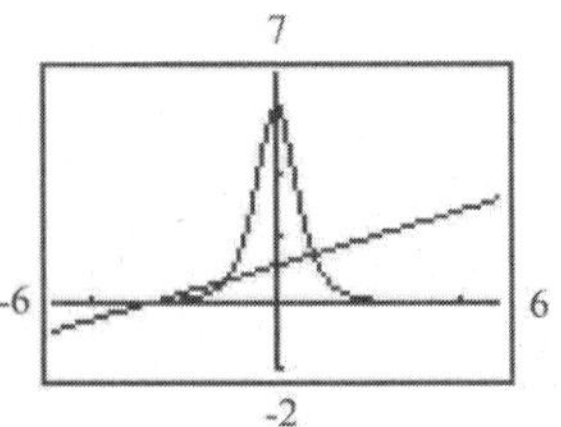

13. $y=(2x-1)^3$; normal line $m=-\frac{1}{24}, x>0$

Therefore, $m_{\tan}=24$

$m_{\tan}=3(2x-1)^2(2)$

$=6(x-1)^2$

$6(2x-1)^2=24$

$(2x-1)^2=4$

$2x-1=\pm 2$

$x=\frac{3}{2}$ or $x=-\frac{1}{2}$ (reject $x<0$)

$x=\frac{3}{2}, y=8$

Eq. of N.L.:

$y-8=-\frac{1}{24}\left(x-\frac{3}{2}\right)$

$24y-192=-x+\frac{3}{2}$

$2x+48y-387=0$

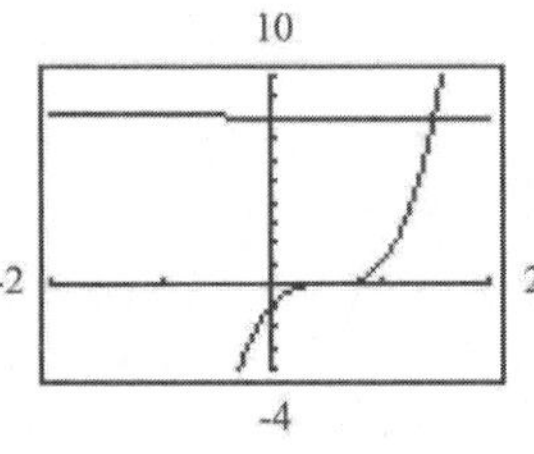

17. $y=x+2x^2-x^4$

$y'=1+4x-4x^3$

At $(1, 2)$, $y'=1+4(1)-4(1)^3=1$

Eq. of T.L. at $(1, 2)$: $y-2=1(x-1)$

$y=x+1$

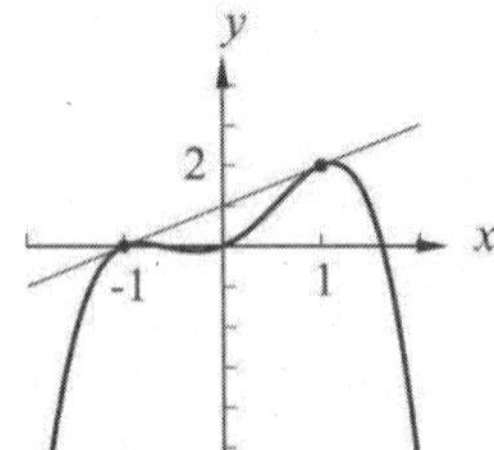

At $(-1, 0)$, $y' = 1 + 4(-1) - 4(-1)^3 = 1$

Eq. of TL at $(-1, 0)$:

$$y - 0 = 1(x - (-1))$$
$$y = x + 1$$

The tangent lines are the same: $y = x + 1$

21. $x^2 + y^2 = a^2$

$$2x + 2yy' = 0$$
$$y'\Big|_{(x_1, y_1)} = \frac{-x}{y}\Big|_{(x_1, y_1)} = \frac{-x_1}{y_1}$$
$$y - y_1 = \frac{-x_1}{y_1}(x - x_1)$$
$$y_1 y - y_1^2 = -x_1 x + x_1^2$$
$$x_1 x + y_1 y = x_1^2 + y_1^2$$
$$x_1 x + y_1 y = a^2$$

25. The parabola has axis along the y-axis, and passes through the points $(-40, 10)$ and $(40, 10)$. Its equation is:

$$x^2 = 4py$$

Substituting $(40, 10)$:

$$40^2 = 4p(10)$$
$$4p = 160$$

Therefore, $x^2 = 160y$

Eq. of T.L. at $(40, 10)$:

$$2x = 160\frac{dy}{dx}$$
$$\frac{dy}{dx} = \frac{x}{80}$$
$$m_{\text{TL}}\Big|_{x=40} = \frac{40}{80} = \frac{1}{2}$$
$$y - 10 = \frac{1}{2}(x - 40)$$
$$x - 2y - 20 = 0$$

29. $y = \frac{4}{x^2 + 1}$; $(-2 < x < 2)$

Supports at $(-1, 2)$, $(0, 4)$, $(1, 2)$

$$y = 4(x^2 + 1)^{-1}$$
$$\frac{dy}{dx} = -4(x^2 + 1)^{-2}(2x)$$
$$m_{\tan} = \frac{-8x}{(x^2 + 1)^2}$$
$$m_{\tan}\Big|_{x=-1} = 2;$$
$$m_{\tan}\Big|_{x=0} = 0;$$
$$m_{\tan}\Big|_{x=1} = -2$$

Eq. of N.L.: at $(-1, 2)$:

$$m_{\text{NL}} = -\frac{1}{2}$$
$$y - 2 = -\frac{1}{2}(x + 1)$$
$$2y - 4 = -x - 1;$$
$$x + 2y - 3 = 0$$

Eq. of N.L.: at $(0, 4)$

$x = 0$; m_{NL} is undefined (vertical).

Eq. of N.L.: at $(1, 2)$:

$$m_{\text{NL}} = \frac{1}{2}$$
$$y - 2 = \frac{1}{2}(x - 1)$$
$$2y - 4 = x - 1$$
$$x - 2y + 3 = 0$$

24.2 Newton's Method for Solving Equations

1. $x^2 - 5x + 1 = 0$, $0 < x < 1$, $f(x) = x^2 - 5x + 1$

$f(0) = 1$, $f(1) = -3$, choose $x_1 = 0.5$

$$f'(x) = 2x - 5$$
$$x_2 = x_1 - \frac{f(x_1)}{f'(x_1)} = 0.5 - \frac{0.5^2 - 5(0.5) + 1}{2(0.5) - 5} = 0.1875$$
$$x_3 = 0.1875 - \frac{(0.1875)^2 - 5(0.1875) + 1}{2(0.1875) - 5} = 0.208\,614\,864\,9$$

Using the quadratic formula,

$$x = \frac{(-5) - \sqrt{(-5)^2 - 4(1)(1)}}{2(1)} = 0.208\,712\,152\,5$$

Therefore, x_3 is correct to 3 decimal places.

5. $x^3-6x^2+10x-4=0$ (between 0 and 1)

$f(x)=x^3-6x^2+10x-4;$

$f'(x)=3x^2-12x+10;$

$f(0)=-4;\ f(1)=1$

Let $x_1=0.7$

n	x_n	$f(x_n)$	$f'(x_n)$	$x_n-\frac{f(x_n)}{f'(x_n)}$
1	0.7	0.403	3.07	0.568 729 6
2	0.568 729 6	−0.069 466 5	4.145 604 5	0.585 486 3
3	0.585 486 3	−0.001 200 9	4.002 547 0	0.585 786 3
4	0.585 786 3	−0.000 000 5	4.000 001 2	0.585 786 4

$x_4=x_5=0.585\ 786$ to 6 decimal places

9. $x^4-x^3-3x^2-x-4=0;$ (between 2 and 3)

$f(x)=x^4-x^3-3x^2-x-4;$

$f'(x)=4x^3-3x^2-6x-1$

$f(2)=-10;\ f(3)=20.$

Let $x_1=2.3$

n	x_n	$f(x_n)$	$f'(x_n)$	$x_n-\frac{f(x_n)}{f'(x_n)}$
1	2.3	−6.3529	17.998	2.652 978 1
2	2.652 978 1	3.097 272 7	36.657 000 1	2.568 484 8
3	2.568 484 8	0.217 499 3	31.576 096	2.561 596 7
4	2.561 596 7	0.001 367 1	31.179 597	2.561 552 8

$x_4=x_5=2.561\ 6$ to 4 decimal places

13. $2x^2=\sqrt{2x+1}$ (the positive real solution)

$f(x)=2x^2-\sqrt{2x+1}$

$f'(x)=4x-\frac{1}{\sqrt{2x+1}}$

Let $x_1=0.8$, since we can see that the graphs of $2x^2$ and $\sqrt{2x-1}$ intersect near $x=0.8$.

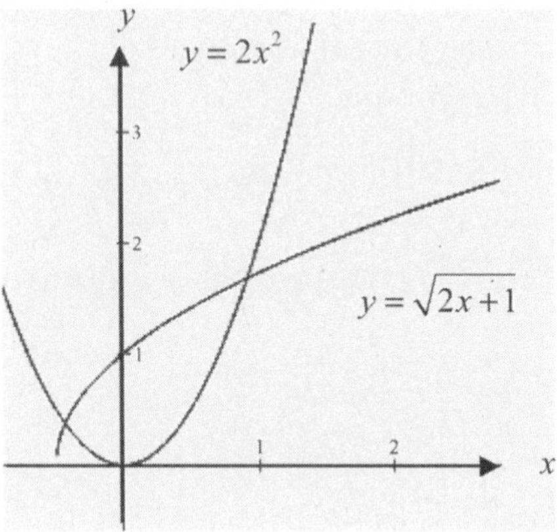

n	x_n	$f(x_n)$	$f'(x_n)$	$x_n-\frac{f(x_n)}{f'(x_n)}$
1	0.8	−0.332 451 5	2.579 826 3	0.928 865 9
2	0.928 865 9	0.035 100 9	3.123 916 4	0.917 629 7
3	0.917 629 7	2.656×10^{-4}	3.076 632 0	0.917 543 3
4	0.917 543 3	1.569×10^{-8}	3.076 268 6	0.917 543 3

The positive root is approximately 0.917 543 3 to 7 decimal places.

17. $f(x)=x^3-2x^2-5x+4$. From the graph, one root lies between -1 and -2, a second between 0 and 1, and a third between 3 and 4

$f'(x)=3x^2-4x-5$

Let $x_1=-1.7$

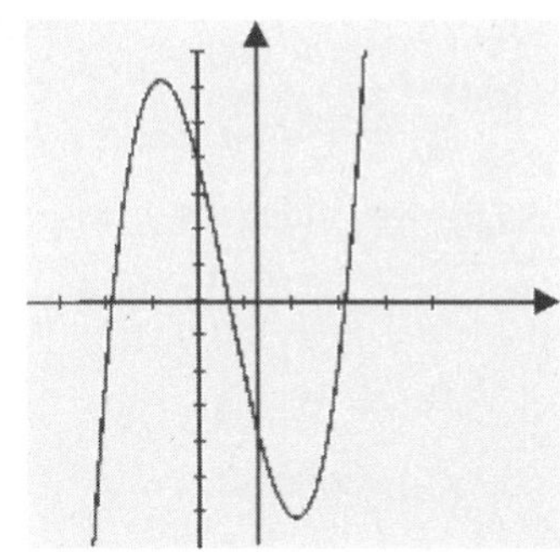

n	x_n	$f(x_n)$	$f'(x_n)$	$x_n=\frac{f(x_n)}{f'(x_n)}$
1	−1.7	1.807	10.47	−1.872 588 3
2	−1.872 588 3	−0.216 626 7	13.010 114 8	−1.855 937 7
3	−1.855 937 7	−0.002 107 4	12.757 265 2	−1.855 772 5
4	−1.855 772 5	$-2.065\ 005\ 0\times10^{-7}$	12.754 765 1	−1.855 772 5

Let $x_1=0.7$

1	0.7	−0.137	−6.33	0.678 357 0
2	0.678 357 0	$3.670\ 385\ 5\times10^{-5}$	−6.332 923 3	0.678 362 8

Let $x_1=3.1$

1	3.1	−0.929	11.43	3.181 277 3
2	3.118 127 73	0.048 760 8	12.636 467 2	3.177 418 6
3	3.177 418 6	$1.122\ 688\ 9\times10^{-4}$	12.578 292 6	3.177 409 7

The roots are $-1.855\ 8$, 0.678 4, and 3.177 4.

21. $f(x) = x^2 - a$

$f'(x) = 2x$

$$x_2 = x_1 - \frac{f(x_1)}{f'(x_1)} = x_1 - \frac{x_1^2}{2x_1} = x_1 - \frac{x_1}{2} + \frac{a}{2x_1}$$

$x_2 = \frac{x_1}{2} + \frac{a}{2x_1}$. Similarly, $x_3 = \frac{x_2^{\,2}}{2} + \frac{a}{2x_2}$ which

generalizes to $x_{n+1} = \frac{x_n}{2} + \frac{a}{2x_n}$.

25. $h = 0.009\,26x^3 - 0.0833x^2 + 0.786$

$h' = 0.02778x^2 - 0.1666x$

From the graph $h = 0$ near 4. Let $x_1 = 4.5$.

Newton's method gives

$x_2 = 4.195\,399\,535$

$x_3 = 4.212\,582\,549$

$x_4 = 4.212\,629\,78$

$x_5 = 4.212\,629\,781$

The root is $x = 4.212\,629\,78$

For $s = 0.786$, $h = 4.21$ cm

29. $V_1 = (2.00)(2.00)(4.00) = 16.00 \text{ cm}^3$

$V_2 = 2V_1 = 32.00 \text{ cm}^3$

Let each dimension increase by x cm as a result of the expansion.

Therefore, $32 = (4.00 + x)(2.00 + x)^2$

$32.00 = 16.00 + 20.00x + 8.00x^2 + x^3$

$f(x) = x^3 + 8.00x^2 + 20.00x - 16.00$

$f'(x) = 3x^2 + 16.00x + 20.00$

$f(0) < 0$; $f(1) > 0$; let $x_1 = 0.5$

n	x_n	$f(x_n)$	$f'(x_n)$	$x_n - \frac{f(x_n)}{f'(x_n)}$
1	0.5	−3.875	28.75	0.634 782 6
2	0.634 782 6	0.175 028 8	31.365 368 6	0.629 202 2
3	0.629 202 2	0.000 308 3	31.254 923 2	0.629 192 4

$x_3 = 0.629\,20$

Therefore, each edge increases by 0.629 cm.

24.3 Curvilinear Motion

1. $x = 4t^2$ $\qquad y = 1 - t^2$

$v_x = \frac{dx}{dt} = 8t\big|_{t=2} = 16 \qquad \frac{dy}{dt} = -2t\big|_{t=2} = -4$

$v = \sqrt{16^2 + (-4)^2} = 16.5$

$\tan\theta = \frac{-4}{16}$, $\theta = -14.0^\circ$

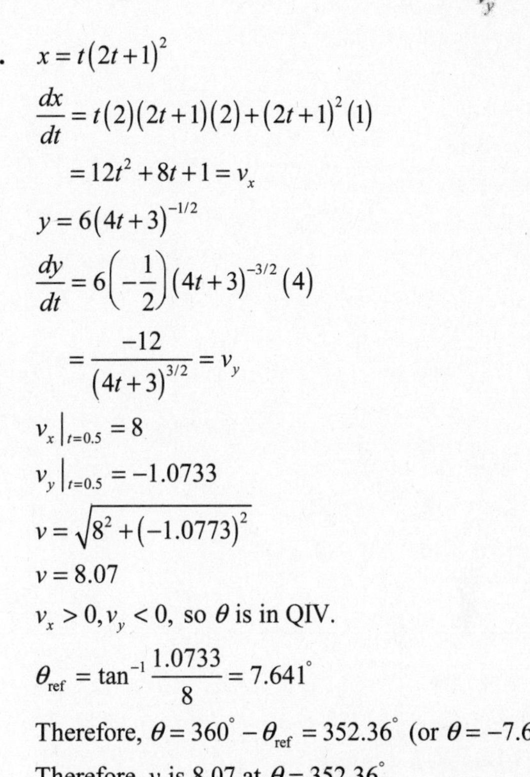

5. $x = t(2t+1)^2$

$\frac{dx}{dt} = t(2)(2t+1)(2) + (2t+1)^2(1)$

$= 12t^2 + 8t + 1 = v_x$

$y = 6(4t+3)^{-1/2}$

$\frac{dy}{dt} = 6\left(-\frac{1}{2}\right)(4t+3)^{-3/2}(4)$

$= \frac{-12}{(4t+3)^{3/2}} = v_y$

$v_x\big|_{t=0.5} = 8$

$v_y\big|_{t=0.5} = -1.0733$

$v = \sqrt{8^2 + (-1.0773)^2}$

$v = 8.07$

$v_x > 0, v_y < 0$, so θ is in QIV.

$\theta_{\text{ref}} = \tan^{-1}\frac{1.0733}{8} = 7.641^\circ$

Therefore, $\theta = 360^\circ - \theta_{\text{ref}} = 352.36^\circ$ (or $\theta = -7.64^\circ$)

Therefore, v is 8.07 at $\theta = 352.36^\circ$

To sketch the curve we tabulate a few points:

t	x	y
0	0	3.464
0.5	2	2.683
1	9	2.268

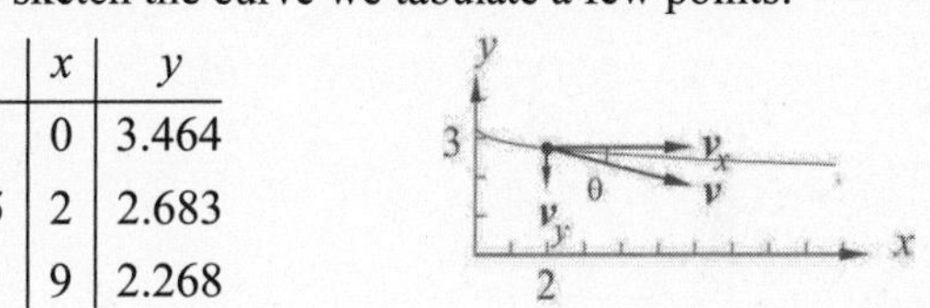

9. $x = t(2t+1)^2$

$v_x = 12t^2 + 8t + 1$

$a_x = 24t + 8;$

$a_x\big|_{t=0.5} = 20.0$

$y = \frac{6}{\sqrt{4t+3}}$

$v_y = -12(4t+3)^{-3/2}$

$a_y = \frac{72}{(4t+3)^{5/2}}$

$a_y\big|_{t=0.5} = 1.288$

$a = \sqrt{20.0^2 + 1.288^2}$

$a = 20.0$

$\theta = \tan^{-1}\dfrac{1.288}{20.0}$

$\theta = 3.68^\circ$

Therefore, a is 20.0 at $\theta = 3.68^\circ$.

13. $x = 0.2t^2$

$v_x = 0.4t$

$a_x = 0.4$

$a_x\big|_{t=2.0} = 0.4$

$y = -0.1t^3$

$v_y = -0.3t^2$

$a_y = -0.6t$

$a_y\big|_{t=2.0} = -1.2$

$a = \sqrt{(0.4)^2 + (-1.2)^2}$

$= 1.3$

$a_x > 0, a_y < 0,$ so θ is in QIV

$\theta_{\text{ref}} = \tan^{-1}\left(\dfrac{1.2}{0.4}\right)$

$= 71.6^\circ$

$\theta = 360^\circ - \theta_{\text{ref}} = 288.4^\circ$

Therefore, a is 1.3 m/min^2 at $\theta = 288^\circ$ (or $\theta = -71.6^\circ$)

17. $x = 10\left(\sqrt{1+t^4} - 1\right)$

$v_x = \dfrac{20t^3}{\sqrt{1+t^4}}$

$y = 40t^{3/2}$

$v_y = 60t^{1/2}; \ (0 \le t \le 100 \text{ s})$

For $t = 10$ s:

$v_x = \dfrac{20t^3}{\sqrt{1+t^4}}\bigg|_{t=10.0 \text{ s}} = 200.0$ m/s

$v_y = 60t^{1/2}\big|_{t=10.0 \text{ s}} = 189.7$ m/s

$v = \sqrt{(200)^2 + (189.7)^2}$

$v = 276$ m/s

$\theta = \tan^{-1}\dfrac{189.7}{200} = 43.5^\circ$

For $t = 100$ s:

$v_x\big|_{t=100 \text{ s}} = 2000.0$ m/s

$v_y\big|_{t=100 \text{ s}} = 600$ m/s

$v = \sqrt{(2000)^2 + (600)^2}$

$v = 2088$ m/s

$\theta = \tan^{-1}\dfrac{600}{2000} = 16.7^\circ$

At 10 s v is 276 m/s at $\theta = 43.5^\circ$, and at 100 s v is 2090 m/s at $\theta = 16.7^\circ$.

21. $x = 10\left(\sqrt{1+t^4} - 1\right) = 10\left(1+t^4\right)^{1/2} - 10;$

$y = 40t^{3/2}$

$\dfrac{dx}{dt} = 5\left(1+t^4\right)^{-1/2}\left(4t^3\right) = \dfrac{20t^3}{\sqrt{1+t^4}}$

$\dfrac{dy}{dt} = 60t^{1/2} = 60\sqrt{t}$

$$a_x = \frac{\left(1+t^4\right)^{1/2}\left(60t^2\right) - \left(20t^3\right)\left(\frac{1}{2}\right)\left(1+t^4\right)^{-1/2}\left(4t^3\right)}{\left(1+t^4\right)}$$

$$= \frac{\left(1+t^4\right)^{-1/2}\left[60t^2\left(1+t^4\right) - 40t^6\right]}{\left(1+t^4\right)}$$

$$= \frac{60t^2\left(1+t^4\right) - 40t^6}{\left(1+t^4\right)^{3/2}}$$

$a_y = 30t^{-1/2}$

$a_x\big|_{t=10.0} = \dfrac{6000(10{,}000) - 40{,}000{,}000}{1{,}000{,}000} = 20.0$

$a_y\big|_{t=10.0} = 30(10)^{-1/2} = 9.49$

$a = \sqrt{(20)^2 + (9.5)^2} = 22.1$ m/s^2

$\tan\theta = \dfrac{9.5}{20.0} = 0.475$

$\theta = 25.4^\circ$

$a_x\big|_{t=100} = \dfrac{60\left(10^4\right)\left(10^8\right) - 40\left(10^{12}\right)}{\left(10^8\right)^{3/2}}$

$= \dfrac{6\times10^{13} - 4\times10^{13}}{10^{12}}$

$= \dfrac{2\times10^{13}}{10^{12}} = 20.0$

$a_y\big|_{t=100} = 30\left(10^2\right)^{-1/2} = 3.0$

$a = \sqrt{(20.0)^2 + (3.0)^2} = 20.2$ m/s^2

$\tan\theta = \dfrac{3.0}{20.0} = 0.150$

$\theta = 8.5^\circ$

25. $d = 88.9$ mm; $r = 44.45$ mm; $x^2 + y^2 = 44.45^2$;

$$2x\frac{dx}{dt} + 2y\frac{dy}{dt} = 0$$

$$xv_x + yv_y = 0$$

$$\frac{v_y}{v_x} = -\frac{x}{y}$$

3600 r/min = 7200π rad/min = ω

$$v = \omega r$$
$$= 7200\pi(44.45)$$
$$= 320\ 000\pi \text{ mm/min}$$
$$= 320\pi \text{ m/min}$$

For $x = 30.5$ mm

$$y = \sqrt{44.45^2 - 30.5^2}$$
$$= 32.335 \text{ mm}$$

$$\frac{v_y}{v_x} = -\frac{x}{y}$$
$$= -0.943\ 25$$

$$v_y = -0.943\ 25 v_x$$

$$v = \sqrt{v_x^2 + v_y^2}$$
$$= \sqrt{v_x^2 + (-0.943\ 25 v_x)^2} = 1.375 v_x$$

$$v_x = \frac{320\pi}{1.3747} = 731 \text{ m/min}$$

$$v_y = -0.943\ 25(731) = -690 \text{ m/min}$$

Alternatively,

$$\tan\theta = \frac{v_y}{v_x}$$
$$= -\frac{x}{y}$$
$$= -0.943\ 25$$

$$\theta = -43.33^\circ$$

$$v_x = 320\pi\cos(-43.33^\circ)$$
$$= 731 \text{ m/min}$$

$$v_y = 320\pi\sin(-43.33^\circ)$$
$$= -690 \text{ m/min}$$

24.4 Related Rates

1. $E = 2.800T + 0.012T^2$

$$\frac{dE}{dt} = 2.800\frac{dT}{dt} + 0.024T\frac{dT}{dt}$$

$$\left.\frac{dE}{dt}\right|_{T=100^\circ\text{C}} = 2.800(1.00) + 0.024(100)(1.00)$$

$$\left.\frac{dE}{dt}\right|_{T=100^\circ\text{C}} = 5.20 \text{ V/min}$$

5. $x^2 + 3y^2 + 2y = 10$

$$2x\frac{dx}{dt} + 6y\frac{dy}{dt} + 2\frac{dy}{dt} = 0$$

$$\frac{dy}{dt} = \frac{-x\frac{dx}{dt}}{3y+1}$$

$$\left.\frac{dy}{dt}\right|_{(3,-1)} = \frac{(-3)(2)}{3(-1)+1}$$
$$= 3$$

9. $v = 18\sqrt{T}$

$$\frac{dv}{dt} = \frac{18}{2\sqrt{T}}\frac{dT}{dt}$$

$$\left.\frac{dv}{dt}\right|_{T=25} = \frac{9}{\sqrt{25}}(0.2)$$
$$= 0.36 \text{ m/s}^2$$

13. $T = \pi\sqrt{\frac{L}{245}} = \frac{\pi}{\sqrt{245}}L^{1/2}$

$$\frac{dT}{dt} = \frac{\pi}{2\sqrt{245}}L^{-1/2}\frac{dL}{dT}$$

$$\left.\frac{dT}{dt}\right|_{L=16.0} = \frac{\pi}{2\sqrt{245}}(16.0)^{-1/2}(-0.100)$$
$$= -0.002\ 51 \text{ s/s}$$

17. $r = \sqrt{0.40\lambda}$; $\frac{d\lambda}{dt} = 0.10\times10^{-7}$;

$$r = (0.40\lambda)^{1/2}$$

$$\frac{dr}{dt} = \frac{1}{2}(0.40\lambda)^{-1/2}(0.40)\frac{d\lambda}{dt}$$
$$= 0.20(0.40\lambda)^{-1/2}\frac{d\lambda}{dt}$$

$$\left.\frac{dr}{dt}\right|_{\lambda=6.0\times10^{-7}} = 0.20\left[0.40(6.0\times10^{-7})\right]^{-1/2}(0.10\times10^{-7})$$
$$= 4.1\times10^{-6} \text{ m/s}$$

21. $y = \frac{1.5}{18}x$

$$12y = x$$

$$V = \frac{1}{2}xy(12) = 6xy, \quad 0 < x < 18,$$
$$0 < y < 1.5$$

$$V = 6(12y)y$$
$$= 72y^2$$

$$\frac{dV}{dt} = 144y\frac{dy}{dt}$$

$$\frac{dy}{dt} = \frac{\frac{dV}{dt}}{144y}$$

$$\left.\frac{dy}{dt}\right|_{y=1} = \frac{0.80}{144(1)}$$

$$= 0.0056 \text{ m/min}$$

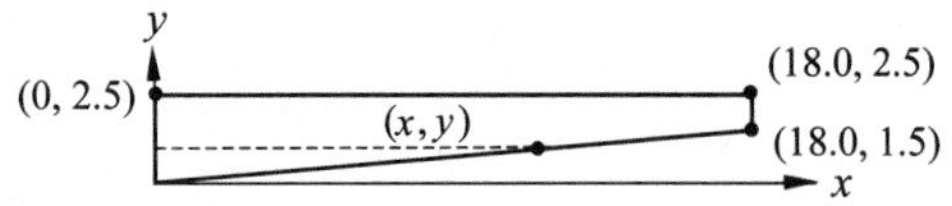

y
(0, 2.5)
(x, y)
(18.0, 2.5)
(18.0, 1.5)
x

25. $V = x^3; \frac{dx}{dt} = -0.50$ mm/min

$$\frac{dV}{dt} = 3x^2\frac{dx}{dt}$$

$$\left.\frac{dV}{dt}\right|_{x=8.20} = 3(8.20)^2(-0.50)$$

$$= -101 \text{ mm}^3\text{ / min}$$

29. $p = \frac{k}{V}; \frac{dV}{dt} = 20.0 \text{ cm}^3\text{ / min}; V = 810 \text{ cm}^3$

To find k, substitute $p = 230, V = 650$:

$$230 = \frac{k}{650}$$

$$k = 1.495\times10^5 \text{ kPa}\times\text{cm}^3$$

$$p = \frac{149\ 500}{V}$$

$$= 149\ 500V^{-1}$$

$$\frac{dp}{dt} = -149\ 500V^{-2}\frac{dV}{dt}$$

$$\left.\frac{dp}{dt}\right|_{V=810} = -149\ 500(810)^{-2}(20)$$

$$= -4.6 \text{ kPa/min}$$

33.

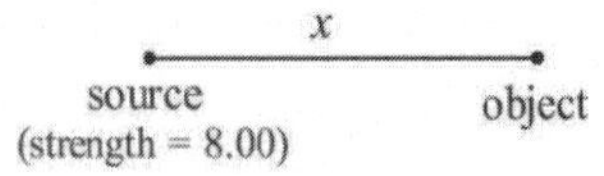

x
source
(strength = 8.00)
object

$$I = \frac{8.00k}{x^2}$$

$$\frac{dI}{dt} = \frac{-2(8.00)k}{x^3}\frac{dx}{dt}$$

$$\left.\frac{dI}{dt}\right|_{x=100} = \frac{-2(8.00)k}{100^3}(-50.0)$$

$$\frac{dI}{dt} = 0.000800k \text{ units/s}$$

37.

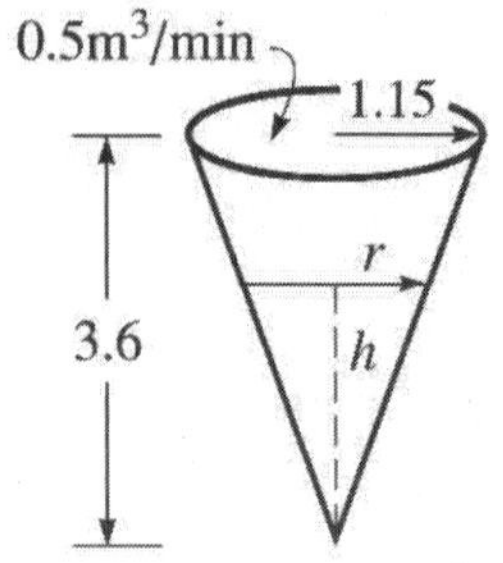

0.5m^3/min
1.15
r
3.6
h

Using similar triangles,

$$\frac{1.15}{3.6} = \frac{r}{h}$$

$$r = \frac{1.15h}{3.6}$$

$$V = \frac{1}{3}\pi r^2 h$$

$$= \frac{1}{3}\pi\left(\frac{1.15}{3.6}h\right)^2 h$$

$$= \frac{1}{3}\pi\left(\frac{1.15}{3.6}\right)^2 h^3$$

$$\frac{dV}{dt} = \pi\left(\frac{1.15}{3.6}\right)^2 h^2\frac{dh}{dt}$$

$$\frac{dh}{dt} = \frac{1}{\pi}\frac{dV}{dt}\left(\frac{3.6}{1.15h}\right)^2$$

$$\left.\frac{dh}{dt}\right|_{h=1.8} = \frac{1}{\pi}(0.50)\left(\frac{3.6}{1.15(1.8)}\right)^2$$

$$= 0.481 \text{ m/min}$$

41. $z^2 = 5^2 + x^2; \frac{dz}{dt} = -2.5$ m/s; $z = 13$ m,

$x = 12.0$ m

$$2z\frac{dz}{dt} = 2x\frac{dx}{dt}$$

$$\frac{dx}{dt} = \frac{z}{x}\frac{dz}{dt}$$

$$\left.\frac{dx}{dt}\right|_{z=13.0} = \frac{13.0}{12.0}(-2.50)$$

$$\frac{dx}{dt} = -2.71 \text{ m/s}$$

The negative sign indicates the boat is approaching the wharf.

24.5 Using Derivatives in Curve Sketching

1. $f(x) = x^3 - 6x^2$

$f'(x) = 3x^2 - 12x = 3x(x-4)$ with $x = 0$, $x = 4$ as critical values.

If $x < 0$, $f'(x) = 3x(x-4) > 0$. $f(x)$ is increasing

If $0 < x < 4$, $f'(x) = 3x(x-4) > 0$. $f(x)$ is decreasing

If $x > 4$, $f'(x) = 3x(x-4) > 0$. $f(x)$ is increasing

inc. $x < 0$, $x > 4$; dec. $0 < x < 4$

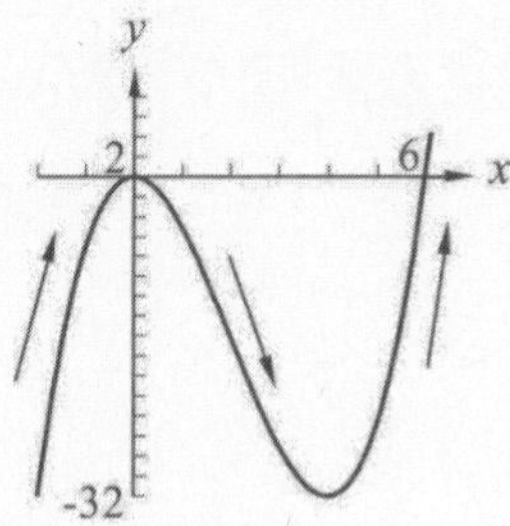

5. $y = x^2 + 2x$

$y' = 2x + 2$

$2x + 2 > 0$

$2x > -2$

$x > -1$; $f(x)$ increasing.

$2x + 2 < 0$

$2x < -2$

$x < -1$; $f(x)$ decreasing.

9. $y = x^2 + 2x$

$y' = 2x + 2$

$y' = 0$ at $x = -1$

$y'' = 2 > 0$ at $x = -1$ and $(-1, -1)$ is a relative minimum. Note also that y decreases for $x < -1$ and increases for $x > -1$.

13. $y = x^2 + 2x$

$y' = 2x + 2$

$y'' = 2$

Thus, $y'' > 0$ for all x. The graph is concave up for all x and has no points of inflection.

17. $y = x^2 + 2x$

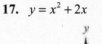

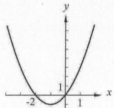

21. $y = 12x - 2x^2$

$y' = 12 - 4x$

$y' = 0$ at $x = 3$

For $x = 3$, $y = 12(3) - 2(3)^2 = 18$ and $(3, 18)$ is a critical point.

$12 - 4x < 0$ for $x > 3$ and the function decreases; $12 - 4x > 0$ for $x < 3$ and the function increases. Hence $(3, 18)$ is a maximum.

$y'' = -4$; thus $y'' < 0$ for all x. There are no inflection points.

The graph is concave down for all x, and $(3, 18)$ is a maximum point.

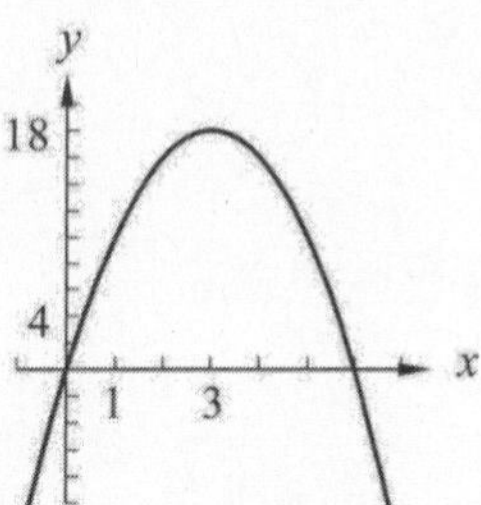

25. $y = x^3 + 3x^2 + 3x + 2$

$y' = 3x^2 + 6x + 3$

$= 3(x^2 + 2x + 1)$

$= 3(x+1)^2$

$= 0$ for $x = -1$

$(-1, 1)$ is a critical point.

$3(x+1)^2 > 0$ for all x so the function is always increasing

$y'' = 6x + 6$

$= 0$ for $x = -1$

$6x + 6 < 0$ for $x < -1$ and the graph is concave down.

$6x+6>0$ for $x>-1$ and the graph is concave up.
Therefore $(-1, 1)$ is an inflection point.
Since there is no change in slope from positive to negative or vice versa, there are no maximum or minimum points.

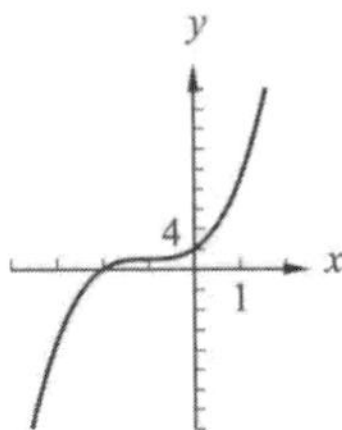

29. $y=4x^3-3x^4+6$
$y'=12x^2-12x^3=12x^2(1-x)=0$
$12x^2(1-x)=0$ for $x=0$ and $x=1$
$(0, 6)$ and $(1, 4)$ are critical points.
$12x^2-12x^3>0$ for $x<0$ and the slope is positive.
$12x-12x^3>0$ for $0<x<1$ and the slope is positive.
$12x-12x^3<0$ for $x>1$ and the slope is negative.
$y''=24x-36x^2$
$=24x-36x^2$
$=12x(2-3x)$
$=0$ for $x=0$, $x=\frac{2}{3}$
$(0, 6)$ and $\left(\frac{2}{3}, \frac{178}{27}\right)$ are possible inflection points.
$24x-36x^2<0$ for $x<0$ and the graph is concave down.
$24x-36x^2>0$ for $0<x<\frac{2}{3}$ and the graph is concave up.
$24x-36x^2<0$ for $x>\frac{2}{3}$ and the graph is concave down.
$(1, 7)$ is a relative maximum point since $y'=0$ at $(1, 7)$ and the slope is positive for $x<1$ and negative for $x>1$. $(0, 6)$ and $\left(\frac{2}{3}, \frac{178}{27}\right)$ are inflection pts since there is a concavity change.

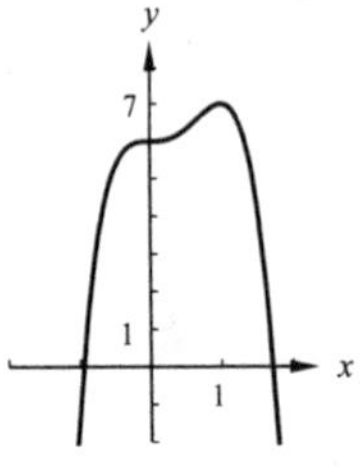

33. $y=x^3-12x$
$y'=3x^2-12$
$y''=6x.$
On a graphing calculator with $x_{\min}=-5$, $x_{\max}=5$, $y_{\min}=-20$, $y_{\max}=20$, enter $y_1=x^3-12x$, $y_2=3x^2-12$, $y_3=6x$. From the graph it is observed that the maximum and minimum values of y occur when y' is zero. A maximum value for y occurs when $x=-2$, and a minimum value occurs when $x=2$. An inflection point (change in curvature) occurs when y'' is zero. x is also zero at this point. Where $y'>0$, y inc.; $y'<0$, y dec. $y''>0$, y concave up;
$y''<0$, y concave down, $y''=0$, y has inflection.

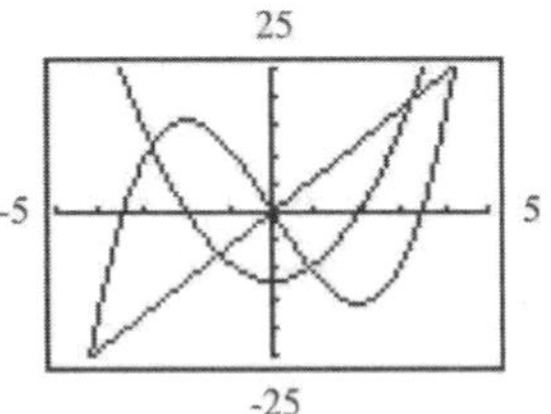

37. The left relative max is above the left relative min and below the right relative min.

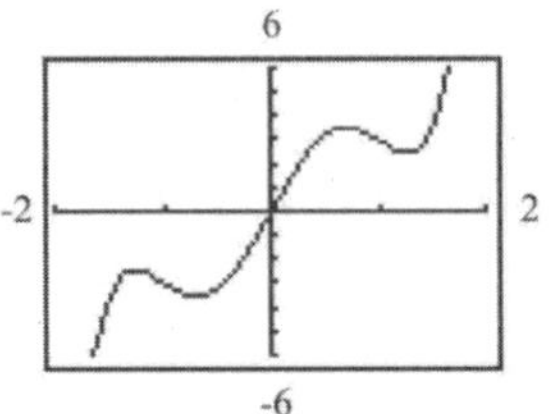

41. $P=4i-0.5i^2$
$P'=4-1.0i$
$=0$ for $i=4.0$
$P''=-1.0<0$, conc. down everywhere
$P'>0$ for $x<4$, P inc.
$P'<0$ for $x>4$, P dec.
$P(4)=8$, $(4, 8)$ max.
$4i-0.5i^2=0$ for $i=0, i=8$

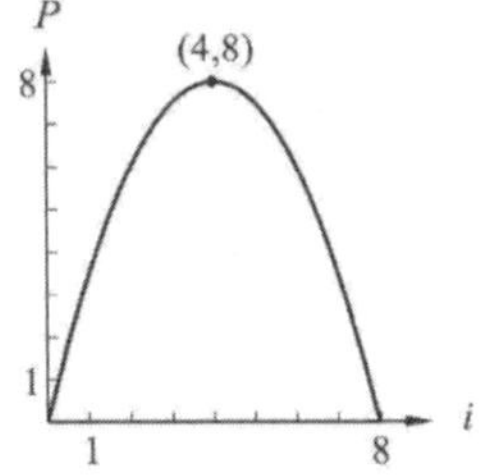

45. $R = 75 - 18i^2 + 8i^3 - i^4$

$R' = -36i + 24i^2 - 4i^3$

$= -4i\left(i^2 - 6i + 9\right)$

$= -4i(i-3)^2$

$R' = 0$ for $i = 0$ and $i = 3$

$(0, 75)$ and $(3, 48)$ are critical points.

$R' > 0$ for $i < 0$, $R' < 0$ for $0 < i < 3$

$R' < 0$ for $i > 3$

Max. at (0, 75), no max. or min. at (3, 48)

$R'' = -36 + 48i - 12i^2$

$= -12(i-1)(i-3)$

$(1, 64)$ and $(3, 48)$ are possible inflection points.

$R'' < 0$ for $i < 1$, concave down

$R'' > 0$ for $1 < i < 3$, concave up

$R'' < 0$ for $i > 3$, concave down

$(1, 64)$ and $(3, 48)$ are inflection points.

(From calculator graph, $R = 0$ for $i = -1.5$ and $i = 5.0$)

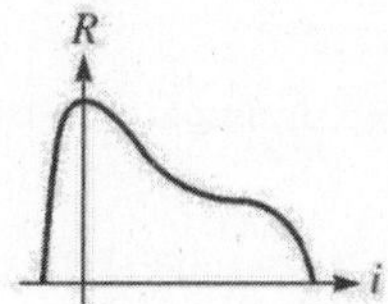

49. $V = x(8-2x)(12-2x), 0 < x < 4$

$V = 4\left(24x - 10x^2 + x^3\right)$

$= 4x^3 - 40x^2 + 96x$

$f'(x) = 4\left(24 - 20x + 3x^2\right)$

$= 0$

By quadratic solution:

$x = 1.57$ and $x = 5.10$ (reject)

$f''(x) = 4(-20 + 6x)$

$f''(1.57) < 0$, rel max. $(1.57, 67.6)$

$f''(5.10) > 0$, rel min. $(5.10, -20.2)$

$f''(x) = 0;\ x = \frac{20}{6}$, infl. $\left(\frac{10}{3}, 23.7\right)$

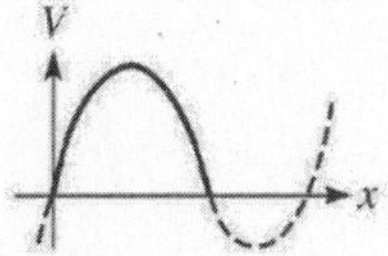

53. $f(-1) = 0$, root at $(-1, 0)$

$f(2) = 2$, point on curve $(2, 2)$

$f'(x) < 0$ for $x < -1$

Therefore, $f(x)$ decreasing for $x < -1$.

$f'(x) > 0$ for $x > -1$

Therefore, $f(x)$ increasing for $x > -1$.

$f''(x) < 0$ for $0 < x < 2$

Therefore, $f(x)$ concave down for $0 < x < 2$.

$f''(x) > 0$ for $x < 0$ or $x > 2$

Therefore, $f(x)$ concave up for $x < 0$.

Therefore, $f(x)$ concave up for $x > 2$.

Summary: $(-1, 0)$ min. point

Inflection point at y-intercept

$(2, 2)$ inflection point

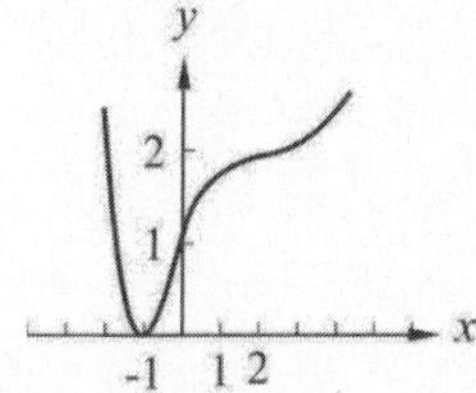

24.6 More on Curve Sketching

1. $y = x - \frac{4}{x}$

(1) Intercepts:

For $x = 0$, y is undefined which means curve is not continuous at $x = 0$ and there are no y-intercepts.

For $y = 0$, $0 = x - \frac{4}{x}$, $x^2 - 4 = 0$, $x = \pm 2$ are the x-intercepts.

(2) Symmetry: symmetric to the origin since $-y = -x - \frac{4}{-x}$ is the same as $y = x - \frac{4}{x}$

(3) Behaviour as x becomes large:

As $x \to \pm\infty$, $\frac{4}{x} \to 0$ and $y \to x$. $y = x$ is an asymptote.

(4) Vertical asymptotes:

y is undefined for $x = 0$. As $x \to 0^-$, $y \to \infty$ and as $x \to 0^+$, $y \to -\infty$. The x-axis is a vertical asymptote.

(5) Domain and range:
domain: $x \neq 0$, range: $-\infty < y < \infty$
(6) Derivatives:

$y' = 1 + \frac{4}{x^2} > 0$, y inc. for $x \neq 0$

$y'' = -\frac{8}{x^3} > 0$ for $x < 0$, y conc. up

$y'' = -\frac{8}{x^3} < 0$ for $x > 0$, y conc. down

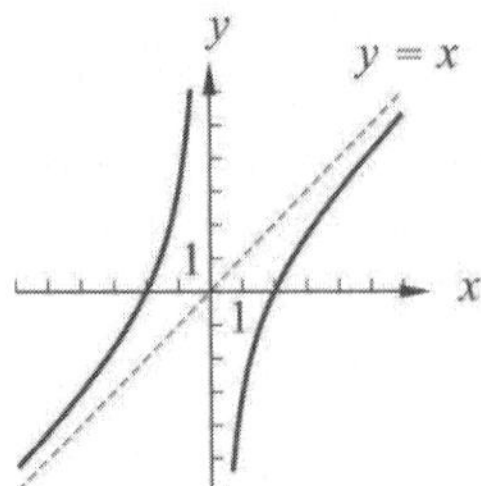

5. $y = x^2 + \frac{2}{x} = \frac{x^3 + 2}{x}$

(1) $\frac{2}{x}$ is undefined for $x = 0$, so the graph is not continuous at the y-axis; i.e., no y-intercept exists.

(2) $\frac{x^3 + 2}{x} = 0$ at $x = \sqrt[3]{-2} = -\sqrt[3]{2}$. There is an x-intercept at $\left(-\sqrt[3]{2}, 0\right)$.

(3) As $x \to \infty$, $x^2 \to \infty$ and $\frac{2}{x} \to 0$, so $x^2 + \frac{2}{x} \to \infty$.

(4) As $x \to 0$ through positive x, $x^2 \to 0$ and $\frac{2}{x} \to \infty$, so $x^2 + \frac{2}{x} \to \infty$.

(5) As $x \to -\infty$, $x^2 \to \infty$ and $\frac{2}{x} \to 0$ so $x^2 + \frac{2}{x} \to \infty$.

$x = 0$ is a vertical asymptote

(6) As $x \to 0$ through negative numbers, $x^2 \to 0$ and $\frac{2}{x} \to -\infty$, so $x^2 + \frac{2}{x} \to -\infty$.

(7) $y' = 2x - 2x^{-2} = 0$ at $x = 1$ and the slope is zero at $(1, 3)$.

(8) $y'' = 2 + 4x^{-3} = 0$ at $x = -\sqrt[3]{2}$ and $\left(-\sqrt[3]{2}, 0\right)$ is an inflection point.

(9) $y'' > 0$ at $x = 1$, so the graph is concave up and $(1, 3)$ is a relative minimum.

(10) Since $\left(-\sqrt[3]{2}, 0\right)$ is an inflection pt, $f''(x) < 0$ for $-\sqrt[3]{2} < x < 0$ and the graph is concave down.
$f''(x) > 0$ for $x < -\sqrt[3]{2}$ and the graph is concave up.
$f''(x) > 0$ for $x > 0$ and the graph is concave up.
(11) Not symmetrical about the x or y-axis nor about the origin.

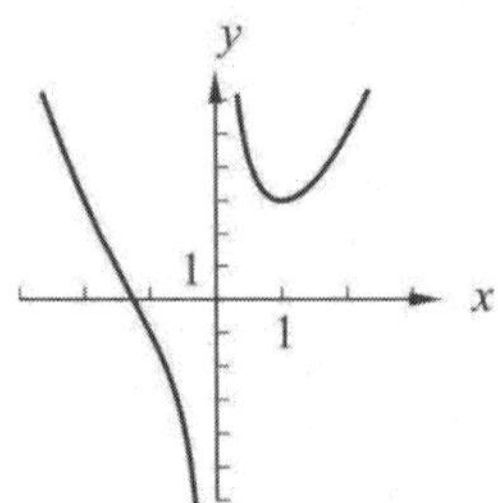

9. $y = \frac{x^2}{x+1}$

$= \frac{x^2 - 1 + 1}{x+1}$

$= x - 1 + \frac{1}{x+1}$

Intercepts:
(1) Function undefined at $x = -1$; not continuous at $x = -1$.
(2) At $x = 0$, $y = 0$. The origin is the only intercept.
(3) Behaviour as x becomes large: As $x \to \pm\infty$, $y \to x - 1$, so $y = x - 1$ is a slant asymptote.
Vertical asymptotes:
(4) As $x \to -1$ from the left, $x + 1 \to 0$ through negative values and $\frac{x^2}{x+1} \to -\infty$ since $x^2 > 0$ for all x. As $x \to -1$ from the right, $x + 1 \to 0$ through positive values and $\frac{x^2}{x+1} \to +\infty$. $x = -1$ is an asymptote.
Symmetry:
(5) The graph is not symmetrical about the y-axis or the x-axis or the origin.
Derivatives:

(6) $y' = \frac{(x+1)(2x) - x^2(1)}{(x+1)^2}$

$= \frac{x(x+2)}{(x+1)^2}$

$y' = 0$ at $x = -2$, $x = 0$.

$(-2, -4)$ and $(0, 0)$ are critical points. Checking the derivative at $x = -3$, the slope is positive, and at $x = -1.5$ the slope is negative. $(-2, -4)$ is a relative maximum point. Checking the derivative at $x = -0.5$, the slope is negative and at $x = 1$ the slope is positive, so $(0, 0)$ is a relative minimum point.

$$y'' = \frac{(x+1)^2(2x+2) - (x^2+2x)(2(x+1))}{(x+1)^4}$$
$$= \frac{2(x+1)(x^2+2x+1-x^2-2x)}{(x+1)^4}$$
$$= \frac{2}{(x+1)^3}$$

$y'' \neq 0$, so no inflection points.
$y'' > 0$ for $x > -1$, concave up
$y'' < 0$ for $x < -1$, concave down
Int. $(0, 0)$, max. $(-2, -4)$, min $(0, 0)$, asym. $x = -1$

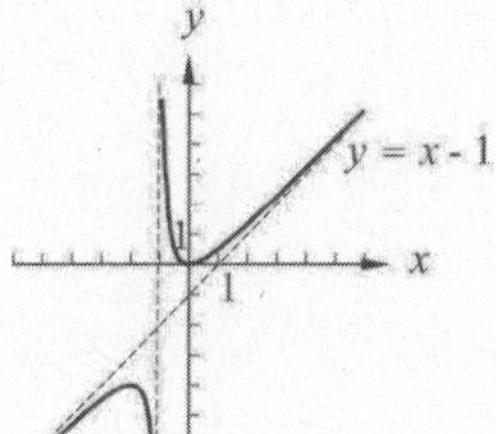

13. $y = \frac{4}{x} - \frac{4}{x^2}$

Intercepts:

(1) There are no y intercepts since $x = 0$ is undefined.

(2) $y = 0$ when $x = 1$ so $(1, 0)$ is an x-intercept,

Asymptotes:

(3) $x = 0$ is an asymptote; the denominator is 0.
$y = 0$ is an asymptote because as $x \to \pm\infty$, $y \to 0$.

Symmetry:

(4) Not symmetrical about the y-axis since $\frac{4}{x} - \frac{4}{x^2}$ is different from $\frac{4}{(-x)} - \frac{4}{(-x)^2}$

(5) Not symmetrical about the x-axis since $y = \frac{4}{x} - \frac{4}{x^2}$ is different from $-y = \frac{4}{x} - \frac{4}{x^2}$

(6) Not symmetrical about the origin since $y = \frac{4}{x} - \frac{4}{x^2}$ is different from $-y = \frac{4}{-x} - \frac{4}{(-x)^2}$

Derivatives:

(7) $y' = -4x^{-2} + 8x^{-3}$
$= 0$ at $x = 2$

(8) $y'' = 8x^{-3} - 24x^{-4}$
$= 0$ at $x = 3$

$y''(2) < 0$ so $(2, 1)$ is a relative maximum.
$y'' < 0$ (concave down) for $x < 3$ and $y'' > 0$ (concave up) for $x > 3$ so $(3, \frac{8}{9})$ is an inflection.

Behaviour as x becomes large:

(9) As $x \to \infty$ or $-\infty$, $\frac{4}{x}$ and $-\frac{4}{x^2}$ each approach 0, so the x-axis is a horizontal asymptote.
As $x \to 0$, $\frac{4}{x} - \frac{4}{x^2} = \frac{4x-4}{x^2}$ approaches $-\infty$, through positive or negative values of x, so the y-axis is a vertical asymptote.

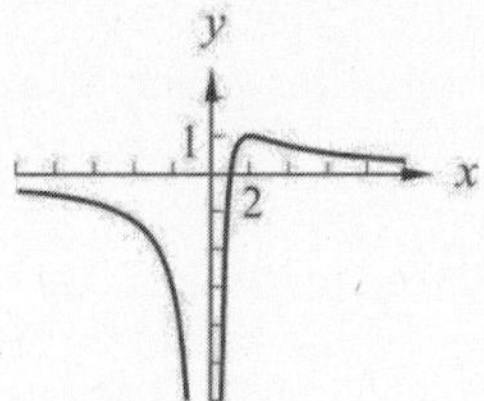

17. $y = \frac{9x}{9 - x^2}$

(1) Intercept, $(0, 0)$

(2) Vertical asymptotes at $x = -3$, $x = 3$

(3) $y' = \frac{(9-x^2)(9) - (9x)(-2x)}{(9-x^2)^2}$
$$= \frac{9(x^2+9)}{(9-x^2)^2}$$
$\neq 0$ for all $x \neq \pm 3$

In fact, $y' > 0$ for all $x \neq \pm 3$, so the graph is always increasing.

(4)
$$y'' = \frac{9\left[(9-x^2)^2(2x) - (x^2+9)(2)(9-x^2)(-2x)\right]}{(9-x^2)^4}$$
$$= \frac{9(9-x^2)(2x)\left[9 - x^2 + 2x^2 + 18\right]}{(9-x^2)^4}$$
$$= \frac{18x(x^2+27)}{(9-x^2)^3}$$
$= 0$ for $x = 0$

Concavity changes at $x = 0, x = \pm 3$:
$y'' > 0$ for $x < -3$ and $0 < x < 3$ (concave up)
$y'' < 0$ for $-3 < x < 0$ and $x > 3$ (concave down)
Therefore $(0,0)$ is an inflection point.
(5) Symmetry: There is symmetry to the origin.
(6) As $x \to +\infty$ and as $x \to -\infty$, $y \to 0$. Therefore, $y = 0$ is an asymptote.

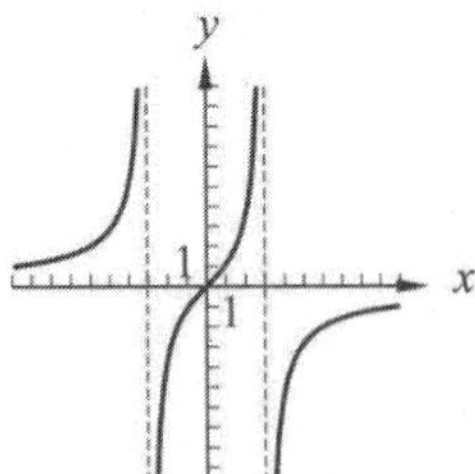

21. $\dfrac{9x^2}{x^2+9} = 9 - \dfrac{81}{x^2+9}$ which simplifies the graphing.

25. $R = \dfrac{200}{\sqrt{t^2 + 40\,000}}$

Intercepts:
(1) if $t = 0$, $R = 1$; R-intercept at $(0, 1)$
(2) No t-intercept since R is never 0
Symmetry:
(3) The function is symmetric with respect to the R-axis. However, R has no physical significance for $t < 0$ so we have no symmetry.
Derivatives:
(4) $R = 200\left(t^2 + 40\,000\right)^{-1/2}$

$$R' = -100\left(t^2 + 40\,000\right)^{-3/2}(2t)$$
$$= -200t\left(t^2 + 40\,000\right)^{-3/2}$$
$$= 0 \text{ for } t = 0$$

(5)

$$R'' = \frac{\left(t^2+40\,000\right)^{3/2}(-200) - (-200t)\left(\frac{3}{2}\right)\left(t^2+40\,000\right)^{1/2}(2t)}{\left[\left(t^2+40\,000\right)^{3/2}\right]^2}$$
$$= \frac{200\left(t^2+40\,000\right)^{1/2}\left[-t^2 - 40\,000 + 3t^2\right]}{\left(t^2+40\,000\right)^3}$$
$$= \frac{400\left(t^2 - 20\,000\right)}{\left(t^2+40\,000\right)^{5/2}}$$
$$= 0 \text{ for } t = \sqrt{20\,000} = 141$$

$R''(0) < 0$ so $t = 0$ is a rel. max.
R'' changes sign at $t = 141$, so there is an inflection point at $(141, 0.82)$
As x becomes large:
(6) As $x \to \infty$, $\frac{200}{\sqrt{t^2+40\,000}}$ approaches zero, so $R = 0$ is a horizontal asymptote.

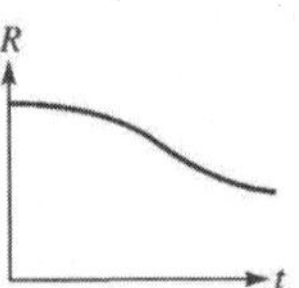

29. $V = \pi r^2 h = 20$

Write height as a function of radius: $h = \dfrac{20}{\pi r^2}$

$$A_T = \text{top} + \text{bottom} + \text{cylinder}$$
$$= \pi r^2 + \pi r^2 + 2\pi r h$$
$$= 2\pi r^2 + 2\pi r\left(\frac{20}{\pi r^2}\right)$$
$$= 2\pi r^2 + \frac{40}{r}$$

There are no intercepts.
As $r \to \infty$, A_T grows without bound, and it grows as the curve $2\pi r^2$ (since $\frac{40}{r}$ approaches 0).
Vertical asymptote at $r = 0$.

$$\frac{dA_T}{dr} = 4\pi r - \frac{40}{r^2}$$
$$= 0 \text{ for } r = \sqrt[3]{\frac{10}{\pi}} = 1.47$$
$$\frac{d^2A}{dr^2} = 4\pi + \frac{80}{r^3}$$
$$= 0 \text{ for } r = -\sqrt[3]{\frac{20}{\pi}} \text{ (reject)}$$

$f''(1.47) > 0$ min. at $(1.47, 40.8)$

$\dfrac{d^2A}{dr^2} > 0$ for $r > 0$

Therefore, A_T is concave up for $r > 0$.

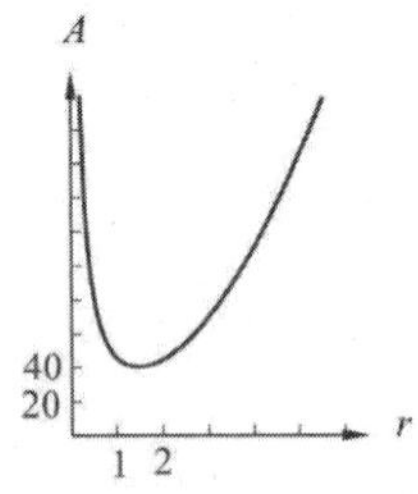

24.7 Applied Maximum and Minimum Problems

1. $A = xy,\ 2x + 2y = 2400 \Rightarrow x + y = 1200$

$A = (1200 - y)y$

$= 1200y - y^2$

$A' = 1200 - 2y$

$= 0$ for $y = 600$

$A'' = -2 < 0$, so $y = 600$ is max.

$x + 600 = 1200$

$x = 600$

$A_{\max} = 600(600) = 360\ 000\ \text{m}^2$

5. $P = EI - RI^2$

$\frac{dP}{dI} = E - 2RI$

$= 0$ for $I = \frac{E}{2R}$

$\frac{d^2P}{dI^2} = -2R < 0$ for all I, therefore max. power at $I = \frac{E}{2R}$

9. $S = 360A - 0.1A^3$, find maximum S.

$S' = 360 - 0.3A^2$

$= 0$

$A^2 = 1200$

$A = 35\ \text{m}^2$

$S'' = -0.6A < 0$ for all valid (positive) A so the graph is concave down and $A = 35\ \text{m}^2$ is a max.

Maximum savings are

$S = 360(35) - 0.1(35)^3 = \$8300.$

13. l = distance to origin

$= \sqrt{x^2 + y^2}$

Substutute $y = x^2 - 4$:

$l = \sqrt{x^2 + (x^2 - 4)^2}$

Minimize $L = l^2$

$L = x^2 + x^4 - 8x^2 + 16$

$= x^4 - 7x^2 + 16$

$L' = 4x^3 - 14x$

$= 0$

$x = 0, x = \pm\sqrt{3.5} = \pm 1.87$

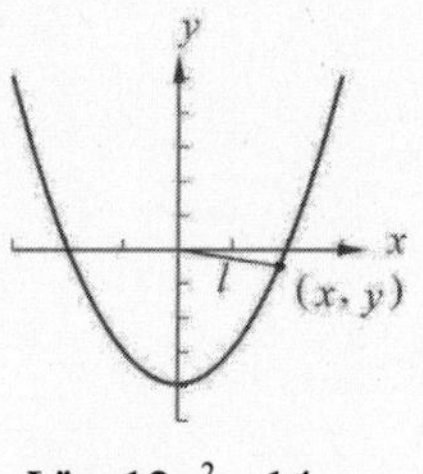

$L'' = 12x^2 - 14$

$L''\big|_{x=0} < 0$, so 0 is a rel. maximum

$L''\big|_{x=1.87} > 0$, $x = 1.87$ is a rel. minimum

$L''\big|_{x=-1.87} > 0, x = -1.87$ is a rel. minimum

$y = \sqrt{3.5}^2 - 4 = -0.5$

$l = \sqrt{3.5 + (-0.5)^2} = 1.94$ units is the closest that the particle comes to the origin.

17.

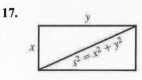

y, x, $s^2 = x^2 + y^2$

$P = 2x + 2y = 48$

$x + y = 24$

Diagonal will be a minimum if $l = s^2$ is a minimum.

$l = x^2 + y^2$

$l = x^2 + (24 - x)^2$

$l = x^2 + 24^2 - 48x + x^2$

$l = 2x^2 - 48x + 24^2$

$\frac{dl}{dx} = 4x - 48$

$= 0$ for $x = 12,$ from which $y = 12$

Dimensions are 12 cm by 12 cm, so a square will minimize the diagonal.

21. $A = xy$

Substitute $y = \frac{A}{x}$ into $P = 2x + 2y$:

$P = 2x + \frac{2A}{x}$

$P' = 2 - \frac{2A}{x^2}$

$= 0$ for $x = \sqrt{A}$

$y = A/\sqrt{A}$

$= \sqrt{A}$

$P'' = \frac{4A}{x^3}\Big|_{x=\sqrt{A}} > 0$, so P is a maximum for $x = y = \sqrt{A}$, a square.

25. $12.0^2 = x^2 + y^2$

$y^2 = 144 - x^2$

Substitute $y = \sqrt{144 - x^2}$ into $A = \frac{1}{2}xy$:

$$A = \frac{1}{2}x\sqrt{144 - x^2}$$

$$\frac{dA}{dx} = \frac{1}{2}\left[x \cdot \frac{1}{2}(144 - x^2)^{-1/2}(-2x) + \sqrt{144 - x^2}\right]$$

$$= \frac{1}{2}\left[\frac{-x^2}{\sqrt{144 - x^2}} + \sqrt{144 - x^2}\right]$$

$$= \frac{1}{2}\left[\frac{-x^2 + 144 - x^2}{\sqrt{144 - x^2}}\right] = \frac{1}{2}\frac{(144 - 2x^2)}{\sqrt{144 - x^2}}$$

$$\frac{dA}{dx} = 0$$

$$= 144 - 2x^2$$

$$x = \sqrt{72} = 8.49$$

Test $\sqrt{72}: f'\sqrt{71} > 0; f'\sqrt{73} < 0$

Therefore, max. $(\sqrt{72}, \sqrt{72})$.

Therefore, legs of triangle will be equal at 8.49 cm for max. area.

29. $lw = 384$

$$A = 384 + 4(24) + 2(6.00w) + 2(4.00l)$$

$$A = 408 + 12.0w + 8.00\frac{384}{w}$$

$$\frac{dA}{dw} = 12.0 - \frac{8.00(384)}{w^2}$$

$$= 0$$

$w = 16.0$ cm, $l = \frac{384}{16.0} = 24.0$ cm

$\frac{d^2A}{dw^2} = \frac{6144}{w^3} > 0$, so the area is minimized if $w = 16.0$ cm

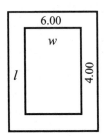

Dimensions: $w + 8.00 = 24.0$ cm

$l + 12.0 = 36.0$ cm

6.00
w
l
4.00

33. $V = (10 - 2x)\left(\frac{15 - 2x}{2}\right)x, \ 0 < x < 5$

$$V = 2x^3 - 25x^2 + 75x$$

$$V' = 6x^2 - 50x + 75 = 0$$

$x = 1.96$, $6.37 > 5$, reject

$V'' = 12x - 50\big|_{x=1.96} < 0$,

so volume is maximized at $x = 1.96$

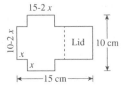

15-2 x
10-2 x
x
x
Lid
10 cm
15 cm

V is a maximum for $x = 2.0$ cm.

37. $y = k(2x^4 - 5Lx^3 + 3L^2x^2)$

$$= 2kx^4 - 5kLx^3 + 3kL^2x^2$$

$$y' = 8kx^3 - 15kLx^2 + 6kL^2x$$

$$= 0$$

$$kx(8x^2 - 15Lx + 6L^2) = 0$$

$kx = 0$; $x = 0$

$$8x^2 - 15Lx + 6L^2 = 0$$

$$x = \frac{-(-15L) \pm \sqrt{(-15L)^2 - 4(8)(6L^2)}}{2(8)}$$

$$= \frac{15L \pm L\sqrt{33}}{16}$$

$$= \frac{15L \pm 5.74L}{16}$$

$x = 0.58L$, $1.3L$ (not valid; this distance is greater than L, the length of the beam)

$$y'' = 24x^2 - 30Lx + 6L^2$$

$y''\big|_{=0} > 0$, so $x = 0$ is a minimum.

$y''\big|_{x=0.58L} < 0$, so $x = 0.58L$ is a maximum.

41. Let x be the distance from factory A.

$$n(x) = \frac{k}{x^2} + \frac{8k}{(8 - x)^2}$$

$$n'(x) = \frac{-2k}{x^3} + \frac{8k \cdot (-2)(-1)}{(8 - x)^3} = 0$$

$$8x^3 = (8 - x)^3$$

$$2x = 8 - x$$

$$3x = 8$$

$$x = \frac{8}{3}$$

$$n''(x)=\frac{6k}{x^4}+\frac{48k}{(8-x)^4}$$

$n''\left(\frac{8}{3}\right)>0 \Rightarrow n(x)$ is min at $x=\frac{8}{3}$ km from A.

45. $2x+\pi d=400$

$$\pi d=400-2x$$

$$d=\frac{400-2x}{\pi}$$

$$A=xd$$

$$=x\left(\frac{400-2x}{\pi}\right)$$

$$=\frac{400x-2x^2}{\pi}$$

$$A'=\frac{400-4x}{\pi}$$

$=0$ for $x=100$ m

$A''=\frac{-4}{\pi}<0$, so the area is maximized when $x=100$ m.

49. Let C = total cost, and let x be the vertical distance from the loading area to the point P. Then,

$$C=50\ 000(10-x)+80\ 000\sqrt{x^2+2.5^2}$$

$$=500\ 000-50\ 000x+80\ 000\left(x^2+6.25\right)^{1/2}$$

$$C'=-50\ 000+40\ 000\left(x^2+6.25\right)^{-1/2}(2x)$$

$$=-50\ 000+80\ 000x\left(x^2+6.25\right)^{-1/2}$$

$$=-50\ 000+\frac{80\ 000x}{\sqrt{x^2+6.25}}$$

$$=\frac{-50\ 000\sqrt{x^2+6.25}+80\ 000x}{\sqrt{x^2+6.25}}=0$$

$$-50\ 000\sqrt{x^2+6.25}+80\ 000x=0$$

$$\sqrt{x^2+6.25}=\frac{-80\ 000x}{-50\ 000}$$

$\sqrt{x^2+6.25}=\frac{8x}{5}$; square both sides

$$x^2+6.25=\frac{64}{25}x^2$$

$$6.25=\frac{64}{25}x^2-x^2$$

$$6.25=\frac{39}{25}x^2$$

$$x^2=6.25\left(\frac{25}{39}\right)=4.00$$

$x=2.00$ km; $10-x=8.00$ km

The pipeline should turn 8.00 km downstream from the refinery.

53. $L(p)=1140p^3(1-p)^{17}, 0<p<1$

$$L'(p)=1140\left[-17p^3(1-p)^{16}+3p^2(1-p)^{17}\right]$$

$$=1140p^2(1-p)^{16}(-17p+3(1-p))$$

$$=1140p^2(1-p)^{16}(3-20p)$$

$=0$ for $p=0, p=1, p=\frac{3}{20}$.

$L'(p)>0$ for $0<p<\frac{3}{20}$,

$L'(p)>0$ for $\frac{3}{20}<p<1$,

so the likelihood is maximized when $p=\frac{3}{20}$.

24.8 Differentials and Linear Approximations

1. $s=\frac{4t}{t^3+4}$

$$ds=\frac{\left(t^3+4\right)(4)-4t\left(3t^2\right)}{\left(t^3+4\right)^2}dt$$

$$ds=\frac{-8t^3+16}{\left(t^3+4\right)^2}dt$$

5. $y=x^5+x$

$$\frac{dy}{dx}=5x^4+1$$

$$dy=\left(5x^4+1\right)dx$$

9. $s=2\left(3t^2-5\right)^4$

$$\frac{ds}{dt}=8\left(3t^2-5\right)^3(6t)$$

$$ds=8\left(3t^2-5\right)^3(6t)\,dt$$

$$ds=48t\left(3t^2-5\right)^3dt$$

13. $y=x^2(1-x)^3$

$$dy=\left[x^2\cdot 3(1-x)^2(-1)+(1-x)^3\cdot 2x\right]dx$$

$$dy = \left(-3x^2(1-x)^2 + 2x(1-x)^3\right)dx$$
$$dy = (1-x)^2\left(-3x^2 + 2x(1-x)\right)dx$$
$$dy = (1-x)^2\left(-5x^2 + 2x\right)dx$$
$$dy = x(1-x)^2(-5x+2)dx$$

17. $y = f(x) = 7x^2 + 4x,$
$$dy = f'(x)dx$$
$$= (14x+4)dx$$
$$\Delta y = f(x+\Delta x) - f(x)$$
$$= 7(4.2)^2 + 4(4.2) - \left(7\cdot 4^2 + 4\cdot 4\right)$$
$$= 12.28$$
$$dy = (14\cdot 4 + 4)(0.2) = 12$$

21. $f(x) = x^2 + 2x;\ f'(x) = 2x + 2$
$$L(x) = f(a) + f'(a)(x-a)$$
$$= f(0) + f'(0)(x-0)$$
$$L(x) = 0^2 + 2\cdot 0 + (2\cdot 0 + 2)(x-0)$$
$$L(x) = 2x$$

25. $C = 2\pi r,\ r = 6370,\ dr = 250$
$$dC = 2\pi dr$$
$$= 2\pi(250)$$
$$= 1570 \text{ km}$$

29. $\lambda = \dfrac{k}{f},$
$$685 = \frac{k}{4.38\times 10^{14}};$$
$$k = 3.00\times 10^{17} \text{ nm} \cdot \text{H}_z$$
$$\frac{d\lambda}{df} = \frac{-k}{f^2}$$
$$d\lambda = \frac{-k}{f^2}df$$
$$d\lambda = \frac{-3.00\times 10^{17}}{\left(4.38\times 10^{14}\right)^2}\cdot\left(0.20\times 10^{14}\right)$$
$$d\lambda = -31.3 \text{ nm}$$

33. $A = \pi r^2$
$$dA = 2\pi r dr$$
$$\frac{dA}{A} = \frac{2\pi r dr}{\pi r^2}$$
$$= 2\frac{dr}{r}$$
$$= 2(2\%) = 4\%$$

37. $L(x) = f'(0)(x-0) + f(0)$
$$f(x) = (1+x)^k,$$
$$f'(x) = k(1+x)^{k-1}$$
$$f(0) = 1$$
$$f'(0) = k$$
$$L(x) = kx + 1$$

41. $C(V) = \dfrac{3.6}{(1+2V)^{1/2}},\ C(4) = 1.2$
$$C'(V) = \frac{-3.6}{(1+2V)^{3/2}},\ C''(4) = \frac{-2}{15}$$
$$L(V) = -\frac{2}{15}(V-4) + 1.2$$
$$= -0.133V + 1.73$$

Review Exercises

1. $y = 3x - x^2$ at $(-1, -4);$
$$y' = 3 - 2x$$
$$y'\big|_{x=-1} = 3 - 2(-1)$$
$$= 5$$
$m = 5$ for tangent line
$$y - y_1 = 5(x - x_1)$$
$$y - (-4) = 5[x - (-1)]$$
$$y + 4 = 5x + 5$$
$$5x - y + 1 = 0$$
To graph:
$$y - \tfrac{9}{4} = -\left(x^2 - 3x + \tfrac{9}{4}\right)$$
This is a parabola opening downward, with vertex at $\left(\frac{3}{2}, \frac{9}{4}\right)$, with intercepts $(0,0)$ and $(0,3)$.

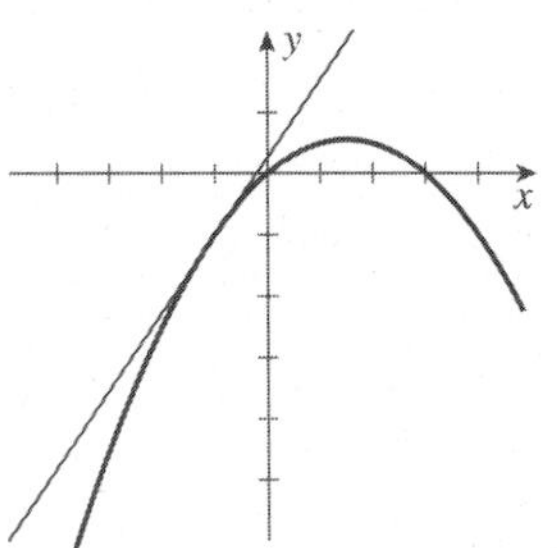

5. $y=\sqrt{x^2+3};\ m_{\tan}=\dfrac{1}{2}$

$y=\left(x^2+3\right)^{1/2}$

$\dfrac{dy}{dx}=\dfrac{1}{2}\left(x^2+3\right)^{-1/2}(2x)$

$=\dfrac{x}{\sqrt{x^2+3}}$

$m_{\tan}=\dfrac{dy}{dx}$

$\dfrac{1}{2}=\dfrac{x}{\sqrt{x^2+3}}$

Squaring both sides,

$4x^2=x^2+3$

$3x^2=3$

$x^2=1$

$x=1$ and $x=-1$.

$x=1$ checks in the equation, but $x=-1$ does not (it is an extraneous root), so the only point where the slope is $\frac{1}{2}$ is $(1,2)$.

Therefore,

$y-2=\dfrac{1}{2}(x-1)$

$y=\dfrac{1}{2}x+\dfrac{3}{2}$ is the equation of the tangent line.

9. $y=0.5x^2+x$

$v_y=\dfrac{dy}{dt}=\dfrac{x\,dx}{dt}+\dfrac{dx}{dt}$

Substitute $v_x=0.5\sqrt{x}$:

$v_y=x\left(0.5\sqrt{x}\right)+0.5\sqrt{x}$

Find v_y at $(2,4)$:

$v_y\big|_{x=2}=2\left(0.5\sqrt{2}\right)+0.5\sqrt{2}$

$=1.5\sqrt{2}=2.12$

13. $x^3-3x^2-x+2=0$ (between 0 and 1)

$f(x)=x^3-3x^2-x+2$

$f'(x)=3x^2-6x-1$

$f(0)=0^3-3\left(0^2\right)-0+2=2$

$f(1)=1^3-3\left(1^2\right)-1+2=-1$

The root is possibly closer to 1 than 0. Let $x_1=0.6$:

n	x_n	$f(x_n)$	$f'(x_n)$	$x_n-\frac{f(x_n)}{f'(x_n)}$
1	0.6	0.536	−3.52	0.752 272 7
2	0.752 272 7	−0.024 293 6	−3.815 893 6	0.745 906 3
3	0.745 906 3	−0.000 030 4	−3.806 309 2	0.745 898 3

$x_4=x_3=0.745\,9$

17. $y=4x^2+16x$

(1) The graph is continuous for all x.

(2) The intercepts are $(0,0)$ and $(-4,0)$.

(3) As $x\to+\infty$ and $-\infty$, $y\to+\infty$.

(4) The graph is not symmetrical about either axis or the origin.

(5) $y'=8x+16$

$y'=0$ at $x=-2$. $(-2,-16)$ is a critical point.

(6) $y''=8>0$ for all x; the graph is concave up and $(-2,-16)$ is a minimum.

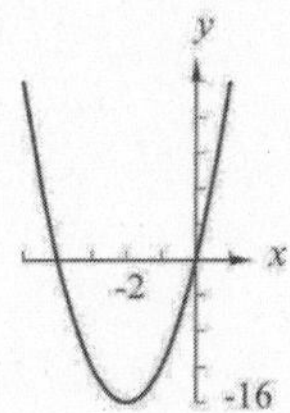

21. $y=x^4-32x$

(1) The graph is continuous for all x.

(2) The intercepts are $(0,0)$ and $\left(2\sqrt[3]{4},0\right)$.

(3) As $x\to-\infty$, $y\to+\infty$; as $x\to+\infty$, $y\to+\infty$.

(4) The graph is not symmetrical about either axis or the origin.

(5) $y'=4x^3-32=0$ for $x=2$

(6) $y''=12x^2$; $y''=0$ at $x=0$; $(0,0)$ is a possible point of inflection. Since $f''(x)>0$; the graph is concave up everywhere (except 0) and (0, 0) is not an inflection point. $(2,-48)$ is a minimum

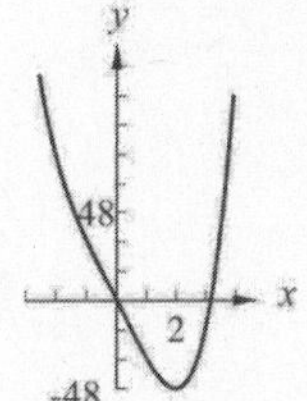

25. $y = f(x)$

$= 4x^3 + \frac{1}{x}$

$dy = f'(x)dx$

$= \left(12x^2 - \frac{1}{x^2}\right)dx$

29. $y = f(x) = 4x^3 - 12,\ x = 2,\ \Delta x = 0.1$

$\Delta y - dy = f(x+\Delta x) - f(x) - f'(x)dx$

$= 4(2.1)^3 - 12 - \left(4(2)^3 - 12\right) - 12(2)^2(0.1)$

$= 0.244$

33. $V = f(r) = \frac{4}{3}\pi r^3,\ r = 3.500,\ \Delta r = 0.012$

$dV = f'(r)dr$

$= 4\pi r^2 dr$

$= 4\pi(3.500)^2(0.012)$

$dV = 1.85\ \text{m}^3$

37. $Z = \sqrt{R^2 + X^2}$

$= \left(R^2 + X^2\right)^{1/2}$

$dZ = \frac{1}{2}\left(R^2 + X^2\right)^{-1/2}(2R)dR$

$dZ = \frac{RdR}{\sqrt{R^2 + X^2}}$

$\text{relative error} = \frac{dZ}{Z} = \frac{RdR}{\sqrt{R^2+X^2}} \cdot \frac{1}{Z}$

$= \frac{RdR}{\sqrt{R^2+X^2}} \cdot \frac{1}{\sqrt{R^2+X^2}}$

$= \frac{RdR}{R^2+X^2}$

41. $y = x^2 + 2$ and $y = 4x - x^2$

$y' = 2x;\qquad y' = 4 - 2x$

To find the point of intersection:

$2x = 4 - 2x$

$4x = 4$

$x = 1$

$y = 1 + 2 = 3$

The point (1, 3) belongs to both graphs; the slope of the tangent line is 2.

$y - y_1 = 2(x - x_1)$

$y - 3 = 2(x - 1)$

$y - 3 = 2x - 2$

$2x - y + 1 = 0$ is the equation of the tangent line.

45. $x = 8t \qquad y = -0.15t^2$

$\frac{dx}{dt} = 8 \qquad \frac{dy}{dt} = -0.30t$

$v_x\big|_{t=12} = 8 \qquad v_y\big|_{t=12} = -3.6$

$v\big|_{t=12} = \sqrt{8^2 + (-3.6)^2}$

$= \sqrt{64 + 12.96}$

$= \sqrt{76.96}$

$= 8.8\ \text{m/s}$

$\theta_{\text{ref}} = \tan^{-1}\frac{3.6}{8}$

$= 24.2^\circ$

$\theta = 360^\circ - \theta_{\text{ref}}$

$= 336^\circ$

49. $d = \frac{1000}{\sqrt{T^2 - 400}}$

$\frac{dd}{dt} = \frac{dd}{dT}\frac{dT}{dt}$

$= -\frac{-1000T}{\left(T^2 - 400\right)^{3/2}}\frac{dT}{dt}$

$\frac{dd}{dt}\bigg|_{T=28.0,\ \frac{dT}{dt}=2.00} = \frac{-1000(28.0)}{\left(28.0^2 - 400\right)^{3/2}}(2.00)$

$= -7.44\ \text{cm/s}$

53. $P = 0.030r^3 - 2.6r^2 + 71r - 200,\ 6 \le r \le 30\ \text{m}^3/\text{s}$

$\frac{dP}{dr} = 0.09r^2 - 5.2r + 71$

$= 0$

$r = 22.1$, $r = 35.6$ (reject), $(6 \le r \le 30)$

$\frac{d^2P}{dr^2} = 0.18r - 5.2$

$\frac{d^2P}{dr^2}\bigg|_{r=22.1} = -1.22 < 0$ maximum

P is a maximum when rate is $22.1\ \text{m}^3/\text{s}$

57.

$$x^2 + y^2 = 2.00^2$$

$$w = 2x,\ h = 2.00 + y$$
$$w = 2\sqrt{2.00^2 - y^2}$$
$$= 2\cdot\sqrt{2.00^2 - (h-2.00)^2}$$
$$= 2\sqrt{-h^2 + 4.00h}$$
$$\frac{dw}{dt} = \left(-h^2 + 4.00h\right)^{-1/2}(-2h + 4.00)\frac{dh}{dt}$$
$$\frac{dh}{dt} = -0.0500$$
$$\left.\frac{dw}{dt}\right|_{h=0.500} = \left(-0.500^2 + 4.00(0.500)\right)^{-1/2}$$
$$\times(-2(0.500) + 4.00)(-0.0500)$$
$$= -0.113 \text{ m/min}$$

61. Let x = distance from home plate to player, then

$$x^2 = 90.0^2 + (18.0t)^2$$
$$= 90.0^2 + 18.0^2t^2$$
$$2x\frac{dx}{dt} = 2(18.0)^2 t$$
$$\frac{dx}{dt} = \frac{18.0^2 t}{x}$$

The player is 40 ft from first base after

$$t = \frac{40}{18}$$
$$= 2.22 \text{ s}$$
$$\left.\frac{dx}{dt}\right|_{t=2.22} = \frac{18.0^2(2.22)}{\sqrt{90^2 + 18.0^2(2.22)^2}}$$
$$= 7.30 \text{ ft/s}$$

65. Let y = altitude of the plane

z = distance from plane to radar station

x = distance from a point on the ground directly below the plane to the radar station

$$z^2 = x^2 + y^2$$
$$2z\frac{dz}{dt} = 2x\frac{dx}{dt} + 2y\frac{dy}{dt}$$

The plane is moving horizontally, so $\frac{dy}{dt} = 0.$

$$z\frac{dz}{dt} = x\frac{dx}{dt} + y(0)$$
$$\sqrt{x^2 + y^2}\,\frac{dz}{dt} = x\frac{dx}{dt}$$

When $x = 8.00$ km, $y = 2400$ m $= 2.400$ km,

$$\sqrt{8.00^2 + 2.40^2}\,(-1110) = 8.00\frac{dx}{dt}$$
$$\frac{dx}{dt} = -1160$$

Actual speed of the plane $= |-1160| = 1160$ km/h

69. $V = x^2y$

To write y in terms of x we use the constraint on the area:

$$x^2 + 4xy = 27$$

Substitute $y = \frac{27 - x^2}{4x}$ into $V = x^2y$:

$$V = x^2\cdot\frac{27 - x^2}{4x}$$
$$= \frac{27x - x^3}{4}$$
$$V' = \frac{27}{4} - \frac{3x^2}{4}$$
$$= 0 \text{ for } x = 3$$

$V'' = \frac{-6}{4} < 0$ for $x > 0$, V is concave down

$$V = \frac{27(3) - 3^3}{4} = 13.5 \text{ dm}^3$$

is the maximum volume.

73. $V = (36 - x)(30 - 2x)(x),\ 0 < x < 15$

$$V(x) = 1080x - 102x^2 + 2x^3$$
$$V'(x) = 1080 - 204x + 6x^2$$
$$= 0 \text{ for } x = 6.56,\ 27.4$$

using the quadratic formula. 27.4 is rejected since $0 < x < 15$. Since $V''(x) = -204 + 12x|_{x=6.56} < 0$ the volume of the drawer is a maximum for $x = 6.56$ cm

77. Let s = cost per cm^2 of stainless steel

$10s$ = cost per cm^2 of silver

$$V = \pi r^2 h = 314$$

$$h = \frac{314}{\pi r^2}$$

$$\text{Cost} = c = s\left(\pi r^2 + 2\pi rh\right) + 10s\left(\pi r^2\right)$$

$$c = 11s\pi r^2 + 2\pi srh$$

$$c = 11s\pi r^2 + 2\pi sr\left(\frac{314}{\pi r^2}\right)$$

$$c(r) = 11s\pi r^2 + \frac{2s(314)}{r} > 0$$

$$c'(r) = 22s\pi r - \frac{2s(314)}{r^2}$$

$$= 0 \text{ for } r = 2.09$$

which is a minimum since

$$c''(r) = 22s\pi + \frac{4s(314)}{r^3} > 0$$

$$h = \frac{314}{\pi r^2} = 23.0$$

The most economical dimensions of the container are $r = 2.09$ cm, $h = 23.0$ cm

81.

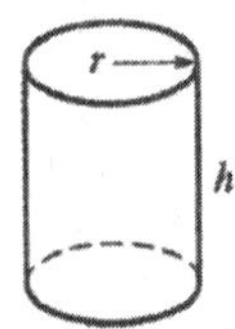

The amount of plastic used is determined by the surface area $S = 2\cdot\pi r^2 + 2\pi r\cdot h.$

$V = \pi r^2 h =$ constant, from which

$$h = \frac{\text{constant}}{\pi r^2}.$$

$$S = 2\pi r^2 + 2\pi r\cdot\frac{\text{constant}}{\pi r^2}$$

$$S(r) = 2\pi r^2 + \frac{2\cdot\text{constant}}{r}$$

$$\frac{dS}{dr} = 4\pi r - \frac{2\cdot\text{constant}}{r^2} = 0 \text{ for minimum}$$

$$4\pi r^3 = 2\cdot\text{constant}$$

$$r^3 = \frac{\text{constant}}{2\pi}$$

$$\frac{h}{r} = \frac{\text{constant}}{\pi\cdot r^3} = \frac{\text{constant}}{\pi\frac{\text{constant}}{2\pi}} = 2$$

The height should be twice the radius to minimize the surface area. In finding the $\frac{h}{r}$ ratio, the constant volume cancels out, so it is not necessary to specify the volume.

CHAPTER 25

INTEGRATION

25.1 Antiderivatives

1. $f(x)=12x^3$; power of x required is 4, $F(x)=ax^4$

$F'(x)=4ax^3=12x^3$

$4a=12$

$a=3$

$F(x)=3x^4$

5. $f(x)=3x^2$

The power of x required in the antiderivative is 3. Therefore, we must multiply by $\frac{1}{3}$. An antiderivative of $3x^2$ is $\frac{1}{3}(3x^3)=x^3$, or $a=1$.

9. The power of x required in the antiderivative of $f(x)=9\sqrt{x}$ is $\frac{3}{2}$.

$\frac{d}{dx}x^{3/2}=\frac{3}{2}x^{1/2}$

Write $f(x)=9\cdot\frac{2}{3}\left(\frac{3}{2}x^{1/2}\right)$

The antiderivative of $9\sqrt{x}$ is $\frac{2}{3}\cdot 9x^{3/2}$, or $a=6$.

13. $f(x)=\frac{5}{2}x^{3/2}$

The power of x required in the antiderivative is $\frac{5}{2}$.

$\frac{d}{dx}ax^{5/2}=\frac{5}{2}ax^{3/2}$

$\frac{5}{2}a=\frac{5}{2}, a=1$

An antiderivative of $\frac{5}{2}x^{3/2}$ is $x^{5/2}$.

17. $f(x)=2x^2-x$

$F(x)=ax^3+bx^2$

$\frac{d}{dx}(ax^3+bx^2)=3ax^2+2bx$

$3a=2;\ a=\frac{2}{3}$

$2b=-1, b=-\frac{1}{2}$

An antiderivative of $2x^2-x$ is $\frac{2}{3}x^3-\frac{1}{2}x^2$.

21. $f(x)=\frac{-7}{x^6}=-7x^{-6}$

$F(x)=ax^{-5}$

$\frac{d}{dx}ax^{-5}=-5ax^{-6}$

$-5a=-7$

$a=\frac{7}{5}$

An antiderivative of $-7x^{-6}$ is $\frac{7}{5}x^{-5}$.

25. $f(x)=x^2-4+x^{-2}$

$F(x)=ax^3+bx+cx^{-1}$

$\frac{d}{dx}(ax^3+bx+cx^{-1})=3ax^2+b-cx^{-2}$

$3a=1, a=\frac{1}{3}$

$b=-4$

$-c=1$

$c=-1$

An antiderivative of x^2-4+x^{-2} is $\frac{1}{3}x^3-4x-\frac{1}{x}$.

29. $f(p)=4(p^2-1)^3(2p)$

$F(p)=a(p^2-1)^4$

$\frac{d}{dp}\left(a(p^2-1)^4\right)=4a(p^2-1)^3(2p)$

This is precisely $f(p)$ with $a=1$, so an antiderivative of $4(p^2-1)^3(2p)$ is $(p^2-1)^4$.

33. $f(x)=\frac{3}{2}(6x+1)^{1/2}(6)$

$F(x)=a(6x+1)^{3/2}$

$\frac{d}{dx}\left(a(6x+1)^{3/2}\right)=\frac{3}{2}a(6x+1)^{1/2}(6)$

This is precisely $f(x)$ with $a=1$, so an antiderivative of $\frac{3}{2}(6x+1)^{1/2}(6)$ is $(6x+1)^{3/2}$.

37. $f(x)=\frac{-2}{(2x+1)^2}=-2(2x+1)^{-2}$

$F(x)=a(2x+1)^{-1}$

$\frac{d}{dx}\left(a(2x+1)^{-1}\right)=-a(2x+1)^{-2}(2)$

$-2a=-2$

$a=1$

An antiderivative of $-2(2x+1)^{-2}$ is $(2x+1)^{-1}$.

25.2 The Indefinite Integral

1. $\int 8x\,dx = 8\int x^1 dx = 8\frac{x^{1+1}}{1+1} + C = 4x^2 + C$

5. $\int 2x\,dx = 2\int x\,dx$

$u = x;\ du = dx;\ n = 1$

$2\int x\,dx = 2\left(\frac{x^{1+1}}{1+1}\right) + C$

$= x^2 + C$

9. $\int 8x^{3/2}dx = 8\int x^{3/2}dx$

$u = x;\ du = dx;\ n = \frac{3}{2}$

$\int 8x^{3/2}dx = \frac{8x^{(3/2)+1}}{\frac{3}{2}+1} + C$

$= \frac{8x^{5/2}}{\frac{5}{2}} + C$

$= \frac{16}{5}x^{5/2} + C$

13. $\int\left(x^2 - x^5\right)dx = \int x^2 dx - \int x^5 dx$

$= \frac{x^3}{3} - \frac{x^6}{6} + C$

$= \frac{1}{3}x^3 - \frac{1}{6}x^6 + C$

17. $\int\left(\frac{t^2}{2} - \frac{2}{t^2}\right)dt = \frac{t^{2+1}}{2(2+1)} - \frac{2t^{-2+1}}{-2+1} + C$

$= \frac{t^3}{6} + \frac{2}{t} + C$

21. $\int\left(2x^{-2.3} + 3^{-2}\right)dx = \int 2x^{-2/3}dx + \int 3^{-2}dx$

$= 2\int x^{-2/3}dx + 3^{-2}\int dx$

$= \frac{2}{\frac{1}{3}}\left(x^{1/3}\right) + 3^{-2}\left(x^1\right) + C$

$= 6x^{1/3} + \frac{1}{9}x + C$

25. $\int\left(x^2 - 1\right)^5(2x\,dx);$

$u = x^2 - 1;\ du = 2x\,dx;\ n = 5$

$\int\left(x^2 - 1\right)^5(2x\,dx) = \frac{\left(x^2 - 1\right)^6}{6} + C$

$= \frac{1}{6}\left(x^2 - 1\right)^6 + C$

29. $\int\left(2\theta^5 + 5\right)^7\theta^4 d\theta = \frac{1}{10}\int\left(2\theta^5 + 5\right)^7\cdot\left(10\theta^4\right)d\theta$

$= \frac{1}{10}\int u^7 du$

$= \frac{1}{10}\frac{u^8}{8} + C$

$= \frac{1}{10}\cdot\frac{\left(2\theta^5 + 5\right)^8}{8} + C$

$= \frac{\left(2\theta^5 + 5\right)^8}{80} + C$

$u = 2\theta^5 + 5, du = 10\theta^4, n = 7$

33. $\int\frac{x\,dx}{\sqrt{6x^2 + 1}} = \int\left(6x^2 + 1\right)^{-1/2}x\,dx$

$u = 6x^2 + 1;\ du = 12x\,dx;\ n = -\frac{1}{2}$

$\int\left(6x^2 + 1\right)^{-1/2}x\,dx = \frac{1}{12}\int\left(6x^2 + 1\right)^{-1/2}(12x\,dx)$

$= \frac{1}{12}\frac{\left(6x^2 + 1\right)^{1/2}}{1/2} + C$

$= \frac{1}{6}\sqrt{6x^2 + 1} + C$

37. $\frac{dy}{dx} = 6x^2$

$dy = 6x^2 dx$

$y = \int 6x^2 dx = 6\int x^2 dx = \frac{6x^3}{3} + C = 2x^3 + C$

The curve passes through $(0,\ 2)$:

$2 = 2\left(0^3\right) + C$

$C = 2$

$y = 2x^3 + 2$

41. $\int 3x^2 dx = x^3 + C$

$\int 3x^2 dx \neq x^3$ since the constant of integration must be included.

45. $\int 3(2x+1)^2 dx = \frac{3}{2}\int(2x+1)^2(2dx)$

$= \frac{3}{2}\frac{(2x+1)^3}{3} + C$

$= \frac{(2x+1)^3}{2} + C$

$\int 3(2x+1)^2 dx \neq (2x+1)^3 + C$ because the factor of $\frac{1}{2}$ is missing.

49. $f''(x)=\sqrt{x}=x^{1/2}$

$$f'(x)=\frac{x^{1/2+1}}{\frac{1}{2}+1}+C=\frac{2}{3}x^{3/2}+C_1$$

$$f(x)=\frac{2}{3}\frac{x^{3/2+1}}{\frac{3}{2}+1}+C_1x+C_2$$

$$f(x)=\frac{4}{15}x^{5/2}+C_1x+C_2$$

53.

$$\frac{di}{dt}=4t-0.6t^2$$

$$di=\left(4t-0.6t^2\right)dt$$

$$i=\int\left(4t-0.6t^2\right)dt$$

$$=2t^2-0.2t^3+C$$

$i=2A$ when $t=0$ s:

$$2=2(0)^2-0.2(0)^3+C$$

$$C=2$$

$$i=2t^2-0.2t^3+2$$

57. $\dfrac{df}{dA}=\dfrac{0.005}{\sqrt{0.01A+1}}=0.005(0.01A+1)^{-1/2}$

$$f(A)=\int 0.005(0.01A+1)^{-1/2}\,dA+C$$

$$=\frac{1}{2}\int(0.01A+1)^{-1/2}(0.01dA)$$

$$=(0.01A+1)^{1/2}+C$$

$f=0$ for $A=0$ m^2:

$$0=(0.01(0)+1)^{1/2}+C$$

$$C=-1$$

$$f(A)=(0.01A+1)^{1/2}-1$$

$$=\sqrt{0.01A+1}-1$$

25.3 The Area Under a Curve

1. (a)

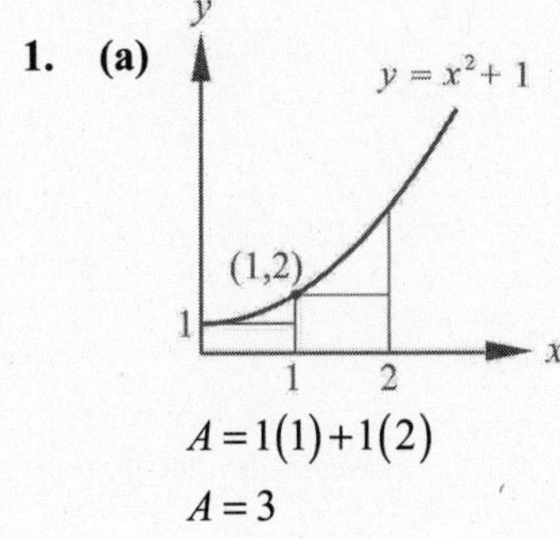

$$A=1(1)+1(2)$$

$$A=3$$

(b)

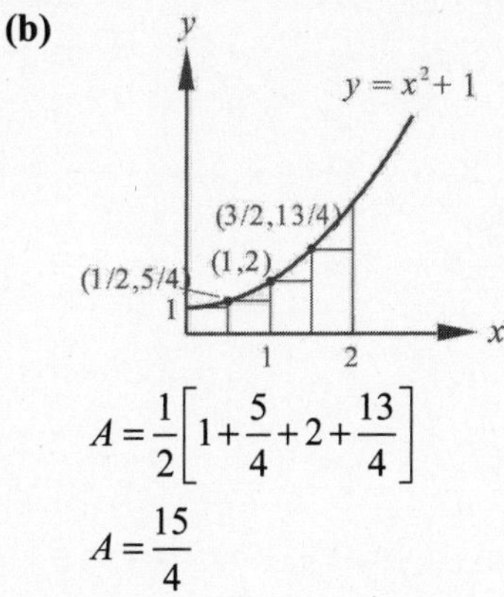

$$A=\frac{1}{2}\left[1+\frac{5}{4}+2+\frac{13}{4}\right]$$

$$A=\frac{15}{4}$$

5. $y=3x$, between $x=0$ and $x=3$

(a)

x	y
0	0
1	3
2	6
3	9

$n=3$; $\Delta x=1$

$$A=1(0+3+6)=9$$

(b)

x	y
0	0
0.3	0.9
0.6	1.8
0.9	2.7
1.2	3.6
1.5	4.5
1.8	5.4
2.1	6.3
2.4	7.2
2.7	8.1
3	9

$n=10$; $\Delta x=0.3$

$$A=0.3(0+0.9+1.8+2.7+3.6+4.5+5.4+6.3+7.2+8.1)$$

$$=0.3(40.5)=12.2$$

9. $y=4x-x^2$, between $x=1$ and $x=4$

(a) $n=6$, $\Delta x=0.5$

$$A=0.5(3.00+3.75+3.75+3.00+1.75+0.00)$$

$$A=7.62$$

x	y
1.0	3.00
1.5	3.75
2.0	4.00
2.5	3.75
3.0	3.00
3.5	1.75
4.0	0.00

(b) $n = 10,\ \Delta x = 0.3$

$$A = 0.3(3.00 + 3.51 + 3.84 + 3.96 + 3.75 + 3.36 + 2.79 + 2.04 + 1.11)$$

$$A = 8.21$$

x	y
1.0	3.00
1.3	3.51
1.6	3.84
1.9	3.99
2.2	3.96
2.5	3.75
2.8	3.36
3.1	2.79
3.4	2.04
3.7	1.11
4.0	0.00

13. $y = \dfrac{1}{\sqrt{x+1}}$, between $x = 3$ and $x = 8$

(a) $n = 5,\ \Delta x = \dfrac{8-3}{5} = 1$

$$A = \sum_{i=1}^{5} A_i = \sum_{i=1}^{5} y_i \Delta x$$

$$y_1 = f(4)$$

$$A = (0.447 + 0.408 + \cdots + 0.354 + 0.333)(1)$$

$$A = 1.92$$

x	y
3	0.5
4	0.447
5	0.408
6	0.378
7	0.355
8	0.333

(b) $n = 10,\ \Delta x = \dfrac{8-3}{10} = 0.5$

$$A = \sum_{i=1}^{10} A_i = \sum_{i=1}^{10} y_i \Delta x$$

$$y_1 = f(3.5)$$

$$A = (0.471 + 0.447 + \cdots + 0.343 + 0.333)(0.5)$$

$$A = 1.96$$

x	y
3	0.5
3.5	0.471
4	0.447
4.5	0.426
5	0.408
5.5	0.392
6	0.378
6.5	0.365
7	0.354
7.5	0.343
8	0.333

17. $y = x^2$, between $x = 0$ and $x = 2$

$$\int x^2 dx = \frac{x^3}{3} + C$$

$$F(x) = \frac{x^3}{3}$$

$$A_{0,\,2} = \left[\int x^2 dx\right]_0^2 = F(2) - F(0) = \frac{8}{3} - 0 = \frac{8}{3}$$

21. $y = \dfrac{1}{x^2} = x^{-2}$, between $x = 1$ and $x = 5$

$$\int x^{-2} dx = \frac{x^{-1}}{-1} + C$$

$$F(x) = \frac{x^{-1}}{-1}$$

$$A_{1,5} = \left[\int x^{-2} dx\right]_1^5 = F(5) - F(1) = -\frac{1}{5} - (-1) = \frac{4}{5} = 0.8$$

25. $y = 3x,\ x = 0$ to $x = 3,\ n = 10,\ \Delta x = 0.3$

Using the table in 5(b) and $y_1 = f(0.3)$,

$$A = 0.3(0.9 + 1.8 + 2.7 + 3.6 + 4.5 + 5.4 + 6.3 + 7.2 + 8.1 + 9.0)$$

$$A = 14.85$$

$A_{inscribed}$	<	A_{exact}	<	$A_{circumscribed}$
12.15	<	13.5	<	14.85

$\frac{12.15+14.85}{2} = 13.5$ because the extra area above $y = 3x$ using circumscribed rectangles is the same as the omitted area under $y = 3x$ using inscribed rectangles.

25.4 The Definite Integral

1. $$\int_1^4 \left(x^{-2} - 1\right) dx = -\frac{1}{x} - x\Big|_1^4 = -\frac{1}{4} - 4 - \left(-\frac{1}{1} - 1\right)$$

$$\int_1^4 \left(x^{-2} - 1\right) dx = -\frac{9}{4}$$

5. $$\int_1^4 x^{5/2} dx = \frac{2}{7} x^{7/2}\Big|_1^4 = \frac{256}{7} - \frac{2}{7} = \frac{254}{7}$$

9. $u = 1 - x;\ du = -dx$

$$\int_{-1.6}^{0.7} (1 - x)^{1/3} dx = -\int_{-1.6}^{0.7} (1 - x)^{1/3} (-dx)$$

$$= -\frac{3}{4}(1 - x)^{4/3}\Big|_{-1.6}^{0.7}$$

$$= -\frac{3}{4}(0.2008 - 3.5752)$$

$$= 2.53$$

13. $$\int_{0.5}^{2.2} \left(\sqrt[3]{x} - 2\right) dx = \int_{0.5}^{2.2} x^{1/3} dx - 2\int_{0.5}^{2.2} dx$$

$$= \left(\frac{3}{4}x^{4/3} - 2x\right)\Big|_{0.5}^{2.2}$$

$$= (2.1460 - 4.4) - (0.2976 - 1)$$

$$= -1.55$$

17. $u = 4 - x^2;\ du = -2x\,dx$

$$\int_{-2}^{-1} 2x\left(4 - x^2\right)^3 dx = -\int_{-2}^{-1} \left(4 - x^2\right)^3 (-2x\,dx)$$

$$= -\frac{\left(4 - x^2\right)^4}{4}\Bigg|_{-2}^{-1}$$

$$= -\left(\frac{81}{4} - 0\right) = -\frac{81}{4}$$

21. $u = 6x + 1;\ du = 6dx$

$$\int_{2.75}^{3.25} \frac{dx}{\sqrt[3]{6x + 1}} = \int_{2.75}^{3.25} (6x + 1)^{-1/3} dx$$

$$= \frac{1}{6}\int_{2.75}^{3.25} (6x + 1)^{-1/3} (6dx)$$

$$= \frac{1}{6} \cdot \frac{3}{2}(6x + 1)^{2/3}\Big|_{2.75}^{3.25}$$

$$= \frac{1}{4}(6x + 1)^{2/3}\Big|_{2.75}^{3.25}$$

$$= \frac{1}{4}(7.4904 - 6.7405) = 0.188$$

25. $$\int_3^7 \sqrt{16t^2 + 8t + 1}dt = \int_3^7 \sqrt{(4t + 1)^2}\, dt$$

$$= \int_3^7 (4t + 1)dt$$

$$= \frac{4t^2}{2} + t\Big|_3^7$$

$$= 2(49) + 7 - 2(9) - 3 = 84$$

29. $u = 2x^2 - x + 1; du = (4x - 1)dx$

$$\int_{-1}^{2} \frac{8x - 2}{\left(2x^2 - x + 1\right)^3} dx$$

$$= \int_{-1}^{2} (8x - 2)\left(2x^2 - x + 1\right)^{-3} dx$$

$$= 2\int_{-1}^{2} (4x - 1)\left(2x^2 - x + 1\right)^{-3} dx$$

$$= \frac{2\left(2x^2 - x + 1\right)^{-2}}{-2}\Bigg|_{-1}^{2}$$

$$= -\left(2x^2 - x + 1\right)^{-2}\Big|_{-1}^{2}$$

$$= -\frac{1}{7^2} - \left(-\frac{1}{4^2}\right) = 0.0421$$

33. $u = z^2 + 4; du = 2zdz$

$$\int_{\sqrt{5}}^{3} 2z\sqrt[4]{z^4 + 8z^2 + 16}dz$$

$$= \int_{\sqrt{5}}^{3} 2z\left(\left(z^2+4\right)^2\right)^{1/4} dz$$

$$= \int_{\sqrt{5}}^{3} \left(z^2+4\right)^{1/2} 2zdz$$

$$= \frac{2}{3}\left(z^2+4\right)^{3/2}\Big|_{\sqrt{5}}^{3} = 13.2$$

37. $y^2 = 4x, \quad y > 0$

For $x = 1, \quad y^2 = 4(1) \Rightarrow y = 2$

For $x = 4, \quad y^2 = 4(4) \Rightarrow y = 4$

$$\int_{x=1}^{x=4} ydx = \int_1^4 2\sqrt{x}dx$$

$$= 2 \cdot \frac{2}{3}x^{3/2}\Big|_1^4$$

$$= \frac{4}{3}\left(4^{3/2} - 1^{3/2}\right) = \frac{28}{3}$$

41. $\int_{-1}^{1} t^{2k}dt = \frac{t^{2k+1}}{2k+1}\Big|_{-1}^{1} = \frac{1^{2k+1}}{2k+1} - \frac{(-1)^{2k+1}}{2k+1}$

$$= \frac{1}{2k+1} + \frac{1}{2k+1}$$

$$\int_{-1}^{1} t^{2k}dt = \frac{2}{2k+1}$$

45. $W = \int_0^{80} (1000 - 5x)\,dx$

$$= \left(1000x - \frac{5}{2}x^2\right)\Big|_0^{80}$$

$$= 1000(80) - \frac{5}{2}(80)^2 - [0]$$

$$= 80\,000 - 16\,000 = 64\,000 \text{ N}\cdot\text{m}$$

49. $\frac{3N}{2E_F^{\ 3/2}} \int_0^{E_F} E^{3/2}dE = \frac{3N}{2E_F^{\ 3/2}} \cdot \frac{E^{5/2}}{\frac{5}{2}}\Big|_0^{E_F}$

$$\frac{3N}{2E_F^{\ 3/2}} \int_0^{E_F} E^{3/2}dE = \frac{3N}{2E_F^{\ 3/2}} \cdot \frac{2E_F^{\ 5/2}}{5} = \frac{3NE_F}{5}$$

25.5 Numerical Integration: The Trapezoidal Rule

1. $\int_1^3 \frac{1}{x}dx, \; n = 2, \; h = \frac{b-a}{n} = \frac{3-1}{2} = 1$

x	y
1	1
2	$\frac{1}{2}$
3	$\frac{1}{3}$

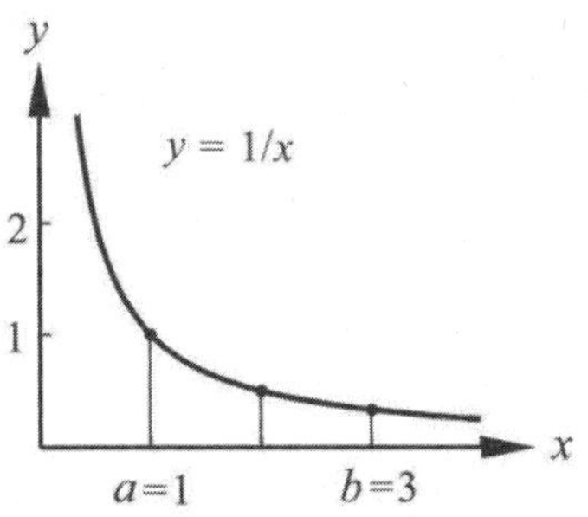

$$A = \frac{1}{2}\left(1 + 2\left(\frac{1}{2}\right) + \frac{1}{3}\right) = \frac{7}{6}$$

5. $\int_1^4 \left(1 + \sqrt{x}\right)dx; \; n = 6; \; h = \frac{4-1}{6} = \frac{1}{2}; \; \frac{h}{2} = \frac{1}{4}$

n	x_n	y_n
0	1	2
1	1.5	2.22
2	2	2.41
3	2.5	2.58
4	3	2.73
5	3.5	2.87
6	4	3

$$A_T = \frac{1}{4}[2 + 2(2.22) + 2(2.41) + 2(2.58) + 2(2.73) + 2(2.87) + 3]$$

$$A = \frac{1}{4}(30.62) = 7.66$$

$$A = \int_1^4 \left(1 + x^{1/2}\right)dx = \left(x + \frac{2}{3}x^{3/2}\right)\Big|_1^4$$

$$A = 4 + \frac{16}{3} - \left(1 + \frac{2}{3}\right) = \frac{23}{3} = 7.67$$

9. $\int_0^5 \sqrt{25-x^2}\,dx;\ n=5;\ h=\frac{5}{5}=1;\ \frac{h}{2}=\frac{1}{2}$

n	x_n	y_n
0	0	5
1	1	4.90
2	2	4.58
3	3	4
4	4	3
5	5	0

$$A_T=\frac{1}{2}\left[5+2(4.90)+2(4.58)+2(4)+2(3)+0\right]$$
$$=19.0$$

13. $\int_0^4 2^x\,dx;\ n=12;\ h=\frac{4}{12}=\frac{1}{3};\ \frac{h}{2}=\frac{1}{6}$

x	y
0	1
$\frac{1}{3}$	1.260
$\frac{2}{3}$	1.587
1	2
$1\frac{1}{3}$	2.520
$1\frac{2}{3}$	3.175
2	4
$2\frac{1}{3}$	5.040
$2\frac{2}{3}$	6.350
3	8
$3\frac{1}{3}$	10.080
$3\frac{2}{3}$	12.699
4	16

$$A=\frac{1}{6}\left[1+2(1.260)+2(1.587)+2(2)+2(.520)\right.$$
$$+2(3.175)+2(4)+2(5.040)+2(6.350)$$
$$\left.+2(8)+2(10.080)+2(12.699)+16\right]$$
$$A=21.7$$

17. The approximate value is less than the exact value because the tops of all the trapezoids are below the curve.

21. $L=2\int_0^{50}\sqrt{6.4\times10^{-7}x^2+1}\,dx;\ n=10;\ h=\frac{50}{10}=5;$

$\frac{h}{2}=\frac{5}{2}$

x	y
0	2
5	2.000 016
10	2.000 063 999
15	2.000 143 955
20	2.000 255 984
25	2.000 399 96
30	2.000 575 917
35	2.000 783 846
40	2.001 037 38
45	2.001 295 58
50	2.001 599 361

$$L=\frac{5}{2}[2+2(2.000\,016+2.000\,063\,999+2.000\,143\,995$$
$$+2.000\,255\,984+2.000\,399\,96+2.000\,575\,917$$
$$+2.000\,783\,846+2.001\,023\,838+2.001\,295\,58)$$
$$+2.001\,599\,361]$$
$$L=100.026\,791\,7\text{ m}$$
$$L=100.027\text{ m}$$

25.6 Simpson's Rule

1. $\int_0^1 \frac{dx}{x+2},\ n=2,\ h=\frac{1-0}{2}=\frac{1}{2}$

x	y
0	$\frac{1}{2}$
$\frac{1}{2}$	$\frac{2}{5}$
1	$\frac{1}{3}$

$$\int_0^1 \frac{dx}{x+2}=\frac{\frac{1}{2}}{3}\left(\frac{1}{2}+4\left(\frac{2}{5}\right)+\frac{1}{3}\right)=0.406$$

5. $\int_1^4 \left(2x+\sqrt{x}\right)dx;\ n=6;\ h=\frac{4-1}{6}=\frac{1}{2};\ \frac{h}{3}=\frac{1}{6}$

$$A_S=\frac{1}{6}\left[3+4(4.22)+2(5.41)+4(6.58)+2(7.73)\right.$$
$$\left.+4(8.87)+10\right]$$
$$A_S=\frac{1}{6}(117.96)=19.7$$

n	x_n	y_n
1	1	3
2	1.5	4.22
3	2	5.41
4	2.5	6.58
5	3	7.73
6	3.5	8.87
7	4	10

$$A = \int_1^4 \left(2x + \sqrt{x}\right) dx = 2\int_1^4 x dx + \int_1^4 x^{1/2} dx$$
$$= \left(x^2 + \frac{2}{3}x^{3/2}\right)\Big|_1^4 = 16 + \frac{16}{3} + -\left(1 + \frac{2}{3}\right)$$
$$= \frac{59}{3} = 19.7$$

9. $\int_1^5 \frac{dx}{x^2 + x}$; $n = 10$; $h = 0.4$; $\frac{h}{3} = \frac{0.4}{3}$

$$A_S = \frac{0.4}{3}[0.5000 + 4(0.2976) + 2(0.1984)$$
$$+ 4(0.1420) + 2(0.1068) + 4(0.0833)$$
$$+ 2(0.0668) + 4(0.0548) + 2(0.0458)$$
$$+ 4(0.0388) + 0.0333]$$
$$= \frac{0.4}{3}(3.8349) = 0.511$$

13. $h = 2$; $\frac{h}{3} = \frac{2}{3}$

x	y
2	0.670
4	2.34
6	4.56
8	3.67
10	3.56
12	4.78
14	6.87

$$\int_2^{14} y dx = \frac{2}{3}[0.67 + 4(2.34) + 2(4.56) + 4(3.67)$$
$$+ 2(3.56) + 4(4.78) + 6.87] = 44.6$$

17. $\bar{x} = 0.9129\int_0^3 x\sqrt{0.3 - 0.1}dx$; $n = 12$; $h = \frac{3}{12} = \frac{1}{4}$; $\frac{h}{3} = \frac{1}{12}$

n	x_n	y_n
1	0	0
2	0.25	0.131
3	0.50	0.25
4	0.75	0.356
5	1	0.447
6	1.25	0.523
7	1.50	0.581

n	x_n	y_n
8	1.75	0.619
9	2	0.632
10	2.25	0.616
11	2.5	0.559
12	2.75	0.435
13	3	0

$$\bar{x} = A_S = \frac{1}{12}[0 + 4(0.131) + 2(0.25) + 4(0.356)$$
$$+ 2(0.447) + 4(0.523) + 2(0.581)$$
$$+ 4(0.619) + 2(0.632) + 4(0.616)$$
$$+ 2(0.559) + 4(0.435) + (0)](0.9129)$$
$$= \frac{1}{12}(15.658)(0.9129) = 1.191$$
$$\bar{x} = 1.19 \text{ cm}$$

Review Exercises

1. $\int(4x^3 - x)dx = \int 4x^3 dx - \int x dx$

$$= \frac{4x^4}{4} - \frac{x^2}{2} + C$$
$$= x^4 - \frac{1}{2}x^2 + C$$

5. $\int_1^4 \left(\frac{\sqrt{x}}{2} + \frac{2}{\sqrt{x}}\right)dx = \frac{1}{2}\int_1^4 x^{1/2}dx + 2\int_1^4 x^{-1/2}dx$

$$= \frac{1}{2}\frac{x^{3/2}}{\frac{3}{2}} + \frac{2x^{1/2}}{\frac{1}{2}}\Bigg|_1^4$$
$$= \frac{1}{3}x^{3/2} + 4x^{1/2}\Big|_1^4$$
$$= \left[\frac{1}{3}(4)^{3/2} + 4(4)^{1/2}\right]$$
$$- \left[\frac{1}{3}(1)^{3/2} + 4(1)^{1/2}\right] = \frac{19}{3}$$

9. $\int\left(5 + \frac{6}{x^3}\right)dx = \int 5dx + \int \frac{6}{x^3}dx$

$$= \int 5dx + \int 6x^{-3}dx$$
$$= (5x) + \left(-\frac{6}{2}x^{-2}\right) + C$$
$$= 5x - \frac{3}{x^2} + C$$

13. $u = 9 - 5n; du = -5dn$

$$\int \frac{dn}{(9-5n)^3} = \int (9-5n)^{-3}\, dn$$
$$= -\frac{1}{5}\int (9-5n)^{-3}(-5\, dn)$$
$$= -\frac{1}{5}\cdot\frac{(9-5n)^{-2}}{-2} + C$$
$$= \frac{1}{10}\cdot\frac{1}{(9-5n)^2} + C$$
$$= \frac{1}{10(9-5n)^2} + C$$

17. $\int_0^2 \frac{3x\,dx}{\sqrt[3]{1+2x^2}} = \int_0^2 (1+2x^2)^{-1/3}(3x)\,dx$

$u = 1+2x^2;\ du = 4x\,dx;\ n = -\frac{1}{3}$

$$\frac{3}{4}\int_0^2 (1+2x^2)^{-1/3}(4x)\,dx = \frac{3}{4}\cdot\frac{(1+2x^2)^{2/3}}{\frac{2}{3}}\Bigg|_0^2$$
$$= \frac{9}{8}(1+2x^2)^{2/3}\Big|_0^2$$
$$= \frac{9}{8}\left[1+2(2^2)\right]^{2/3}$$
$$-\frac{9}{8}\left[1+2(0)^2\right]^{2/3}$$
$$= \frac{9}{8}(9)^{2/3} - \frac{9}{8}(1)^{2/3}$$
$$= \frac{9}{8}\left(\sqrt[3]{81}-1\right)$$
$$= \frac{9}{8}\left(3\sqrt[3]{3}-1\right)$$

21. $\int \frac{(2-3x^2)\,dx}{(2x-x^3)^2} = \int (2x-x^3)^{-2}(2-3x^2)dx;$

$u = 2x-x^3;\ du = (2-3x^2)\,dx;\ n = -2$

$$\int (2x-x^3)^{-2}(2-3x^2)dx = \frac{(2x-x^3)^{-1}}{-1} + C$$
$$= -\frac{1}{(2x-x^3)} + C$$

25. $\frac{dy}{dx} = 3 - x^2$

$$y = 3x - \frac{x^3}{3} + C$$

Substitute $(-1,3)$:

$$3 = 3(-1) - \frac{(-1)^3}{3} + C$$
$$C = \frac{17}{3}$$
$$y = 3x - \frac{x^3}{3} + \frac{17}{3}$$

29. $\int_3^8 F(v)\,dv - \int_4^8 F(v)\,dv = \int_3^4 F(v)\,dv$

The area under F from 3 to 8 minus the area under F from 4 to 8 leaves the area under F from 3 to 4.

33. $\int_0^1 x^3 dx = \frac{x^4}{4}\Big|_0^1 = \frac{1}{4}$

$$\int_1^2 (x-1)^3\, dx = \frac{(x-1)^4}{4}\Bigg|_1^2$$
$$= \frac{(2-1)^4}{4} - \frac{(1-1)^4}{4} = \frac{1}{4}$$

which shows $\int_0^1 x^3 dx = \int_1^2 (x-1)^3\, dx$

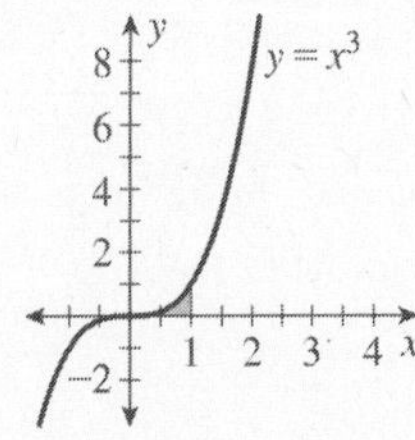

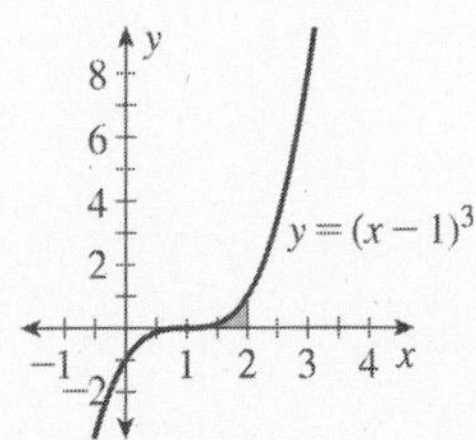

$y = (x-1)^3$ is $y = x^3$ shifted right one unit, so the areas are the same.

37. Since $f(x) > 0$, the graph is above the x-axis.

Since $f''(x) < 0$, for $a \le x < b$, f is concave down for $a \le x \le b$.

$$\int_a^b f(x)\,dx > A_{\text{trapezoid}}$$

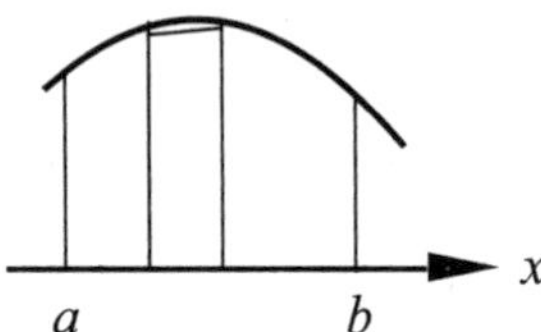

If a function is concave, the tops of the trapezoids are all below the curve. Therefore, the area under the curve is greater than the approximation by the trapezoidal rule.

41. $h = \frac{3-1}{4} = \frac{1}{2}; \frac{h}{3} = \frac{1}{6}$

$x_0 = 1,\ f(1) = \frac{1}{2(1)-1} = 1,$

$x_1 = 1.5,\ f(1.5) = \frac{1}{2(1.5)-1} = \frac{1}{2},$

$x_2 = 2.0,\ f(2.0) = \frac{1}{2(2.0)-1} = \frac{1}{3},$

$x_3 = 2.5, f(2.5) = \frac{1}{2(2.5)-1} = \frac{1}{4},$

$x_4 = 3.0, f(3.0) = \frac{1}{2(3.0)-1} = \frac{1}{5}$

$$\int_1^3 \frac{dx}{2x-1} \approx \frac{h}{3}[y_0 + 4y_1 + 2y_2 + 4y_3 + y_4]$$

$$= \frac{1}{6}\left[1 + \frac{4}{2} + \frac{2}{3} + \frac{4}{4} + \frac{1}{5}\right]$$

$$= \frac{73}{90} = 0.811$$

45. $y = x\sqrt[3]{2x^2+1},\ a = 1,\ b = 4,\ n = 3.$

$h = \frac{b-a}{n} = \frac{4-1}{3} = 1, \frac{h}{2} = \frac{1}{2}$

$x_0 = 1,\ y_0 = 1\sqrt[3]{2\cdot 1^2 + 1} = \sqrt[3]{3},$

$x_1 = 2,\ y_1 = 2\sqrt[3]{2\cdot 2^2 + 1} = 2\sqrt[3]{9},$

$x_2 = 3,\ y_2 = 3\sqrt[3]{2\cdot 3^2 + 1} = 3\sqrt[3]{19},$

$x_3 = 4,\ y_3 = 4\sqrt[3]{2\cdot 4^2 + 1} = 4\sqrt[3]{33}$

$$\int_1^4 x\sqrt[3]{2x^2+1}dx \approx \frac{1}{2}[y_0 + 2y_1 + 2y_2 + y_3]$$

$$= \frac{1}{2}\left[\sqrt[3]{3} + 4\sqrt[3]{9} + 6\sqrt[3]{19} + 4\sqrt[3]{33}\right] = 19.30$$

49. $y(x) = 4 + \sqrt{1 + 8x - 2x^2}, \Delta x = \frac{4}{8} = \frac{1}{2}$

x	y	x	y
0	5	$\frac{5}{2}$	6.915
$\frac{1}{2}$	6.121	3	6.646
1	6.646	$\frac{7}{2}$	6.121
$\frac{3}{2}$	6.915	4	5
2	7		

$$A = \frac{1}{2}\left[y(0) + y\left(\frac{1}{2}\right) + y(1) + y\left(\frac{3}{2}\right) + y(2) + y\left(\frac{5}{2}\right)\right.$$

$$\left. + y(3) + y\left(\frac{7}{2}\right)\right]$$

$A = 25.7\text{ m}^2$

53. Velocity function: $v(t) = \frac{ds}{dt} = -9.8t + 16$

Position function: $s(t) = -4.9t^2 + 16t + s_0$

$s_0 = \text{initial position} = s(0) = 48$

Position function: $s(t) = -4.9t^2 + 16t + 48$

Position at $t = 4.0$:

$s(4.0) = -4.9(4.0)^2 + 16(4.0) + 48$

$= 33.6$

Displacement $= s(4.0) - s(0)$

$= 33.6 - 48$

$= -14.4$ m

57. $s = \int_0^4 t\sqrt{4 + 9t^2}\,dt$

$$= \frac{1}{18}\int_0^4 (4 + 9t^2)^{1/2}\,18t\,dt$$

$$= \frac{1}{18}\left(\frac{2}{3}\right)(4 + 9t^2)^{3/2}\Big|_0^4$$

$$s = \frac{1}{27}\left(148^{3/2} - 4^{3/2}\right)$$

$s = 66.4$ cm

$u = 4 + 9t^2,\ du = 18t\,dt$

CHAPTER 26

APPLICATIONS OF INTEGRATION

26.1 Applications of the Indefinite Integral

1. $s=\int(v_0-9.8t)\,dt=v_0t-4.9t^2+C$

$24.5=v_0(0)-4.9(0)^2+C$

$24.5=C$

$s=v_0t-4.9t^2+24.5$

$0=v_0(1.0)-4.9(1.0)^2+24.5$

$v_0=-19.6\text{ m/s}$

top -19.6 m/s

24.5 m time = 1.0 s

ground

5. $\frac{ds}{dt}=-0.25\text{ m/s}$

$ds=-0.25dt$

$s=-0.25\int dt=-0.25t+C_1$

$t=0,\ s=8;\text{therefore, } C_1=8$

$s=-0.25t+8.00$

$=8.00-0.25t$

9. At impact: $t=0,\ a=-250\text{ m/s}^2, s_{\text{impact}}=0$

$v_{\text{impact}}=96\frac{\text{km}}{\text{h}}\cdot\frac{1000\text{m}}{1\text{km}}\cdot\frac{1\text{h}}{3600\text{s}}$

$=26.7\text{ m/s},$

At stop: $t=t_{\text{stop}},\ v=0$

$a=-250$

$v=-250t+v_{\text{impact}}$

$=-250t+26.7$

$s=-125t^2+26.7t+s_{\text{impact}}$

$s=-125t^2+26.7t$

At stop:

$v=0$

$=-250t_{\text{stop}}+26.7$

$t_{\text{stop}}=0.107$

$s_{\text{stop}}=-125(0.107)^2+26.7(0.107)$

$=1.4\text{ m}$

13. $v=\int -9.8dt$

$=-9.8t+C$

$v=v_0;\ t=0$

$C=v_0$

$v=-9.8t+v_0$

$s=\int(-9.8t+v_0)dt$

$=-4.9t^2+v_0t+C_1$

$s=0,\ t=0$

$C_1=0$

$s=-4.9t^2+v_0t$

$s=30\text{ when } v=0$

$30=-4.9t^2+v_0t$

$0=-9.8t+v_0$

Substitute $t=\frac{v_0}{9.8}$ into $30=-4.9t^2+v_0t$:

$$30=-4.9\left(\frac{v_0}{9.8}\right)^2+v_0\left(\frac{v_0}{9.8}\right)$$

$$\frac{4.9v_0^2}{9.8^2}=30$$

$$v_0=\sqrt{\frac{30(9.8)^2}{4.9}}=24\text{ m/s}$$

17. $i=0.230\mu\text{A}=0.230\times10^{-6}\text{ A}$

$t=1.50\text{ ms}=1.50\times10^{-3}\text{ s}$

$q=\int i\,dt$

$=\int 0.230\times10^{-6}dt$

$=0.230\times10^{-6}t+C_1$

$q=0,\ t=0$

$C_1=0;$

$q=0.230\times10^{-6}t$

Find q for $t = 1.50\times10^{-3}$ s:

$$q = 0.230\times10^{-6}\left(1.50\times10^{-3}\right)$$
$$= 0.345\times10^{-9} = 0.345 \text{ nC}$$

21. $C = 2.5\ \mu\text{F} = 2.5\times10^{-6}$ F

$i = 25$ mA $= 25\times10^{-3}$A

$t = 12$ ms $= 12\times10^{-3}$ s

$$V_c = \frac{1}{C}\int i\,dt$$
$$= \frac{1}{2.5\times10^{-6}}\int 0.025dt$$
$$= \frac{1}{2.5\times10^{-6}}(0.025)t + C_1$$
$$= 1.0\times10^{4}t + C_1$$

$V_c = 0$ at $t = 0$ so $C_1 = 0$

$V_c = 1.0\times10^{4}t$

Find V_c for $t = 0.012$ s:

$$V_c = 1.0\times10^{4}(0.012)$$
$$= 120 \text{ V}$$

25. $\omega = \dfrac{d\theta}{dt} = 16t + 0.5t^2$

$$d\theta = \left(16t + 0.50t^2\right)dt$$
$$\theta = 8t^2 + \frac{0.50t^3}{3} + C$$

$\theta = 0,\ t = 0$ so $C = 0$

$$\theta = 8t^2 + \frac{0.50t^3}{3}$$

Find θ for $t = 10.0$ s:

$$\theta = 8(10.0)^2 + \frac{0.50}{3}(10.0)^3$$
$$= 970 \text{ rad}$$

29. $\dfrac{dV}{dx} = \dfrac{-k}{x^2}$

$$V = -\int\frac{k}{x^2}dx$$
$$= kx^{-1} + C$$
$$= \frac{k}{x} + C$$

$$0 = \lim_{x\to\infty} V = \lim_{x\to\infty}\frac{k}{x} + C$$
$$C = 0$$
$$V = \frac{k}{x}$$
$$V\big|_{x=x_1} = \frac{k}{x_1}$$

26.2 Areas by Integration

1. $A = \int_1^3 ydx = \int_1^3 x^2dx$

$$= \frac{x^3}{3}\bigg|_1^3$$
$$A = \frac{(3)^3}{3} - \frac{(1)^3}{3}$$
$$A = \frac{26}{3}$$

5. $y = 6 - 4x;\ x = 0,\ y = 0,\ y = 3$

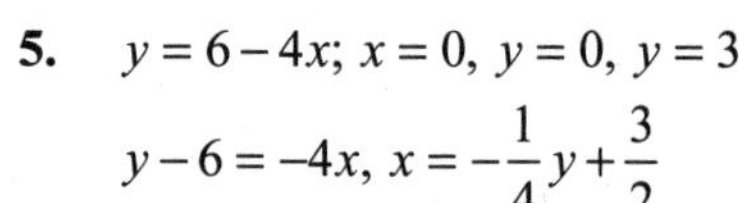

$$y - 6 = -4x,\ x = -\frac{1}{4}y + \frac{3}{2}$$

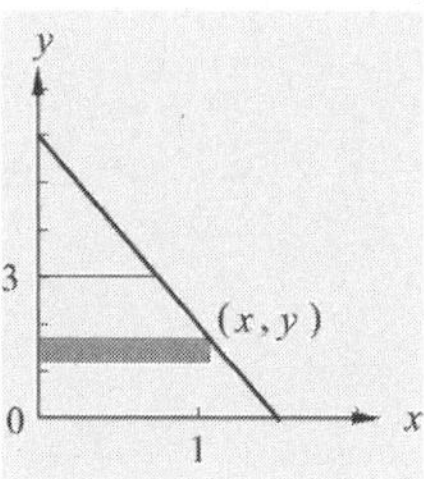

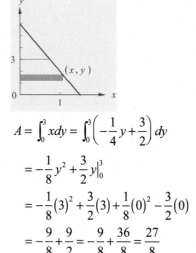

$$A = \int_0^3 xdy = \int_0^3\left(-\frac{1}{4}y + \frac{3}{2}\right)dy$$
$$= -\frac{1}{8}y^2 + \frac{3}{2}y\Big|_0^3$$
$$= -\frac{1}{8}(3)^2 + \frac{3}{2}(3) + \frac{1}{8}(0)^2 - \frac{3}{2}(0)$$
$$= -\frac{9}{8} + \frac{9}{2} = -\frac{9}{8} + \frac{36}{8} = \frac{27}{8}$$

9. $y = x^{-2};\ y = 0,\ x = 2,\ x = 3$

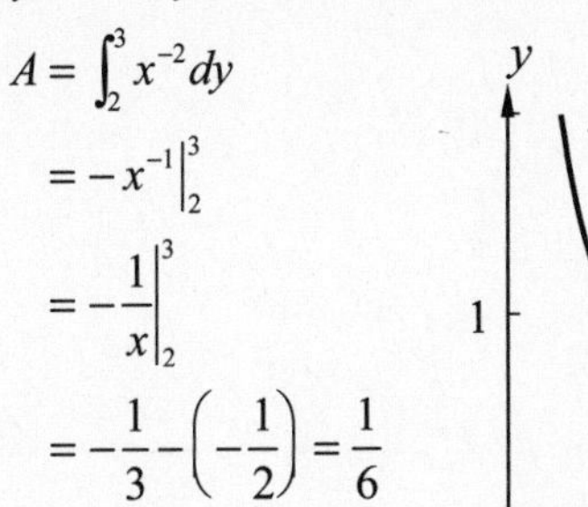

$$A = \int_2^3 x^{-2}dy$$
$$= -x^{-1}\Big|_2^3$$
$$= -\frac{1}{x}\Bigg|_2^3$$
$$= -\frac{1}{3} - \left(-\frac{1}{2}\right) = \frac{1}{6}$$

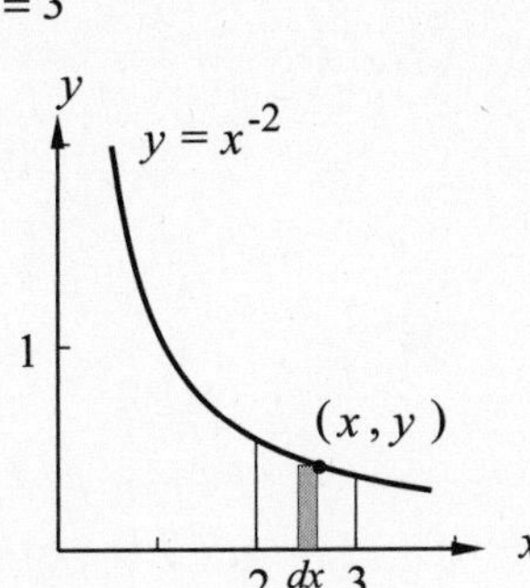

13. $y = \frac{2}{\sqrt{x}};\ x = 0,\ y = 1,\ y = 4$

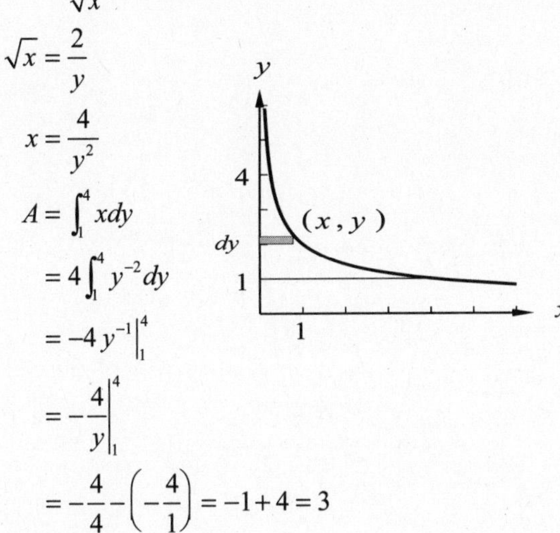

$$\sqrt{x} = \frac{2}{y}$$
$$x = \frac{4}{y^2}$$
$$A = \int_1^4 x dy$$
$$= 4\int_1^4 y^{-2}dy$$
$$= -4y^{-1}\Big|_1^4$$
$$= -\frac{4}{y}\Bigg|_1^4$$
$$= -\frac{4}{4} - \left(-\frac{4}{1}\right) = -1 + 4 = 3$$

17. Insersections:

$$x - 2\sqrt{x} = 0$$
$$x\left(1 - \frac{2}{\sqrt{x}}\right) = 0$$
$$x = 0 \quad \text{or} \quad x = 4$$
$$A = \int_0^4 \left(0 - \left(x - 2\sqrt{x}\right)\right)dx$$
$$A = -\frac{x^2}{2} + \frac{2x^{3/2}}{\frac{3}{2}}\Bigg|_0^4$$
$$A = \frac{8}{3}$$

21. $y = x^4 - 8x^2 + 16,\ y = 16 - x^4$

Intersections:

$$x^4 - 8x^2 + 16 = 16 - x^4$$
$$2x^4 - 8x^2 = 0$$
$$2x^2\left(x^2 - 4\right) = 0$$
$$x = 0 \text{ or } x = 2 \text{ or } x = -2$$

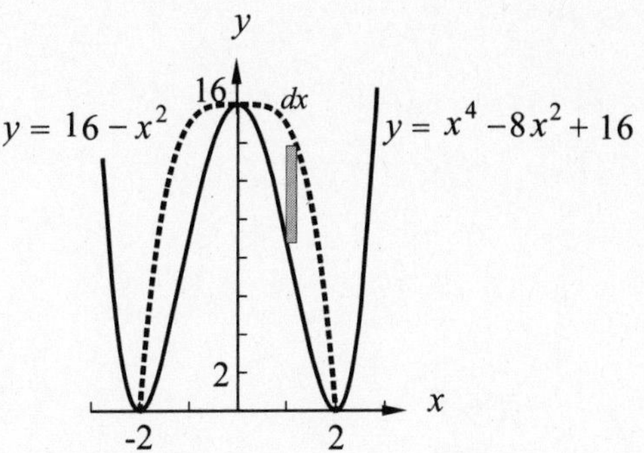

$$A = 2\int_0^2 \left(16 - x^4 - \left(x^4 - 8x^2 + 16\right)\right)dx$$
$$A = 2\int_0^2 \left(8x^2 - 2x^4\right)dx$$
$$= 2\left(\frac{8}{3}x^3 - \frac{2}{5}x^5\right)\Bigg|_0^2 = \frac{256}{15}$$

$\int_{-2}^2 y_2 - y_1 dx = 2\int_0^2 y_2 - y_1 dx$ because of symmetry with respect to the y-axis.

25. $y = x^5;\ x = -1,\ x = 2,\ y = 0$

Between -1 and 0, x^5 is below $y = 0$.

Between 0 and 2, x^5 is above $y = 0$.

We compute two integrals and sum.

$$A = \int_{-1}^0 (0 - y)dx + \int_0^2 (y - 0)dx$$
$$= -\int_{-1}^0 x^5 dx + \int_0^2 x^5 dx$$
$$= -\frac{x^6}{6}\Bigg|_{-1}^0 + \frac{x^6}{6}\Bigg|_0^2$$
$$= 0 - \left(-\frac{1}{6}\right) + \frac{64}{6} - 0 = \frac{65}{6}$$

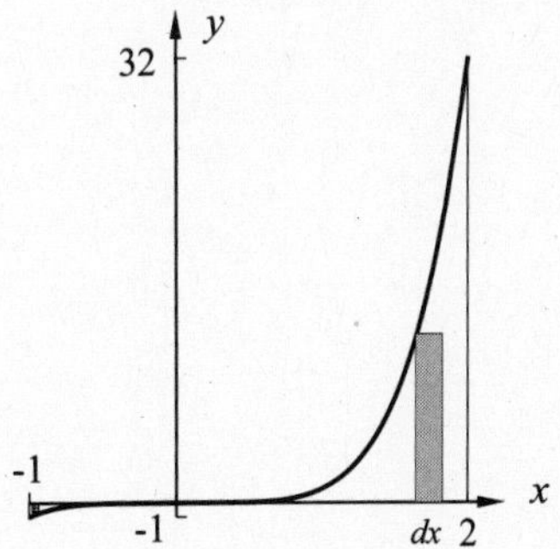

29. $y = 4 - x^2, y = 4x - x^2, x = 0, x = 2$

$$A = 2\int_1^2 \left(4x - x^2 - \left(4 - x^2\right)\right)dx$$

$$= 2\int_1^2 (4x-4)\,dx = 2\left(2x^2 - 4x\right)\Big|_1^2 = 4$$

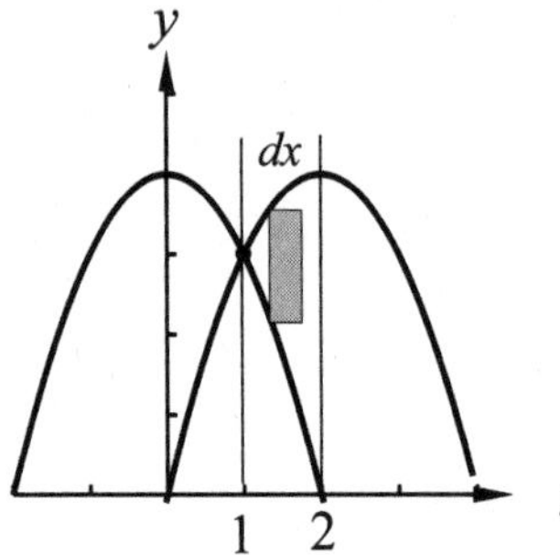

33.

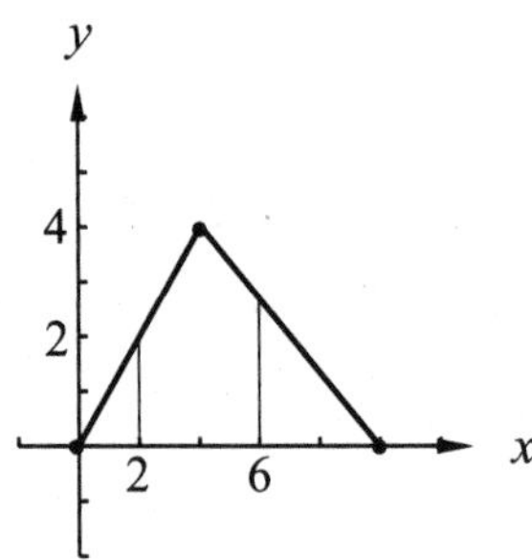

$$A = \int_0^4 x\,dx + \int_4^{10}\left(-\frac{2}{3}(x-4)+4\right)dx = \int_4^{10}\left(-\frac{2}{3}x + \frac{20}{3}\right)dx$$

$$A = \frac{x^2}{2}\Bigg|_0^4 + \left(-\frac{x^2}{3} + \frac{20x}{3}\right)\Bigg|_4^{10}$$

$$= \frac{4^2}{2} + \left(-\frac{10^2}{3} + \frac{20(10)}{3} + \frac{4^2}{3} - \frac{20(4)}{3}\right)$$

$$A = 20$$

37. $\int_0^2 \left(4 - x^2\right)dx = 2\int_0^{\sqrt{c}} \left(c - x^2\right)dx$

$$4x - \frac{x^3}{3}\Bigg|_0^2 = 2\left(cx - \frac{x^3}{3}\right)\Bigg|_0^{\sqrt{c}}$$

$$8 - \frac{8}{3} = 2\left(c\sqrt{c} - \frac{c\sqrt{c}}{3}\right)$$

$$\frac{16}{3} = \frac{4}{3}c^{3/2}$$

$$c = 4^{2/3}$$

41. **(a)** $\int_0^2 \left(8x - x^4\right)dx = 4x^2 - \frac{x^5}{5}\Bigg|_0^2$

$$= \frac{48}{5}$$

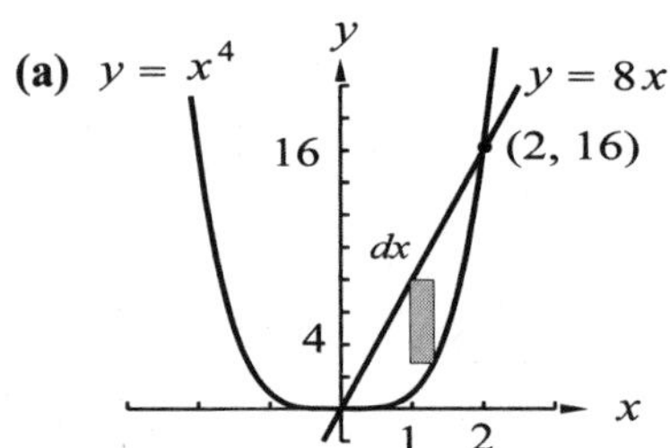

(b) $\int_0^{16}\left(\sqrt[4]{y} - \frac{y}{8}\right)dy = \frac{4}{5}y^{5/4} - \frac{y^2}{16}\Bigg|_0^{16}$

$$= \frac{48}{5}$$

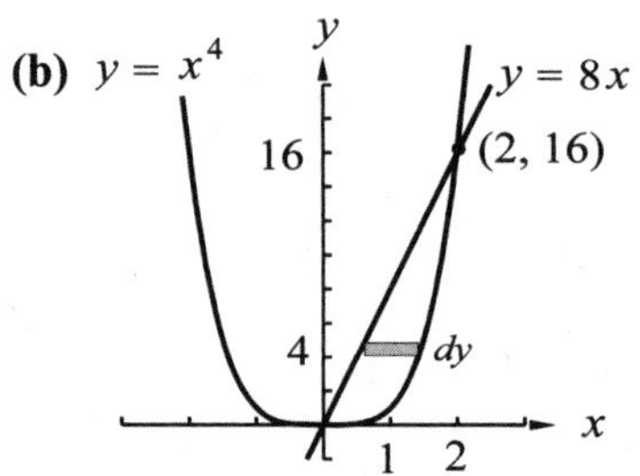

45. $v = 1 - 0.01\sqrt{2t+1}$; $t = 10$ s to $t = 100$ s

$$\Delta s = \int_{10}^{100}\left(1 - 0.01\sqrt{2t+1}\right)dt$$

$$= \int_{10}^{100} dt - 0.01\int_{10}^{100}(2t+1)^{1/2}\,dt$$

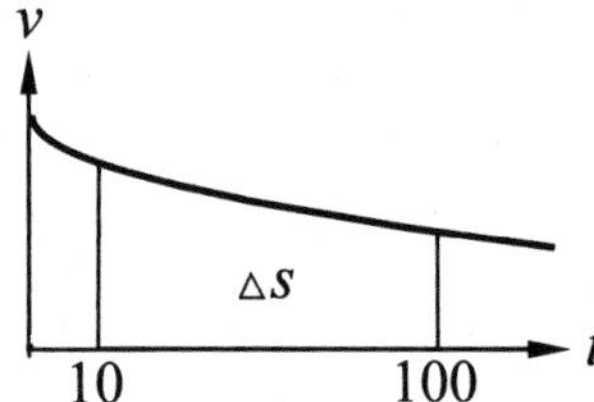

$$\Delta s = \int_{10}^{100} dt - \frac{0.01}{2}\int (2t+1)^{1/2}\,2\,dt$$

$$= t - \frac{0.01}{2}\cdot\frac{2}{3}(2t+1)^{3/2}\Bigg|_{10}^{100}$$

$$= \left[t - \frac{0.01}{3}(2t+1)^{3/2}\right]\Bigg|_{10}^{100}$$

$$= 90.501 - 9.679$$

$\Delta s = 80.8$ km

change in position from $t = 10$ s to $t = 100$ s

49. $y = 0.25x^4$ and $y = 12 - 0.25x^4$ intersect when

$$0.25x^4 = 12 - 0.25x^4$$
$$0.50x^4 = 12$$
$$x^4 = 24$$
$$x = \sqrt[4]{24}$$
$$y = 0.25\left(\sqrt[4]{24}\right)^4 = 6$$

y $y = 0.25x^4$

dx

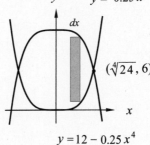

$(\sqrt[4]{24}, 6)$

x

$y = 12 - 0.25x^4$

$$A = 2\int_0^{\sqrt[4]{24}} \left(12 - 0.25x^4 - 0.25x^4\right) dx$$
$$A = 2\int_0^{\sqrt[4]{24}} \left(12 - 0.50x^4\right) dx$$
$$A = 2\left(12x - \frac{0.50x^5}{5}\right)\Bigg|_0^{\sqrt[4]{24}}$$
$$A = 42.5 \text{ dm}^2$$

26.3 Volumes by Integration

1. $y = x^3, x = 2, y = 0$ about the x-axis.

$$V = \int_0^2 \pi y^2 dx = \pi \int_0^2 \left(x^3\right)^2 dx$$
$$V = \pi \frac{x^7}{7}\Bigg|_0^2 = \frac{128\pi}{7}$$

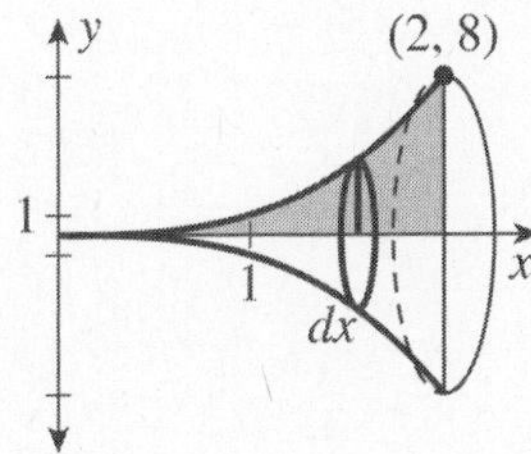

5. Shell: $dV = 2\pi(\text{radius})(\text{height})(\text{thickness})$

radius $= x$, height $= y$, thickness $= dx$

$$dV = 2\pi xy dx$$
$$V = \int_0^2 2\pi x(2 - x) dx$$
$$= 2\pi\left[x^2 - \frac{x^3}{3}\right]\Bigg|_0^2$$
$$= \frac{8\pi}{3}$$

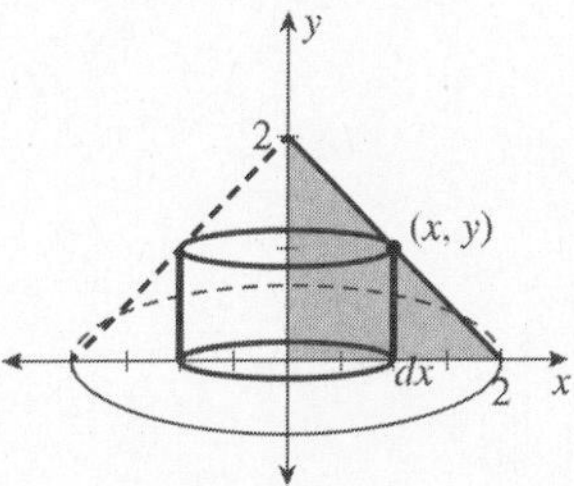

9. $y = 3\sqrt{x}, y = 0, x = 4$

Disc: $dV = \pi(\text{radius})^2(\text{thickness})$

radius $= y$, thickness $= dx$

$$dV = \pi y^2 dx$$
$$V = \pi\int_0^4 y^2 dx = \pi\int_0^4 9x dx$$
$$= \pi\left(\frac{9}{2}x^2\right)\Bigg|_0^4$$
$$= \pi\left[\frac{9}{2}(4)^2 - 0\right] = 72\pi$$

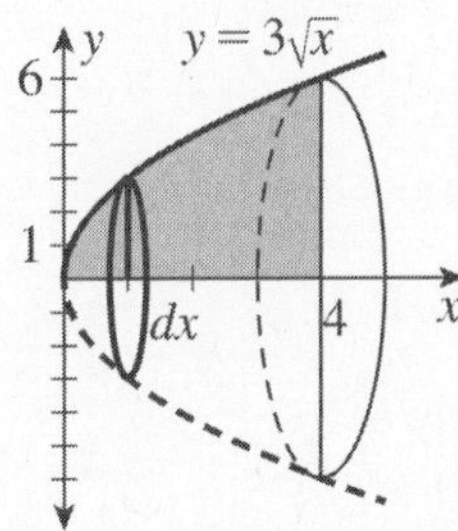

13. $y = x^2 + 1, x = 0, x = 3, y = 0$

Disc: $dV = \pi(\text{radius})^2(\text{thickness})$

radius $= y$, thickness $= dx$

$$dV = \pi y^2 dx$$

$$V = \pi\int_0^3 (x^2+1)^2\,dx$$
$$= \pi\int_0^3 (x^4+2x^2+1)\,dx$$
$$= \pi\left(\frac{1}{5}x^5+\frac{2}{3}x^3+x\right)\Big|_0^3$$
$$= \pi\left[\frac{1}{5}(3)^5+\frac{2}{3}(3)^3+3-0\right]$$
$$= \frac{348}{5}\pi$$

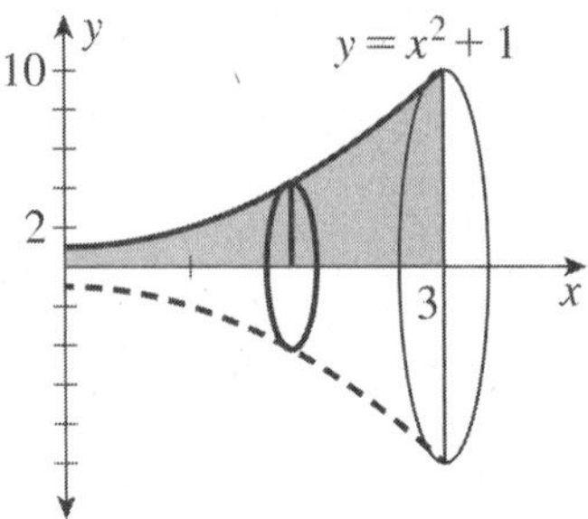

17. $y = x^{1/3}, x = 0, y = 2$

$x = y^3; x^2 = y^6$

Disc: $dV = \pi(\text{radius})^2(\text{thickness})$

radius $= x$, thickness $= dy$

$dV = \pi x^2 dy$

$$V = \pi\int_0^2 y^6 dy$$
$$= \frac{\pi}{7}y^7\Big|_0^2 = \frac{128\pi}{7}$$

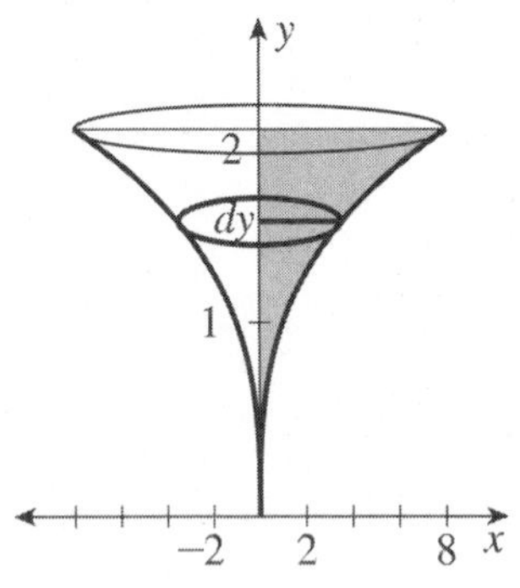

21. $x^2 - 4y^2 = 4, x = 3$

Shell: $dV = 2\pi(\text{radius})(\text{height})(\text{thickness})$

radius $= x$, height $= 2y$, thickness $= dx$

$dV = 2\pi x(2y)dx$

$x^2 - 4y^2 = 4, y = \sqrt{\dfrac{x^2-4}{4}}$

Intercept: $\dfrac{x^2-4}{4} = 0$

$x = 2$

$$V = 4\pi\int_2^3 x\sqrt{\frac{x^2-4}{4}}\,dx$$
$$= \frac{4\pi}{2}\int_2^3 (x^2-4)^{1/2}\,2xdx$$

$u = x^2 - 4, du = 2xdx$

$$V = \pi\cdot\frac{2}{3}(x^2-4)^{3/2}\Big|_2^3$$
$$= \frac{2\pi}{3}(5^{3/2}) - 0$$
$$= \frac{10\sqrt{5}}{3}\pi$$

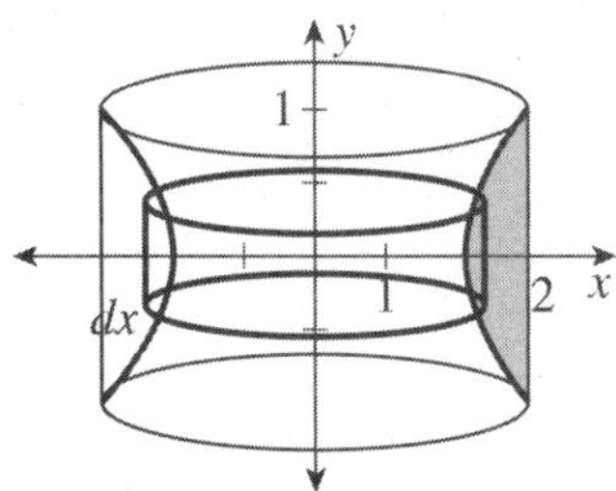

25. $y = \sqrt{4-x^2}$, Quad I

Shell: $dV = 2\pi(\text{radius})(\text{height})(\text{thickness})$

radius $= x$, height $= y$, thickness $= dx$

$dV = 2\pi xydx$

$$V = 2\pi\int_0^2 x\sqrt{4-x^2}\,dx$$

$u = 4 - x^2, du = -2xdx$

$$V = -\pi\int_0^2 (4-x^2)^{1/2}(-2xdx)$$
$$= -\pi\frac{2}{3}(4-x^2)^{3/2}\Big|_0^2$$
$$= -\frac{2\pi}{3}(0-8) = \frac{16\pi}{3}$$

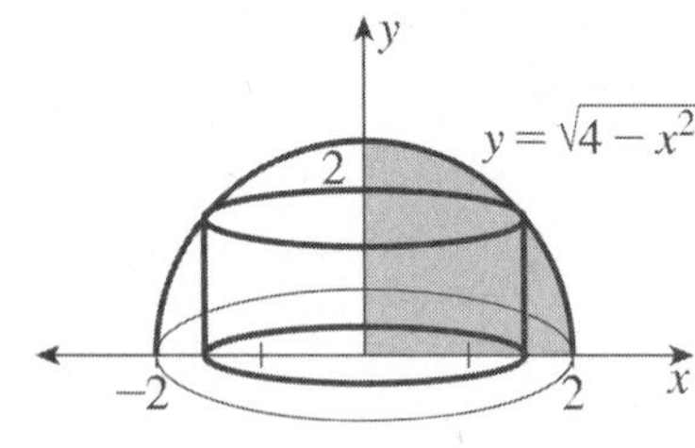

29. $y = 4x - x^2, y = 0$, rotated around $x = 4$

Shell: $dV = 2\pi(\text{radius})(\text{height})(\text{thickness})$

radius $= 4 - x$, height $= y$, thickness $= dx$

$dV = 2\pi(4-x)ydx$

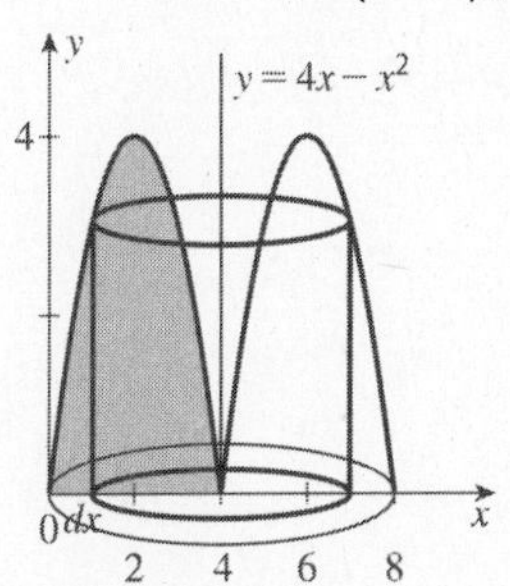

$$V = 2\pi\int_0^4 (4-x)(4x-x^2)dx$$

$$= 2\pi\int_0^4 (16x - 8x^2 + x^3)dx$$

$$= 2\pi\left[8x^2 - \frac{8}{3}x^3 + \frac{1}{4}x^4\right]\Bigg|_0^4$$

$$= \frac{128}{3}\pi$$

33.

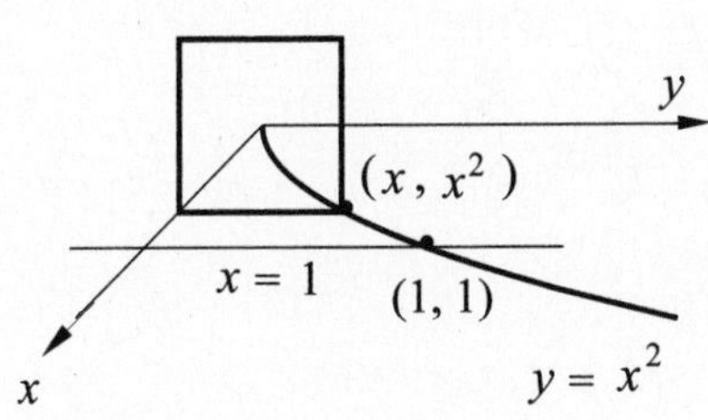

The volume of the solid is the sum of elements of volume as we vary x. Each element has width dx and length and height y, so

$dV = y^2dx$

As we sum all the elements of volume we get

$$V = \int_0^1 (x^2)^2 dx = \frac{x^5}{5}\Bigg|_0^1 = \frac{1}{5}$$

37. $x^2 + y_2^2 = 9;\ y_1 = 1$

$y_1 = 1, y_2^2 = 9 - x^2$

Volume of lead removed:

V = volume of cylinder + spherical caps

Interesections:

$x^2 + 1^2 = 9$

$x = \pm 2\sqrt{2}$

Because of symmetry, we calculate only the volume from rotation in the first quadrant.

$dV = \pi y_1^2 dx + \pi y_2^2 dx$

$$V = 2\left[\pi\int_0^{2\sqrt{2}} (1)dx\right] + 2\pi\int_{2\sqrt{2}}^3 (9 - x^2)dx$$

$$= 2\pi x\Big|_0^{2\sqrt{2}} + 2\pi\left[9x - \frac{x^3}{3}\right]\Bigg|_{2\sqrt{2}}^3$$

$$= 2\pi(2\sqrt{2}) + 2\pi\left(27 - 9 - \left[18\sqrt{2} - \frac{16\sqrt{2}}{3}\right]\right)$$

$$= 2\pi\left[2\sqrt{2} + 18 - 18\sqrt{2} + 16\frac{\sqrt{2}}{3}\right]$$

$V = 18.3\text{ cm}^3$

26.4 Centroids

1. $y = |x|$ is symmetric with respect to y-axis $\Rightarrow \bar{x} = 0$

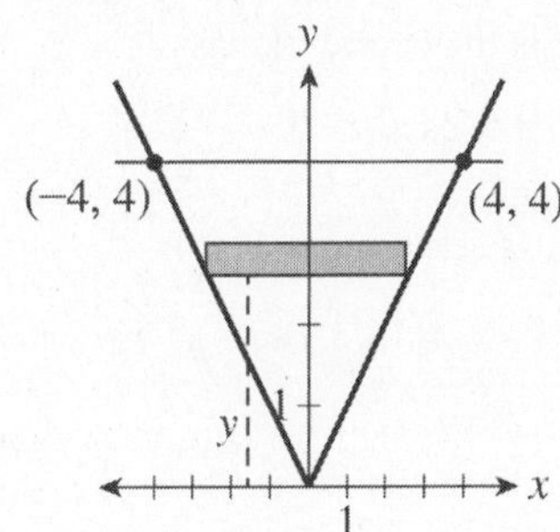

$$\bar{y} = \frac{\int_0^4 y(2x)dy}{\int_0^4 2xdy}$$

$$\bar{y} = \frac{\int_0^4 2y^2dy}{\int_0^4 2ydy} = \frac{\frac{y^3}{3}\Big|_0^4}{\frac{y^2}{2}\Big|_0^4}$$

$$= \frac{\frac{4^3}{3}}{\frac{4^2}{2}} = \frac{8}{3}$$

$$(\bar{x}, \bar{y}) = \left(0, \frac{8}{3}\right)$$

5. $M\overline{x} = m_1x_1 + m_2x_2 + m_3x_3 + m_4x_4$

$(42+24+15+84)\overline{x} = 42(-3.5)+24(0)+15(2.6)+84(3.7)$

$\overline{x} = 1.2$ cm

42 g 24 g 15 g 84 g

-4 -3 -2 -1 0 1 2 3 4

-3.5 0 2.6 3.7

9. Break area into three ractangles.

First: centre $(-1.00, -1.00)$; $A_1 = 4.00$

Second: centre $(0, 0.50)$; $A_2 = 4.00$

Third: centre $(2.50, 1.50)$; $A_3 = 3.00$

Taking moments with respect to the y-axis:

$4.00(-1.00)+4.00(0)+3.00(2.50)$

$= (4.00+4.00+3.00)\overline{x}$

$\overline{x} = 0.32$

Taking moments with respect to the x-axis:

$4.00(-1.00)+4.00(0.50)+3.00(1.50) = 11.00\overline{y}$

$\overline{y} = 0.23$

$(0.32 \text{ cm}, 0.23 \text{ cm})$ is the centre of mass.

13. $y = 4 - x$, and axes

$x = 4 - y$

$$\overline{x} = \frac{\int_0^4 xy\,dy}{\int_0^4 y\,dx} = \frac{\int_0^4 x(4-x)dx}{\int_0^4 (4-x)dx} = \frac{\int_0^4 (4x-x^2)dx}{\int_0^4 (4-x)dx}$$

$$= \frac{(2x^2 - \frac{1}{3}x^3)\big|_0^4}{(4x-\frac{1}{2}x^2)\big|_0^4} = \frac{\frac{32}{3}}{8} = \frac{4}{3}$$

$$\overline{y} = \frac{\int_0^4 y(x)\,dy}{\int_0^4 x\,dx} = \frac{\int_0^4 y(4-y)dy}{\int_0^4 (4-y)dy} = \frac{\int_0^4 (4y-y^2)dy}{\int_0^4 y(4-y)dy}$$

$$= \frac{(2y^2 - \frac{1}{3}y^3)\big|_0^4}{(4y-\frac{1}{2}y^2)\big|_0^4} = \frac{4}{3}$$

The centre of mass is at $(\overline{x}, \overline{y}) = \left(\frac{4}{3}, \frac{4}{3}\right)$.

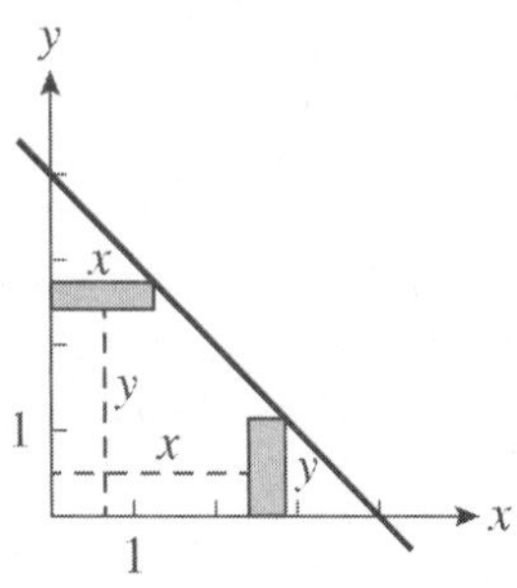

17. $A = \int_0^2 x\,dx + \int_2^3 (6-2x)dx$

$$= \frac{x^2}{2}\Big|_0^2 + \left(6x - x^2\right)\Big|_2^3$$

$= 3$

$$\overline{x} = \frac{\int_0^2 x(3x-2x)dx + \int_2^3 x(6-2x)dx}{3}$$

$$= \frac{\frac{x^3}{3}\Big|_0^2 + 3x^2 - \frac{2x^3}{3}\Big|_2^3}{3}$$

$$= \frac{5}{3}$$

$$\overline{y} = \frac{\int_0^6 y(\frac{y}{2}-\frac{y}{3})dy}{3}$$

$$= \frac{\frac{y^3}{18}\Big|_0^6}{3}$$

$$= \frac{12}{3} = 4$$

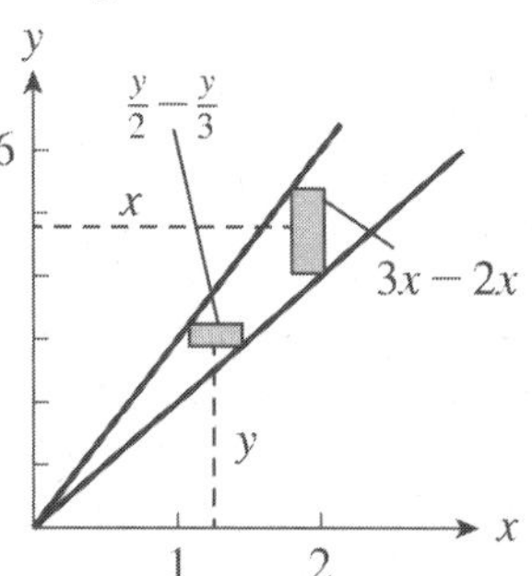

The centre of mass is at $(\overline{x}, \overline{y}) = \left(\frac{5}{3}, 4\right)$.

21. Curves intersect when $\frac{x^2}{4p} = a$

$x = \pm 2\sqrt{pa}$

$y = a$

Region is symmetric with respect to $y-$axis, so $\overline{x} = 0$.

$$\overline{y} = \frac{\int y(2x)dy}{\int 2x\,dy}$$

$$= \frac{\int_0^a 2y(2\sqrt{py})dy}{\int_0^a 2(2\sqrt{py})dy}$$

$$\bar{y} = \frac{4\sqrt{p}\int_0^a y^{3/2}dy}{4\sqrt{p}\int_0^a y^{1/2}dy}$$

$$= \frac{\frac{2}{5}y^{5/2}\Big|_0^a}{\frac{2}{3}y^{3/2}\Big|_0^a}$$

$$= \frac{3}{5}\frac{a^{5/2}}{a^{3/2}} = \frac{3}{5}a$$

$$(\bar{x}, \bar{y}) = \left(0, \frac{3}{5}a\right)$$

25. $y^2 = 4x,\ y = 0,\ x = 1,$

$y^2 = 4x,\ x = \dfrac{y^2}{4}$

Rotate about y-axis, $\bar{x} = 0$

$$\bar{y} = \frac{\int_0^2 y\left(-\left(\frac{y^2}{4}\right)^2 + 1\right)dy}{\int_0^2\left(-\left(\frac{y^2}{4}\right)^2 + 1\right)dy}$$

$$\bar{y} = \frac{\int_0^2\left(-\frac{1}{16}y^5 + y\right)dy}{\int_0^2\left(-\frac{1}{16}y^4 + 1\right)dy} = \frac{-\frac{1}{96}y^6 + \frac{1}{2}y^2\Big|_0^2}{-\frac{1}{80}y^5 + y\Big|_0^2}$$

$$= \frac{-\frac{64}{96} + 2}{-\frac{32}{80} + 2} = \frac{\frac{128}{96}}{\frac{128}{80}} = \frac{5}{6}$$

$$(\bar{x}, \bar{y}) = \left(0, \frac{5}{6}\right)$$

29. Triangle is area bounded by $y = \dfrac{a}{b}x$, x-axis, $x = b$

$$\bar{x} = \frac{\int_0^b xy\,dx}{\int_0^b y\,dx} = \frac{\int_0^b x\frac{a}{b}x\,dx}{\int_0^b \frac{a}{b}x\,dx}$$

$$= \frac{\frac{a}{b}\int_0^b x^2dx}{\frac{a}{b}\int_0^b xdx} = \frac{\frac{x^3}{3}\Big|_0^b}{\frac{x^2}{2}\Big|_0^b} = \frac{\frac{b^3}{3}}{\frac{b^2}{2}} = \frac{2}{3}b$$

$$\bar{y} = \frac{\int_0^b y(b-x)\,dy}{\int_0^b (b-x)\,dy} = \frac{\int_0^b y\left(b - \frac{b}{a}y\right)dy}{\int_0^b\left(b - \frac{b}{a}y\right)dy}$$

$$= \frac{\int_0^b\left(by - \frac{b}{a}y^2\right)dy}{\int_0^b\left(b - \frac{b}{a}y\right)dy} = \frac{\frac{b}{2}y^2 - \frac{b}{3a}y^3\Big|_0^a}{by - \frac{b}{2a}y^2\Big|_0^a}$$

$$= \frac{\frac{a^2b}{2} - \frac{a^2b}{3}}{ab - \frac{ab}{2}} = \frac{a}{3}$$

$$(\bar{x}, \bar{y}) = \left(\frac{2}{3}b, \frac{1}{3}a\right)$$

33. Bounded area: $y = -4x + 80, y = 60, y = 0,$
$x = 0, \bar{x} = 0$

$y = -4x + 80; 4x = 80 - y; x = 20 - \dfrac{1}{4}y$

$$\bar{y} = \frac{\int_0^{60}\left(400y - 10y^2 + \frac{1}{16}y^3\right)dy}{\int_0^{60}\left(400 - 10y + \frac{1}{16}y^2\right)dy}$$

$$= \frac{200y^2 - \frac{10}{3}y^3 + \frac{1}{64}y^4\Big|_0^{60}}{400y - 5y^2 + \frac{1}{48}y^3\Big|_0^{60}}$$

$$= \frac{720\,000 - 720\,000 + 202\,500}{24\,000 - 18\,000 + 4\,500}$$

$= 19.3$ cm from larger base

$$(\bar{x}, \bar{y}) = (0, 19.3)$$

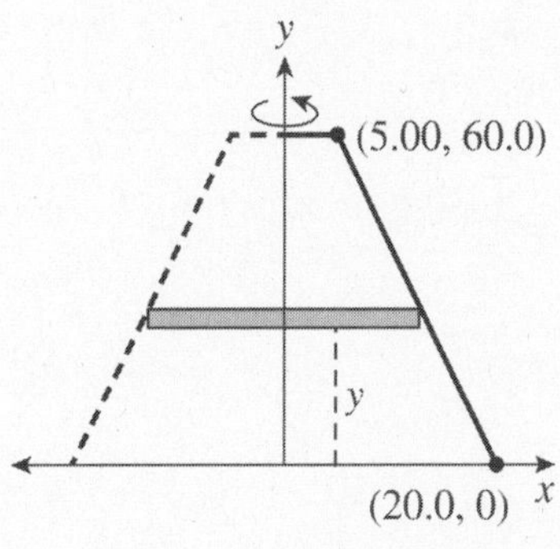

26.5 Moments of Inertia

1. 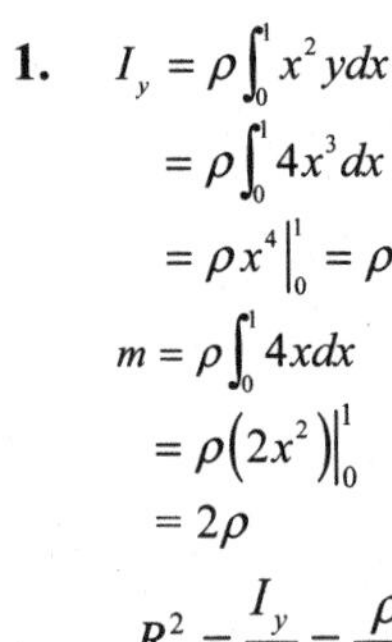

$$I_y = \rho\int_0^1 x^2 y\,dx = \rho\int_0^1 4x^3\,dx = \rho x^4\Big|_0^1 = \rho$$

$$m = \rho\int_0^1 4x\,dx = \rho\left(2x^2\right)\Big|_0^1 = 2\rho$$

$$R_y^2 = \frac{I_y}{m} = \frac{\rho}{2\rho}$$

$$R_y = \frac{\sqrt{2}}{2}$$

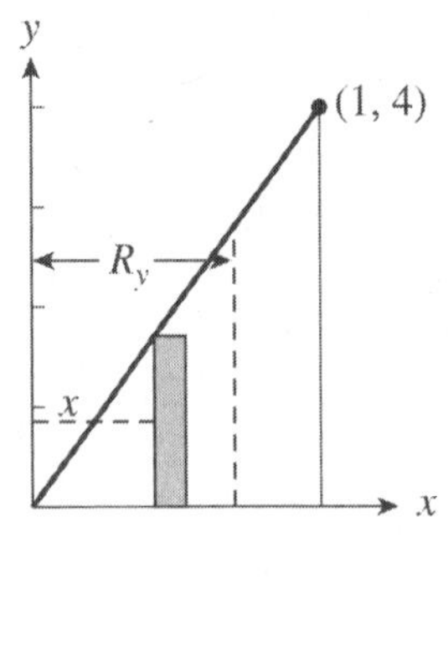

5. $I = m_1x_1^2 + m_2x_2^2 + m_3x_3^2$

$I = 45.0(-3.80)^2 + 90.0(0.00)^2 + 62.0(5.50)^2$

$I = 2530\text{g}\cdot\text{cm}^2$

$I = MR^2$

$2530 = (45.0 + 90.0 + 62.0)R^2$

$R = 3.58\text{cm}$

45.0 g 90.0 g 62.0 g

-4 -3 -2 -1 0 1 2 3 4 5 6

-3.80 0.00 5.50

9. $y^2 = x,\ x = 9$, x-axis, with respect to the x-axis

Intersection: $(9, 3)$

$$I_x = \rho\int_0^3 y^2\left(9 - y^2\right)dy = \rho\int_0^3\left(9y^2 - y^4\right)dy = \rho\left(3y^3 - \frac{1}{5}y^5\right)\Big|_0^3 = \rho\left(81 - \frac{243}{5}\right) = \frac{162}{5}\rho$$

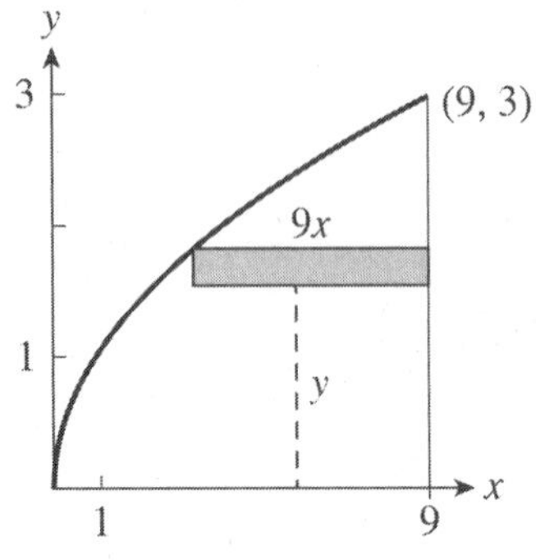

13. $y = \frac{b}{a}x;\ x = a,\ y = 0$

$$I_x = \rho\int_0^b y^2(a - x)\,dy = \rho\int_0^b y^2\left(a - \frac{ay}{b}\right)dy = \rho a\int_0^b\left(y^2 - \frac{y^3}{b}\right)dy = \rho a\left(\frac{1}{3}y^3 - \frac{1}{4b}y^4\right)\Big|_0^b = \frac{\rho ab^3}{12}$$

The mass of the plate is $m = \frac{\rho ab}{2}$, so

$$I_x = \left(\frac{\rho ab}{2}\right)\left(\frac{b^2}{6}\right) = \frac{1}{6}mb^2$$

17. $y^2 = x^3,\ y = 8$, y-axis with respect to x-axis

Intersection: $(4, 8)$

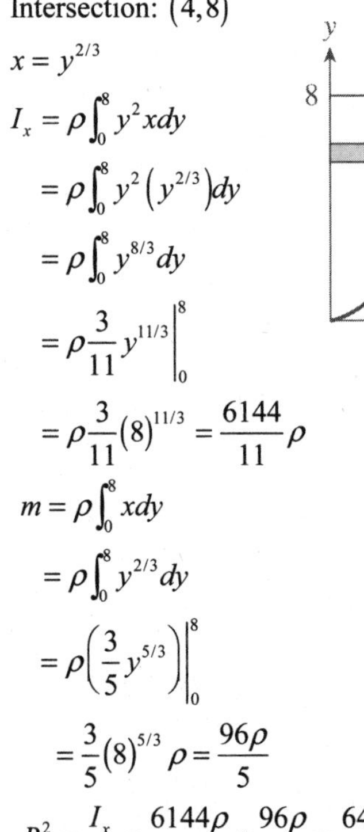

$x = y^{2/3}$

$$I_x = \rho\int_0^8 y^2x\,dy = \rho\int_0^8 y^2\left(y^{2/3}\right)dy = \rho\int_0^8 y^{8/3}\,dy = \rho\frac{3}{11}y^{11/3}\Big|_0^8 = \rho\frac{3}{11}(8)^{11/3} = \frac{6144}{11}\rho$$

$$m = \rho\int_0^8 x\,dy = \rho\int_0^8 y^{2/3}\,dy = \rho\left(\frac{3}{5}y^{5/3}\right)\Big|_0^8 = \frac{3}{5}(8)^{5/3}\rho = \frac{96\rho}{5}$$

$$R^2 = \frac{I_x}{m} = \frac{6144\rho}{11} \div \frac{96\rho}{5} = \frac{64(5)}{11}$$

$$R = \sqrt{\frac{64(5)}{11}} = \frac{8}{11}\sqrt{55}$$

21. $y = 4x - x^2$, $y = 0$, rotated about y-axis

$$I_y = 2\pi\rho\int_0^4 (4x - x^2)x^3 dx$$
$$= 2\pi\rho\int_0^4 (4x^4 - x^5)dx$$
$$= 2\pi\rho\left[\frac{4}{5}x^5 - \frac{1}{6}x^6\right]_0^4$$
$$= 2\pi\rho\left[\frac{4}{5}(4)^5 - \frac{1}{6}(4)^6\right] = \frac{4096\pi\rho}{15}$$
$$m = 2\pi\rho\int_0^4 (4x - x^2)(x)dx$$
$$= 2\pi\rho\int_0^4 (4x^2 - x^3)dx$$
$$= 2\pi\rho\left[\frac{4}{3}x^3 - \frac{1}{4}x^4\right]_0^4$$
$$= 2\pi\rho\left[\frac{4}{3}(4)^3 - \frac{1}{4}(4)^4\right] = \frac{128\pi\rho}{3}$$
$$R_y^2 = \frac{I_y}{m} = \frac{\frac{4096\pi\rho}{15}}{\frac{128\pi\rho}{3}} = \frac{32}{5}$$
$$R_y = \sqrt{\frac{32}{5}\left(\frac{5}{5}\right)} = \frac{4\sqrt{10}}{5}$$

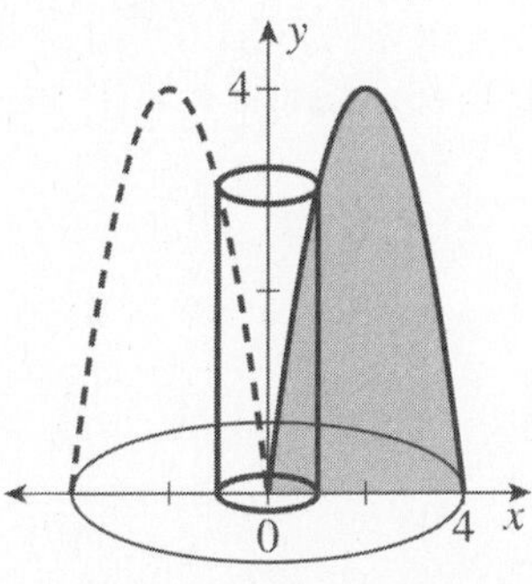

25. $r = 0.600$ cm, $h = 0.800$ cm, $m = 3.00$ g

$$y = \frac{0.600}{0.800}x = 0.750x;\ x = 1.333y$$
$$I_x = 2\pi\rho\int_0^{0.600} (0.800 - 1.333y)y^3 dy$$
$$= 2\pi\rho(0.200y^4 - 0.2667y^5)\Big|_0^{0.600}$$
$$= 2\pi\rho(0.005\,181)$$
$$m = \frac{\rho}{3}\pi r^2 h,\ 2\pi\rho = \frac{6m}{r^2 h} = \frac{6(3.00)}{(0.600^2)(0.800)}$$
$$= 62.5 \text{ g/cm}^3$$
$$I_x = (62.5)(0.005\,181) = 0.324 \text{ g}\cdot\text{cm}^2$$

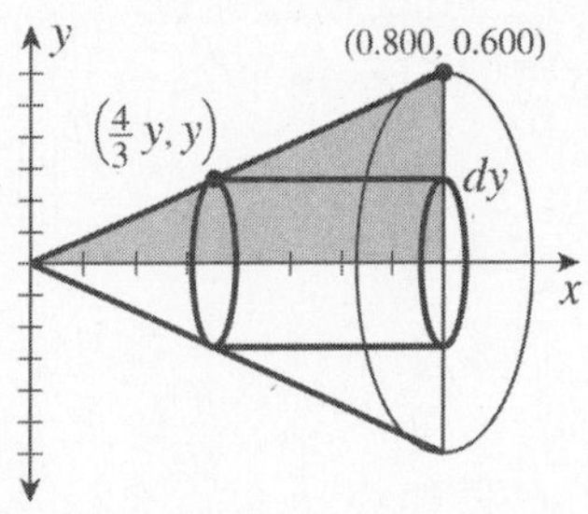

26.6 Other Applications

1. $f(x) = kx$

$$6.0 = k(2.0)$$
$$k = 3.0 \text{ N/cm}$$

Since the spring (of natural length 12.0) is to be stretched from a length of 15.0 to 18.0, $a = 3.0$ and $b = 6.0$

$$W = \int_{3.0}^{6.0} 3.0x dx$$
$$= 1.5x^2\Big|_{3.0}^{6.0}$$
$$W = 40.5 \text{ N}\cdot\text{cm}$$

5. $f(x) = kx$

$$6.0 = k(1.5)$$
$$k = 4.0 \text{ N/cm}$$
$$W = \int_0^{2.0} 4.0x\, dx$$
$$= 2.0x^2\Big|_0^{2.0}$$
$$= 2.0(2.0^2) - 0$$
$$W = 8.0 \text{ N}\cdot\text{cm}$$

9. $f(x) = \frac{kq_1q_2}{x^2}$, $1.0\,\text{pm} = 1.0\times10^{-12}$, $4.0\,\text{pm} = 4.0\times10^{-12}\,\text{m}$

$$W = \int_{1.0\times10^{-12}}^{4.0\times10^{-12}} \frac{9.0\times10^9(1.6\times10^{-19})^2}{x^2}dx$$
$$= 9.0\times10^9(1.6\times10^{-19})^2\left(-\frac{1}{x}\right)\Big|_{1.0\times10^{-12}}^{4.0\times10^{-12}}$$
$$= -23.04\times10^{-29}[0.25\times10^{12} - 10^{12}]$$
$$= -23.04\times10^{-29}(0.75\times10^{12})$$
$$= 1.7\times10^{-16}\text{ J}$$

13. $W = 6000(8.0) + \int_0^8 48(8-x)\,dx$

$$W = 48\ 000 + 48\left(8x - \frac{x^2}{2}\right)\Bigg|_0^8$$

$$W = 50\ 000\ \text{N}\cdot\text{m}$$

17.

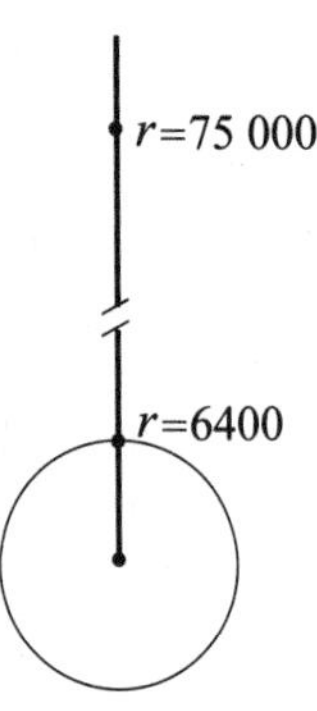

$$F = \frac{Gm_1m_2}{r^2}$$

$$160 = \frac{Gm_1m_2}{6400^2}$$

$$Gm_1m_2 = 160\left(6400^2\right)$$

$$W = \int_{75\ 000}^{6400} -\frac{Gm_1m_2}{r^2}\,dr = \frac{Gm_1m_2}{r}\Bigg|_{75\ 000}^{6400}$$

$$W = 160\left(6400^2\right)\left(\frac{1}{6400} - \frac{1}{75\ 000}\right)$$

$$W = 9.37\times 10^5\ \text{N}\cdot\text{km}$$

21. $F = \gamma\int_a^b lh\,dh$

$$= 9800\int_0^{0.80}(4.0x)\,dx$$

$$= 9800\left(2.0x^2\right)\Big|_0^{0.80}$$

$$= 9800(1.28 - 0) = 12.5\ \text{kN}$$

25.

surface (15, 0) x
(0, 1) (15, 1)
(30 − 15y, y)
(0, 2)
y

$$\gamma = 9800$$

$$F = 9800\int_0^1 y(15dy) + 9800\int_1^2 y(30-15y)\,dy$$

$$F = 9800(15)\frac{y^2}{2}\Bigg|_0^1 + 9800\left(15y^2 - 5y^3\right)\Big|_1^2$$

$$F = 172\ 000\ \text{N}$$

29. $F_{\text{TOP}} = 9800(2.00)^2(1.00) = 3.92\times 10^4\ \text{N}$

$F_{\text{BOTTOM}} = 9800(2.00)^2(3.00) = 1.18\times 10^5\ \text{N}$

The difference in the two forces is the buoyant force.

33. $\eta = 0.768s - 0.000\ 04s^3$, $s_1 = 30.0$ km/h, $s_2 = 90.0$ km/h

$$\eta_{av} = \int_{30.0}^{90.0}\frac{(0.768s - 0.000\ 04s^3)}{60.0}\,ds$$

$$\eta = \frac{1}{60.0}\left(\frac{0.768s^2}{2} - 0.000\ 01s^4\right)\Big|_{30.0}^{90.0}$$

$$= \frac{1}{60.0}\left\{\frac{0.768}{2}(90.0)^2 - 0.000\ 01(90.0)^4 - \left(\frac{0.768}{2}(30.0)^2 - 0.000\ 01(30.0)^4\right)\right\}$$

$$= \frac{1}{60.0}(2.454\times 10^3 - 3.375\times 10^2)$$

$$= 35.3\%$$

37. $y = \frac{r}{h}x$

$$\frac{dy}{dx} = \frac{r}{h};\left(\frac{dy}{dx}\right)^2 = \frac{r^2}{h^2}$$

$$S = 2\pi\int_0^h \frac{r}{h}x\sqrt{1 + \frac{r^2}{h^2}}\,dx$$

$$= 2\pi\frac{r}{h}\frac{1}{h}\sqrt{h^2 + r^2}\int_0^h x\,dx$$

$$= \frac{2\pi r}{h^2}\sqrt{h^2 + r^2}\,\frac{x^2}{2}\Big|_0^h$$

$$= \frac{\pi r}{h^2}\sqrt{h^2 + r^2}x^2\Big|_0^h$$

$$= \frac{\pi r}{h^2}\sqrt{h^2 + r^2}h^2$$

$$= \pi r\sqrt{h^2 + r^2}$$

Review Exercises

1. Travelling horizontally: $t = \frac{s}{v} = \frac{17.1}{45.0} = 0.38 \text{ s}$

Travelling vertically:

$$v = v_0 - 9.8t = -9.8t$$
$$s = \int v dt$$
$$= \int -9.8t dt$$
$$= -4.9t^2 + s_0$$
$$= -4.9t^2$$

Evaluating at $t = 0.38$ s,

$$s = -4.9(0.38)^2$$
$$= -0.708 \text{ m}$$

The ball drops 0.708 m.

5. $a = -0.750$

$$v = -0.750t + 2.50$$
$$= 0$$
$$t = \frac{2.50}{0.750}$$
$$= \frac{10}{3} \text{ s}$$
$$s = -\frac{0.750}{2}t^2 + 2.50t$$

Evaluating at $t = \frac{10}{3}$:

$s = 4.17 < 4.20$

The ball does not make it to the hole.

9. $V_c = \frac{1}{C}\int i\, dt$

$$V_c = \frac{i}{C}\int dt = \frac{it}{C} + V_0 = \frac{it}{C} + 0$$
$$V_c = \frac{12\times10^{-3}\left(25\times10^{-6}\right)}{5.5\times10^{-9}}$$
$$= 55 \text{ V}$$

13. Intercept: $(9,0)$

$$A = \int_0^9 y\, dx = \int_0^9 \sqrt{9-x}\, dx$$
$$A = -\int_0^9 (9-x)^{1/2}(-dx)$$
$$A = -\frac{2}{3}(9-x)^{3/2}\Big|_0^9 = 18$$

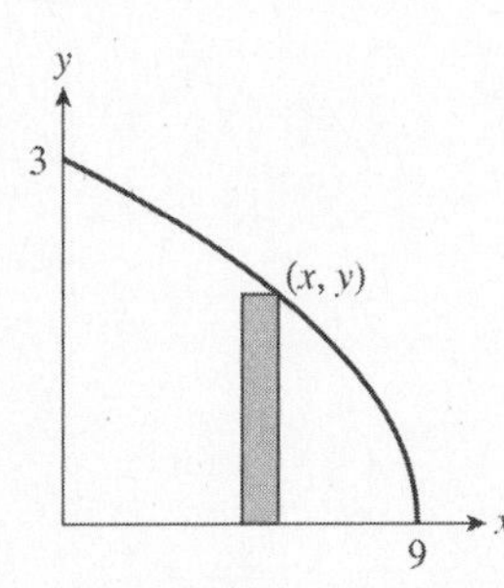

17. $A = \int_0^3 \left(x^2 - \left(x^3 - 2x^2\right)\right)dx$

$$A = \int_0^3 \left(3x^2 - x^3\right)dx$$
$$A = x^3 - \frac{x^4}{4}\Big|_0^3$$
$$A = \frac{27}{4}$$

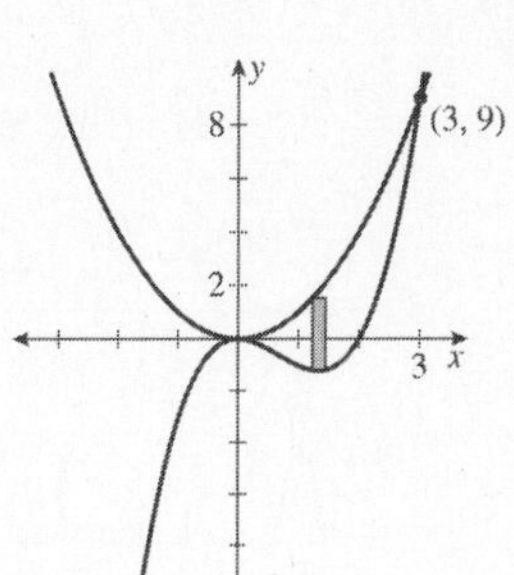

21.

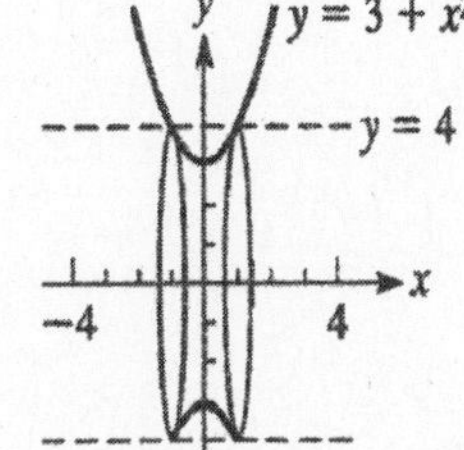

$$V = \pi\int_{-1}^{1} 4^2\, dx - \pi\int_{-1}^{1} y^2 dx = \pi\int_{-1}^{1}\left(4^2 - y^2\right)dx$$
$$= \pi\int_{-1}^{1}\left[16 - \left(3+x^2\right)^2\right]dx$$
$$= \pi\int_{-1}^{1}\left(16 - 9 - 6x^2 - x^4\right)dx$$
$$= \pi\int_{-1}^{1}\left(7 - 6x^2 - x^4\right)dx$$
$$= \pi\left(7x - 2x^3 - \frac{1}{5}x^5\right)\Big|_{-1}^{1}$$
$$= \pi\left(7 - 2 - \frac{1}{5}\right) - \pi\left(-7 + 2 + \frac{1}{5}\right) = \frac{24\pi}{5} + \frac{24\pi}{5}$$
$$= \frac{48\pi}{5}$$

25.

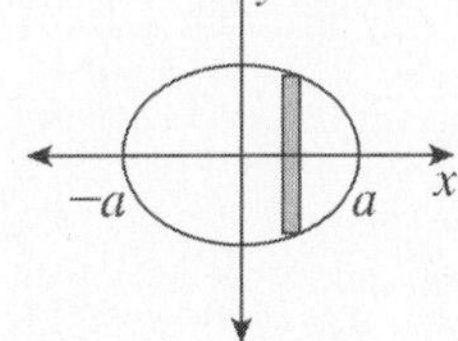

$$V = \pi\int_{-a}^{a} y^2 dx = \pi\int_{-a}^{a}\left(\frac{a^2b^2 - b^2x^2}{a^2}\right)dx = \pi\int_{-a}^{a}\left(b^2 - \frac{b^2}{a^2}x^2\right)dx$$
$$= \pi\left(b^2x - \frac{b^2}{3a^2}x^3\right)\Big|_{-a}^{a}$$
$$= \pi\left[\left(ab^2 - \frac{ab^2}{3}\right) - \left(-ab^2 + \frac{ab^2}{3}\right)\right] = \pi\left(2ab^2 - \frac{2ab^2}{3}\right)$$
$$= \pi\left(\frac{4ab^2}{3}\right) = \frac{4\pi ab^2}{3} = \frac{4}{3}\pi ab^2$$

29. $y^2 = x^3$ and $y = 3x$ intersect at (0, 0) and (9, 27).

$$y = x^{3/2}, x = y^{2/3}, x = \frac{y}{3}$$

$$\bar{x} = \frac{\int_0^9 x\left(3x - x^{3/2}\right)dx}{\int_0^9 \left(3x - x^{3/2}\right)dx}$$

$$= \frac{\int_0^9 \left(3x^2 - x^{5/2}\right)dx}{\int_0^9 \left(3x - x^{3/2}\right)dx}$$

$$\bar{x} = \frac{x^3 - \frac{2}{7}x^{7/2}\Big|_0^9}{\frac{3}{2}x^2 - \frac{2}{5}x^{5/2}\Big|_0^9} = \frac{30}{7}$$

$$\bar{y} = \frac{\int_0^{27} y\left(y^{2/3} - \frac{1}{3}y\right)dy}{\frac{243}{10}} = \frac{\int_0^{27}\left(y^{5/3} - \frac{y^2}{3}\right)dy}{\frac{243}{10}}$$

$$\bar{y} = \frac{\frac{3}{8}y^{8/3} - \frac{1}{9}y^3\Big|_0^{27}}{\frac{243}{10}} = \frac{45}{4}$$

$$(\bar{x}, \bar{y}) = \left(\frac{30}{7}, \frac{45}{4}\right)$$

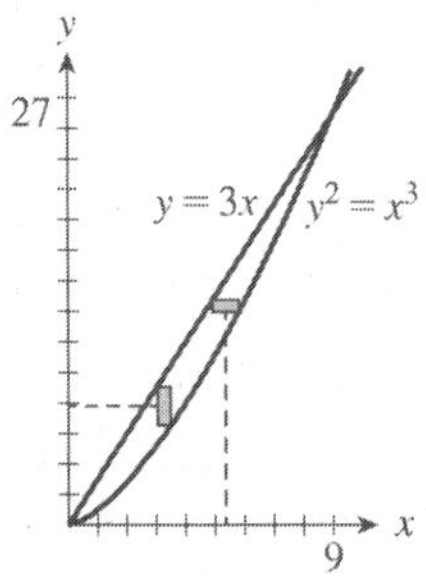

33. $I_y = \rho\int_0^2 x^2\left(\left(3x - x^2\right) - x\right)dx$

$$I_y = \rho\int_0^2 \left(2x^3 - x^4\right)dx$$

$$I_y = \rho\left(\frac{x^4}{2} - \frac{x^5}{5}\right)\Bigg|_0^2$$

$$I_y = \frac{8\rho}{5}$$

37. rope has a weight per unit length $= \frac{20}{30}$ N/m

$$W = W_{\text{bucket}} + W_{\text{rope}}$$

$$W = 80(30) + \int_0^{30} \frac{2}{3}(30 - x)\,dx$$

$$= 2400 + \frac{2}{3}(30x) - \frac{x^2}{3}\Bigg|_0^{30}$$

$$= 2700 \text{ N}\cdot\text{m}$$

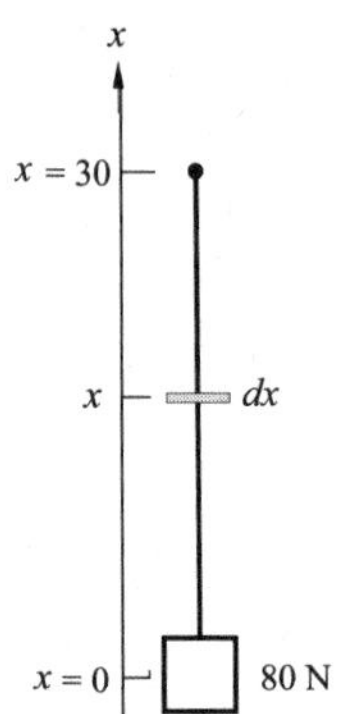

41. $a = -6$

$$v(t) = 12 - \int 6\,dt$$

$$= 12 - 6t$$

$$v(5) = 12 - 6t\big|_{t=5} = -18 \text{ m/s}$$

After 5.0 s the velocity of the rock is 18 m/s down the slope.

45. Using discs and $\frac{x^2}{1.5^2} + \frac{y^2}{10^2} = 1$ as the equation of the ellipse, we have

$$V = \int_0^{10} \pi x^2\,dy$$

$$V = \pi\int_0^{10} 1.5^2\left(1 - \frac{y^2}{10^2}\right)dy$$

$$V = \pi(1.5)^2\left(y - \frac{1}{3}\frac{y^3}{10^2}\right)\Bigg|_0^{10}$$

$$V = \pi(1.5)^2\left(10 - \frac{1}{3}\frac{10^3}{10^2}\right) = 47 \text{ m}^3$$

49. The circumference of the bottom, $c = 2\pi r = 9\pi$, is the length of the vertical surface area l for Eq. (26.27).

$$F = \rho\int_a^b lh\,dh$$
$$= 10.6\int_0^{3.25}(9\pi)h\,dh$$
$$= 10.6\left[4.50\pi h^2\right]_0^{3.25}$$
$$= 10.6\left[4.50\pi(3.25)^2 - 0\right] = 1580 \text{ kN}$$

53. $a = 5\times10^{14}$

$v = 0 + 5\times10^{14}t$

$s = 2.5\times10^{14}t^2$

When $s = 2.5$ cm $= 0.025$ m,

$0.025 = 2.5\times10^{14}t^2$

$t = 1\times10^{-8}$

$$v = 5\times10^{14}\left(1\times10^{-8}\right)$$
$$= 5\times10^{6} \text{ m/s}$$

57. We place a coordinate system at the centre of the top of the float, with the y-axis perpendicular to the fixed diameter. Therefore, slices parallel to the y-axis have cross sections that are squares. We take those as our elements of volume.
An element of volume at x of thickness dx has area $(2y)^2 = 4\left(a^2 - x^2\right)$, so its volume is $4\left(a^2 - x^2\right)dx$.
The volume is the sum of all the elements of volume from $x = -a$ to $x = a$. Because of symmetry, we need only compute the integral from 0 to a.

$$V = 2\int_0^a 4\left(a^2 - x^2\right)dx$$
$$= 8\left(a^2x - \frac{x^3}{3}\right)\Bigg|_0^a$$
$$= 8\left(\frac{2}{3}a^3\right) = \frac{16}{3}a^3$$

CHAPTER 27

DIFFERENTIATION OF TRANSCENDENTAL FUNCTIONS

27.1 Derivatives of the Sine and Cosine Functions

1. $r = \sin^2 2\theta^2$

$$\frac{dr}{d\theta} = 2\sin 2\theta^2(2(2\theta))\cos 2\theta^2$$

$$\frac{dr}{d\theta} = 8\theta\sin 2\theta^2\cos 2\theta^2$$

$$\frac{dr}{d\theta} = 4\theta\sin 4\theta^2$$

5. $y = 2\sin(2x^3 - 1)$

$$\frac{dy}{dx} = 2\cos(2x^3 - 1)(6x^2)$$

$$= 12x^2\cos(2x^3 - 1)$$

9. $y = 2\cos(3x - \pi)$

$$\frac{dy}{dx} = 2[-\sin(3x - \pi)(3)]$$

$$= -6\sin(3x - \pi)$$

13. $y = 3\cos^3(5x + 2)$

$$\frac{dy}{dx} = 3(3)\cos^2(5x + 2)[-\sin(5x + 2)(5)]$$

$$\frac{dy}{dx} = -45\cos^2(5x + 2)\sin(5x + 2)$$

17. $y = 3x^3\cos 5x$

$$\frac{dy}{dx} = 3[x^3(-5\sin 5x) + \cos 5x(3x^2)]$$

$$\frac{dy}{dx} = 9x^2\cos 5x - 15x^3\sin 5x$$

21. $y = \sqrt{1 + \sin 4x} = (1 + \sin 4x)^{1/2}$

$$\frac{dy}{dx} = \frac{1}{2}(1 + \sin 4x)^{-1/2}(4\cos 4x)$$

$$\frac{dy}{dx} = \frac{2\cos 4x}{\sqrt{1 + \sin 4x}}$$

25. $y = \dfrac{2\cos x^2}{3x - 1}$

$$\frac{dy}{dx} = \frac{(3x-1)(2)(-\sin x^2)(2x) - 2\cos x^2(3)}{(3x-1)^2}$$

$$\frac{dy}{dx} = \frac{-4x(3x-1)\sin x^2 - 6\cos x^2}{(3x-1)^2}$$

$$\frac{dy}{dx} = \frac{4x(1-3x)\sin x^2 - 6\cos x^2}{(3x-1)^2}$$

29. $s = \sin(\sin 2t)$

$$\frac{ds}{dt} = \cos(\sin 2t)\cos(2t)(2)$$

$$\frac{ds}{dt} = 2\cos 2t\cos(\sin 2t)$$

33. $p = \dfrac{1}{\sin s} + \dfrac{1}{\cos s} = (\sin s)^{-1} + (\cos s)^{-1}$

$$\frac{dp}{ds} = \frac{-\cos s}{\sin^2 s} + \frac{\sin s}{\cos^2 s}$$

37. **(a)** $\cos 1.0000 = 0.5403023059$, calculator

Represents $\dfrac{d}{dx}(\sin x)$ at $x = 1$ (derivative, slope of T.L. to sine curve at $x = 1$).

(b) $(\sin 1.0001 - \sin 1.0000)/0.0001$
$= 0.5402602315$, calculator = slope of secant line through the two points on sine curve where $x = 1.000$ and $x = 1.0001$.

41. $\sin(xy) + \cos 2y = x^2$

$$\cos(xy)\left(x\frac{dy}{dx} + y\right) + (-\sin 2y)\left(2\frac{dy}{dx}\right) = 2x$$

$$x\frac{dy}{dx}\cos(xy) + y\cos(xy) - 2\frac{dy}{dx}\sin 2y = 2x$$

$$\frac{dy}{dx}(x\cos(xy) - 2\sin 2y) = 2x - y\cos(xy)$$

$$\frac{dy}{dx} = \frac{2x - y\cos xy}{x\cos xy - 2\sin 2y}$$

45. $$\cos 2x = 2\cos^2 x - 1$$
$$-\sin 2x(2) = 2(2)\cos x(-\sin x) - 0$$
$$-2\sin 2x = -4\sin x\cos x$$
$$\sin 2x = 2\sin x\cos x$$

49. $$y = \frac{2\sin 3x}{x};\ x = 0.15$$
$$\frac{dy}{dx} = 2\left(\frac{x(3)\cos 3x - \sin 3x}{x^2}\right)$$
$$m_{\text{TL}} = \frac{6x\cos 3x - 2\sin 3x}{x^2}$$
$$m_{\text{TL}}\big|_{x=0.15} = -2.65$$

53. $y = 1.85\sin 36\pi t$
$$v = \frac{dy}{dx} = (1.85\cos 36\pi t)(36\pi)$$
$$v\big|_{t=0.0250} = [1.85\cos(36\pi\cdot 0.025)][36\pi]$$
$$= -199\ \text{cm/s}$$

27.2 Derivatives of the Other Trigonometric Functions

1. $y = 3\sec^2 x^2$
$$\frac{dy}{dx} = 3(2)(\sec x^2)\frac{d}{dx}(\sec x^2)$$
$$\frac{dy}{dx} = 6\sec x^2 \sec x^2 \tan x^2 (2x)$$
$$\frac{dy}{dx} = 12x\sec^2 x^2 \tan x^2$$

5. $y = 5\cot(0.25\pi - \theta)$
$$\frac{dy}{d\theta} = -5\csc^2(0.25\pi - \theta)\cdot(-1)$$
$$= 5\csc^2(0.25\pi - \theta)$$

9. $y = -3\csc\sqrt{2x+3}$
$$\frac{dy}{dx} = -3[-\csc\sqrt{2x+3}\cot\sqrt{2x+3}\cdot\frac{1}{2}(2x+3)^{-1/2}(2)]$$
$$\frac{dy}{dx} = \frac{3\csc\sqrt{2x+3}\cot\sqrt{2x+3}}{\sqrt{2x+3}}$$

13. $y = 2\cot^4 \frac{1}{2}x$
$$\frac{dy}{dx} = 2(4)\cot^3\frac{1}{2}x\left[-\csc^2\frac{1}{2}x\left(\frac{1}{2}\right)\right]$$
$$= -4\cot^3\frac{1}{2}x\csc^2\frac{1}{2}x$$

17. $y = 3\csc^4 7x$
$$\frac{dy}{dx} = 3\cdot 4\csc^3 7x[-\csc 7x\cot 7x(7)]$$
$$= -84\csc^4 7x\cot 7x$$

21. $y = 4\cos x\csc x^2$
$$\frac{dy}{dx} = 4[\cos x(-\csc x^2\cot x^2\cdot 2x) + \csc x^2(-\sin x)]$$
$$\frac{dy}{dx} = -4\csc x^2(2x\cos x\cot x^2 + \sin x)$$

25. $y = \dfrac{2\cos 4x}{1+\cot 3x}$
$$\frac{dy}{dx} = \frac{(1+\cot 3x)[-2\sin 4x(4)] - 2\cos 4x(-\csc^2 3x)(3)}{(1+\cot 3x)^2}$$
$$\frac{dy}{dx} = \frac{-8\sin 4x(1+\cot 3x) + 6\cos 4x\csc^2 3x}{(1+\cot 3x)^2}$$
$$\frac{dy}{dx} = \frac{2(-4\sin 4x - 4\sin 4x\cot 3x + 3\cos 4x\csc^2 3x)}{(1+\cot 3x)^2}$$

29. $r = \tan(\sin 2\pi\theta)$
$$\frac{dr}{d\theta} = \sec^2(\sin 2\pi\theta)\cos(2\pi\theta)(2\pi)$$
$$= 2\pi\cos 2\pi\theta\sec^2(\sin 2\pi\theta)$$

33. $x\sec y - 2y = \sin 2x$
$$x\sec y\tan y\frac{dy}{dx} + \sec y - 2\frac{dy}{dx} = 2\cos 2x$$
$$x\sec y\tan y\frac{dy}{dx} - 2\frac{dy}{dx} = 2\cos 2x - \sec y$$
$$\frac{dy}{dx}(x\sec y\tan y - 2) = 2\cos 2x - \sec y$$
$$\frac{dy}{dx} = \frac{2\cos 2x - \sec y}{x\sec y\tan y - 2}$$

37. $y = \tan 4x\sec 4x$
$$\frac{dy}{dx} = \tan 4x\cdot 4\sec 4x\tan 4x + \sec 4x\cdot 4\sec^2 4x$$
$$dy = 4\sec 4x(\tan^2 4x + \sec^2 4x)dx$$

41. (a)

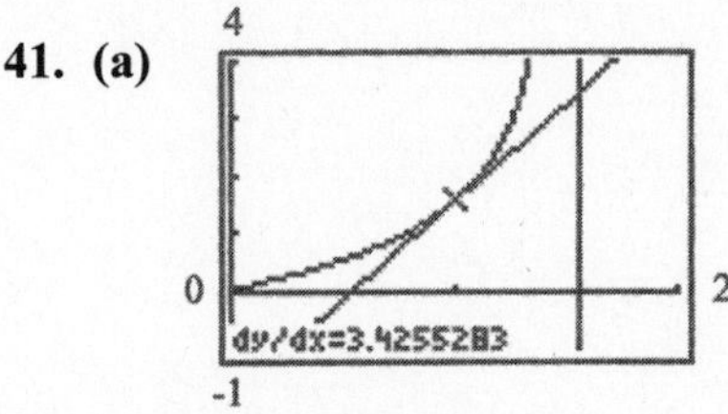

(b)

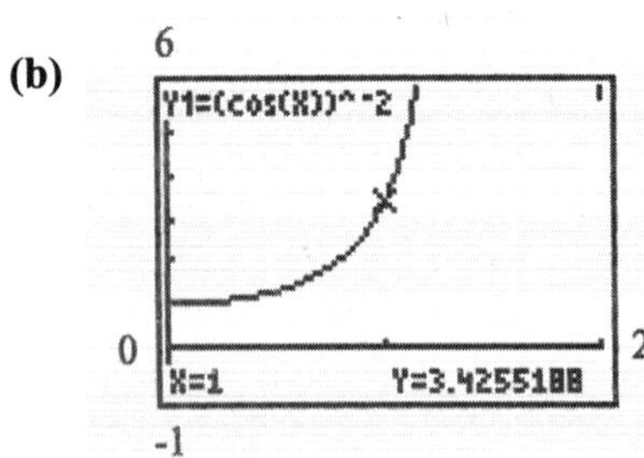

The values are the same to four decimal places.

45. $y = 2\cot 3x; x = \dfrac{\pi}{12};$

$$\frac{dy}{dx} = 2(-\csc^2 3x)(3)$$
$$= -6\csc^2 3x;$$
$$\left.\frac{dy}{dx}\right|_{x=\pi/12} = -6\csc^2\frac{\pi}{4}$$
$$= -6(\sqrt{2})^2$$
$$= -12$$

49. $y = 2t^{1.5} - \tan 0.1t$

$$v = \frac{dy}{dt} = 3t^{0.5} - 0.1\sec^2 0.1t$$
$$v|_{t=15} = 3(15)^{0.5} - 0.1\sec^2[0.1(15)]$$
$$= -8.4\,\text{cm/s}$$

27.3 Derivatives of the Inverse Trigonometric Functions

1. $y = \sin^{-1} x^2$

$$\frac{dy}{dx} = \frac{1}{\sqrt{1-(x^2)^2}}(2x) = \frac{2x}{\sqrt{1-x^4}}$$

5. $y = 2\sin^{-1} 3x^3$

$$\frac{dy}{dx} = 2\frac{1}{\sqrt{1-9x^6}}(9x^2)$$
$$= \frac{18x^2}{\sqrt{1-9x^6}}$$

9. $y = 2\cos^{-1}\sqrt{2-x}$

$$\frac{dy}{dx} = 2\cdot\frac{-1}{\sqrt{1-(2-x)}}\cdot\frac{1}{2}(2-x)^{-1/2}(-1)$$

$$\frac{dy}{dx} = \frac{1}{\sqrt{x-1}\sqrt{2-x}}$$
$$= \frac{1}{\sqrt{(x-1)(2-x)}}$$

13. $y = 6\tan^{-1}\left(\dfrac{1}{x}\right)$

$$\frac{dy}{dx} = 6\frac{1}{1+\frac{1}{x^2}}\left(-\frac{1}{x^2}\right)$$
$$= \frac{-\frac{6}{x^2}}{(x^2+1)/x^2}$$
$$= \frac{-6}{x^2+1}$$

17. $v = 0.4u\tan^{-1} 2u$

$$\frac{dv}{du} = 0.4u\cdot\frac{2}{1+4u^2} + 0.4\tan^{-1} 2u$$
$$\frac{dv}{du} = \frac{0.8u}{1+4u^2} + 0.4\tan^{-1} 2u$$

21. $y = \dfrac{\sin^{-1} 2x}{\cos^{-1} 2x}$

$$\frac{dy}{dx} = \frac{\cos^{-1} 2x\dfrac{1}{\sqrt{1-4x^2}}(2) - \sin^{-1} 2x\cdot\dfrac{-1}{\sqrt{1-4x^2}}(2)}{(\cos^{-1} 2x)^2}$$
$$\frac{dy}{dx} = \frac{2}{\sqrt{1-4x^2}}\left[\frac{\cos^{-1} 2x + \sin^{-1} 2x}{(\cos^{-1} 2x)^2}\right]$$
$$\frac{dy}{dx} = \frac{2(\cos^{-1} 2x + \sin^{-1} 2x)}{\sqrt{1-4x^2}(\cos^{-1} 2x)^2}$$

25. $u = \left[\sin^{-1}(4t+3)\right]^2$

$$\frac{du}{dt} = 2\left[\sin^{-1}(4t+3)\right]\frac{4}{\sqrt{1-(4t+3)^2}}$$
$$= \frac{2\sqrt{2}\sin^{-1}(4t+3)}{\sqrt{-2t^2-3t-1}}$$

29. $y = \dfrac{1}{1+4x^2} - \tan^{-1} 2x$

$$= (1+4x^2)^{-1} - \tan^{-1} 2x$$
$$\frac{dy}{dx} = -(1+4x^2)^{-2}(8x) - \frac{1}{1+4x^2}(2)$$

$$=\frac{-8x}{(1+4x^2)^2}-\frac{2}{1+4x^2}$$
$$=\frac{-8x-2(1+4x^2)}{(1+4x^2)^2}$$
$$=\frac{-8x-2-8x^2}{(1+4x^2)^2}$$
$$=\frac{-2(1+4x+4x^2)}{(1+4x^2)^2}$$
$$\frac{dy}{dx}=\frac{-2(1+2x)^2}{(1+4x^2)^2}$$

33. $2\tan^{-1}xy+x=3$

$$2\frac{1}{1+x^2y^2}\left(x\frac{dy}{dx}+y\right)+1=0$$
$$\frac{2x}{1+x^2y^2}\frac{dy}{dx}+\frac{2y}{1+x^2y^2}=-1$$
$$\frac{2x}{1+x^2y^2}\frac{dy}{dx}=-1-\frac{2y}{1+x^2y^2}$$
$$\frac{2x}{1+x^2y^2}\frac{dy}{dx}=\frac{-1-x^2y^2-2y}{1+x^2y^2}$$
$$2x\frac{dy}{dx}=-1-x^2y^2-2y$$
$$\frac{dy}{dx}=\frac{-(x^2y^2+2y+1)}{2x}$$

37. $y=(\sin^{-1}x)^3$

$$\frac{dy}{dx}=3(\sin^{-1}x)^2\frac{1}{\sqrt{1-x^2}}$$
$$dy=\frac{3(\sin^{-1}x)^2\,dx}{\sqrt{1-x^2}}$$

41. $y=x\tan^{-1}x$

$$\frac{dy}{dx}=\frac{x}{1+x^2}+\tan^{-1}x$$
$$\frac{d^2y}{dx^2}=\frac{1+x^2-x(2x)}{\left(1+x^2\right)^2}+\frac{1}{1+x^2}$$
$$\frac{d^2y}{dx^2}=\frac{1-x^2}{\left(1+x^2\right)^2}+\frac{1}{1+x^2}\times\frac{1+x^2}{1+x^2}$$
$$\frac{d^2y}{dx^2}=\frac{1-x^2+1+x^2}{\left(1+x^2\right)^2}$$
$$=\frac{2}{\left(1+x^2\right)^2}$$

45. $y=\tan^{-1}2x$

$$\frac{dy}{dx}=\frac{1}{1+(2x)^2}(2)$$
$$=\frac{2}{1+4x^2}$$
$$=2\left(1+4x^2\right)^{-1}$$
$$\frac{d^2y}{dx^2}=2(-1)\left(1+4x^2\right)^{-2}(8x)$$
$$=\frac{-16x}{\left(1+4x^2\right)^2}$$

49. $t=\frac{1}{\omega}\sin^{-1}\frac{A-E}{mE}$

$$=\frac{1}{\omega}\sin^{-1}\left(\frac{A-E}{E}\right)\left(\frac{1}{m}\right)$$
$$=\frac{1}{\omega}\sin^{-1}\left(\frac{A-E}{E}\right)m^{-1}$$
$$u=\left(\frac{A-E}{E}\right)m^{-1};\ \frac{du}{dm}=-\left(\frac{A-E}{E}\right)m^{-2}$$
$$\frac{dt}{dm}=\frac{1}{\omega\sqrt{1-\left(\frac{A-E}{E}\right)^2m^{-2}}}\left(\frac{-A+E}{Em^2}\right)$$
$$=\frac{E-A}{\omega Em^2\sqrt{1-\frac{(A-E)^2}{E^2m^2}}}$$
$$=\frac{E-A}{\omega Em^2\sqrt{\frac{E^2m^2-(A-E)^2}{E^2m^2}}}$$
$$\frac{dt}{dm}=\frac{E-A}{\omega m\sqrt{E^2m^2-(A-E)^2}}$$

53. $\tan\theta=\frac{h}{x}$

$$\theta=\tan^{-1}\frac{h}{x}$$
$$\frac{d}{dx}\frac{h}{x}=\frac{d}{dx}hx^{-1}=-hx^{-2}=-\frac{h}{x^2}$$
$$\frac{d\theta}{dx}=\frac{1}{1+\frac{h^2}{x^2}}\left(-\frac{h}{x^2}\right)$$
$$\frac{d\theta}{dx}=\frac{-h}{\frac{x^2+h^2}{x^2}\left(x^2\right)}$$
$$\frac{d\theta}{dx}=\frac{-h}{h^2+x^2}$$

27.4 Applications

1. Sketch the curve $y=\sin x-\frac{x}{2}, 0\le x\le 2\pi$.

$x=0\Rightarrow y=0,(0,0)$ is both x-intercept and y-intercept. Using Newton's method or zero feature on a graphing calculator, (1.90,0) is the only x-intercept for $0\le x\le 2\pi$. For $x=2\pi$,

$y=\sin 2\pi-\frac{2\pi}{2}=-\pi\Rightarrow(2\pi,-\pi)$ is the right – hand endpoint.

No vertical asymptotes (no denominators that become zero).

$$\frac{dy}{dx}=\cos x-\frac{1}{2}=0 \text{ for } x=\frac{\pi}{3},\frac{5\pi}{3}$$

$$\frac{d^2y}{dx^2}=-\sin x$$

$$\left.\frac{d^2y}{dx^2}\right|_{x=\frac{\pi}{3}}=\frac{-\sqrt{3}}{2}, \text{ so max}\left(\frac{\pi}{3},0.34\right)$$

$$\left.\frac{d^2y}{dx^2}\right|_{x=\frac{5\pi}{3}}=\frac{\sqrt{3}}{2}, \text{ so min}\left(\frac{5\pi}{3},-3.48\right)$$

$$\left.\frac{d^2y}{dx^2}\right|_{x=0,\pi}=0 \text{ for } x=0,\pi$$

$$\frac{d^2y}{dx^2}<0 \text{ for } 0<x<\pi$$

$$\frac{d^2y}{dx^2}>0 \text{ for } x<0, \pi<x<2\pi$$

$(0,0)$ and $\left(\pi,-\frac{\pi}{2}\right)$ are inflection points

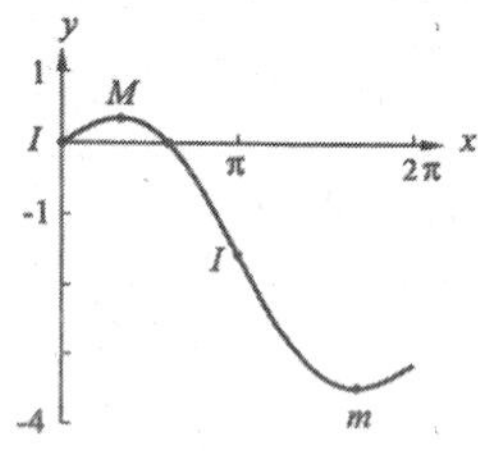

5. $y=\tan^{-1}x$

$\frac{dy}{dx}=\frac{1}{1+x^2}>0$ for all x. Therefore, the graph is always increasing.

9. $y=x\sin^{-1}x$

$$y|_{x=0.5}=0.26179939$$

$$\frac{dy}{dx}=\frac{x}{\sqrt{1-x^2}}+\sin^{-1}x$$

$$\left.\frac{dy}{dx}\right|_{x=0.5}=1.1009497$$

$$y-0.26179939=1.1009497(x-0.5)$$

$$y=1.1x-0.29$$

13. $y=6\cos x-8\sin x$; minimum value occurs when $f'(x)=0$.

$$f'(x)=-6\sin x-8\cos x=0$$

$$\sin x=-\frac{8}{6}\cos x$$

$$\tan x=-\frac{8}{6}=-\frac{4}{3}$$

$\theta_{\text{ref}}=0.927$ (a 3, 4, 5 triangle)

In QII: $\theta=\pi-0.927=2.21$

In QIII: $\theta=\pi+0.927=4.07$

$f''(x)=-6\cos x+8\sin x$

$f''(2.21)>0$, min; $f''(4.07)<0$, max.

Minimum occurs where $x=2.21$ rad.

$$f(2.21)=6\left(\frac{-3}{5}\right)-8\left(\frac{4}{5}\right)$$

$$=\frac{-18}{5}+\frac{-32}{5}=\frac{-50}{5}=-10$$

Minimum value of $y=-10$.

17. $y=0.50\sin 2t+0.30\cos t$;

$$v=\frac{dy}{dt}=1.00\cos 2t-0.30\sin t$$

$$v|_{t=0.40}=1.00\cos 0.80-0.30\sin 0.40$$

$$=0.58 \text{ m/s}$$

$$a=\frac{d^2y}{dt^2}=-2.00\sin 2t-0.30\cos t$$

$$a|_{t=0.40}=-2.00\sin 0.80-0.30\cos 0.40$$

$$=-1.7 \text{ m/s}^2$$

21. $v_x=\frac{dx}{dt}$

$$=-19(6\pi)\sin 6\pi t$$

$$=341$$

$$v_y = \frac{dy}{dt}$$
$$= 19(6\pi)\cos 6\pi t$$
$$= 111$$
$$v = \sqrt{v_x^2 + v_y^2}$$
$$v = \sqrt{(-19(6\pi)\sin 6\pi t)^2 + (19(6\pi)\cos 6\pi t)^2}$$
$$v|_{t=0.600} = 358 \text{ cm/s}$$
$$\tan\theta = \frac{19(6\pi)\cos 6\pi t}{-19(6\pi)\sin 6\pi t}\bigg|_{t=0.600}$$
$$\theta = 0.314 \text{ rad}$$
$$= 0.314 \text{ rad}\cdot\frac{180^\circ}{\pi \text{ rad}}$$
$$= 18.0^\circ$$

25. $s = 4.9t^2$

$$\tan\theta = \frac{60.0 - s}{40.0}$$
$$= \frac{60.0 - 4.9t^2}{40.0}$$
$$\theta = \tan^{-1}\left(\frac{60.0 - 4.9t^2}{40.0}\right)$$
$$\frac{d\theta}{dt} = \frac{-39\,200t}{5.2\times10^5 - 58\,800t^2 + 2400t^4}$$
$$\frac{d\theta}{dt}\bigg|_{t=1.0\text{ s}} = -0.085 \text{ rad/s}$$

29. $F = \dfrac{0.25w}{0.25\sin\theta + \cos\theta}$

$$\frac{dF}{d\theta} = \frac{-0.25w(0.25\cos\theta - \sin\theta)}{(0.25\sin\theta + \cos\theta)^2}$$
$$= 0$$
$$0.25\cos\theta - \sin\theta = 0$$
$$\tan\theta = 0.25$$
$$\theta = 0.245 \text{ rad}$$
$$= 0.245\cdot\frac{180}{\pi}$$
$$= 14.0^\circ$$
$$\frac{d^2F}{d\theta^2} = -0.25w\left[\frac{(0.25\sin\theta + \cos\theta)^2(-0.25\sin\theta - \cos\theta)}{(0.25\sin\theta + \cos\theta)^4} - \frac{(0.25\cos\theta - \sin\theta)^2(0.25\sin\theta + \cos\theta)}{(0.25\sin\theta + \cos\theta)^4}\right]$$
$$\frac{d^2F}{d\theta^2}\bigg|_{\theta=0.245} > 0, \text{ so } F \text{ is a minimum.}$$

33. $V = 0.48(1.2 - \cos 1.26t)$

$$\frac{dV}{dt} = 0.48(1.26)\sin 1.26t$$
$$= 0$$
$$1.26t = 0, \pi$$
$$t = 0, \frac{\pi}{1.26}$$
$$\frac{d^2V}{dt^2} = 0.48(1.26)^2\cos 1.26t$$
$$\frac{d^2V}{dt^2}\bigg|_{t=\frac{\pi}{1.26}} < 0, \text{ max at } t = \frac{\pi}{1.26}$$
$$V_{\max} = 0.48\left(1.2 - \cos 1.26\frac{\pi}{1.26}\right) = 1.056$$
$$V_{\max} = 1.06 \text{ L/s}$$

37. Let a be the distance between the end of the pole touching the ground and the wall. Then

$\tan\theta = \dfrac{1.8}{a}$, so $a = 1.8\cot\theta$

Also,

$\cos\theta = \dfrac{a + 1.2}{y}$, so $y = \dfrac{a + 1.2}{\cos\theta}$

Substituting $a = 1.8\cot\theta$:

$$y = \frac{1.8\cot\theta + 1.2}{\cos\theta}$$
$$= 1.8\csc\theta + 1.2\sec\theta$$
$$\frac{dy}{d\theta} = 1.8(-\csc\theta\cot\theta) + 1.2(\sec\theta\tan\theta)$$
$$= -1.8\csc\theta\cot\theta + 1.2\sec\theta\tan\theta$$
$$= -\frac{1.8}{\sin\theta}\cdot\frac{\cos\theta}{\sin\theta} + \frac{1.2}{\cos\theta}\cdot\frac{\sin\theta}{\cos\theta}$$
$$= -\frac{1.8\cos\theta}{\sin^2\theta} + \frac{1.2\sin\theta}{\cos^2\theta}$$
$$= -\frac{1.8\cos\theta}{\sin^2\theta}\cdot\frac{\cos^2\theta}{\cos^2\theta} + \frac{1.2\sin\theta}{\cos^2\theta}\cdot\frac{\sin^2\theta}{\sin^2\theta}$$
$$= \frac{-1.8\cos^3\theta + 1.2\sin^3\theta}{\sin^2\theta\cos^2\theta}$$
$$-1.8\cos^3\theta + 1.2\sin^3\theta = 0$$
$$1.2\sin^3\theta = 1.8\cos^3\theta$$
$$\tan\theta = \sqrt[3]{\frac{3}{2}}$$

$$\theta = \tan^{-1}\sqrt[3]{\frac{3}{2}} = 0.853 \text{ rad}$$

$$y_{\theta=0.853} = 4.21 \text{ m}$$

$$\frac{dA}{d\theta} < 0 \text{ for } \theta < 0.853$$

$$\frac{dA}{d\theta} > 0 \text{ for } \theta > 0.853,$$

so A is a minimum at $\theta = 0.853$.

27.5 Derivative of the Logarithmic Function

1. $y = \ln\cos 4x$

$$\frac{dy}{dx} = \frac{1}{\cos 4x}(-\sin 4x)(4)$$

$$\frac{dy}{dx} = -4\tan 4x$$

5. $y = 4\log_5(3-x)$

$$\frac{dy}{dx} = 4\frac{1}{3-x}(\log_5 e)(-1)$$

$$= \frac{4}{x-3}\log_5 e$$

9. $y = 2\ln\tan 2x$

$$\frac{dy}{dx} = 2\frac{1}{\tan 2x}\sec^2 2x(2)$$

$$= \frac{4\sec^2 2x}{\tan 2x}$$

$$= \frac{4\sec^2 2x}{\frac{\sec 2x}{\csc 2x}}$$

$$\frac{dy}{dx} = 4\sec 2x\csc 2x$$

13. $y = \ln\left(x-x^2\right)^3$

$$\frac{dy}{dx} = \frac{1}{\left(x-x^2\right)^3}\cdot 3\left(x-x^2\right)^2(1-2x)$$

$$= \frac{-6x+3}{x-x^2}$$

17. $y = 3x\ln(6-x)$

$$\frac{dy}{dx} = 3x\cdot\frac{-1}{6-x} + 3\ln(6-x)$$

$$= \frac{3x}{x-6} + 3\ln(6-x)$$

21. $r = 0.5\ln[\cos(\pi\theta^2)]$

$$\frac{dr}{d\theta} = 0.5.\frac{1}{\cos(\pi\theta^2)}\cdot(-\sin(\pi\theta^2))\cdot 2\pi\theta$$

$$\frac{dr}{d\theta} = -\pi\theta\tan(\pi\theta^2)$$

25. $u = 3v\ln^2 2v$

$$\frac{du}{dv} = 3v\cdot 2\ln 2v\cdot\frac{1}{2v}\cdot 2 + 3\ln^2 2v$$

$$\frac{du}{dv} = 6\ln 2v + 3\ln^2 2v$$

29. $r = \ln\frac{v^2}{v+2}$

$$= \ln v^2 - \ln(v+2)$$

$$= 2\ln v - \ln(v+2)$$

$$\frac{dr}{dv} = \frac{2}{v} - \frac{1}{v+2}$$

$$= \frac{2v+4-v}{v(v+2)}$$

$$\frac{dr}{dv} = \frac{v+4}{v(v+2)}$$

33. $y = x - \ln^2(x+y)$

$$\frac{dy}{dx} = 1 - 2\ln(x+y)\frac{1}{x+y}\left(1+\frac{dy}{dx}\right)$$

$$\frac{dy}{dx} = 1 - \frac{2\ln(x+y)}{x+y} - \frac{2\ln(x+y)}{x+y}\frac{dy}{dx}$$

$$\frac{dy}{dx}\left[1+\frac{2\ln(x+y)}{x+y}\right] = \frac{x+y-2\ln(x+y)}{x+y}$$

$$\frac{dy}{dx}\left[\frac{x+y+2\ln(x+y)}{x+y}\right] = \frac{x+y-2\ln(x+y)}{x+y}$$

$$\frac{dy}{dx} = \frac{x+y-2\ln(x+y)}{x+y+2\ln(x+y)}$$

37. $y = (1+x)^{(1/x)}$

(a)

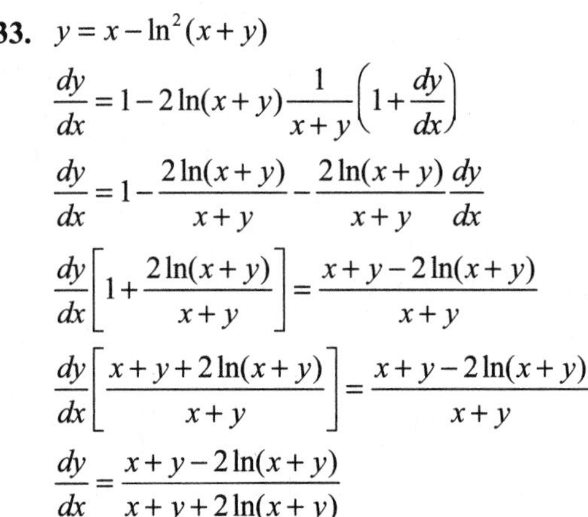

(b)

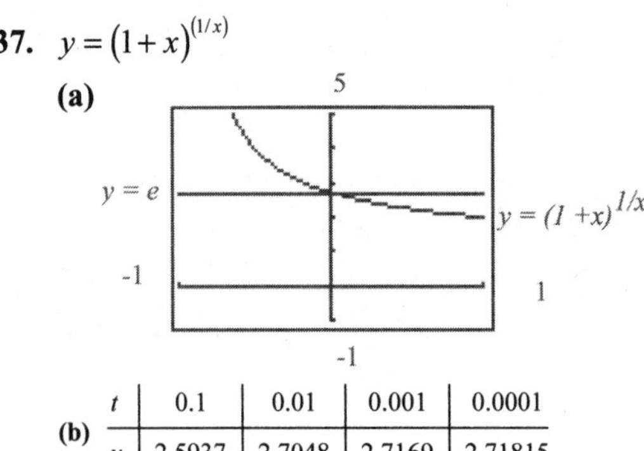

t	0.1	0.01	0.001	0.0001
y	2.5937	2.7048	2.7169	2.71815

41. $y = \sin^{-1} 2x + \ln\sqrt{1-4x^2};\ x = 0.250;$

$$= \sin^{-1} 2x + \frac{1}{2}\ln\left(1-4x^2\right)$$

$$\frac{dy}{dx} = \frac{2}{\sqrt{1-(2x)^2}} + \frac{-8x}{2\left(1-4x^2\right)}$$

$$= \frac{2\sqrt{1-4x^2} - 4x}{1-4x^2}$$

$$\left.\frac{dy}{dx}\right|_{x=0.250} = \frac{2\sqrt{1-4(0.25)^2} - 4(0.25)}{1-4(0.25)^2}$$

$$= 0.976$$

45. $y = \tan^{-1} 2x + \ln(4x^2+1);$

$$m_{\tan} = \frac{dy}{dx}$$

$$= \frac{1}{1+4x^2}(2) + \frac{1}{4x^2+1}(8x)$$

$$= \frac{2}{1+4x^2} + \frac{8x}{4x^2+1}$$

$$= \frac{2+8x}{1+4x^2}$$

When $x = 0.625$,

$$m_{\tan} = \frac{2+8(0.625)}{1+4(0.625)^2}$$

$$= 2.73$$

49. $y_1 = \ln(x^2), x^2 \neq 0$

$$\frac{dy_1}{dx} = \frac{1}{x^2}(2x)$$

$$= \frac{2}{x}$$

$$\left.\frac{dy_1}{dx}\right|_{x=-1} = -2$$

$$y_2 = 2\ln x, x > 0$$

$$\frac{dy_2}{dx} = 2\left(\frac{1}{x}\right)$$

$= \left.\frac{2}{x}\right|_{x=-1}$ is not defined since

$-1 < 0$

For values of x where both functions are defined, the derivatives are the same.

53. $y = \ln\left(\frac{1+\sqrt{1+x^2}}{x}\right) - \sqrt{1-x^2}$

$$\frac{dy}{dx} = \frac{1}{\frac{1+\sqrt{1+x^2}}{x}}\left(\frac{x\left(0+\frac{2x}{2\sqrt{1+x^2}}\right) - (1+\sqrt{1+x^2})(1)}{x^2}\right) - \frac{-2x}{2\sqrt{1-x^2}}$$

$$\frac{dy}{dx} = \frac{x}{1+\sqrt{1+x^2}}\left(\frac{\frac{x^2}{\sqrt{1+x^2}} - 1 - \sqrt{1+x^2}}{x^2}\right) + \frac{x}{\sqrt{1-x^2}}$$

$$\frac{dy}{dx} = \frac{x^2 - \sqrt{1+x^2} - 1 - x^2}{x(1+\sqrt{1+x^2})\sqrt{1+x^2}} + \frac{x}{\sqrt{1-x^2}}$$

$$\frac{dy}{dx} = \frac{-1-\sqrt{1+x^2}}{x\sqrt{1+x^2} + x + x^3} + \frac{x}{\sqrt{1-x^2}}$$

$$\frac{dy}{dx} = \frac{x^2 - x^2 - 1 - \sqrt{1+x^2}}{x(x^2+1+\sqrt{1+x^2})} + \frac{x}{\sqrt{1-x^2}}$$

$$\frac{dy}{dx} = \frac{x^2 - (x^2+1+\sqrt{1+x^2})}{x(x^2+1+\sqrt{1+x^2})} + \frac{x}{\sqrt{1-x^2}}$$

$$\frac{dy}{dx} = \frac{x}{x^2+1+\sqrt{1+x^2}} - \frac{1}{x} + \frac{x}{\sqrt{1-x^2}}$$

27.6 Derivative of the Exponential Function

1. $y = \ln\sin e^{2x}$

$$\frac{dy}{dx} = \frac{1}{\sin e^{2x}}(\cos e^{2x})(2e^{2x})$$

$$\frac{dy}{dx} = 2e^{2x}\cot e^{2x}$$

5. $y = 6e^{\sqrt{x}}$

$$\frac{dy}{dx} = 6e^{\sqrt{x}} \cdot \frac{1}{2\sqrt{x}}$$

$$= \frac{3e^{\sqrt{x}}}{\sqrt{x}}$$

9. $R = Te^{-T}$

$$\frac{dR}{dT} = T(e^{-T})(-1) + (1)(e^{-T})$$
$$= e^{-T} - Te^{-T}$$
$$= e^{-T}(1-T)$$

13. $r = \frac{2(e^{2s} - e^{-2s})}{e^{2s}}$

$$r = 2(1 - e^{-4s})$$
$$\frac{dr}{ds} = 2(4e^{-4s})$$
$$\frac{dr}{ds} = 8e^{-4s}$$

17. $y = \frac{2e^{3x}}{4x+3}$

$$\frac{dy}{dx} = \frac{(4x+3)(2e^{3x})(3) - (2e^{3x})(4)}{(4x+3)^2}$$
$$\frac{dy}{dx} = \frac{(12x+9)(2e^{3x}) - 8e^{3x}}{(4x+3)^2}$$
$$= \frac{2e^{3x}(12x+5)}{(4x+3)^2}$$

21. $y = (2e^{2x})^3 \sin x^2$

$$= 8e^{6x} \sin x^2$$
$$\frac{dy}{dx} = 8e^{6x}(\cos x^2)(2x) + \sin x^2(8e^{6x})(6)$$
$$= 16e^{6x}(x\cos x^2 + 3\sin x^2)$$

25. $y = xe^{xy} + \sin y$

$$\frac{dy}{dx} = x(e^{xy})\left(x\frac{dy}{dx} + y\right) + (1)e^{xy} + \cos y\frac{dy}{dx}$$
$$= x^2e^{xy}\frac{dy}{dx} + xy(e^{xy}) + e^{xy} + \cos y\frac{dy}{dx}$$
$$\frac{dy}{dx} - x^2e^{xy}\frac{dy}{dx} - \cos y\frac{dy}{dx} = xye^{xy} + e^{xy}$$
$$\frac{dy}{dx}(1 - x^2e^{xy} - \cos y) = xye^{xy} + e^{xy}$$
$$\frac{dy}{dx} = \frac{e^{xy}(xy+1)}{1 - x^2e^{xy} - \cos y}$$

29. $I = \ln \sin 2e^{6t}$

$$\frac{dI}{dt} = \frac{1}{\sin 2e^{6t}} \cdot \cos 2e^{6t} \cdot 12e^{6t}$$
$$= 12e^{6t} \cot 2e^{6t}$$

33. **(a)** $e = e^x = 2.7182818$ when $x = 1.0000$. This is the slope of a tangent line to the curve $f(x) = e^x$ when $x = 1.0000$. It is the curve $f'(x) = e^x$, since $\frac{de^x}{dx} = e^x$.

(b) $\frac{e^{1.0001} - e^{1.0000}}{0.0001} = 2.7184178$ This is the slope of a secant line through the curve.
$f(x) = e^x$at $x = 1.0000$, where $\Delta x = 0.0001$.

$$\lim_{\Delta x \to 0} \frac{e(x + \Delta x - e^x)}{\Delta x} = \frac{de^x}{dx} = e^x$$

For $\Delta x = 0.0001$, the slope of the tangent line is approximately equal to the slope of the secant line.

37. $y = e^{-x/2} \cos 4x$

$$\frac{dy}{dx} = -4(\sin 4x)\left(e^{-x/2}\right) - \frac{1}{2}e^{-x/2}\cos 4x$$
$$\left.\frac{dy}{dx}\right|_{x=0.625} = -1.46$$

41.

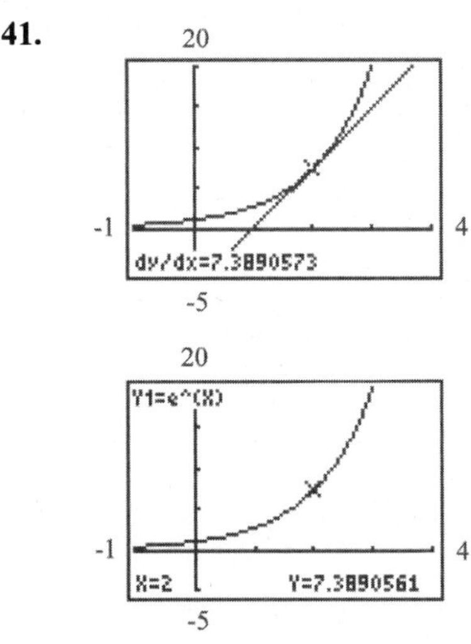

45. To show:

$$y = \frac{e^{2x} - 1}{e^{2x} + 1} \text{ satisfies } \frac{dy}{dx} = 1 - y^2$$
$$\frac{dy}{dx} = \frac{(e^{2x}+1)2e^{2x} - (e^{2x}-1)2e^{2x}}{(e^{2x}+1)^2}$$
$$\frac{dy}{dx} = \frac{2e^{2x}(e^{2x} + 1 - e^{2x} + 1)}{(e^{2x}+1)^2}$$
$$= \frac{4e^{2x}}{(e^{2x}+1)^2}$$
$$1 - y^2 = 1 - \frac{(e^{2x}-1)^2}{(e^{2x}+1)^2}$$
$$= \frac{(e^{2x}+1)^2 - (e^{2x}-1)^2}{(e^{2x}+1)^2}$$

$$= \frac{e^{4x} + 2e^{2x} + 1 - e^{4x} + 2e^{2x} - 1}{(e^{2x}+1)^2}$$

$$= \frac{4e^{2x}}{(e^{2x}+1)^2}$$

Therefore,

$$\frac{dy}{dx} = 1 - y^2$$

49. $y = Ae^{kx} + Be^{-kx}$

$$y' = Ake^{kx} - Bke^{-kx}$$
$$y'' = Ak^2e^{kx} + Bk^2e^{-kx}$$
$$= k^2\left(Ae^{kx} + Be^{-kx}\right)$$
$$= k^2y$$

53. $i = 4.42e^{-66.7t}\sin(226t)$

$$\frac{di}{dt} = 4.42e^{-66.7t}\cos(226t)\cdot 226 - 294.8e^{-66.7t}\sin(226t)$$
$$\frac{di}{dt} = e^{-66.7t}(999\cos(226t) - 295\sin(226t))$$

57. $\frac{d}{dx}\sinh u = \frac{d}{dx}\frac{1}{2}\left(e^u - e^{-u}\right)$

$$= \frac{1}{2}\left(e^u\frac{du}{dx} - e^{-u}\cdot -\frac{du}{dx}\right)$$
$$= \frac{1}{2}e^u\frac{du}{dx} + \frac{1}{2}e^{-u}\frac{du}{dx}$$
$$= \left[\frac{1}{2}\left(e^u + e^{-u}\right)\right]\frac{du}{dx}$$

$$\frac{d}{dx}\sinh u = \cosh u\frac{du}{dx}$$

$$\frac{d}{dx}\cosh u = \frac{d}{dx}\frac{1}{2}\left(e^u + e^{-u}\right)$$
$$= \frac{1}{2}\left(e^u\frac{du}{dx} + e^{-u}\cdot -\frac{du}{dx}\right)$$
$$= \frac{1}{2}e^u\frac{du}{dx} - \frac{1}{2}e^{-u}\frac{du}{dx}$$
$$= \left[\frac{1}{2}\left(e^u - e^{-u}\right)\right]\frac{du}{dx}$$

$$\frac{d}{dx}\cosh u = \sinh u\frac{du}{dx}$$

27.7 L'Hospital's Rule

1. $\lim_{x\to\infty}\frac{3e^{2x}}{5\ln x} = \lim_{x\to\infty}\frac{\frac{d}{dx}3e^{2x}}{\frac{d}{dx}5\ln x}$

$$= \lim_{x\to\infty}\frac{6e^{2x}}{\frac{5}{x}}$$
$$= \lim_{x\to\infty}\frac{6xe^{2x}}{5}$$
$$= \infty$$

5. $\lim_{x\to\infty}\frac{x\ln x}{x+\ln x} = \lim_{x\to\infty}\frac{x\frac{1}{x}+\ln x}{1+\frac{1}{x}}$

$$= \infty$$

9. $\lim_{x\to 0}\frac{\ln x}{x^{-1}} = \lim_{x\to 0}\frac{\frac{1}{x}}{-\frac{1}{x^2}}$

$$= -\lim_{x\to 0}x$$
$$= 0$$

13. $\lim_{x\to 1}\frac{\sin\pi x}{x-1} = \lim_{x\to 1}\frac{\pi\cos\pi x}{1}$

$$= -\pi$$

17. $\lim_{x\to 0^+}x\ln\sin x = \lim_{x\to 0^+}\frac{\ln\sin x}{\frac{1}{x}}$

$$= \lim_{x\to 0^+}\frac{\frac{1}{\sin x}\cos x}{-\frac{1}{x^2}}$$
$$= -\lim_{x\to 0^+}\frac{x\cos x}{\frac{\sin x}{x}}$$
$$= 0$$

21. $\lim_{x\to +\infty}\frac{1+e^{2x}}{2+\ln x} = \lim_{x\to +\infty}\frac{2e^{2x}}{\frac{1}{x}}$

$$= \lim_{x\to +\infty}2xe^{2x}$$
$$= \infty$$

25. $\lim_{x\to 0^+}(\sin x)(\ln x) = \lim_{x\to 0^+}\frac{\ln x}{\frac{1}{\sin x}}$

$$= \lim_{x\to 0^+}\frac{\frac{1}{x}}{\frac{-\cos x}{\sin^2 x}}$$
$$= -\lim_{x\to 0^+}\frac{\sin x}{x}\cdot\frac{\sin x}{\cos x}$$
$$= 0$$

29. $\lim_{x\to 0}\frac{\sqrt{1-x}-\sqrt{1+x}}{x} = \lim_{x\to 0}\frac{\frac{-1}{2\sqrt{1-x}} - \frac{1}{2\sqrt{1+x}}}{1}$

$$= -\frac{1}{2} - \frac{1}{2}$$
$$= -1$$

33. $$y = x^x$$
$$\ln y = \ln x^x = x\ln x$$
$$\lim_{x\to 0}\ln y = \lim_{x\to 0} x\ln x$$
$$= \lim_{x\to 0}\frac{\ln x}{\frac{1}{x}}$$
$$= \lim_{x\to 0}\frac{\frac{1}{x}}{-\frac{1}{x^2}}$$
$$= -\lim_{x\to 0} x = 0$$
$$e^{\ln \lim_{x\to 0} y} = e^0 = 1 \Rightarrow \lim_{x\to 0} y = 1$$

37. $\sin x$ varies between -1 and 1 as $x\to\infty$.

27.8 Applications

1. $y = e^{-x}\sin x,\ 0\le x\le 2\pi$

$y = 0$ for $x = 0, \pi, 2\pi$. Intercepts: (0, 0), $(\pi,0),(2\pi,0)$

No symmetry to the axes or the origin, and no vertical asymptotes.

$$\frac{dy}{dx} = e^{-x}\cos x - e^{-x}\sin x$$
$$= 0$$
$$\sin x = \cos x$$
$$x = \frac{\pi}{4}, \frac{5\pi}{4}$$
$$\frac{d^2y}{dx^2} = e^{-x}(-\sin x - \cos x) - e^{-x}(\cos x - \sin x)$$

$$\frac{d^2y}{dx^2} = e^{-x}(-2\cos x)$$
$$= 0 \text{ for } x = \frac{\pi}{2}, \frac{3\pi}{2}$$
$$\left.\frac{d^2y}{dx^2}\right|_{x=\frac{\pi}{4}} = -2e^{-\pi/4}\cos\frac{\pi}{4}$$
$$= -0.64 < 0 \Rightarrow \left(\frac{\pi}{4}, 0.322\right) \text{ is max}$$
$$\left.\frac{d^2y}{dx^2}\right|_{\frac{5\pi}{4}} = -2e^{-5\pi/4}\cos\frac{5\pi}{4}$$
$$= 0.03 > 0 \Rightarrow \left(\frac{5\pi}{4}, -0.014\right) \text{ is a min}$$

$\frac{d^2y}{dx^2}$ changes from negative to positive at $x = \frac{\pi}{2}$

$\left(\frac{\pi}{2}, 0.208\right)$ is an inflection point

$\frac{d^2y}{dx^2}$ changes from positive to negative at $x = \frac{3\pi}{2}$

$\left(\frac{3\pi}{2}, -0.009\right)$ is an inflection point

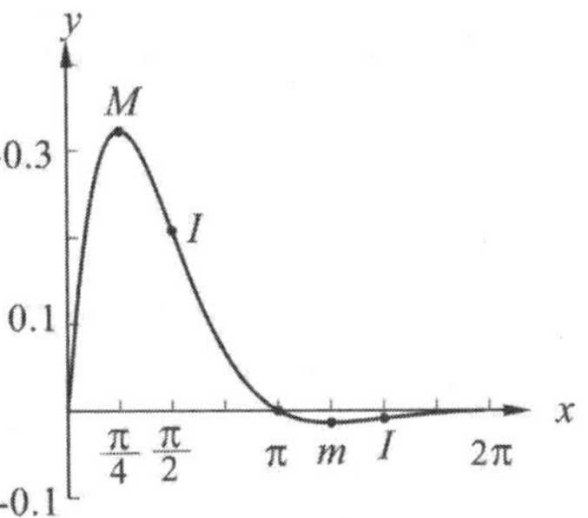

5. $$y = 3xe^{-x} = \frac{3x}{e^x}$$
$$\frac{dy}{dx} = (3x)(-e^{-x}) + 3e^{-x}$$
$$= (3-3x)e^{-x}$$
$$\frac{d^2y}{dx^2} = (-3)(e^{-x}) + (3-3x)(-e^{-x})$$
$$= -3e^{-x} - 3e^{-x} + 3xe^{-x}$$
$$= (3x-6)e^{-x}$$

(1) Intercept: $x = 0, y = 0$ (origin)

(2) Symmetry: None

(3) As $x\to+\infty$, $y = 0$, horizontal asymptote $y = 0$

As $x\to-\infty$, $y\to-\infty$

(4) Vertical asymptote: none

(5) Domain: all x, range to be determined

(6) Set $\frac{dy}{dx} = 0$

$e^{-x}(3-3x) = 0$ for $x = 1$

$f''(1) < 0$, so Max $\left(1, \frac{1}{2}\right)$

Set $\frac{d^2y}{dx^2}=0$

$e^{-x}(3x-6)=0$ for $x=2,$

so inflection point at $\left(2,\frac{2}{e^2}\right)$

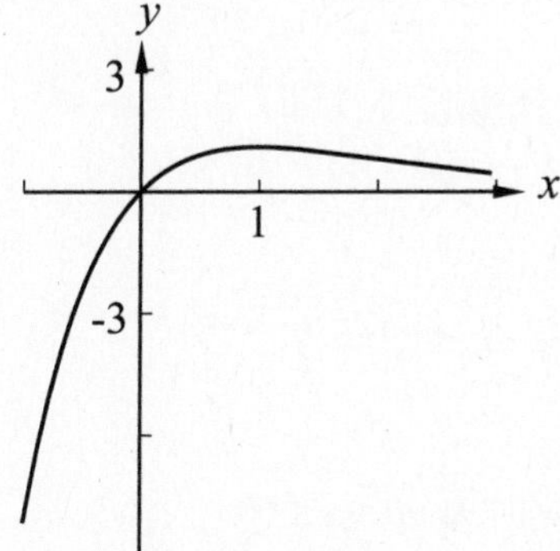

9. $y=4e^{-x^2}$

$\frac{dy}{dx}=4e^{-x^2}(-2x)=\frac{-8x}{e^{x^2}}$

$\frac{d^2y}{dx^2}=\frac{e^{x^2}(-8)+8x(2xe^{x^2})}{e^{2x^2}}=8e^{-x^2}\left(2x^2-1\right)$

(1) Intercepts: $x=0, y=4, (0,4)$ intercept.

(2) Symmetry: yes, with respect to y-axis.

(3) As $x\to\pm\infty, y\to 0$ positively; x-axis is a horizontal asymptote.

(4) No vertical asymptote.

(5) Domain: all x; range to be determined.

(6) $\frac{dy}{dx}=\frac{-8x}{e^{x^2}}=0$ for $x=0$

$\left.\frac{d^2y}{dx^2}\right|_{x=0}<0,$ so max. at $(0,4)$

Range: $0<y\le 4$

$\frac{d^2y}{dx^2}=2x^2-1=0$ for $x=\pm\sqrt{\frac{1}{2}}=\pm\frac{\sqrt{2}}{2};$

$\left(-\frac{\sqrt{2}}{2},\frac{4}{\sqrt{e}}\right),\left(\frac{\sqrt{2}}{2},\frac{4}{\sqrt{e}}\right)$ are inflection points.

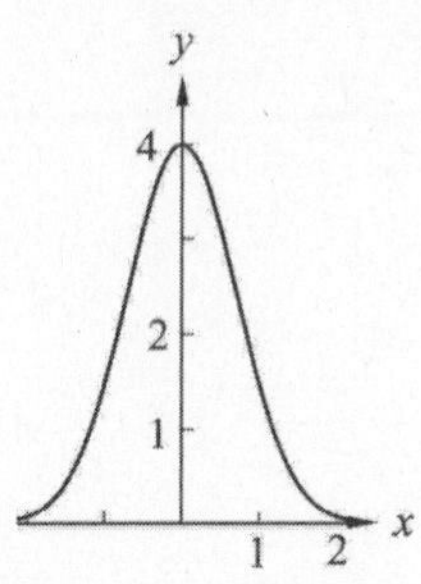

13. $y=\frac{1}{2}(e^x-e^{-x})$

$\frac{dy}{dx}=\frac{1}{2}(e^x+e^{-x})$

$\frac{d^2y}{dx^2}=\frac{1}{2}(e^x-e^{-x})$

(1) Intercepts: $x=0, y=0, (0,0)$

(2) Symmetry with respect to the origin.

(3) $x\to+\infty, y\to+\infty, x\to-\infty, y\to-\infty$

(4) No vertical asymptote.

(5) Domain: all x; range: all y.

(6) $\frac{dy}{dx}=\frac{e^x+e^{-x}}{2}>0$ for all x, so the function is always increasing and there are no maxima or minima.

$\frac{d^2y}{dx^2}=\frac{e^x-e^{-x}}{2}$

$=0$

$e^x=e^{-x}$

$x=-x$

$x=0,$ therefore inflection point at (0,0)

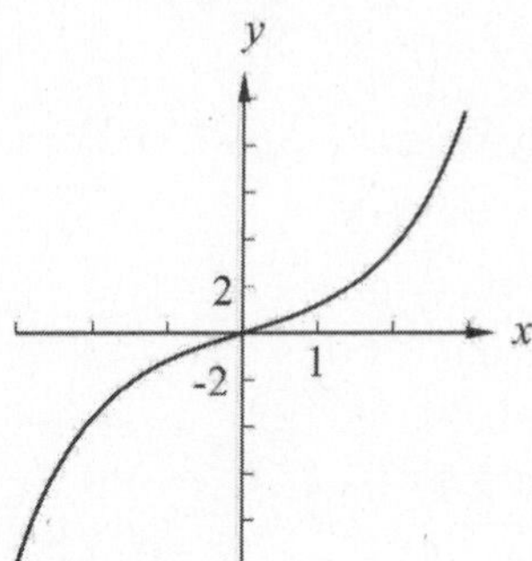

17. $y=x^2\ln x$

$\frac{dy}{dx}=x^2\left(\frac{1}{x}\right)+(\ln x)(2x)$

$=x+2x\ln x;$

$\left.\frac{dy}{dx}\right|_{x=1}=1+2\ln 1$

$=1+2(0)=1$

Slope is $1, x=1, y=0;$ using slope intercept form of the equation and substituting gives $0=1(1)+b$ or $b=1.$ The equation is $y=(1)x-1$ or $y=x-1.$

21. $f(x)=x^2-3+\ln 4x$

$$f'(x)=2x+\frac{1}{x}$$

$$f(1)=-0.6137$$

$$f(2)=3.079$$

Choose $x_1=1.5$

n	x_n	$f(x_n)$	$f'(x_n)$	$x_n-\frac{f(x_n)}{f'(x_n)}$
1	1.5	1.041 759 4	3.666 666 7	1.215 883 8
2	1.215 883 8	0.060 138 9	3.254 214 6	1.197 403 4
3	1.197 403 5	0.000 224 8	3.229 947 3	1.197 333 8
4	1.197 333 8	3.156×10^{-9}	3.229 856 6	1.197 333 8

The root is 1.197 333 8.

25. $P=100e^{-0.005t}; t=100$ days;

$$\frac{dP}{dt}=100e^{-0.005t}(-0.005)$$

$$=-0.05e^{-0.005t}$$

$$\left.\frac{dP}{dt}\right|_{t=100}=-0.303 \text{ W/day}$$

29. $\ln p=\frac{a}{T}+b\ln T+c$

$$\frac{1}{p}\frac{dp}{dT}=-\frac{a}{T^2}+\frac{b}{T}$$

$$\frac{dp}{dT}=\left(-\frac{a}{T^2}+\frac{b}{T}\right)p$$

33. $y=\ln\sec x; -1.5\le x\le 1.5;$

$$\frac{dy}{dx}=\tfrac{1}{\sec x}\cdot\sec x\tan x$$

$=\tan x=0$ at $x=0$;

$x=0$ is a critical value

Multiples of 2π are also critical values, but they are not part of the path of the roller mechanism.

$\frac{d^2y}{dx^2}=\sec^2 x$

$\sec^2(0)=1$ so the curve is concave up and there is a minimum point at $x=0, y=\ln\sec 0=\ln 1=0$ recurring at multiples of

$x=2\pi; (2\pi,0), (4\pi,0), \ldots$

$(0,0)$ is an intercept.

Asymptotes occur where $\frac{dy}{dx}=\tan x$ is undefined. These values are odd multiples of $\frac{\pi}{2}; -\frac{\pi}{2}, \frac{\pi}{2}, \frac{3\pi}{2}\cdots$ The graph is periodic, but the path of the roller mechanism corresponds only to the part of the first cycle, with $-1.5\le x\le 1.5$,

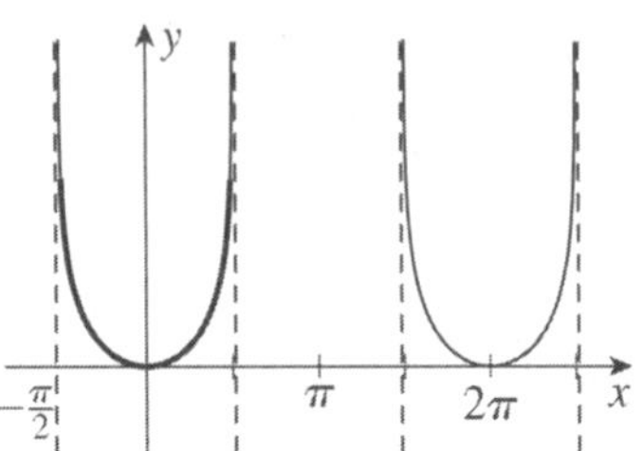

37. $y=6.0e^{-0.020x}\sin(0.20x), 0\le x\le 60$

$$\frac{dy}{dx}=e^{-0.020x}\left[\frac{6\cos(0.2x)}{5}-\frac{3\sin(0.2x)}{25}\right]=0$$

$\tan 0.2x=10$

$$0.2x=\tan^{-1}10+k\pi$$

$$x=5\tan^{-1}10+5k\pi$$

$k=0 \quad x_1=7.355638372, y_1=5.153476505$

$k=1 \quad x_2=23.06360164, y_2=-3.764113107$

$k=2 \quad x_3=38.77156491, y_3=2.749318343$

$k=3 \quad x_4=54.47952818, y_4=-2.008109516$

$(x_1,y_1)=117.6°\text{W}50.2°\text{N}$, maximum

$(x_2,y_2)=101.9°\text{W}41.2°\text{N}$, minimum

$(x_3,y_3)=86.2°\text{W } 47.7°\text{N}$, maximum

$(x_4,y_4)=70.5°\text{W}43.0°\text{N}$, maximum

41. $s=kx^2\ln\frac{1}{x}=k[x^2(\ln 1-\ln x)]=-kx^2\ln x$

$$\frac{ds}{dx}=-k\left(x^2\frac{1}{2}+\ln x2x\right)=-k(x+2x\ln x)$$

$$\frac{ds}{dx}=-kx(1+2\ln x)=0$$

For max., min.:

$$x=0; \ln x=-\frac{1}{2}; x=e^{-1/2}=\frac{1}{\sqrt{e}}=0.607$$

Review Exercises

1. $y=3\cos(4x-1);$

$$\frac{dy}{dx}=\left[-3\sin(4x-1)\right][4]$$

$$=-12\sin(4x-1)$$

5. $y=\csc^2(3x+2);$

$$\frac{dy}{dx}=2\csc(3x+2)[-\csc(3x+2)\cot(3x+2)](3)$$
$$=-6\csc^2(3x+2)\cot(3x+2)$$

9. $y=\left(e^{x-3}\right)^2$

$$\frac{dy}{dx}=2\left(e^{x-3}\right)\left(e^{x-3}\right)(1)$$
$$=2e^{2(x-3)}$$

13. $y=10\tan^{-1}\left(\frac{x}{5}\right)$

$$\frac{dy}{dx}=10\left[\frac{1}{1+\left(\frac{x}{5}\right)^2}\right]\frac{1}{5}$$
$$=\frac{2}{1+\left(\frac{x}{5}\right)^2}$$
$$=\frac{50}{25+x^2}$$

17. $y=\sqrt{\csc 4x+\cot 4x}=(\csc 4x+\cot 4x)^{1/2}$

$$\frac{dy}{dx}=\frac{1}{2}(\csc 4x+\cot 4x)^{-1/2}$$
$$\times\left(-4\csc 4x\cot 4x-4\csc^2 4x\right)$$
$$=\frac{1}{2}(\csc 4x+\cot 4x)^{-1/2}(-4\csc 4x)$$
$$\times(\csc 4x+\cot 4x)$$
$$=-2\csc 4x(\csc 4x+\cot 4x)^{1/2}$$
$$=(-2\csc 4x)\sqrt{\csc 4x+\cot 4x}$$

21. $y=\dfrac{\cos^2 x}{e^{3x}+\pi^2}$

$$\frac{dy}{dx}=\frac{\left(e^{3x}+\pi^2\right)[2\cos x(-\sin x)]-\left(\cos^2 x\right)\left(e^{3x}\right)(3)}{\left(e^{3x}+\pi^2\right)^2}$$
$$=\frac{\left(e^{3x}+\pi^2\right)[-2\sin x\cos x]-3e^{3x}\cos^2 x}{\left(e^{3x}+\pi^2\right)^2}$$
$$=\frac{-\cos x\left[\left(e^{3x}+\pi^2\right)(2\sin x)+3e^{3x}\cos x\right]}{\left(e^{3x}+\pi^2\right)^2}$$
$$=\frac{-\cos x\left[2e^{3x}\sin x+2\pi^2\sin x+3e^{3x}\cos x\right]}{\left(e^{3x}+\pi^2\right)^2}$$

25. $y=\ln\left(\csc x^2\right)$

$$\frac{dy}{dx}=\frac{1}{\csc x^2}\left(-\csc x^2\cot x^2\right)(2x)$$
$$=-2x\cot x^2$$

29. $L=0.1e^{-2t}\sec(\pi t)$

$$\frac{dL}{dt}=0.1e^{-2t}\sec(\pi t)\tan(\pi t)\cdot\pi+\sec(\pi t)\cdot 0.1(-2)e^{-2t}$$
$$\frac{dL}{dt}=0.1\pi e^{-2t}\sec(\pi t)\tan(\pi t)-0.2e^{-2t}\sec(\pi t)$$
$$\frac{dL}{dt}=e^{-2t}\sec(\pi t)\cdot[0.1\pi\tan(\pi t)-0.2]$$

33. $\tan^{-1}\frac{y}{x}=x^2e^y$

$$u=\frac{y}{x}=yx^{-1};\ \frac{du}{dx}=-yx^{-2}+x^{-1}\frac{dy}{dx}$$
$$\frac{1}{1+\left(yx^{-1}\right)^2}\left(-yx^{-2}+x^{-1}\frac{dy}{dx}\right)=x^2e^y\frac{dy}{dx}+2xe^y$$
$$\frac{\frac{-y}{x^2}+\frac{1}{x}\frac{dy}{dx}}{1+y^2x^{-2}}=x^2e^y\frac{dy}{dx}+2xe^y$$
$$\frac{-y}{x^2}+\frac{1}{x}\frac{dy}{dx}=\left(x^2e^y\frac{dy}{dx}+2xe^y\right)\left(1+y^2x^{-2}\right)$$
$$\frac{-y}{x^2}+\frac{1}{x}\frac{dy}{dx}=x^2e^y\frac{dy}{dx}+2xe^y+y^2e^y\frac{dy}{dx}$$
$$+2x^{-1}y^2e^y$$
$$\frac{1}{x}\frac{dy}{dx}-x^2e^y\frac{dy}{dx}-y^2e^y\frac{dy}{dx}=2xe^y+2x^{-1}y^2e^y+\frac{y}{x^2}$$
$$\frac{dy}{dx}\left(\frac{1}{x}-x^2e^y-y^2e^y\right)=2xe^y+\frac{2y^2ey}{x}+\frac{y}{x^2}$$
$$\frac{dy}{dx}\left(\frac{1-x^3e^y-xy^2e^y}{x}\right)=\frac{2x^3e^y+2xy^2e^y+y}{x^2}$$
$$\frac{dy}{dx}=\frac{2x^3e^y+2xy^2e^y+y}{x-x^4e^y-x^2y^2e^y}$$

37. $\ln xy+ye^{-x}=1$

Using implicit differentiation,

$$\frac{d\ln xy}{dx}+\frac{d\ ye^{-x}}{dx}=\frac{d(1)}{dx}$$
$$\frac{1}{xy}\left(x\frac{dy}{dx}+y\right)+ye^{-x}(-1)+e^{-x}\frac{dy}{dx}=0$$
$$\frac{1}{y}\frac{dy}{dx}+\frac{1}{x}-ye^{-x}+e^{-x}\frac{dy}{dx}=0$$

$$\frac{1}{y}\frac{dy}{dx}+e^{-x}\frac{dy}{dx}=ye^{-x}-\frac{1}{x}$$

$$\frac{dy}{dx}\left(\frac{1}{y}+e^{-x}\right)=ye^{-x}-\frac{1}{x}$$

$$\frac{dy}{dx}\left(\frac{1+ye^{-x}}{y}\right)=\frac{x\,ye^{-x}-1}{x}$$

$$\frac{dy}{dx}=\left(\frac{x\,ye^{-x}-1}{x}\right)\left(\frac{y}{1+y^{e-x}}\right)$$

$$=\frac{y\left(xye^{-x}-1\right)}{x\left(1+ye^{-x}\right)}$$

41. $y=x-\cos 0.5x$

$y=x-\cos 0.5x\big|_{x=0}=-1\Rightarrow y\text{-int: }(0,\,-1)$

$0=x-\cos 0.5x\Rightarrow x=0.9\Rightarrow x\text{-int: }(0,\,0.9)$

$\frac{dy}{dx}=1+0.5\sin 0.5x$

$=0$

$\sin x=-2$, no solution $\Rightarrow$ no critical points

$\frac{dy}{dx}>0$ for all x, so the function is always increasing

$\frac{d^2y}{dx^2}=0.25\cos 0.5x$

$=0$ for $x=\pi+k(2\pi)$

$\frac{d^2y}{dx^2}=$ changes sign at $x=\pi+k(2\pi)$

$\Rightarrow x=\pi+k(2\pi)$ are inflection points.

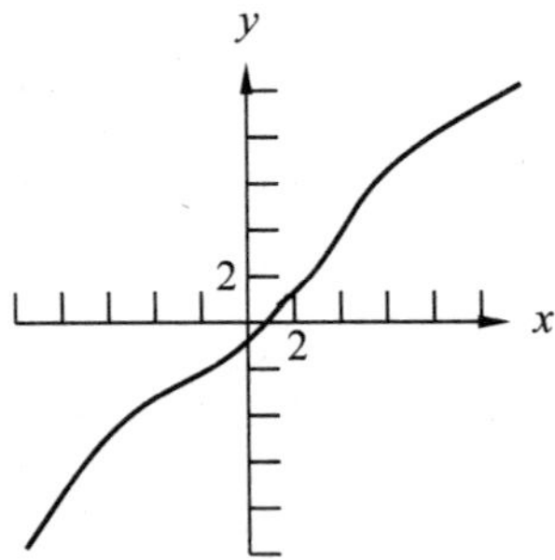

45. $y=4\cos^2\left(x^2\right)$

$\frac{dy}{dx}=2\left[4\cos\left(x^2\right)\right]\left[-\sin\left(x^2\right)\right](2x)$

$=-16x\cos x^2\sin x^2$

$\left.\frac{dy}{dx}\right|_{x=1}=-16\cos\left(1^2\right)\sin\left(1^2\right)$

$=-16(0.5403)(0.8415)$

$=-7.27$

$f(1)=4\cos^2\left(1^2\right)=4(0.5403)^2=1.168$

$y=-7.27x+b$

$1.168=-7.27(1)+b$

$b=8.44$

$y=-7.27x+8.44$

$7.27x+y-8.44=0$

49. $\lim\limits_{x\to 0}\frac{\sin 2x}{\sin 3x}=\lim\limits_{x\to 0}\frac{2\cos 2x}{3\cos 3x}$

$=\frac{2}{3}$

53. $\lim\limits_{x\to\infty}\frac{\ln x}{\sqrt[3]{x}}=\lim\limits_{x\to\infty}\frac{\frac{1}{x}}{\frac{1}{3}x^{-\frac{2}{3}}}$

$=3\lim\limits_{x\to\infty}\frac{1}{\sqrt[3]{x}}$

$=0$

57. $y=\sin 3x$

$\frac{dy}{dx}=3\cos 3x$

$\frac{d^2y}{dx^2}=-9\sin 3x$

$=-9y$

61. $y=e^x-2e^{-x}$

$\frac{dy}{dx}=e^x+2e^{-x}$

$\frac{d^2y}{dx^2}=e^x-2e^{-x}>0$

$e^{2x}>2$

$2x>\ln 2$

$x>\frac{1}{2}\ln 2=0.347$

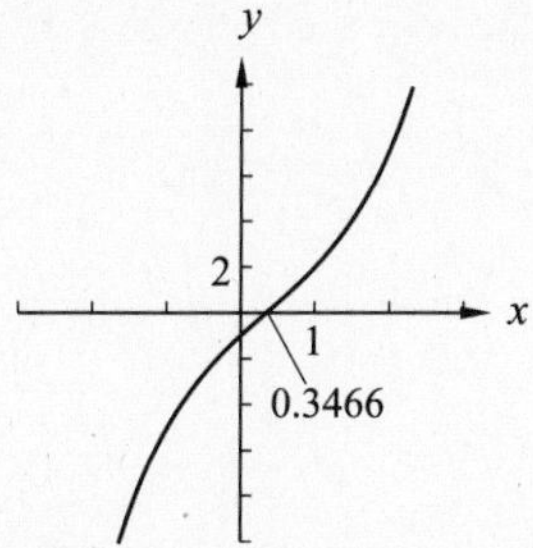

$y = e^x - 2e^{-x}$ is concave up for $x > \frac{1}{2}\ln 2$

65. $F = \dfrac{200\mu}{\mu\sin\theta + \cos\theta}$

$$\frac{dF}{d\theta} = \left.\frac{-200\mu(\mu\cos\theta - \sin\theta)}{(\mu\sin\theta + \cos\theta)^2}\right|_{\substack{\mu=0.20\\ \theta=15^\circ}}$$

$$\frac{dF}{d\theta} = 2.5\text{ N}$$

69. $T = 17.2 + 5.2\cos\left[\dfrac{\pi}{6}(x - 0.50)\right]$

$$\frac{dT}{dt} = \frac{dT}{dx}\frac{dx}{dt} = -5.2\sin\left[\frac{\pi}{6}(x-0.50)\right]\left(\frac{\pi}{6}\right)\frac{dx}{dt}$$

$$\left.\frac{dT}{dt}\right|_{x=2} = -5.2\sin\left[\frac{\pi}{6}(2-0.50)\right]\left(\frac{\pi}{6}\right)(0.033)$$

$$\frac{dT}{dt} = -0.064^\circ\text{C/day}$$

73. $n = xN\log_x N = 8x\log_x 8$ for $N = 8$

$n = \dfrac{8x\ln 8}{\ln x}$, $1 < x \le 10 \Rightarrow x = 1$ is a vertical asymptote and no y-intercept.

$n = \dfrac{8x\ln 8}{\ln x} = 0$ has no solution for $1 < x \le 10 \Rightarrow$ no x-intercept.

$$\frac{dn}{dx} = 8\ln 8\frac{\ln x - x\left(\frac{1}{x}\right)}{(\ln x)^2}$$

$$= 0$$

$$\ln x = 1$$

$$x = e$$

$(e, 8e\ln 8)$ is a minimum since $\dfrac{d^2n}{dx^2} > 0$

$$\frac{d^2n}{dx^2} = 8\ln 8\frac{(\ln x)^2\left(\frac{1}{x}\right) - (\ln x - 1)(2\ln x)\left(\frac{1}{x}\right)}{(\ln x)^2} = 0$$

$$\ln x(2 - \ln x) = 0$$

$\ln x = 0$ or $\ln x = 2$

$x = 1$, reject $\quad x = e^2$

$\left(e^2, \dfrac{8e^2\ln 8}{2}\right)$ is an inflection point since $\dfrac{d^2n}{dx^2}$ changes sign.

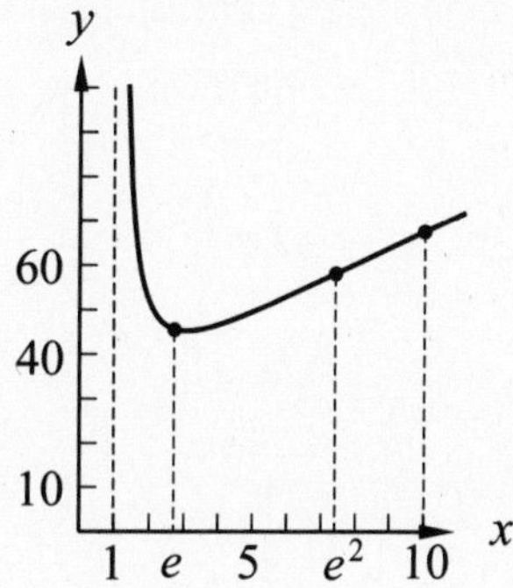

77. $\theta = \sin^{-1}\left(\dfrac{Ff}{R}\right)$

$$\frac{d\theta}{dF} = \frac{1}{\sqrt{1 - \frac{F^2f^2}{R^2}}}\left(\frac{f}{R}\right)$$

$$= \frac{1}{\sqrt{\frac{R^2 - F^2f^2}{R^2}}}\left(\frac{f}{R}\right)$$

$$\frac{d\theta}{dF} = \frac{f}{\sqrt{R^2 - F^2f^2}}$$

81. $T = 30 + 60(0.5)^{0.200t}$

$$T|_{t=5.00} = 60$$

$$\frac{dT}{dt} = 60(0.5)^{0.200t}\ln 0.5(0.200)$$

$$\left.\frac{dT}{dt}\right|_{t=5.00} = -4.16$$

$$L(t) = -4.16(t-5) + 60$$

$$= -4.16t + 80.8$$

85. $y = 3.00e^{-0.500x^2}$

$$A = 2xy = 6.00xe^{-0.500x^2}$$

$$\frac{dA}{dx} = 6.00e^{-0.500x^2} - 6.00xe^{-0.500x^2}$$

$$= 6.00(1-x)e^{-0.500x^2}$$

$$= 0 \text{ for } x = 1.00$$

$$y(1.00) = 1.82$$

$$\frac{d^2A}{dx^2} = 6.00\left[(1-x)e^{-0.500x^2}(-x) - e^{-0.500x^2}\right]$$

$$= 6.00(-x + x^2 - 1)e^{-0.500x^2}$$

$\left.\frac{d^2A}{dx^2}\right|_{x=1} < 0$, so the area is a maximum, and 2.00 in. wide, 1.82 in. high are the dimensions of the rectangular passage with the largest area.

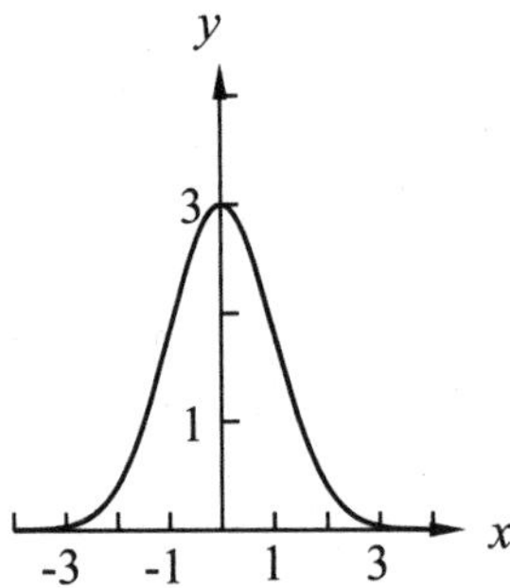

89. $S = 640 + 81\pi\left(\csc\theta - \frac{2}{3}\cot\theta\right),\quad 0 < \theta < 90^\circ$

$$\frac{dS}{d\theta} = 81\pi\left(-\csc\theta\cot\theta + \frac{2}{3}\csc^2\theta\right) = 0$$

$$\csc\theta\left(\frac{2}{3}\csc\theta - \cot\theta\right) = 0$$

$\csc\theta = 0$, no solution or $\frac{2}{3}\csc\theta - \cot\theta = 0$

$$\frac{2}{3\sin\theta} = \frac{\cos\theta}{\sin\theta}$$

$$\cos\theta = \frac{2}{3}$$

$$\theta = 48.2^\circ$$

93. $I = \frac{k\cos\theta}{r^2},\ \cos\theta = \frac{h}{r}$

$$I = \frac{kh}{r^3}$$

To write r in terms of h, we use the Pythagorean theorem:

$$r^2 = h^2 + 10.0^2$$

$$r^3 = \left(h^2 + 100\right)^{3/2}$$

$$I = \frac{kh}{\left(h^2 + 100\right)^{3/2}}$$

$$\frac{dI}{dh} = \frac{2k\left(50 - h^2\right)}{\left(h^2 + 100\right)^{5/2}}$$

$$= 0 \text{ for } h^2 = 50,\ h = 7.07$$

$$\frac{dI}{dh} > 0 \text{ for } h < 7.07$$

$$\frac{dI}{dh} < 0 \text{ for } h > 7.07$$

The illuminance at the circumference will be a max for a light placed 7.07 cm above the centre of the circle.

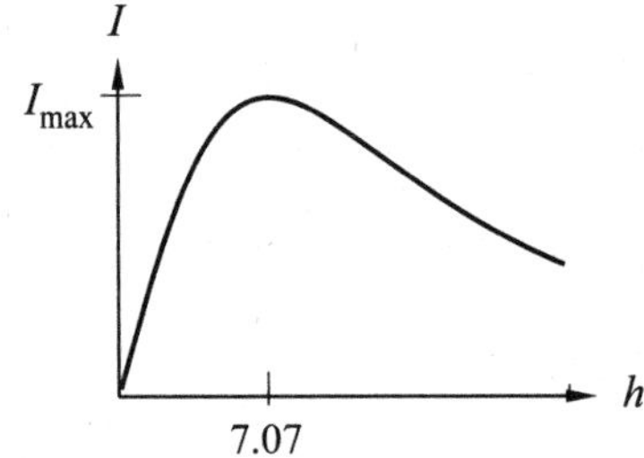

97. $y = \frac{H}{w}\cosh\frac{wx}{H}$

$$\frac{dy}{dx} = \sinh\frac{wx}{H}$$

$$\frac{d^2y}{dx^2} = \frac{w}{H}\cosh\frac{wx}{H}$$

$$\frac{w}{H}\sqrt{1 + \left(\frac{dy}{dx}\right)^2} = \frac{w}{H}\sqrt{1 + \left(\sinh\frac{wx}{H}\right)^2}$$

$$= \frac{w}{H}\sqrt{\left(\cosh\frac{wx}{H}\right)^2}$$

$$= \frac{w}{H}\cosh\frac{wx}{H}$$

$$= \frac{d^2y}{dx^2}$$

CHAPTER 28

METHODS OF INTEGRATION

28.1 The General Power Formula

1. $u = \cos x, du = -\sin x dx$

In order to be able to integrate, $\cos x$ must be changed to $-\sin x$:

$$\int \cos^3 x(-\sin x\, dx) = \frac{1}{4}\cos^4 x + C$$

5. $u = \cos\theta;\ n = \frac{1}{2};\ du = -\sin\theta\, d\theta$

$$0.4\int\sqrt{\cos\theta}\sin\theta\, d\theta = -0.4\int(\cos\theta)^{1/2}(-\sin\theta\, d\theta)$$

$$= -\frac{0.8}{3}(\cos\theta)^{3/2} + C$$

9. $u = \cos 2x;\ n = 1;\ du = -2\sin 2x\, dx$

$$\int_0^{\pi/8} \cos 2x \sin 2x\, dx$$

$$= -\frac{1}{2}\int_0^{\pi/8}(\cos 2x)^1(-2\sin 2x\, dx)$$

$$= -\frac{1}{2}\frac{(\cos 2x)^2}{2}\bigg|_0^{\pi/8}$$

$$= -\frac{1}{4}(\cos^2 2x)\Big|_0^{\pi/8}$$

$$= -\frac{1}{4}\left(\cos^2\frac{\pi}{4} - \cos^2 0\right)$$

$$= -\frac{1}{4}\left(\frac{1}{2} - 1\right) = \frac{1}{8}$$

13. $u = \tan^{-1} 5x;\ du = \frac{1}{1+25x^2}(5dx);\ n = 1$

$$\int\frac{5\tan^{-1}5x}{1+25x^2}dx = \int(\tan^{-1}5x)^1\frac{5dx}{1+25x^2}$$

$$= \frac{1}{2}(\tan^{-1}5x)^2 + C$$

17. $u = \ln(2x+3);\ du = \frac{1}{2x+3}(2dx);\ n = 1$

$$\int_0^{1/2}\frac{\ln(2x+3)}{2x+3}dx = \frac{1}{2}\int_0^{1/2}[\ln(2x+3)]\frac{2dx}{2x+3}$$

$$= \frac{1}{2}\frac{[\ln(2x+3)]^2}{2}\bigg|_0^{1/2}$$

$$= \frac{1}{4}\ln^2(2x+3)\Big|_0^{1/2}$$

$$= \frac{1}{4}(\ln^2 4 - \ln^2 3)$$

$$= 0.179$$

21. $u = 1 - e^{2t},\ du = -2e^{2t}dt$

$$\int\frac{e^{2t}}{(1-e^{2t})^3}dt = -\frac{1}{2}\int(1-e^{2t})^{-3}(-2e^{2t}dt)$$

$$= -\frac{1}{2}\int u^{-3}du$$

$$= -\frac{1}{2}\cdot\frac{u^{-3+1}}{-3+1} + C$$

$$= -\frac{1}{2}\cdot\frac{1}{-2(u^2)} + C$$

$$= \frac{1}{4(1-e^{2t})^2} + C$$

25. $u = 1 + \cot x, du = -\csc^2 x dx$

$$\int_{\pi/6}^{\pi/4}(1+\cot x)^2\csc^2 x\, dx)$$

$$= -\int_{\pi/6}^{\pi/4}(1+\cot x)^2(-\csc^2 x\, dx)$$

$$= -\frac{(1+\cot x)^3}{3}\bigg|_{\pi/6}^{\pi/4}$$

$$= -\frac{(1+\cot\frac{\pi}{4})^3}{3} + \frac{(1+\cot\frac{\pi}{6})^3}{3}$$

$$= 2\sqrt{3} + \frac{2}{3}$$

$$= 4.13$$

29. $\int\frac{dx}{x\ln^2 x} = \int\ln^{-2}x\left(\frac{1}{x}dx\right) = \int u^{-2}du$

with $u = \ln x,\ du = \frac{1}{x}dx,\ n = -2$

33. $u = 1 + \tan^{-1} 2x; du = \dfrac{2dx}{1+(2x)^2}$

$$A = \int_0^2 \frac{1+\tan^{-1} 2x}{1+4x^2} dx$$
$$= \frac{1}{2}\int_0^2 \left(1+\tan^{-1} 2x\right)\left(\frac{2dx}{1+4x^2}\right)$$
$$= \frac{1}{4}\left(1+\tan^{-1} 2x\right)^2\Big|_0^2$$
$$= \frac{1}{2}[2.326^2 - 1] = 1.10$$

37. $P = mnv^2 \int_0^{\pi/2} \sin\theta\cos^2\theta\, d\theta;\ n = 2;$

$u = \cos\theta;\ du = -\sin\theta\, d\theta$

$$P = -mnv^2 \int_0^{\pi/2} \cos^2\theta(-\sin\theta\, d\theta)$$
$$= -mnv^2\left[\frac{\cos^3\theta}{3}\right]_0^{\pi/2}$$
$$= -mnv^2\left[\frac{1}{3}\left(\cos^3\frac{\pi}{2} - \cos^3 0\right)\right]$$
$$= -mnv^2\left[\frac{1}{3}(0-1)\right]$$
$$= -mnv^2\left(-\frac{1}{3}\right) = \frac{1}{3}mnv^2$$

28.2 The Basic Logarithmic Form

1. Add $2x$ to the integrand, then $u = x^2+1, du = 2xdx,$

$$\int \frac{2x\,dx}{x^2+1} = \ln\left|x^2+1\right| + C$$

5. $\displaystyle\int \frac{2x\,dx}{4-3x^2};\ u = 4-3x^2; du = -6x\,dx$

$$-\frac{1}{3}\int \frac{-6x\,dx}{4-3x^2} = -\frac{1}{3}\ln\left|4-3x^2\right| + C$$

9. $\dfrac{0.4}{2}\displaystyle\int \frac{\csc^2 2\theta\, d\theta}{\cot 2\theta};u = \cot 2\theta;\ du = -2\csc^2 2\theta\, d\theta$

$$\frac{0.4}{2}\int \frac{2\csc^2 2\theta\, d\theta}{\cot 2\theta} = -0.2\ln\left|\cot 2\theta\right| + C$$

13. $u = 1 - e^{-x}; du = e^{-x}dx$

$$\int \frac{e^{-x}dx}{1-e^{-x}} = \ln\left|1-e^{-x}\right| + C$$

17. $u = 1 + 4\sec x; du = 4\sec x\tan x\,dx$

$$\int \frac{\sec x\tan x\,dx}{1+4\sec x} = \frac{1}{4}\int \frac{4\sec x\tan x\,dx}{1+4\sec x}$$
$$= \frac{1}{4}\ln\left|1+4\sec x\right| + C$$

21. $u = \ln r; du = \dfrac{dr}{r}$

$$0.5\int \frac{dx}{r\ln r} = 0.5\int \frac{1}{\ln r}\cdot\frac{dr}{r}$$
$$= 0.5\ln\left|\ln r\right| + C$$

25. $n = -\dfrac{1}{2};\ u = 1-2x;\ du = -2dx$

$$\int \frac{6dx}{\sqrt{1-2x}} = -3\int (1-2x)^{-1/2}(2dx)$$
$$= -3(1-2x)^{1/2}(2) + C$$
$$= -6\sqrt{1-2x} + C$$

29. $u = 4 + \tan 3x; du = 3\sec^2 3x\,dx$

$$\int_0^{\pi/12} \frac{\sec^2 3x}{4+\tan 3x}dx = \frac{1}{3}\int_0^{\pi/12} \frac{3\sec^2 3x\,dx}{(4+\tan 3x)}$$
$$= \frac{1}{3}\ln\left|4+\tan 3x\right|_0^{\pi/12}$$
$$= \frac{1}{3}(\ln 5 - \ln 4)$$
$$= \frac{1}{3}\ln\frac{5}{4}$$
$$= 0.0744$$

33.

$$\begin{array}{r} 1 \\ x+4\overline{)x-4} \\ -x-4 \\ -8 \end{array}$$

$$\int \frac{x-4}{x+4}dx = \int dx - \int \frac{8}{x+4}dx$$
$$= x - 8\ln\left|x+4\right| + C$$

37. $m=\frac{dx}{dx}=\frac{\sin x}{3+\cos x}; y=\int\frac{1}{3+\cos x}\times\sin x\,dx$

$y=-\int\frac{1}{3+\cos x}(-\sin x)\,dx;$

$u=3+\cos x; du=-\sin x\,dx$

$y=-\int\frac{1}{u}du$

$=-\ln|u|+C$

$=-\ln(3+\cos x)+C$

Substitute $x=\frac{\pi}{3}, y=2$:

$2=-\ln\left(3+\cos\frac{\pi}{3}\right)+C$

$2=-\ln(3+0.5)+C$

$C=2+\ln 3.5$

$y=-\ln(3+\cos x)+\ln 3.5+2$; substituting for C

$y=\ln\frac{3.5}{3+\cos x}+2$

41. $a(t)=(t+4)^{-1}$

$v(t)=\ln|t+4|+c$

$v(0)=0$

$=\ln|0+4|+c$

$c=-\ln 4$

$v(t)=\ln|t+4|-\ln 4$

$v(4)=\ln|4+4|-\ln 4$

$=\ln 8-\ln 4$

$=\ln\frac{8}{4}$

$v(4)=\ln 2$ m/s

45. $t=L\int\frac{di}{E-iR};$

$u=E-iR; du=-Rdi$

$t=-\frac{L}{R}\int\frac{-Rdi}{E-iR}$

$=\frac{-L}{R}\ln|E-iR|+C$

Substitute $t=0, i=0$:

$0=-\frac{L}{R}\ln E+C$

$C=\frac{L}{R}\ln|E|$

$t=\frac{L}{R}(-\ln|E-iR|+\ln|E|)$

$t=\frac{L}{R}\ln\left|\frac{E}{E-iR}\right|$

$\frac{R}{L}t=\ln\left|\frac{E}{E-iR}\right|$

$e^{Rt/L}=\frac{E}{E-iR}$

$i=\frac{E}{R}-\frac{E}{R}e^{-Rt/L}$

$i=\frac{E}{R}\left(1-e^{-Rt/L}\right)$

28.3 The Exponential Form

1. $u=x^3, du=3x^2dx$

Multiply the integrand by $3x$.

$\int 3x^2e^{x^3}dx=\int e^{x^3}(3x^2dx)$

$=e^{x^3}+C$

5. $u=2x+5; du=2dx$

$\int e^{2x+5}dx=\frac{1}{2}\int e^{2x+5}(2dx)$

$=\frac{1}{2}e^{2x+5}+C$

9. $y=x^3; du=3x^2dx$

$\int 6x^2e^{x^3}dx=6\int e^{x^3}(x^2dx)$

$=\frac{6}{3}\int e^{x^3}(3x^2dx)$

$=2e^{x^3}+C$

13. $u=2\sec\theta; du=2\sec\theta\tan\theta\,d\theta$

$\int 4(\sec\theta\tan\theta)\,e^{2\sec\theta}d\theta=\frac{4}{2}\int e^{2\sec\theta}2\sec\theta\tan\theta\,d\theta$

$=2e^{2\sec\theta}+C$

17. $u=2x; du=2dx$

$\int_1^3 3e^{2x}(e^{-2x}-1)=3\int(e^0-e^{2x})dx$

$=3\int dx-\frac{3}{2}\int e^{2x}(2dx)$

$=3x-\frac{3}{2}e^{2x}\Big|_1^3$

$=\left[9-\frac{3}{2}e^6-\left(3-\frac{3}{2}e^2\right)\right]$

$=6-\frac{3}{2}(e^6-e^2)=-588$

21. $u = \tan^{-1} 2x;\ du = \dfrac{2}{1+4x^2}dx$

$$\int \frac{e^{\tan^{-1} 2x}dx}{4x^2+1} = \frac{1}{2}\int e^{\tan^{-1} 2x}\frac{2dx}{4x^2+1}$$
$$= \frac{1}{2}e^{\tan^{-1} 2x} + C$$

25. $u = \cos^2 x;$

$$du = 2\cos x(-\sin x)dx$$
$$= -2\sin x \cos x\ dx$$
$$= -2\sin 2x\ dx$$

$$\int_0^{\pi} (\sin 2x)e^{\cos^2 x}dx = -\frac{1}{2}\int_0^{\pi} e^{\cos^2 x}(-2\sin 2x\ dx)$$
$$= -\frac{1}{2}e^{\cos^2 x}\Big|_0^{\pi}$$
$$= -\frac{1}{2}[e^{(\cos\pi)^2} - e^{(\cos 0)^2}]$$
$$= -\frac{1}{2}[e - e] = 0$$

29. $A = \int_0^2 3e^x dx$

$$= 3e^x\Big|_0^2$$
$$= 3e^2 - 3e^0 = 19.2$$

33. $1+e^x\overline{)1}$ with quotient 1; $\underline{1+e^x}$; remainder $-e^x$ $\Rightarrow \dfrac{1}{1+e^x} = 1 - \dfrac{e^x}{1+e^x}$

$u = 1+e^x,\ du = e^x dx$

$$\int \frac{dx}{1+e^x} = \int dx - \int \frac{e^x}{1+e^x}dx$$
$$= x - \ln\left|1+e^x\right| + C \text{ and since } 1+e^x > 0$$
$$= x - \ln\left(1+e^x\right) + C$$

37.

$$u = \frac{x}{2}, du = \frac{dx}{2}$$
$$y_{av} = \frac{\int_a^b y dx}{b-a}$$
$$= \frac{\int_0^4 4e^{x/2}dx}{4-0}$$
$$= 2\int_0^4 e^{x/2}\left(\tfrac{1}{2}dx\right)$$
$$= 2e^{x/2}\Big|_0^4$$
$$= 2\left(e^2 - 1\right) = 12.8$$

41. $qe^{t/RC} = \dfrac{E}{R}\int e^{t/RC}dt$

$$u = \frac{t}{RC};\ du = \frac{1}{RC}dt$$
$$qe^{t/RC} = RC\cdot\frac{E}{R}\int e^{t/RC}\left(\frac{1}{RC}\right)dt$$
$$qe^{t/RC} = EC(e^{t/RC}) + C_1$$

Substitute $q = 0, t = 0:$

$$0 = EC + C_1$$
$$C_1 = -EC$$
$$qe^{t/RC} = EC(e^{t/RC}) - EC$$
$$q = EC - \frac{EC}{e^{t/RC}}$$
$$q = EC(1 - e^{-t/RC})$$

45. $p = \int_0^{100} (0.05)x^{-0.5}e^{-0.1x^{0.5}}dx$

$$u = -0.1x^{0.5}, du = -0.05x^{-0.5}dx$$
$$p = -\int_0^{100} e^{-0.1x^{0.5}}\left(-0.05x^{-0.5}dx\right)$$
$$= -e^{-0.1x^{0.5}}\Big|_0^{100}$$
$$= 0.632$$

28.4 Basic Trigonometric Forms

1. $u = x^3, du = 3x^2dx$

If we change $x\ dx$ to $3x^2dx$, then

$$\int \sec^2 x^3(3x^2dx) = \tan x^3 + C$$

5. $u = 3\theta;\ du = 3\ d\theta$

$$\int 0.3\sec^2 3\theta\ d\theta = \frac{0.3}{3}\int \sec^2 3\theta(3d\theta)$$
$$= 0.1\tan 3\theta + C$$

9. $u = x^3;\ du = 3x^2dx$

$$\int_{0.5}^{1} x^2\cot x^3 dx = \frac{1}{3}\int_{0.5}^{1}\cot x^3(3x^2dx)$$
$$= \frac{1}{3}\ln\left|\sin x^3\right|\Big|_{0.5}^{1}$$
$$= \frac{1}{3}\left(\ln\left|\sin 1\right| - \ln\left|\sin\frac{1}{8}\right|\right)$$
$$= 0.6365$$

13. $u = \frac{1}{x} = x^{-1};\ du = -1x^{-2}dx = -\frac{dx}{x^2}$

$$\int \frac{\sin(\frac{1}{x})}{x^2}dx = -\int \sin\left(\frac{1}{x}\right)\left(-\frac{dx}{x^2}\right)$$
$$= -\left[-\cos\left(\frac{1}{x}\right)\right] + C$$
$$= \cos\left(\frac{1}{x}\right) + C$$

17. $u = \frac{x}{2},\ du = \frac{1}{2}dx$

$$\int \sqrt{\frac{1-\cos x}{2}}dx = \int \sin\frac{x}{2}dx$$
$$= 2\int \sin\frac{x}{2}\left(\frac{1}{2}dx\right)$$
$$= -2\cos\frac{x}{2} + C$$

21. $u = 2T,\ du = 2dT$

$$\int \frac{2\tan T}{1-\tan^2 T}dT = \frac{1}{2}\int \tan 2T\,(2dT)$$
$$= -\frac{1}{2}\ln|\cos 2T| + C$$

25. $\sin 3x\left(\frac{1}{\sin 3x} + \frac{1}{\cos 3x}\right) = 1 + \tan 3x;$

$u = 3x;\ du = 3\ dx$

$$\int_0^{\pi/9} \sin 3x(\csc 3x + \sec 3x)dx$$
$$= \int_0^{\pi/9} (1 + \tan 3x)dx$$
$$= \int_0^{\pi/9} dx + \int_0^{\pi/9} \tan 3x\ dx$$
$$= \int_0^{\pi/9} dx + \frac{1}{3}\int_0^{\pi/9} \tan 3x(3\ dx)$$
$$= \left(x - \frac{1}{3}\ln|\cos 3x|\right)\Big|_0^{\pi/9}$$
$$= \frac{\pi}{9} - \frac{1}{3}\ln\left|\cos\frac{\pi}{3}\right| - \left(0 - \frac{1}{3}\ln|\cos 0|\right)$$
$$= \frac{\pi}{9} - \frac{1}{3}\ln\left(\frac{1}{2}\right) = \frac{\pi}{9} + \frac{1}{3}\ln 2$$
$$= 0.580$$

29. $\int \frac{dx}{1+\sin x} = \int \frac{dx}{1+\sin x}\cdot\frac{1-\sin x}{1-\sin x}$

$$= \int \frac{(1-\sin x)}{1-\sin^2 x}dx$$
$$= \int \frac{1-\sin x}{\cos^2 x}dx$$
$$= \int \frac{1}{\cos^2 x}dx + \int \frac{-\sin x\ dx}{\cos^2 x}$$
$$= \int (\sec^2 x\ dx - \sec x \tan x)dx$$
$$= \tan x - \sec x + C$$

33. $y = \sec x;\ x = 0;\ x = \frac{\pi}{3}$

$y = 0$; rotated about x-axis, discs.

$dV = \pi r^2 dx;\ r = y$

$$V = \int_0^{\pi/3} \pi y^2 dx$$
$$= \pi\int_0^{\pi/3} \sec^2 x\ dx$$
$$= \pi \tan x\Big|_0^{\pi/3}$$
$$= \pi\sqrt{3} = 5.44$$

37. $y = \tan x^2;\ y = 0,\ x = 1$

$$\overline{x} = \frac{\int_0^1 xy\ dx}{0.3984} = \frac{\int_0^1 (\tan x^2)x\ dx}{0.3984}$$

$u = x^2;\ du = 2x\ dx$

$$\overline{x} = \frac{\frac{1}{2}\int_0^1 \tan x^2(2x\ dx)}{0.3984}$$
$$= \frac{-\frac{1}{2}\ln\left|\cos x^2\right|\Big|_0^1}{0.3984}$$
$$= \frac{-\frac{1}{2}(\ln \cos 1 - \cos 0)}{0.3984} = \frac{0.3078}{0.3984}$$
$$= 0.7726\text{ m}$$

28.5 Other Trigonometric Forms

1. $u = \sin 2x, du = 2\cos 2x dx$

Change dx to $\cos 2x\ dx$, then

$$\int \sin^2 2x(\cos 2x\ dx) = \frac{1}{2}\int \sin^2 2x(2\cos 2x\ dx)$$
$$= \frac{1}{2}\cdot\frac{\sin^3 2x}{3} + C$$
$$= \frac{\sin^3 2x}{6} + C$$

5. $u = 2x; du = 2dx; u = \cos 2x; du = -2\sin 2x\, dx$

$$\begin{aligned}\int \sin^3 2x\,dx &= \int \sin^2 2x \sin 2x\,dx \\ &= \int (1-\cos^2 2x)\sin 2x\,dx \\ &= \int \sin 2x\,dx - \int \cos^2 2x \sin 2x\,dx \\ &= \frac{1}{2}\int \sin 2x(2dx) + \frac{1}{2}\int \cos^2 2x(-2\sin 2x)dx \\ &= -\frac{1}{2}\cos 2x + \frac{1}{2}\frac{\cos^3 2x}{3} + C \\ &= -\frac{1}{2}\cos 2x + \frac{1}{6}\cos^3 2x + C\end{aligned}$$

9. $u = \cos x;\ du = -\sin x\,dx$

$$\begin{aligned}\int_0^{\pi/4} 5\sin^5 x\,dx &= 5\int_0^{\pi/4} \sin x \sin^4 x\,dx \\ &= 5\int_0^{\pi/4} \sin x(1-\cos^2 x)^2 dx \\ &= 5\int_0^{\pi/4} \sin x(1-2\cos^2 x+\cos^4 x)dx \\ &= 5\left(\int_0^{\pi/4} \sin x\,dx - 2\int_0^{\pi/4} \cos^2 x \sin x\,dx \right. \\ &\qquad \left. + \int_0^{\pi/4} \cos^4 x \sin x\,dx\right) \\ &= \left[5(-\cos x) + 10\frac{\cos^3 x}{3} - 5\frac{\cos^5 x}{5}\right]_0^{\pi/4} \\ &= 5\left(-\frac{1}{\sqrt{2}}\right) + \frac{10}{3}\left(-\frac{1}{\sqrt{2}}\right)^3 - \left(\frac{1}{\sqrt{2}}\right)^5 \\ &\qquad -\left[5(-1) + \frac{10}{3}(1) - 1\right] \\ &= \frac{64-43\sqrt{2}}{24} = 0.133\end{aligned}$$

13.
$$\begin{aligned}\int 2(1+\cos 3\phi)^2 d\phi &= \int 2(1+2\cos 3\phi + \cos^2 3\phi)d\phi \\ &= \int (2+4\cos 3\phi + 2\cos^2 3\phi)d\phi \\ &= 2\phi + \frac{4\sin 3\phi}{3} + \int (1+\cos 6\phi)d\phi \\ &= 2\phi + \frac{4\sin 3\phi}{3} + \phi + \frac{1}{6}\int \cos 6\phi(6d\phi) \\ &= 3\phi + \frac{4\sin 3\phi}{3} + \frac{\sin 6\phi}{6} + C \\ &= \frac{1}{3}(9\phi + 4\sin 3\phi + \sin 3\phi \cos 3\phi) + C\end{aligned}$$

17. $u = \tan x; du = \sec^2 x\,dx$

$$\begin{aligned}\int_0^{\pi/4} \tan x \sec^4 x\,dx &= \int_0^{\pi/4} \tan x \sec^2 x(1+\tan^2 x)dx \\ &= \int_0^{\pi/4} (\tan x)^1 \sec^2 x\,dx \\ &\qquad + \int_0^{\pi/4} \tan^3 x \sec^2 x\,dx \\ &= \left[\frac{1}{2}(\tan x)^2 + \frac{1}{4}(\tan x)^4\right]_0^{\pi/4} \\ &= \frac{1}{2}(1)^2 + \frac{1}{4}(1)^4 = \frac{3}{4}\end{aligned}$$

21.
$$\begin{aligned}\int 0.5\sin s \sin 2s\,ds &= \int \sin^2 s \cos s\,ds \\ &= \frac{\sin^3 s}{3} + C\end{aligned}$$

25.
$$\begin{aligned}&\int \frac{1-\cot x}{\sin^4 x}dx \\ &= \int (1-\cot x)\csc^4 x\,dx \\ &= \int (1-\cot x)\csc^2 x(1+\cot^2 x)dx \\ &= \int (1-\cot x+\cot^2 x-\cot^3 x)\csc^2 x\,dx \\ &= \int \csc^2 x\,dx - \int \cot x \csc^2 x\,dx + \int \cot^2 x \csc^2 x\,dx \\ &\quad - \int \cot^3 x \csc^2 x\,dx \\ &= \int \csc^2 x\,dx + \int \cot x(-\csc^2 x\,dx) \\ &\quad - \int \cot^2 x(-\csc^2 x\,dx) + \int \cot^3 x(-\csc^2 x\,dx) \\ &= -\cot x + \frac{\cot^2 x}{2} - \frac{\cot^3 x}{3} + \frac{\cot^4 x}{4} + C \\ &= \frac{1}{4}\cot^4 x - \frac{1}{3}\cot^3 x + \frac{1}{2}\cot^2 x - \cot x + C\end{aligned}$$

29.
$$\begin{aligned}\int \sec^6 x\,dx &= \int \sec^4 x(1+\tan^2 x)dx \\ &= \int \sec^4 x\,dx + \int \sec^4 x \tan^2 x\,dx \\ &= \int \sec^2 x(1+\tan^2 x)dx \\ &\qquad + \int \sec^2 x(1+\tan^2 x)\tan^2 x dx \\ &= \int \sec^2 x\,dx + \int \tan^2 x \sec^2 x\,dx \\ &\qquad + \int \tan^2 x \sec^2 x\,dx + \int \tan^4 x \sec^2 x\,dx \\ &= \tan x + \frac{2}{3}\tan^3 x + \frac{1}{5}\tan^5 + C\end{aligned}$$

33. $u = e^{-x}, du = -e^{-x}$

$$\int \frac{\sec e^{-x}}{e^x} dx = -\int \sec u du$$
$$= -\ln|\sec u + \tan u|$$
$$= -\ln|\sec e^{-x} + \tan e^{-x}| + C$$

37. Rotate about x-axis, using discs

$$V = \pi \int_0^{\pi} y^2 dx$$
$$= \pi \int_0^{\pi} \sin^2 x\, dx$$
$$= \pi \int_0^{\pi} \frac{1}{2}(1-\cos 2x) dx$$
$$= \frac{\pi}{2}\int_0^{\pi} dx - \frac{\pi}{2}\int_0^{\pi} \cos 2x\, dx$$
$$= \frac{\pi}{2}x - \frac{\pi}{2} \times \frac{1}{2}\sin 2x \Big|_0^{\pi}$$
$$= \frac{\pi^2}{2} - \frac{\pi}{4}\sin 2\pi - 0 + \frac{\pi}{4}\sin 0$$
$$= \frac{1}{2}\pi^2 = 4.93$$

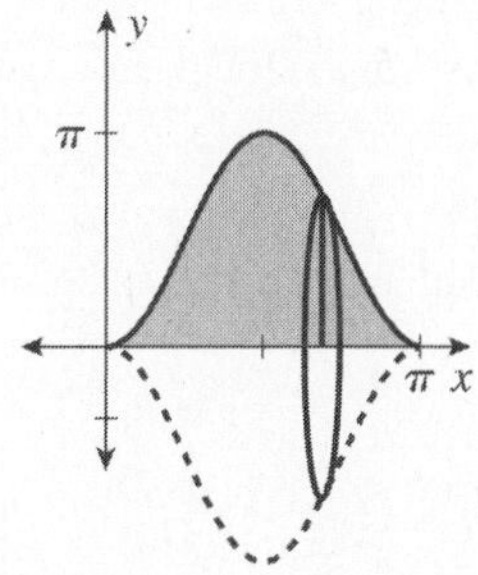

41. $\int \sin x \cos x\, dx; u = \sin x; du = \cos x\, dx$

$$\int u\, du = \frac{1}{2}u^2 + C = \frac{1}{2}\sin^2 x + C_1$$

Let $u = \cos x; du = -\sin x\, dx$

$$-\int \cos x(-\sin x) dx = -\frac{1}{2}\cos^2 x + C_2$$
$$\frac{1}{2}\sin^2 x + C_1 = \frac{1}{2}(1-\cos^2 x + C_2)$$
$$= \frac{1}{2} - \frac{1}{2}\cos^2 x + C_1$$
$$-\frac{1}{2}\cos^2 x + C_2 = \frac{1}{2} - \frac{1}{2}\cos^2 x + C_1$$
$$C_2 = C_1 + \frac{1}{2}$$

45. $v = \frac{\sin^3 t}{3} + 6$

$$s = \int \frac{\sin^3 t}{3} dt + \int 6dt$$
$$s = \frac{1}{3}\int \sin^2 t \sin t dt + \int 6dt$$
$$s = \frac{1}{3}\int (1-\cos^2 t)\sin t dt + 6\int dt$$
$$s = \frac{1}{3}\int \sin t dt - \frac{1}{3}\int \cos^2 t \sin t dt + 6\int dt$$
$$s = \frac{1}{3}\cos t + \frac{1}{9}\cos^3 t + 6t + s_0$$
$$s(0) = 0 = -\frac{1}{3}\cos 0 + \frac{1}{9}\cos^3 0 + 6(0) + s_0 \Rightarrow s_0 = \frac{2}{9}$$
$$s = -\frac{1}{3}\cos t + \frac{1}{9}\cos^3 t + 6t + \frac{2}{9}$$

49. $V_{\text{rms}} = \sqrt{\frac{1}{T}\int_0^T y^2 dt}$

$$V_{\text{rms}} = \sqrt{\frac{1}{1/60.0}\int_0^{1/60.0} (340 \sin 120\pi t)^2 dt}$$
$$= \sqrt{60}\sqrt{\int_0^{1/60.0} 340^2 \frac{1-\cos 240\pi t}{2} dt}$$
$$= \sqrt{60}\sqrt{\frac{340^2}{2}\left(t - \frac{1}{240\pi}\sin 240\pi t\right)\Bigg|_0^{1/60.0}}$$
$$= 240 \text{ V}$$

28.6 Inverse Trigonometric Forms

1. Change dx to $-x\,dx$, then

$$\int \frac{-x\,dx}{\sqrt{9-x^2}} = \frac{1}{2}\int (9-x^2)^{-1/2}(-2x\,dx)$$
$$= \frac{1}{2}\frac{1(9-x^2)^{1/2}}{\frac{1}{2}} + C$$
$$= \sqrt{9-x^2} + C$$

5. $a = 8; u = x; du = dx;$

$$\int \frac{dx}{64+x^2} = \frac{1}{8}\tan^{-1}\frac{x}{8} + C$$

9. $\int_0^2 \frac{3e^{-t}dt}{1+9e^{-2t}}$

$= -\int_0^2 \frac{-3e^{-t}dt}{1+(3e^{-t})^2}$ which has form $\int \frac{dx}{1+x^2}$

$= -\tan^{-1}\left(3e^{-t}\right)\Big|_0^2$

$= -\tan^{-1}\frac{3}{e^2} + \tan^{-1} 3$

$= 0.863$

13.

$u = 9x^2 + 16; du = 18x\,dx$

$\int \frac{8x\,dx}{9x^2+16} = \frac{8}{18}\int \frac{18x\,dx}{9x^2+16}$

$= \frac{4}{9}\ln\left|9x^2+16\right| + C$

17. $a = 1; u = e^x; du = e^x dx$

$\int \frac{e^x dx}{\sqrt{1-e^{2x}}} = \sin^{-1}(e^x) + C$

21. $a = 2; u = x+2; du = dx$

$\int \frac{4\,dx}{\sqrt{-4x-x^2}} = \int \frac{4\,dx}{\sqrt{4-(x+2)^2}}$

$= 4\int \frac{dx}{\sqrt{4-(x+2)^2}}$

$= 4\sin^{-1}\left(\frac{x+2}{2}\right) + C$

25. $a = 2; u = x; du; = dx; n = \frac{1}{2}; u = 4-x^2;$

$du = -2x\,dx$

$\int \frac{2-x}{\sqrt{4-x^2}}dx = \int \frac{2\,dx}{\sqrt{4-x^2}} - \int \frac{x\,dx}{\sqrt{4-x^2}}$

$= 2\int \frac{dx}{\sqrt{4-x^2}} + \frac{1}{2}\int \frac{(-2x\,dx)}{\left(4-x^2\right)^{1/2}}$

$= 2\sin^{-1}\frac{x}{2} + \frac{1}{2}\left(4-x^2\right)^{1/2}\cdot 2 + C$

$= 2\sin^{-1}\frac{x}{2} + \sqrt{4-x^2} + C$

29. $\int \frac{x^2+3x^5}{1+x^6}dx = \int \frac{x^2}{1+x^6}dx + \int \frac{3x^5}{1+x^6}dx$

$\int \frac{x^2}{1+x^6}dx = \frac{1}{3}\int \frac{3x^2}{1+\left(x^3\right)^2}dx$

$= \frac{1}{3}\tan^{-1} x^3$

$\int \frac{3x^5}{1+x^6}dx = \frac{1}{2}\int \frac{6x^5 dx}{1+x^6}$

$= \frac{1}{2}\ln\left(1+x^6\right)$

$\int \frac{x^2+3x^5}{1+x^6}dx = \frac{1}{3}\tan^{-1} x^3 + \frac{1}{2}\ln\left(1+x^6\right) + C$

33. **(a)** General power, $\int u^{-1/2} du$ where $u = 4-9x^2$.

$du = -18x\,dx$; numerator can fit du of denominator.

Square root becomes $-1/2$ power.

Does not fit inverse sine form.

(b) Inverse sine; $a = 2; u = 3x; du = 3\,dx$

(c) Logarithmic; $u = 4-9x; du = -9\,dx$

37. $y = \frac{1}{1+x^2};$

$A = \int_0^2 \frac{1}{1+x^2}dx;$

$a = 1; u = x; du = dx$

$A = \frac{1}{1}\tan^{-1}\frac{x}{1}\Big|_0^2$

$= \tan^{-1} 2 - \tan^{-1} 0$

$= 1.11$

41. $\int \frac{dx}{\sqrt{A^2-x^2}} = \int \sqrt{\frac{k}{m}}dt$

Integrate both sides:

$\sin^{-1}\frac{x}{A} = \sqrt{\frac{k}{m}}t + C$

Solve for C by letting $x = x_0$ and $t = 0$.

$\sin^{-1}\frac{x_0}{A} = \sqrt{\frac{k}{m}}(0) + C$

$C = \sin^{-1}\frac{x_0}{A};$

Therefore,

$\sin^{-1}\frac{x}{A} = \sqrt{\frac{k}{m}}t + \sin^{-1}\frac{x_0}{A}$

28.7 Integration by Parts

1. $u = \sqrt{1-x}, dv = x dx$

$du = \frac{-1}{2\sqrt{1-x}}dx, v = \frac{x^2}{2}$

$$\int x\sqrt{1-x}dx = \frac{x^2}{2}\sqrt{1-x} - \int \frac{x^2}{2}\frac{-dx}{2\sqrt{1-x}}$$
$$= \frac{x^2}{2}\sqrt{1-x} + \frac{1}{4}\int \frac{x^2 dx}{\sqrt{1-x}}$$

No, the substitution $u = \sqrt{1-x}, dv = x\,dx$ does not work since $\int \frac{x^2 dx}{\sqrt{1-x}}$ is more complex than $\int x\sqrt{1-x}dx$.

5. $\int 4xe^{2x}dx;$

$u = x;\ du = dx;\ dv = e^{2x}dx$

$v = \frac{1}{2}\int e^{2x}(2dx) = \frac{1}{2}e^{2x}$

$$\int 4xe^{2x}dx = \frac{4}{2}xe^{2x} - \frac{4}{2}\int e^{2x}dx$$
$$= 2xe^{2x} - \int e^{2x}(2\,dx)$$
$$= 2xe^{2x} - e^{2x} + C$$

9. $\int 2\tan^{-1}x\,dx;$

$u = \tan^{-1}x; du = \frac{1}{1+x^2}dx;$

$dv = dx; v = x$

$$2\int \tan^{-1}x\,dx = 2\left(x\tan^{-1}x - \int \frac{x\,dx}{1+x^2}\right)$$

$u = x^2;\ du = 2xdx$

$$= 2\left(x\tan^{-1}x - \frac{1}{2}\int \frac{2x\,dx}{1+x^2}\right)$$
$$= 2x\tan^{-1}x - \ln\left(1+x^2\right) + C$$

13. $\int x\ln x\,dx;$

$u = \ln x; du = \frac{dx}{x}; dv = x\,dx; v = \frac{x^2}{2}$

$$\int x\ln x\,dx = \frac{1}{2}x^2\ln x - \frac{1}{2}\int x^2\frac{dx}{x}$$
$$= \frac{1}{2}x^2\ln x - \frac{1}{2}\int x\,dx$$
$$= \frac{1}{2}x^2\ln x - \frac{1}{2}\cdot\frac{x^2}{2} + C$$
$$= \frac{1}{2}x^2\ln x - \frac{1}{4}x^2 + C$$

17. $\int_0^{\pi/2} e^x\cos x\,dx;$

We perform the integral as an indefinite integral first.

$u = e^x; = e^z dx; dv = \cos x\,dx; v = \sin x$

$\int e^x\cos x\,dx = e^x\sin x - \int \sin xe^x\,dx$

$u = e^x; du = e^x dx; dv = \sin x\,dx; v = -\cos x$

$\int e^x\cos x\,dx = e^x\sin x - \left(-e^x\cos x + \int e^x\cos dx\right)$

$\int e^x\cos dx = e^x\sin x + e^x\cos x - \int e^x\cos x\,dx$

$2\int e^x\cos x\,dx = e^x\sin x + e^x\cos x + C$

Now we evaluate the definite integral:

$$\int_0^{\pi/2} e^x\cos xdx = \frac{1}{2}e^x(\sin + \cos x)\Big|_0^{\pi/2}$$
$$= \frac{1}{2}e^{\pi/2}(1+0) - \frac{1}{2}e^0(0+1)$$
$$= \frac{1}{2}\left(e^{\pi/2} - 1\right) = 1.91$$

21. $u = \cos(\ln x), \qquad dv = dx$

$v = x$

$du = -\sin(\ln x)\cdot\frac{1}{x}dx$

$\int \cos(\ln x)dx = x\cos(\ln x) + \int x\sin(\ln x)\frac{1}{x}dx$

$u = \sin(\ln x), \qquad dv = dx$

$v = x$

$du = \cos(\ln x)\cdot\frac{1}{x}dx$

$$\int \cos(\ln x)dx = x\cos(\ln x) + x\sin(\ln x) - \int x\cos(\ln x)\cdot\frac{1}{x}dx$$

$2\int \cos(\ln x)dx = x\cos(\ln x) + x\sin(\ln x)$

$\int \cos(\ln x)dx = \frac{x}{2}\left(\cos(\ln x) + \sin(\ln x)\right) + C$

25. $A = \int_0^2 xe^{-x}dx;$

$u = x;\ du = dx;\ dv = e^{-x}dx; v = \int e^{-x}dx = -e^{-x}$

$$A = -xe^{-x}\Big|_0^2 - \int_0^2 -e^{-x}dx$$
$$= -xe^{-x} - e^{-x}\Big|_0^2$$
$$= 2e^{-2} - e^{-2} - (0-1)$$
$$= 1 - \frac{3}{e^2} = 0.594$$

29. $\bar{x}=\dfrac{\int_a^b xydx}{\int_a^b ydx}$

$=\dfrac{\int_0^{\pi/2} x(\cos x)dx}{\int_0^{\pi/2}\cos xdx}$

Let $u = x$; $du = dx$; $dv = \cos x$; $v = \sin x$

$\bar{x}=\dfrac{x\sin x|_0^{\pi/2}-\int_0^{\pi/2}\sin xdx}{\sin x|_0^{\pi/2}}$

$=\dfrac{x\sin x|_0^{\pi/2}-(-\cos x)|_0^{\pi/2}}{1}$

$=x\sin x+\cos x|_0^{\pi/2}$

$=\dfrac{\pi}{2}-1=0.571$

33. $v=\dfrac{ds}{dt}=\dfrac{t^3}{\sqrt{t^2+1}};$

$s=\int\dfrac{t^3dt}{\sqrt{t^2+1}}$

Let $u=t^2$; $du=2t\,dt$; $dv=\dfrac{t\,dt}{(t^2+1)^{1/2}}$

$v=\dfrac{1}{2}\int\dfrac{2t\,dt}{(t^2+1)^{1/2}}$

$=\dfrac{1}{2}(2)(t^2+1)^{1/2}$

$=(t^2+1)^{1/2}$

$s=t^2(t^2+1)^{1/2}-\int(t^2+1)^{1/2}(2t\,dt)$

$=t^2(t^2+1)^{1/2}-\dfrac{2}{3}(t^2+1)^{3/2}+C$

Substitute $s=0, t=0$:

$0=-\dfrac{2}{3}+C;\ C=\dfrac{2}{3}$

$s=\dfrac{1}{3}\left[3t^2(t^2+1)^{1/2}-2(t^2+1)^{3/2}+2\right]$

$=\dfrac{1}{3}\left[(t^2-2)(t^2+1)^{1/2}+2\right]$

28.8 Integration by Trigonometric Substitution

1. Delete the x^2 before the radical in the denominator.

$\int\dfrac{dx}{\sqrt{1-x^2}}=\sin^{-1}x+C$

5. $\int\dfrac{dx}{x^2\sqrt{x^2+1}}.$

Let $x=\tan\theta$, $dx=\sec^2\theta d\theta$, $\sqrt{x^2+1}=\sec\theta$

$\int\dfrac{dx}{x^2\sqrt{x^2+1}}=\int\dfrac{\sec^2\theta d\theta}{\tan^2\theta\sec\theta}$

$=\int\dfrac{d\theta}{\tan\theta}$

$=\int\cot\theta\csc\theta d\theta$

9. Let $x=\sin\theta$; $dx=\cos\theta d\theta$, $\sqrt{1-x^2}=\cos\theta$

$\int\dfrac{\sqrt{1-x^2}}{x^2}dx=\int\dfrac{\cos\theta}{\sin^2\theta}\cos\theta d\theta$

$=\int\dfrac{\cos^2\theta}{\sin^2\theta}d\theta$

$=\int\cot^2\theta d\theta$

$=\int(\csc^2\theta-1)d\theta$

$=\int\csc^2\theta d\theta-\int d\theta$

$=-\cot\theta-\theta+C$

$=\dfrac{-\sqrt{1-x^2}}{x}-\sin^{-1}x+C$

13. Let $z=3\tan\theta$; $dz=3\sec^2\theta\,d\theta, \sqrt{z^2+9}=\sec\theta$

$\int\dfrac{6dz}{z^2\sqrt{z^2+9}}=6\int\dfrac{3\sec^2\theta\,d\theta}{9\tan^2\theta(3\sec\theta)}$

$=\dfrac{6}{9}\int\dfrac{\sec\theta\,d\theta}{\tan^2\theta}$

$=\dfrac{6}{9}\int\dfrac{\cos\theta\,d\theta}{\sin^2\theta}$

$=\dfrac{6}{9}\int\csc\theta\cot\theta\,d\theta$

$=-\dfrac{6}{9}\csc\theta+C$

$\tan\theta=\dfrac{z}{3};\ \csc\theta=\dfrac{\sqrt{z^2+9}}{z}$

$\dfrac{-6}{9}\csc\theta+C=-\dfrac{2\sqrt{z^2+9}}{3z}+C$

17. $\int_0^{0.5}\dfrac{x^3dx}{\sqrt{1-x^2}},$

$x=\sin\theta$; $dx=\cos\theta\,d\theta, \sqrt{1-x^2}=\cos\theta$

$\int\dfrac{\sin^3\theta\cos\theta\,d\theta}{\cos\theta}=\int\sin^3\theta\,d\theta$

$$= \int \sin\theta \sin^2\theta\, d\theta$$
$$= \int \sin\theta(1-\cos^2\theta)d\theta$$
$$= \int \sin\theta\, d\theta - \int \cos^2\theta \sin\theta\, d\theta$$
$$= -\cos\theta + \frac{\cos^3\theta}{3} + C$$
$$\int_0^{0.5} \frac{x^3dx}{\sqrt{1-x^2}} = -\sqrt{1-x^2} + \frac{1}{3}\left(\sqrt{1-x^2}\right)^3\Big|_0^{0.5}$$
$$= -\sqrt{1-0.5^2} + \frac{1}{3}\left(\sqrt{1-0.5^2}\right)^3 + \sqrt{1} - \frac{1}{3}\sqrt{1}$$
$$= 0.017$$

21. $\int \frac{dy}{y\sqrt{4y^2-9}}$;

$$2y = 3\sec\theta; y = \frac{3}{2}\sec\theta, dy = \frac{3}{2}\sec\theta\tan\theta\, d\theta$$
$$\sqrt{4y^2-9} = 3\tan\theta$$
$$\int \frac{\frac{3}{2}\sec\theta\tan\theta\, d\theta}{\frac{3}{2}\sec\theta\cdot 3\tan\theta} = \frac{1}{3}\int d\theta$$
$$= \frac{1}{3}\theta + C$$
$$= \frac{1}{3}\sec^{-1}\frac{2}{3}y + C$$
$$\int_{2.5}^{3} \frac{dy}{y\sqrt{4y^2-9}} = \frac{1}{3}\sec^{-1}\left(\frac{2y}{3}\right)\Big|_{2.5}^{3}$$
$$= \frac{1}{3}\cos^{-1}\left(\frac{3}{2y}\right)\Big|_{2.5}^{3}$$
$$= 0.0400$$

25. (a) $u = 1-x^2, du = -2xdx$

$$\int x\sqrt{1-x^2}dx = -\frac{1}{2}\int\left(1-x^2\right)^{1/2}(-2xdx)$$
$$= -\frac{1}{2}\frac{\left(1-x^2\right)^{3/2}}{3/2} + C$$
$$= -\frac{1}{3}\left(1-x^2\right)^{3/2} + C$$

(b) $x = \sin\theta$

$$dx = \cos\theta d\theta$$
$$\sqrt{1-x^2} = \cos\theta$$

$u = \cos\theta, du = -\sin\theta d\theta$

$$\int x\sqrt{1-x^2}dx = \int \sin\theta\cos\theta\cos\theta d\theta$$
$$= -\int \cos^2\theta(-\sin\theta d\theta)$$
$$= -\frac{\cos^3\theta}{3} + C$$
$$= -\frac{\left(1-x^2\right)^{3/2}}{3} + C$$

29. $x^2 + y^2 = a^2; y = \sqrt{a^2-x^2}$

$$I_y = \rho\int_0^a x^2y\, dx$$
$$= \rho\int_0^a x^2\sqrt{a^2-x^2}dx$$
$$x = a\sin\theta; dx = a\cos\theta\, d\theta, \sqrt{a^2-x^2} = a\cos\theta$$
$$\rho\int a^2\sin^2\theta\cdot a\cos\theta\cdot a\cos\theta\, d\theta$$
$$= \rho a^4\int \sin^2\theta\cos^2\theta\, d\theta$$
$$= \rho a^4\int \sin^2\theta(1-\sin^2\theta)d\theta$$
$$= \rho a^4\int (\sin^2\theta - \sin^4\theta)d\theta$$
$$= \rho a^4\left[\frac{1}{2}\int(1-\cos 2\theta)d\theta - \int\frac{1^2}{2}(1-\cos\theta)^2 d\theta\right]$$
$$= \rho a^4\left[\frac{1}{2}\int d\theta - \frac{1}{2}\int\cos 2\theta\, d\theta - \int\frac{1}{4}(1-2\cos 2\theta + \cos^2 2\theta)d\theta\right]$$
$$= \rho a^4\left[\frac{1}{2}\int d\theta - \frac{1}{2}\int\cos 2\theta\, d\theta - \frac{1}{4}\int d\theta\frac{1}{2}\cos 2\theta - \frac{1}{4}\cos^2 2\theta\, d\theta\right]$$
$$= \rho a^4\left[\frac{1}{2}\int d\theta - \frac{1}{4}d\theta - \frac{1}{4}\int\frac{1}{2}(1+\cos 4\theta)d\theta\right]$$
$$= \rho a^4\left[\frac{1}{4}\int d\theta - \frac{1}{8}\int d\theta - \frac{1}{4}\int\cos 4\theta\, d\theta\right]$$
$$= \rho a^4\left[\frac{1}{8}\int d\theta - \frac{1}{32}\int\cos 4\theta\, d\theta\right]$$
$$= \rho a^4\left[\frac{1}{8}\theta - \frac{1}{32}\sin 4\theta\right]$$
$$(\sin 4A = 2\sin 2A\cos 2A = 2[2\sin A\cos A(1-2\sin^2 A)])$$
$$= \rho a^4\left[\frac{1}{8}\sin^{-1}\frac{x}{a} - \frac{1}{32}(4\sin\theta\cos\theta)(1-2\sin^2\theta)\right] + C$$

$$I_y = \rho a^4\left[\frac{1}{8}\sin^{-1}\frac{x}{a} - \frac{1}{8}\cdot\frac{x}{a}\cdot\frac{\sqrt{a^2-x^2}}{a}\left(1-2\frac{x^2}{a^2}\right)\right]_0^a$$

$$= \rho a^4\left[\frac{1}{8}\sin^{-1}1\right] = \frac{1}{8}\rho a^4\frac{\pi}{2}$$

$$I_y = \frac{1}{16}\pi\rho a^4$$

$$m = \rho\int_0^c y\,dx = \rho\int_0^c \sqrt{a^2-x^2}\,dx$$

$$\rho\int\sqrt{a^2-x^2}\,dx = \rho\int a^2\cos^2\theta\,d\theta$$

$$= \rho a^2\int\frac{1}{2}(1+\cos 2\theta)d\theta$$

$$= \frac{\rho a^2}{2}\int d\theta - \frac{\rho a^2}{2}\int\cos 2\theta\,d\theta$$

$$= \frac{\rho a^2}{2}\theta - \frac{\rho a^2}{4}\sin 2\theta + C$$

$$= \frac{\rho a^2}{2}\theta - \frac{\rho a^2}{2}\sin\theta\cos\theta + C$$

$$m = \frac{\rho a^2}{2}\left(\sin^{-1}\frac{x}{a} - \frac{x}{a}\frac{\sqrt{a^2-x^2}}{a}\right)\Bigg|_0^a$$

$$= \frac{\rho a^2}{2}(\sin^{-1}1)$$

$$m = \frac{1}{4}\pi\rho a^2$$

$$I_y = \frac{1}{16}\pi\rho a^4$$

$$= \frac{1}{4}\pi\rho a^2\left(\frac{1}{4}a^2\right) = \frac{1}{4}ma^2$$

33. The centroid of the boat rudder is $\overline{x} = 0$ by symmetry. The centroid of one side of the boat rudder is

$$\overline{x} = \frac{\int_0^2 x\left(0.5x^2\sqrt{4-x^2}\,dx\right)}{\pi/2} = 1.36$$

37. $u = \sqrt{x+1} \Rightarrow u^2 = x+1 \Rightarrow x = u^2-1$

$$du = \frac{1}{2\sqrt{x+1}}dx$$

$$2u\,du = dx$$

$$\int x\sqrt{x+1}\,dx = \int\left(u^2-1\right)u\left(2u\,du\right)$$

$$= \int\left(2u^4-2u^2\right)du$$

$$= \frac{2u^5}{5} - \frac{2u^3}{3} + C$$

$$= \frac{2(x+1)^{5/2}}{5} - \frac{2(x+1)^{3/2}}{3} + C$$

28.9 Integration by Partial Fractions: Nonrepeated Linear Factors

1. $\dfrac{10-x}{x^2+x-2} = \dfrac{10-x}{(x-1)(x+2)} = \dfrac{A}{x-1} + \dfrac{B}{x+2}$

$$10-x = A(x+2) + B(x-1)$$

for $x = -2$:

$$10-1 = A(1+2) + B(1-1)$$

$$9 = 3A$$

$$A = 3$$

for $x = 1$:

$$10-1 = A(1+2) + B(1-1)$$

$$9 = 3A$$

$$A = 3$$

$$\frac{10-x}{x^2+x-2} = \frac{3}{x-1} + \frac{-4}{x+2}$$

5. $\dfrac{x^2-6x-8}{x^3-4x} = \dfrac{x^2-6x-8}{x(x^2-4)}$

$$= \frac{x^2-6x-8}{x(x+2)(x-2)}$$

$$= \frac{A}{x} + \frac{B}{x+2} + \frac{C}{x-2}$$

9. $\displaystyle\int\frac{dx}{x^2-4} = \int\frac{dx}{(x+2)(x-2)}$

$$= \int\frac{-\frac{1}{4}}{x+2}dx + \int\frac{\frac{1}{4}}{x-2}dx$$

$$= -\frac{1}{4}\ln|x+2| + \frac{1}{4}\ln|x-2| + C$$

$$= \frac{1}{4}\ln\left|\frac{x-2}{x+2}\right| + C$$

13. $\int_0^1 \frac{2t+4}{3t^2+5t+2}dt$

$= \int_0^1 \frac{8}{3t+2}dt - \int_0^1 \frac{2}{t+1}dt$

$= \frac{8\ln(3t+2)}{3}\Big|_0^1 - 2\ln(t+1)\Big|_0^1$

$= \frac{8\ln 5}{3} - \frac{8\ln 2}{3} - 2\ln 2 + 2\ln 1$

$= 1.06$

17. $\int \frac{6x^2-2x-1}{4x^3-x}dx$

$= \int \frac{6x^2-2x-1}{x(4x^2-1)}dx = \int \frac{6x^2-2x-1}{x(2x+1)(2x-1)}dx$

$= \int \frac{dx}{x} + \int \frac{\frac{3}{2}}{2x+1}dx - \frac{\frac{1}{2}}{2x-1}dx$

$= \ln|x| + \frac{3\ln|2x+1|}{4} - \frac{\ln|2x-1|}{4} + C$

$= \frac{4\ln|x|}{4} + \frac{3\ln|2x+1|}{4} - \frac{\ln|2x-1|}{4} + C$

$= \frac{1}{4}\ln\left|\frac{x^4(2x+1)^3}{2x-1}\right| + C$

21. $\int \frac{dV}{(V^2-4)(V^2-9)}$

$= \int \frac{dV}{(V-2)(V+2)(V+3)(V-3)}$

$= \int \frac{\frac{1}{30}}{V-3}dV - \int \frac{\frac{1}{30}dV}{V+3} - \int \frac{\frac{1}{20}dV}{V-2} + \int \frac{\frac{1}{20}dV}{V+2}$

$= \frac{1}{30}\ln|V-3| - \frac{1}{30}\ln|V+3|$

$-\frac{1}{20}\ln|V-2| + \frac{1}{20}\ln|V+2| + C$

$= \frac{2}{60}\ln|V-3| - \frac{2}{60}\ln|V+3| - \frac{3}{60}\ln|V-2|$

$+\frac{3}{60}\ln|V+2| + C$

$= \frac{1}{60}\ln\left|\frac{(V+2)^3(V-3)^2}{(V-2)^3(V+3)^2}\right| + C$

25. $\frac{1}{u(a+bu)} = \frac{A}{u} + \frac{B}{a+bu}$

$1 = A(a+bu) + Bu$

for $u = 0$: $1 = aA$

$A = \frac{1}{a}$

for $u = \frac{a}{b}$: $1 = B\left(-\frac{a}{b}\right)$

$B = -\frac{b}{a}$

$\int \frac{du}{u(a+bu)}$

$= \int \frac{\frac{1}{a}}{u}du + \int \frac{-\frac{b}{a}}{a+bu}du$

$= \frac{1}{a}\ln|u| - \frac{1}{a}\int \frac{b\,du}{a+bu}$

$= \frac{1}{a}\ln|u| - \frac{1}{a}\ln|a+bu| + C$

$= -\frac{1}{a}(-\ln|u| + \ln|a+bu| + C$

$= -\frac{1}{a}\ln\frac{|a+bu|}{|u|} + C$

$= \frac{1}{a}\ln\left|\frac{a+bu}{u}\right| + C$

29. $V = \int_1^3 \frac{2\pi x\,dx}{(x^3+3x^2+2x)}$

$= \int_1^3 \frac{2\pi\,dx}{x+1} - \int_1^3 \frac{2\pi\,dx}{x+2}$

$V = 2\pi\ln(x+1)\Big|_1^3 - 2\pi\ln(x+2)\Big|_1^3$

$= 2\pi[\ln 4 - \ln 2 - \ln 5 + \ln 3]$

$V = 2\pi\ln\frac{4\cdot 3}{2\cdot 5} = 2\pi\ln\frac{6}{5}$

$= 1.15$

33. $W = \int_0^{0.5} \frac{4x\,dx}{x^2+3x+2}$

$= \int_0^{0.5} \frac{8dx}{x+2} - \int_0^{0.5} \frac{4dx}{x+1}$

$= 8\ln|x+2| - 4\ln|x+1|\Big|_0^{0.5}$

$= 0.163\ \text{N}\cdot\text{cm}$

28.10 Integration by Partial Fractions: Other Cases

1. $\frac{2}{x(x+3)^2} = \frac{A}{x} + \frac{B}{x+3} + \frac{C}{(x+3)^2}$

5. $\int \frac{x-8}{x^3-4x^2+4x} = \int \frac{-2}{x}dx + \int \frac{2}{x-2}dx - \int \frac{3}{(x-2)^2}dx$

$$= -2\ln|x| + 2\ln|x-2| + \frac{3}{x-2} + C$$

$$= 2\ln\left|\frac{x-2}{x}\right| + \frac{3}{x-2} + C$$

9. $\frac{2s}{(s-3)^3} = \frac{A}{s-3} + \frac{B}{(s-3)^2} + \frac{C}{(s-3)^3}$

$$= \frac{A(s-3)^2 + B(s-3) + C}{(s-3)^3}$$

$$2s = A(s-3)^2 + B(s-3) + C$$

$$s = 3,\ 6 = C$$

$$2s = A(s-3)^2 + B(s-3) + 6$$

$$\left.\begin{array}{l} s=1,\quad 2 = 4A - 2B + 6 \\ s=2,\quad 4 = A - B + 6 \end{array}\right\} A = 0,\ B = 2$$

$$\int_1^2 \frac{2s}{(s-3)^3}ds = \int_1^2 \frac{2}{(s-3)^2}ds + \int_1^2 \frac{6}{(s-3)^3}dx$$

$$= 2\cdot\frac{(s-3)^{-2+1}}{-2+1} + 6\cdot\left.\frac{(s-3)^{-3+1}}{-3+1}\right|_1^2$$

$$= \left.\frac{-2}{(s-3)} - \frac{3}{(s-3)^2}\right|_1^2 = -\frac{5}{4}$$

13. $\frac{x^2+x+5}{(x+1)(x^2+4)} = \frac{A}{x+1} + \frac{Bx+C}{x^2+4}$

$$= \frac{A(x^2+4) + (Bx+C)(x+1)}{(x+1)(x^2+4)}$$

$$x^2 + x + 5 = A(x^2+4) + (Bx+C)(x^2+1)$$

$$\left.\begin{array}{ll} x=0, & 5 = 4A + \quad C \\ x=1, & 1, = 7 = 5A + 2B + 2C \\ x=2, & 11 = 8A + 6B + 3C \end{array}\right\} A = 1, B = 0, C = 1$$

$$\int_0^2 \frac{x^2+x+5}{(x+1)(x^2+4)}dx$$

$$= \int_0^2 \frac{dx}{x+1} + \int_1^2 \frac{1}{x^2+4}dx$$

$$= \ln|x+1| + \frac{1}{2}\tan^{-1}\left.\frac{x}{2}\right|_0^2 = 1.49$$

17. $\frac{10x^3+40x^2+22x+7}{(4x^2+1)(x^2+6x+10)} = \frac{Ax+B}{4x^2+1} + \frac{Cx+D}{x^2+6x+10}$

$$10x^3 + 40x^2 + 22x + 7$$

$$= (Ax+B)(x^2+6x+10) + (Cx+D)(4x^2+1)$$

$$= Ax^3 + 6Ax^2 + 10Ax + Bx^2 + 6Bx + 10B + 4Cx^3$$

$$+ Cx + 4Dx^2 + D$$

$$= (A+4C)x^3 + (6A+B+4D)x^2$$

$$+ (10A+6B+C)x + 10B + D$$

$$\left.\begin{array}{ll} (1)\ A + 4C & = 10 \\ (2)\ 6A + B + 4D & = 40 \\ (3)\ 10A + 6B + C & = 22 \\ (4)\ 10B + D & \quad 7 \end{array}\right\} A = 2, B = 0, C = 2, D = 7$$

$$\int \frac{10x^3+40x^2+22x+7}{(4x^2+1)(x^2+6x+10)}dx = \int \frac{2x}{4x^2+1}dx$$

$$+ \int \frac{2x+7}{x^2+6x+10}dx$$

$$\int \frac{2x}{4x^2+1}dx = \frac{1}{4}\int \frac{8x}{4x^2+1}dx$$

$$= \frac{\ln(4x^2+1)}{4}$$

$$\int \frac{2x+7}{x^2+6x+10}dx = \int \frac{2x+7}{x^2+6x+9+1}dx$$

$$= \int \frac{2x+7}{(x+3)^2+1}dx$$

$$u = x+3, du = dx, x = u-3$$

$$\int \frac{2x+7}{x^2+6x+10}dx = \int \frac{2(u-3)+7}{u^2+1}du$$

$$= \int \frac{2u-6+7}{u^2+1}du$$

$$= \int \frac{2u}{u^2+1}du$$

$$= \int \frac{2u}{u^2+1}du + \int \frac{1}{u^2+1}du$$

$$= \ln(u^2+1) + \tan^{-1}u + C$$

$$= \ln(x^2+6x+10) + \tan^{-1}(x+3) + C$$

$$\int \frac{10x^3+40x^2+22x+7}{(4x^2+1)(x^2+6x+10)}dx$$

$$= \frac{\ln(4x^2+1)}{4} + \ln(x^2+6x+10) + \tan^{-1}(x+3) + C$$

21. $\frac{x}{(x-2)^3}=\frac{A}{(x-2)}+\frac{B}{(x-2)^2}+\frac{C}{(x-2)^3}$

$$=\frac{A(x-2)^2+B(x-2)+C}{(x-2)^3}$$

$$x=A(x^2-4x+4)+B(x-2)+C$$

$$x=Ax^2-4Ax+4A+Bx-2B+C$$

x^2: $A=0$

x: $1=-4A+B\Rightarrow B=1$

Constant: $0=4A-2B+C$

$$0=-2+C$$

$$C=2$$

$$\int\frac{xdx}{(x-2)^3}=\int\frac{dx}{(x-2)^2}+\int\frac{2dx}{(x-2)^3}$$

$$=\frac{(x-2)^{-2+1}}{-2+1}+2\frac{(x-2)^{-3+1}}{-3+1}+C$$

$$=\frac{-1}{(x-2)}-\frac{1}{(x-2)^2}+C$$

$$=\frac{-1(x-2)-1}{(x-2)^2}+C$$

$$=\frac{1-x}{(x-2)^2}+C$$

25. $V=\int_0^2 2\pi x\cdot\frac{4}{x^4+6x^2+5}dx$

$$\frac{4x}{x^4+6x^2+5}=\frac{Ax+B}{x^2+5}+\frac{Cx+D}{x^2+1}$$

$$4x=(A+C)x^3+(B+D)x^2+(A+5C)x+B+5D$$

$$A+C=0;B+D=0;A+5C=4;B+5D=0$$

$$A=-1,B=0,C=1,D=0$$

$$\int\frac{4x}{x^4+6x^2+5}dx=\int\frac{-xdx}{x^2+5}dx+\int\frac{xdx}{x^2+1}dx$$

$$=-\frac{1}{2}\ln(x^2+5)+\frac{1}{2}\ln(x^2+1)+C$$

$$V=2\pi\int_0^2\frac{4x}{x^4+6x^2+5}dx$$

$$=2\pi\left[-\frac{1}{2}\ln(x^2+5)+\frac{1}{2}\ln(x^2+1)\right]\Bigg|_0^2$$

$$=2\pi\left[-\frac{1}{2}\ln 9+\ln 5\right]=3.21$$

29. $\bar{x}=\frac{\int_1^2 x\cdot\frac{4}{x^3+x}dx}{\int_1^2\frac{4}{x^3+x}dx}=\frac{\pi-4\tan^{-1}\frac{1}{2}}{2\ln\frac{8}{5}}$

$$\bar{x}=1.370$$

28.11 Integration by Use of Tables

1. $\int\frac{xdx}{(2+3x)^2}$ is formula 3 with $u=x$, $du=dx$, $a=2$, $b=3$.

5. $u=x^2$, $du=2xdx$

$$\int\frac{xdx}{(4-x^4)^{3/2}}=\frac{1}{2}\int\frac{2xdx}{\left(2^2-(x^2)^2\right)^{3/2}}=\frac{1}{2}\int\frac{du}{(2^2-u^2)^{3/2}}$$

which is formula 25 with $a=2$.

9. Formula 1; $u=x$; $a=2$; $b=5$; $du=dx$

$$\int\frac{3xdx}{2+5x}=3\int\frac{xdx}{2+5x}$$

$$=3\left\{\frac{1}{25}(2+5x)-2\ln|2+5x|\right\}+C$$

$$=\frac{3}{25}\left[2+5x-2\ln|2+5x|\right]+C$$

13. Formula 24; $u=y$, $du=dy, a=2$

$$\int\frac{dy}{(y^2+4)^{3/2}}=\frac{y}{4\sqrt{y^2+4}}+C$$

17. Formula 17; $u=2x$; $du=2dx$; $a=3$

$$\int\frac{\sqrt{4x^2-9}}{x}dx=\int\frac{\sqrt{(2x)^2-3^2}}{2x}2dx$$

$$=\sqrt{4x^2-9}-3\sec^{-1}\left(\frac{2x}{3}\right)+C$$

21. Formula 52; $u=r^2$; $du=2r\,dr$

$$6\int\tan^{-1}r^2(r\,dr)$$

$$=3\int\tan^{-1}r^2(2r\,dr)$$

$$=3\left[r^2\tan^{-1}r^2-\frac{1}{2}\ln(1+r^4)\right]+C$$

$$=3r^2\tan^{-1}r^2-\frac{3}{2}\ln(1+r^4)+C$$

25. Formula 11; $u = 2x$; $du = 2dx$; $a = 1$

$$\int \frac{dx}{x\sqrt{4x^2+1}} = \int \frac{2dx}{2x\sqrt{(2x)^2+1^2}}$$

$$= -\ln\left(\frac{1+\sqrt{4x^2+1}}{2x}\right) + C$$

29. Formula 40; $a=1$; $u=x$; $du = dx$; $b = 5$

$$\int_0^{\pi/12} \sin\theta \cos 5\theta\, d\theta = -\frac{\cos(-4\theta)}{2(-4)} - \frac{\cos 6\theta}{12}\Bigg|_0^{\pi/12}$$

$$= \frac{1}{8}\cos 4\theta - \frac{1}{12}\cos 6\theta\Bigg|_0^{\pi/12}$$

$$= 0.0208$$

33. $u = x^2, du = 2x\,dx, u^2 = x^4$

$$\int \frac{2xdx}{(1-x^4)^{3/2}} = \int \frac{du}{(1-u^2)^{3/2}}$$

Formula 25; $a = 1$

$$\int \frac{2xdx}{(1-x^4)^{3/2}} = \frac{u}{\sqrt{1-u^2}} + C;$$

$$\int \frac{2x\,dx}{(1-x^4)^{3/2}} = \frac{x^2}{\sqrt{1-x^4}} + C$$

37. Formula 46; $u = x^2$; $du = 2x\,dx$; $n = 1$

$$\int x^3 \ln x^2 dx = \frac{1}{2}\int x^2 \ln x^2 (2x\,dx)$$

$$= \frac{1}{2}\left[(x^2)^2\left(\frac{\ln x^2}{2} - \frac{1}{4}\right)\right] + C$$

$$= \frac{1}{2}\left[\frac{x^4}{2}\left(\ln x^2 - \frac{1}{2}\right)\right] + C$$

$$= \frac{1}{4}x^4\left(\ln x^2 - \frac{1}{2}\right) + C$$

41. Let $u = t^3$, $du = 3t^2 dt$, formula 19, $a = 1$

$$\int t^2\left(t^6+1\right)^{3/2} dt = \int \frac{1}{3}\left(u^2+1\right)^{3/2} du,$$

$$= \frac{1}{3}\left[\frac{u}{4}\left(u^2+1\right)^{3/2} + \frac{3u}{8}\sqrt{u^2+1} + \frac{3}{8}\ln\left(u+\sqrt{u^2+1}\right)\right]$$

$$= \frac{1}{12}t^3\left(t^6+1\right)^{3/2} + \frac{t^3}{8}\sqrt{t^6+1} + \frac{1}{8}\ln\left(t^3+\sqrt{t^6+1}\right) + C$$

45. From Exercise 35 of Section 26.6;

$$s = \int_a^b \sqrt{1+\left(\frac{dy}{dx}\right)^2}\, dx;$$

$$y = x^2; \frac{dy}{dx} = 2x$$

$$s = \int_0^1 \sqrt{1+(2x)^2}\, dx$$

$$= \frac{1}{2}\int_0^1 \sqrt{(2x)^2+1}(2dx)$$

Formula 14; $u = 2x$; $du = 2\,dx, a = 1$

$$s = \frac{1}{2}\left[\frac{2x}{2}\sqrt{4x^2+1} + \frac{1}{2}\ln(2x+\sqrt{4x^2+1})\right]\Bigg|_0^1$$

$$= \frac{1}{2}\left[\left(1\sqrt{5} + \frac{1}{2}\ln(2+\sqrt{5})\right) - \frac{1}{2}\ln 1\right]$$

$$= \frac{1}{4}[2\sqrt{5} + \ln(2+\sqrt{5})] = 1.48$$

49. $F = \gamma\int_0^3 lhdh = \gamma\int_0^3 x(3-y)dy = \gamma\int_0^3 \frac{3-y}{\sqrt{1+y}}dy$

Power rule and formula #6 with $a = 1, b = 1, u = y$

$$\int \frac{3-y}{\sqrt{1+y}}dy = 3\int \frac{dy}{\sqrt{1+y}} - \int \frac{ydy}{\sqrt{1+y}}$$

$$= 3\frac{(1+y)^{1/2}}{\frac{1}{2}} - \left[\frac{-2(2-y)\sqrt{1+y}}{3(1)^2}\right] + C$$

$$F = \gamma\int_0^3 \frac{3-y}{\sqrt{1+y}}dy$$

$$= \gamma\left[6(1+y)^{1/2} + \frac{2}{3}(2-y)(1+y)^{1/2}\right]\Bigg|_0^3$$

$$= \gamma\left[6(2) + \frac{2}{3}(-1)(2) - 6(1) - \frac{2}{3}(2)(1)\right]$$

$$F = \lambda\left(12 - \frac{4}{3} - 6 - \frac{4}{3}\right)$$

$$= \frac{10\lambda}{3}$$

$$= \frac{10(9800)}{3}$$

$$= 32.7\text{ kN}$$

Review Exercises

1. $u = -8x$, $du = -8dx$

$$\int e^{-8x}dx = -\frac{1}{8}\int e^{-8x}(-8dx)$$

$$= -\frac{1}{8}e^{-8x} + C$$

5. $\int_0^{\pi/2}\frac{4\cos\theta\, d\theta}{1+\sin\theta}=4\int_0^{\pi/2}\frac{\cos\theta\, d\theta}{1+\sin\theta};$

$u=1+\sin\theta, du=\cos\theta$

$=4\ln(1+\sin\theta)\Big|_0^{\pi/2}$

$=2.77$

9. $\int_0^{\pi/2}\cos^3 2\theta\, d\theta=\int_0^{\pi/2}\cos^2 2\theta\cos 2\theta\, d\theta$

$=\int_0^{\pi/2}(1-\sin^2 2\theta)\cos 2\theta\, d\theta$

$=\int_0^{\pi/2}\cos 2\theta\, d\theta-\int_0^{\pi/2}\sin^2 2\theta\cos 2\theta\, d\theta$

$=\frac{1}{2}\int_0^{\pi/2}\cos 2\theta(2d\theta)-\frac{1}{2}\int_0^{\pi/2}\sin^2 2\theta\cos 2\theta(2d\theta)$

$=\frac{1}{2}\left[\sin 2\theta-\frac{1}{3}\sin^3 2\theta\right]_0^{\pi/2}$

$=\frac{1}{2}\left[\left(\sin\pi-\frac{1}{3}\sin^3\pi\right)-\left(\sin 0-\frac{1}{3}\sin^3 0\right)\right]$

$=\frac{1}{2}(0)=0$

13. $\int(\sin t+\cos t)^2\cdot\sin t\, dt$

$=\int(\sin^2 t+2\sin t\cos t+\cos^2 t)\cdot\sin t dt$

$=\int(1+2\sin t\cos t)\cdot\sin t dt$

$=\int(\sin t+2\sin^2 t\cos t)dt$

$=\int\sin t\, dt+2\int\sin^2 t(\cos t\, dt)$

$=-\cos t+\frac{2\sin^3 t}{3}+C$

$(u=\sin t, du=\cos t dt)$

17. $\int\sec^4 3x dx=\int\sec^2 3x\sec^2 3x dx$

$=\int(1+\tan^2 3x)\sec^2 3x dx$

$=\frac{1}{3}\int\sec^2 3x(3dx)+\frac{1}{3}\int\tan^2 3x\sec^2 3x(3dx)$

$=\frac{1}{3}\tan 3x+\frac{1}{3}\frac{\tan^3 3x}{3}+C$

$=\frac{1}{9}\tan^3 3x+\frac{1}{3}\tan 3x+C$

21. $u=x^2, du=2xdx$

$\int\frac{3xdx}{4+x^4}=3\int\frac{xdx}{4+x^4}=3\int\frac{1}{2^2+(x^2)^2}xdx$

$=\frac{3}{2}\int\frac{1}{2^2+(x^2)^2}2xdx$

$=\frac{3}{2}\left(\frac{1}{2}\tan^{-1}\frac{x^2}{2}\right)+C$

$=\frac{3}{4}\tan^{-1}\frac{x^2}{2}+C$

25. $u=e^{2x}+1, du=e^{2x}(2dx)$

$\int\frac{e^{2x}dx}{\sqrt{e^{2x}+1}}=\frac{1}{2}\int(e^{2x}+1)^{-1/2}e^{2x}(2dx)$

$=\frac{1}{2}(e^{2x}+1)^{1/2}(2)+C$

$=\sqrt{e^{2x}+1}+C$

29. $\int_0^{\pi/6}3\sin^2 3\phi d\phi=\int_0^{\pi/6}3\cdot\frac{(1-\cos 6\phi)}{2}d\phi$

$=\int_0^{\pi/6}\frac{3}{2}d\phi-\frac{1}{4}\int_0^{\pi/6}\cos 6\phi(6d\phi)$

$=\frac{3}{2}\phi\Big|_0^{\pi/6}-\frac{1}{4}\sin 6\phi\Big|_0^{\pi/6}$

$=\frac{3}{2}\left[\frac{\pi}{6}-0\right]-\frac{1}{4}[\sin\pi-\sin 0]=\frac{\pi}{4}$

33. $\frac{3u^2-6u-2}{u^2(3u+1)}=\frac{Au+B}{u^2}+\frac{C}{3u+1}$

$=\frac{(Au+B)(3u+1)+Cu^2}{u^2(3u+1)}$

$3u^2-6u-2=3Au^2+Au+3Bu+B+Cu^2$

$3u^2-6u-2=(3A+C)u^2+(A+3B)u+B$

(1) $3A+C=3, 3(0)+C=3, C=3$

(2) $A+3B=-6; A+3(-2)=-6, A=0$

(3) $B=-2$

$\int\frac{3u^2-6u-2}{u^2(3u+1)}du=\int\frac{-2}{u^2}du+\int\frac{3}{3u+1}du$

$=\frac{2}{u}+\ln|3u+1|+C$

37. $u=\ln x, du=\frac{dx}{x}$

$\int_1^e 3\cos(\ln x)\cdot\frac{dx}{x}=3\sin(\ln x)\Big|_1^e$

$$= 3\sin(\ln e) - 3\sin(\ln 1)$$
$$= 3\sin(1) - 3\sin(0)$$
$$= 3\sin 1 = 2.52$$

41. $u = \cos x,\ du = -\sin x dx$

$$\int \frac{\sin x\cos^2 x}{5+\cos^2 x}dx = -\int \frac{u^2 du}{5+u^2}$$
$$= -\int\left(1 - \frac{5}{5+u^2}\right)du$$
$$= -u + \frac{5}{\sqrt{5}}\tan^{-1}\frac{u}{\sqrt{5}} + C$$
$$= -\cos x + \frac{5}{\sqrt{5}}\tan^{-1}\frac{\cos x}{\sqrt{5}} + C$$

45. $\int e^{\ln 4x}dx = \int 4x dx = 2x^2 + C$

$\int \ln e^{4x}dx = \int 4x\, dx = 2x^2 + C$

The integrals are the same because the functions are the same.

49. Use the general power formula.

Let $u = e^x + 1, du = e^x dx,\ n = 2.$

$$\int e^x(e^x+1)^2 dx = \int u^2 du$$
$$= \frac{u^3}{3} + C_1$$
$$= \frac{(e^x+1)^3}{3} + C_1$$
$$= \frac{e^{3x}}{3} + e^{2x} + e^x + \frac{1}{3} + C_1$$

$$\int e^x(e^x+1)^2 dx = \int e^x(e^{2x} + 2e^x + 1)dx$$
$$= \int(e^{3x} + 2e^{2x} + e^x)dx$$
$$= \int e^{3x}dx + \int 2e^{2x}dx + \int e^x dx$$
$$= \frac{1}{3}\int e^{3x}(3dx) + 2\left(\frac{1}{2}\right)\int e^{2x}(2dx) + \int e^x dx$$
$$= \frac{1}{3}e^{3x} + e^{2x} + e^x + C_2$$

The two integrals differ by a constant, with

$C_2 = C_1 + \frac{1}{3}$

53. (a) $u = x^2 + 4$

$du = 2xdx$

$$\int \frac{x}{\sqrt{x^2+4}}dx = \int \frac{\frac{1}{2}du}{\sqrt{u}}$$
$$= u^{1/2} + C$$
$$= \sqrt{x^2+4} + C$$

(b) $x = 2\tan\theta$

$dx = 2\sec^2\theta d\theta$

$\sqrt{x^2+4} = \sec\theta$

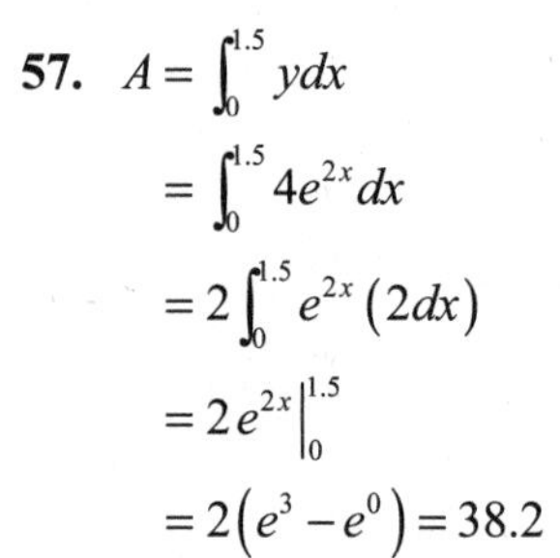

$$\int \frac{xdx}{\sqrt{x^2+4}} = \int \frac{2\tan\theta(2\sec^2\theta d\theta)}{\sec\theta}$$
$$= \int 2\tan\theta\sec\theta d\theta$$
$$= 2\sec\theta + C$$
$$= 2\frac{\sqrt{x^2+4}}{2} + C$$
$$= \sqrt{x^2+4} + C$$

(a) is simpler

57. $A = \int_0^{1.5} ydx$

$$= \int_0^{1.5} 4e^{2x}dx$$
$$= 2\int_0^{1.5} e^{2x}(2dx)$$
$$= 2e^{2x}\Big|_0^{1.5}$$
$$= 2(e^3 - e^0) = 38.2$$

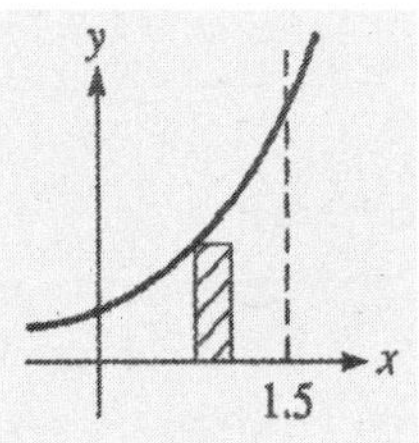

61. $\int \tan^{-1} 2xdx$

$u = \tan^{-1} 2x, \qquad dv = dx$

$du = \frac{1}{1+4x^2}(2dx) \qquad v = x$

$$\int \tan^{-1}(2x)dx = x\tan^{-1}2x - \int x\frac{2dx}{1+4x^2}$$

$$= x\tan^{-1}2x - \frac{1}{4}\int\frac{8x\,dx}{1+4x^2}$$
$$= x\tan^{-1}2x - \frac{1}{4}\ln\left(1+4x^2\right)$$
$$A = \int_0^2 \tan^{-1}2x\,dx = x\tan^{-1}2x - \frac{1}{4}\ln\left(1+4x^2\right)\Big|_0^2$$
$$A = 2\tan^{-1}4 - \frac{1}{4}\ln 17 = 1.94$$

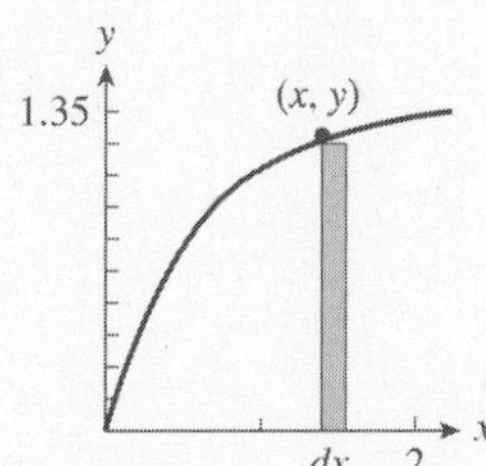

65. $V = \int_0^{\pi}\pi\left(e^x\sin x\right)^2 dx$
$$= \pi\int_0^{\pi}e^{2x}\sin^2 x\,dx$$
$$= \frac{\pi}{2}\int_0^{\pi}e^{2x}\left(1-\cos 2x\right)dx$$
$$= \frac{\pi}{4}e^{2x}\Big|_0^{\pi} - \frac{\pi}{2}\int_0^{\pi}e^{2x}\cos 2x\,dx$$
$$= \frac{\pi}{4}\left(e^{2\pi}-1\right) - \frac{\pi e^{2x}}{8}\left(\cos 2x+\sin 2x\right)\Big|_0^{\pi}$$
$$= \frac{\pi}{4}\left(e^{2\pi}-1\right) - \frac{\pi}{8}\left(e^{2\pi}-1\right)$$
$$= \frac{\pi}{8}\left(e^{2\pi}-1\right) = 210$$

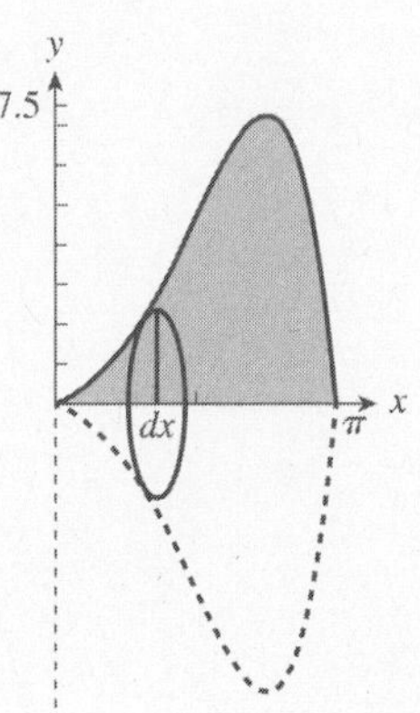

69. $I = \frac{1}{2}\int_0^{0.25}\frac{5}{1+2t}(2dt)$
$$= \frac{5}{2}\ln\left(1+2t\right)\Big|_0^{0.25} = 1.01\text{ N}\cdot\text{s}$$

73. $y = 16.0\left(e^{x/32}+e^{-x/32}\right)$, $x = -25.0$ to $x = 25.0$
$$L = \sqrt{1+\left(\frac{dy}{dx}\right)^2}\,dx$$
$$\frac{dy}{dx} = \frac{1}{2}\left(e^{x/32}-e^{-x/32}\right)$$
$$1+\left(\frac{dy}{dx}\right)^2 = \frac{e^{x/16}-e^{-x/16}+2}{4}$$
$$\sqrt{1+\left(\frac{dy}{dx}\right)^2} = \frac{e^{x/32}-e^{-x/32}}{2}$$
$$L = 2\int_0^{25}\frac{e^{x/32}-e^{-x/32}}{2}dx = 32\,e^{x/32}-e^{-x/32}\Big|_0^{25}$$
$$L = 32\left(e^{25/32}-e^{-25/32}\right) = 55.24375836$$
$$L = 55.2\text{ m}$$

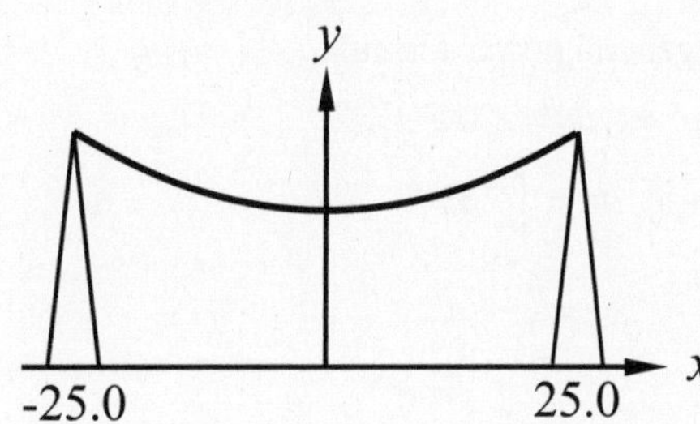

77. $y_{rms} = \sqrt{\frac{1}{T}\int i^2\,dt} = \sqrt{\frac{1}{T}\int\left(2\sin t\right)^2 dt}$
$$\int_0^{2\pi}\left(2\sin t\right)^2 dt = \int_0^{2\pi}4\sin^2 t\,dt = 4\int_0^{2\pi}\sin^2 t\,dt$$
$$= 4\left(\frac{t}{2}-\frac{1}{2}\sin t\cos t\right)\Big|_0^{2\pi}$$

Formula 29 in table of integrals.
$$= 2t - 2\sin t\cos t\Big|_0^{2\pi} = 4\pi;\ T = 2\pi$$
$$y_{rms} = \sqrt{\frac{1}{2\pi}(4\pi)} = \sqrt{2}$$

81. $V=\pi\int_{2.00}^{4.00} y^2\,dx=\pi\int_{2.00}^{4.00} e^{-0.2x}\,dx$; and

$u=-0.2x\; du=-0.2dx$

$$=-\frac{\pi}{0.2}\int_{2.00}^{4.00} e^{-0.2x}(-0.2)\,dx$$

$$=\frac{-\pi}{0.2}e^{-0.2x}\Big|_{2.00}^{4.00}$$

$$=-\frac{\pi}{2}\left[e^{-0.8}-e^{-0.4}\right]=3.47\text{ cm}^3$$

85. (a) Integrate using the double angle formula:

$$A=\int_0^{\pi}(4-y)\,dx$$

$$=\int_0^{\pi}\left(4-4\cos^2 x\right)dx$$

$$=\int_0^{\pi}4\left(1-\cos^2 x\right)dx$$

$$=\int_0^{\pi}\left(4-4\frac{1+\cos 2x}{2}\right)dx$$

$$=\left[2x-\sin 2x\right]\Big|_0^{\pi}$$

$$=2\pi$$

(b) Integrate using a trigonometric identity and then formula 29

$$A=\int_0^{\pi}(4-y)\,dx$$

$$=\int_0^{\pi}4\left(1-\cos^2 x\right)dx$$

$$=\int_0^{\pi}4\sin^2 x\,dx$$

$$=4\left[\frac{x}{2}-\frac{1}{2}\sin x\cos x\right]\Bigg|_0^{\pi}$$

$$A=2\pi$$

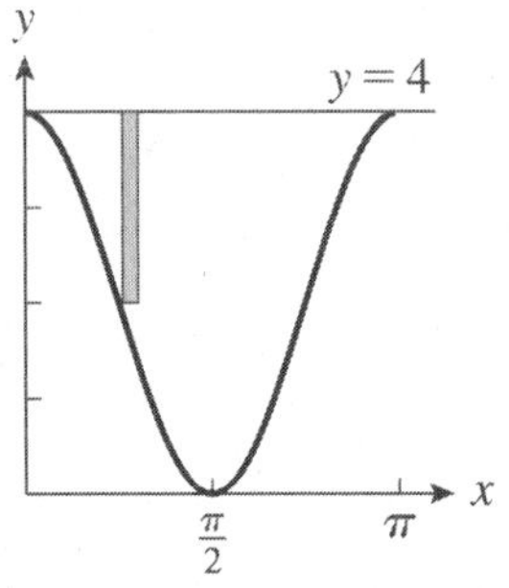

CHAPTER 29

PARTIAL DERIVATIVES AND DOUBLE INTEGRALS

29.1 Functions of Two Variables

1. $f(x, y) = 3x^2 + 2xy - y^3$

$f(-2, 1) = 3(-2)^2 + 2(-2)(1) - (1)^3$

$= 7$

5. From geometry:

$A = 2\pi rh + 2\pi r^2;$

$V = \pi r^2 h$ so write h as a function of V:

$h = \dfrac{V}{\pi r^2}$

Substitute in A:

$A = 2\pi r\left(\dfrac{V}{\pi r^2}\right) + 2\pi r^2$

$= \dfrac{2V}{r} + 2\pi r^2$

9. $f(x, y) = 2x - 6y$

$f(0, -4) = 2(0) - 6(-4) = 24$

13. $Y(y,t) = \dfrac{2-3y}{t-1} + 2y^2 t$

$Y(y,2) = \dfrac{2-3y}{2-1} + 2y^2(2)$

$= 2 - 3y + 4y^2$

17. $H(p,q) = p - \dfrac{p - 2q^2 - 5q}{p+q}$

$H(p,q+k)$

$= p - \dfrac{p - 2(q+k)^2 - 5(q+k)}{p+q+k}$

$= \dfrac{p(p+q+k) - p + 2(q^2 + 2kq + k^2) + 5(q+k)}{p+q+k}$

$= \dfrac{p^2 + pq + pk - p + 2q^2 + 4kq + 2k^2 + 5q + 5k}{p+q+k}$

21. $f(x,y) = xy + x^2 - y^2$

$f(x,x) - f(x,0)$

$= x(x) + x^2 - x^2 - [x(0) + x^2 - 0^2]$

$= x^2 + x^2 - x^2 - x^2$

$= 0$

25. $f(x,y) = \frac{\sqrt{y}}{2x}$; considering $\sqrt{y}, y \ge 0$ for real values of $f(x,y)$; considering $2x, x \ne 0$ to avoid division by zero. Thus, $y < 0$ and $x = 0$ are not permissible.

29. $V = iR;\ i = 3\ A,\ R = 6\ \Omega$

$V = 3(6) = 18$ V

33. For a, b, with the same sign: circle if $a = b$, ellipse if $a \ne b$.

For a and b with different signs: hyperbola.

37. $p = 2\ell + 2w; \ell = \dfrac{p-2w}{2}$

$A = \ell w = \dfrac{p-2w}{2}w = \dfrac{pw - 2w^2}{2}$

$p = 250$ cm, $w = 55$ cm

$A = \dfrac{250(55) - 2(55)^2}{2} = 3850\ \text{cm}^2$

29.2 Curves and Surfaces in Three Dimensions

1. $3x - y + 2z + 6 = 0$

Intercepts: $(0, 0, -3), (0, 6, 0), (-2, 0, 0)$

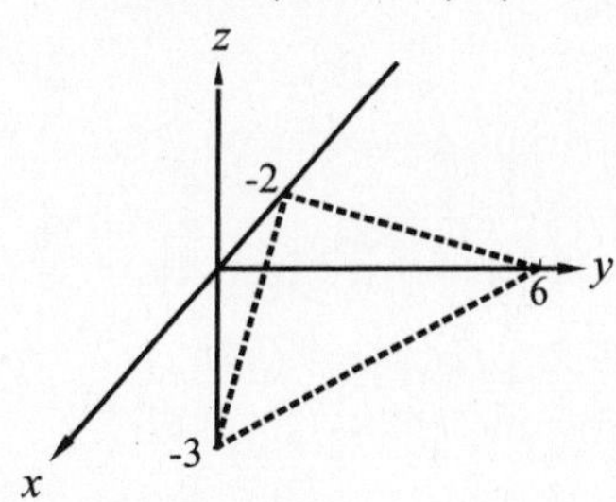

5. $x + y + 2z - 4 = 0$; plane

Intercepts: $(4, 0, 0), (0, 4, 0), (0, 0, 2)$

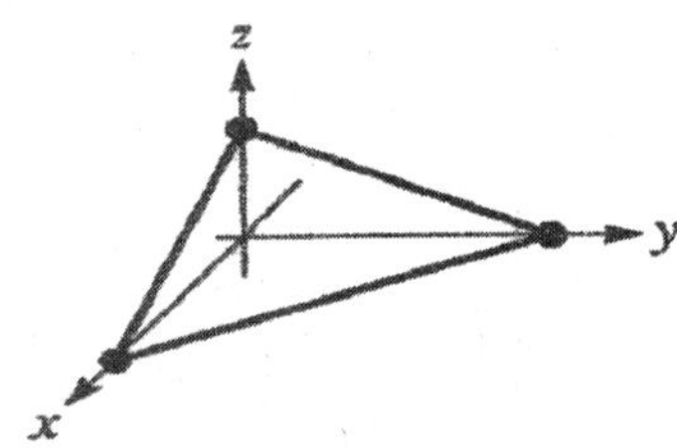

9. $z = y - 2x - 2$; plane

Intercepts: $(-1, 0, 0), (0, 2, 0), (0, 0, -2)$

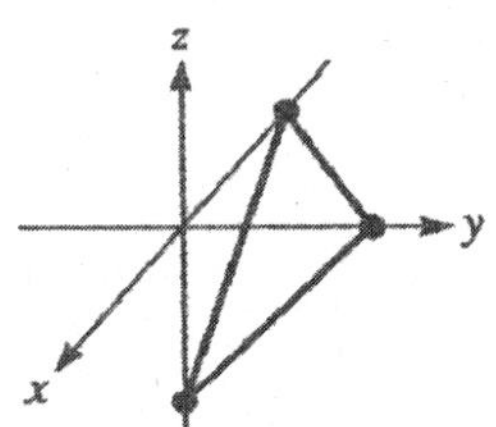

13. $x^2 + y^2 + z^2 = 4$

Intercepts: $(\pm 2, 0, 0), (0, \pm 2, 0), (0, 0, \pm 2)$

Traces:

yz-plane: $y^2 + z^2 = 4$, circle, $r = 2$

xz-plane: $x^2 + z^2 = 4$, circle, $r = 2$

xy-plane: $x^2 + y^2 = 4$, circle, $r = 2$

The surface is a sphere with radius 2.

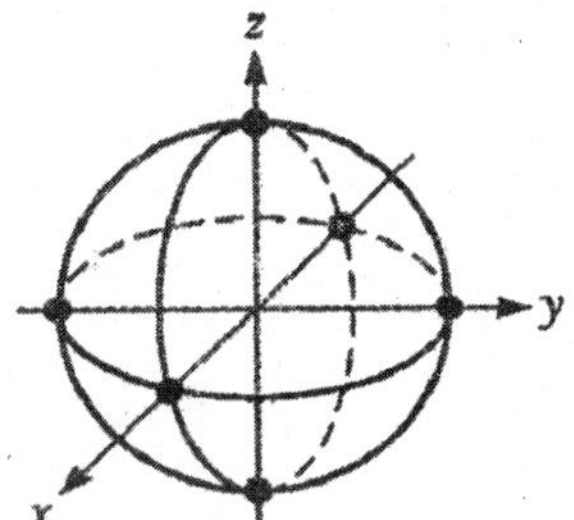

17. $z = 2x^2 + y^2 + 2$

Intercepts: No *x*-intercept, no *y*-intercept, (0,0,2)

Traces:

yz-plane:: $z = y^2 + 2$; parabola, $V(0, 0, 2)$

xz-plane: $z = 2x^2 + 2$, parabola, $V(0, 0, 2)$

xy-plane: No trace, $(2x^2 + y^2 + 2 \neq 0)$

Section: For $z = 4, 2x^2 + y^2 = 2$, ellipse

The surface is an elliptical paraboloid.

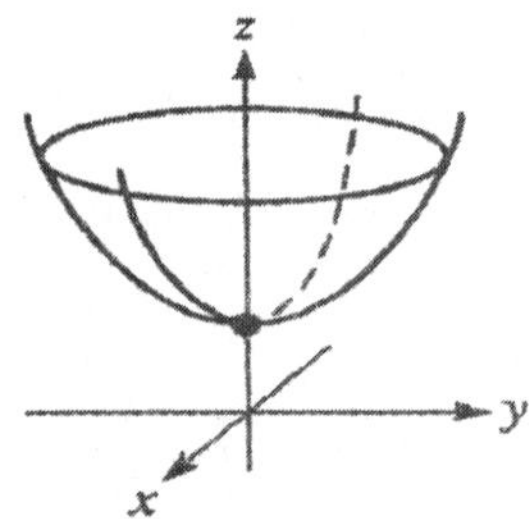

21. $x^2 + y^2 = 16$

Intercepts: $(\pm 4, 0, 0), (0, \pm 4, 0)$, no *z*-intercept.

Traces and sections:

Since *z* is not present in the equation, the trace and sections are circles $x^2 + y^2 = 16$, with $r = 4$, for all *z*. This is a cylindrical surface.

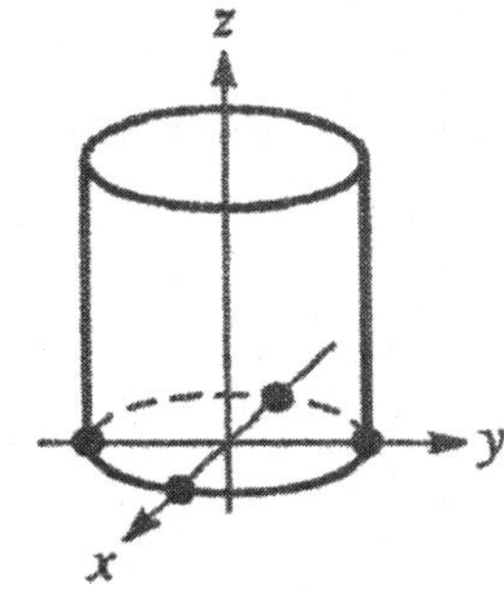

25. **(a)** $\left(3, \frac{\pi}{4}, 5\right)$

$$x = r\cos\theta = 3\cos\frac{\pi}{4} = \frac{3\sqrt{2}}{2}$$

$$y = r\sin\theta = 3\cos\frac{\pi}{4} = \frac{3\sqrt{2}}{2}$$

$$z = z = 5$$

$$\left(\frac{3\sqrt{2}}{2}, \frac{3\sqrt{2}}{2}, 5\right)$$

(b) $\left(2, \frac{\pi}{2}, 3\right)$

$$x = r\cos\theta = 2\cos\frac{\pi}{2} = 0$$

$$y = r\sin\theta = 2\sin\frac{\pi}{2} = 2$$

$$z = z = 3$$

$$(0, 2, 3)$$

(c) $\left(4, \frac{\pi}{3}, 2\right)$

$$x = r\cos\theta = 4\cos\frac{\pi}{3} = 4\left(\frac{1}{2}\right) = 2$$

$$y = r\sin\theta = 4\sin\frac{\pi}{3} = 4\left(\frac{\sqrt{3}}{2}\right) = 2\sqrt{3}$$

$$z = z = 2$$

$$\left(2, 2\sqrt{3}, 2\right)$$

29. $r^2 = 4z$

$$x^2 + y^2 = 4z$$

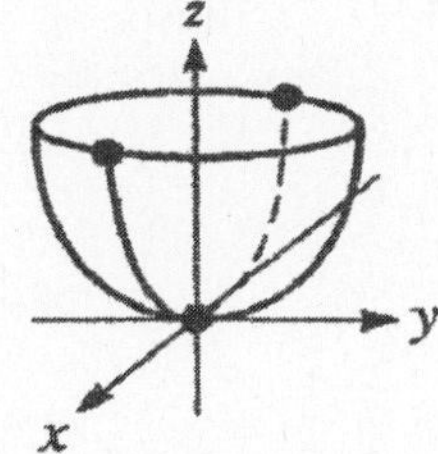

33. $2x^2 + 2y^2 + 3z^2 = 6$

Intercepts: $(\pm\sqrt{3},0,0),(0,\pm\sqrt{3},0),(0,0,\pm\sqrt{2})$

Traces:

yz-plane: $2y^2 + 3z^2 = 6$,

ellipse, $a = \sqrt{3}, b = \sqrt{2}$

xz-plane: $2x^2 + 3z^2 = 6$,

ellipse, $a = \sqrt{3}, b = \sqrt{2}$

xy-plane: $2x^2 + 2y^2 = 6$,

circle, $x^2 + y^2 = 3, r = \sqrt{3}$

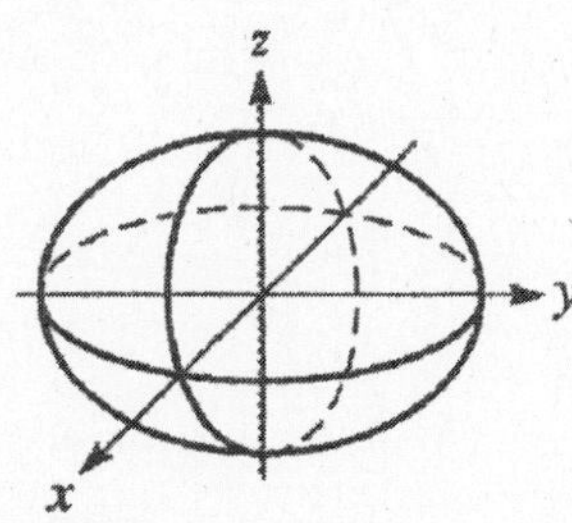

37. $x^2 + y^2 - 2y = 0$

Since z does not appear in the equation, all the traces and sections are circles,

$$x^2 + y^2 - 2y + 1 = 1$$

$$x^2 + (y-1)^2 = 1$$

with centre $(0, 1, z)$ for z and $r = 1$. This is a cylindrical surface.

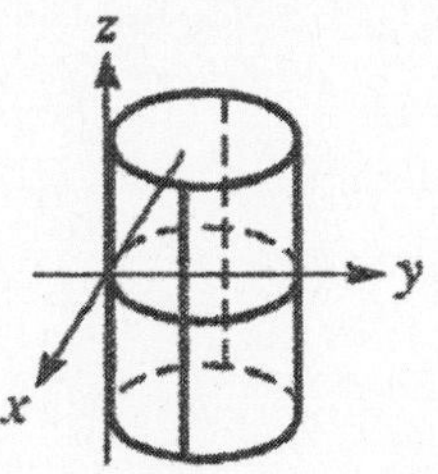

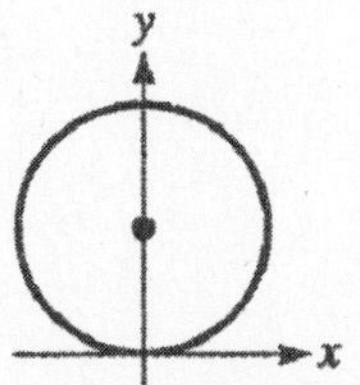

29.3 Partial Derivatives

1. $z = \dfrac{x\ln y}{y^2+1}$

$$\frac{\partial z}{\partial x} = \frac{\ln y}{y^2+1}$$

$$\frac{\partial z}{\partial y} = \frac{x\left(y^2+1\right)1/y - 2xy\ln y}{\left(y^2+1\right)^2}$$

$$\frac{\partial z}{\partial y} = \frac{x\left(y^2+1\right) - 2xy^2\ln y}{y\left(y^2+1\right)^2}$$

5. $f(x,y) = xe^{2y}$

$$\frac{\partial f}{\partial x} = e^{2y}$$

$$\frac{\partial f}{\partial y} = xe^{2y}(2) = 2xe^{2y}$$

9. $\phi = r\sqrt{1+2rs}$

$$\frac{\partial \phi}{\partial r} = r\left(\frac{1}{2}\right)(1+2rs)^{-1/2}(2s) + (1+2rs)^{1/2}$$

$$= \frac{rs}{(1+2rs)^{1/2}} + (1+2rs)^{1/2}$$

$$= \frac{1+3rs}{\sqrt{1+2rs}}$$

$$\frac{\partial\phi}{\partial s} = r\left(\frac{1}{2}\right)(1+2rs)^{-1/2}(2r)$$
$$= \frac{r^2}{\sqrt{1+2rs}}$$

13. $z = \sin xy$

$$\frac{\partial z}{\partial x} = (\cos xy)(y)$$
$$= y\cos xy$$
$$\frac{\partial z}{\partial y} = (\cos xy)(x)$$
$$= x\cos xy$$

17. $f(x,y) = \dfrac{2\sin^3 2x}{1-3y}$

$$\frac{\partial f}{\partial x} = \frac{2(3)(\sin^2 2x)(\cos 2x)(2)}{1-3y}$$
$$= \frac{12\sin^2 2x\cos 2x}{1-3y}$$
$$\frac{\partial f}{\partial y} = -(2\sin^3 2x)(1-3y)^{-2}(-3)$$
$$= \frac{6\sin^3 2x}{(1-3y)^2}$$

21. $z = \sin x + \cos xy - \cos y$

$$\frac{\partial z}{\partial x} = \cos x - (\sin xy)(y)$$
$$= \cos x - y\sin xy$$
$$\frac{\partial z}{\partial y} = -(\sin xy)(x) + \sin y$$
$$= -x\sin xy + \sin y$$

25. $z = 3xy - x^2$

$$\frac{\partial z}{\partial x} = 3y - 2x$$
$$\left.\frac{\partial z}{\partial x}\right|_{(1,-2,-7)} = 3(-2) - 2(1) = -8$$

29. $z = 2xy^3 - 3x^2y$

$$\frac{\partial z}{\partial x} = 2y^3 - 6xy$$
$$\frac{\partial z}{\partial y} = 6xy^2 - 3x^2$$
$$\frac{\partial^2 z}{\partial x^2} = -6y, \frac{\partial^2 z}{\partial x^2} = 12xy$$
$$\frac{\partial^2 z}{\partial x\partial y} = \frac{\partial^2 z}{\partial x\partial y} = 6y^2 - 6x$$

33. $z = 9 - x^2 - y^2$

$$\frac{\partial z}{\partial y} = -2y$$
$$\left.\frac{\partial z}{\partial y}\right|_{(1,2,4)} = -4$$
$$\left.\frac{\partial z}{\partial y}\right|_{(2,2,1)} = -4$$

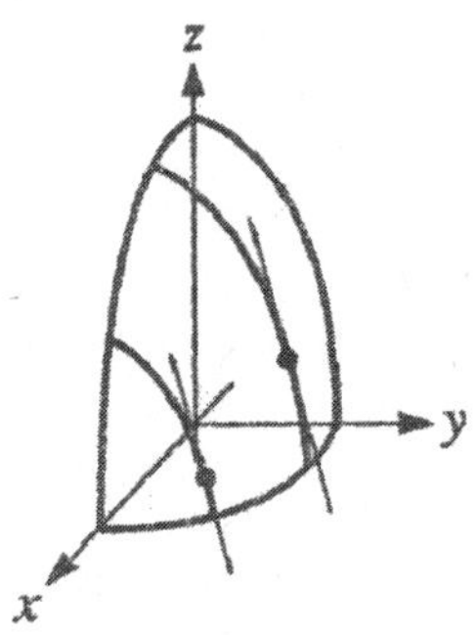

37. $V = \pi r^2 h + \dfrac{1}{2}\left(\dfrac{4}{3}\pi r^3\right)$

$$\frac{\partial V}{\partial r} = 2\pi rh + 2\pi r^2$$
$$\left.\frac{\partial V}{\partial r}\right|_{r=2.65,h=4.20} = 2\pi(2.65)(4.20) + 2\pi(2.65)^2$$
$$\frac{\partial V}{\partial r} = 114\text{ cm}^2$$

41. $i_b = 50(e_b + 5e_c)^{1.5}$

$$\frac{\partial i_b}{\partial e_c} = 50(1.5)(e_b + 5e_c)^{0.5}(5)$$
$$= 375(e_b + 5e_c)^{0.5}$$

For $e_b = 200$ V and $e_c = -20$ V

$$\frac{\partial i_b}{\partial e_c} = 375(200 - 100)^{0.5}$$
$$= 3750\ \mu\text{A/V} = 3.75\,10^{-3}1/\Omega$$

45. Laplace's equation is $\dfrac{\partial^2 u}{\partial x^2} + \dfrac{\partial^2 u}{\partial y^2} = 0.$

Let $u(x,y) = e^{-x}\sin y$. Then

$$\frac{\partial u}{\partial x} = -e^{-x}\sin y \qquad \frac{\partial u}{\partial y} = e^{-x}\cos y$$

$$\frac{\partial^2 u}{\partial x^2} = e^{-x}\sin y \qquad \frac{\partial^2 u}{\partial y^2} = -e^{-x}\sin y$$

Substituting into Laplace's equation:

$$\frac{\partial^2 u}{\partial x^2} + \frac{\partial^2 u}{\partial y^2} = e^{-x}\sin y - e^{-x}\sin y = 0,$$

so the equation is satisfied.

29.4 Double Integrals

1. $$\int_0^1\int_{x^2}^{x}(x+y)\,dy\,dx = \int_0^1\left(xy+\frac{y^2}{2}\right)\Bigg|_{x^2}^{x}dx$$

$$= \int_0^1\left(x^2+\frac{x^2}{2}-\left(x^3+\frac{x^4}{2}\right)\right)dx$$

$$= \int_0^1\left(-\frac{x^4}{2}-x^3+\frac{3x^2}{2}\right)dx$$

$$= -\frac{x^5}{10}-\frac{x^4}{4}+\frac{3x^2}{6}\Bigg|_0^1 = \frac{3}{20}$$

5. $$\int_1^2\int_0^{y^2} xy^2\,dx\,dy = \int_1^2 y^2\left(\frac{1}{2}x^2\right)\Bigg|_0^{y^2}dy$$

$$= \int_1^2 y^2\left(\frac{1}{2}y^4\right)dy$$

$$= \frac{1}{2}\int_1^2 y^6\,dy$$

$$= \frac{1}{14}y^7\Bigg|_1^2$$

$$= \frac{127}{14}$$

9. $$\int_0^{\pi/6}\int_{\pi/3}^{y}\sin x\,dx\,dy$$

$$= \int_0^{\pi/6}(-\cos x)\Big|_{\pi/3}^{y}dy$$

$$= -\int_0^{\pi/6}\left(\cos y-\cos\frac{\pi}{3}\right)dy$$

$$= -\int_0^{\pi/6}\left(\cos y-\frac{1}{2}\right)dy$$

$$= -\sin y+\frac{1}{2}y\Bigg|_0^{\pi/6}$$

$$= -\sin\frac{\pi}{6}+\frac{\pi}{12}$$

$$= \frac{\pi}{12}-\frac{1}{2}$$

$$= \frac{\pi-6}{12}$$

13. $$\int_1^2\int_0^x yx^3e^{xy^2}\,dy\,dx = \frac{1}{2}\int_1^2\int_0^x x^2\left(2xye^{xy^2}\right)dy\,dx$$

$$= \frac{1}{2}\int_1^2 x^2\left(e^{xy^2}\right)\Big|_0^x dx$$

$$= \frac{1}{2}\int_1^2 x^2\left(e^{x\cdot x^2}-1\right)dx$$

$$= \frac{1}{6}\int_1^2 3x^2e^{x^3}dx-\frac{1}{2}\int_1^2 x^2dx$$

$$= \frac{1}{6}e^{x^3}-\frac{1}{6}x^3\Bigg|_1^2$$

$$= \frac{1}{6}\left[e^8-8-(e-1)\right]$$

$$= 495$$

17. The trace of the surface in the xy plane is $y = 4-x$, so y goes from 0 to $4-x$. The x-intercept of the trace is $x = 4$, so x goes from to 0 to 4:

$$V = \int_0^4\int_x^{4-x} z\,dy\,dx = \int_0^4\int_x^{4-x}(4-x-y)\,dy\,dx$$

$$= \int_0^4\left(4y-xy-\frac{1}{2}y^2\right)\Bigg|_0^{4-x}dx$$

$$= \int_0^4\left[4(4-x)-x(4-x)-\frac{1}{2}(4-x)^2\right]dx$$

$$= \int_0^4\left(8-4x+\frac{1}{2}x^2\right)dx$$

$$= 8x-2x^2+\frac{1}{6}x^3\Bigg|_0^4$$

$$= 32-32+\frac{64}{6}=\frac{32}{3}$$

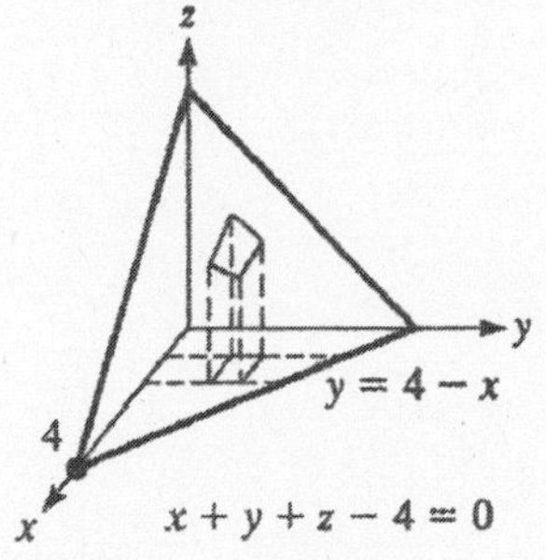

21. y varies between x and 2, and x varies between 0 and 2.

$$V = \int_0^2 \int_x^2 z\,dy\,dx$$
$$= \int_0^2 \int_x^2 \left(2+x^2+y^2\right)dy\,dx$$
$$= \int_0^2 \left(2y+x^2y+\frac{1}{3}y^3\right)\Big|_x^2 dx$$
$$= \int_0^2 \left(4+2x^2+\frac{8}{3}-2x-x^3-\frac{1}{3}x^3\right)dx$$
$$= \int_0^2 \left(\frac{20}{3}-2x+2x^2-\frac{4}{3}x^3\right)dx$$
$$= \frac{20}{3}x-x^2+\frac{2}{3}x^3-\frac{1}{3}x^4\Big|_0^2$$
$$= \frac{40}{3}-4+\frac{16}{3}-\frac{16}{3}=\frac{28}{3}$$

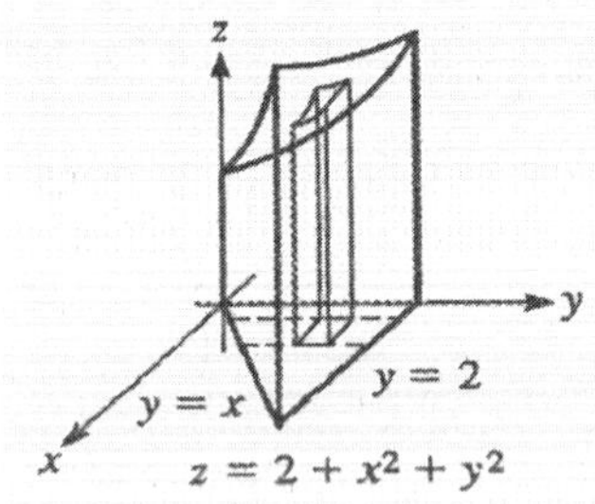

25. Set up the origin in the left-hand back corner of the wedge. Then the volume of interest is the volume under the plane $z=10-2y$. (This is the trace on the zy-plane, and it is a cylindrical surface.) It is bounded by $x=0$, $x=12$, $y=0$, $y=5$. We have

$$\int_0^5 \int_0^{12} z\,dx\,dy = \int_0^5 \int_0^{12} (10-2y)\,dx\,dy$$
$$= \int_0^5 (10-2y)x\Big|_0^{12} dy$$
$$= 12\int_0^5 (10-2y)dy$$
$$= 12\left(10y-y^2\right)\Big|_0^5$$
$$= 12(50-25)$$
$$= 300 \text{ cm}^2$$

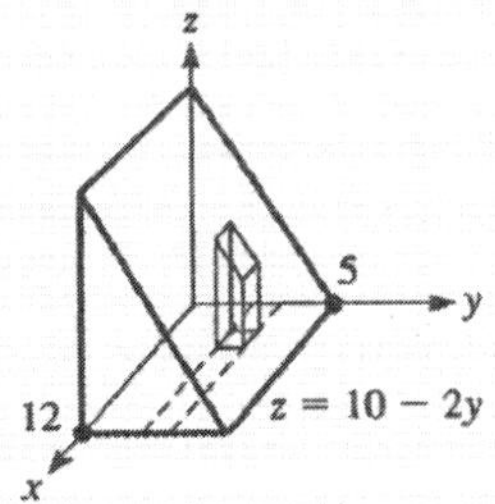

Review Exercises

1. $f(x, y) = 3x^2y - y^3$

$f(-1, 4) = 3(-1)^2(4) - 4^3 = -52$

5. $x - y + 2z - 4 = 0$, plane

intercepts: $(0, 0, 2)$, $(0, -4, 0)$, $(4, 0, 0)$

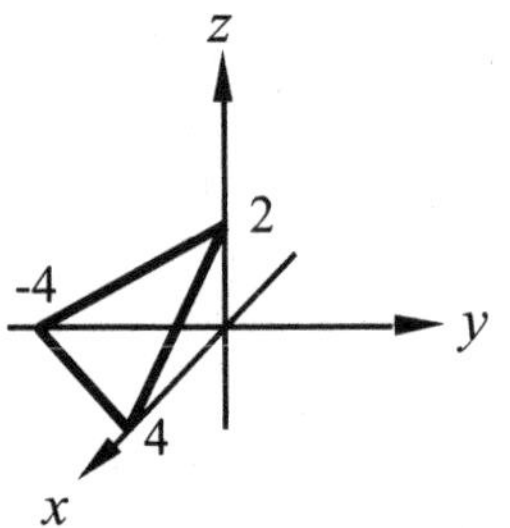

9. $z = 5x^3y^2 - 2xy^4$

$$\frac{\partial z}{\partial x} = 5y^2\left(3x^2\right) - 2y^4(1) = 15x^2y^2 - 2y^4$$
$$\frac{\partial z}{\partial y} = 5x^3(2y) - 2x\left(4y^3\right) = 10x^3y - 8xy^3$$

13. $z = \dfrac{2x-3y}{x^2y+1}$

$$\frac{\partial z}{\partial x} = \frac{\left(x^2y+1\right)(2)-(2x-3y)(2xy)}{\left(x^2y+1\right)^2}$$
$$= \frac{2-2x^2y+6xy^2}{\left(x^2y+1\right)^2}$$
$$\frac{\partial z}{\partial y} = \frac{\left(x^2y+1\right)(-3)-(2x-3y)\left(2x^2\right)}{\left(x^2y+1\right)^2}$$
$$= \frac{-\left(3+2x^3\right)}{\left(x^2y+1\right)^2}$$

17. $z = \sin^{-1}\sqrt{x+y}$

$$\frac{\partial z}{\partial x} = \frac{1}{\sqrt{1-(x+y)}}\left(\frac{1}{2}\right)(x+y)^{-1/2}(1)$$
$$= \frac{1}{2\sqrt{(x+y)(1-x-y)}}$$
$$\frac{\partial z}{\partial y} = \frac{1}{\sqrt{1-(x+y)}}\left(\frac{1}{2}\right)(x+y)^{-1/2}(1)$$
$$= \frac{1}{2\sqrt{(x+y)(1-x-y)}}$$

21. $\int_0^2 \int_1^2 (3y+2xy)\,dx\,dy = \int_0^2 \left(3xy+x^2y\right)\Big|_1^2\,dy$

$= \int_0^2 (6y+4y-3y-y)\,dy$

$= \int_0^2 6y\,dy$

$= 3y^2\Big|_0^2 = 12$

25. $\int_0^1 \int_0^{2x} x^2 e^{xy}\,dy\,dx = \int_0^1 \int_0^{2x} x\left(e^{xy}x\,dy\right)dx$

$= \int_0^1 x e^{xy}\Big|_0^{2x}\,dx$

$= \int_0^1 x\left(e^{2x^2}-1\right)dx$

$= \frac{1}{4}e^{2x^2} - \frac{1}{2}x^2\Big|_0^1$

$= \frac{1}{4}e^2 - \frac{1}{2} - \frac{1}{4}$

$= \frac{1}{4}\left(e^2-3\right) = 1.10$

29. $z = \sqrt{x^2+4y^2}$

Intercepts: $(0, 0, 0)$

Traces:

$z \geq 0$ for all x and y

(defined by positive square root)

In yz-plane: $z = \pm 2y$

In xz-plane: $z = \pm x$

In xy-plane: $(0, 0, 0)$

Section: For $z = 2$

$x^2+4y^2 = 4$ (ellipse)

The surface is an elliptical cone.

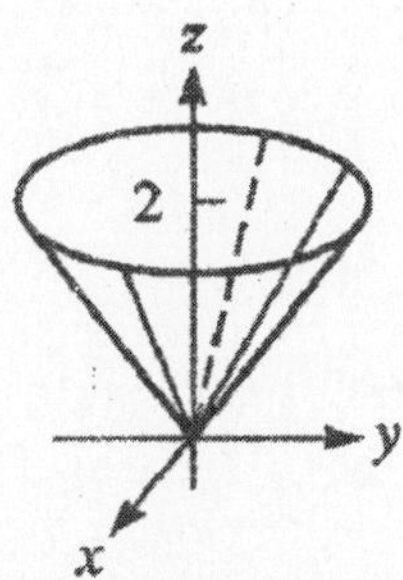

33. **(a)** $\theta = 3$ is a vertical half-plane

(b) $z = r^2 = x^2+y^2$

Intercepts: $(0, 0, 0)$

Traces are parabolas, sections are circles

The surface is a circular paraboloid.

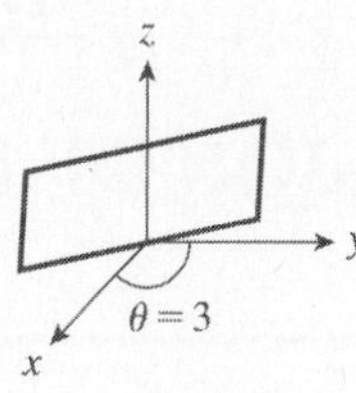

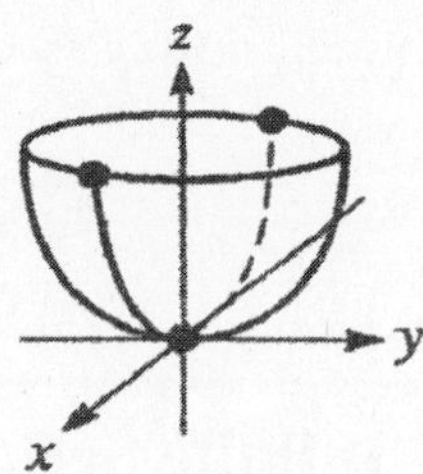

37. $i_c = i_e\left(1-e^{-2V_c}\right)$

$\alpha = \frac{\partial i_c}{\partial i_e} = 1-e^{-2V_c}$

For $V_c = 2$ V,

$\alpha = 1-e^{-4} = 0.982$

41. $L = L_0 + k_1F + k_2T + k_3FT^2$

$\alpha = \frac{1}{L}\frac{\partial L}{\partial T}$

$\alpha = \frac{k_2+2k_3FT}{L_0+k_1F+k_2T+k_3FT^2}$

45. The region is bounded by the vertical plane $y = 0$, the vertical plane $x = 0$, the cylinder $x^2+y^2 = 16$, and is under the plane $z = 8-x$.

$V = \int_0^4 \int_0^{\sqrt{16-x^2}} z\,dy\,dz$

$= \int_0^4 \int_0^{\sqrt{16-x^2}} (8-x)\,dy\,dz$

$= \int_0^4 (8-x)y\Big|_0^{\sqrt{16-x^2}}\,dx$

$= \int_0^4 (8-x)\sqrt{16-x^2}\,dx$

$= 8\int_0^4 \sqrt{16-x^2}\,dx - \int_0^4 x\sqrt{16-x^2}\,dx$

(formula 15 and the power rule)

$= 8\left[\frac{x}{2}\sqrt{16-x^2} + \frac{16}{2}\sin^{-1}\frac{x}{4}\right] + \frac{1}{3}\left(16-x^2\right)^{3/2}\Big|_0^4$

$= 8\left[8\sin^{-1}1\right] - \frac{1}{3}(16)^{3/2}$

$= 64\sin^{-1}1 - \frac{64}{3}$

$= 64\left(\frac{\pi}{2}\right) - \frac{64}{3} = 32\left(\pi - \frac{2}{3}\right)$

$= 79.2$

CHAPTER 30

EXPANSION OF FUNCTIONS IN SERIES

30.1 Infinite Series

1. $\sum_{u=1}^{\infty} 0.5^n = 0.5 + 0.5^2 + 0.5^3 + 0.5^4 + \cdots + 0.5^n + \cdots$

$S_1 = 0.5, S_2 = 0.75, S_3 = 0.875, S_4 = 0.9375.$

The series now converges

5. $a_n = \dfrac{1}{n+2}; n = 0, 1, 2, 3, \ldots$

$a_0 = \dfrac{1}{0+2} = \dfrac{1}{2}$ $\quad a_2 = \dfrac{1}{2+2} = \dfrac{1}{4}$

$a_1 = \dfrac{1}{1+2} = \dfrac{1}{3}$ $\quad a_3 = \dfrac{1}{3+2} = \dfrac{1}{5}$

9. $a_n = \cos\dfrac{n\pi}{2}, n = 0, 1, 2, 3, \ldots$

$a_0 = \cos\dfrac{0\cdot\pi}{2} = 1$

$a_1 = \cos\dfrac{1\cdot\pi}{2} = 0$

$a_2 = \cos\dfrac{2\cdot\pi}{2} = -1$

$a_3 = \cos\dfrac{3\cdot\pi}{2} = 0$

(a) $1, 0, -1, 0$

(b) $1 + 0 - 1 + 0 \ldots$

13. $\dfrac{1}{2\times 3} - \dfrac{1}{3\times 4} + \dfrac{1}{4\times 5} - \dfrac{1}{5\times 6} + \cdots$

$n = 1,\ a_1 = \dfrac{1}{(1+1)(1+2)} = \dfrac{1}{2\times 3}$

$n = 2,\ a_2 = \dfrac{-1}{(2+1)(2+2)} = \dfrac{-1}{3\times 4}$

$a_n = \dfrac{(-1)^{n+1}}{(n+1)(n+2)}$

17. $1 + \dfrac{1}{2} + \dfrac{2}{3} + \dfrac{3}{4} + \dfrac{4}{5} + \ldots$

$S_0 = 1;\ S_1 = 1 + \dfrac{1}{2} = \dfrac{3}{2} = 1.5$

$S_2 = 1 + \dfrac{1}{2} + \dfrac{2}{3} = \dfrac{13}{16} = 2.1666667$

$S_3 = 1 + \dfrac{1}{2} + \dfrac{2}{3} + \dfrac{3}{4} = \dfrac{35}{12} = 2.9166667$

$S_4 = 1 + \dfrac{1}{2} + \dfrac{2}{3} + \dfrac{3}{4} + \dfrac{4}{5} = \dfrac{223}{60} = 3.7166667$

The partial sums do not seem to converge, so the series appears to be divergent.

21. $\sum_{n=1}^{\infty} \dfrac{2n+1}{n^2(n+1)^2}$

First five terms:

$a_1 = \dfrac{3}{4}; a_2 = \dfrac{5}{36}; a_3 = \dfrac{7}{144}; a_4 = \dfrac{9}{400}; a_5 = \dfrac{11}{900}$

First five partial sums:

$S_1 = 0.75;$

$S_2 = \dfrac{3}{4} + \dfrac{5}{36} = 0.8888889$

$S_3 = \dfrac{3}{4} + \dfrac{5}{36} + \dfrac{7}{144} = 0.9375000$

$S_4 = \dfrac{3}{4} + \dfrac{5}{36} + \dfrac{7}{144} + \dfrac{9}{400} = 0.9600000$

$S_5 = \dfrac{3}{4} + \dfrac{5}{36} + \dfrac{7}{144} + \dfrac{9}{400} + \dfrac{11}{900} = 0.9722222$

Convergent, converging to 1 (approx. sum)

25. $1 + 2 + 4 + \cdots + 2^n + \cdots; n = 0, 1, 2, 3, \cdots$

$S_0 = 1$

$S_1 = 3$

$S_2 = 7$

$S_3 = 15$

$S_n = 2^{n+1} - 1$

$\lim_{n\to\infty} Sn = \lim_{n\to\infty}(2^{n+1} - 1) = \infty$, divergent

Also, it is a geometric series with $r = 2 > 1$, so the series is divergent.

29. $10 + 9 + 8.1 + 7.29 + 6.561 + \cdots + 10(0.9)^n + \cdots;$

$n = 0, 1, 2, 3, \ldots$

$a = 10, r = 0.9 < 1$, so the series is convergent.

$S = \dfrac{10}{1 - 0.9} = 100$

33. $\sum_{n=0}^{\infty}(x-4)^n$ is a GS with $a_1=1$, $r=x-4$ which, converges for

$|x-4|<1$

$-1<x-4<1$

$3<x<5$

37. $S_n=\dfrac{a_1(1-r^n)}{(1-r)}$; $r\neq 1$; geometric series

Series: $\frac{1}{2}+\frac{1}{4}+\frac{1}{8}+\cdots$; $a_n=\frac{1}{2^n}, a=\frac{1}{2}, r=\frac{1}{2}$

$f(x)=\dfrac{a_1(1-r^x)}{(1-r)}$; $f(x)=\dfrac{(1-r^x)}{(1-\frac{1}{2})}=(1-r^x)$

x	y
0	0
1	$\frac{1}{2}$
2	$\frac{3}{4}$
3	$\frac{7}{8}$
4	$\frac{15}{16}$
5	$\frac{31}{32}$

The infinite series approaches 1.

41. $\sum_{n=0}^{\infty}x^n=1+x+x^2+\cdots+x^n+\cdots$

For $|x|<1$, $a_1=1, r=x$, and the series converges because $|r|<1$.

$S=\dfrac{1}{1-x}$

$\sum_{n=0}^{\infty}x^n=\dfrac{1}{1-x}$

30.2 Maclaurin Series

1. $f(x)=\dfrac{2}{2+x}$, $f(0)=1$

$f'(x)=\dfrac{-2}{(2+x)^2}$, $f'(0)=-\dfrac{1}{2}$

$f''(x)=\dfrac{4}{(2+x)^3}$, $f''(0)=\dfrac{1}{2}$

$f'''(x)=\dfrac{-12}{(2+x)^4}$, $f'''(0)=-\dfrac{3}{4}$

$f(x)=\dfrac{2}{2+x}=1-\dfrac{1}{2}x+\dfrac{1}{4}x^2-\dfrac{1}{8}x^3+\cdots$

5. $f(x)=\cos x \quad f(0)=1$

$f'(x)=-\sin x \quad f'(0)=0$

$f''(x)=-\cos x \quad f''(0)=-1$

$f'''(x)=\sin x \quad f'''(0)=0$

$f^{iv}(x)=\cos x \quad f^{iv}(0)=1$

$f(x)=\cos x=f(0)+f''(0)\dfrac{x^2}{2!}+f^{iv}(0)\dfrac{x^4}{4!}-\cdots$

$\cos x=1-1\dfrac{x^2}{2}+1\dfrac{x^4}{24}-\cdots$

$\cos x=1-\dfrac{1}{2}x^2+\dfrac{1}{24}x^4-\cdots$

9. $f(x)=e^{-2x} \quad f(0)=1$

$f'(x)=-2e^{-2x} \quad f'(0)=-2$

$f''(x)=4e^{-2x} \quad f''(0)=4$

$e^{-2x}=1-2x+4\dfrac{x^2}{2}-\cdots$

$=1-2x+2x^2-\cdots$

13. $f(x)=\dfrac{1}{(1-x)} \quad f(0)=1$

$f'(x)=\dfrac{1}{(1-x)^2} \quad f'(0)=1$

$f''(x)=\dfrac{2}{(1-x)^3} \quad f''(0)=2$

$\dfrac{1}{(1-x)}=1+x+\dfrac{2x^2}{2}+\cdots$

$=1+x+x^2+\cdots$

17. $f(x)=\cos^2 x \quad f(0)=1$

$f'(x)=-2\sin x\cos x \quad f'(0)=0$

$f''(x)=2-4\cos^2 x \quad f''(0)=-2$

$f'''(x)=8\sin x\cos x \quad f'''(0)=0$

$f^{iv}(x)=16\cos^2 x-8 \quad f^{iv}(0)=8$

$\cos^2 x=1-2\dfrac{x^2}{2!}+8\dfrac{x^4}{4!}+\cdots$

$=1-x^2+\dfrac{1}{3}x^4+\cdots$

21. $f(x)=\tan^{-1}x$

$$f'(x)=\frac{1}{1+x^2}=(1+x^2)^{-1}$$
$$f''(x)=-(1+x^2)^{-2}2x=-2x(1+x^2)^{-2}$$
$$f'''(x)=-2x[-2(1+x^2)^{-3}(2x)]+(1+x^2)^{-2}(-2)$$
$$f(0)=0$$
$$f'(0)=1$$
$$f''(0)=0$$
$$f'''(0)=-2$$
$$f(x)=0+1x+\frac{0x^2}{2!}-\frac{2x^3}{3!}+\cdots$$
$$=x-\frac{1}{3}x^3+\cdots$$

25. $f(x)=\ln\cos x$

$$f'(x)=-\frac{1}{\cos x}\sin x=-\tan x$$
$$f''(x)=-\sec^2 x$$
$$f'''(x)=-2\sec x\sec x\tan x=-2\sec^2 x\tan x$$
$$f^{iv}(x)=-2\sec^2 x\sec^2 x-2\tan x(2\sec x\sec x\tan x)$$
$$f(0)=\ln 1=0$$
$$f'(0)=0$$
$$f''(0)=-1$$
$$f'''(0)=0$$
$$f^{iv}(0)=-2-0=-2$$
$$f(x)=0+0x-\frac{1x^2}{2!}+\frac{0x^3}{3!}-\frac{2x^4}{4!}+\cdots$$
$$=-\frac{1}{2}x^2-\frac{1}{12}x^4-\cdots$$

29. **(a)** It is not possible to find a Maclaurin's expansion for $f(x)=\csc x$ since $\csc x$ is not defined when $x=0$.

(b) It is not possible to find a Maclauri's expansion for $f(x)=\ln x$ since $\ln x$ is not defined when $x=0$.

33. The Maclaurin's expansion of $f(x)=e^{3x}$ is

$f(x)=1+3x+\frac{9}{2}x^2+\frac{9}{2}x^3+\cdots$. The linearization is $L(x)=1+3x$, the first two terms of the expansion.

37. $f(x)=x^2, f'(x)=2x, f''(x)=2, f^{(n)}(x)=0, n\ge 3$

$$f(0)=0, f'(0)=0, f''(0)=2, f^{(n)}(x)=0, n\ge 3$$
$$x^2=0+0(x)+2\frac{x^2}{2!}+0\frac{x^3}{3!}+\cdots$$
$$x^2=x^2$$

41. $0\le R\le 1; R=e^{-0.001t}$

$$\frac{dR}{dt}=-0.001e^{-0.001t};\frac{d^2R}{dt^2}=1\times10^{-6}e^{-0.001t}$$
$$f(0)=1; f'(x)=-0.001; f''(0)=1\times10^{-6}$$
$$e^{-0.001t}=1-0.001t+1\times10^{-6}\frac{t^2}{2!}-\cdots$$
$$=1-0.001t+10\times10^{-7}\frac{t^2}{2}-\cdots$$
$$=1-0.001t+(5\times10^{-7})t^2-\cdots$$

30.3 Operations with Series

1. $e^x=1+x+\frac{x^2}{2!}+\frac{x^3}{3!}+\cdots$

$$e^{2x^2}=1+2x^2+\frac{(2x^2)^2}{2!}+\frac{(2x^2)^3}{3!}+\cdots$$
$$e^{2x^2}=1+2x^2+2x^4+\frac{4}{3}x^6+\cdots$$

5. $f(x)=\sin\left(\frac{1}{2}x\right);$

$$g(x)=\sin x=x-\frac{x^3}{3!}+\frac{x^5}{5!}-\frac{x^7}{7!}+\cdots$$
$$f(x)=g\left(\frac{1}{2}x\right)$$
$$=\frac{1}{2}x-\frac{\left(\frac{1}{2}x\right)^3}{3!}+\frac{\left(\frac{1}{2}x\right)^5}{5!}-\frac{\left(\frac{1}{2}x\right)^7}{7!}+\cdots$$
$$=\frac{1}{2}x-\frac{x^3}{2^3 3!}+\frac{x^5}{2^5 5!}-\frac{x^7}{2^7 7!}+\cdots$$

9. $f(x)=\ln\left(1+x^2\right)$

$$g(x)=\ln(1+x)$$
$$=x-\frac{x^2}{2}+\frac{x^3}{3}-\frac{x^4}{4}+\cdots$$

$$\ln\left(1+x^2\right)=g(x^2)$$
$$=x^2-\frac{\left(x^2\right)^2}{2}+\frac{\left(x^2\right)^3}{3}-\frac{\left(x^2\right)^4}{4}+\cdots$$
$$=x^2-\frac{1}{2}x^4+\frac{1}{3}x^6-\frac{1}{4}x^8+\cdots$$

13. $\int_0^{0.2}\cos\sqrt{x}dx=\int_0^{0.2}\left(1-\frac{\left(\sqrt{x}\right)^2}{2}+\frac{\left(\sqrt{x}\right)^4}{24}\right)dx$
$$=\int_0^{0.2}\left(1-\frac{1}{2}x+\frac{1}{24}x^2\right)dx$$
$$=\left(x-\frac{1}{4}x^2+\frac{1}{72}x^3\right)\Bigg|_0^{0.2}$$
$$=0.2-\frac{1}{4}(0.2)^2+\frac{1}{72}(0.2)^3$$
$$=0.190$$

17. $e^x\sin x=f(x)$

$$e^x=1+x+\frac{x^2}{2!}+\frac{x^3}{3!}+\cdots \quad (1)$$
$$\sin x=x-\frac{x^3}{3!}+\frac{x^5}{5!}-\cdots \quad (2)$$

Multiply (1) by x from (2): $x+x^2+\frac{x^3}{2!}+\frac{x^4}{3!}+\cdots$

Multiply (1) by $-\frac{x^3}{3!}$ from (2): $-\frac{x^3}{3!}-\frac{x^4}{3!}-\cdots$

Combine:
$$e^x\sin x=x+x^2+\frac{x^3}{2!}+\frac{x^4}{3!}-\frac{x^3}{3!}-\frac{x^4}{3!}+\cdots$$
$$=x+x^2+\frac{1}{2}x^3-\frac{1}{6}x^3+\cdots$$
$$=x+x^2+\frac{2}{6}x^3+\cdots$$
$$=x+x^2+\frac{1}{3}x^3+\cdots$$

21. Replacing x in $\ln\left(1+x\right)=x-\frac{1}{2}x^2+\frac{1}{3}x^3-\frac{1}{4}x^4+\ldots$ with $\sin x$ gives

$$\ln\left(1+\sin x\right)=\sin x-\frac{1}{2}\sin^2 x+\frac{1}{3}\sin^3 x-\frac{1}{4}\sin^4 x+\ldots$$

from which, replacing $\sin x$ with

$$\sin x=x-\frac{1}{6}x^3+\frac{1}{120}x^5-\ldots,$$
$$\ln(1+\sin x)=x-\frac{1}{6}x^3+\frac{1}{120}x^5-\ldots$$
$$-\frac{1}{2}\left(x-\frac{1}{6}x^3+\frac{1}{120}x^5-\ldots\right)^2$$
$$+\frac{1}{3}\left(x-\frac{1}{6}x^3+\frac{1}{120}x^5-\ldots\right)^3-\ldots$$

Considering through the x^4 terms:
$$\ln(1+\sin x)=x-\frac{1}{6}x^3-\frac{1}{2}x^2+\frac{1}{6}x^4+\frac{1}{3}x^3+\ldots$$
$$\ln(1+\sin x)=x-\frac{1}{2}x^2+\frac{1}{6}x^3+\frac{1}{6}x^4+\ldots$$

25.
$$\cos x=1-\frac{x^2}{2!}+\frac{x^4}{4!}-\frac{x^6}{6!}+\cdots$$
$$\int\cos x\,dx$$
$$=\int dx-\frac{1}{2}\int x^2dx+\frac{1}{4!}\int x^4dx$$
$$-\frac{1}{6!}\int x^6dx+\cdots$$
$$=x-\frac{1}{2}\frac{x^3}{3}+\frac{1}{4!}\frac{x^5}{5}-\frac{1}{6!}-\frac{1}{6!}\frac{x^7}{7}+\cdots$$
$$=x-\frac{x^3}{3!}+\frac{x^5}{5!}-\frac{x^7}{7!}+\cdots=\sin x$$

29. $\int_0^1 e^x dx=e^x\Big|_0^1=e-e^0=e-1$
$$=2.7182818-1=1.7182818$$
$$=1.72$$
$$e^x=1+x+\frac{x^2}{2!}+\frac{x^3}{3!}+\cdots$$
$$\int_0^1\left(1+x+\frac{x^2}{2}+\frac{x^3}{6}\right)dx=x+\frac{x^2}{2}+\frac{x^3}{6}+\frac{x^4}{24}\Bigg|_0^1$$
$$=1+\frac{1}{2}+\frac{1}{6}+\frac{1}{24}$$
$$=1.7083333=1.71$$

The approximation is correct to 2 significant digits.

33. $y=x^2e^x; x=0.2, x\text{-axis}$
$$A_{0,0.2}=\int_0^{0.2}x^2e^xdx$$
$$=\int_0^{0.2}x^2\left(1+x+\frac{x^2}{2}\right)dx$$

$$= \int_0^{0.2}\left(x^2 + x^3 + \frac{1}{2}x^4\right)dx$$

$$= \frac{x^3}{3} + \frac{x^4}{4} + \frac{x^5}{10}\Big|_0^{0.2}$$

$$= \frac{1}{3}(0.2)^3 + \frac{1}{4}(0.2)^4 + \frac{1}{10}(0.2)^5$$

$$= 0.00310$$

37. $K = \left[\left(1 - \frac{v^2}{c^2}\right)^{-1/2} - 1\right]mc^2$

$$= \left[1 - \frac{1}{2}\left(\frac{-v^2}{c^2}\right) + \frac{-\frac{1}{2}\left(-\frac{1}{2} - 1\right)}{2!}\left(\frac{-v^2}{c^2}\right)^2 + \cdots - 1\right]mc^2$$

$$= \left[\frac{1}{2}\frac{v^2}{c^2} + \frac{3}{8}\frac{v^4}{c^4} + \cdots\right]mc^2$$

$$= \frac{1}{2}mv^2 + \frac{3}{8}\frac{mv^4}{c^2} + \ldots$$

$= \frac{1}{2}mv^2$ for v much smaller than c.

41. $y_1 = e^x, y_2 = 1, y_3 = 1 + x, y_4 = 1 + x + \frac{1}{2}x^2;$

$x_{\min} = -5, x_{\max} = 5, y_{\min} = -1, y_{\max} = 3$

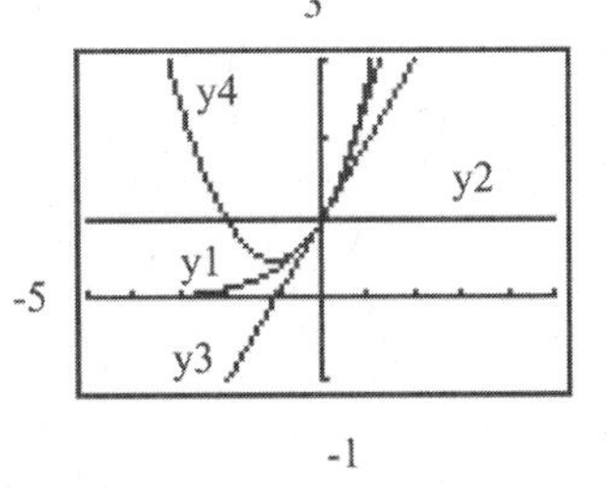

30.4 Computations by Use of Series Expansions

1. $e^x = 1 + x + \frac{x^2}{2!} + \cdots$

$$e^{-0.1} = 1 + (-0.1) + \frac{(-0.1)^2}{2!} + \cdots$$

$$e^{-0.1} = 0.905$$

5. $\sin 0.1, (2 \text{ terms});$

$$\sin x = x - \frac{x^3}{3!}$$

$$\sin 0.1 = 0.1 - \frac{(0.1)^3}{6}$$

$$= 0.0998333$$

$(\sin 0.1 = 0.0998334$ from calculator$)$

9. $\cos \pi^\circ$, (2 terms); $\pi^\circ = \frac{\pi^2}{180}$ radians

$$\cos x = 1 - \frac{x^2}{2!}$$

$$\cos \pi^\circ = 1 - \frac{\left(\frac{\pi^2}{180}\right)^2}{2}$$

$$= 0.9984967733$$

$(\cos \pi^\circ = 0.9984971499$ from calculator$)$

13. $\sin 0.3625, (3 \text{ terms})$

$$\sin x = x - \frac{x^3}{3!} + \frac{x^5}{5!}$$

$$\sin 0.3625 = 0.3625 - \frac{(0.3625)^3}{6} + \frac{(0.3625)^5}{5!}$$

$$= 0.35461303$$

$(\sin 0.3625 = 0.35461287$ from calculator$)$

17. 1.032^6, 3 terms

$$(1+x)^6 = 1 + 6x + 15x^2$$

$$(1.032) = 1 + 6(0.032) + 15(0.032)^2$$

$$= 1.20736$$

$((1.032)^6 = 1.20803,$ from calculator$)$

21. $\sqrt{1+x} = (1+x)^{1/2}$

$$(1+x)^n = 1 + nx + \frac{n(n-1)}{2!}x^2$$

$$(1+x)^{1/2} = 1 + \frac{1}{2}x + \frac{\frac{1}{2}\left(-\frac{1}{2}\right)}{2!}x^2$$

$$= 1 + \frac{1}{2}x - \frac{1}{8}x^2$$

$$\sqrt{1.1076} = \sqrt{1 + 0.1076} = (1 + 0.1076)^{1/2}$$

$$(1 + 0.1076)^{1/2} = 1 + \frac{1}{2}(0.1076) - \frac{1}{8}(0.1076)^2$$

$$= 1.052353$$

$(\sqrt{1.1076} = 1.052426$ from calculator$)$

25. From Exercise 5, $\sin(0.1) = 0.1 - \dfrac{0.1^3}{6} = 0.0998$

The maximum possible error is the value of the first term omitted,

$$\frac{x^5}{5!} = \left|\frac{0.1^5}{120}\right| = 8.3\times10^{-8}$$

29. $(1+x)^n = 1 + nx + \dfrac{n(n-1)}{2!}x^2$

$$\sqrt{3.92} = 2\left(1+(-0.02)\right)^{1/2}$$
$$= 2\left[1+\frac{1}{2}(-0.02)+\frac{\frac{1}{2}\left(\frac{1}{2}-1\right)}{2!}(-0.02)^2\right]$$
$$= 1.9799$$

33. $e^x = 1+x+\dfrac{x^2}{2}+\dfrac{x^3}{3!}+\dfrac{x^4}{4!}+\cdots > 1+x+\dfrac{x^2}{2}$

for $x > 0$ since the terms of the expansion for e^x after those on right-hand side of the inequality have a positive value.

37. $f(t) = \dfrac{E}{R}\left(1-e^{-Rt/L}\right);$

$$e^x = 1+x+\frac{x^2}{2}+\cdots$$
$$e^{-Rt/L} = 1-\frac{Rt}{L}+\frac{R^2t^2}{2L^2}+\cdots$$
$$i = \frac{E}{R}\left[1-\left(1-\frac{Rt}{L}+\frac{R^2t^2}{2L^2}\right)\right] = \frac{E}{L}\left(t-\frac{Rt^2}{2L}\right)$$

The approximation will be valid for small values of t.

30.5 Taylor Series

1. $f(x) = x^{1/2}, f(1) = 1$

$$f'(x) = \frac{1}{2x^{1/2}}, f'(1) = \frac{1}{2}$$
$$f''(x) = -\frac{1}{4x^{3/2}}, f''(1) = -\frac{1}{4}$$
$$f'''(x) = \frac{3}{8x^{5/2}}, f'''(1) = \frac{3}{8}$$
$$\sqrt{x} = 1+\frac{1}{2}(x-1)+\frac{-\frac{1}{4}(x-1)^2}{2!}+\frac{\frac{3}{8}(x-1)^3}{3!}+\cdots$$
$$\sqrt{x} = 1+\frac{1}{2}(x-1)-\frac{1}{8}(x-1)^2+\frac{1}{16}(x-1)^3-\cdots$$

5. $\sqrt{4.2}$

$$\sqrt{x} = 2+\frac{(x-4)}{4}-\frac{(x-4)^2}{64}+\frac{(x-4)^3}{512}$$
$$\sqrt{4.2} = 2+\frac{(4.2-4)}{4}-\frac{(4.2-4)^2}{64}+\frac{(4.2-4)^3}{512}$$
$$= 2.049;$$

($\sqrt{4.2} = 2.04939$ from calculator)

9. $\sin x = \dfrac{1}{2}+\dfrac{\sqrt{3}}{2}\left(x-\dfrac{\pi}{6}\right)-\dfrac{1}{4}\left(x-\dfrac{\pi}{6}\right)^2$

$$x = 29.53^\circ$$
$$= \frac{1}{2}+\frac{\sqrt{3}}{2}\left(\frac{29.53\pi}{180}-\frac{\pi}{6}\right)-\frac{1}{4}\left(\frac{29.53\pi}{180}-\frac{\pi}{6}\right)^2$$
$$= 0.49288;$$

($\sin 29.53^\circ = 0.4928792$ from calculator)

13. $\sin x; a = \dfrac{\pi}{3}$

$$f(x) = \sin x \qquad f\left(\frac{\pi}{3}\right) = \frac{\sqrt{3}}{2}$$
$$f'(x) = \cos x \qquad f'\left(\frac{\pi}{3}\right) = \frac{1}{2}$$
$$f''(x) = -\sin x \qquad f''\left(\frac{\pi}{3}\right) = -\frac{\sqrt{3}}{2}$$
$$\sin x = \frac{\sqrt{3}}{2}+\frac{1}{2}\left(x-\frac{\pi}{3}\right)-\frac{\sqrt{3}}{2}\cdot\frac{1}{2!}\left(x-\frac{\pi}{3}\right)^2-\cdots$$
$$= \frac{1}{2}\left[\sqrt{3}+\left(x-\frac{\pi}{3}\right)-\frac{\sqrt{3}}{2!}\left(x-\frac{\pi}{3}\right)^2+\cdots\right]$$

17. $\tan x;\ a = \dfrac{\pi}{4}$

$$f(x) = \tan x \qquad f\left(\frac{\pi}{4}\right) = 1$$
$$f'(x) = \sec^2 x \qquad f'\left(\frac{\pi}{4}\right) = (\sqrt{2})^2 = 2$$
$$f''(x) = 2\sec x\sec x\tan x = 2\sec^2 x\tan x$$
$$f''(x) = 2(\sqrt{2})^2(1) = 4$$
$$\tan x = 1+2\left(x-\frac{\pi}{4}\right)+\frac{4\left(x-\frac{\pi}{4}\right)^2}{2!}+\cdots$$
$$= 1+2\left(x-\frac{\pi}{4}\right)+2\left(x-\frac{\pi}{4}\right)^2+\cdots$$

21. $f(x)=\dfrac{1}{x+2},\ f(3)=\dfrac{1}{5}$

$$f'(x)=-\frac{1}{(x+2)^2},\ f'(3)=-\frac{1}{25}$$

$$f''(x)=\frac{2}{(x+2)^3},\ f''(3)=\frac{2}{125}$$

$$\frac{1}{x+2}=\frac{1}{5}-\frac{1}{25}(x-3)+\frac{1}{125}(x-3)^2$$

25. $\sqrt{9.3};\ a=9$

$$f(x)=\sqrt{x} \qquad f'(9)=3$$

$$f'(x)=\frac{1}{2\sqrt{x}} \qquad f'(9)=\frac{1}{6}$$

$$f''(x)=-\frac{1}{4x^{3/2}} \qquad f''(9)=-\frac{1}{108}$$

$$\sqrt{x}=3+\frac{1}{6}(x-9)-\frac{1}{108}\frac{(x-9)^2}{2!}$$

$$\sqrt{9.3}=3+\frac{1}{6}(0.3)-\frac{1}{108}\frac{(0.3)^2}{2}=3.0496$$

29. $\sin x=\dfrac{1}{2}\left[\sqrt{3}+\left(x-\dfrac{\pi}{2}\right)-\dfrac{\sqrt{3}}{2}\left(x-\dfrac{\pi}{3}\right)^2\right];\ a=\dfrac{\pi}{3}$

$$61^\circ=60^\circ+1^\circ=\frac{\pi}{3}+\frac{\pi}{180}$$

$$\sin 61^\circ=\frac{1}{2}\left[\sqrt{3}+\frac{\pi}{180}-\frac{\sqrt{3}}{2}\left(\frac{\pi}{180}\right)^2\right]=0.87462$$

33. Expand $f(x)=2x^3+x^2-3x+5$ about $x=1$

$$f'(x)=6x^2+2x-3$$

$$f''(x)=12x+2$$

$$f'''(x)=12$$

$$f(x)=f(1)+f'(1)(x-1)\frac{f''(1)(x-1)^2}{2!}+\frac{f'''(1)(x-1)^2}{3!}$$

$$=2(1)^3+1^2-3(1)+5+\left(6(1)^2+2(1)-3\right)(x-1)$$

$$+\frac{(12(1)+2)(x-1)^2}{2!}+\frac{12(x-1)^3}{3!}$$

$$=5+5(x-1)+7(x-1)^2+2(x-1)^3$$

37. $i=6\sin\pi t, \qquad i(\pi/2)=6\sin\pi^2/2$

$$i'=6\pi\cos\pi t, \qquad i'(\pi/2)=6\pi\cos\pi^2/2$$

$$i''=-6\pi^2\sin\pi t, \quad i''(\pi/2)=-6\pi^2\sin\pi^2/2$$

$$i=6\sin\frac{\pi^2}{2}+6\pi\cos\frac{\pi^2}{2}\left(t-\frac{\pi}{2}\right)$$

$$-3\pi^2\sin\frac{\pi^2}{2}\left(t-\frac{\pi}{2}\right)^2+\cdots$$

41. $f(x)=\dfrac{1}{x};\ x=0\ to\ x=4$

(a) $y_1=\dfrac{1}{x}$

(b) $y_2=\dfrac{1}{2}-\dfrac{1}{4}(x-2)$

Graph in part (b) will fit the graph in part (a) well for values of x close to $x=2$.

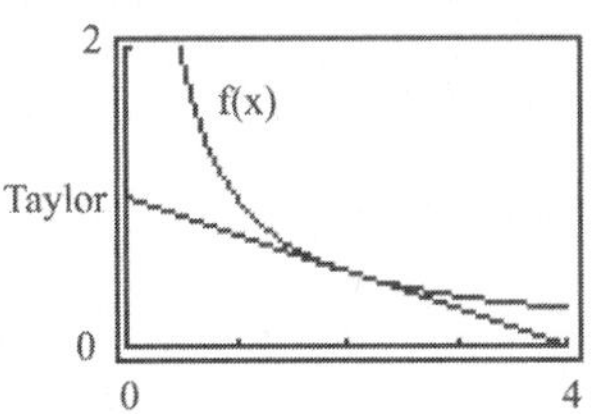

30.6 Introduction to Fourier Series

1. $f(x)=\begin{cases}-2, -\pi\le x<0\\ 2, \quad 0\le x<\pi\end{cases}$

$$a_0=\frac{1}{2\pi}\int_{-\pi}^{0}(-2)dx+\frac{1}{2\pi}\int_0^{\pi}2dx=0$$

$$a_n=\frac{1}{\pi}\int_{-\pi}^{0}-2\cos nx\,dx+\frac{1}{\pi}\int_0^{\pi}2\cos nx\,dx=0$$

$$b_n=\frac{1}{\pi}\int_{-\pi}^{0}-2\sin nx\,dx+\frac{1}{\pi}\int_0^{\pi}2\sin nx\,dx$$

$$=\frac{4}{\pi}\left(\frac{1}{n}-\frac{\cos(\pi n)}{n}\right)\Bigg|_0^{\pi}$$

$$b_n=\frac{4}{\pi}(1-\cos\pi n)\Big|_0^{\pi}=\begin{cases}\dfrac{8}{n\pi}, n\text{ odd}\\ 0, n\text{ neven}\end{cases}$$

$$b_1=\frac{8}{\pi}, b_3=\frac{8}{3\pi}, b_5=\frac{8}{5\pi}$$

$$f(x)=\frac{8}{\pi}\sin x+\frac{8}{3\pi}\sin 3x+\frac{8}{5\pi}\sin 5x+\cdots$$

$$f(x)=\frac{8}{\pi}\left(\sin x+\frac{1}{3}\sin 3x+\frac{1}{5}\sin 5x+\cdots\right)$$

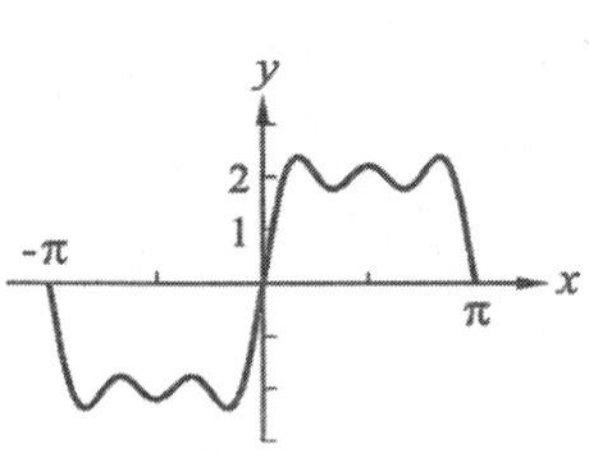

y 2 1 -π π x

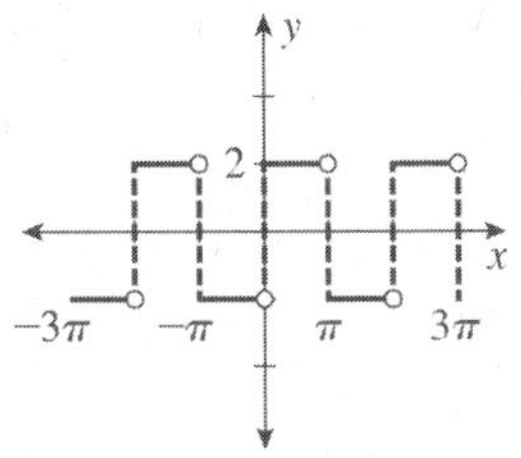

y 2 x −3π −π π 3π

5. $f(x)=\begin{cases}1 & -\pi\le x<0\\ 2 & 0\le x<\pi\end{cases}$

$$a_0=\frac{1}{2\pi}\int_{-\pi}^{0}1\,dx+\frac{1}{2\pi}\int_{0}^{\pi}2\,dx$$
$$=\frac{x}{2\pi}\Big|_{-\pi}^{0}+\frac{2x}{2\pi}\Big|_{0}^{\pi}$$
$$=0+\frac{\pi}{2\pi}+\frac{2\pi}{2\pi}-0=\frac{1}{2}+1=\frac{3}{2}$$
$$a_1=\frac{1}{\pi}\int_{-\pi}^{0}1\cos x\,dx+\frac{1}{\pi}\int_{0}^{\pi}2\cos x\,dx$$
$$=\frac{1}{\pi}\sin x\Big|_{-\pi}^{0}+\frac{2}{\pi}\sin x\Big|_{0}^{\pi}$$
$$=\frac{1}{\pi}(0-0)+\frac{2}{\pi}(0-0)=0$$
$a_n=0$ since $\sin n\pi=0$
$$b_1=\frac{1}{\pi}\int_{-\pi}^{0}1\sin x\,dx+\frac{1}{\pi}\int_{0}^{\pi}2\sin x\,dx$$
$$=-\frac{1}{\pi}\cos x\Big|_{-\pi}^{0}-\frac{2}{\pi}\cos x\Big|_{0}^{\pi}$$
$$=-\frac{1}{\pi}(1+1)-\frac{2}{\pi}(-1-1)$$
$$=-\frac{2}{\pi}+\frac{4}{\pi}=\frac{2}{\pi}$$
$$b_2=\frac{1}{\pi}\int_{-\pi}^{0}1\sin 2x\,dx+\frac{1}{\pi}\int_{0}^{\pi}2\sin x\,dx$$
$$=-\frac{1}{2\pi}\cos 2x\Big|_{-\pi}^{0}-\frac{1}{\pi}\cos 2x\Big|_{0}^{\pi}$$
$$=-\frac{1}{2\pi}(1-1)-\frac{1}{\pi}(1-1)=0$$
$$b_3=\frac{1}{\pi}\int_{-\pi}^{0}\sin 3x\,dx+\frac{1}{\pi}\int_{0}^{\pi}2\sin 3x\,dx$$
$$=-\frac{1}{3\pi}\cos 3x\Big|_{-\pi}^{0}-\frac{2}{3\pi}\cos 3x\Big|_{0}^{\pi}$$
$$=-\frac{1}{3\pi}(1+1)-\frac{2}{3\pi}(-1-1)=\frac{2}{3\pi}$$

Therefore, $b_n=0$ for n even; $b_n=\dfrac{2}{n\pi}$ for n odd.

Therefore, $f(x)=\dfrac{3}{2}+\dfrac{2}{\pi}\sin x+\dfrac{2}{3\pi}\sin 3x+\cdots$

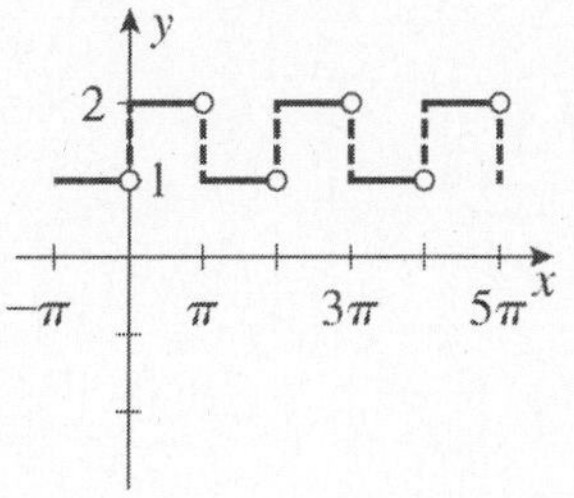

9. $f(x)=\begin{cases}-1 & -\pi\le x<0\\ 0 & 0\le x<\dfrac{\pi}{2}\\ 1 & \dfrac{\pi}{2}\le x<\pi\end{cases}$

$$a_0=\frac{1}{2\pi}\int_{-\pi}^{0}-dx+\frac{1}{2\pi}\int_{\pi/2}^{\pi}dx$$
$$=-\frac{1}{2\pi}x\Big|_{-\pi}^{0}+\frac{1}{2\pi}x\Big|_{\pi/2}^{\pi}$$
$$=-\frac{1}{2\pi}\left[\pi-\left(\pi-\frac{\pi}{2}\right)\right]=-\frac{1}{4}$$
$$a_1=\frac{1}{\pi}\int_{-\pi}^{0}-\cos x\,dx+\frac{1}{\pi}\int_{\pi/2}^{\pi}\cos x\,dx$$
$$=-\frac{1}{\pi}\sin x\Big|_{-\pi}^{0}+\frac{1}{\pi}\sin x\Big|_{\pi/2}^{\pi}$$
$$=-\frac{1}{\pi}(\sin x\,|_{-\pi}^{0}-\sin x\,|_{\pi/2}^{\pi})=-\frac{1}{\pi}$$
$$a_2=\frac{1}{\pi}\int_{-\pi}^{0}-\cos 2x\,dx+\frac{1}{\pi}\int_{\pi/2}^{\pi}\cos 2x\,dx$$
$$=-\frac{1}{2\pi}\sin 2x\Big|_{-\pi}^{0}+\frac{1}{2\pi}\sin 2x\Big|_{\pi/2}^{\pi}$$
$$=-\frac{1}{2\pi}\left(\sin 2x\,|_{-\pi}^{0}-\sin 2x\,|_{\pi/2}^{\pi}\right)=0$$
$$a_3=\frac{1}{\pi}\int_{-\pi}^{0}-\cos 3x\,dx+\frac{1}{\pi}\int_{\pi/2}^{\pi}\cos 3x\,dx$$
$$=-\frac{1}{3\pi}\sin 3x\Big|_{-\pi}^{0}+\frac{1}{3\pi}\sin 3x\Big|_{\pi/2}^{\pi}$$
$$=-\frac{1}{3\pi}\left(\sin 3x\Big|_{-\pi}^{0}-\sin 3x\Big|_{\pi/2}^{\pi}\right)=\frac{1}{3\pi}$$

Therefore,

$$a_n = \pm\frac{1}{n\pi} \text{ for } n \text{ odd; } a_n = 0 \text{ for } n \text{ even}$$

$$b_1 = \frac{1}{\pi}\int_{-\pi}^{0} -\sin x\,dx + \frac{1}{\pi}\int_{\pi/2}^{\pi} \sin x\,dx$$

$$= \frac{1}{\pi}\cos x\Big|_{-\pi}^{0} - \frac{1}{\pi}\cos x\Big|_{\pi/2}^{\pi}$$

$$= \frac{1}{\pi}\left(\cos x\Big|_{-\pi}^{0} - \cos x\Big|_{\pi/2}^{\pi}\right) = \frac{3}{\pi}$$

$$b_2 = \frac{1}{\pi}\int_{-\pi}^{0} -\sin 2x\,dx + \frac{1}{\pi}\int_{\pi/2}^{\pi} \sin 2x\,dx$$

$$= \frac{1}{2\pi}\cos 2x\Big|_{-\pi}^{0} - \frac{1}{2\pi}\cos 2x\Big|_{\pi/2}^{\pi}$$

$$= \frac{1}{2\pi}\left(\cos x\Big|_{-\pi}^{0} - \cos 2x\Big|_{\pi/2}^{\pi}\right) = -\frac{1}{\pi}$$

$$b_3 = \frac{1}{\pi}\int_{-\pi}^{0} -\sin 3x\,dx + \frac{1}{\pi}\int_{\pi/2}^{\pi} \sin 3x\,dx$$

$$= \frac{1}{3\pi}\cos 3x\Big|_{-\pi}^{0} - \frac{1}{3\pi}\cos 3x\Big|_{\pi/2}^{\pi}$$

$$= \frac{1}{3\pi}\left(\cos 3x\Big|_{-\pi}^{0} - \cos 3x\Big|_{\pi/2}^{\pi}\right)$$

$$= \frac{1}{\pi}$$

$$b_n = (-1)^n\frac{1}{\pi} \text{ for } n > 1$$

$$f(x) = -\frac{1}{4} - \frac{1}{\pi}\cos x + \frac{1}{3\pi}\cos 3x - \cdots$$
$$+ \frac{3}{\pi}\sin x - \frac{1}{\pi}\sin 2x + \frac{1}{\pi}\sin 3x - \cdots$$

13. $f(x) = e^x, -\pi \le x < \pi$

$$a_0 = \frac{1}{2\pi}\int_{-\pi}^{\pi} e^x\,dx = \frac{e^{\pi} - e^{-\pi}}{2\pi}$$

$$a_1 = \frac{1}{\pi}\int_{-\pi}^{\pi} e^x\cos x\,dx = -\frac{e^{\pi} - e^{-\pi}}{2\pi}$$

$$a_2 = \frac{1}{\pi}\int_{-\pi}^{\pi} e^x\cos 2x\,dx = \frac{e^{\pi} - e^{-\pi}}{5\pi}$$

$$b_1 = \frac{1}{\pi}\int_{-\pi}^{\pi} e^x\sin x\,dx = -\frac{e^{\pi} - e^{-\pi}}{2\pi}$$

$$b_2 = \frac{1}{\pi}\int_{-\pi}^{\pi} e^x\sin 2x\,dx = -\frac{2(e^{\pi} - e^{-\pi})}{5\pi}$$

$$e^x = \frac{e^{\pi} - e^{-\pi}}{2\pi} - \frac{e^{\pi} - e^{-\pi}}{2\pi}\cos x + \frac{e^{\pi} - e^{-\pi}}{5\pi}\cos 2x + \cdots$$
$$+ \frac{e^{\pi} - e^{-\pi}}{2\pi}\sin x - \frac{2(e^{\pi} - e^{-\pi})}{5\pi}\sin 2x + \cdots$$

$$e^x = \frac{e^{\pi} - e^{-\pi}}{\pi}\left(\frac{1}{2} - \frac{1}{2}\cos x + \frac{1}{5}\cos 2x + \cdots + \frac{1}{2}\sin x\right.$$
$$\left. - \frac{2}{5}\sin 2x + \cdots\right)$$

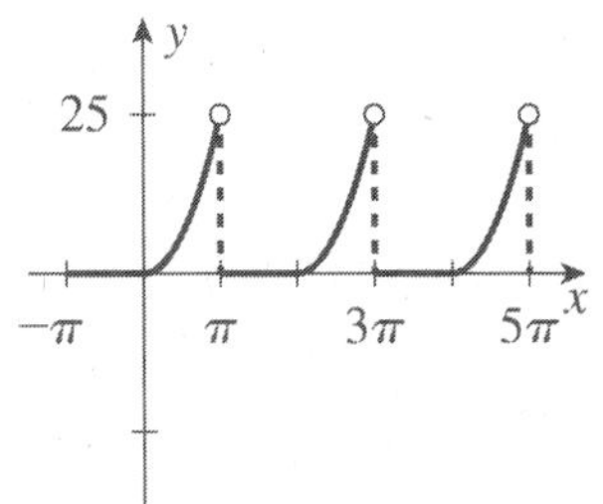

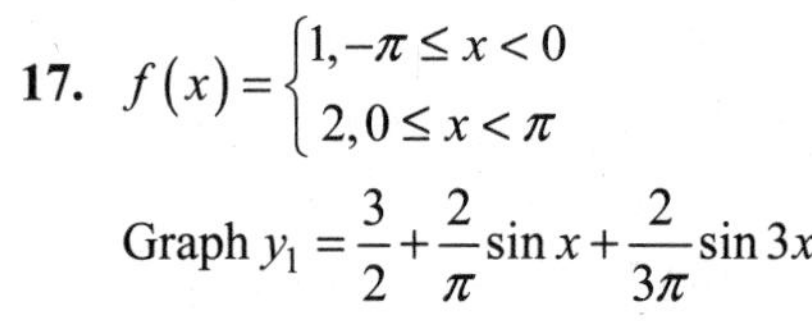

17. $f(x) = \begin{cases} 1, -\pi \le x < 0 \\ 2, 0 \le x < \pi \end{cases}$

Graph $y_1 = \frac{3}{2} + \frac{2}{\pi}\sin x + \frac{2}{3\pi}\sin 3x$

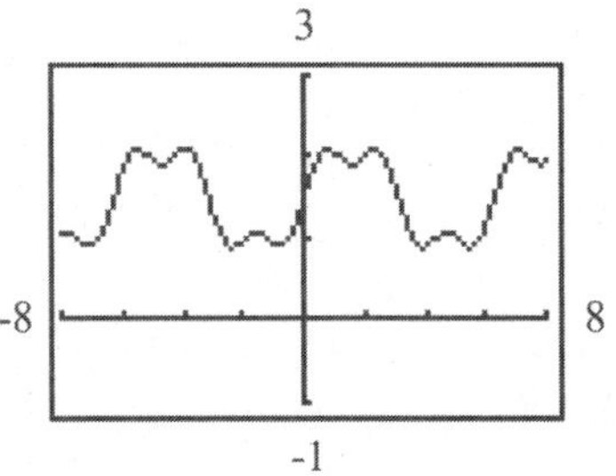

21. $F(t) = \begin{cases} 0, & -\pi \le t < 0 \\ t^2 + t, & 0 < t < \pi \end{cases}$

$$a_0 = \frac{1}{2\pi}\int_0^{\pi}\left(t^2 + t\right)dt = \frac{\pi^2}{6} + \frac{\pi}{4}$$

$$a_1 = \frac{1}{\pi}\int_0^{\pi}\left(t^2 + t\right)\cos t\,dt = -\frac{2}{\pi} - 2$$

$$a_2 = \frac{1}{\pi}\int_0^{\pi}\left(t^2 + t\right)\cos 2t\,dt = \frac{1}{2}$$

$$a_3 = \frac{1}{\pi}\int_0^{\pi}\left(t^2 + t\right)\cos 3t\,dt = \frac{-2 - 2\pi}{9\pi}$$

$$b_1 = \frac{1}{\pi}\int_0^{\pi}\left(t^2+t\right)\sin t\,dt = \pi - \frac{4}{\pi}+1$$
$$b_2 = \frac{1}{\pi}\int_0^{\pi}\left(t^2+t\right)\sin 2t\,dt = \frac{-\pi-1}{2}$$
$$b_3 = \frac{1}{\pi}\int_0^{\pi}\left(t^2+t\right)\sin 3t\,dt = \frac{\pi}{3}-\frac{4}{27\pi}+\frac{1}{3}$$
$$F(t) = \frac{\pi^2}{6}+\frac{\pi}{4}-\left(\frac{2}{\pi}+2\right)\cos t+\frac{1}{2}\cos 2t$$
$$-\left(\frac{2+2\pi}{9\pi}\right)\cos 3t+\cdots+\left(\pi-\frac{4}{\pi}+1\right)\sin t$$
$$-\left(\frac{\pi+1}{2}\right)\sin 2t+\left(\frac{\pi}{3}-\frac{4}{27\pi}+\frac{1}{3}\right)\sin 3t+\cdots$$

30.7 More About Fourier Series

1. $f(x)\begin{cases} 2 & -\pi\le x-\frac{\pi}{2},\frac{\pi}{2}\le x<\pi \\ 3 & -\frac{\pi}{2}\le x<\frac{\pi}{2}\end{cases}$

From Example 3,

$$f_1(x) = \frac{1}{2}+\frac{2}{\pi}\left(\cos x - \frac{\cos 3x}{3}+\frac{\cos 5x}{5}\right)-\cdots$$
$$f(x) = f_1(x)+2$$
$$= \frac{5}{2}+\frac{2}{\pi}\left(\cos x - \frac{\cos 3x}{3}+\frac{\cos 5x}{5}-\cdots\right)$$

5. $f(x)=\begin{cases} 5 & -3\le x<0 \\ 0 & 0\le x<3\end{cases}$

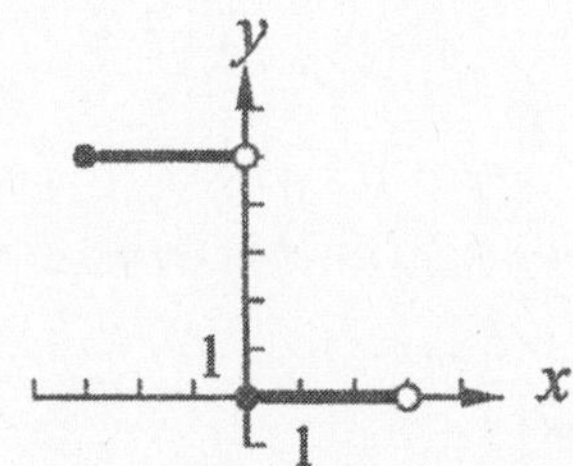

from the graph $f(x)$ is neither odd nor even. (It is, however, a shift of an odd function.)

9. $f(x)=|x| \quad -4\le x<4$

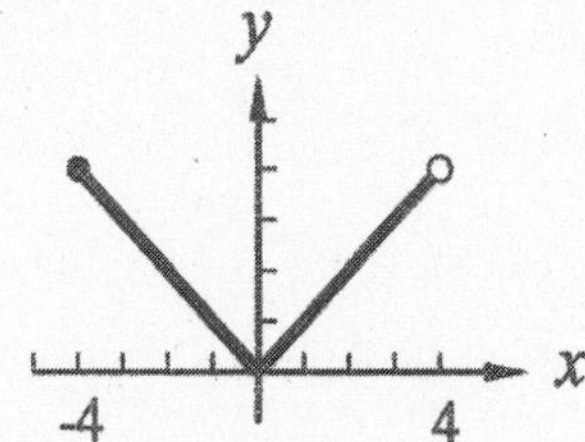

is even from the graph.

13. From the graph, $f(x)=2-x$, $-4\le x<4$ is not odd or even. However, $f(x)$ is the same as the function $f_1(x)=-x$ shifted up 2 units. Since $f_1(x)$ is odd, its expansion contains only sine terms. Therefore, the expansion of $f(x)$ contains only sine terms and the constant term 2.

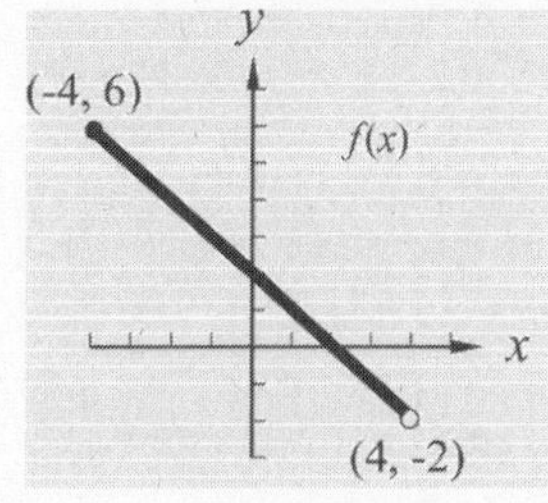

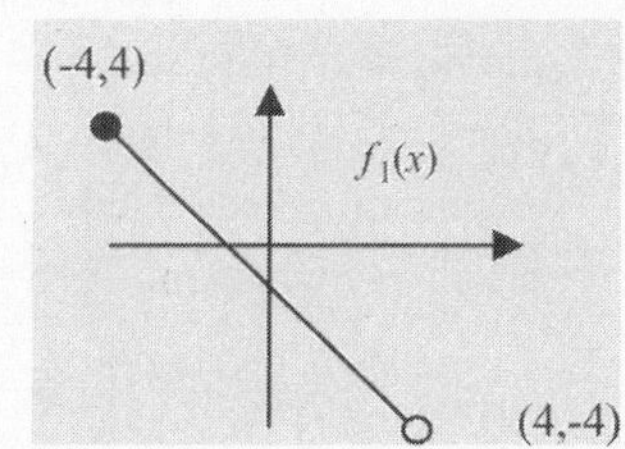

17. $f(x)=\begin{cases} 5 & -3\le x<0 \\ 0 & 0\le x<3\end{cases}$

period $= 6 = 2L$, $L = 3$

Although $f(x)$ is neither odd nor even, it is the same as the function

$$f_1(x)=\begin{cases} \frac{5}{2} & -3\le x<0 \\ -\frac{5}{2} & 0\le x<3\end{cases}$$

shifted up $\frac{5}{2}$ units. Since $f_1(x)$ is odd, the expansion of $f(x)$ will have only sine terms plus the constant $\frac{5}{2}$. Therefore,

$$a_0 = \frac{5}{2}$$
$$a_n = 0$$
$$b_n = \frac{1}{L}\int_{-L}^{L} f(x)\sin\frac{n\pi x}{L}\,dx$$
$$= \frac{1}{3}\int_{-3}^{0} 5\sin\frac{n\pi x}{3}\,dx+\frac{1}{3}\int_0^3 0\cdot\sin\frac{2\pi x}{3}\,dx$$
$$b_n = \frac{5\cos(n\pi)-5}{n\pi} = \frac{5}{\pi}\left(\frac{\cos(n\pi)-1}{n}\right)$$

n	b_n
1	$\frac{5}{\pi}\cdot(-2) = \frac{-10}{\pi}$
2	0
3	$\frac{5}{\pi}\left(-\frac{2}{3}\right) = \frac{-10}{3\pi}$
4	0
5	$\frac{5}{\pi}\left(-\frac{2}{5}\right) = \frac{-10}{5\pi}$

$$f(x) = a_0 + a_1 \cos\frac{\pi x}{L} + a_2 \cos\frac{2\pi x}{L} + a_3 \cos\frac{3\pi x}{L} + \cdots$$
$$+ b_1, \sin\frac{\pi x}{L} + b_2\frac{2\pi x}{L} + b_3\frac{3\pi x}{L} + \cdots$$
$$f(x) = \frac{5}{2} - \frac{10}{\pi}\left(\sin\frac{\pi x}{3} + \frac{1}{3}\sin\frac{3\pi x}{3} + \frac{1}{5}\sin\frac{5\pi}{3} + \cdots\right)$$

21. $f(x) = \begin{cases} -x & -4 \le x < 0 \\ x & 0 \le x < 4 \end{cases}$

period $= 2L = 8$, so $L = 4$

The function is even, and therefore, its Fourier expansion contains only cosine terms and possibly a constant. Therefore $b_n = 0$ and

$$a_0 = \frac{1}{8}\int_{-4}^{0} -x dx + \frac{1}{8}\int_0^4 x dx$$
$$= -\frac{1}{16}x^2\Big|_{-4}^{0} + \frac{1}{16}x^2\Big|_0^4$$
$$= 2$$
$$a_n = \frac{1}{4}\int_{-4}^{0} -x\cos\frac{n\pi x}{4}dx + \frac{1}{4}\int_0^4 x\cos\frac{n\pi x}{4}dx$$
$$= 2\left[\frac{1}{4}\int_0^4 x\cos\frac{n\pi x}{4}dx\right]$$
$$= \frac{1}{2}\frac{16}{(n\pi)^2}\int_0^4 \frac{n\pi}{4}x\cos\frac{n\pi}{4}x\frac{n\pi}{4}dx$$

(formula 48 from appendix)

$$= \frac{8}{(n\pi)^2}\left(\cos\frac{n\pi x}{4} + \frac{n\pi x}{4}\sin\frac{n\pi x}{4}\right)\Big|_0^4$$
$$= \frac{8}{(n\pi)^2}(\cos n\pi + n\pi\sin n\pi - 1)$$
$$= -\frac{8}{(n\pi)^2}(1 - \cos n\pi)$$
$$a_1 = -\frac{16}{\pi^2}; a_2 = 0; a_3 = -\frac{16}{9\pi^2}$$

Therefore,

$$f(x) = 2 - \frac{16}{\pi^2}\cos\frac{\pi x}{4} - \frac{16}{9\pi^2}\cos\frac{3\pi x}{4} + \cdots$$
$$= 2 - \frac{16}{\pi^2}\left(\cos\frac{\pi x}{4} + \frac{1}{9}\cos\frac{3\pi x}{4}\cdots\right)$$

25. Expand $f(x) = x^2$ in a half-range cosine series for $0 \le x < 2$.

$$a_0 = \frac{1}{L}\int_0^L f(x)dx = \frac{1}{2}\int_0^2 x^2 dx$$
$$= \frac{1}{2}\frac{x^3}{3}\Big|_0^2$$
$$= \frac{1}{6}(2^3 - 0) = \frac{4}{3}$$

$$a_n = \frac{2}{L}\int_0^L f(x)\cos\frac{n\pi x}{L}dx, (n = 1, 2, 3, \ldots)$$
$$a_n = \frac{2}{2}\int_0^2 x^2\frac{n\pi x}{2}dx$$
$$= \frac{2x}{\frac{n^2\pi^2}{4}}\cos\frac{n\pi x}{2} + \left(\frac{x^2}{\frac{n\pi}{2}} - \frac{2}{\frac{n^3\pi^3}{8}}\right)\sin\frac{n\pi x}{2}\Big|_0^2$$
$$a_n = \frac{8x}{n^2\pi^2}\cos\frac{n\pi x}{2} + \left(\frac{2x^2}{n\pi} - \frac{16}{n^3\pi^3}\right)\sin\frac{n\pi x}{2}\Big|_0^2$$
$$a_n = \frac{16}{n^2\pi^2}\cos n\pi + \left(\frac{8}{n\pi} - \frac{16}{n^3\pi^3}\right)\sin n\pi$$
$$a_n = \frac{16}{n^2\pi^2}\cos n\pi$$
$$a_1 = \frac{16}{1^2\pi^2}\cos\pi = \frac{-16}{\pi^2}$$
$$a_2 = \frac{16}{2^2\pi^2}\cos 2\pi = \frac{4}{\pi^2}$$
$$a_3 = \frac{16}{3^2\pi^2}\cos 3\pi = \frac{-16}{9\pi^2}$$
$$f(x) = \frac{4}{3} - \cos\frac{\pi x}{2} + \frac{4}{\pi^2}\cos\frac{2\pi x}{2}$$
$$- \frac{16}{9\pi^2}\cos\frac{3\pi x}{2} + \cdots$$
$$f(x) = \frac{4}{3} - \frac{16}{\pi^2}\left(\cos\frac{\pi x}{2} - \frac{1}{4}\cos\pi x + \frac{1}{9}\cos\frac{3\pi x}{2} - \cdots\right)$$

Review Exercises

1.
$$f'(x) = \frac{1}{1+e^x} = (1+e^x)^{-1}$$
$$f'(x) = -(1+e^x)^{-2}(e^x) = -e^x(1+e^x)^{-2}$$
$$f''(x) = -(1+e^x)^{-2}e^x + e^x(2)(1+e^x)^{-3}(e^x)$$
$$f'''(x) = -(1+e^x)^{-2}e^x + e^x(2)(1+e^x)^{-3}(e^x)$$
$$+ 2e^{2x}(-3)(1+e^x)^{-4}e^x$$
$$+ (1+e^x)^{-3}(2e^{2x})(2)$$
$$f(0) = \frac{1}{1+1} = \frac{1}{2}$$
$$f'(0) = -1(2^{-2}) = -\frac{1}{4}$$
$$f''(0) = -\frac{1}{4} + \frac{1}{4} = 0$$
$$f'''(0) = -\frac{1}{4} + \frac{1}{4} - \frac{3}{8} + \frac{1}{2} = \frac{1}{8}$$
$$f(x) = \frac{1}{2} - \frac{1}{4}x + \frac{0x^2}{2!} + \left(\frac{1}{8}\right)\frac{x^3}{3!} + \cdots$$
$$= \frac{1}{2} - \frac{1}{4}x + \frac{1}{48}x^3 - \cdots$$

5. $f(x) = (x+1)^{1/3} \qquad f(0) = 1$

$f'(x) = \frac{1}{3}(x+1)^{-2/3} \qquad f'(0) = \frac{1}{3}$

$f''(x) = -\frac{2}{9}(x+1)^{-5/3} \qquad f''(0) = -\frac{2}{9}$

$$f(x) = 1 + \frac{1}{3}x - \frac{2x^2}{9(2)} + \cdots$$
$$= 1 + \frac{1}{3}x - \frac{1}{9}x^2 + \cdots$$

The series can also be found using the binomial series formula directly with $n = \frac{1}{3}$:

$$(1+x)^{1/3} = 1 + \frac{1}{3}x + \frac{\frac{1}{3}\left(-\frac{2}{3}\right)}{2!}x^2 + \cdots$$
$$= 1 + \frac{x}{3} - \frac{x^2}{9} + \cdots$$

9. $f(x) = \cos(a+x), \qquad f(0) = \cos a$

$f'(x) = -\sin(a+x), \qquad f'(0) = -\sin a$

$f''(x) = -\cos(a+x), \qquad f''(0) = -\cos a$

$f'''(x) = \sin(a+x), \qquad f'''(0) = \sin a$

$$f(x) = \cos(a+x)$$
$$= \cos a - \sin a \cdot x - \frac{\cos a}{2!}x^2 + \sin a \cdot \frac{x^3}{3!} - \cdots$$

The series can also be found through the cosine of the sum:

$$\cos(a+x) = \cos a \cos x - \sin a \sin x$$
$$= \cos a\left(1 - \frac{x^2}{2!} + \cdots\right) - \sin a\left(x - \frac{x^3}{3!} + \cdots\right)$$
$$= \cos a - \sin a \cdot x - \frac{\cos a}{2!}x^2 + \sin a \cdot \frac{x^3}{3!} - \cdots$$

13. See Exercise 5.

$$\sqrt[3]{1+x} = 1 + \frac{1}{3}x - \frac{1}{9}x^2 + \cdots$$
$$\sqrt[3]{1+0.3} = 1 + \frac{1}{3}(0.3) - \frac{1}{9}(0.3)^2$$
$$\sqrt[3]{1.3} = 1.09$$

17. $\ln(1+x) = x - \frac{x^2}{2} + \frac{x^3}{3} - \cdots$

$$\ln[1+(-0.1828)] = -0.1828 - \frac{(-0.1828)^2}{2} + \frac{(-0.1828)^3}{3} - \cdots$$
$$\ln 0.8172 = -0.202$$

21. $f(x) = \sqrt{x}, a = 144$

$f(x) = \sqrt{x}, f(144) = 12$

$f'(x) = \frac{1}{2}x^{-1/2}, f'(144) = \frac{1}{24}$

$f''(x) = -\frac{1}{4}x^{-3/2}, f''(144) = -\frac{1}{6912}$

$$\sqrt{x} = 12 + \frac{1}{24}(x-144) - \frac{1}{6912}\frac{(x-144)^2}{2} + \cdots$$
$$\sqrt{148} = 12 + \frac{1}{24}(4) - \frac{1}{6912}\frac{(4)^2}{2}$$
$$= 12.16551$$
$$\sqrt{148} = 12.2$$

25. $f(x) = \cos x, a = \frac{\pi}{3}$

$f(x) = \cos x, f\left(\frac{\pi}{3}\right) = \frac{1}{2}$

$f'(x) = -\sin x, f'\left(\frac{\pi}{3}\right) = \frac{-1}{2}\sqrt{3}$

$f''(x) = -\cos x, f''\left(\frac{\pi}{3}\right) = \frac{-1}{2}$

$$f(x) = \frac{1}{2} - \frac{1}{2}\sqrt{3}\left(x - \frac{\pi}{3}\right) - \frac{1}{4}\left(x - \frac{\pi}{3}\right)^2 + \cdots$$

29. $f(x) = \begin{cases} \pi - 1, & -4 \le x < 0 \\ \pi + 1, & 0 \le x < 4 \end{cases}$ is Example 7 in 30.7 shifted up $\pi - 1$ units

$$f(x) = \pi - 1 + 1 + \frac{4}{\pi}\sin\frac{\pi x}{4} + \frac{4}{3\pi}\sin\frac{3\pi x}{4} + \cdots$$
$$f(x) = \pi + \frac{4}{\pi}\left(\sin\frac{\pi x}{4} + \frac{1}{3}\sin\frac{3\pi x}{4} + \cdots\right)$$

33. $f(x) = x$

$-2 \le x < 2$, period $= 4, L = 2$;

The function is odd, so the series consist only of sine terms.

$$b_n = \frac{1}{2}\int_{-2}^{2} x \sin\frac{n\pi x}{2}\,dx$$
$$= \frac{1}{2}\left(\frac{2}{n\pi}\right)^2\left(\sin\frac{n\pi x}{2} - \frac{n\pi x}{2}\cos\frac{n\pi x}{2}\right)\Bigg|_{-2}^{2}$$
$$= \frac{2}{n^2\pi^2}[\sin n\pi - n\pi\cos n\pi - \sin(-n\pi) - n\pi\cos(-n\pi)]$$
$$= \frac{2}{n^2\pi^2}(-n\pi\cos n\pi - n\pi\cos n\pi)$$
$$= \frac{-4}{n\pi}\cos n\pi;$$

$$b_1 = -\frac{4}{\pi}\cos\pi = \frac{4}{\pi},$$
$$b_2 = -\frac{4}{2\pi}\cos 2\pi = -\frac{2}{\pi},$$
$$b_3 = \frac{-4}{3\pi}\cos 3\pi = \frac{4}{3\pi}$$
$$f(x) = \frac{4}{\pi}\left(\sin\frac{\pi x}{2} - \frac{1}{2}\sin\pi x + \frac{1}{3}\sin\frac{3\pi x}{2} - \cdots\right)$$

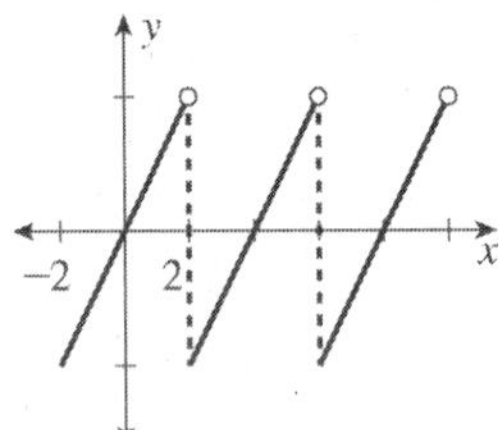

37. It is a geometric series for which $|r| < 1 = 0.75$. Therefore, the series converges.

$$S = \frac{64}{1-0.75} = 256$$

41. $f(x) = \tan x, \qquad f\left(\frac{\pi}{4}\right) = 1$

$f'(x) = 1 + \tan^2 x, \qquad f'\left(\frac{\pi}{4}\right) = 2$

$f''(x) = 2\tan x\left(+\tan^2 x\right), \qquad f''(x) = 4$

$$f(x) = \tan x = 1 + 2\left(x - \frac{\pi}{4}\right) + \frac{4\left(x-\frac{\pi}{4}\right)^2}{2!} + \cdots$$
$$= 1 + 2\left(x - \frac{\pi}{4}\right) + 2\left(x - \frac{\pi}{4}\right)^2 + \cdots$$

45. $\sin x = x - \dfrac{x^3}{3!} + \cdots$

$\sin(x+h) - \sin(x-h)$

$$= (x+h) - \frac{(x+h)^3}{3!} + \cdots - (x-h) + \frac{(x-h)^3}{3!} - \cdots$$
$$= x + h - \frac{x^3 + 3x^2h + 3xh^2 + h^3}{3!} + \cdots$$
$$-x + h + \frac{x^3 - 3x^2h + 3xh^2 - h^3}{3!} - \cdots$$
$$= 2h - \frac{6x^2h}{3!} - \frac{2h^3}{3!} + \cdots$$
$$= 2h\left(1 - \frac{x^2}{2} + \cdots\right) - \frac{2h^3}{3!} + \cdots$$
$= 2h\cos x$ for small h

49. $\sin^2 x = \dfrac{1}{2}(1 - \cos 2x)$

$$= \frac{1}{2}\left(1 - \left(1 - \frac{(2x)^2}{2!} + \frac{(2x)^4}{4!} - \frac{(2x)^6}{6!} + \cdots\right)\right)$$
$$= \frac{1}{2}\left(1 - 1 + 2x^2 - \frac{2}{3}x^4 + \frac{4}{45}x^6 - \cdots\right)$$
$$= x^2 - \frac{1}{3}x^4 + \frac{2}{45}x^6 - \cdots$$

53. $f(x) = \dfrac{1}{1-x} = 1 + x + x^2 + \cdots$

$$\frac{1}{1+x} = \frac{1}{1-(-x)} = 1 + (-x) + (-x)^2 + \cdots$$
$$= 1 - x + x^2 - \cdots$$

57. $e^x = 1 + x + \dfrac{x^2}{2!} + \dfrac{x^3}{3!}$

$$e^{0.9} = 1 + (0.9) + \frac{0.9^2}{2} + \frac{0.9^3}{6} = 2.4265$$
$$e^x = e\left[1 + (x-1) + \frac{(x-1)^2}{2}\right]$$
$$e^{0.9} = e\left[1 + (0.9-1) + \frac{(0.9-1)^2}{2}\right] = 2.4600$$

$e^{0.9} = 2.459603$ directly from the calculator.

61. $\tan^{-1} x = \displaystyle\int \frac{1}{1+x^2}dx$

$$= \int\left(1 - x^2 + x^4 - x^6 + \cdots\right)dx$$
$$\tan^{-1} x = x - \frac{x^3}{3} + \frac{x^5}{5} - \frac{x^7}{7} + \cdots$$

65. $N = N_0 e^{-\lambda t} \cdot N$

$$= N_0\left[1 + (-\lambda t) + \frac{(-\lambda t)^2}{2!} + \frac{(-\lambda t)^3}{3!} + \cdots\right]$$
$$= N_0\left[1 - \lambda t + \frac{\lambda^2 t^2}{2} - \frac{\lambda^3 t^3}{6} + \cdots\right]$$

69. $f(x) = \ln\frac{1+x}{1-x}$,

$$f(0) = \ln\frac{1+0}{1-0} = \ln 1 = 0$$

$$f'(x) = -\frac{2}{x^2-1}, f'(0) = 2$$

$$f''(x) = \frac{4x}{(x+1)^2(x-1)^2}, f''(0) = 0$$

$$f'''(x) = \frac{-4(3x^2+1)}{(x+1)^3(x-1)^3}, f'''(0) = 4$$

$$f^{iv}(x) = \frac{48x(x^2+1)}{(x+1)^4(x-1)^4}, f^{iv}(0) = 0$$

$$f^{(5)}(x) = \frac{-48(5x^4+10^2+1)}{(x+1)^5(x-1)5}, f^{(5)}(0) = 48$$

$$f^{(6)}(x) = \frac{480x(3x^4+10x^2+3)}{(x+1)^6(x-1)^6}, f^{(6)}(0) = 0$$

$$f^{(7)}(x) = \frac{-1440(7x^6+35x^4+21x^2+1)}{(x+1)^7(x-1)^7}, f^{(7)}(0)$$

$$= 1440$$

$$V = \ln\frac{1+x}{1-x} = 2x + \frac{4x^3}{3!} + \frac{48x^5}{5!} + \frac{1440x^7}{7!}$$

$$V = \ln\frac{1+x}{1-x} = 2x + \frac{2}{3}x^3 + \frac{2}{5}x^5 + \frac{2}{7}x^7$$

Alternatively, using the properties of logarithms and the known expansion for $\ln(1+x)$:

$$\ln(1+x) = x - \frac{x^2}{2} + \frac{x^3}{3} - \frac{x^4}{4} + \frac{x^5}{5} - \frac{x^6}{6} + \frac{x^7}{7} - \cdots$$

$$\ln(1-x) = -x - \frac{x^2}{2} - \frac{x^3}{3} - \frac{x^4}{4} - \frac{x^5}{5} - \frac{x^6}{6} - \frac{x^7}{7} - \cdots$$

$$V = \ln(1+x) - \ln(1-x)$$

$$= x - \frac{x^2}{2} + \frac{x^3}{3} - \frac{x^4}{4} + \frac{x^5}{5} - \frac{x^6}{6} + \frac{x^7}{7} \cdots$$

$$-\left[-x - \frac{x^2}{2} - \frac{x^3}{3} - \frac{x^4}{4} - \frac{x^5}{5} - \frac{x^6}{6} - \frac{x^7}{7} - \cdots\right]$$

$$= 2x + \frac{2}{3}x^3 + \frac{2}{5}x^5 + \frac{2}{7}x^7 + \cdots$$

73. $f(x) = \begin{cases} 0, & -\pi \le x < 0 \\ \sin t, & 0 \le x < \pi/2 \\ 0, & \pi/2 \le x < \pi \end{cases}$

$$a_0 = \frac{1}{2\pi}\int_{-\pi}^{\pi} f(x)\,dx$$

$$= \frac{1}{2\pi}\int_{-\pi}^{0} 0\,dx + \frac{1}{2\pi}\int_0^{\pi/2} \sin t\,dt + \frac{1}{2\pi}\int_{\pi/2}^{\pi} 0\,dx$$

$$= \frac{-1}{2\pi}\cos t\Big|_0^{\pi/2}$$

$$= \frac{1}{2\pi}\left[-\cos\frac{\pi}{2} + \cos 0\right] = \frac{1}{2\pi}$$

$$a_1 = \frac{1}{\pi}\int_0^{\pi/2} \sin t\cos t\,dt$$

$$= \frac{1}{\pi}\cdot\frac{\sin^2 t}{2}\Bigg|_0^{\pi/2}$$

$$= \frac{1}{2\pi}$$

$$a_2 = \frac{1}{\pi}\int_0^{\pi/2} \sin t\cos 2t\,dt$$

$$= \frac{1}{\pi}\left[-\frac{\cos(-t)}{2(-1)} - \frac{\cos(3t)}{2(3)}\right]_0^{\pi/2}$$

$$= \frac{1}{\pi}\left[-\frac{1}{2} + \frac{1}{6}\right] = -\frac{1}{3\pi}$$

$$b_n = \frac{1}{\pi}\int_{-\pi}^{\pi} f(t)\sin nt\,dt = \frac{1}{\pi}\int_0^{\pi/2} \sin t\sin nt\,dt$$

$$b_1 = \frac{1}{\pi}\int_0^{\pi/2} \sin^2 t dt = \frac{1}{\pi}\left[\frac{t}{2} - \frac{\sin t\cos t}{2}\right]_0^{\pi/2}$$

$$= \frac{1}{4}$$

$$b_2 = \pi\int_0^{\pi/2} \sin t\sin 2t\,dt$$

$$= \frac{1}{\pi}\left[\frac{\sin t}{2} - \frac{\sin 3t}{6}\right]_0^{\pi 2}$$

$$= \frac{1}{\pi}\left[\frac{\sin\pi/2}{2} - \frac{\sin 3\pi/2}{6}\right]$$

$$b_2 = \frac{1}{\pi}\left[\frac{1}{2} - \frac{-1}{6}\right] = \frac{2}{3\pi}$$

$$f(x) = \frac{1}{2\pi} + \frac{1}{\pi}\left(\frac{1}{2}\cos t - \frac{1}{3}\cos 2t + \cdots\right)$$

$$+ \frac{1}{4}\sin t + \frac{2}{3\pi}\sin 2t + \cdots$$

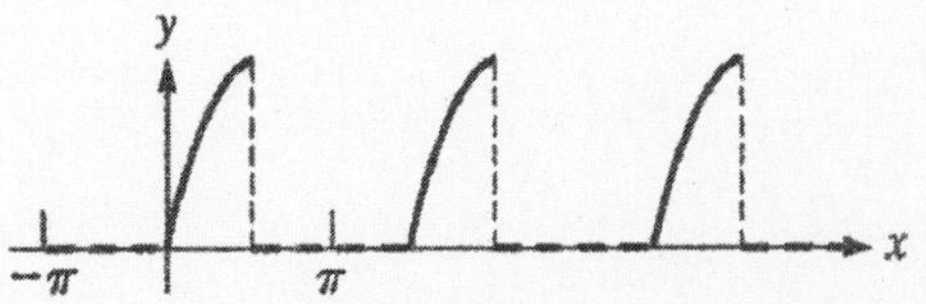

CHAPTER 31

DIFFERENTIAL EQUATIONS

31.1 Solutions of Differential Equations

1. $y = c_1e^{-x} + c_2e^{2x}$

$$\frac{dy}{dx} = -c_1e^{-x} + 2c_2e^{2x}$$

$$\frac{d^2y}{dx^2} = c_1e^{-x} + 4c_2e^{2x}$$

$$\begin{aligned}\frac{d^2y}{dx^2} - \frac{dy}{dx} &= c_1e^{-x} + 4c_2e^{2x} - (-c_1e^{-x} + 2c_2e^{2x}) \\ &= c_1e^{-x} + 4c_2e^{2x} + c_1e^{-x} - 2c_2e^{2x} \\ &= 2c_1e^{-x} + 2c_2e^{2x} \\ &= 2(c_1e^{-x} + c_2e^{2x}) \\ &= 2y\end{aligned}$$

$y = 4e^{-x}$

$$\frac{dy}{dx} = -4e^{-x}$$

$$\frac{d^2y}{dx^2} = 4e^{-x}$$

$$\begin{aligned}\frac{d^2y}{dx^2} - \frac{dy}{dx} &= 4e^{-x} - (-4e^{-x}) \\ &= 8e^{-x} \\ &= 2(4e^{-x}) \\ &= 2y\end{aligned}$$

5. $y'' + 3y' - 4y = 3e^x$

$$y = c_1e^x + c_2e^{-4x} + \frac{3}{5}xe^x$$

$$\begin{aligned}y' &= c_1e^x - 4c_2e^{-4x} + \frac{3}{5}(xe^x + e^x) \\ &= \left(c_1 + \frac{3}{5}\right)e^x - 4c_2e^{-4x} + \frac{3}{5}xe^x\end{aligned}$$

$$\begin{aligned}y'' &= \left(c_1 + \frac{3}{5}\right)e^x - 16c_2e^{-4x} + \frac{3}{5}(xe^x + e^x) \\ &= \left(c_1 + \frac{6}{5}\right)e^x + 16c_2e^{-4x} + \frac{3}{5}xe^x\end{aligned}$$

Substitute y, y', y'' into the differential equation.

$$\left(c_1 + \frac{6}{5}\right)e^x + 16c_2e^{-4x} + \frac{3}{5}xe^x$$
$$+3\left(\left(c_1 + \frac{3}{5}\right)e^x - 4c_2e^{-4x} + \frac{3}{5}xe^x\right)$$
$$-4\left(c_1e^x + c_2e^{-4x} + \frac{3}{5}xe^x\right) = 3e^x$$

$$\left(c_1 + \frac{6}{5} + 3c_1 + \frac{9}{5} - 4c_1\right)e^x + (16c_2 - 12c_2 - 4c_2)e^{-4x}$$
$$+\left(\frac{3}{5} + \frac{9}{5} - \frac{12}{5}\right)xe^x = 3e^x$$

$$3e^x = 3e^x$$

Since y is a solution to the second-order equation, and there are two arbitrary constants, it is the general solution.

9. $y = 3\cos 2x, y' = -6\sin 2x,\ y'' = -12\cos 2x$

Substitute y and y' into the differential equation.

$$y'' + 4y = 0$$
$$-12\cos 2x + 4(3\cos 2x) = 0$$
$$0 = 0$$

$y = c_1\sin 2x + c_2\cos 2x; y' = 2c_1\cos 2x - 2c_2\sin 2x$

$y'' = -4c_1\sin 2x - 4c_2\cos 2x$

Substitute y and y' into the differential equation.

$$-4c_1\sin 2x - 4c_2\cos 2x + 4(c_1\sin 2x + c_2\cos 2x) = 0$$
$$0 = 0$$

Both functions are solutions to the differential equation.

13. $y = 2 + x - x^3$

$$\frac{dy}{dx} = 1 - 3x^2$$

This is the original equation, so the function is a solution to the differential equation.

17. $y'' + 9y = 4\cos x;$

$2y = \cos x;\ y = \frac{1}{2}\cos x;$

$y' = -\frac{1}{2}\sin x;\ y'' = -\frac{1}{2}\cos x$

Substitute y and y''.

$$-\frac{1}{2}\cos x+9\left(\frac{1}{2}\cos x\right)=4\cos x$$
$$-\frac{1}{2}\cos x+9\left(\frac{1}{2}\cos x\right)=4\cos x$$
$$\frac{8}{2}\cos x=4\cos x$$
$$4\cos x=4\cos x \text{ identity}$$

The function is a solution to the differential equation.

21. $x\frac{d^2y}{dx^2}+\frac{dy}{dx}=0;$

$$y=c_1\ln x+c_2$$
$$\frac{dy}{dx}=\frac{c_1}{x}=c_1x^{-1}$$
$$\frac{d^2y}{dx^2}=-c_1x^{-2}=-\frac{c_1}{x^2}$$

Substitute $\frac{dy}{dx}$ and $\frac{d^2y}{dx^2}$.

$$x\left(\frac{-c_1}{x^2}\right)+c_1x^{-1}=0$$
$$-\frac{c_1}{x}+\frac{c_1}{x}=0$$
$$0=0 \text{ identity}$$

The function is a solution to the differential equation.

25. $y=c_1e^x+c_2e^{2x}+\frac{3}{2}$

$$y'=c_1e^x+2c_2e^{2x}$$
$$y''=c_1e^x+4c_2e^{2x}$$
$$y''-3y'+2y=3$$
$$c_1e^x+4c_2e^{2x}-3\left(c_1e^x+2c_2e^{2x}\right)$$
$$+2\left(c_1e^x+c_2e^{2x}+\frac{3}{2}\right)=3$$
$$c_1e^x+4c_2e^{2x}-3c_1e^x-6c_2e^{2x}$$
$$+2c_1e^x+2c_2e^{2x}+3=3$$
$$3=3 \text{ identity}$$

The function is a solution to the differential equation.

29. $(y')^2+xy'=y$

$$y=cx+c^2$$
$$y'=c$$

Substitute y'.

$$(c)^2+x(c)=y$$
$$c^2+cx=y$$
$$y=y \text{ identity}$$

The function is a solution to the differential equation.

33. $y=x^3+c_1x^2+c_2$

Substitute $x=0, y=-4$.

$$-4=0^3+c_1(0)^2+c_2$$
$$c_2=-4$$

$y=x^3+c_1x^2-4$ is the particular solution.

31.2 Separation of Variables

1. $2xy\,dx+(x^2+1)dy=0$

$$\frac{2x\,dx}{x^2+1}+\frac{dy}{y}=0$$
$$\ln(x^2+1)+\ln y=\ln c$$
$$\ln(y(x^2+1))=\ln c$$
$$y(x^2+1)=c$$

5. $y^2dx+dy=0$

Divide by y^2 and integrate.

$$dx+\frac{dy}{y^2}=0$$
$$x+\frac{y^{-1}}{-1}=c$$
$$x-\frac{1}{y}=c$$

9. $x^2+(x^3+5)y'=0$

$$(x^3+5)\frac{dy}{dx}=-x^2$$

Multiply by $\frac{dx}{x^3+5}$ and integrate ($u=x^3+5$).

$$dy=\frac{-x^2dx}{x^3+5}$$
$$y=-\frac{1}{3}\ln(x^3+5)+c$$
$$3y+\ln(x^3+5)=3c_1$$

Since $3c_1$ is constant, $3y+\ln(x^3+5)=c$

13. $e^{x^2}dy=x\sqrt{1-y}\,dx$

Divide by $e^{x^2}\sqrt{1-y}$ and integrate.

$$\frac{dy}{\sqrt{1-y}}=\frac{x\,dx}{e^{x^2}}$$
$$\frac{dy}{(1-y)^{1/2}}=e^{-x^2}x\,dx$$
$$-\frac{(1-y)^{1/2}}{\frac{1}{2}}=-\frac{1}{2}e^{-x^2}+c$$

$$-2\sqrt{1-y} = -\frac{1}{2}e^{-x^2} + c$$
$$4\sqrt{1-y} = e^{-x^2} - 2c_1$$
Write $-2c_1 = c.$
$$4\sqrt{1-y} = e^{-x^2} + c$$

17. $y' - y = 4$
$$\frac{dy}{dx} = 4 + y$$
Multiply by $\frac{dx}{4+y}$ and integrate.
$$\frac{dy}{4+y} = dx$$
$$\ln(4+y) = x + c$$

21. $y \tan x\, dx + \cos^2 x\, dy = 0$
Divide by $y \cos^2 x$ and integrate.
$$\frac{\tan x\, dx}{\cos^2 x} + \frac{dy}{y} = 0$$
$$\tan x \sec^2 x\, dx + \frac{dy}{y} = 0$$
$$\frac{1}{2}\tan^2 x + \ln y = c_1$$
Let $2c_1 = c$
$$\tan^2 x + 2\ln y = c$$

25. $e^{\cos\theta} \tan\theta d\theta + \sec\theta dy = 0$
Multiply by $\cos\theta$ and integrate.
$$e^{\cos\theta} \sin\theta d\theta + dy = 0$$
$$-e^{\cos\theta} + y = c$$

29. $2\ln t\, dt + t\, di = 0$
Divide by t and integrate.
$$2\ln t \frac{dt}{t} + di = 0$$
$$\frac{2(\ln t)^2}{2} + i = c$$
$$(\ln t)^2 + i = c$$
$$i = c - (\ln t)^2$$

33. $\frac{dy}{dx} + yx^2 = 0$
Multiply by $\frac{dx}{y}$ and integrate.
$$\frac{dy}{y} + x^2 dx = 0$$
$$\ln y + \frac{x^3}{3} = c$$

Substitute $x = 0, y = 1$
$$\ln 1 = c$$
$$c = 0$$
$$3\ln y + x^3 = 0$$

37. $y^2 e^x dx + e^{-x} dy = y^2 dx; x = 0, y = 2$
$$y^2\left(1 - e^x\right)dx = e^{-x} dy$$
Multiply by $\frac{e^x}{y^2}$ and integrate.
$$\left(e^x - e^{2x}\right)dx = \frac{dy}{y^2}$$
$$e^x - \frac{1}{2}e^{2x} = -\frac{1}{y} + c$$
Substitute $x = 0, y = 2$
$$e^0 - \frac{1}{2}e^0 = -\frac{1}{2} + c$$
$$c = 1$$
$$e^x - \frac{1}{2}e^{2x} = -\frac{1}{y} + 1$$
$$2e^x - e^{2x} = -\frac{2}{y} + 2$$
$$e^{2x} - \frac{2}{y} = 2(e^x - 1)$$

41. $dT + 0.15(T - 10)dt = 0, \qquad T(0) = 40$
Divide by $T - 10$ and integrate.
$$\frac{dT}{T-10} + 0.15dt = 0$$
$$\ln(T - 10) + 0.15t = \ln c$$
$$\ln\left(c(T-10)\right) = -0.15t$$
$$T - 10 = ce^{-0.15t}$$
$$T = ce^{-0.15t} + 10$$
Substitute $t = 0, T = 40.$
$$40 = c + 10$$
$$c = 30$$
$$T(t) = 30e^{-0.15t} + 10$$

31.3 Integrating Combinations

1. $x\, dy + y\, dx + 2xy^2 dy = 0$
$$\frac{x\, dy + y\, dx}{xy} + 2y\, dy = 0$$
$$\frac{d(xy)}{xy} + 2y\, dy = 0$$
$$\ln xy + y^2 = c$$

5. $$y\,dx - x\,dy + x^3dx = 2dx$$
$$x\,dy - y\,dx - x^3dx = -2dx$$
$$\frac{(x\,dy - y\,dx)}{x^2} - x\,dx = -\frac{2dx}{x^2}$$
$$d\left(\frac{y}{x}\right) - xdx = -\frac{2dx}{x^2}$$
$$\frac{y}{x} - \frac{1}{2}x^2 = 2x^{-1} + c_1$$
$$y - \frac{1}{2}x^3 = 2 + c_1x$$
$$2y - x^3 = 4 + 2c_1x$$
$$x^3 - 2y = cx - 4$$

9. $$\sin x\,dy = (1 - y\cos x)\,dx$$
$$\sin x\,dy + y\cos x\,dx = dx$$
$$d(y\sin x) = dx$$
$$y\sin x = x + c$$

13. $\tan(x^2 + y^2)dy + x\,dx + y\,dy = 0$
$$dy + \frac{x\,dx + y\,dy}{\tan(x^2 + y^2)} = 0$$
Multiply both sides by 2 so that the numerator of the quotient is the differential of $x^2 + y^2$.
$$2dy + \cot(x^2 + y^2)(2x\,dx + 2y\,dy) = 0$$
$$2dy + \cot(x^2 + y^2)d(x^2 + y^2) = 0$$
$$2y + \ln\sin(x^2 + y^2) = c$$

17. $10xdy + 5ydx + 3ydy = 0$
$5(2xdy + ydx) + 3ydy = 0$
Multiply by y.
$$5(2xydy + y^2dx) + 3y^2dy = 0$$
$$5d(xy^2) + 3y^2dy = 0$$
$$5xy^2 + y^3 = c$$

21. $ydx - xdy = y^3dx + y^2xdy; x = 2, y = 4$
$$\frac{ydx - xdy}{y^2} = ydx + xdy$$
$$d\left(\frac{x}{y}\right) = d(xy)$$
$$\frac{x}{y} = xy + c$$
Substitute $x = 2, y = 4$:
$$\frac{2}{4} = 2(4) + c$$
$$c = -\frac{15}{2}$$
$$\frac{x}{y} = xy - \frac{15}{2}$$
$$2x = 2xy^2 - 15y$$

25. $$e^{-x}dy - 2y\,dy = ye^{-x}\,dx$$
$$e^{-x}dy - ye^{-x}\,dx = 2y\,dy$$
$$d\left(ye^{-x}\right) = d\left(y^2\right)$$
$$ye^{-x} = y^2 + c$$

31.4 The Linear Differential Equation of the First Order

1. $$dy + \left(\frac{2}{x}\right)ydx = 3dx$$
$$e^{\int\frac{2}{x}dx} = e^{2\ln x} = e^{\ln x^2} = x^2$$
$$yx^2 = \int 3x^2dx + c$$
$$yx^2 = x^3 + c$$
$$y = x + cx^{-2}$$

5. $dy + 2ydx = e^{-4x}dx; P = 2, Q = e^{-4x}$
$$e^{\int 2dx} = e^{2x}$$
$$ye^{2x} = \int e^{-4x}e^{2x}dx$$
$$= -\frac{1}{2}\int e^{-2x}(-2dx)$$
$$= -\frac{1}{2}e^{-2x} + c$$
$$y = -\frac{1}{2}e^{-2x}e^{-2x} + ce^{-2x}$$
$$y = -\frac{1}{2}e^{-4x} + ce^{-2x}$$

9.
$$dy = 3x^2(2 - y)dx$$
$$dy = 6x^2dx - 3x^2ydx$$
$$dy + 3x^2y = 6x^2dx; P = 3x^2, Q = 6x^2$$
$$e^{\int 3x^2dx} = e^{x^3}$$
$$ye^{x^3} = \int 6x^2e^{x^3}dx + c$$
$$= 2e^{x^3} + c$$
$$y = ce^{-x^3} + 2$$
Note: The equation can also be solved by separation of variables.

13. $dr + r\cot\theta d\theta = d\theta$

$dr + \cot\theta r d\theta = d\theta; P = \cot\theta, Q = 1$

$e^{\int \cot\theta d\theta} = e^{\ln\sin\theta} = \sin\theta$

$r\sin\theta = \int \sin\theta d\theta + c$

$r\sin\theta = -\cos\theta + c$

$r = -\dfrac{\cos\theta}{\sin\theta} + \dfrac{c}{\sin\theta}$

$r = -\cot\theta + c\csc\theta$

17. $y' + y = x + e^x$

$dy + y\,dx = \left(x + e^x\right)dx; P = 1, Q = x + e^x$

$ye^{\int dx} = \int \left(x + e^x\right)e^{\int dx}dx$

$ye^x = \int \left(xe^x + e^{2x}\right)dx$

$ye^x = xe^x - e^x + \dfrac{1}{2}e^{2x} + c$

$y = x - 1 + \dfrac{1}{2}e^x + ce^{-x}$

21. $y' = x^3(1 - 4y)$

$\dfrac{dy}{dx} = x^3 - 4x^3 y$

$dy + 4x^3 y dx = x^3 dx; P = 4x^3, Q = x^3$

$e^{4\int x^3 dx} = e^{x^4}$

$ye^{x^4} = \int x^3 e^{x^4} dx + c$

$= \dfrac{1}{4}\int e^{x^4} 4x^3 dx + c$

$= \dfrac{1}{4}e^{x^4} + c$

$y = \dfrac{1}{4} + ce^{-x^4}$

25. $\sqrt{1+x^2}\,dy + x(1+y)\,dx = 0$

$\sqrt{1+x^2}\,dx + xdx + xy\,dx = 0$

$dy + \dfrac{x}{\sqrt{1+x^2}}y\,dx = -\dfrac{x}{\sqrt{1+x^2}}dx, \quad P = \dfrac{x}{\sqrt{1+x^2}}$

$e^{\int \frac{x}{\sqrt{1+x^2}}dx} = e^{\sqrt{1+x^2}}$

$ye^{\sqrt{1+x^2}} = \int \dfrac{-x}{\sqrt{1+x^2}}e^{\sqrt{1+x^2}}dx + c$

$ye^{\sqrt{1+x^2}} = -e^{\sqrt{1+x^2}} + c$

$y = -1 + ce^{-\sqrt{1+x^2}}$

29. $y' = 2(1 - y)$

(1) Solve by separation of variables.

$\dfrac{dy}{1-y} = 2dx$

$-\ln(1-y) = 2x - \ln c$

$\ln\dfrac{c}{1-y} = 2x$

$c = (1-y)e^{2x}$

$1 - y = ce^{-2x}$

$y = 1 - ce^{-2x}$

(2) Solve as a first-order equation.

$dy = 2dx - 2ydx$

$dy + 2ydx = 2dx; P = 2, Q = 2$

$e^{\int 2dx} = e^{2x}$

$ye^{2x} = \int 2e^{2x}dx + c$

$ye^{2x} = e^{2x} + c$

$y = 1 + ce^{-2x}$

33. $\dfrac{dy}{dx} + 2y\cot x = 4\cos x; x = \dfrac{\pi}{2}, y = \dfrac{1}{3};$

$dy + 2y\cot xdx = 4\cos xdx; P = 2\cot x; Q = 4\cos x$

$e^{\int 2\cot xdx} = e^{2\ln\sin x}$

$= e^{\ln(\sin x)^2}$

$= \sin^2 x$

$y\sin^2 x = \int 4\cos x(\sin x)^2 dx + c$

$= \dfrac{4(\sin x)^3}{3} + c$

$y = \dfrac{4}{3}\sin x + c(\csc^2 x)$

Substitute $x = \dfrac{\pi}{2}, y = \dfrac{1}{3}$

$\dfrac{1}{3} = \dfrac{4}{3}\sin\dfrac{\pi}{2} + c$

$c = -1$

$y = \dfrac{4}{3}\sin x - \csc^2 x$

37. $y' + P(x)y = Q(x)y^2$

$dy + P(x)y\,dx = Q(x)y^2\,dx$, which is not linear because $Q(x)y^2$ is not a function of x only.

Let $u = \dfrac{1}{y}$, so $dy = -y^2\,du$, then

$dy + P(x)y\,dx = Q(x)y^2 dx$ is

$-y^2 du + P(x)y\,dx = Q(x)y^2\,dx$

$du - P(x)u\,dx = -Q(x)\,dx$, which is linear.

31.5 Numerical Solutions of First-order Equations

1. $\frac{dy}{dx} = x+1$

x	y	$x+1$	dy	y(correct)
0.0	1.00	1.0	0.20	1.00
0.2	1.20	1.2	0.24	1.22
0.4	1.44	1.4	0.28	1.48
0.6	1.72	1.6	0.32	1.78
0.8	2.04	1.8	0.36	2.12
1.0	2.40	2.0	0.40	2.50

$y = \frac{1}{2}x^2 + x + c$

$y = 1$ when $x = 0$

$c = 1 \qquad y = \frac{1}{2}x^2 + x + 1$

5.

x	y Approximate	y Exact
0.0	1	1
0.1	$1.0+(0.0+1)(0.1)=1.10$	1.105
0.2	$1.10+(0.1+1)(0.1)=1.21$	1.220
0.3	$1.21+(0.2+1)(0.1)=1.33$	1.345
0.4	$1.33+(0.3+1)(0.1)=1.46$	1.480
0.5	$1.46+(0.4+1)(0.1)=1.60$	1.625
0.6	$1.60+(0.5+1)(0.1)=1.75$	1.780
0.7	$1.75+(0.6+1)(0.1)=1.91$	1.945
0.8	$1.91+(0.7+1)(0.1)=2.08$	2.120
0.9	$2.08+(0.8+1)(0.1)=2.26$	2.305
1.0	$2.26+(0.9+1)(0.1)=2.45$	2.500

9. $\frac{dy}{dx} = xy + 1$, $x = 0$ to $x = 0.4$, $\Delta x = 0.1$, $(0, 0)$

x	y	y to 4 Places
0	0	0
0.1	0.1003339594	0.1003
0.2	0.20268804	0.2027
0.3	0.3091639819	0.3092
0.4	0.42203172548	0.4220

13. $\frac{dy}{dx} = \cos(x+y)$, $x = 0$ to $x = 0.6$, $\Delta x = 0.1$, $\left(0, \frac{\pi}{2}\right)$

x	y
0	$\frac{\pi}{2} = 1.5708$
0.1	1.5660
0.2	1.5521
0.3	1.5302
0.4	1.5011
0.5	1.4656
0.6	1.4244

17. $\frac{di}{dt} = 2i = \sin t$

t	i	$\sin t - 2i$	di
0.0	0.0000	0.0000	0.0000
0.1	0.0000	0.0998	0.0100
0.2	0.0100	0.1787	0.0179
0.3	0.0279	0.2398	0.0240
0.4	0.0518	0.2857	0.0286
0.5	0.0804	0.3186	0.0319

$i = 0.0804$ A for $t = 0.5$ s

$di + 2i\,dt = \sin t\,dt;\ e^{\int 2dt} = e^{2t}$

$ie^{2t} = \int e^{2t}\sin t\,dt = \frac{e^{2t}(2\sin t - \cos t)}{4+1} + c$

$i = \frac{1}{5}(2\sin t - \cos t) + ce^{-2t}$ (Formula 49)

$i = 0$ for $t = 0$, $0 = \frac{1}{5}(0-1) + c$, $c = \frac{1}{5}$

$i = \frac{1}{5}\left(2\sin t - \cos t + e^{-2t}\right)$

$i = 0.0898$ A for $t = 0.5t$

31.6 Elementary Applications

1. $y^2 = kx$

$2y\frac{dy}{dx} = k = \frac{y^2}{x}$

$\frac{dy}{dx} = \frac{y}{2x}$ for slope of any member of family

$\frac{dy}{dx} = -\frac{2x}{y}$ for slope of orthogonal trajectories.

$y\,dy = -2x\,dx$

$\frac{y^2}{2} = -x^2 + \frac{c}{2}$

$y^2 + 2x^2 = c$

5. $\frac{dy}{dx}=\frac{2x}{y};\ y\,dy=2x\,dx$

$\frac{1}{2}y^2=x^2+c$

Substitute $x=2, y=3$:

$\frac{1}{2}(9)=4+c$

$c=0.5$

$\frac{1}{2}y^2=x^2+0.5$

$y^2=2x^2+1$

$y=\pm\sqrt{2x^2+1}$

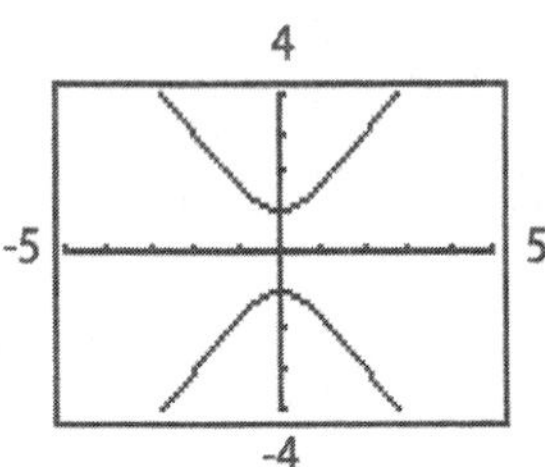

9. $\frac{dy}{dx}=ke^x$

$y=ke^x$, so $k=\frac{y}{e^x}$

Substitute for k in the equation for the derivative.

$\frac{dy}{dx}=\frac{y}{e^x}e^x=y$

$\left.\frac{dy}{dx}\right|_{OT}=-\frac{1}{y}$

$y\,dy=-dx$

Integrating,

$\frac{y^2}{2}=-x+\frac{c}{2}$

$y^2=c-2x$

The orthogonal trajectories to the exponential family are parabolas.

13. From Example 3,

$N=N_0(0.5)^{t/5.27}$

When $t=2.00$:

$N=N_0(0.5)^{2.00/5.27}$

$=N_0(0.769)$

76.9% of the initial amount remains

17. $\frac{dN}{dt}=r-kN$

Solving by separation of variables:

$\frac{dN}{r-kN}=dt$

$-\frac{1}{k}\ln(r-kN)=t+c$

$N=0$ for $t=0$

$c=-\frac{1}{k}\ln r$

$-\frac{1}{k}\ln(r-kN)=t-\frac{1}{k}\ln r$

$\ln\frac{r-kN}{r}=-kt$

$r-kN=re^{-kt}$

$N=\frac{r}{k}\left(1-e^{-kt}\right)$

21. $r\frac{dS}{dr}=2(a-S)$

$r\,dS=2a\,dr-2S\,dr$

$dS+\frac{2}{r}S\,dr=2a\frac{dr}{r}$

$P=\frac{2}{r}, Q=\frac{2a}{r}$

$e^{\int P\,dr}=e^{\int(2/r)dr}=e^{2\ln r}=e^{\ln r^2}=r^2$

$Sr^2=\int\left(2a\frac{dr}{r}\cdot r^2\right)=\int 2ar\,dr$

$Sr^2=ar^2+c$

$S=a+\frac{c}{r^2}$

25. $\frac{dT}{dt}=k(T-25)$

$\frac{dT}{T-25}=k\,dt$

Solve by separation of variables:

$\ln(T-25)=kt+\ln c$

$T=25+ce^{kt}$

Substitute $T=90, t=0$:

$90=25+c$

$c=65$

$T=25+65e^{kt}$

Substitute $T=60, t=5$:

$60=25+65e^{5k}$

$$e^{5k} = \frac{35}{65}$$
$$5k = \ln\left(\tfrac{35}{65}\right)$$
$$k = -0.124$$
$$T = 25 + 65e^{-0.124t}$$
Find t for $T = 40$:
$$40 = 25 + 65e^{-0.124t}$$
$$e^{-0.124t} = \tfrac{15}{65}$$
$$-0.124t = \ln\left(\tfrac{15}{65}\right)$$
$$t = 12 \text{ min}$$

29. $\frac{dc}{dt} = k(c_0 - c)$

Solving by separation of variables:
$$\frac{dc}{c - c_0} = -k dt$$
$$\ln(c - c_0) = -kt + \ln c_1$$
$$\ln\frac{c - c_0}{c_1} = -kt$$
Substitute $t = 0, c = 0$:
$$\ln\frac{0 - c_0}{c_1} = 0$$
$$c_1 = -c_0$$
$$c - c_0 = -c_0 e^{-kt}$$
$$c = c_0\left(1 - e^{-kt}\right)$$

33. $V = E\sin\omega t; t = 0, i = 0$
$$L\frac{di}{dt} + Ri = E\sin\omega t$$
$$L\,di + Ri\,dt = E\sin\omega t\,dt$$
$$di + \frac{R}{L}i\,dt = \frac{E}{L}\sin\omega t\,dt$$
$$P = \frac{R}{L}; Q = \frac{E}{L}\sin\omega t\,dt$$
$$e^{\int(R/L)dt} = e^{Rt/L}$$
$$ie^{Rt/L} = \int\frac{E}{L}\sin\omega t\,e^{Rt/L}dt = \frac{E}{L}\int e^{Rt/L}\sin\omega t\,dt$$
Integration by parts or formula 49 with $a = R/L$, $b = \omega$:
$$ie^{Rt/L} = \frac{E}{L}\cdot\frac{e^{Rt/L}\left(\frac{R}{L}\sin\omega t - \omega\cos\omega t\right)}{\frac{R^2}{L^2} + \omega^2} + c$$
$$= \frac{E}{R^2 + L^2\omega^2}e^{Rt/L}(R\sin\omega t - \omega L\cos\omega t) + c$$
Substitute $t = 0, i = 0$:
$$0 = \frac{E}{R^2 + L^2\omega^2}(-\omega L) + c$$
Therefore, $c = \frac{\omega LE}{R^2 + L^2\omega^2}$
$$ie^{Rt/L} = \frac{E}{R^2 + L^2\omega^2}\left[e^{Rt/L}(R\sin\omega t - \omega L\cos\omega t) + \omega L\right]$$
$$i = \frac{E}{R^2 + L^2\omega^2}\left(R\sin\omega t - \omega L\cos\omega t + \omega Le^{-Rt/L}\right)$$

37. $\frac{dv}{dt} = 9.8 - v$

Solving by separation of variables:
$$\frac{dv}{9.8 - v} = dt$$
$$-\ln(9.8 - v) = t - \ln c$$
$$\ln\frac{9.8 - v}{c} = -t$$
$$9.8 - v = ce^{-t}$$
$$v = 9.8 - ce^{-t}$$
Starting from rest means $v = 0$ when $t = 0$;
$$0 = 9.8 - c$$
$$c = 9.8$$
Therefore,
$$v = 9.8 - 9.8e^{-t} = 9.8\left(1 - e^{-t}\right)$$
$$\lim_{t\to\infty} 9.8\left(1 - e^{-t}\right) = 9.8$$

41. $\frac{dx}{dt} = 6t - 3t^2$
$$dx = \left(6t - 3t^2\right)dt$$
$$x = \frac{6t^2}{2} - \frac{3t^3}{3} + c$$
$$x = 3t^2 - t^3 + c$$
Substitute $x = 0$ for $t = 0$:

$0 = c$, so
$$x = 3t^2 - t^3$$
Substitute x into y:
$$y = 2\left(3t^2 - t^3\right) - \left(3t^2 - t^3\right)^2$$
$$y = -t^6 + 6t^5 - 9t^4 - 2t^3 + 6t^2$$

45. $\frac{dv}{dt} = kv$
$$\frac{dv}{v} = k\,dt$$

$\ln v = kt + \ln c$

Let $v = 33\ 000$ when $t = 0$; $33\ 000 = c$;

$\ln v = kt + \ln 33\ 000$

$\ln v - \ln 33\ 000 = kt$

$\ln \frac{v}{33\ 000} = kt$

$\frac{v}{33\ 000} = e^{kt}$

$v = 33\ 000 e^{kt}$

Let $v = 19\ 700$ when $t = 3$:

$19\ 700 = 33\ 000 e^{3k}$

$e^k = (0.597)^{1/3}$

$v = 33\ 000(0.597)^{t/3}$

After 11 years, $v = 33\ 000(0.597)^{11/3} = \$4977$

49. $y = e^{x/2} + k$, isobars

$\frac{dy}{dx} = \frac{1}{2}e^{x/2}$

The slope must be the negative reciprocal for the orthogonal trajectories,

$\left.\frac{dy}{dx}\right|_{\text{OT}} = -2e^{-x/2}$

$y = 4e^{-x/2} + c$, wind direction

We graph the following isobars and orthogonal trajectories:

isobars: $y = e^{x/2}$, $y = e^{x/2} + 3$

wind direction: $y = 4e^{-x/2} + 1$, $y = 4e^{-x/2} - 2$

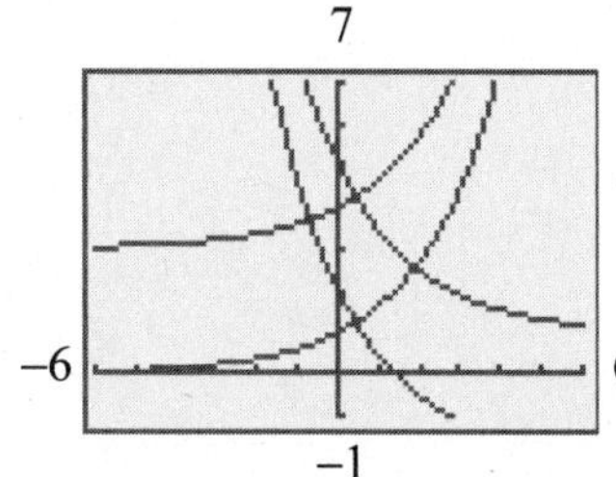

31.7 Higher-order Homogeneous Equations

1. $D^2y - 5Dy = 0$

$m^2 - 5m = 0$

$m(m-5) = 0$

$m = 0, m = 5$

$y = c_1 + c_2e^{5x}$

5. $3\frac{d^2y}{dx^2} + 4\frac{dy}{dx} + y = 0$

$3D^2y + 4Dy + y = 0$

$3m^2 + 4m + 1 = 0$

$(3m+1)(m+1) = 0$

$m_1 = -\frac{1}{3}, m_2 = -1$

$y = c_1e^{-(1/3)x} + c_2e^{-x}$

9. $2D^2y - 3y = Dy$

$2D^2y - Dy - 3y = 0$

$2m^2 - m - 3 = 0$

$(2m-3)(m+1) = 0$

$m = \frac{3}{2}, m = -1$

$y = c_1e^{\frac{3x}{2}} + c_2e^{-x}$

13. $3D^2y + 8Dy - 3y = 0$

$3m^2 + 8m - 3 = 0$

$(3m-1)(m+3) = 0$

$m_1 = \frac{1}{3}$ and $m_2 = -3$

$y = c_1e^{x/3} + c_2e^{-3x}$

17. $2\frac{d^2y}{dx^2} - 4\frac{dy}{dx} + y = 0$

$2D^2y - 4Dy + y = 0$

$2m^2 - 4m + 1 = 0$

Quadratic formula: $m = \frac{4 \pm \sqrt{16-8}}{4}$;

$m_1 = 1 + \frac{\sqrt{2}}{2},\ m_2 = 1 - \frac{\sqrt{2}}{2}$

$y = c_1e^{\left(1+\left(\sqrt{2}/2\right)\right)x} + c_2e^{\left(1-\left(\sqrt{2}/2\right)\right)x}$

21. $y'' = 3y' + y$

$D^2y - 3Dy - y = 0$

$m^2 - 3m - 1 = 0$

Quadratic formula:

$m = \frac{3 \pm \sqrt{9+4}}{2}; m_1 = \frac{3}{2} + \frac{\sqrt{13}}{3}; m_2 \frac{3}{2} - \frac{\sqrt{13}}{2}$

$y = c_1e^{\left((3/2)+\left(\sqrt{13}/2\right)\right)x} + c_2e^{\left((3/2)-\left(\sqrt{13}/2\right)\right)x}$

$y = e^{3x/2}\left(c_1e^{x\sqrt{13}/2} + c_2e^{-x(\sqrt{13}/2)}\right)$

25. $2D^2y+5aDy-12a^2y=0,\ a>0$

$2m^2+5am-12a^2=0$

$(2m-3a)(m+4a)=0$

$m=\dfrac{3a}{2},\qquad m=-4a$

$y=c_1e^{\frac{3a}{2}x}+c_2e^{-4ax}$

29. $D^2y-Dy=12y$

$D^2y-Dy-12y=0$

$y=0$ when $x=0$; $y=1$ when $x=1$

$m^2-m-12=0$

$(m-4)(m+3)=0$

$m_1=4; m_2=-3$

$y=c_1e^{4x}+c_2e^{-3x}$

Substituting given values:

$0=c_1+c_2$; therefore, $c_1=-c_2$;

$1=c_1e^4+c_2e^{-3}$

$1=c_1e^4+c_2e^{-3}=-c_2e^4+\dfrac{c_2}{e^3}=\dfrac{-c_2e^7+c_2}{e^3}$

$e^3=c_1\left(1-e^7\right)$; therefore,

$c_2=-\dfrac{e^3}{\left(1-e^7\right)},c_1=\dfrac{e^3}{\left(1-e^7\right)}$

$y=-\dfrac{e^3}{\left(1-e^7\right)}e^{4x}+\dfrac{e^3}{\left(1-e^7\right)}e^{-3x}$

$=\dfrac{e^3}{e^7-1}e^{4x}-\dfrac{e^3}{e^7-1}e^{-3x}$

$y=\dfrac{e^3}{e^7-1}\left(e^{4x}-e^{-3x}\right)$

33. $D^4y-5D^2y+4y=0$

$m^4-5m^2+4=0$

$(m^2-4)(m^2-1)=0$

$m_1=2,m_2=-2,m_3=1,m_4=-1$

$y=c_1e^x+c_2e^{-2x}+c_3e^x+c_4e^{-x}$

31.8 Auxiliary Equation with Repeated or Complex Roots

1. $\dfrac{d^2y}{dx^2}+10\dfrac{dy}{dx}+25y=0$

$D^2y+10Dy+25y=0$

$m^2+10m+25=0$

$(m+5)^2=0$

$m=-5,-5$

$y=e^{-5x}\left(c_1+c_2x\right)$

5. $D^2y-2Dy+y=0$

$m^2-2m+1=0$

$(m-1)^2=0$

$m=1,1$

$y=e^x\left(c_1+c_2x\right)$

9. $D^2y+9y=0$

$m^2+9=0$

$m_1=3j$ and $m_2=-3j$

$\alpha=0,\beta=3$

$y=e^{0x}\left(c_1\sin 3x+c_2\cos 3x\right)$

$y=c_1\sin 3x+c_2\cos 3x$

13. $D^4y-y=0$

$m^4-1=0$

$\left(m^2-1\right)\left(m^2+1\right)=0$

$(m-1)(m+1)\left(m^2+1\right)=0$

$m=\pm1, m=\pm j\ (\alpha=0,\beta=1)$

$y=c_1e^x+c_2e^{-x}+c_3\sin x+c_4\cos x$

17. $16D^2y-24Dy+9y=0$

$16m^2-24m+9=0$

$(4m-3)^2=0$

$m=\dfrac{3}{4},\dfrac{3}{4}$

$y=e^{3x/4}\left(c_1+c_2x\right)$

21. $2D^2y+5y=4Dy$

$2D^2y+5y-4Dy=0$

$2m^2-4m+5=0$

Quadratic formula:

$m=\frac{4\pm\sqrt{16-40}}{4}=\frac{4\pm2\sqrt{-6}}{4}$

$m_1=1+\frac{\sqrt{6}}{2}j; m_2=1-\frac{\sqrt{6}}{2}j; \alpha=1, \beta=\frac{1}{2}\sqrt{6}$

$y=e^x\left(c_1\cos\frac{1}{2}\sqrt{6}x+c_2\sin\frac{1}{2}\sqrt{6}x\right)$

25. $2D^2y-3Dy-y=0$

$2m^2-3m-1=0$

By the quadratic formula,

$m=\frac{3\pm\sqrt{9+8}}{4}$

$m_1=\frac{3}{4}+\frac{\sqrt{17}}{4}, m_2=\frac{3}{4}-\frac{\sqrt{17}}{4}$

$y=c_1e^{((3/4)+(\sqrt{17}/4))x}+c_2e^{((3/4)+(\sqrt{17}/4))x};$

$y=e^{(3/4)x}(c_1e^{x(\sqrt{17}/4)}+c_2e^{-x(\sqrt{17}/4)})$

29. $D^3y-6D^2y+12Dy-8y=0$

$m^3-6m^2+12m-8=0$

$(m-2)(m^2-4m+4)=0$

$(m-2)(m-2)(m-2)=0$

$m=2,2,2$ repeated root

$y=e^{2x}(c_1+c_2x+c_3x^2)$

33. $D^2y+2Dy+10y=0$

$m^2+2m+10=0$

By the quadratic formula,

$m=\frac{-2\pm\sqrt{4-40}}{2}$

$m_1=-1+3j; m_2=-1-3j, \alpha=-1, \beta=3;$

$y=e^{-x}(c_1\sin3x+c_2\cos3x)$

Substituting $y=0$ when $x=0$:

$0=e^0(c_1\sin0+c_2\cos0); c_2=0$

Substituting $y=e^{-\pi/6}, x=\frac{\pi}{6}$:

$e^{-\pi/6}=e^{-\pi/6}\left(c_1\sin\frac{\pi}{2}\right)$

$e^{-\pi/6}=e^{-\pi/6}c_1$

$c_1=1$

$y=e^{-x}\sin3x$

37. $y=c_1e^{3x}+c_2e^{-3x}$

The roots are $m=3, m=-3$, so the auxiliary equation is

$(m-3)(m+3)=0$

$m^2-9=0$

Therefore, $D^2y-9y=0$

41. **(a)** For $a=0$, $D^2y+ay=0$ is $D^2y=0$ from which $m^2=0$, and $m=0$ is a double root

Therefore $y=c_1+c_2x$

Substitute $y=0, x=0$:

$0=c_1+c_2(0)$

$c_1=0$ and $y=c_2x$

Substitute $y=0, x=1$:

$0=c_2(1)$

$c_1=0$ and $y=0$ is the solution.

(b) For $a<0$, $-a>0$ and $D^2y+ay=0$ has $m^2+a=0\Rightarrow m=\pm\sqrt{-a}$ from which

$y=c_1e^{\sqrt{-a}x}+c_2e^{-\sqrt{-a}x}$

Substitute $y=0, x=0$:

$0=c_1+c_2$, so $c_1=-c_2$

Substitute $y=0, x=1$:

$0=-c_2e^{\sqrt{-a}}+c_2e^{-\sqrt{-a}}$

$c_2=0$ and $y=0$ is the solution.

31.9 Solutions of Nonhomogeneous Equations

1. b: x^2+2x+e^{-x}

We choose the particular solution to be of the form

$y_p=A+Bx+Cx^2+Ee^{-x}$

5. $D^2y-Dy-2y=4$

$m^2-m-2=0$

$(m-2)(m+1)=0$

$m_1=2, m_2=-1$

$y_c=c_1e^{2x}+c_2e^{-x}$

Assume a particular solution of the form $y_p=A$

$Dy_p=0; D^2y_p=0$

Substituting in the differential equation:

$0-0-2A=4$

$A=-2$

Therefore, $y_p=-2$ and the complete solution is

$y=c_1e^{2x}+c_2e^{-x}-2$

9. $y''-3y'=2e^x+xe^x$

$D^2y-3Dy=2e^x+xe^x$

$m^2-3m=0$

$m(m-3)=0; m_1=0, m_2=3$

$y_c=c_1e^0+c_2e^{3x}=c_1+c_2e^{3x}$

Let $y_p=Ae^x+Bxe^x$

Then $Dy_p=Ae^x+B(xe^x+e^x)=Ae^x+Bxe^x+Be^x$

$D^2y_p=Ae^x+Be^x+B(xe^x+e^x)$

$=Ae^x+Be^x+Bxe^x+Be^x$

$D^2y_p=Ae^x+2Be^x+Bxe^x$

Substituting in the differential equation:

$Ae^x+2Be^x+Bxe^x-3(Ae^x+Bxe^x+Be^x)$

$=Ae^x+2Be^x+Bxe^x-3Ae^x-3Bxe^x-3Be^x$

$=-2Ae^x-Be^x-2Bxe^x$

$=e^x(-2A-B)+xe^x(-2B)$

$=2e^x+xe^x$

Equating coefficients of similar terms:

$-2A-B=2$ and $-2B=1$

$B=-\frac{1}{2}$ and $A=-\frac{3}{4}$

The complete solution is

$y=c_1+c_2e^{3x}-\frac{3}{4}e^x-\frac{1}{2}xe^x$

13. $\frac{d^2y}{dx^2}-2\frac{dy}{dx}+y=2x+x^2+\sin 3x$

$D^2y-2Dy+y=2x+x^2+\sin 3x$

$m^2-2m+1=$

$(m-1)^2=0$

$m=1$ (double root)

$y_c=e^x(c_1+c_2x)$

Let $y_p=A+Bx+Cx^2+E\sin 3x+F\cos 3x$

Then $Dy_p=B+2Cx+3E\cos 3x-3F\sin 3x$

$D^2y_p=2C-9E\sin 3x-9F\cos 3x$

Substituting in the differential equation:

$2C-9E\sin 3x-9F\cos 3x$

$-2(B+2Cx+3E\cos 3x-3F\sin x)$

$+A+Bx+Cx^2+E\sin 3x+F\cos 3x$

$=2C-9E\sin 3x-9F\cos 3x-2B-4Cx-6E\cos 3x$

$+6F\sin x+A+Bx+Cx^2+E\sin 3x+F\cos 3x$

$=(2C-2B+A)+\sin 3x(6F-8E)+$

$\cos 3x(-8F-6E)$

$+x(B-4)+Cx^2$

$=2x+x^2+\sin 3x$

Equating the coefficients of similar terms:

$2C-2B+A=0; 6F-8E=1; -8F-6E=0;$

$B-4=2; C=1$

Therefore, $C=1, B=6, A=10, E=-\frac{2}{25}, F=\frac{3}{5}$

The complete solution is

$y=e^x(c_1+c_2x)+10+6x+x^2-\frac{2}{25}\sin 3x+\frac{3}{50}\cos 3x$

17. $D^2y-Dy-30y=10$

$m^2-m-30=0$

$(m-6)(m+5)=0$

$m_1=-5, m_2=6$

$y_c=c_1e^{-5x}+c_2e^{6x}$

Let $y_p=A$

Then $Dy_p=0, D^2y_p=0$

Substituting in the differential equation:

$0-0-30A=10$

$A=-\frac{1}{3}$

The complete solution is $y=c_1e^{-5x}+c_2e^{6x}-\frac{1}{3}$

21. $D^2y-4y=\sin x+2\cos x$

$D^2y-4y=0$

$m^2-4=0$

$m_1=2; m_2=-2$

$y_c=c_1e^{2x}+c_2e^{-2x}$

Let $y_p=A\sin x+B\cos x$

Then $Dy_p=A\cos x-B\sin x$

$D^2y_p=-A\sin x-B\cos x$

Substituting into the differential equation:

$-A\sin x-B\cos x-4A\sin x-4B\cos x$

$=-5A\sin x-5B\cos x$

$=\sin x+2\cos x$

Equating the coefficients of similar terms:

$-5A=1$, so $A=-\frac{1}{5}$; $-5B=2$, so $B=-\frac{2}{5}$

The complete solution is

$y=c_1e^{2x}+c_2e^{-2x}-\frac{1}{5}\sin x-\frac{2}{5}\cos x$

25. $D^2y+5Dy+4y=xe^x+4$

$D^2y+5Dy+4y=0$

$m^2+5m+4=0$

$(m+1)(m+4)=0$

$m_1=-1, m_2=-4$

$y_c=c_1e^{-x}+c_2e^{-4x}$

Let $y_p=Ae^x+Bxe^x+C$

$Dy_p=Ae^x+Bxe^x+Be^x$

$D^2y_p=Ae^x+B(xe^x+e^x)+Be^x$

$\quad =Ae^x+2Be^x+Bxe^x$

Substituting in the differential equation:

$Ae^x+2Be^x+Bxe^x+5(Ae^x+Bxe^x+Be^x)$

$\quad +4(Ae^x+Bxe^x+C)$

$=(10A+7B)e^x+10Bxe^x+4C$

$=xe^x+4$

Equating the coefficients of similar terms:

$10A+7B=0; 10B=1; 4C=4$

Therefore, $B=\frac{1}{10}, C=1, A=-\frac{7}{100}$

The complete solution is

$y=c_1e^{-x}+c_2e^{-4x}-\frac{7}{100}e^x+\frac{1}{10}xe^x+1$

29. $D^2y+y=\cos x$

$D^2y+y=0$

$m^2+1=0$

$m=\pm i$

$y_c=c_1\sin x+c_2\cos x$

Let $y_p=x(A\sin x+B\cos x)$

Then $Dy_p=x(A\cos x-B\sin x)+A\sin x+B\cos x$

$D^2y_p=x(-A\sin x-B\cos x)+A\cos x$

$\quad -B\sin x+A\cos x-B\sin x$

Substituting in the differential equation:

$-x(A\sin x+B\cos x)+2A\cos x-2B\sin x$

$+x(A\sin x+B\cos x)=\cos x$

$2A\cos x-2B\sin x=\cos x$

$2A=1, B=0$

$A=\frac{1}{2}$

$y_p=\frac{1}{2}x\sin x$

The complete solution is

$y=c_1\sin x+c_2\cos x+\frac{1}{2}x\sin x$

33. $D^2y-Dy-6y=5-e^x$

$D^2y-Dy-6y=0$

$m^2-m-6=0$

$(m-3)(m+2)=0$

$m_1=3, m_2=-2$

$y_c=c_1e^{3x}+c_2e^{-2x}$

Let $y_p=A+Be^x$

Then $Dy_p=Be^x$ and $D^2y_p=Be^x$

Substituting in the differential equation:

$Be^x-Be^x-6(A+Be^x)=5-e^x$

$-6A-6Be^x=5-e^x$

$-6A=5$, so $A=-\frac{5}{6}$

$-6B=-1$ so $B=\frac{1}{6}$

The complete solution is $y=c_1e^{3x}+c_2e^{-2x}-\frac{5}{6}+\frac{1}{6}e^x$

Substituting $x=0, y=2$:

$2=c_1+c_2-\frac{5}{6}+\frac{1}{6}$

$c_1=\frac{8}{3}-c_2$

$D_y=3c_1e^{3x}-2c_2e^{-2x}+\frac{1}{6}e^x$

Substituting $Dy=4, x=0$:

$4=3c_1-2c_2+\frac{1}{6}$

Substituting c_1 :

$4=8-3c_2-2c_2+\frac{1}{6}$

$c_2=\frac{5}{6}, c_1=\frac{11}{6}$

The solution is

$y=\frac{11}{6}e^{3x}+\frac{5}{6}e^{-2x}+\frac{1}{6}e^x-\frac{5}{6}$

$=\frac{1}{6}\left(11e^{3x}+5e^{-2x}+e^x-5\right)$

37. $Dy-y=x^2$

$m-1=0$

$m=1$

$y_c=c_1e^x$

Let $y_p=A+Bx+Cx^2$

Then $Dy_p=B+2Cx$

Substituting in the differential equation:

$B+2Cx-A-Bx-Cx^2=x^2$

Equating the coefficients of similar terms:

$-C=1$, so $C=-1$

$2C-B=0$, so $B=-2$

$A-B=0$, so $A=-2$

The solution is $y=c_1e^x-2-2x-x^2$

Solving with an integrating factor would require integration by parts, so this method is simpler.

31.10 Applications of Higher-order Equations

1. $x=c_1\sin 4t+c_2\cos 4t; x=0, Dx=2$ for $t=0$

Substituting $x=0, t=0$:

$0=c_1\sin(4(0))+c_2\cos(4(0))$

$c_2=0$

$x=c_1\sin 4t$

$Dx=4c_1\cos 4t$

Substituting $Dx=2, t=0$:

$2=4c_1\cos(4(0))$

$c_1=\frac{1}{2}$

The solution is $x=\frac{1}{2}\sin 4t$

5. $D^2\theta+\frac{g}{l}\theta=0; g=9.8m/s^2, l=1.0m;$

$D^2\theta+9.8\theta=0$

$m^2+9.8=0$

$m_1=j\sqrt{9.8}, m_2=-j\sqrt{9.8}; \alpha=0, \beta=\sqrt{9.8}$

$\theta=c_1\sin\sqrt{9.8}t+c_2\cos\sqrt{9.8}t$

Substituting $\theta=0.1$ when $t=0$;

$0.1=c_1\sin 0+c_2\cos 0$

$0.1=c_2$

$D\theta=c_1\sqrt{9.8}\cos\sqrt{9.8}t-0.1\sqrt{9.8}\sin\sqrt{9.8}t$

Substituting $D\theta=0$ when $t=0$;

$0=c_1\sqrt{9.8}\cos 0-c_2\sqrt{9.8}\sin 0$

$c_1=0$

$\theta=0.1\cos\sqrt{9.8}t=0.1\cos 3.1t$

To graph, we consider the key values of the cosine function:

$\sqrt{9.8}t$	t	$\cos\sqrt{9.8}t$	$0.1\cos\sqrt{9.8}t$
0	0	1	0.1
$\frac{\pi}{2}$	0.50	1	0
π	1.00	−1	−0.1
$\frac{3\pi}{2}$	1.50	0	0
2π	2.00	1	0.1

The period is $\frac{2\pi}{\sqrt{9.8}}=2.00$, and the amplitude is 0.1.

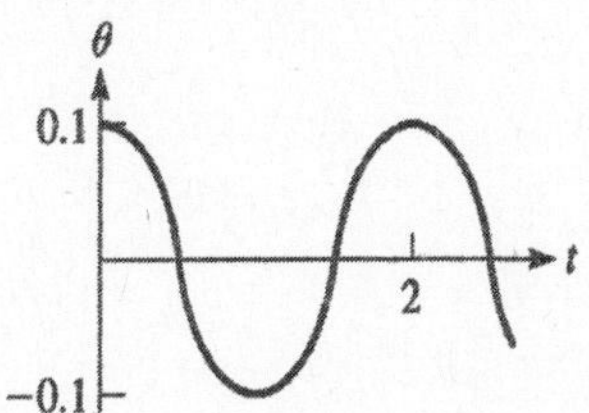

9. $m\frac{d^2s}{dt^2}=-\frac{mg}{e}(s-L)$

$\frac{d^2s}{dt^2}+\frac{g}{e}s=\frac{g}{e}L$

$m^2+\frac{g}{e}=0$

$m=\pm j\sqrt{\frac{g}{e}}$

$s_c=c_1\sin\sqrt{\frac{g}{e}}t+c_2\cos\sqrt{\frac{g}{e}}t$

Let $s_p=A+Bt$

$Ds_p=B$

$D^2s_p=0$

Substituting in the differential equation:

$0+\frac{g}{e}(A+Bt)=\frac{g}{e}L$

Equating the coefficients of similar terms:

$\frac{g}{e}A=\frac{g}{e}L$, so $A=L$

$\frac{g}{e}B=0$, so $B=0$

$s=c_1\sin\sqrt{\frac{g}{e}}t+c_2\cos\sqrt{\frac{g}{e}}t+L$

Substituting $t=0, s=s_0$:

$s_0=c_1\sin 0+c_2\cos 0+L$

$c_2=s_0-L$

$$\frac{ds}{dt}=c_1\sqrt{\frac{g}{e}}\cos\sqrt{\frac{g}{e}}t-(s_0-L)\sqrt{\frac{g}{e}}\sin\sqrt{\frac{g}{e}}t$$

Substituting $t=0, \frac{ds}{dt}=0$:

$$0=c_1\sqrt{\frac{g}{e}}\cos(0)-(s_0-L)\sqrt{\frac{g}{e}}\sin(0)$$

$$c_1=0$$

$$s=(s_0-L)\cos\sqrt{\frac{g}{e}}t+L$$

13. $mD^2x+kx=4\sin 2t$

$$\frac{4.00}{9.80}m^2+80.0=0$$

$$m=\pm 14j$$

$$x_c=c_1\sin 14t+c_2\cos 14$$

Let $x_p=A\sin 2t+B\cos 2t$

Then $Dx_p=2A\cos 2t-2B\sin 2t$

$$D^2x_p=-4A\sin 2t-4B\cos 2t$$

Substituting in the differential equation:

$$\frac{4.00}{9.80}(-4A\sin 2t-4B\cos 2t)$$
$$+80(A\sin 2t+B\cos 2t)=4\sin 2t$$

$$\frac{3840}{49}A\sin 2t+\frac{3840}{49}B\cos 2t=4\sin 2t$$

Equating the coefficients of similar terms:

$$\frac{3840}{49}A=4, \text{ so } A=\frac{49}{960}; B=0$$

$$x_p=\frac{49}{960}\sin 2t$$

The complete solution is

$$x=c_1\sin 14t+c_2\cos 14t+\frac{49}{960}\sin 2t$$

$$Dx=14c_1\cos 14t-14c_2\sin 14t+\frac{98}{960}\cos 2t$$

When $t=0,\ x=0.100,\ Dx=0$

$$0.100=c_1\sin 0+c_2\cos 0+\frac{49}{960}\sin 0$$

$$c_2=0.100$$

$$0=14c_1\cos 0-1.4\sin 0+\frac{98}{960}\cos 0$$

$$c_1=-\frac{7}{960}$$

$$x=-\frac{7}{960}\sin 14t+0.100\cos 14t+\frac{49}{960}\sin 2t$$

17. $L\frac{d^2q}{dt^2}+R\frac{dq}{dt}+\frac{q}{C}=E$

$L=0.100\text{H}, R=0, C=100\mu\text{F}=10^{-4}\text{ F}, E=100\text{ V}$

$$0.100\frac{d^2q}{dt^2}+\frac{q}{10^{-4}}=100$$

$$\frac{d^2q}{dt^2}+10^5q=1000$$

$$\frac{d^2q}{dt^2}+10^5q=0$$

$$m^2+10^5=0$$

$$m=\pm 10^{5/2}j=\pm 316j; \alpha=0, \beta=316$$

$$q_c=c_1\sin 316t+c_2\cos 316t$$

Let $q_p=A$, then $q_p'=0, q_p''=0$

Substituting in the differential equation:

$$0+0+10^5A=1000$$

$$A=0.01$$

The complete solution is

$$q=c_1\sin 316t+c_2\cos 316t+0.01$$

Substituting $q=0, t=0$:

$$0=0+c_2(1)+0.01$$

$$c_2=-0.01$$

$$\frac{dq}{dt}=316c_1\cos 316t+3.16\sin 316t$$

Substituting $\frac{dq}{dt}=0,\ t=0: 0=316c_1+0$, so $c_1=0$

Therefore, $q=0.01(1-\cos 316t)$

21. $L\frac{d^2q}{dt^2}+R\frac{dq}{dt}+\frac{q}{C}=E$

$L=8.00\text{mH}=8.00\times 10^{-3}\text{H}; R=0;$

$C=0.50\mu\text{F}=5.00\times 10^{-7}\text{F}$

$E=20.0e^{-200t}\text{mV}=2.00\times 10^{-2}e^{-200t}\text{V}$

$$8.00\times 10^{-3}\frac{d^2q}{dt^2}+\frac{q}{5.00\times 10^{-7}}=2.00\times 10^{-2}e^{-200t}$$

$$\frac{d^2q}{dt^2}+2.50\times 10^8q=2.50e^{-200t}$$

$$m^2+2.50\times 10^8=0;$$

$$m=\pm 1.58\times 10^4j; \alpha=0, \beta=1.58\times 10^4$$

$$q_c=c_1\sin 1.58\times 10^4t+c_2\cos 1.58\times 10^4t$$

Let $q_p=Ae^{-200t}$, then $q_p'=-200Ae^{-200t}$ and

$q_p''=4.00\times 10^4Ae^{-200t}$

Substituting into the differential equation,

$$4.00\times10^4 Ae^{-200t}+2.50\times10^8 Ae^{-200t}=2.50e^{-200t}$$

Equating the coefficients:

$$4.00\times10^4 A+2.50\times10^8 A=2.50$$

$$A=10^{-8}$$

$$q=c_1\sin1.58\times10^4t+c_2\cos1.58\times10^4+10^{-8}e^{-200t}$$

Substituting $q=0,\ t=0$:

$$0=c_1(0)+c_2(1)+10^{-8}$$

$$c_2=-10^{-8}$$

$$i=\frac{dq}{dt}=1.58\times10^4c_1\cos1.58\times10^4t$$
$$-1.58\times10^4c_2\sin1.58\times10^4t-200\times10^{-8}e^{-200t}$$

Substituting $i=0, t=0$:

$$0=1.58\times10^4c_1(1)-2.00\times10^{-6}(1)$$

$$c_1=\frac{2.00\times10^{-6}}{1.58\times10^4}$$

$$i=1.58\times10^4\times\frac{2\times10^{-6}}{1.58\times10^4}\cos1.58\times10^4t$$
$$-1.58\times10^4\times(-10^{-8})\sin1.58\times10^4t$$
$$-200\times10^{-6}e^{-200t}$$
$$=2.00\times10^{-6}\cos1.58\times10^4t$$
$$+1.58\times10^{-4}\sin1.58\times10^4t$$
$$-2.00\times10^6e^{-200t}$$
$$=10^{-6}[2.00\cos(1.58\times10^4t)]$$
$$+158(1.58\times10^4t)-2.00e^{-200t}]$$

25. $EI\dfrac{d^4y}{dx^4}=w$

$EID^4y=0$

$m^4=0$ and $m=0$ is a quadruple root

$y_c=c_1+c_2+c_3x^2+c_4x^3$

Let $y_p=Ax^4$, then $\dfrac{d^4y_p}{dx^4}=24A$

Substituting in the differential equation,

$24AEI=w$, so $A=\dfrac{w}{24EI}$

The complete solution is

$$y=c_1+c_2+c_3x^2+c_4x^3+\frac{w}{24EI}x^4$$

Substituting $x=0, y=0$ we get $c_1=0$

$$y=c_2x+c_3x^2+c_4x^3+\frac{w}{24EI}x^4$$

$$y'=c_2+2c_3x+3c_4x^2+\frac{4w}{24EI}x^3$$

Substituting $x=0, y'=0$ we get $c_2=0$

$$y=c_3x^2+c_4x^3+\frac{w}{24EI}x^4$$

$$y''=2c_3+6c_4x+\frac{12w}{24EI}x^2$$

Substituting $x=L, y''=0$:

$$0=2c_3+6c_4L+\frac{12w}{24EI}L^2$$

$$y'''=6c_4+\frac{24w}{24EI}x$$

Substituting $x=L, y'''=0$:

$$0=6c_4+\frac{24w}{24EI}\cdot L,\ \text{so } c_4=\frac{-wL}{6EI}$$

$$0=2c_3+6\cdot\frac{-wL}{6EI}\cdot L+\frac{12w}{24EI}L^2,$$

$$\text{so } c_3=\frac{wL^2}{4EI}$$

The solution is

$$y=\frac{wL^2}{4EI}x^2+\frac{-wL}{6EI}x^3+\frac{w}{24EI}x^4$$

$$y=\frac{w}{24EI}(6L^2x^2-4Lx^3+x^4)$$

31.11 Laplace Transforms

1. $f(t)=1, t>0$

$$L(f)=\int_0^\infty e^{-st}\cdot1\,dt$$
$$=\lim_{c\to\infty}\frac{-1}{s}\int_0^c e^{-st}(-s\,dt)$$
$$=-\frac{1}{s}\lim_{c\to\infty}e^{-st}\Big|_0^c$$
$$=-\frac{1}{s}\left[\lim_{c\to\infty}\left(e^{-sc}-1\right)\right]$$
$$=-\frac{1}{s}(0-1)$$
$$L(f)=\frac{1}{s}$$

5. $f(t)=e^{3t}$; from Transform 3 of the table, with $a=-3: L(e^{3t})=\dfrac{1}{s-3}$

9. $f(t)=\cos2t-\sin2t$

$L(f)=L(\cos2t)-L(\sin2t)$

By Transforms 5 and 6 with $a=2$:

$$L(f)=\frac{s}{s^2+4}-\frac{2}{s^2+4}$$

$$L(f)=\frac{s-2}{s^2+4}$$

13. $f''+f'; f(0)=0; f'(0)=0$

$$L[f''+f']$$
$$= L(f'')+L(f')$$
$$= s^2L(f)-sf(0)-f'(0)+sL(f)-f(0)$$
$$= s^2L(f)-s\cdot 0-0+sL(f)-0$$
$$= s^2L(f)+sL(f)$$

17. $F(s)=\dfrac{2}{s^3}$

$$L^{-1}(F)=L^{-1}\left(\frac{2}{s^3}\right)=2L^{-1}\left(\frac{1}{s^3}\right)$$
$$L^{-1}(F)=\frac{2t^2}{2}=t^2 \text{ (Transform 2)}$$

21. $F(s)=\dfrac{1}{s^3+3s^2=3s+1}$

$$L^{-1}(F)=L^{-1}\left[\frac{1}{(s+1)^3}\right]$$
$$=\frac{1}{2}L^{-1}\left[\frac{2!}{(s+1)^3}\right]$$
$$=\frac{1}{2}t^2e^{-t}$$

(Transform 12)

25. $F(s)=\dfrac{4s^2-8}{(s+1)(s-2)(s-3)}=\dfrac{-\frac{1}{3}}{s+1}+\dfrac{-\frac{8}{3}}{s-2}+\dfrac{7}{s-3}$

$$L^{-1}(F)$$
$$=-\frac{1}{3}L^{-1}\left(\frac{1}{s+1}\right)-\frac{8}{3}L^{-1}\left(\frac{1}{s-2}\right)+7L^{-1}\left(\frac{1}{s-3}\right)$$
$$=-\frac{1}{3}e^{-t}-\frac{8}{3}e^{2t}+7e^{3t}$$

29. Take $f(t)=e^{-at}$. Then

$$F(s)=L(f)=\frac{1}{s+a}$$
$$-\frac{d}{ds}F(s)=-\frac{d}{ds}\left(\frac{1}{s+a}\right)$$
$$=-\left(\frac{-1}{(s+a)^2}\right)$$
$$=\frac{1}{(s+a)^2}$$

From Transform 11, $L\{tf(t)\}=L\left(te^{-at}\right)=\dfrac{1}{(s+a)^2}$

Therefore, $L\{tf(t)\}=-\dfrac{d}{ds}F(s)$

31.12 Solving Differential Equations by Laplace Transforms

1. $2y'-y=0; y(0)=2$

$$L(2y')-L(y)=L(0)$$
$$2L(y')-L(y)=0$$
$$2sL(y)-2(2)-L(y)=0$$
$$L(y)=\frac{4}{2s-1}=2\frac{1}{s-\frac{1}{2}}$$

From Transform 3, $y=2e^{\frac{1}{2}t}$

5. $y'+y=0; y(0)=1$

$$L(y')+L(y)=L(0)$$
$$L(y')+L(y)=0$$
$$sL(y)-y(0)+L(y)=0$$
$$sL(y)-1+L(y)=0$$
$$(s+1)L(y)=1$$
$$L(y)=\frac{1}{s+1}$$

Transform 3 with $a=-1$ $\quad y=e^{-t}$

9. $y'+3y=e^{-3t}; y(0)=1$

$$L(y')+L(3y)=L(e^{-3t})$$
$$[sL(y)-1)]+3L(y)=\frac{1}{s+3}$$
$$(s+3)L(y)=\frac{1}{s+3}+1$$
$$L(y)=\frac{1}{(s+3)^2}+\frac{1}{s+3}$$

The inverse is found from Transforms 11 and 3

$$y=te^{-3t}+e^{-3t}=(1+t)e^{-3t}$$

13. $4y''+4y'+5y=0,\ y(0)=1,\ y'(0)=-\dfrac{1}{2}$

$$4L(y'')+4L(y')+5L(y)=0$$
$$4\left(s^2L(y)-sy(0)-y'(0)\right)+4\left(sL(y)-y(0)\right)+5L(y)=0$$
$$4s^2L(y)-4s+2+4sL(y)-4+5L(y)=0$$
$$\left(4s^2+4s+5\right)L(y)=4s+2$$
$$L(y)=\frac{4s+2}{4s^2+4s+5}=\frac{4\left(s+\frac{1}{2}\right)}{4\left(s^2+s+\frac{1}{4}\right)+4}$$
$$L(y)=\frac{s+\frac{1}{2}}{\left(s+\frac{1}{2}\right)^2+1}$$

Use Transform 20, $a=\dfrac{1}{2},\ b=1$

$$y=e^{-1/2t}\cos t$$

17. $y''+y=1; y(0)=1; y'(0)=1$

$L(y'')+L(y)=L(1)$

$s^2L(y)-s-1+L(y)=\frac{1}{s}$

$(s^2+1)L(y)=\frac{1}{s}+s+1$

$L(y)=\frac{1}{s(s^2+1^2)}+\frac{s}{s^2+1^2}+\frac{1}{s^2+1^2}$

By Transforms 7, 5, and 6:

$y=1-\cos t+\cos t+\sin t$

$y=1+\sin t$

21. $y''-4y=10e^{3t}, y(0)=5, y'(0)=0$

$L(y'')-4L(y)=10\cdot L(e^{3y})$

Use Transform 3 with $a=-3$:

$s^2L(y)-sy(0)-y'(0)-4L(y)=\frac{10}{s-3}$

$s^2L(y)-5s-0-4L(y)=\frac{10}{s-3}$

$(s^2-4)L(y)=5s+\frac{10}{s-3}$

$L(y)=\frac{5s}{(s+2)(s-2)}+\frac{10}{(s+2)(s-2)(s-3)}$

Write in partial fractions,

$L(y)=\frac{\frac{5}{2}}{s+2}+\frac{\frac{5}{2}}{s-2}+\frac{\frac{1}{2}}{s+2}+\frac{-\frac{5}{2}}{s-2}+\frac{2}{s-3}$

$L(y)=\frac{3}{s+2}+\frac{2}{s-3}$

Use Transform 3 twice:

$y=3e^{-2t}+2e^{3t}$

25. $2v'=6-v$; since the object starts at rest, $v(0)=0$, $v'(0)=0$

$2L(v')+L(v)=6L(1)$

Use Transform 1:

$2sL(v)-0+L(v)=\frac{6}{s}$

$(2s+1)L(v)=\frac{6}{s}$

$L(v)=\frac{6}{s(2s+1)}=6\left[\frac{\frac{1}{2}}{s(s+\frac{1}{2})}\right]$

By Transform 4 with $a=\frac{1}{2}$: $v=6(1-e^{-t/2})$

29. $L\frac{d^2q}{dt^2}+R\frac{dq}{dt}+\frac{q}{C}=E; q(0)=0$

$0(q'')+50q'+\frac{q}{4\times10^{-6}}=40$

$50q'+\frac{10^6}{4}q=40$

$50L(q')+\frac{10^6}{4}L(q)=L(40)$

$50[sL(q)-q(0)]+\frac{10^6}{4}L(q)=\frac{40}{s}$

$L(q)\left(50s+\frac{10^6}{4}\right)=\frac{40}{s}$

$L(q)=\frac{40}{s\left(50s+\frac{10^6}{4}\right)}$

$L(q)=\frac{40}{50s(s+5000)}$

$=\frac{40}{50(5000)}\cdot\frac{5000}{s(s+5000)}$

$=0.00016\cdot\frac{5000}{s(s+5000)}$

$q=L^{-1}\left[0.00016\cdot\frac{5000}{s(s+5000)}\right]$

$q=1.60\times10^{-4}(1-e^{-5000t})$

(from Transform 4, with $a=5000$)

33. $D^2y+9y=18\sin 3t;\ y=0,\ Dy=0, t=0$

$y''+9y=18\sin 3t$

$L(y'')+9L(y)=18L(\sin 3t)$

Use Transform 6 with $a=3$:

$s^2L(y)-sy(0)-y'(0)+9L(y)=18\cdot\frac{3}{s^2+9}$

$L(y)(s^2+9)=\frac{54}{s^2+9}$

$L(y)=\frac{54}{(s^2+9)^2}$

Use Transform 15, with $a=3, 2a^3=54$

$y=L^{-1}\left[\frac{54}{(s^2+9)^2}\right]$

$y=\sin 3t-3t\cos 3t$

37. $y^{iv} = k$

$L(y^{iv}) = L(k)$

$s^4 L(y) - s^3 \cdot y(0) - s^2 \cdot y'(0) - s \cdot y''(0) - y'''(0) = \frac{k}{s}$

$s^4 L(y) - as^2 - b = \frac{k}{s}$

$s^4 L(y) = as^2 + \frac{k}{s} + b = \frac{as^3 + k + bs}{s}$

$L(y) = \frac{as^3 + bs + k}{s^5} = \frac{a}{s^2} + \frac{b}{s^4} + \frac{k}{s^5}$

$y = ax + \frac{b}{6}x^3 + \frac{k}{24}x^4$

$y' = a + \frac{b}{2}x^2 + \frac{k}{6}x^3$

$y' = bx + \frac{k}{2}x^2$

$y''' = b + kx$

At $x = L, y = 0,$ so

$aL + \frac{bL^3}{6} + \frac{kL^4}{24} = 0$

Also, at $x = L, y = 0,$ so

$bL + \frac{k}{2}L^2 = 0$ and $b = \frac{-kL}{2}$

Therefore,

$aL - \frac{kL}{2} \cdot \frac{L^3}{6} + \frac{kL^4}{24} = 0$

$a = \frac{kL^3}{24}$

$y = \frac{kL^3}{24}x - \frac{kL}{12} + \frac{k}{24x}x^4 = \frac{k}{24}(L^3 x - 2Lx^3 + x^4)$

Since $k = w/EI$,

$y = \frac{w}{24EI}(L^3 x - 2Lx^3 + x^4)$

Review Exercises

1. $4xy^3 dx + (x^2 + 1)dy = 0$

The equation is separable; divide by $y^3(x^2 + 1)$.

$\frac{4x}{x^2 + 1}dx + \frac{dy}{y^3} = 0$

Integrating ($u = x^2 + 1, du = 2xdx$),

$2\ln(x^2 + 1) - \frac{1}{2y^2} = c$

5. $2D^2 y + Dy = 0$

The auxiliary equation is

$2m^2 + m = 0$

$m(2m + 1) = 0$

$m_1 = 0$ and $m_2 = -\frac{1}{2}$

$y = c_1 e^0 + c_2 e^{-x/2}$

$y = c_1 + c_2 e^{-x/2}$

9. $(x + y)dx + (x + y^3)dy = 0$

$x\,dx + y\,dx + x\,dy + y^3\,dy = 0$

$x\,dx + d(xy) + y^3\,dy = 0$

Integrating,

$\frac{1}{2}x^2 + xy + \frac{1}{4}y^4 = c_1$

$2x^2 + 4xy + y^4 = c$

13. $dy = (2y + y^2)dx$

By separation of variables: $\frac{dy}{y(2 + y)} = dx$

Integrating by formula 2:

$-\frac{1}{2}\ln\left(\frac{2 + y}{y}\right) = x + \ln c_1$

$\ln\frac{2 + y}{y} = -2x - \ln c_1^2$

$c_1^2\left(\frac{2 + y}{y}\right) = e^{-2x}$

$y = c_1^2(y + 2)e^{2x}$

$y = c(y + 2)e^{2x}$

$y(1 - ce^{2x}) = 2c$

$y = \frac{2c}{1 - ce^{2x}}$

17. $y' + 4y = 2e^{-2x}$

$\frac{dy}{dx} + 4y = 2e^{-2x}$

$dy + 4 \cdot y\,dx = 2e^{-2x}dx$

$e^{\int 4dx} = e^{4x}$

$ye^{4x} = \int 2e^{-2x}e^{4x} + c$

$= e^{2x} + c$

$y = e^{-2x} + ce^{-4x}$

21. $2D^2s + Ds - 3s = 6$

$2m^2 + m - 3 = 0$

$(m-1)(2m+3) = 0$

$m_1 = 1,\ m_2 = -\dfrac{3}{2}$

$s_c = c_1e^t + c_2e^{-3t/2}$

Let $s_p = A$, then $s_p' = 0;\ s_p'' = 0$

Substituting into the differential equation,

$2(0) + 0 - 3A = 6$

$A = -2$

$s_p = -2$

The complete solution is $s = c_1e^t + c_2e^{-3t/2} - 2$

25. $9D^2y - 18Dy + 8y = 16 + 4x$

$$D^2y - 2Dy + \frac{8}{9}y = \frac{16}{9} + \frac{4}{9}x$$

$$m^2 - 2m + \frac{8}{9} = 0$$

$$m = \frac{2 \pm \sqrt{4 - 4\left(\frac{8}{9}\right)}}{2};$$

$$m_1 = \frac{2}{3};\ m_2 = \frac{4}{3}$$

$y_c = c_1e^{2x/3} + c_2e^{4x/3}$

Let $y_p = A + Bx$. Then $y_p' = B;\ y_p'' = 0$

Substituting into the differential equation.

$$0 - 2B + \frac{8}{9}(A + Bx) = \frac{16}{9} + \frac{4}{9}x$$

$$\left(-2B + \frac{8}{9}A\right) + \frac{8}{9}Bx = \frac{16}{9} + \frac{4}{9}x$$

$$-2B + \frac{8}{9}A = \frac{16}{9};\ \frac{8}{9}B = \frac{4}{9}$$

$$B = \frac{1}{2}$$

$$-2\left(\frac{1}{2}\right) + \frac{8}{9}A = \frac{16}{9}$$

$$A = \frac{25}{8}$$

$$y_p = \frac{1}{2}x + \frac{25}{8}$$

The complete solution is

$$y = c_1e^{2x/3} + c_2e^{4x/3} + \frac{1}{2}x + \frac{25}{8}$$

29. $y'' - 7y' - 8y = 2e^{-x}$

$D^2y - 7Dy - 8y = 0$

$m^2 - 7m - 8 = 0$

$(m+1)(m-8) = 0$

$m = -1,\ m = 8$

$y_c = c_1e^{-x} + c_2e^{8x}$

Let $y_p = Axe^{-x}$

Then $y_p' = Ae^{-x}(1-x)$,

$y_p'' = Ae^{-x}(x-2)$

Substituting in the differential equation:

$$Ae^{-x}(x-2) - 7Ae^{-x}(1-x) - 8Axe^{-x} = 2e^{-x}$$

$$A(x-2) - 7A(1-x) - 8Ax = 2$$

$$-9A = 2$$

$$A = \frac{-2}{9}$$

$$y_p = \frac{-2}{9}xe^{-x}$$

$y = y_c + y_p$

$$y = c_1e^{-x} + c_2e^{8x} + \frac{-2}{9}xe^{-x}$$

33. $3y' = 2y\cot x$

$$\frac{dy}{dx} = \frac{2}{3}y\cot x$$

$$\frac{dy}{y} = \frac{2}{3}\cot x\,dx$$

$$\ln y = \frac{2}{3}\ln\sin x + \ln c$$

$\ln y - \ln\sin^{2/3}x = \ln c;$

$$\ln\frac{y}{\sin^{2/3}x} = \ln c$$

$$\frac{y}{\sin^{2/3}x} = c$$

$$y = c\sin^{2/3}x$$

Substituting $y = 2, x = \dfrac{\pi}{2}$:

$$2 = c\sin^{2/3}\frac{\pi}{2}$$

$c = 2$

Therefore,

$y = 2\sin^{2/3}x = 2\sqrt[3]{\sin^2 x}$

$y^3 = 8\sin^2 x$

37. $D^2v + Dv + 4v = 0$

$m^2 + m + 4 = 0$

$m = \frac{-1 \pm \sqrt{-15}}{2}$

$m = -\frac{1}{2} \pm \frac{\sqrt{15}}{2} j;\ \alpha = -\frac{1}{2},\ \beta = \frac{\sqrt{15}}{2};$

$v = e^{-t/2}\left(c_1 \sin\frac{\sqrt{15}}{2}t + c_2 \cos\frac{\sqrt{15}}{2}t\right)$

$Dv = e^{-t/2}\left(\frac{\sqrt{15}}{2}c_1 \cos\frac{\sqrt{15}}{2}t - \frac{\sqrt{15}}{2}c_2 \sin\frac{\sqrt{15}}{2}t\right)$

$+\left(c_1 \sin\frac{\sqrt{15}t}{2} c_2 \cos\frac{\sqrt{15}}{2}t\right)\left(-\frac{1}{2}e^{-t}\right)$

Substituting $v = 0$ when $t = 0$:

$0 = e^0\left(c_1 \cdot 0 + c_2\right)$

$c_2 = 0$

Substituting $Dv = \sqrt{15}, t = 0$:

$\sqrt{15} = e^0\left(\frac{\sqrt{15}}{2}c_1\right)$

$c_1 = 2$

Therefore, $y = 2e^{-t/2}\sin\left(\frac{\sqrt{15}}{2}t\right)$

41. $4y' - y = 0; y(0) = 1$

$L(4y') - L(y) = 0$

$4L(y') - L(y) = 0$

$4sL(y) - 1 - L(y) = 0$

by Eq. (30.24)

$(4s - 1)L(y) = 1$

$L(y) = \frac{1}{4s - 1} = \frac{1}{4(s - \frac{1}{4})}$

By Transform 3 with $a = -\frac{1}{4}, y = \frac{1}{4}e^{t/4}$.

45. $y'' + y = 0, y(0) = 0, y'(0) = -4$

$L(y'') + L(y) = 0$

$s^2L(y) + -sy(0) - y'(0) + L(y) = 0$

$s^2L(y) + 4 + L(y) = 0$

$(s^2 + 1)L(y) = -4$

$L(y) = -4\left(\frac{1}{s^2 + 1}\right)$

By Transform 6 with $a = 1$,

$y = -4\sin t$.

49. $\frac{dy}{dx} = 1 + y^2$, $x = 0$ to $x = 0.4$, $\Delta x = 0.1$, $(0, 0)$

x	y	y
0	0	0
0.1	$0 + (1 + 0^2)(0.1)$	0.1
0.2	$0.1 + (1 + 0.1^2)(0.1)$	0.201
0.3	$0.201 + (1 + 0.201^2)(0.1)$	0.3050401
0.4	$0.305 + (1 + 0.305^2)(0.1)$	0.4143450463

53. $\frac{dx}{dt} = 2t$

$x = t^2 + c$

Substitute $x = 0, t = 0$:

$1 = 0^2 + c$

$c = 1$

Therefore, x in terms of t is given by

$x = t^2 + 1$

Taking derivatives with respect to t on both sides of $xy = 1$:

$x\frac{dy}{dt} + y\frac{dx}{dt} = 0$

Substituting $\frac{dx}{dt} = 2t$ and $x = \frac{1}{y}$:

$\frac{1}{y}\frac{dy}{dt} + y(2t) = 0$

Multiplying by dt / y,

$\frac{1}{y^2}dy + 2t\,dt = 0$

$-\frac{1}{y} + t^2 = c$

Substituting $y = 0, t = 0$:

$-\frac{1}{1} + 0^2 = c$

$c = -1$

$-\frac{1}{y} + t^2 = -1$

$-1 + yt^2 = -y$

$y(t^2 + 1) = 1$

$y = \frac{1}{t^2 + 1}$

In terms of t, $x = t^2 + 1$ and $y = \frac{1}{t^2 + 1}$.

57. $\dfrac{dm}{dt} = km$

Solve by separation of variables:

$\dfrac{dm}{m} = kdt$

$\ln m = kt + c$

Substituting $t = 0, m = m_0 : \ln m_0 = k(0) + c$

$c = \ln m_0$

$\ln m = kt + \ln m_0$

$\ln \dfrac{m}{m_0} = kt$

$\dfrac{m}{m_0} = e^{kt}$

$m = m_0 e^{kt}$

61. $\dfrac{dy}{dx} = \dfrac{y}{y-x}, (-1, 2), y > 0$

$ydy = ydx + xdy = d(xy)$

$\dfrac{y^2}{2} = xy + C$

Substituting $x = -1, y = 2$:

$\dfrac{2^2}{2} = (-1)(2) + C$

$C = 4$

$\dfrac{y^2}{2} = xy + 4$

$y^2 = 2xy + 8$

$y^2 - 2xy - 8 = 0$

To find y explicitly, we use the quadratic formula with $a = 1, b = -2x, c = -8$:

$y = \dfrac{2x \pm \sqrt{4x^2 + 32}}{2}$

$= x \pm \sqrt{x^2 + 8}$

The path is a hyperbola that consists of the two functions:

$y = x + \sqrt{x^2 + 8}$ and

$y = x - \sqrt{x^2 + 8}$

65. $N = N_0 e^{kt}$

Substituting $N = N_0 / 2, t = 1.28 \times 10^9$:

$\dfrac{N_0}{2} = N_0 e^{k(1.28\times10^9)}$

$k = \dfrac{\ln 0.5}{1.28 \times 10^9}$

$N = N_0 e^{-\frac{\ln 0.5}{1.28\times10^9}t}$

Find t for $N = 0.75N_0 : 0.75N_0 = N_0 e^{\frac{\ln 0.5}{1.28\times10^9}t}$

$t = \dfrac{(\ln 0.75)(1.28\times10^9)}{\ln 0.5}$

$t = 5.31\times10^8$ years

69. See Example 2, Section 31.6.

$y = kx^5, \ k = \dfrac{y}{x^5}$

Take the derivative with respect to x: $y' = 5kx^4$

Substitute the expresion for k:

$y' = 5\left(\dfrac{y}{x^5}\right)x^4 = \dfrac{5y}{x}$

The slope of the orthogonal trajectories will be the negative reciprocal of y', therefore,

$\left.\dfrac{dy}{dx}\right|_{OT} = -\dfrac{x}{5y}$

Solve by separating variables:

$5y\,dy = -x\,dx$

$\dfrac{5}{2}y^2 = -\dfrac{1}{2}x^2 + \dfrac{1}{2}c$

$5y^2 + x^2 = c$

The orthogonal trajectories are ellipses.

73. $L\dfrac{di}{dt} + Ri = E$

$2\dfrac{di}{dt} + 40i = 20$

$\dfrac{di}{dt} + 20i = 10$

$di + 20i\,dt = 10\,dt$

$P = 20, Q = 10$

$ie^{\int 20\,dt} = \int 10e^{\int 20\,dt} + c$

$ie^{20t} = \int 10e^{20t} + c$

$= 0.5e^{20t} + c$

Substituting $i = 0, t = 0$:

$0 = 0.5 + c$

$c = -0.5$

$ie^{20t} = 0.5e^{20t} - 0.5$

$= 0.5\left(e^{20t} - 1\right)$

$i = 0.5\left(1 - e^{-20t}\right)$

77. $LD^2q + RDq + \frac{q}{C} = E$

$L = 0.5$ H, $R = 6\Omega$, $C = .02$ F, $E = 24\sin 10t$

$0.5D^2q + 6Dq + 50q = 24\sin 10t$

$D^2q + 12Dq + 100q = 48\sin 10t$

$D^2q + 12Dq + 100q = 0$

$m^2 + 12m + 100 = 0$

$$m = \frac{-12 \pm \sqrt{144 - 400}}{2} = -6 \pm 8j$$

$q_c = e^{-6t}(c_1 \sin 8t + c_2 \cos 8t)$

Let $q_p = A\sin 10t + B\cos 10t$, then

$Dq_p = 10A\cos 10t - 10B\sin 10t$

$D^2q_p = -100A\sin 10t - 100B\cos 10t$

Substituting in the differential equation:

$-100A\sin 10t - 100B\cos 10t +$
$12(10A\cos 10t - 10B\sin 10t)$
$+100(A\sin 10t + B\cos 10t) = 48\sin 10t$

$-120B\sin 10t + 120A\cos 10t = 48\sin 10t$

Equating the coefficients of similar terms:

$-120B = 48; 120A = 0$

Therefore, $B = -0.4$, $A = 0$

The complete solution is

$q = e^{-6t}(c_1 \sin 8t + c_2 \cos 8t) - 0.4\cos 10t$

Substituting $q = 0$, $t = 0$:

$0 = c_2 - 0.4$, so $c_2 = 0.4$

$q = e^{-6t}(c_1 \sin 8t + 0.4\cos 8t) - 0.4\cos 10t$

$Dq = e^{-6t}(-6c_1 \sin 8t - 2.4\cos 8t + 8c_1 \cos 8t - 3.2\sin 8t)$
$+ 4\sin 10t$

Substituting $Dq = 0, t = 0 : 0 = -2.4 + 8c_1$, so $c_1 = 0.3$

The solution is

$q = e^{-6t}(0.3\sin 8t + 0.4\cos 8t) - 0.4\cos 10t$

81. $2\frac{di}{dt} + i = 12,\ i(0) = 0$

$2L(i') + L(i) = 12L(1)$

Use Transform 1: $2[sL(i) - i(0)] + L(i) = \frac{12}{s}$

$(2s+1)L(i) = \frac{12}{s}$

$$L(i) = \frac{12}{s(2s+1)}$$

$$= 12\frac{\frac{1}{2}}{s(s+1/2)}$$

Use Transform 4, $a = \frac{1}{2}$:

$$i = 12\left(1 - e^{-t/2}\right)$$

Evaluate i for $t = 0.3$:

$$i = 12\left(1 - e^{-0.3/2}\right)$$

$$i = 1.67 \text{ A}$$

85. $m = 0.25$ kg,

$k = 16$ N/m

$0.25D^2y + 16y = \cos 8t \quad (D = d/dt)$

$D^2y + 64y = 4\cos 8t,\ y(0) = 0,\ Dy(0) = 0$

$L(y'') + 64L(y) = 4L(\cos 8t)$

Use Transform 5:

$$s^2L(y) - s\cdot 0 - 0 + 64L(y) = 4\left(\frac{s}{s^2+64}\right)$$

$$L(y) = \frac{4s}{(s^2+64)^2}$$

$$= \frac{1}{4}\left[\frac{2(8s)}{(s^2+8^2)^2}\right]$$

Using Transform 16,

$y = 0.25t\sin 8t$

89. $EI\frac{d^2y}{dx^2} = M,\ M = 2000x - 40x^2$

$EID^2y = 2000x - 40x^2$

$$D^2y = \frac{1}{EI}(2000x - 40x^2)$$

$$Dy = \frac{1}{EI}\left(1000x^2 - \frac{40}{3}x^3\right) + c_1$$

$$y = \frac{1}{EI}\left(\frac{1000}{3}x^3 - \frac{10}{3}x^4\right) + c_1x + c_2$$

$y = 0$ for $x = 0$ and $x = L$

$$0 = \frac{1}{EI}(0 - 0) + 0 + c_2,\ c_2 = 0$$

$$0 = \frac{1}{EI}\left(\frac{1000}{3}L^3 - \frac{10}{3}L^4\right) + c_1L$$

$$c_1 = \frac{1}{EI}\left(\frac{10}{3}L^3 - \frac{1000}{3}L^2\right)$$

$$y = \frac{1}{EI}\left(\frac{1000}{3}x^3 - \frac{10}{3}x^4\right) + \frac{1}{EI}\left(\frac{1000}{3}L^3 - \frac{1000}{3}L^2\right)x$$

$$= \frac{10}{3EI}\left(100x^3 - x^4 + L^3x - 100L^2x\right)$$